Lecture Notes in Production Engineering

Lecture Notes in Production Engineering (LNPE) is book series that reports the latest research and developments in Production Engineering, comprising:

- Biomanufacturing
- Control and Management of Processes
- Cutting and Forming
- Design
- Life Cycle Engineering
- Machines and Systems
- Optimization
- Precision Engineering and Metrology
- Surfaces

LNPE publishes authored conference proceedings, contributed volumes and authored monographs that present cutting-edge research information as well as new perspectives on classical fields, while maintaining Springer's high standards of excellence. Also considered for publication are lecture notes and other related material of exceptionally high quality and interest. The subject matter should be original and timely, reporting the latest research and developments in all areas of production engineering. The target audience of LNPE consists of advanced level students, researchers, as well as industry professionals working at the forefront of their fields. Much like Springer's other Lecture Notes series, LNPE will be distributed through Springer's print and electronic publishing channels.

Indexed in Scopus* *Indexed in EI Compendex To submit a proposal or request further information please contact Anthony Doyle, Executive Editor, Springer (anthony.doyle@springer.com).

Thomas Bauernhansl · Alexander Verl ·
Mathias Liewald · Hans-Christian Möhring
Editors

Production at the Leading Edge of Technology

Proceedings of the 13th Congress of the German Academic Association for Production Technology (WGP), Freudenstadt, November 2023

Springer

Editors
Thomas Bauernhansl
Institut für Industrielle Fertigung und Fabrikbetrieb - IFF
Universität Stuttgart
Stuttgart, Germany

Alexander Verl
Institut für Steuerungstechnik und Werkzeugmaschinen - ISW
Universität Stuttgart
Stuttgart, Germany

Mathias Liewald
Institut für Umformtechnik - IFU
Universität Stuttgart
Stuttgart, Germany

Hans-Christian Möhring
Institut für Werkzeugmaschinen - IfW
Universität Stuttgart
Stuttgart, Germany

ISSN 2194-0525 ISSN 2194-0533 (electronic)
Lecture Notes in Production Engineering
ISBN 978-3-031-47393-7 ISBN 978-3-031-47394-4 (eBook)
https://doi.org/10.1007/978-3-031-47394-4

This Springer imprint is published by the registered company Springer Nature Switzerland AG
The registered company address is: Gewerbestrasse 11, 6330 Cham, Switzerland

Paper in this product is recyclable.

Preface

On behalf of the German Academic Association for Production Technology (WGP), Frankfurt/Germany, four member institutes agreed to organize its 13th Annual Plenary Congress entitled "Production at the Leading Edge of Technology" in Freudenstadt from November 20–23, 2023. The Institutes for Industrial Manufacturing and Management (IFF), for Metal Forming Technology (IFU), for Control Engineering of Machine Tools and Manufacturing Units (ISW) and for Machine Tools (IfW) cordially welcome researchers and guests to present recent scientific findings in the areas of production technologies. We are looking forward to exciting and fruitful discussions with experts from industry and research!

Research in production is constantly pushing the boundaries of what is feasible. The presentations show production processes that advance into new areas of production in terms of fundamental findings, methodology, more efficient use of resources or interdisciplinarity. But where does the search for new borders lead to? Which borders do we still have to cross and which ones do we prefer not to cross?

The focus of the congress is on production processes in actually limiting fields related to extreme velocity, size, accuracy, methodology, use of resources and interdisciplinarity. Challenges from the fields of metal forming, cutting processes and machine tools, automated assembly and robotics, process planning and management sciences, as well as interdisciplinary research topics, will be addressed. This year, the key topics of production technologies for the circular economy, environmentally neutral production, resilient value-added systems and digitalization as an enabler for sustainable production are core of the congress. Therefore, the program summarizes peer-reviewed contributions from production science and industrial research. Scheduled sessions provide an overview of current trends in production research and give an insight into ongoing research recently conducted in the German Academic Association for Production Technology.

We wish all participants an interesting and inspiring Annual Plenary Congress and look forward to welcoming you to Freudenstadt!

November 2023

Prof. Thomas Bauernhansl
Universität Stuttgart
Institut für Industrielle Fertigung und Fabrikbetrieb IFF

Prof. Alexander Verl
ISW

Prof. Mathias Liewald
IFU

Prof. Hans-Christian Möhring
IfW

Organization

University of Stuttgart

Institute of Industrial Manufacturing and Management

Thomas Bauernhansl
Victoria Pohl
Florian Roth
Selina Rögele
Marco Maier

University of Stuttgart

Institute for Control Engineering of Machine Tools and Manufacturing Units

Alexander Verl
Xenia Günther

University of Stuttgart

Institute for Metal Forming Technology

Mathias Liewald
Maximilian Bachmann

University of Stuttgart

Institute for Machine Tools

Hans-Christian Möhring
Sibylle Krug

Contents

Digitization as an Enabler for a Sustainable Production

Conceptional Framework to Reduce Misinformation in Production System Engineering

S. Karch[1(✉)], A. Strahilov[1], and A. Lüder[2]

[1] let's dev GmbH & Co. KG, Alter Schlachthof 33, 76131 Karlsruhe, Germany
sabrina.karch@letsdev.de

[2] Chair of Production Systems and Automation, Otto-von-Guericke-University, Universitätspl. 2, 39106 Magdeburg, Germany

Abstract. In addition to the efficient use of materials, the significance of information as a resource increase in digital age. Digital tools are seen as a catalyst in generating and processing of digital data and information. One of the main challenges is to find the right information in the increasing oversupply of information as well as to manage this resource sustainably throughout the entire product life cycle. Misinformation is identified to be one of the main drivers of inefficiency in an engineering activity and is influenced by quantity and quality of data and information.

To further reduce this inefficiency, the study presents a conceptual framework that pursues the approach of demand-orientation as well as the smallest possible and appropriate quantum of data and information. Senior experts confirm the conceptual framework in semi-structured interviews. The outlook refers to use cases for investigation in a multi case study.

Keywords: Production Systems Engineering · Data and Information flow · Lean Digitalization

1 Introduction

The exchange of data and information on demand is the main focus, especially when experts from different disciplines work together collaboratively, such as in production systems engineering. The effectiveness and efficiency of collaboration as well as in each individual engineering activity depends on the logistical principle of information: right information, at the right time, in the right quantity, at the right place, in the right quality [1].

In today's processes, native data of interim or final results are converted to PDF or exchanged as MS Office file using email [2, 30].

In a currently discussed ideal, a process participant receives required data in a usable format by a tool-independent Common Data Model, a multilateral data platform. In this way, many actors have the possibility to collaborate on engineering data simultaneously and across companies based on current and consistent data [3].

T. Bauernhansl et al. (Eds.): WGP 2023, LNPE, pp. 3–13, 2024.
https://doi.org/10.1007/978-3-031-47394-4_1

The requirements for a data unit, named as data quantum, have to be defined and standardized. For this purpose, the study aims to develop and validate a fundamental framework for a specific data quantum and follows the research question:

How to define the smallest possible and appropriate quantum of data from the user's point of view?

First, an overview of the theoretical background and related work is given (Sect. 2). Based on this literature-centered study a conceptional framework is developed (Sect. 3). Section 4 raises the research methodology and Sect. 5 presents the research results validated by the semi-structured interviews. Section 6 discusses the results and limitations. Finally, a summary and an outlook for further work is given (Sect. 7).

2 Theoretical Background and Related Work

2.1 Data Management

Data management is an integral part of working with data quanta. Hanksche [4] considers data management to be "the planning, design, monitoring and control of the use, distribution and communication of information in organizations in order to achieve strategic goals". Data management includes all concepts that enable the organizational and technical implementation to provide data reliably, enable effective data use, and ensure data quality, data protection, and information security [4]. In addition, data management includes, e. g., data security, data distribution and access control [5].

Data are syntactic entities, input to interpretation processes and become information as interpreted data within contextual, semantic relations. Knowledge results from individual interpretation of information, experience and learning processes. Consequently, information is the input and output of a decision, which is finally made based on knowledge [6]. This is also the understanding behind Dippold's definition of data management. Value is created by translating information into decisions and actions. Data management has the function of making data available just in time to every process participant [7].

Relevant sub-aspects are implemented in different use cases, e. g. in the project SemAnz 4.0 with the target to describe data points semantically unambiguously [8]. ECLASS is a data standard for the classification of products and services using standardized ISO-compliant characteristics [9]. The data receiver using ECLASS can easily interpret the received data, but cannot evaluate the quality.

The project management can provide guidance on the question of the right amount of data by analogy with release planning. However, operational solutions are not discussed in this context [10].

The presented use cases cover relevant aspects of data management and working with data quanta, but do not represent a holistic view. The target is to gather significant aspects and transfer them into a generally valid framework.

2.2 Lean Paradigm in Data Flow

Lean is known for the paradigm of eliminating waste and work without added value [11] and addresses the effectiveness and efficiency of activities and processes.

The customer is the starting point for lean systems approach, which is summarized by Womack and Jones in five principles [12].

In the context of data and information, three aspects are of key importance: customer orientation, elimination of waste and the ideal of one-piece flow.

Customer Orientation. Ohno underlines the role of the customer by affirming that all activities, including the processing of information, are only relevant if they result in products, services or an activity, customers want to buy (T. Ohno according to [13]). In reference to this customer centricity, a downstream process step is also to be considered as an internal customer whose needs and requirements must be taken int account [14]. In the context of transformation of data and information, these internal customers have information needs. Consequently, needs orientation is to be seen as a prerequisite for an effective and efficient processing of data and information in accordance with the information logistics principles.

Elimination of Waste. Waste, or muda in Japanese, means "to toil" or "pointless effort", has an immediate impact on effectiveness and efficiency and includes all activities that do not benefit the customer [14]. Misinformation is identified to be one of the main drivers of inefficiency in an engineering activity and is influenced by quantity and quality of data and information [15]. In the increasing availability of data and information, misinformation must be eliminated to ensure an efficient process.

One Piece Flow. An ideal from a lean point of view is the so-called one-piece flow or the supply with the smallest possible units so that the customer's needs are met on demand [14]. This idea is also to pursue in the context of data in order to meet demand orientation and the logistical principles of data and information [1].

The resulting framework should meet these requirements of lean paradigm.

2.3 Data-Driven Engineering

Production System Engineering (PSE) is a model-based and multi-disciplinary process involving various domain experts [16] and considering a Production System of Systems (PSoS). PSE is highly complex, often concerning coevolving products and system components [17]. To handle the complexity and ensure effective and efficient engineering, different engineering process methods have been developed and standardized considering the designed PSoS as a digital value creation chain [18]. Major representatives of engineering guidelines are ISO 18828 [19], VDI/VDE Guidelines 2221 [20], 2206 [21], and 4499 [22].

PSE engineers in different domains take design decisions based on suitable engineering knowledge [23]. The engineering activities form a network, linked by data exchange of externalized engineering knowledge [24]. The interaction of engineers is challenged [25] by the different engineering habits, especially limiting efficient and effective information exchange and utilization. Especially the engineering quality evaluation is impeded by a missing integrated information representation covering the individual knowledge of the experts and their cross-discipline.

3 Conceptional Framework

The study proposes a conceptional framework based on the literature reviewed, to respond the research question. The framework is intended to meet the requirements of the lean paradigm (see Fig. 1) described in Sect. 2.2.

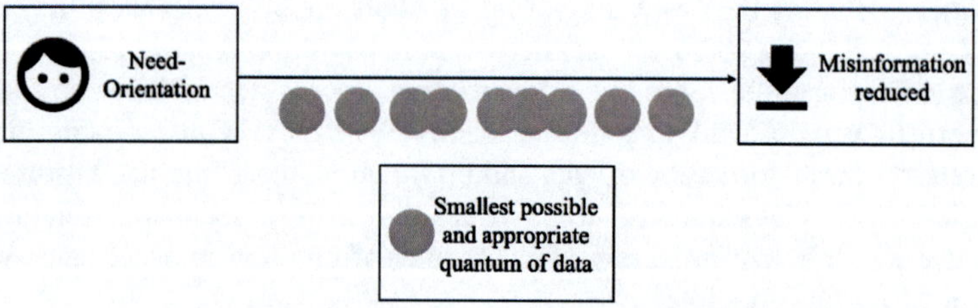

Fig. 1. Correlation of lean paradigm and misinformation.

The framework intends a description of a data quantum through five main aspects. Each data quantum contains information about security, transaction, data type, rights and roles, and data quality (see Fig. 2).

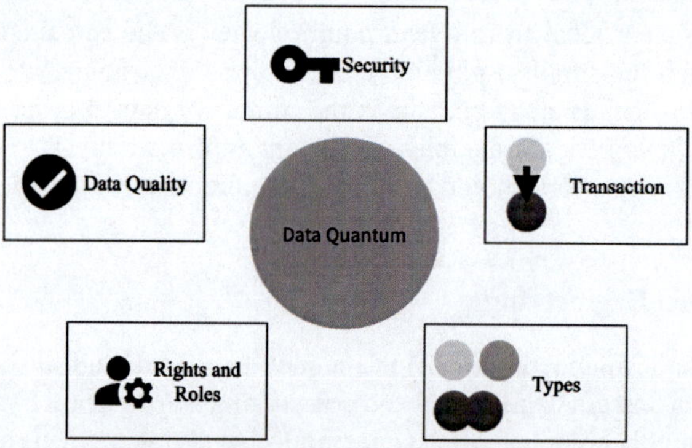

Fig. 2. Main aspects of a data quantum.

Security. The category security includes the technical and organizational protection of data itself as well as of data transmission. This includes, for example, the transmission technology, encryption or the API (Application Programming Interface) interface [26].

Transaction. In a transaction, a dataset is left in a consistent state after error-free execution of an instruction. Changed data can be transmitted via the pull or push principle as well as in a permanent transmission [26].

Data Type. This category gives an overview of how data can be organized to be exchanged. With the data types it is to be distinguished between values, attributes, properties, structure, class, list with values, files and formats. It can also be a set of data in the form of files in various formats.

Rights and Roles. With the rights and roles concept, it is important to distinguish which roles exist and which rights, such as read, write, export and convert, are permitted. The question of data owner also needs to be clarified. Rights and roles can also be assigned to data security, but are listed separately here due to their importance [5].

Data Quality. Data quality includes in particular, the topicality, completeness and consistency of data [5].

4 Method

To address the research question, the qualitative study is based on the Design Science Method and semi-structured interviews with senior experts. Interviews, as data collection method, are used to gather complex information, opinions and experiences. The semi-structured interview allows the respondents to answer in their own words and to contribute their ideas and experience. Expert opinion is a convenient way to reflect the conditions of practice and can be used to validate artifact designs at an early date [27, 28].

The interview guide is designed for an interview of about 30 to 45 min and follows a general structure of three main parts: welcome message with a note about recording, the main part with questions, a statement of gratitude for participation [28].

The interview questions are listed in the Table 1. Questions Q1 to Q5 refer to the categories presented in Sect. 3. In question Q6, the experts have the opportunity to add another category to the framework. Question Q7 and Q8 aims to ask the experts about barriers and what needs to be done to make the framework work.

Table 1. Overview of interview questions.

ID	Question
Q1	Can you confirm the category "security"? What is missing in your expertise?
Q2	Can you confirm the category "transaction"? What is missing in your expertise?
Q3	Can you confirm the category "data types"? What is missing in your expertise?
Q4	Can you confirm the category "roles and rights"? What is missing in your expertise?
Q5	Can you confirm the category "data quality"? What is missing in your expertise?
Q6	Would you like to add a category?
Q7	Why can't the framework be successful from your point of view?
Q8	What to do to make the framework work?

Table 2 provides an overview of the interviews and give background information on the interview partners. These have experience in computer science and software development or consider the framework from a user perspective as an engineer.

The results of the interviews have been transcribed and analyzed using Mayring's qualitative content analysis method. Inductive category development was applied, whereby the categories are formed spontaneously during the analysis of the transcribed interviews [29].

Table 2. Overview of interviews and interview partners.

Professional Activity	Professional Experience (years)	Position	ID
Software Development	22	Head of Development	I1
Software Development	16	CTO	I2
Engineering	28	Managing Director	I3
Engineering	28	Senior Researcher	I4

5 Results

The results of the expert interviews are presented and interpreted below, with a particular focus on the different opinions expressed by the experts, but also on the areas of agreement between them.

5.1 Results for the Content-Related Questions (Q1–Q6)

All experts confirm the importance of the framework, in particular the idea of standardization and the categories defined. As a result of their expertise, the following additions are of particular interest.

Question Q1 Security. In this category, the focus in the interviews is on three aspects, the protection of the data itself, the encryption of data during transmission, and the encrypted transmission path.

Manipulation of data is possible at the sender, on the transmission path as well as at the receiver, therefore an authenticity check should be conducted (I1), (I4). Signed data (I1) as well as copy protection is also a way to protect data (I2). To ensure that data does not get to unauthorized persons, authorized persons can verify themselves, e. g. via a digital ID card. A two-factor authentication can additionally protect data, in the simplest case via e-mail or using TOTP (Time-based On-time Passwords) (I2).

TOTP uses an algorithm to generate a temporary passcode using the current time of day as an authentication factor [34].

The experts agree that cost and benefit of *encrypting data or data packets* must be weighed up and should only be considered for particularly security-critical data in a collaborative use case (I1), (I2). Not separately worth protecting are, e. g. data of purchased parts. The closer a plant or PLC (programmable logic controller) program gets to the final state, the more protective the specific plant data are (I3). In standard cases, the *encryption of devices and transmission* channel should be secured, e. g. with HTTPS (Hypertext transfer protocol secure) encryption (I1), (I2).

HTTPS is based on HTTP and uses an additional protocol to encrypt data and transmission of data [32, 33].

Question Q2 Transaction. Each transaction must be provided with a *unique ID* and an authenticity check must be performed. Third party libraries make it possible to check the scope that has changed, to compare the entire document and to verify the update instead of checking data that has not changed with each modification (I2).

Versioning can be used, where access login can be tracked and meta-information about the changes can be provided (I1).

The advantage of *pull request* is that the data can be retrieved on demand. With a *push transaction*, the relevance must be configured in advance, similar to a PubSub (publishing subscribe) mechanism (I1). The challenge is to find out whether new data is relevant for a user. One approach is that this information must be read by all users and the access is logged, whereby data protection according to GDPR (General Data Protection Regulation) must be taken into account (I1). The GDPR regulates the processing of personal data of natural persons by natural persons, companies or organizations in the EU [31].

Information needs may change over the project duration. If the maintenance or operations department takes over responsibility from the project manager for a new plant, e. g. the project manager will only need information on changes (I3).

Question Q3 Types. In addition to classic data formats, a *stream* could also be considered for live data (I1), (I2). In interview (I4) the object reference was added, i. e. in relation to which artifact data is exchanged.

Question Q4 Rights and Roles. Data construct forms the fundament for the implementation of the *rights and role allocation.* An ACL (Access Control List) can be used to specify users and authorizations in regards to data (I2).

An example of the importance of roles and rights is the programming of the PLC software, especially for safety functions. It must be ensured that only trained employees can make changes. The transfer of responsibilities is also conceivable, e. g. if the PLC source code is transferred to the client (I3). The exchange of a subset of existing data, e. g., on an as-needed basis to external suppliers, can also be relevant and could be regulated via rights and roles (I4).

In the collaboration, it is important to establish *hierarchies or representative* regulations, so that, for example, all parties, engineers or lawyers, can accept modified data or specific parts of it depending on their area of expertise in a contract or operating manual (I2). In principle, the *review process* by the parties involved improves the quality of the changed data (I1).

A *reminder* can be a useful function in this context (I1), (I2) and an automatic reminder up to an automatic escalation is conceivable (I1). The challenge is to find balance between reminders and acceptance by users without forcing them too much into a process flow. Not to forget that an application built on the framework takes on a kind of social function (I1).

A *recommender* or an artificial intelligence in compliance with data protection could help to find data immediately or make suggestions about relevant data (I2).

The importance of *transparency* about given information, if necessary prioritized according to groups of people, is emphasized in the interview (I3), but it is important that changes can only be made by authorized persons.

Question Q5 Data Quality. A data quantum should always have an *identifier*, when and by whom it was created or changed, in the sense of versioning (I1). In general, data quality can be increased with the help of defined *responsibilities* (I2).

The *evaluation of quality* is challenging in the case of machine or automatically generated data, such as e. g. in a simulation (I2). Moreover, data quality is often a subjective perception, which makes it difficult to measure (I1).

In networked systems, a technical failure can also have a negative impact on data quality and *actuality* if, e. g. data has not yet been synchronized at a certain point. Nowadays, data actuality is supported with the help of the PubSub mechanism (I2).

It is essential for data quality and actuality, that all parties involved can access the *same data in one place* (I3). In terms of actuality, it is necessary to provide status information such as draft or final (I3).

Question Q6 Add a Category. In the interview (I2), the importance of *data protection* in accordance with GDPR is emphasized as a separate category for a data quantum.

GDPR compliance must be checked in two steps. First, is there data that is covered by data protection and second, do the technical solutions cover the requirements of the GDPR, such as regulated transfer, expiration dates, and deletion periods.

5.2 Results for the Implementation-Oriented Questions (Q7–Q8)

Question Q7 Why can't the Framework be Successful. From a *technical perspective*, the experts see no reason for the failing of the framework and a demand-driven supply of data quanta (I1), (I2).

Of particular importance, however, is user *acceptance* as a social aspect. If an application does not bring the user any tangible benefits, such as time savings, better information supply, or if user feel overregulated and limited in their competence or importance, the framework can fail (I1), (I2), (I4).

Another reason for failure may be that the application based on the framework becomes too *complex*, so that the solution is not used. Also, unprofessionally maintained *responsibilities*, bring the framework to failure (I2).

Question Q8 What to do to Make the Framework Work. In the first place, it is necessary to create *acceptance* for the framework and related applications, by means of comprehensible documentation and a simple possibility of using the framework. A balance must be struck between technically feasible and sensible solutions, with a view to acceptance by users (I2).

An *MVP* (minimum viable product) must be defined that covers the minimum requirements of the parties and has at least the same acceptance as today's solutions, whereby a structural improvement in the background makes the decisive difference (I1), (I2). The framework can be successful if it is implemented in such a way that it not only meets the requirements of a research project but also takes *critical mass* into account (I1).

6 Discussion and Limitations

The framework was confirmed in technical feasibility. The proposed categories can be retained, supplemented by the category Data protection according to GDPR. The presented framework gives a basic, standardized ordering concept for working with data quanta and thus answers the research question presented at the beginning of this paper. The results of the interviews can be transferred into requirements for working with data quanta as well as applications based on the framework. In addition to technical aspects, social aspects that affect acceptance are also in the foreground, which emphasizes a further design from the user's point of view.

It cannot be excluded that other aspects might be relevant when further experts are interviewed. In addition, the focus of the study is on the design of the framework for a data quantum, the interconnection of multiple data quanta needs to be further investigated. Applying the framework to one or more use cases can also provide further insights.

7 Conclusion and Outlook

The demand-oriented supply of data and information out of the oversupply is one of the main challenges in digital age. In production systems engineering, data and information are the equivalent of material in production. An efficient, sustainable handling of these and the elimination of misinformation have a direct impact on the efficiency of the entire product life cycle. In order to be able to organize data in a needs-oriented manner, the research question of *how to define the smallest possible and appropriate quantum of data from the user's point of view* was investigated and the framework of the data quantum elaborated.

In semi-structured interviews with experts, this framework was confirmed and initial, detailed requirements were developed.

The framework is to be investigated in a multi-case study as part of further research activities. The aim is to analyze its operational applicability and to gain insights into its practicability. Attention will be paid to the provision of data quanta in line with demand-oriented requirements, technical implementation and acceptance by users.

Acknowledgement. Parts of this work were supported by the Federal Ministry of Education and Research (BMBF) within the research project H2Giga – FertiRob.

References

1. Augustin, S.: Information als Wettbewerbsfaktor – Informationslogistik. TÜV Rheinland, Köln (1990)
2. Strahilov, A., Hämmerle, H.: Engineering workflow and software tool chains of automated production systems. In: Biffl, S., Lüder, A., Gerhard, D. (eds.) Multi-Disciplinary Engineering for Cyber-Physical Production Systems, pp. 207–234. Springer, Cham (2017). https://doi.org/10.1007/978-3-319-56345-9_9

3. Lüder, A., Baumann, L., Behnert, A.-K., Rinker, F., Biffl, S.: Paving pathways for digitalization in engineering - common concepts in engineering chains. In: 25th IEEE International Conference on Emerging Technologies and Factory Automation (ETFA), pp. 1401–1404, ETFA, Vienna, Austria (2020)
4. Hanksche, I.: Digitaler Wandel- Lean & Systematisch. Springer Vieweg, Wiesbaden (2021)
5. Hildebrand, K., Gebauer, M., Mielke, M.: Daten- und Informationsqualität – Die Grundlage der Digitalisierung. Springer Vieweg, Wiesbaden (2021)
6. Aamodt, A., Nygård, M.: Different roles and mutual dependencies of data, information, and knowledge - an AI perspective on their integration. North-Holland Elsevier, Data Knowl. Eng. **16**, 191–222 (1995)
7. Dippold, R., Meier, A., Schnider, W., Schwinn, K.: Unternehmensweites Datenmanagement – Von der Datenbankadministration bis zum Informationsmanagement. Springer Vieweg, Wiesbaden (2005)
8. Semantische Allianz für Industrie 4.0.: https://www.hsu-hh.de/aut/forschung/forschungsthemen/semantische-allianz-fuer-industrie-4-0. Accessed 06 June 2023
9. Eclass. https://eclass.eu/. Accessed 06 June 2023
10. Felkai, R., Beiderwieden, A.: Projektmanagement für technische Projekte – Ein Leitfaden für Studium und Beruf. Springer, Heidelberg Dordrecht London New York (2015)
11. Ohno, T.: Toyota Production System – Beyond Large-Scale Production. Productivity Press, New York (1988)
12. Womack, J.P., Jones, and D. T.: Lean Thinking – Ballast abwerfen, Unternehmensgewinne steigern. Campus, Frankfurt am Main (2013)
13. Liker, J.K., Braun, A.: Der Toyota Weg - 14 Managementprinzipien des weltweit erfolgreichsten Automobilkonzerns. FBV, München (2014)
14. Bertagnolli, F.: Lean Management – Introduction and In-Depth Study of Japanese Management Philosophy. Springer, Wiesbaden (2022)
15. Karch, S., et al.: Lean Engineering – Identifying waste in engineering chains. In: 56th CIRP Conference on Manufacturing Systems, CIRP CMS, South Africa (2023). accepted
16. Biffl, S., Lüder, A., Gerhard, D.: Multi-disciplinary Engineering for Cyber-Physical Production Systems: Data Models and Software Solutions for Handling Complex Engineering Projects. Springer, Cham (2017)
17. Mehr, R., Lüder, A.: Managing Complexity within the Engineering of Product and Production Systems. In Biffl, S., Eckhart, M., Lüder, A., Weippl, E. (Eds.): Security and Quality in Cyber-Physical Systems Engineering, pp. 57–79. Springer, Cham (2019)https://doi.org/10.1007/978-3-030-25312-7_3
18. Paetzold, K.: Product and systems engineering/CA* tool chains. In: Biffl, S., Lüder, A., Gerhard, D. (eds.) Multi-Disciplinary Engineering for Cyber-Physical Production Systems, pp. 27–62. Springer, Cham (2017). https://doi.org/10.1007/978-3-319-56345-9_2
19. ISO/TR 18828: Industrial automation systems and integration - Standardized procedures for production systems engineering, Series of 5 parts (2018–2020)
20. VDI/VDE 2221–2019: Design of technical products and systems - Model of product design. Beuth Verlag (2019)
21. VDI/VDE 2206–2021: Development of mechatronic and cyber-physical systems. Beuth Verlag (2021)
22. VDI/VDE 4499 2008–2022: Digital factory. Beuth Verlag. 5 parts (2008–2022)
23. Drath, R., Barth, M., Fay, A.: Offenheitsmetrik für Engineering-Werkzeuge. Atp Edition, **54**(09), 46–55 (2012)
24. Behnert, A.-K., Rinker, F., Luder, A., Biffl, S.: Migrating engineering tools towards an automationml-based engineering pipeline. In: IEEE 19th International Conference on Industrial Informatics (INDIN), pp. 1–7 (2021)

25. Biffl, S., et al.: Experiences with Technical Debt in Parallel Multi-Disciplinary Systems Engineering Christian Doppler Laboratory for Security and Quality Improvement in the Production System Lifecycle. Technical Report (2019)
26. Fink, A., Schneidereit, G., Voß, S.: Grundlagen der Wirtschaftsinformatik. Physica-Verlag, Heidelberg (2005)
27. Wieringa, R. J.: Design Science Methodology for Information Systems and Software Engineering.Springer, Berlin Heidelberg (2014). https://doi.org/10.1007/978-3-662-43839-8
28. Johannesson, P., Perjons, E.: An Introduction to Design Science. Springer, Cham (2014)
29. Mayring, P.: Qualitative Inhaltsanalyse – Grundlagen und Techniken. Beltz, Weinheim (2010)
30. Schleipen, M., Mersch, T., Karch, S.: Engineering und Inbetriebnahme von Produktionsanlagen – Eine Analyse der beteiligten Prozesse und des entstehenden Datenflusses. Atp Magazin (2023) – accepted
31. GDPR. https://eur-lex.europa.eu/eli/reg/2016/679/oj. Accessed 25 Jul 2023
32. RFC Standard 2818. https://datatracker.ietf.org/doc/html/rfc2818. Accessed 25 Jul 2023
33. RFC Standard 2616. https://www.ietf.org/rfc/rfc2616.txt. Accessed 25 Jul 2023
34. RFC Standard 6238. https://datatracker.ietf.org/doc/html/rfc6238. Accessed 25 Jul 2023

Evaluating CAD Usage to Generate Synthetic Images for Machine Learning-Based Object Detection

T. Schmelter(✉), T. Nowak, M. Knott, and B. Kuhlenkötter

Ruhr-University Bochum, Universitaetsstr. 150, 44801 Bochum, Germany
schmelter@lps.rub.de

Abstract. When implementing machine learning (ML) methods, for example, in the context of quality control in production, a major concern for small and medium-sized enterprises (SMEs) is the effort, time, and thus cost required to train a powerful ML model. Since nowadays most manufacturers already have computer-aided design (CAD) models of their products, using these for training an ML algorithm could be beneficial. In this paper, the performance of the YOLO Object Detection Algorithm is evaluated on images directly rendered from a CAD model in Autodesk Inventor and real-world images of the same product. The main goal is to reduce the number of real-world images needed to train the model using CAD-Renderings. This is especially useful before the start of production since real-world images are not available at that point. This will continue to enable faster adaptation to new products and thus drive digitization in companies.

Keywords: CAD Rendering · Artificial Intelligence · Machine Learning · Object Detection · Quality Control

1 Introduction

In the context of advancing digitalization - especially among small and medium-sized enterprises (SMEs) in Germany - the use of artificial intelligence (AI) is becoming increasingly attractive for many to be able to withstand the strong international competitive pressure [1]. This is further promoted by the constantly growing amount of generated data in the companies themselves and machine learning (ML) models – as a subcategory of AI - made available online as open-source software. This offers the advantage that the actual effort of programming can be reduced to a minimum and existing ML models only have to be trained with their company-specific data, which reduces the required know-how in the company. Thus, the application and targeted use of AI are becoming possible for more and more companies [2].

The greatest obstacles to AI usage are the high implementation costs, the lack of expertise in the company, and the generation of the company-specific data basis for training the AI. The generation of this data basis poses many challenges and requires a lot of time and knowledge about the required quality of the data, which reduces the economic efficiency and thus the potential of AI for companies. Especially in the field of image-processing AI, the automated generation of the datasets is particularly complicated and very costly due to the enormous time commitment of an employee to generate

T. Bauernhansl et al. (Eds.): WGP 2023, LNPE, pp. 14–23, 2024.
https://doi.org/10.1007/978-3-031-47394-4_2

a sufficiently large data basis [3]. To lower the inhibition thresholds and reduce the time required, which makes the use of an image-processing AI more economical, the question of using synthetic images to train an AI is coming to the fore, especially since most companies in the manufacturing industry already have their data in digital CAD files.

The goal is to define a procedure that enables SMEs to implement image-processing AI for use in quality control at a low cost and without a great deal of know-how. With the ability to analyze data in real time, AI and ML algorithms are helping manufacturers identify quality issues early, reduce waste, and improve customer satisfaction.

In this study, it will be examined to what extent the construction of an image-processing AI can be reduced by using an open-source network and generating synthetic training images automatically in a commercially available CAD program (Autodesk Inventor). Based on this, the YOLOv7 object detection algorithm is trained and subsequently verified how well the AI correctly classifies components based on the synthetic images in real-world applications. In addition, the training is repeated with a proportional use of synthetic and real images. Thus, the quality of the classification of the AI in real use is compared between different mixing ratios of real to synthetic images to evaluate a proportional reduction of the workload in terms of the results quality.

2 Related Work

The literature has already found approaches for training image recognition ML algorithms with synthetic data. SALAS et al. trained a Convolutional Neural Network (CNN) with virtual images generated from a 3D model in OpenGL. Doing so, they were able to generate the data faster than in conventional ways, but according to their statements they only achieved "similar results to training with few real data" [4]. BLOCK et al. on the other hand, use a small amount of real data in front of a green screen to insert different backgrounds with an image bot and thus virtually extend the data basis [5]. CAD data was used by PLANCHE et al. whereas the focus in this work was on the generation and classification of 3D models [6]. The possibility of generating synthetic images from a CAD program and training a ML model with different mixing ratios to create an optimal ratio of synthetic to real images with a simultaneously high mean Average Precision (mAP) for the simplest possible setup of an optical quality control of components has not been investigated yet.

2.1 Artificial Intelligence and Machine Learning

The German Research Center for Artificial Intelligence defines AI as a property of an IT system to create approximations of a human brain by perceiving, understanding, acting, and learning and thus exhibiting human-like, intelligent behavior. Today's AI applications can handle specific, predefined problems based on computer science and mathematics [7].

ML is a subfield of AI that involves training algorithms to make predictions or decisions based on patterns found in data. It is a process of teaching a computer to recognize and learn from patterns in data, without being explicitly programmed. In ML, models are built and trained on large datasets and then used to make predictions or take actions on new, unseen data. Each dataset is usually divided into different subsets to prevent training and validating on the same data. The focus of this work is the process of supervised learning. In contrast to the processes of unsupervised learning and reinforcement learning, the datasets of the training are classified in advance, which allows the AI to assign its decision to a predefined class. This decision is based on a calculated probability [8]. In this paper, the YOLOv7 algorithm is used.

2.2 YOLOv7

The YOLO algorithm is a popular object detection algorithm that can detect and localize objects in an image or video in real time, using bounding boxes. Such a bounding box limits the minimum rectangular area in an image in which the object to be detected is located. YOLO works by dividing an input image into a grid of cells, and for each cell, it predicts the bounding box and class probabilities of the objects present in that cell. The algorithm uses a single neural network to process the entire image, rather than using multiple networks at different scales like some other object detection algorithms [9, 10], hence the name YouOnlyLookOnce (YOLO). First, it takes an input image and resizes it to a fixed size. The image is divided into a grid of cells for the grid division. Each cell is responsible for detecting objects within that cell. Afterward, the algorithm predicts one or more bounding boxes for each cell while each bounding box is represented by five parameters: the x and y coordinates of the center of the box, the width and height of the box, and the confidence score. Finally, it will assign a class to each bounding box by predicting the probability of each class of object that could be present in the box. YOLO will give the final bounding boxes and their corresponding class probabilities as an output [11].

For this paper, the seventh version of YOLO is used, which (at the time of running the experiments) was the best object detection algorithm in terms of speed and accuracy on the MS COCO dataset[1] [13], which makes it a practical choice for SMEs. The other motivation for choosing this specific algorithm is the complete release of the source code[2] and therefore providing an easy-to-use framework for training and testing [14].

2.3 Evaluation Metric: Mean Average Precision

To evaluate the results, the mAP metric is used. It measures the accuracy of an algorithm in detecting and localizing objects in an image or video. The mAP is computed based on the precision and recall of the algorithm's predictions. Further, we differentiate between mAP@0.5 and mAP@0.5:0.95.

The mAP@0.5 is a standard evaluation metric for object detection algorithms and is computed using a fixed threshold of 0.5 for the intersection over union (IoU) metric. The IoU metric is the ratio of the area of overlap between the predicted bounding box and the manually predefined ground-truth bounding box of an object in an image with the area of union of both boxes (s. Fig. 1).

If the IoU between the predicted and manually generated ground-truth bounding box is greater or equal to 0.5, the detection is considered a true positive for the mAP@0.5.

[1] Comparison dataset on which various ML models are tested and compared with each other.
[2] https://github.com/WongKinYiu/yolov7.

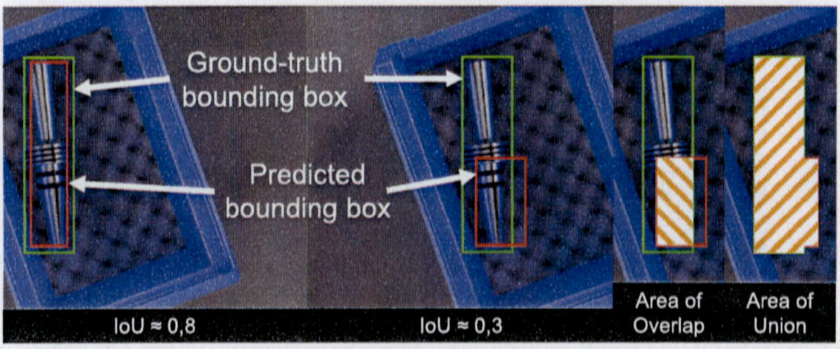

Fig. 1. Manually and automatically created boxes as well as resulting areas for the determination of the IoU

The mAP@0.5:0.95, on the other hand, uses a range of thresholds for the IoU metric, from 0.5 to 0.95, to evaluate the algorithm's accuracy. Consequently, a model needs to predict more accurate bounding boxes to increase mAP@0.5:0.95 making it more comprehensive than mAP@0.5 because it considers the variation in the IoU threshold used in different applications [15].

3 Generation of Synthetic Images (CAD-Renderings)

Autodesk Inventor Professional 2023, which includes an already implemented option for rendering with Inventor Studio, was used to generate synthetic images from CAD files. To generate images as realistic as possible a focus is on the lighting, size of the representation, proportions, and selection of materials. Further parameters were the gloss, reflectivity, and self-exposure of the materials. In Fig. 2 a real image is juxtaposed with a synthetic image of a sample product.

Fig. 2. Comparison between a synthetic image (left) and a real image (right) of the UniLokk in a visual storage box on soft foam for a product-friendly transport

As an application example, an optical quality control with object detection for a bottle cap, the so-called "UniLokk" is examined [16]. The cap can be manufactured in various configurations, whereby in this case the focus is on the colors of the rubber rings as a variant and detecting the product itself. During production and assembly, the Unilokk is transported in visual storage boxes. The different ring colors are shown in Fig. 3 (synthetic images).

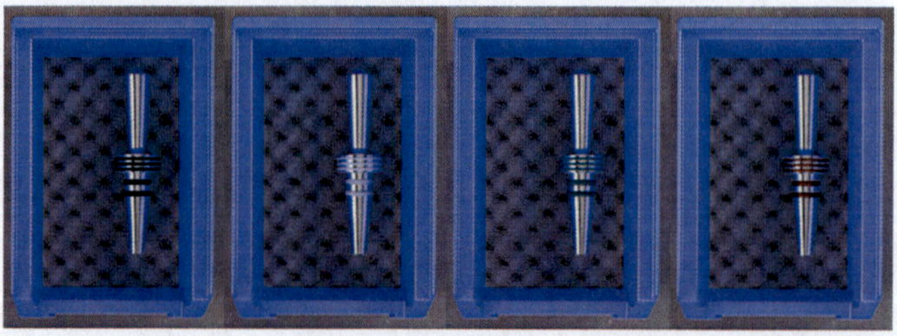

Fig. 3. Display of the different ring colors/ variants of the UniLokk

The goal of using synthetic images is to reduce the amount of effort needed to create a sufficiently large dataset, which is why the generation of synthetic images by rendering was automated in the software. To obtain as many different positions of the component as possible, the camera moves in a meandering pattern over the object and rotates after the traverse path has been completed. In addition, the component rotates and changes position within the transport box. This causes the captured image and the UniLokk to shift, which is equivalent to changing the position of the transport box under the camera. With this technique, any number of images with different positions of the component can be generated within the scope of a rendering.

4 Experimental Setup

In total 2000 images for the real dataset and 2000 images for the synthetic dataset were collected. Both datasets use the same 9 classes which are evenly distributed over the 4 different colors (red, green, white, black). Every color can be split further into 2 classes (upper, lower) for the rings on the UniLokk resulting in 8 different classes. The last class is the UniLokk itself, which marks the entire bottle closure. The manually annotated ground truth bounding boxes and corresponding labels for the upper rings, lower rings, and UniLokk are shown for one example in Fig. 4.

Fig. 4. UniLokk with the marking of different labels for training

To test the ability to generalize from an AI trained with synthetic to real images, differently distributed datasets were created. For every dataset, a YOLO object detector with standard hyperparameters was trained and tested to see the effects of different real and synthetic images ratio. Table 1 gives an overview of the distribution of the images for each dataset.

Table 1. Image distributions of the different datasets

Name	Train (80%)		Valid (20%)	
	Real	Render	Real	Render
Real10	160	0	40	0
Real25	400	0	100	0
Real50	800	0	200	0
Real75	1200	0	300	0
Real90	1440	0	360	0
Real100	1600	0	400	0
Real90-Render10	1440	160	360	40
Real75-Render25	1200	400	300	100
Real50-Render50	800	800	200	200
Real25-Render75	400	1200	100	300
Real10-Render90	160	1440	40	360
Render100	0	1600	0	400

For actual use of the model in a real-world application, the performance on synthetic images does not matter. However, first, it must be evaluated whether the model learned from the data using the validation dataset which contains the same distribution of synthetic and real images. Figure 5 shows the learning process for selected mixture ratios, which show the largest differences over 70 epochs each.

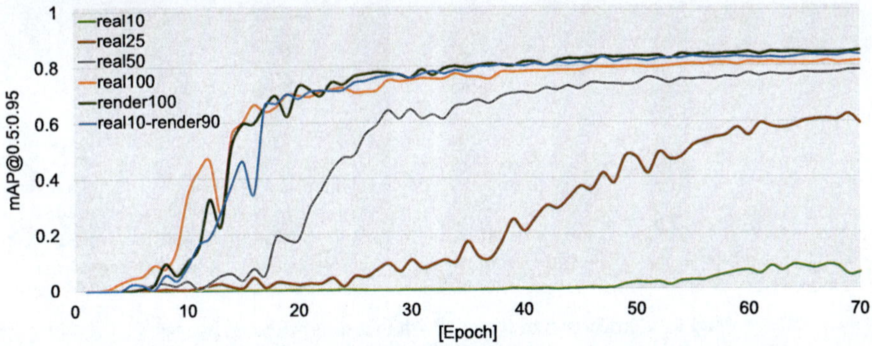

Fig. 5. Achieved mAP of the different distributions over 70 epochs of training

With only 10% of real images (160 images) the model cannot be trained sufficiently well, since only a mAP@0.5:0.95 value of approx. 0.1 is achieved and the first results of the training can only be seen after a period of almost 50 epochs. A clear improvement can be seen with an increase of the training data to 400 real images (real25) or 800 real images (real50). Not only are better values achieved for the mAP, but also the training starts to achieve good values from an earlier epoch. The datasets real100, render100, and mixed-real10-render90, for which 2000 images in total were used in each case, all achieve an equally good mAP@0.5:0.95 of 0.85 at almost the same speed. It is also evident that the mAP@0.5:0.95 stagnates from approx. 40 epochs onwards and that the training could therefore have been terminated earlier. This would have worsened the result only minimally but would have saved training time. Especially in an economic context, fast training and the resulting time savings are particularly important. Thus, it can be stated that the models, which were trained with 2000 samples each, were trained as required and their functionality was verified/ proven.

Since the models so far have only been tested for their functionality with images of their mixture of real and synthetic images, they must be tested with exclusively real images. From this, it can be evaluated to what extent the models, which were trained proportionally with synthetic images, also show correspondingly good decision-making reliability with real images - as they are taken in practice.

The following Fig. 6 shows the mAP of the different trained models testing real images[3]. The shaded area shows the difference between mAP@0.5 (higher) and mAP@0.5:0.95 (lower). To avoid visual clutter only the most relevant experiments are shown. First, the mAP@0.5 turns out better in all cases, which is due to its definition, whereby it allows a higher deviation than the mAP@0.5:0.95. Ultimately, the desired accuracy must be defined for the respective use case before the implementation in the company. In addition, it must be noted that a comparison is drawn here with the manually created bounding boxes, which can also contain deviations and are not drawn 100% accurately.

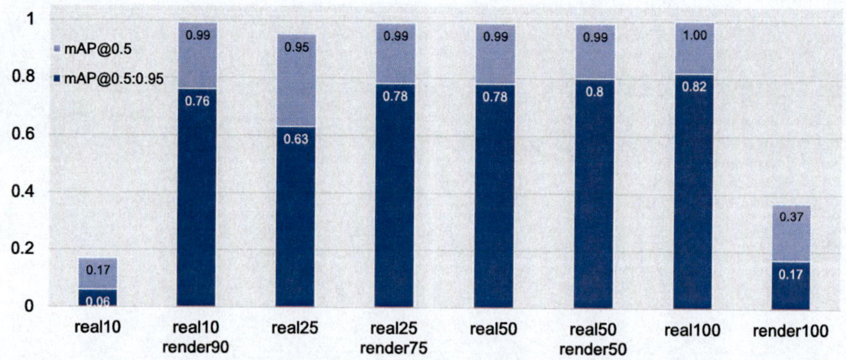

Fig. 6. Comparison of the mAP while testing with real images

[3] Evaluated with 'confidence-threshold' of 0.001 and 'IoU-threshold' of 0.65 for non-max suppression.

The results of the models, which were trained exclusively with real images, match with the results from the training, which are shown in Fig. 5 since these were trained - due to the data basis - exclusively with real images. The model which was trained with 100% synthetic images shows an insufficient result with a mAP of 0.12 and 0.37. However, it is remarkable that already with an addition of 10% of real images to the synthetic images, a mAP@0.5:0.95 of 0.76 can be achieved, which is close to the mAP@0.5:0.95 of the models, which were trained with real images only. Under all conditions, the model trained with 100% real images shows the best results with 0.82 and 1.00. If, however, the accuracy of mAP@0.5 is sufficient for the respective application, all models which were trained with a 100% amount of data – except for the one with 100% synthetic images - show a very good result of 0.99 to 1.00. Since the mAP refers to the adjustment of the bounding box, in the following the confusion matrix is investigated further, which refers to the correct assignment of the classes - and thus to the correct detection. In Fig. 7 four matrices are compared.

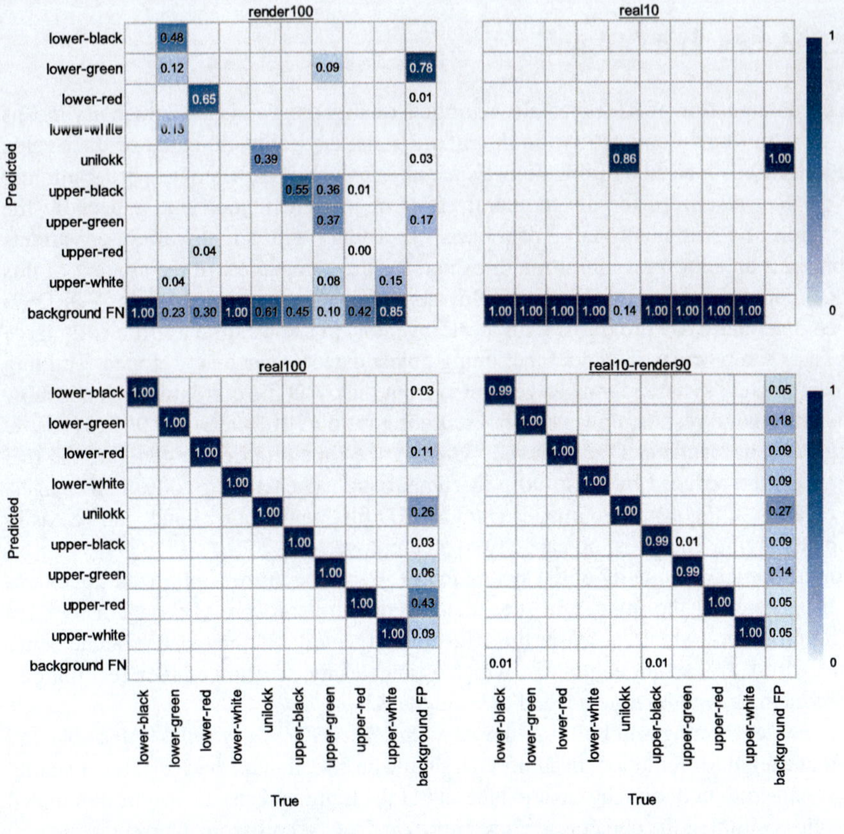

Fig. 7. Comparison of the confusion matrix of different mixing ratios

In a model trained with 100% synthetic images (render100), only a very small percentage is assigned to the correct class. The majority is not classified correctly and is therefore assigned to the background class as a false negative. This means that no UniLokk detected, although one was present. This effect can be seen even more clearly in the upper right corner of the model, which was trained with only 10% real images (real10). Here the database was too small, as can be seen in Fig. 7, which is why almost no classes were recognized at all and almost all values were classified as false negatives in the background class.

In the two lower matrices, a clear trend in the diagonal can be seen, as should be the case in a well-functioning model. The predominant deviations, in this case, are in the false positive class concerning the background, suggesting that the model has detected classes at locations where there are no UniLokks or colored rings. In the mixture of 90% synthetic and 10% real images, a slightly worse result can be seen in comparison to the 100% real images, since there was a mix-up of green and black rings in a proportion of 0.01 and in case of isolated black rings these were not correctly recognized and assigned.

5 Summary and Outlook

The implementation of AI, especially in optical quality control, is an elementary means for SMEs to remain competitive in the international environment, enhance their value creation and advance their digitization. Especially the lack of knowledge in programming and the low resources for the temporal effort of implementation and especially the generation of a sufficiently large data basis - especially with a high variety of variants - represents an enormous inhibition threshold for the companies. In the context of this work, it could be proven that by the employment of open-source ML models (YOLOv7) and commercial CAD programs (Autodesk Inventor) in a short time a sufficiently large data basis can be created, and a functioning object detector can be developed. Training with completely synthetic images generated from the CAD file currently does not show sufficiently good results, however, with a mixing ratio of 10% real and 90% synthetic images decent results can be achieved, whereby the time required to generate the real images can be reduced by up to 90% in comparison to generating 100% real images. The creation of the synthetic images from a CAD file, on the other hand, can be easily automated, tying up almost no capacity of an employee.

In the sequel, the quality of the synthetic images can be further optimized, for example, by generating the images in specialized rendering software. This again requires several programs, which increases the effort in this regard, but may also produce better results with 100% synthetic images, which eliminates the creation of 10% real images, thus reducing again the time needed to build the AI.

Likewise, training and building the AI with YOLOv7 is very quickly feasible and provides a reliable prediction. In addition, it should be investigated to what extent creating the ground-truth bounding boxes and labeling of the training data can also be automated since the position of the object is already known in the CAD program. Currently, most of the time needed to build the AI according to the procedure presented here is for manually labeling the images.

References

1. Seifert, I., et al.: Potentiale der künstlichen Intelligenz im produzierenden Gewerbe in Deutschland (2018). https://www.bmwk.de/Redaktion/DE/Publikationen/Studien/potenziale-kuenstlichen-intelligenz-im-produzierenden-gewerbe-in-deutschland.html
2. Rammer, C., Kompetenzen und Kooperationen zu Künstlicher Intelligenz: Ergebnisse einer Befragung von KI-aktiven Unternehmen in Deutschland (2022). https://www.de.digital/DIGITAL/Redaktion/DE/Digitalisierungsindex/Publikationen/publikation-download-ki-kompetenzen.pdf?__blob=publicationFile&v=1
3. Dukino, C., et al: Künstliche Intelligenz in der Unternehmenspraxis (2019)
4. Salas, A.J.C., Meza-Lovon, G., Fernández, M.E.L., Raposo, A.: Training with synthetic images for object detection and segmentation in real machinery images. In: 2020 33rd SIBGRAPI Conference on Graphics, Patterns and Images (SIBGRAPI), pp. 226–233 (2020)
5. Block, L., Raiser, A., Schön, L., Braun, F., Riedel, O.: Image-bot: generating synthetic object detection datasets for small and medium-sized manufacturing companies. Procedia CIRP **107**, 434–439 (2022). https://doi.org/10.1016/j.procir.2022.05.004
6. Planche, B., et al.: DepthSynth: Real-Time Realistic Synthetic Data Generation from CAD Models for 2.5D Recognition (2017). https://arxiv.org/pdf/1702.08558
7. Bundesregierung, KI-Strategie Bundesregierung (2018). https://www.bundesregierung.de/resource/blob/997532/1550276/3f7d3c41c6e05695741273e78b8039f2/2018-11-15-ki-strategie-data.pdf
8. Kreutzer, R.T., Sirrenberg, M.: Understanding artificial intelligence: fundamentals, use cases and methods for a corporate AI journey. Springer, Cham , (2020). https://ebookcentral.proquest.com/lib/kxp/detail.action?docID=5905235
9. Girshick, R., Donahue, J., Darrell, T., Malik, J.: Rich feature hierarchies for accurate object detection and semantic segmentation (2013). https://arxiv.org/pdf/1311.2524
10. He, K., Zhang, X., Ren, S., Sun, J.: Spatial pyramid pooling in deep convolutional networks for visual recognition (2014). https://arxiv.org/pdf/1406.4729
11. Redmon, J., Divvala, S., Girshick, R., Farhadi, A.: You Only Look Once: Unified, Real-Time Object Detection (2015). https://arxiv.org/pdf/1506.02640
12. Zaidi, S.S.A., Ansari, M.S., Aslam, A., Kanwal, N., Asghar, M., Lee, B.: A survey of modern deep learning based object detection models. Digital Signal Process. **126**, 103514 (2022). https://doi.org/10.1016/j.dsp.2022.103514
13. Lin, T.-Y., et al.: Microsoft COCO: Common Objects in Context (2014). https://arxiv.org/pdf/1405.0312
14. Wang, C.-Y., Bochkovskiy, A., Liao, H.-Y.M.: YOLOv7: Trainable bag-of-freebies sets new state-of-the-art for real-time object detectors (2022). https://arxiv.org/pdf/2207.02696
15. Padilla, R., Passos, W.L., Dias, T.L.B., Netto, S.L., Da Silva, E.A.B.: A comparative analysis of object detection metrics with a companion open-source toolkit. Electronics **10**(3), 279 (2021). https://doi.org/10.3390/electronics10030279
16. Oberc, H., Fahle, S., Prinz, C., Kuhlenkötter, B.: A practical training approach in learning factories to make artificial intelligence tangible. Procedia CIRP **93**, 467–472 (2020). https://doi.org/10.1016/j.procir.2020.04.074

Systematic Lightweight Design of Production Equipment with a Digital Toolchain

Johannes Scholz(✉), Simon Zeidler, Florian Koessler, and Jürgen Fleischer

wbk Institute of Production Science, Karlsruhe Institute of Technology (KIT), 76131 Karlsruhe, Germany

johannes.scholz@kit.edu

Abstract. The importance of sustainable production is growing all the time. Lightweight design of production equipment is a possible measure to increase the sustainability. These potentials are often not used to their full extent due to a high focus on investment costs. To resolve this target conflict, this paper presents the lightweight design of production equipment based on a systematic development process and the corresponding digital toolchain. It includes a newly lightweight design technology planning tool to optimize cost, CO_2 and weight from a production perspective. Based on a rough description of the component and previously modelled requirements an optimized shape, production process and material are recommended. Subsequently, the component can be further detailed with this knowledge. Through the digital toolchain the effects of different designs on the usage of the production equipment can be easily estimated. This enables the developer to identify the potential of lighter but more expensive components.

Keywords: Lightweight design · digitalization · sustainable production

1 Introduction

Legal requirements for companies to report on sustainability indicators increase the importance of sustainable production. This requires sustainable machines and equipment in the production. To achieve this, the machines and systems must be designed to be both energy-efficient in operation and able to be produced in a resource saving manner. This leads to an increasing pressure for the machine and production equipment industry to improve the sustainability. Like in the automotive or aerospace industry where additional costs are accepted for lighter and more efficient products lightweight design has great potential to increase the sustainability. The realization of this potential is limited by the high time pressure in the engineer-to-order market and the customer's unwillingness to pay more for lighter products. To solve the target conflict between costs and a reduced weight, to reduce the CO_2-Footprint of the product, the integration of the production and the material in the early design process as well as the usage of a continuous digital chain for the development is necessary. From the production perspective the center of the digital toolchain is a technology chain planning with focus on the evaluation of cost, CO_2 and weight. In Sect. 2, existing development approaches

T. Bauernhansl et al. (Eds.): WGP 2023, LNPE, pp. 24–33, 2024.
https://doi.org/10.1007/978-3-031-47394-4_3

for an early integration of the production are presented. In Sect. 3, the developed digital toolchain for lightweight design optimized products is presented. Section 4 presents the application of the methodology to production equipment. At the end, a conclusion as well as an outlook is given.

2 Related Work

The development approach of the VDI 2221 [1] is a commonly used development process. It describes a systematic process for problem solving during product development and supports in structuring the development task. However, the development process integrates the production and the material late in the development process. This lead to expensive iterations and loss of time. Furthermore, this approach does not provide any specific methods for lightweight design. Therefore, several lightweight specific approaches were developed [1]. Klein [2] and Krause [3] developed a development process based on the VDI 2221 for lightweight design. During the requirements recording a guideline regarding the lightweight important loads is given. Additionally, lightweight specific optimization methods like topology optimization are integrated in the process [2, 3]. However, the approach does not support an early integration of material and production in the product development process. The development process by Hufenbach and Helms [4] supports an intensive interaction between product, production and material during the whole process and points out the importance of iteration during development of lightweight products. They integrated the interdependencies of product, manufacturing, joining technology and material in the development process and demonstrate the importance of an early integration of material and production in the development process [4]. The weakness of the process is the limitation to fiber-reinforced plastics instead of considering also other materials, production processes and product designs regarding CO_2, costs and resulting product weight. Furthermore, the process focusses on one part and is not considering the interdependencies between components.

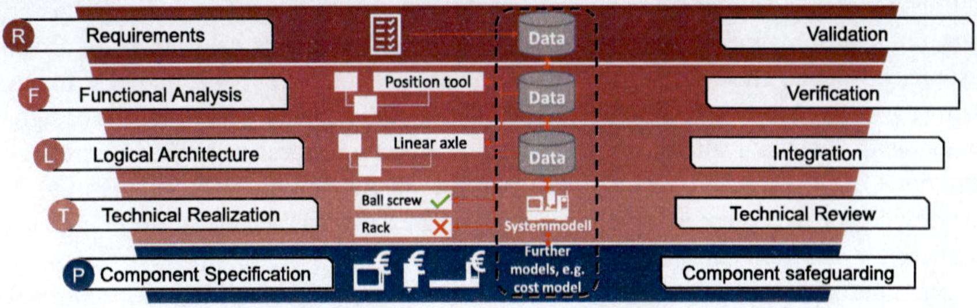

Fig. 1. MBSE approach for the use in production environments

For this purpose, a systematic lightweight development process was developed [5, 6]. The process is based on the concept of Model-Based Systems Engineering (MBSE). After an initialization with the definition of a target system that contains the goals for cost weight and CO_2, requirements are defined. In general, based on the requirements (R), firstly a functional view (F) on the system is build up. This is followed by a logical view (L) and a detailed physical view (P). The authors already extended this by a technical view (T) as the gap between L and P was found to be too significant (Fig. 1). The structure follows the well-known scheme of the V-Model [7] that gets extended by production, material and joining technology which are included along all development levels. The level of detail is increased along the process. The necessary iterations demand the process to being non time-serial while still maintaining the build-up of the development levels. Subsequently, system and process integration follow up before an acceptance test is carried out. For decisions along the development process, it is crucial to provide the necessary data regarding costs, CO_2 and weight and connect it to the product description in the several views.

For optimizing products from a production point of view early in the development process, technology chain planning methods are well known. Klocke et al. [8] define a technology chain as a machine independent combination of manufacturing technologies or manufacturing processes in a defined order for manufacturing a product. For generating technology chains, technologies are described with e.g., manufacturable materials, manufacturable tolerances or number of parts. This technology parameters are compared with product features like required tolerances or complexities which enables selecting suitable manufacturing technologies. The technologies are evaluated regarding the fulfillment of the product requirements. Afterwards, material, geometry or manufacturing technology are changed to improve the evaluation. Based on the fulfillment of product requirements after a manufacturing technology, further manufacturing technologies are added and alternatives of technology chains are generated [8]. This method has its limitations in not supporting the iterative design and in evaluating the manufacturing technology chains regarding factors like costs or CO_2. However, the methodology clearly shows the need of a rough description of product geometry to analyzing interdependencies between material and production. For this reason, the technology chain planning should be included in the technical view.

For handling the target conflicts of product function, costs and other criteria several approaches integrated multi-criterial decision making in the technology chain planning. For example, Müller [9] generates technology chains considering material parameters, number of manufactured parts, flexibility as well as ecological criteria for already designed products. The evaluation of this parameters requires a suitable database for analyzing the manufacturing technologies as well as a high degree of detail for the designed components. For the evaluation the Analytic Hierarchy Process (AHP) combined with Technique of Order Preference Similarity to the Ideal Solution (TOPSIS) is used [9]. These two approaches are used in several approaches to find the best solution in case of conflicting evaluation parameters.

The approaches of Rey et al. [10] and Jacob et al. [11] identified the importance of iterative product design in combination with technology chain planning. Both of the approaches are able to deal with uncertain component parameters and do not require a fully designed component [10, 11]. Rey et al. considers classic manufacturing technologies in their approach. With Jacob et al. it is possible to generate hybrid technology chains with additive and subtractive manufacturing processes. Through the flexibility of additive manufacturing hybrid technology chains also show great potential for lightweight design. Furthermore, this approach uses a structured Unified Modeling Language (UML) - based requirements description model which shows great potential for integration in a systematic development process [10, 11].

All of the presented technology chain planning approaches show great opportunities for integrating the production in the early design process. However, each of the approaches are just able to optimize one component based on the requirements. The approaches do not consider interdependencies between components in case of weight changes of other components which effect the requirements on the optimized component. Furthermore, CO_2-emissions are not considered and the presented approaches are not integrated in an overall development approach which is continuously digitized. This leads to limitations in optimizing a system regarding cost, CO_2 and weight over its whole lifecycle.

3 Digital Toolchain and Technology Chain Planning

To address this limitation, a variety of methods has been implemented to support the systematic development approach. Starting point is a target system where development goals are set. These development goals for example are restrictions for the product e.g., costs and weight, as well as lightweight or sustainability specific requirements that allow the derivation of requirements of product, production and material. The target system sets the overall goal of the development for the whole product and weights the importance of the target figures costs, CO_2 and weight. After this strategy has been specified, requirements are derived and noted in the R-view. Here a functional hierarchy, a functional model and a minimal design propose possible methods that can be applied regarding the structure of the product. Existing reference products or previous product generations can be analysed via a modified function-mass-analysis to identify optimization potentials [12]. Based on the results components with great potential can be prioritized in the further development stages.

The effects of interdependencies between product design, production, material and joining technology can be analysed based on the knowledge from previous development cycles to detect compatibilities of functions and material classes. Knowing the functional level, the L-view can be started with the generation of possible solutions to the generated functional artefacts. The authors already proposed a lightweight creativity process [13] which proved suitable by embodying the iterative nature of product development. Through the breakdown of assemblies via taking a functional view along the paradigm of systems engineering, existing creativity techniques and evaluation methods for idea generation and evaluation have been analyzed and rethought from a lightweight and sustainable design perspective. The methods were tested on the later introduced use case from the field of robotics, which enabled the identification of the potential of LWCM for a lightweight and sustainable design. Following the creativity phase, more detail requiring methods find application. The T-view describes the rough dimensions of the solutions from the L-view. With the resulting rough geometry, a more detailed integration of the production is possible at this point. For this purpose, the later described technology chain planning for specific components was developed. Additionally, the design alternatives can be evaluated by one-dimensional simulation tools to identify the resulting energy consumption during usage.

For the implementation and the transition to the more detailed P-view, a lightweight specific construction catalogue and a method to quantify secondary effects of changes in a subsystem can be applied. In the P-view the level of detail of the solutions is very high. Therefore, very specific and specialized methods like multi criterial decision making, topology optimizations and multi body simulations can be carried out. Especially, during the topology optimization the information from the technology chain planning is used to set the design freedom and reducing the needed time. With this view exact numbers on cost, weight and (lifecycle) CO_2-emissions can be generated. Up until now, all these tools and methods coexisted and got applied uniquely for each task. Therefore, production was always integrated late into the product development process that took place. The reason for this was the strongly limited linkage between the methods and therefore the difficulty of creating a process chain. This showed the need for interfaces between the methods. These interfaces got addressed in the project SyProLei. Here, the whole process from the requirements up to the physical design was digitalized and linked continuously. The continuous link to production hereby represented a large difficulty. Especially the link to and integration of technology chains had not yet been created.

Digital continuity is essential for a flexible, robust and most important weight, CO_2 and production considering product development. The vast variety of methods and their different implementations as well as necessary requirements, inputs and outputs make the connection and continuity of the process difficult to achieve by means of the tools themselves. Therefore, an instance of coordination and data handling on a superior level is necessary. For this purpose, a Product Lifecycle Management (PLM) system like Siemens Teamcenter can be used [14]. In this environment, the systems modelling workbench (SMW) is used to model the functional and the logical view. These views are connected to the 1D-Simulationtool AMESIM, which enables the simulation of the behavior during usage. Furthermore, it is connected with the CAD-System Siemens NX for connecting the functions and logical elements with the geometry information. Connecting the information over the development cycle and saving them in a PLM system supports the knowledge transfer between product generations and the evaluation of solutions. With this extension, the MBSE-character can be implemented in Teamcenter as can be seen in Fig. 2. Furthermore, the definition of interfaces allows for an implementation of environment-foreign tools into the digital toolchain. The systematic development supports the engineer in connecting functions, logical elements and the resulting components. Furthermore, impacts of other components on the currently designed component can easily be identified.

To integrate the production into the digital toolchain, an extended technology chain planning approach was developed and is integrated in the technical design. The considered component is described by requirements regarding loads and tolerances and the needed functions. Through the MBSE approach product features can be generated from the knowledge in the PLM system. Furthermore, the manufacturing technologies and the materials are modelled in a database, which is modelled with an UML-Diagram in Fig. 3. The manufacturing technologies are described by their process costs, manufacturable geometry features and the manufacturable materials. Beside the manufacturable complexity possible specific lightweight design options for each manufacturing technology are stored. Possible options are solid, hollow, sandwich, stiffeners and freeform.

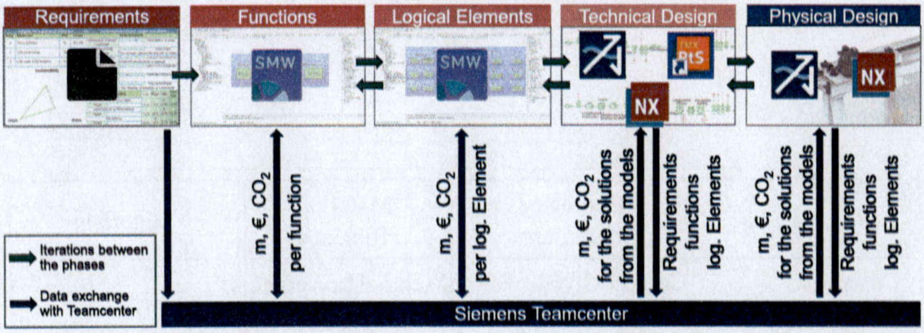

Fig. 2. Phases and methods and their implementation in the Siemens environment

Each of the materials are then described by the manufacturing process depending process energy for one kg material, the parameters of the material itself and information regarding the end-of-life to enable a evalaution over the whole lifecycle.

To generate technology chains, the component description is compared to the material and manufacturing database. Thus, suitable manufacturing technology chains are generated. These technology chains are evaluated regarding costs, resulting component weight, ecologic criteria and lightweight design potential

The costs consider the material costs and the manufacturing costs. The material costs are including the purchasing costs based on the required material mass as well as the recycling costs. These are calculated based on the required recycling energy from the database and a cost factor per energy unit.

The ecologic criteria consist of the CO_2-Emissions for raw material production, production process and the recycling of the component material as well as the material efficiency and the recycling rate of the material. The material efficiency depends on the manufacturing technology and the recycling rate on the material and can be directly taken from a database. The CO_2-Emissions are calculated based on the process energy, raw material energy and recycling energy multiplied by a CO_2-factor.

Additionally, lightweight design is evaluated. The indicators are the specific strength and the specific stiffness of the material as well as the lightweight design indicator of the material. This indicator describes the suitability of a material for different load cases with a value between one (bad) and five (good) [2].

Due to the different units of the indicators all of them are normalized to a value between one (bad) and five (good). The evaluation criteria are resulting in a target conflict. For this purpose, the TOPSIS algorithm is used for the evaluation of the solutions.

Furthermore, the construction options are assigned to specific load cases so that they will be selected based on the requirements. With this connection between manufacturing technology and construction principles, the engineer gets ideas how the component can be improved considering the optimized technology chain. Furthermore, the solutions are directly evaluated regarding their costs and CO_2-emissions. This helps identifying the potential of solutions.

After an initial generation of technology chains, the approach enables two optimization approaches. With the first one, an optimization can be started which varies product parameters like dimensions, material or technologies to find better solutions. The other

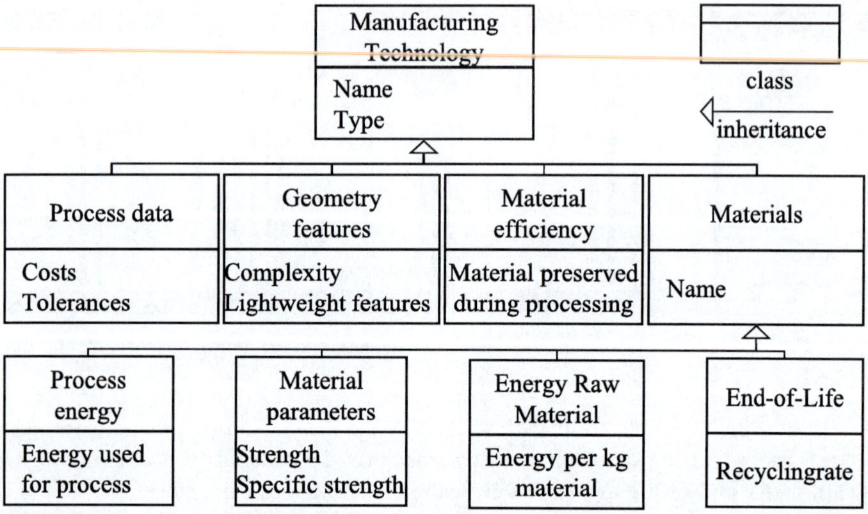

Fig. 3. Description of the manufacturing technologies and the materials for the technology chain planning

optimization approach enables the user to analyze how other components effect the costs, weight or CO_2-emissions of the component. For example, the loads applied to the component can be reduced by 5%, 10% or 15%. To achieve this reduction, changes to other components are required, but this approach allows effects to be identified even during manufacturing. This is enabled by the continuously digitalized toolchain.

The result of the technology chain planning is the optimized description of component dimensions and features as well as the material and the required manufacturing technologies. Furthermore, the component is described with its weight and its costs and CO_2-emissions during production and disposal. The information regarding dimensions and features can be transferred to a parametrized CAD-Model in Siemens NX which can be used for further development steps in the P-Phase like topology optimization.

4 Case Study

The system to be optimized in the case study for a further product generation is a portal robot designed to transport and load a tool between a tool storage and a machining center. Therefore, a robot equipped with a gripping system is mounted on a linear axis portal that spans from the furthest machine to the tool storage. The movement of the system demands energy and therefore produces CO_2-emissions. To minimize these emissions, a system with the optimum of production emissions and weight (emissions of time in use) should be targeted. As of now, the focus will be set on the robot and further downstream systems.

Initially the requirements were investigated and found to be covered fully. On this basis a functional model of the robotic system was build using the SMW. It was found, that the current product featured some parts that are not necessary by means of the functional model. Applying the FMA using the information stored in Siemens Teamcenter, it could be shown, that out of the in-house produced parts especially the gripping system introduces a lot of weight into the total system by requiring adapter flanges to fit onto the robot. As shown in Fig. 4, mass and CO_2 emissions of the function "Connect to the robot" should be decreased (red arrows), while costs can be increased (green arrows) until a target value (black line) is reached. This black line represents the relative amount of the costs permitted in relation to the relative importance of the respective functions.

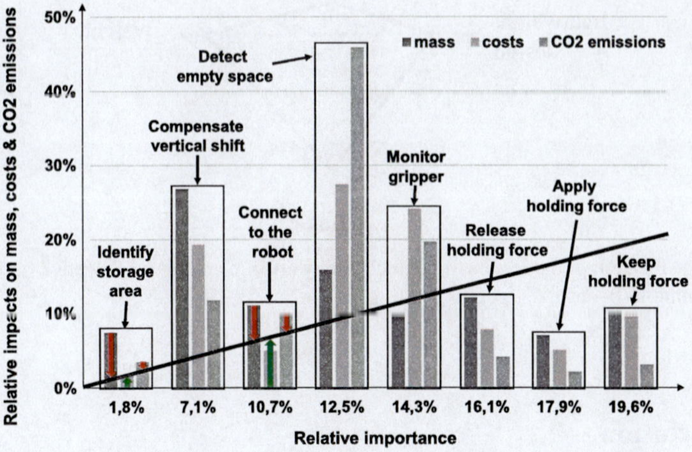

Fig. 4. Extended function mass analysis (example: portal robot) [6]

As a next step to optimize the system for minimal cost and emission, two paths could be followed. Firstly, new gripping ideas could be generated via creative processes. This has been done in [13]. Therefore, the focus of this paper lies more on the optimization of the identified parts and the effects of this process on the total system. As function to optimize "Connect to the robot" of the gripping system was chosen. This is realized with a flange. The requirements of the system are brought down to the specific component as well as the resulting requirements from related components like forces. Through experience knowledge it is identified based on the function and the requirements the number and position of drilling holes and the rough geometry. This information is translated to the requirement model for the flange which consists of number of holes, permissible installation space, required tolerances, applied forces, surface quality and hardness. Furthermore, a parametrized CAD-Model is generated with the important features, so the information of the technology chain can be easily translated to the CAD-model. With this information the design, material and technology chain are generated and evaluated as described in Sect. 3. For a better visualization of the strengths and weakness of different solutions a spider diagram like in Fig. 5 is used. For the presented solution in Fig. 5 the solution is very good regarding ecologic criteria as well as for the weight and the lightweight design. The solution lacks in costs. The selection of a component needs to be made with the weights of the target system. Based on the generated CAD-model, a topology optimization for example is carried out with considering the resulting design restrictions of the technology chain which leads to faster development times and better results.

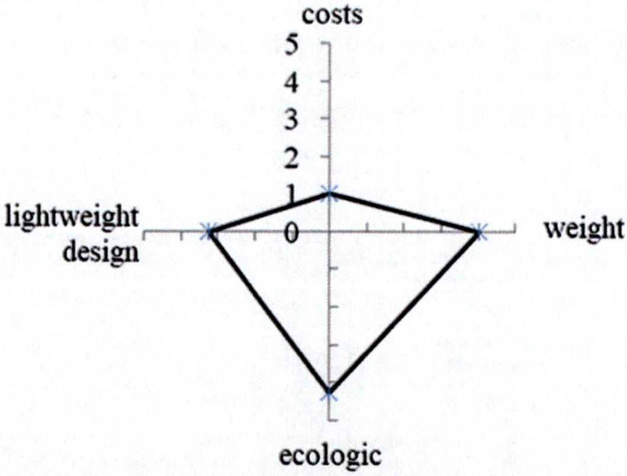

Fig. 5. Visualization of the results of the technology chain planning for Selective Laser Melting with Aluminium (0 – bad, 5 – good)

5 Conclusion

To overcome the target conflict of CO_2, costs and weight especially in industries with high cost pressure like the machine and plant engineering a digital toolchain which enables an early integration of the production in the development process. For this purpose, the paper shows a digital toolchain based on the Siemens Teamcenter with a technology chain planning which enables the user to analyze interdependencies between components and taking into account effects on the whole lifecycle of a product. To improve the technology chain planning in the digital toolchain it will be necessary to build up a database which includes a connection between product features and functions to support the early integration of the production.

References

1. Verein Deutscher Ingenieure (2019) VDI 2221 Part 1: Design of technical products and systems - Model of product design
2. Klein, B.: Leichtbau-Konstruktion. Springer, Wiesbaden (2013)
3. Krause, D.: Leichtbau. In: Rieg, F., Steinhilper, R. (eds.) Handbuch Konstruktion, pp. 463–484. Hanser, München (2012)
4. Hufenbach, W., Helms, O.: Zum methodischen Konstruieren von Leichtbaustrukturen aus kohlenstofffaserverstärkten Kunststoffen. Konstruktion – Zeitschrift für Produktentwicklung und Ingenieur-Werkstoffe 62:68–74 (2010)

5. Scholz, J., Kaspar, J., Quirin, S., et al.: Concept of a systemic development process to realize lightweight potentials. Zeitschrift für wirtschaftlichen Fabrikbetrieb **116**, 797–800 (2021). https://doi.org/10.1515/zwf-2021-0182
6. Kaspar, J., König, K., Scholz, J., et al.: SyProLei - A systematic product development process to exploit lightweight potentials while considering costs and CO2 emissions. Procedia CIRP **109**, 520–525 (2022). https://doi.org/10.1016/j.procir.2022.05.288
7. Verein Deutscher Ingenieure (2021) VDI/VDE 2206:2021–11 Development of mechatronic and cyber-physical systems
8. Klocke, F., Fallböhmer, M., Kopner, A., et al.: Methods and tools supporting modular process design. Robot. Comput. Integr. Manuf. **16**, 411–423 (2000). https://doi.org/10.1016/S0736-5845(00)00024-7
9. Müller, J.: Integrative Gestaltung von Fertigungsprozess- und Prüffolgen für sicherheitskritische Bauteile. Dissertation (2018)
10. Rey, J., Apelt, S., Trauth, D., et al.: Highly iterative technology planning: processing of information uncertainties in the planning of manufacturing technologies. Prod. Eng. Res. Devel. **13**, 361–371 (2019). https://doi.org/10.1007/s11740-019-00882-7
11. Jacob, A., Windhuber, K., Ranke, D., et al.: Planning, evaluation and optimization of product design and manufacturing technology chains for new product and production technologies on the example of additive manufacturing. Procedia CIRP **70**, 108–113 (2018). https://doi.org/10.1016/j.procir.2018.02.049
12. Albers, A., Revfi, S., Spadinger, M.: Extended target weighing approach - Identification of lightweight design potential for new product generations. In: Maier A, Kim H, Oehmen J et al. (eds) Design methods and tools. Curran Associates Inc, Red Hook, NY (2018)
13. König, K., Zeidler, S., Walter, R., et al.: Lightweight creativity methods for idea generation and evaluation in the conceptual phase of lightweight and sustainable design
14. Kleiner, S., Mai, K., Kaspar, J., et al.: SyProLei – Systematischer Produktentwicklungsprozess zur Hebung von Leichtbau-Potentialen. In: Wilke D, Koch W, Kaffenberger R et al. (eds) Tag des Systems Engineering 2022: Tagungsband Paderborn, 16.-18. November 2022, 1. Auflage. Gesellschaft für Systems Engineering, Bremen (2022)

Collaborative Tasks in Construction: A Model for Human-Exoskeleton Interaction to Minimize Muscle Exertion

E. Bances[1(✉)], U. Schneider[1,2], B. Garcia[1], J. Siegert[1], and T. Bauernhansl[1,2]

[1] University of Stuttgart - IFF, 70569 Stuttgart, Germany
nelson-enrique.bances-purizaca@iff.uni-stuttgart.de
[2] Fraunhofer IPA, 70569 Stuttgart, Germany

Abstract. In the EU, 75% of buildings are energy inefficient, emphasizing the priority of their renovations or upgrades to reach the EU's decarbonization goals. However, renovating existing buildings frequently entails physically demanding tasks, which can result in worker injuries, reduced productivity, and health issues.

This document highlights the potential of using exoskeleton technology in collaborative tasks. These approaches' effectiveness was assessed using two methods: electromyography (EMG) to assess muscular activity levels and a real-time motion tracking system to observe object manipulation.

A muscle activation assessment was conducted on ten participants, analyzing four muscles across two distinct phases of the collaborative task. The results indicate that 55% of these cases displayed positive changes marked by a reduction in muscle activity when exoskeletons were employed. In 27.5% of the cases, no significant changes were observed; while 17.5% resulted in unfavorable outcome.

Keywords: Exoskeleton · Collaborative Task · Electromyography

1 Introduction

In 2020, the European Commission (EC) revealed that nearly 75% of existing EU buildings don't meet current energy performance standards. This is alarming because buildings contribute to around 40% of the EU's overall energy consumption and 36% of its energy-related greenhouse gas emissions [1]. Promoting renewables is vital for cutting emissions, but the most effective energy efficiency strategy is to reduce overall consumption [2]. The EC indicates that building upgrades could cut EU energy use by 5–6%, leading to a 5% drop in CO2 emissions. Despite renovation benefits, less than 1% of EU buildings are upgraded yearly [1]. Renovating existing structures is key to achieving a 55% emission cut by 2030 (vs. 1990) and becoming climate-neutral by 2050 [3].

While modern buildings are increasingly being manufactured modularly in factories with automatic machines, building renovation remains a labor-intensive process that heavily relies on manual techniques, requiring significant physical and cognitive effort from workers. Unfortunately, a shortage of skilled construction workers has been reported in 15 European countries [4]. The European Centre for the Development

T. Bauernhansl et al. (Eds.): WGP 2023, LNPE, pp. 34–43, 2024.
https://doi.org/10.1007/978-3-031-47394-4_4

of Vocational Training (CEDEFOP) estimates that Germany alone will need roughly 200,000 new and replacement workers by 2035 [5]. This shortage of skilled workers could pose substantial challenges to the timely and effective renovation of existing buildings. Furthermore, the physical demands of construction, combined with the high prevalence rates of musculoskeletal disorders, can affect the ability of workers to sustain work until retirement age [6].

1.1 Opportunities and Challenges of Exoskeleton and Other Technologies in Construction

Exoskeletons are designed to augment and enhance the physical capabilities of workers engaged in physically demanding industries. By reducing the physical exertion needed to perform tasks, exoskeletons enable workers to complete their jobs with reduced fatigue and a lower risk of musculoskeletal injuries [7].

Recent studies have highlighted the potential benefits that exoskeleton technology can bring to the construction sector [8]. For instance, experts have recommended using exoskeletons with back and shoulder support for tasks like brick masonry or drywall installation. These exoskeletons are designed to effectively address challenges such as repetitive forward bending, uncomfortable positions for the back and shoulders, and heavy lifting [10]. Nevertheless, there are hurdles to overcome in order to facilitate broader adoption and enhance the usability of exoskeletons in construction. One major challenge is ensuring adequate motion flexibility while wearing exoskeletons and addressing concerns related to their bulky size [9]. These factors can significantly impact the comfort and overall effectiveness of exoskeletons for workers on construction sites. Therefore, it is crucial to provide workers with a training period to familiarize themselves with the specific exoskeleton model used [11].

Other technologies like virtual and augmented reality made significant strides in the construction industry. A recent review has found that numerous augmented reality (AR) systems have been proven as efficient, usable, applicable, and accurate approaches in hazard identification, and safety training. However, it is important to acknowledge the challenges and limitations associated with implementing AR in construction. One notable challenge is the limited field of view, which may affect the visualization of crucial information within the work environment. Additionally, occlusion is another challenge where physical objects obstruct the view of virtual elements, potentially affecting the accuracy and usability of AR systems [12].

Finally, combining these technologies could present potential benefits in construction. Nevertheless, limited research has been conducted on the integration of AR with exoskeletons in real industrial tasks. Some studies have explored the use of AR and a passive exoskeleton in related fields, such as ergonomic shoulder control for video game applications, demonstrating promising results in reducing muscle activity and discomfort during experiments using AR in static positions [13].

2 Method

The study focuses on exploring exoskeletons' potential to enhance worker safety and improve coordination during physically demanding collaborative tasks and object manipulation. In particular, the study aims to assess the effectiveness of shoulder support exoskeletons in minimizing muscle activity levels and helping to manipulate heavy loads. This goal will be accomplished by analyzing electromyography (EMG) data. By providing empirical evidence through the EMG study, the research aims to reinforce the idea that incorporating exoskeletons into collaborative work environments can significantly contribute to a safer working environment for individuals.

2.1 Participants

Ten healthy adults (five couples) were recruited for the study. All participants exhibited right-handedness. None of the participants had prior experience with exoskeletons. The selection criteria encompassed: 1) an age of 18 years or above; 2) the absence of upper extremity musculoskeletal pain; 3) no previous record of upper extremity musculoskeletal ailments. These criteria were designed to mitigate the potential for musculoskeletal discomfort during the experimental tasks.

2.2 Experiment Setup

The experiment extends a prior live study on collaborative tasks involving exoskeletons in the construction sector. Here, two collaborative tasks were carried out: firstly, the installation of wooden beams (similar to the focus of this investigation), followed by the positioning of 3-m lightweight wooden strips on ceiling panels. These collaborative tasks were carried out without exoskeletons and with exoskeletons—each lasting for an hour, separated by a one-hour break. This study evaluated the participants' range of motion, gathered user feedback, assessed system usability, and quantified the level of bodily discomfort experienced while utilizing an exoskeleton. Importantly, this ongoing research received ethical approval from the University of Stuttgart on September 21, 2021, as detailed in [14]. For the new study, two participants work together to transport and position a heavy metallic beam. The beam weighs 30 kg and measures 2.5 m in length. The task begins with the beam placed on a sturdy stool at the starting point (SP). The main objective is to accurately position the beam at an endpoint (EP) located approximately six meters away and within a structure 2.1 m high. In addition, an obstacle along the path adds complexity to the task, requiring enhanced coordination between the two individuals to minimize physical exertion. Once the beam is in position, the participants lift it and return it to its original location (SP). However, this time, the beam is transported at a higher height to overcome the initial obstacle. The entire activity is divided into two phases as illustrated in Fig. 1.

- Phase Forward (PF): During this stage, the couple initiates the process by elevating the beam from the SP and successfully navigating it to the EP, while evading an obstacle with a variable height and a width of 1.5 m positioned midway.
- Phase Backward (PB): After reaching the EP on the metallic structure, the beam is dismounted and transported in an elevated position back to the SP.

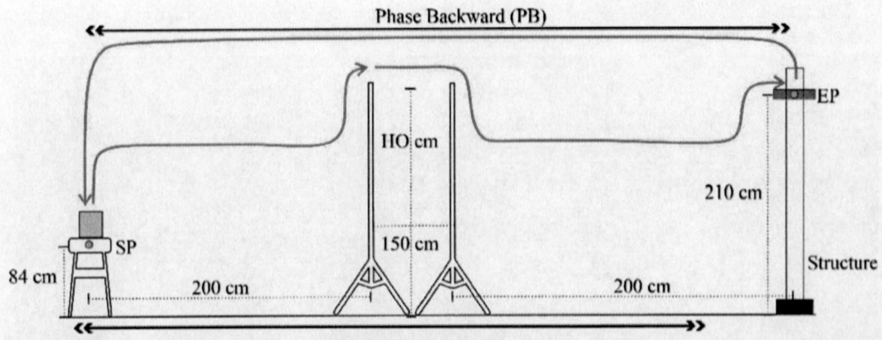

Fig. 1. The experimental configuration consists of a designated SP, positioned 84 cm above the ground level, serving as the initial placement for the beam. The EP is affixed to a metallic framework at an elevation of 210 cm. Positioned 200 cm away from the SP, a 150 cm width obstacle, referred to as the "high obstacle" (HO), is situated. The dimensions of the HO are determined based on the individual participants' body height and arm length, ensuring a personalized approach to the obstacle's placement.

2.3 Evaluation Scenarios

The experiment is structured into cycles and two scenarios. A cycle is considered complete when the beam has been lifted, transported, and positioned through the entire trajectory, which includes the PF and PB. The total experiment comprises five complete cycles to gather comprehensive data and reinforce the findings. The participants had a 60 s pause between the cycles.

In the first scenario, both individuals perform the five cycles without any support (No Exo). This scenario serves as a baseline for comparison. In the second scenario, exoskeletons (Exo) provide physical support for the participants.

An EMG system measured the activity of the following muscles in each participant. The *right upper trapezius* (TrRg), *left upper trapezius* (TrLf), *right middle deltoid* (DtRg) and *left middle deltoid* (DtLf). These muscles play a vital role, especially when raising the arms overhead. The Maximum Voluntary Contraction (MVC) technique assesses each muscle's maximum activation and strength. These MVC values serve as a reference for monitoring any changes in muscle strength during the two phases of the study. The experimental setup is visualized in Fig. 2.

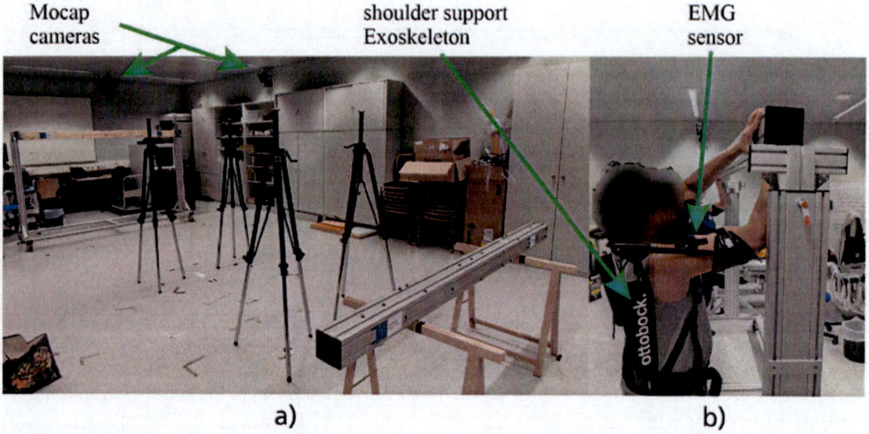

Fig. 2. Graphic representation of the experimental setup: a) The workspace is enclosed by a Vicon® motion capture system (mocap) based on ten cameras with a spatial separation of approximately six meters between the start point (SP) and end point (EP). b) Each participant was equipped with a Paexo® passive exoskeletons designed to provide shoulder support. To precisely monitor muscle activation, an EMG System Delsys® Avanti was employed.

3 Data Analysis and Results

The collaborative task raises several points for analysis, particularly regarding the impact of using an exoskeleton, which can either restrict free movement or enhance flexibility. In a collaborative context, the movement of each individual is influenced by their partner's actions, creating a leader-follower dynamic. However, if there is insufficient coordination during the transportation of the beam, it can pose safety hazards for the workers. Abrupt or forceful movements, combined with delayed reactions, have the potential to cause injuries. It is, therefore, crucial to examine the coordination and synchronization between participants wearing exoskeletons to ensure a safe working environment.

The analysis comprises two fundamental components:

- Identification of Leadership Roles: The initial facet involves the identification of leadership roles and follower roles within each phase. Our hypothesis contends that leaders impose external forces upon followers during movement, potentially affecting muscle activity and, subsequently, triggering sensations of discomfort or even heightening the potential for injury.
- Muscle Activity Comparison (Exoskeleton vs. No Exoskeleton): Our hypothesis posits that despite assuming the follower's role, the exoskeleton serves to optimize their movements, resulting in enhanced efficiency of muscle engagement. This optimization could lead to a reduction in overall muscular exertion.

3.1 Identification of Leadership Roles

Ideally, the couple should work synchronized and coordinated to transport the beam, with both individuals acting as mirrors of each other without compromising their partner's movements. However, achieving this desired synchronization may not always be feasible in practice. Often, one person assumes the role of a leader and initiates the movements by starting to move first, while the other person becomes a follower. In both scenarios, the rotation of the beam during the trajectory is examined to analyze leadership in the task and synchronization in movement. The beam can rotate on its axes if the participant on the right (PR) or the participant on the left (PL) exerts a force on their corresponding ends (see Fig. 3). All rotations could cause discomfort or induce muscular overload in the couple. Exoskeleton technology is employed to restrict movements in each participant and mitigate these rotational motions. Ultimately, each phase of the task is subject to individual analysis.

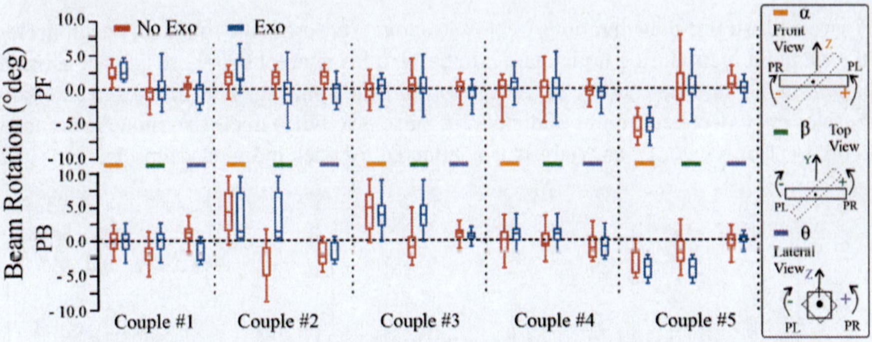

Fig. 3. Comparison of beam axis rotation throughout both phases of the trajectory.

The PF involves dynamic and short-duration movements, starting from a resting position and then lifting it to overcome the obstacle before returning to its resting state. Figure 3 depicts the rotations of each couple during each phase of the task. In the dynamic phase or PF, the rotation around the Y-axis (alpha), the angles for all couples are positive, indicating that PL is the one lifting the beam above the initial value. This could exert an additional force on PR. In the case of exoskeletons, it is observed that it slightly helps to reduce or keep the angle as with No Exo. Lastly, with respect to rotation around the Z-axis (beta), there is a tendency towards negative angle values, indicating that PL is a few steps ahead of PR. Similar to the previous case, when using the exoskeleton, the values are not significantly affected. During this phase, it becomes evident that PL assumed the role of the leader, while PR took on the role of the follower, except couple #5, where PR took the initiative of the movement.

During the static phase or PB, the exoskeleton assumes a more prominent role due to the necessity of raising the beam with fully extended arms. Notably, distinguishing the leader role for couples #1 and #4 proves challenging, given the negligible changes in rotation angles across both scenarios, where mean values remain below 1° degree. For couple #5, PR continues to hold the leadership role over PF. However, couples #2 and #3 exhibit divergent outcomes. Without the exoskeleton, the beta angle displays negative values, implying a slight inclination towards PL leading over PR. In contrast, with the exoskeleton engaged, PR emerges as the leader, suggesting a potential role reversal between the two modes. Moreover, both PL configurations in each mode feature an elevated beam (positive alpha angle), facilitating the application of pushing or pulling forces, potentially exerted on PR. This intricate interplay underscores the dynamics of leadership and force distribution within these couples across distinct modes.

To summarize, there is a multitude of potential scenarios where either PL or PR may assume the roles of leader or follower. However, the leadership classification for this study is shown in Table 1. In addition, these scenarios will be carefully compared based on their muscle activity patterns across both modes, a comparison that will be elaborated upon in the subsequent chapter.

3.2 Muscle Activity Comparison

Figure 4 illustrates the alterations in muscle activation across the four analyzed muscles in each participant during both phases. Initially, it is apparent that there is no consistent pattern in the variation across all muscles within each pairing. This implies that certain muscles display enhancement (reduction in muscle activity) in the exo mode in contrast to others. Consequently, an analysis is conducted for each individual muscle.

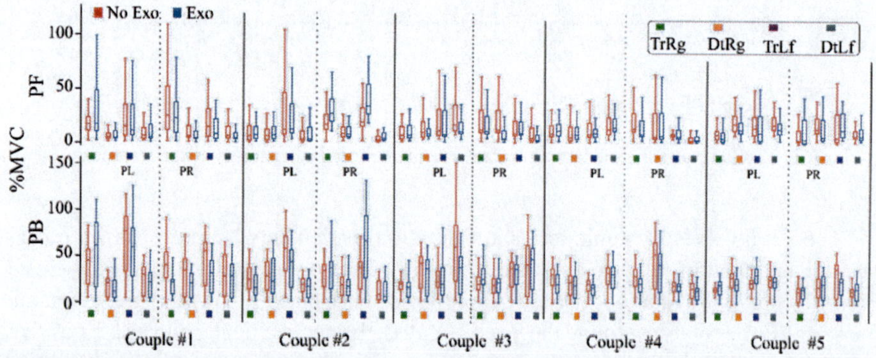

Fig. 4. Comparison of muscle activity (%) of left and right middle deltoid during the PF and PB of both participants showing the difference in both scenarios with no support (NS) and wearing an exoskeleton (OE).

To evaluate the influence of the exoskeleton on individual muscles during collaborative tasks, the study conducted a comprehensive analysis across 80 cases, encompassing four muscles, ten participants, and two phases, aimed at determining the extent to which muscles benefited from the use of exoskeletons during tasks. The findings presented in Table 1 unveiled 27 instances of a "good" impact, 17 instances showed a "positive" effect, and 21 instances were categorized as "identical," indicating the absence of significant bias. There were only five instances of a "bad" impact and ten instances classified as "negative". Furthermore, another factor to consider is whether there is a genuine reduction in muscle activity improvement when participants assume the roles of leaders or followers while utilizing the exoskeleton. For example, when participants take on the role of a leader, in most instances (18 times), the use of an exoskeleton shows a "good" impact, indicating an enhancement ranging from 10% to 25%. A similar trend is observed when participants act as followers, with eleven instances showing "positive" impacts (>25%) and nine instances of "good" impacts compared to scenarios where an exoskeleton is not used.

Table 1. Resume of the impact of the exoskeleton in each muscle activation level (%) between the no exo and exo scenarios.

C#	Phase	Lead/Follow	TrRg	DtRg	TrLf	DtLf
			PL/PR	PL/PR	PL/PR	PL/PR
1	PF	PL/PR	neg/pos	iden/pos	good/pos	iden/pos
	PB	PL/PR	neg/pos	good/pos	good/pos	good/good
2	PF	PL/PR	good/neg	good/neg	good/neg	iden/bad
	PB	PR/PL	pos/bad	iden/iden	pos/neg	good/iden
3	PF	PL/PR	pos/good	iden/pos	iden/iden	pos/good
	PB	PL/PR	good/neg	good/com	iden/iden	pos/iden
4	PF	PL/PR	iden/pos	iden/good	good/iden	iden/good
	PB	PL/PR	pos/pos	good/good	good/iden	iden/good
5	PF	PR/PL	good/neg	good/good	good/bad	good/good
	PB	PR/PL	neg/bad	good/iden	neg/pos	iden/bad

• Δ (%) = a reduction/increase between the mean values of the No-Exo and Exo mode.
• negative: $-25 > \Delta$; bad: $-25 > \Delta > -10$; identical: $-10 > \Delta > 10$; good: $10 > \Delta > 25$; positive: $\Delta > 25$
• # Cases: 80 total (4 muscles, 10 participants, 2 phases)
• Leader: 4 "negative", 0 "bad", 11 "identical", 18 "good", 6 "positive"
• Follower: 6 "negative", 5 "bad", 10 "identical", 9 "good", 11 "positive"

4 Discussion and Conclusion

The presented study aims to examine the influence of exoskeletons on cooperative tasks. In such tasks, the range of movement is constrained by the partners' actions, which may result in the users adopting unexpected positions or exerting excessive force on the manipulated object, leading to muscle strain or potential injuries. This study focuses on evaluating two key aspects of the task.

The first aspect pertains to the complexity of coordinating and synchronizing movements in a collaborative task. Typically, in object manipulation tasks, one user assumes the role of the leader while the other acts as the follower. In an ideal scenario, both users would have similar movements mirroring each other. However, this scenario is not realistic, as factors such as body height or the users' physical condition can influence object manipulation. Considering these factors, the method employed involved analyzing changes in rotation around the object axis during the task to determine who takes the role of leader or follower.

The second aspect examined in this study focused on the variations in muscle activation levels during the task. EMG technology monitored each participant's activity of the right and left upper trapezius and middle deltoid muscles. The MVC normalization method was employed to compare the results. It analyzed the reduction or increase of the muscle activation of each muscle and each participant independently in each phase. This gave a total of 80 cases to analyze whether the muscle experience benefits from exoskeletons or not during the task. The results showed that in 44 cases, the exoskeleton presented benefits in its use. In another 21 instances, it could not be determined whether it was better or worse, which was classified as having no significant impact. Only 15 showed adverse outcomes.

Here, the initial conclusion suggests that the exoskeleton does not yield an enhancement in movement synchronization or coordination among participants, or it might necessitate further complementary studies to validate this observation. Nonetheless, researchers argue that employing AR technology could enhance the synchronization of movements within couples during tasks. This prospect could offer insights into coordinating motions and subsequently mitigating the exertion of forces and reducing the muscle activation of both participants even more during the collaborative task. That is why the experimental setup is being extended to AR data input, all this seeking new know-how for a future smart and highly integrated assist system. Despite the analysis being limited to only five couples in this study, the results furnish compelling evidence of the exoskeleton's efficacy in lowering muscle activation levels during collaborative tasks. Looking ahead, the study aims to enhance its scope by including a larger participant pool for more experimental testing.

In conclusion, an exoskeleton does indeed hold significant potential for reducing muscle effort during collaborative tasks. However, other external factors should be evaluated in future studies, such as users' lack of experience in using exoskeletons. A trial and testing period for the technology is recommended. Another aspect to consider is the customization of exoskeletons. In this study, only an average measurement of exoskeletons was used, but not all participants have the same anthropometric characteristics, which may cause discomfort during their use. Additionally, it is important to consider extended trial periods with elevated muscle fatigue levels to ensure a deeper comprehension and more precise outcomes.

Acknowledgements. Supported by the Deutsche Forschungsgemeinschaft (DFG, German Research Foundation) under Germany's Excellence Strategy – EXC 2120/1 – 390831618.

Ethical Standards. This study is being conducted under ethical approval Az 21-018 and Az. 23-027 by University of Stuttgart.

References

1. European Commission: Communication from the Commission to the European Parliament, the Council, the European Economic and Social Committee and the Committee of the Regions. A Renovation Wave for Europe - greening our buildings, creating jobs, improving lives. COM 662 (2020)

2. Federal Ministry for Economic Affairs and Energy: Energy solutions made in Germany. BMWi, Berlin (2018)
3. European Commission: Communication from the Commission to the European Parliament, the Council, the European Economic and Social Committee and the Committee of the Regions. Fit for 55': delivering the EU's 2030 Climate Target on the way to climate neutrality. COM 550 (2021)
4. European Labour Authority: Eures, report on labour shortages and surpluses. ELA, Luxembourg (2022)
5. European Centre for the Development of Vocational Training (CEDEFOP). Skills Forecast. https://www.cedefop.europa.eu/en/publications-and-resources/data-visualisations/skills-forecast. Accessed 10 May 2023
6. Merkus, S.L., et al.: Physical capacity, occupational physical demands, and relative physical strain of older employees in construction and healthcare. Int. Arch. Occup. Environ. Health **92**, 295–307 (2019)
7. Baldassarre, A., et al.: Industrial exoskeletons from bench to field: human-machine interface and user experience in occupational settings and tasks. Front. Public Health **10**, 4375 (2022)
8. Zhu, Z., Dutta, A., Dai, F.: Exoskeletons for manual material handling–a review and implication for construction applications. Autom. Constr. **122**, 103493 (2021)
9. Andrade, C. and Dan N.-R.: Occupational exoskeleton adoption and acceptance in construction and industrial work: a scoping review. In: Proceedings of the Human Factors and Ergonomics Society Annual Meeting, vol. 66, no. 1. Sage CA: Los Angeles, CA: SAGE Publications, Los Angeles (2022)
10. Golabchi, A., et al.: Impact of passive back-support exoskeletons on manual material handling postures in construction. In: ISARC. Proceedings of the International Symposium on Automation and Robotics in Construction, vol. 39, pp. 359–366. IAARC Publications, Canada (2022)
11. Bances, E., et al.: Applicability of exoskeletons in timber prefabrication: actions for exoskeleton research. Procedia CIRP **107**, 1210–1215 (2022)
12. El Asmar, P.G., et al.: Contextualizing benefits and limitations reported for augmented reality in construction research. J. Inf. Technol. Constr. **26**, 720–738 (2021)
13. Kong, Y.-K., et al.: A passive upper-limb exoskeleton reduced muscular loading during augmented reality interactions. Appl. Ergon. **109**, 103982 (2023)
14. Kopp, V., et al.: Exoworkathlon: a prospective study approach for evaluation of industrial exoskeletons. Wearable Technol. **3**, E22 (2022)

Modeling of an Edge Computing Cluster for Optimized Distribution of Tasks in Production Environments

B. Brockhaus, F. Gast(✉), and M. Weigold

Institute of Production Management, Technology, and Machine Tools (PTW), Technical University of Darmstadt, Darmstadt, Germany
f.gast@ptw.tu-darmstadt.de

Abstract. Due to the digitalization of production, a vast amount of data is being generated that, through analysis and processing, offers new opportunities for implementing efficiency improvements. To process the data in real time, existing computers at the edge of production networks can be interconnected to form an edge computing cluster to distribute additional tasks among these devices. On the shop floor, satisfying latency requirements is necessary. A distribution model to optimally distribute tasks to existing computing resources is presented. The distribution model relies on a model of the network formed by the distributed edge computing nodes and their connections. Both computing nodes and their connections are characterized with various attributes. A metaheuristic approach is used to solve the optimization problem of the best distribution of tasks regarding latency and reliability.

Keywords: Edge Computing · Service-Placement · Metaheuristic

1 Introduction

Enterprises are facing the twin transition, requiring them to reduce their global ecologic footprint as well as compete in an increasingly digitalized market.

One way to reduce the ecologic footprint is to increase resource efficiency in their production systems. This can be achieved by using fewer operating supplies or reducing rejected parts, thereby reducing waste. This classical approach is known as process optimization, but there are other ways of reducing waste in production systems.

A typical value stream contains many non-value-adding steps, such as quality control or logistics. Especially quality control can benefit from digitalization and the implementation of artificial intelligence (AI). E.g. Fertig et al. demonstrate a machine learning pipeline for predictive quality in milling processes based on machine tool internal data [1]. Other use cases for digitalization include the development of a pay-per-stress model for machine tools from data collected during the production process [2] and the real-time monitoring of a motor spindle [3]. Many data-driven use cases in the production domain have requirements, such as a real-time integration [4] or high necessary data volumes, hence they are usually deployed on edge computing devices [5].

T. Bauernhansl et al. (Eds.): WGP 2023, LNPE, pp. 44–53, 2024.
https://doi.org/10.1007/978-3-031-47394-4_5

In Industry 4.0, new edge devices are often installed for each application, which is wasteful due to underutilization. Sharing devices between tasks, a common practice in the cloud-native world, has not been widely adopted in mechanical engineering. The authors of this paper aim to combine existing, underutilized edge devices on the shop floor into a computing cluster, considering production-specific requirements like hard-real-time capabilities. This paper presents a modeling approach to depict computing capabilities and network infrastructure, including communication path availability, between the nodes. The model is supposed to find the optimal distribution of a set of tasks onto the nodes in this computing cluster.

Edge Computing – if following cloud-native approaches – is often called fog computing, although the terms can be differentiated as in [6, 7]. For the purpose of this article, the terms will be used interchangeably.

2 Related Works

Most publications about fog or edge computing from the production engineering domain focus on use cases [8–10]. Only a fraction addresses the technical requirements of production engineering.

Wang et al. use machine learning to estimate latency in combined edge-cloud computing [11]. Zhang and Wei determine the placement of services in edge cloud computing depending on WiFi-Latencies [12]. Jiang et al. present a k-means clustering algorithm to determine the optimal number of fog servers on the shopfloor regarding latency [13]. Bayer et al. present an approach that places software either on the cloud or in the fog-layer based on the container orchestration engine Kubernetes [14]. The decision is determined by the available information regarding the essential hardware components connected to the node, in addition to the overall layout of the network.

The distribution of tasks to underutilized nodes with suboptimal connections within production environments remains largely unexplored in the manufacturing literature. However, there are some studies in the context of industry 4.0 and smart factories that explore task distribution to edge/fog or cloud environments.

Yin et al.'s paper, introduces the use of fog computing in smart manufacturing. Real-time data analysis is required at assembly line terminals to detect exceptions and diagnose device faults. When the resources of the fog network are insufficient, task containers are offloaded to the cloud. The approach presented assumes the use of monolithic tasks, each contained in a separate container [15].

Li et al. present a distribution algorithm for a manufacturing edge layer network with respect to deadlines. The so-called SAE scheduling algorithm selects the appropriate edge server (ES) from the edge server set (ESS) according to the task load, communication time, and computing power. In addition, computing units can cooperate for tasks with low latency requirements when an ES is insufficient to guarantee low latency [16].

In the paper of Wan et al. an energy-aware model for distributing tasks among robots in an automated production facility is presented. To optimize the energy-aware model, a dynamic scheduling system for the intelligent factory based on multi-agent technology is developed. According to the optimal solution calculated by an improved Particle Swarm Optimization (PSO) algorithm, a task-related priority based on the energy intensity of smart equipment is created [17].

Similarly, Karamoozian et al. use a metaheuristic, namely the Gravitational Search Algorithm (GSA), to find latency minimal distributions of IoT applications on a fog cluster. The applications, which consist of dependent processing elements (PE), are modeled as a directed acyclic graph (DAG) [18].

Wang and Li consider the static distribution of heterogeneous tasks in the context of smart factories on an equally heterogeneous fog network. Among other issues, the real-time requirements of fault detection tasks are considered. A hybrid metaheuristic is proposed, which consists of the combination of an improved PSO and Ant Colony Optimization (ACO). The energy consumption and the delay are considered as evaluation components of the found solutions. The proposed approach outperforms both the separate meta-heuristics PSO and ACO in terms of completion time of multiple independent tasks [19].

Outside the manufacturing domain, many different algorithms have been proposed in the literature to solve task distribution problems [20]. Metaheuristics, which mainly mimic natural phenomena or rules, are often utilized in this context because they have good self-organizing, learning, and adaption capabilities and efficiently achieve optimal solutions for large and complex problems within a relatively low time [19, 21].

All in all, the presented approaches address some of the necessary components for task distribution in manufacturing edge clusters, but no publication offers a satisfying combination of heterogeneous edge devices, network influences, and tasks comprised of individual dependent subtasks.

3 Modeling of the Edge Cluster

This section explains the model for the computational cluster and the network connections between the devices. A simple resource graph RG = (V, R) represents the network's topology in the production environment. The nodes $v_i \in V$ each represent a computing resource or a network switch and the edges $r_{ij} \in R$ describe network connections between the nodes.

3.1 Computing Nodes

The edge computers are modeled with 13 attributes that identify the computer, describe the processor and graphics card performance as well as the RAM and hard disk memory capacity.

In addition to the other parameters, the performance of the central processing unit (CPU) must be specified for the edge device. In the context of this work, the performance is measured by the time required to process a task. The higher the performance, the shorter the time the computer takes to calculate a task. Quantifying the performance of a processor and predicting the time it will take to complete a task before executing it presents a challenging issue, mainly because processors and tasks are highly complex.

Therefore a simplified "iron law of processor performance" [22] (Eq. (1)) is used to estimate the computation time t_{comp} for a program using the performance indicator million instructions per seconds (MIPS). Standardized tasks with fixed amounts of different instruction types, called benchmarks, are often used to determine the MIPS_i of a processing node v_i and we presume that our tasks have a similar ratio of these types. The number of instructions n^k_{inst} is generated with the aid of a profiler.

$$t^{kv_i}_{\text{comp}} = n^k_{\text{inst}}/MIPS_i \quad (1)$$

It is assumed that all computers require the same set of instructions n^k_{inst} for a given task k, making comparability between processors possible. Furthermore, due to poor quantifiability, it is assumed that the simultaneous execution of several tasks on one processor or processor core by hyperthreading does not reduce the processor's performance, although this does not correspond to actual experiments [23]. Even though the computing time can only be estimated, this is no problem since the values are only compared relative to each other. Despite the mostly different functionalities and architectures of graphics processors, the model described above is for simplification purposes also applied analogously for graphics cards.

3.2 Network Connections

The network connections between the nodes can be both wired and wireless. The connections are modeled with six attributes regardless of the type. These are the bandwidth B, the average usage, the used protocols for Layer 1 and 2 of the TCP/IP-Reference Model (Link and Internet Layer), and a packet loss probability for that connection. The bandwidth describes the amount of data that can be transferred per second per direction between two nodes over the connection. The packet loss probability is especially relevant for the transport protocol, which regulates the possible retransmission of packets. The loss probability is therefore specified for modeling these resends and is considered in the optimization problem.

The transfer time of a packet between two nodes v_s and v_i with regards to the packet loss probability $P_{r_{si}}(\text{pl})$ of the connection r_{si} is estimated via an expectancy value E as displayed in Eq. (2). The transfer times for the two cases of no packet loss (npl) and packet loss (pl) are calculated with Eq. (3). Here the amount of data to be transmitted d_{trans} is comprised of the net amount of data, overheads, and protocol dependent retransmissions if a packet loss occurs.

$$E\left[t^{v_s v_i}_{\text{trans}}\right] = \left(1 - P_{r_{si}}(\text{pl})\right) \cdot t^{v_s v_i}_{\text{transnpl}} + P_{r_{si}}(\text{pl}) \cdot t^{v_s v_i}_{\text{transpl}} \quad (2)$$

$$t_{\text{trans}}(r_{si}) = d_{\text{trans}}/B_{r_{si}} \quad (3)$$

4 Modelling of Tasks

For modeling tasks, in the literature they are either represented as individual, independent tasks or as connected subtasks in form of a directed acyclic graph (DAG). The vertices of the DAG represent subtasks to be completed in this context, which can be distributed individually, and edges represent the communication between the subtasks. To leverage the possibilities of distributed computing resources in edge/fog and cloud environments, many papers use DAGs to model their tasks [18, 24–30].

In our case, the edges contain attributes, e.g., the necessary data quantity being communicated from the source to the sink subtask. Depending on whether the entire task is executed continuously or once, the unit of data quantity is MB or MB/s. Another attribute for continuous tasks is the amount of data in MB, which must be transferred to start the calculation of the subtask.

Additional vertices are data sources modelled similarly as tasks, but with no computational requirements associated. Similar to the source nodes, the possibility of sinks is also implemented in the model. A subtask can send its results to one or more specific nodes in the network for further processing or storage. Since no task is executed in the sink itself, the specification of the required data for sinks is not necessary.

5 Task Placement as an Optimization Problem

This section presents the developed methodology for placing tasks in the edge computing cluster. The underlying optimization problem and the GSA, which was used to solve this problem, are described. Additionally, the GSA-metaheuristic is validated with exemplary tasks in the edge cluster of the TEC-Lab at the Institute PTW of the Technical University Darmstadt.

5.1 The Optimization Problem

First, an evaluation criterion must be selected to determine which task distribution among the computers in the production environment is optimal. Based on this evaluation criterion, different distributions can be quantified and compared. The minimization or maximization of the criterion forms the optimization goal to be achieved.

In the literature, energy consumption, response time, cost, processing time, and latency are often considered essential parameters in task distribution [20]. In production, latency and real-time requirements are becoming increasingly important due to digitalization. Therefore, the latency was chosen as the main optimization objective and is calculated in Eq. (4). The latency L for a subtask k on a computing node v_i is comprised of the processing time and the maximum of the recursively calculated total latency of the set of preceding subtasks ($P(k)$) plus the expectancy value of the corresponding transmission latency. The processing time directly follows from the processor capacity of the node v_i and the number of instructions for task k according to Eq. (1), whereas the transmission expectancy value is calculated in Eq. (2).

$$L(k, v_i) = t_{\text{comp}}^{kv_i} + \max_{p \epsilon P(k)} \left(L(p, v_s) + E\left[t_{\text{trans}}^{v_s v_i}\right]\right) \tag{4}$$

The optimization function is consequently created by adding the with the factor w_L weighted total latency of the last node Ω in the task graph to a with the factor w_C weighted cost function C, which increases by a specific cost for every overrun of the resource constraints. The final optimization function can therefore be written as follows:

$$\textbf{\textit{Minimize}} w_L \cdot L(\Omega, v_i) + w_C \cdot C \tag{5}$$

5.2 Metaheuristic for Solving the Optimization Problem

As the related works show, metaheuristics are often chosen to optimize task allocations. Primarily, PSO and ACO are widely used to distribute tasks in a production environment. In this work, the GSA was chosen as a metaheuristic. GSA is a recent metaheuristic but has shown high potential in terms of convergence rate and minimization results [18, 31, 32].

The GSA was developed by Rashedi et al. in 2009 and is based on Newton's law of gravity and the interaction of masses [31]. A set of agents is created, with each agent being treated as an object and the mass of each agent regarded as a performance metric. All these objects attract each other depending on their masses and the distance between them via gravitation. The gravitational force results in a global movement towards the most massive objects, which represent the best solutions to a problem.

To not get stuck in a local optimum and to improve the algorithm's exploration ability a random component is added both to the force that acts on a mass and the resulting velocity of the mass. Additionally, the proposed performance optimization where only a linearly decreasing amount of agents with the highest masses (Kbest) attracts the other agents is used, which offers a good compromise of exploration and exploitation [31].

For the application of GSA, it must be encoded to our model by defining the search space and the interpretation of the solution. In the context of allocating tasks on computing resources, each mass represents a possible distribution of the tasks. The dimension of the agents corresponds to the number of subtasks T of a task and the position of the masses is therefore displayed in T coordinates. The resulting allocation for the solution vector of an agent can be extracted by rounding down the values for each coordinate, where the resulting integers represent the identifier of computing resources on which the subtasks are to be executed.

5.3 Model Evaluation

To test the developed distribution model described in this paper the edge cluster within the TEC-Lab at the Institute for Production Management, Technology and Machine Tools (PTW) at the Technical University of Darmstadt is used. Specifically, the computer network around two machine tools is considered for this work. The network formed between these two production machines in the TEC-Lab consists of twelve edge devices, five switches and 16 connections. For each device, the required resource capabilities and average usages are collected to create a graph of the cluster and network according to the developed model.

An exemplary continuous condition monitoring task is used for the validation and test of the model (see Fig. 1). The task is comprised of four subtasks (red), which get data from one source node (blue). Additionally, one subtask sends its results to a sink node (green) in the cluster. The four subtasks are an AI Model that calculates the condition of the workpiece, a Material Removal Detection, which decerns whether the tool is currently removing material, an OEE Calculator, calculating the overall equipment effectiveness for the machine, and an IoT Gateway subtask, which saves the data from the machine for training the AI Model with this data.

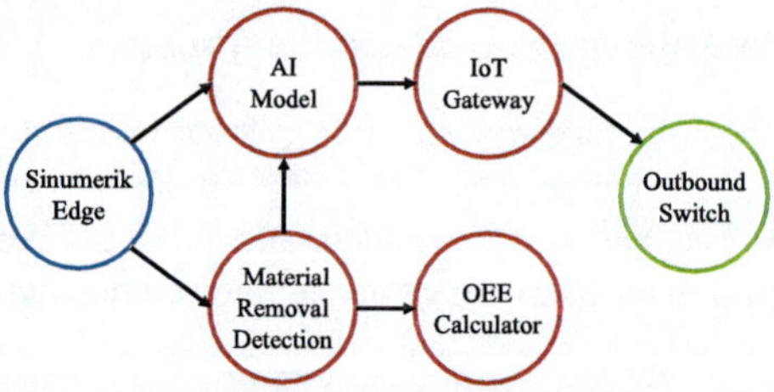

Fig. 1. Graph Visualization of the validation task

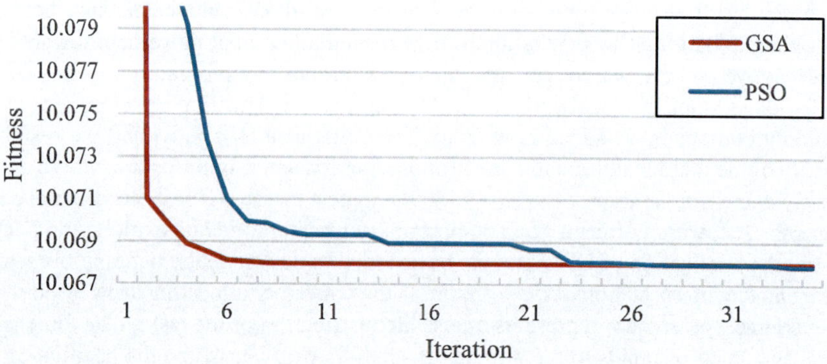

Fig. 2. Average best per iteration for Gravitational Search Algorithm and Particle Swarm Optimization

To evaluate the GSA approach, it is compared with the PSO metaheuristic on the TEC-Lab cluster with the described exemplary task. For both metaheuristics, 50 runs with 100 iterations are carried out. The best solution for the developed optimization function, which is comprised of latency and cost for non-compliance with constraints, is averaged over the 50 runs for every iteration. The detailed view of the results for the first 35 iterations for both GSA and PSO is shown in Fig. 2.

As can be seen in the figure, the GSA converges faster than the PSO algorithm, but PSO finally reaches a minimally lower average value. Thus, the GSA gets stuck in a local minimum more often. However, the difference in the average final solution is in the range of only one millisecond for the achieved latency and can therefore be considered insignificant. PSO reaches the global optimum more consistently than GSA, but the execution time of GSA is 19,6% faster, which is crucial in a latency sensitive production environment.

In conclusion, the GSA chosen for this work has a fast execution time, which can be further reduced with fewer iterations due to the high convergence rate and provides near optimal results for the use case. However, in contrast to the analysis of other works, no better results could be obtained for the experimental network of the TEC-Lab and the modeled task compared to PSO [18, 31].

6 Summary and Outlook

In this paper a method to model processor, memory, and additional capacities of heterogeneous edge devices in a production environment to which tasks are to be distributed was developed. The necessary attributes of these tasks were also specified and the ability to model dependencies of subtasks and additional data transfers to and from computer nodes (sources and sinks) as a DAG was implemented.

Based on low latency and reliability requirements, a specialized optimization function that minimizes the tasks latency while considering the packet loss probability of the network connections was proposed. This optimization problem is then solved with the metaheuristic GSA. The results of a real-world scenario were evaluated against the established PSO metaheuristic. GSA shows comparatively high convergence rate and low execution time but does not reach the best results as often as the PSO.

Since this work is primarily concerned with latency minimization and fast response in the production environment, the GSA metaheuristic is preferable to PSO despite the minimally worse optimization results due to faster execution and convergence.

It can be assumed that the GSA results can be further improved by adjusting search parameters. However, if a more precise solution is required, the PSO algorithm can also be used with the developed cluster, network, and task models. In addition, more detailed modeling of processors, graphics cards, and task requirements could be integrated.

Besides optimization of GSA, the presented approach shall be integrated into the Kubernetes Scheduling Framework. Additionally, the developed approach could be expanded by the dynamic distribution and reallocation during runtime of tasks. The implementation with Kubernetes will allow for the adoption of the developed solution with the use cases of our research group like the quality prediction of produced parts or the real-time monitoring of motor-spindle in a production environment.

References

1. Fertig, A., Preis, C., Weigold, M.: Quality prediction for milling processes: automated parametrization of an end-to-end machine learning pipeline. Prod. Eng. Res. Devel. **17**(2), 237–245 (2023). https://doi.org/10.1007/s11740-022-01173-4
2. Stanula, P., Praetzas, C., Kohn, O., Metternich, J., Weigold, M., Buchwald, A.: Stress-oriented, data-based payment model for machine tools. Procedia CIRP **93**, 1526–1531 (2020). https://doi.org/10.1016/j.procir.2020.03.080
3. Berchtenbreiter, V., von Elling, M., Tepper, C., Weber, M., Weigold, M.: Datengetriebene Produktion zum Anfassen. Zeitschrift für wirtschaftlichen Fabrikbetrieb **117**(10), 618–622 (2022). https://doi.org/10.1515/zwf-2022-1140
4. Lukas, M., Stock, D., Csiszar, A.: FabOS: Towards an open, distributed, real-time-capable, and secure operating system for production. Procedia CIRP **104**, 962–967 (2021). https://doi.org/10.1016/j.procir.2021.11.162
5. Um, J., Gezer, V., Wagner, A., Ruskowski, M.: Edge computing in smart production. In: Berns, K., Görges, D. (eds.) RAAD 2019. AISC, vol. 980, pp. 144–152. Springer, Cham (2020). https://doi.org/10.1007/978-3-030-19648-6_17
6. Bermbach, D., Pallas, F., Pérez, D.G., Plebani, P., Anderson, M., Kat, R., Tai, S.: A Research Perspective on Fog Computing. In: Braubach, L., Murillo, J.M., Kaviani, N., Lama, M., Burgueño, L., Moha, N., Oriol, M. (eds.) ICSOC 2017. LNCS, vol. 10797, pp. 198–210. Springer, Cham (2018). https://doi.org/10.1007/978-3-319-91764-1_16

7. Iorga, M., Feldman, L., Barton, R., Martin, M.J., Goren, N.S., Mahmoudi, C.: Fog computing conceptual model. Gaithersburg, MD, NIST Special Publication, pp. 500–325 (2018)
8. Guo, J., Martínez-García, M.: Key technologies towards smart manufacturing based on swarm intelligence and edge computing. Comput. Electr. Eng. **92**, 1–6 (2021). https://doi.org/10.1016/j.compeleceng.2021.107119
9. Kubiak, K., Dec, G., Stadnicka, D.: Possible applications of edge computing in the manufacturing industry: systematic literature review. Sensors **22**(7), 2445 (2022). https://doi.org/10.3390/s22072445
10. Merino, R., Bediaga, I., Iglesias, A., Munoa, J.: Hybrid edge–cloud-based smart system for chatter suppression in train wheel repair. Appl. Sci. **9**(20), 4283 (2019). https://doi.org/10.3390/app9204283
11. Wang, Y., et al.: MPCSM: microservice placement for edge-cloud collaborative smart manufacturing. IEEE Trans. Ind. Inf. **17**(9), 5898–5908 (2021). https://doi.org/10.1109/TII.2020.3036406
12. Zhang, Y., Wei, H.-Y.: Risk-aware cloud-edge computing framework for delay-sensitive industrial IoTs. IEEE Trans. Netw. Serv. Manage. **18**(3), 2659–2671 (2021). https://doi.org/10.1109/TNSM.2021.3092790
13. Jiang, C., Wan, J., Abbas, H.: An edge computing node deployment method based on improved k-means clustering algorithm for smart manufacturing. IEEE Syst. J. **15**(2), 2230–2240 (2021). https://doi.org/10.1109/JSYST.2020.2986649
14. Bayer, T., Moedel, L., Reich, C.: A fog-cloud computing infrastructure for condition monitoring and distributing industry 4.0 services. In: Proceedings of the 9th International Conference on Cloud Computing and Services Science, Heraklion, Crete, Greece, pp. 233–240 (2019)
15. Yin, L., Luo, J., Luo, H.: Tasks scheduling and resource allocation in fog computing based on containers for smart manufacturing. IEEE Trans. Ind. Inf. **14**(10), 4712–4721 (2018). https://doi.org/10.1109/TII.2018.2851241
16. Li, X., Wan, J., Dai, H.-N., Imran, M., Xia, M., Celesti, A.: A hybrid computing solution and resource scheduling strategy for edge computing in smart manufacturing. IEEE Trans. Ind. Inf. **15**(7), 4225–4234 (2019). https://doi.org/10.1109/TII.2019.2899679
17. Wan, J., Chen, B., Wang, S., Xia, M., Li, D., Liu, C.: Fog computing for energy-aware load balancing and scheduling in smart factory. IEEE Trans. Ind. Inf. **14**(10), 4548–4556 (2018). https://doi.org/10.1109/TII.2018.2818932
18. Karamoozian, A., Hafid, A., Aboulhamid, E.M.: On the fog-cloud cooperation: how fog computing can address latency concerns of IoT applications. In: 2019 Fourth International Conference on Fog and Mobile Edge Computing (FMEC), Rom, Italien, pp. 166–172 (2019)
19. Wang, J., Li, D.: Task scheduling based on a hybrid heuristic algorithm for smart production line with fog computing. Sensors (Basel, Switzerland) **19**(5), 1023 (2019). https://doi.org/10.3390/s19051023
20. Bansal, S., Aggarwal, H., Aggarwal, M.: A systematic review of task scheduling approaches in fog computing. Trans. Emerg. Telecommun. Technol. 33, e4523 (2022). https://doi.org/10.1002/ett.4523
21. Talbi, E.-G.: Metaheuristics: From Design to Implementation. Wiley, Hoboken (2009)
22. Shen, J.P., Lipasti, M.H.: Modern Processor Design: FUNDAMENTALS of Superscalar Processors. Waveland Press, Long Grove (2013)
23. Eeckhout, L.: Computer architecture performance evaluation methods. Synth. Lect. Comput. Architect. **5**(1), 1–145 (2010). https://doi.org/10.2200/S00273ED1V01Y201006CAC010
24. Convolbo, M.W., Chou, J.: Cost-aware DAG scheduling algorithms for minimizing execution cost on cloud resources. J. Supercomput. **72**(3), 985–1012 (2016). https://doi.org/10.1007/s11227-016-1637-7

25. Liu, L., et al.: Dependent task placement and scheduling with function configuration in edge computing. In: 2019 IEEE/ACM International Symposium on Quality of Service (IWQoS), Phoenix, Arizona, USA, pp. 1–10 (2019)
26. Madhura, R., Elizabeth, B.L., Uthariaraj, V.R.: An improved list-based task scheduling algorithm for fog computing environment. Computing **103**(7), 1353–1389 (2021). https://doi.org/10.1007/s00607-021-00935-9
27. Pandit, M.K., Mir, R.N., Chishti, M.A.: Adaptive task scheduling in IoT using reinforcement learning. IJICC **13**(3), 261–282 (2020). https://doi.org/10.1108/IJICC-03-2020-0021
28. Sakellariou, R., Zhao, H.: A hybrid heuristic for DAG scheduling on heterogeneous systems. In: 2004 18th International Parallel and Distributed Processing Symposium, Santa Fe, NM, USA, pp. 111–123 (2004)
29. Yadav, A.M., Tripathi, K.N., Sharma, S.C.: A bi-objective task scheduling approach in fog computing using hybrid fireworks algorithm. J. Supercomput. **78**(3), 4236–4260 (2022). https://doi.org/10.1007/s11227-021-04018-6
30. Zheng, W., Wang, C., Chen, Z., Zhang, D.: A priority-based level heuristic approach for scheduling DAG applications with uncertainties. In: 2022 IEEE 25th International Conference on Computer Supported Cooperative Work in Design (CSCWD), Hangzhou, China, pp. 1022–1027 (2022)
31. Rashedi, E., Nezamabadi-pour, H., Saryazdi, S.: GSA: a gravitational search algorithm. Inf. Sci. **179**(13), 2232–2248 (2009). https://doi.org/10.1016/j.ins.2009.03.004
32. Rashedi, E., Rashedi, E., Nezamabadi-pour, H.: A comprehensive survey on gravitational search algorithm. Swarm Evol. Comput. **41**, 141–158 (2018). https://doi.org/10.1016/j.swevo.2018.02.018

Interaction of Artificial Intelligence and Machining Processes Regarding Industry 4.0 Production Systems

D. Hinzmann[1(✉)], D. Hasselder[1], S. Lezama[1], O. Kirik[1], V. Pandey[1], E. Bosler[1], D. Spitta[1], E. Uhlmann[1,2], J. Krüger[1], V. Markl[1], D. Oberschmidt[1], and M. Rötting[1]

[1] Technical University of Berlin, Pascalstr. 8-9, 10587 Berlin, Germany
daniel.hinzmann@tu-berlin.de

[2] Fraunhofer Institute for Production Systems and Design Technology IPK, Pascalstr. 8-9, 10587 Berlin, Germany

Abstract. Artificial intelligence (AI) has already been the subject of extensive scientific and industrial research in order to solve current challenges in intelligent manufacturing. However, these approaches often do not yet reach industrial small and medium-sized enterprises (SMEs). To meet the demand for resilient and sustainable production processes, the aim is to develop concepts to transfer applications into cyber physical production systems. Therefore, AI methods can be introduced as part of a smart manufacturing chain. In this work, an infrastructure to efficiently handle production data which is independent of the individual production process such as machining, grinding or electrical discharge machining is presented. Subsequently, an exemplary roadmap for SMEs is given to establish AI applications within the scope of manufacturing industry 4.0. As a result, an increase in efficiency and profitability of individual production processes as well as the overall production chain can be achieved.

Keywords: Artificial intelligence · industry 4.0 · smart manufacturing · cyber physical systems

1 Introduction

AI technologies, especially machine learning and deep learning, have developed rapidly in recent years. This is mainly due to the increasing availability of large amounts of data and high computing power, which are a basic prerequisite for complex calculations with AI technologies [1]. An increase in research regarding the application of AI can also be observed in production technology research [2–4]. Despite this development, applications into industrial use-cases still pose a challenge within the industry sector [5–7].

In order to establish methodical approaches to the application-specific transfer of AI in production, basic conditions that predominate in production need to be addressed [8]:

- lack of prior knowledge in the field of AI

T. Bauernhansl et al. (Eds.): WGP 2023, LNPE, pp. 54–63, 2024.
https://doi.org/10.1007/978-3-031-47394-4_6

- skepticism about AI-controlled processes
- concern for complex AI-based interrelationship

Therefore, the possibility to apply current AI approaches regarding an exemplary milling process as the industrially relevant use-case is presented within this paper.

2 Data Infrastructure

One of the key ideas in Industry 4.0 is to harness big data and analytics to fully automate the manufacturing process, enhance production performance, and boost productivity. In modern smart manufacturing, a variety of built-in and external sensors and video cameras are used to initially capture the manufacturing process. The sensed data are gathered, analyzed, and data-driven strategies are implemented to optimize the manufacturing process, such as adjusting the machine parameters. The modified manufacturing procedure is repeated with improved parameters, and new data is generated as the foundation for the next iteration of an optimization cycle. This is also referred to as closed-loop manufacturing. The goal is to continuously extract insights from data in order to realize data-driven improvements in the manufacturing process [9].

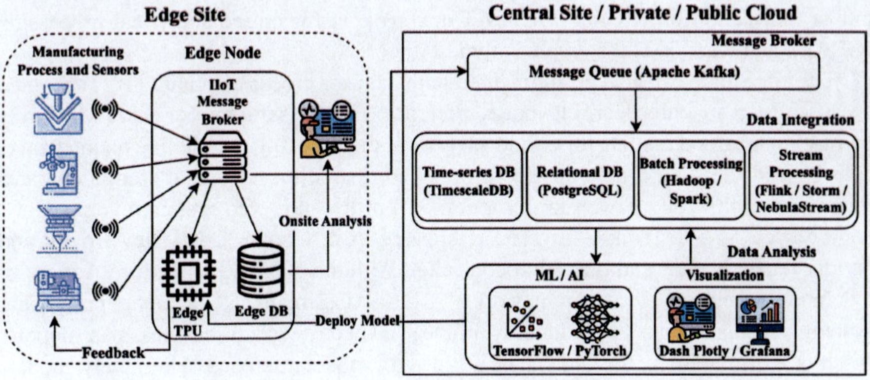

Fig. 1. A conceptual architecture of the data pipeline for Industrial IoT.

Figure 1 depicts the conceptual architecture of the data infrastructure that was established to realize a continuous data-driven manufacturing process. The primary advantage of the layered architecture is that users can match and combine different components based on their requirements and scale each component independently. In the subsequent sub-sections individual components of the architecture are described.

2.1 Edge Site/Edge Node

The manufacturing process is captured using various sensors, such as temperature, pressure etc. Many use-cases in manufacturing also capture the process using video sensors. The captured or sensed data can be transmitted to an edge node using a *light-weight*

IIoT protocol such as Message Queuing Telemetry Transport (MQTT) [10], Open Platform Communications Unified Architecture (OPC-UA) [11], Constrained Application Protocol (CoAP) [12], etc. OPC-UA is favored as recent research has shown that it is the communication standard in industrial automation and offers competitive performance in terms of transmission latency [13]. This also applies in applications where resource-constrained devices connect with one another over low-bandwidth and unstable wireless networks [14]. Many use cases generate high-frequency process data, and the latency to access machine parameters via these protocols in these instances is too high for real-time analytics and model inference [15]. In such cases, industrial grade edge devices that directly interface with the machine control system, such as the Siemens Sinumerik Edge device, are necessary. The captured data from this layer can also be forwarded to a central site which can be a public or private cloud or company's own datacenter, see Sect. 2.2.

Many manufacturing processes require human interaction to monitor the production process. This can be facilitated by connecting to the lightweight message broker and displaying the time-series data as well as the captured video stream using a popular open-source data visualization tool such as Dash Plotly [16], Grafana [17], etc. Additionally, it may be necessary to store or process the acquired data on the edge device for visualization and interactive data analysis. Some of the popular choice of systems for edge data analysis include NebulaStream [18] for stream processing, SQLite [19] and PostgreSQL [20] for online transaction processing (OLTP), DuckDB [21] for online analytical processing (OLAP), and Kùzu [22] for graph analytics.

The edge node can also be augmented with a tensor processing unit (TPU) designed specifically for machine learning model inference. This is useful for use-cases which rely on machine learned models for online analysis at the edge, e.g., predictive maintenance, control feedback to the machine for process correction etc. A recent study compares the inference time and power consumption of four low-powered system on chip (SoC) edge devices - Asus Tinker Edge R, Raspberry Pi 4, Google Coral Dev Board, and Nvidia Jetson Nano - and one microcontroller, Arduino Nano 33 BLE, for a variety of machine learned models and frameworks [23]. NebulaStream [18], a stream processing system which is designed to efficiently utilize such heterogeneous devices, also supports machine learning inference and can be utilized for model inference. Alternately, models can be served and inferred using frameworks for model serving, such as TensorFlow Serving [24], TorchServe [25], etc.

Finally, many manufacturing companies might have several such production plants or assembly lines located across a wide-geographic area [26]. For such scenarios, the message broker can be deployed on a high-end server or a cluster of machines at each edge site or manufacturing plant. The edge node augmented with edge TPU, however, can be machine or line specific depending on the requirements of the use-case.

2.2 Central Site/Public/Private Cloud

The central site can be a public cloud provider or a privately hosted cloud and datacenter. The central site is made up of three layers: a scalable message broker, a data integration layer and an analysis layer. Each will be presented below:

Message Broker (Apache Kafka). The data transmitted from various edge nodes and edge sites is delivered to a scalable message broker especially designed for storing massive amounts of event-driven data. For this the industry-grade open-source distributed event streaming platform Apache Kafka [27] is used. This is a highly scalable publish-subscribe messaging system. At its core, it is a distributed commit log. It was initially developed by LinkedIn to work as the company's central platform for event pipelining and online data integration tasks [27]. Due to its ability to manage high-throughput, fault-tolerant, real-time data streams, Kafka has acquired immense popularity over the years and is currently used in a variety of applications, including real-time analytics, log aggregation, event sourcing, and stream processing [28]. Kafka's durability and scalability also make it appropriate for Industrial IoT. In the following data pipeline, Apache Kafka is largely responsible for providing a method to publish and subscribe to a stream of records to be sent to potentially *multiple* systems and other layers as well as storing these records in a fault-tolerant manner.

Data Integration Layer. Depending on the use-cases, the data integration layer consists of one or more components. The following subsections discuss each component:

Time-Series Database. Time-series databases, often known as TSDBs, are a distinctive type of database that is designed with the specific goal of managing enormous amounts of time-series data. The data captured from machine is also time-series in nature, and further many use-cases require the data to be accessed based on time for a variety of analysis such as outlier detection. Time-series database can also facilitate replaying of the stream desired in a variety of domains. Some of the widely used time-series databases today include TimescaleDB [29], InfluxDB [30], Apache IoTDB [31] and Apache Druid [32].

Relational Database. Relational database management systems (RDBMS) are software systems that help designing and managing relational databases. A relational database stores the data in a *structured* format in tables represented by rows, and columns. Each row in a table represents an individual record, while each column defines the attributes of that record. Structured query language (SQL) is commonly used to query, modify, and retrieve data from RDBMS. Some of the widely used time-series databases today include PostgreSQL [20], MySQL [33], Oracle [34], SQLite [19], etc.

Batch Processing. This layer is responsible for processing data in a batch. In batch processing, a collection of data is processed as a single unit. Data is accumulated over time and stored in a storage system until there is a sufficient quantity to process. Once the batch has been produced, the batch is processed in its entirety. Batch processing is typically employed when data does not require hard real-time processing. Apache Spark [35] and Apache Hadoop [36] are a few popular batch processing systems.

Stream Processing. Stream processing, as opposed to batch processing, involves processing data in real-time as it is generated or received. In stream processing, data is processed as individual events or records instead of a batch of records. Consequently, stream processing aims to minimize processing delays and provide real-time insights. Among the most prominent stream processing frameworks are Apache Flink [37] and Apache Storm [38]. NebulaStream [18] is a next-gen streaming system designed to bring computation closer to data sources for IoT-related applications.

Data Analysis Layer. This layer is largely responsible for developing machine learning/deep learning models and visualizing data from the previous layers. Popular open-source frameworks for data visualization, such as Dash Plotly [16] or Grafana [17], can be used. Popular machine learning frameworks such as PyTorch [39] and TensorFlow [40] and can be used to train machine learning models. PyTorch and TensorFlow are two well-known open-source deep learning frameworks that are often used in the areas of artificial intelligence and machine learning. They provide a set of tools and libraries that make it easier to build, train, and deploy neural networks.

3 Use-Case Application

By implementing the utilized data infrastructure outlined in Sect. 2, the requirement for a data-driven manufacturing process is given. Thus, data acquisition regarding the applied use-case as well as the application of AI models is shown.

3.1 Experimental Setup and Data Acquisition

Based on the previously described conceptual data pipeline, the deployed data pipeline is presented below. In this scenario, the CNC machine tool type D-600 by Stepcraft GmbH & Co. KG, Menden, Germany which comes equipped with UCCNC [41] software for controlling the machining parameters is used. Additionally, an RGB Camera type Ace 2 by Basler AG, Ahrensburg, Germany and the industrial XDK100 IoT device are mounted on the D-600 machine tool to gather data.

The tool head is equipped with a camera that is capable of capturing images of both the cutting tool and the machined surface of the workpiece. It is connected to the edge node using an USB 3.0 cable and is operated using Pylon [42]. Collected images are transmitted to the edge node using the TCP/IP protocol. The XDK110 IoT device is equipped with a range of built-in sensors, including a BMA280 accelerometer (2000 Hz), a BMG160 gyroscope (2000 Hz), a BMM150 magnetometer (300 Hz), a BME280 pressure sensor, as well as acoustic, temperature, and humidity sensors (182 Hz). These sensors enable the device to provide various parameters that can be used for condition monitoring and predictive maintenance. In order to transfer the collected sensor data from the IoT device to the edge node, the MQTT protocol is employed. This implementation takes the available sensors and their methods to interact with the core element Apache Kafka as a basis that may vary on the sensor setup.

Gathered data by the camera and IoT device at the edge node is subsequently transmitted to an Apache Kafka instance for storage, streaming, and processing capabilities. In the current setup, the Kafka instance is also deployed at the edge site and the data is transmitted over LAN. As illustrated in Fig. 2, a bridge between MQTT and Apache Kafka is established for this purpose.

To demonstrate a milling application, the CNC machine is operated with a rotational speed of $v_r = 15{,}000$ 1/min, a feed rate of $v_f = 50$ mm/sec a radial and axial depth of cut of $a_p = a_e = 1$ mm on exemplary wooden workpiece material. The image and sensor data from the milling process, as visualized in Fig. 3, are captured and stored for the subsequent training of AI models to enable advanced data analysis such as chatter detection and surface roughness analysis.

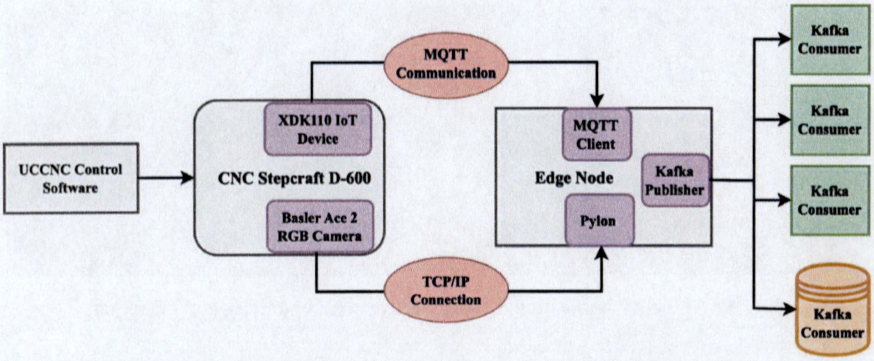

Fig. 2. Overview of the data acquisition and data streaming pipeline.

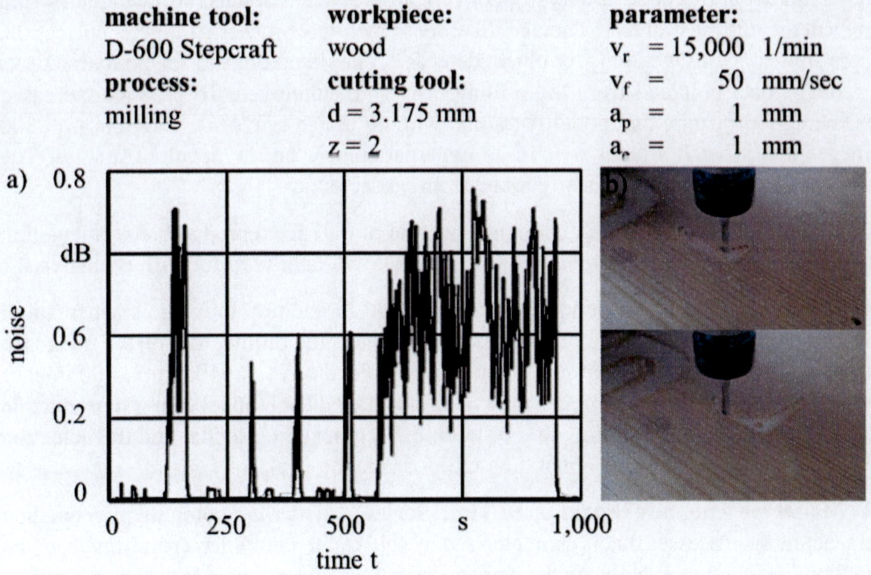

Fig. 3. Data acquisition within experimental setup, a) Sample acoustic data; b) image data

3.2 Application of AI Models in CNC Machining

After acquiring image and time-series data as mentioned in Sect. 3.1, several state-of-the-art AI models are employed for various machine learning tasks. This includes using a pre-trained YOLOv8n model for segmentation and object detection and utilizing LSTM Autoencoder for anomaly detection was used on time-series data.

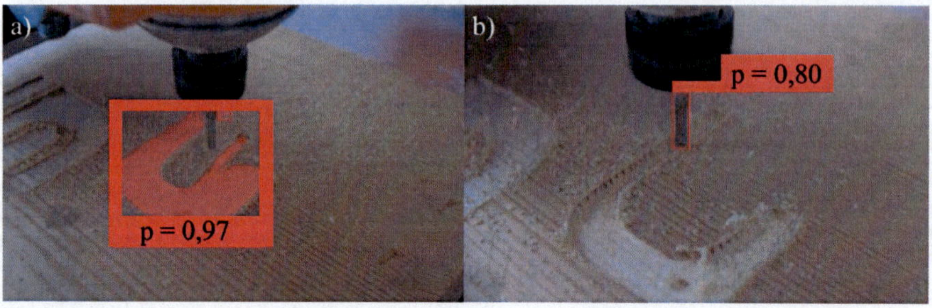

Fig. 4. AI model application; a) surface detection; b) cutting tool detection

AI Model for Semantic Segmentation on Image Data. CNC machining relies on calibrated process parameters, including feed rate v_f, rotational speed v_r, as well as radial and axial depth of cut a_e and a_p respectively, to produce a defined surface quality and prevent the milling tool from damage. In order to facilitate process parameter correction, a pre-trained YOLOv8n [43] for object detection, classification, and segmentation tasks on image data obtained from the milling process is employed. To produce increased segmentation results, the model is trained with an image size of $A = 640$ p, $n = 300$ images and a batch size of $s = 16$ as hyperparameters on the acquired images. The remaining hyperparameters are retained from preset values.

The AI model achieved a confidence interval of 0.97 for reproducible segmentation of the machined workpiece area, see Fig. 4A, that will later used for surface analysis.

AI Model for Object Detection on Image Data. In addition to image segmentation, the pre-trained YOLOv8n model is employed to detect the milling tool on the images to monitor the condition of the tool. The model is trained on $A = 640$ p, $n = 100$ images with 10 epochs. With a confidence interval of $p = 0.80$ the AI model shows reproducible cutting tool detection, see Fig. 4b. For both, the segmentation model and the detection model, CVAT is used as the annotation tool for labeling [44].

AI Model for Anomaly Detection in Time-Series. Chatter is a common phenomenon in machining processes that is characterized by self-excited vibrations resulting from the continuous interaction between the workpiece and the cutting tool. Accurately identifying or forecasting the occurrence of chatter is essential as it can mitigate the impact on the quality of the component [45]. By analyzing sensor and process data, for instance by employing AI models to detect anomalies, it is possible to identify and prevent chatter from causing significant damage to the machine tool and the workpiece. Autoencoders are widely used in the context of anomaly detection, while LSTM models are mainly applied for analyzing time-series data. Therefore, a LSTM Autoencoder [46] model on acoustic data with a sample rate of 40 Hz was employed. The model architecture includes an encoder and decoder. The encoder comprises two LSTM layers with 128 and 64 units, while the decoder has two LSTM layers with 64 and 128 units. Training utilizes the Adam optimizer with 2,000 epochs.

Autoencoders learn efficient representations of the unlabeled data by minimizing the reconstruction error. As illustrated in Fig. 5, the model optimally fits the acoustic data

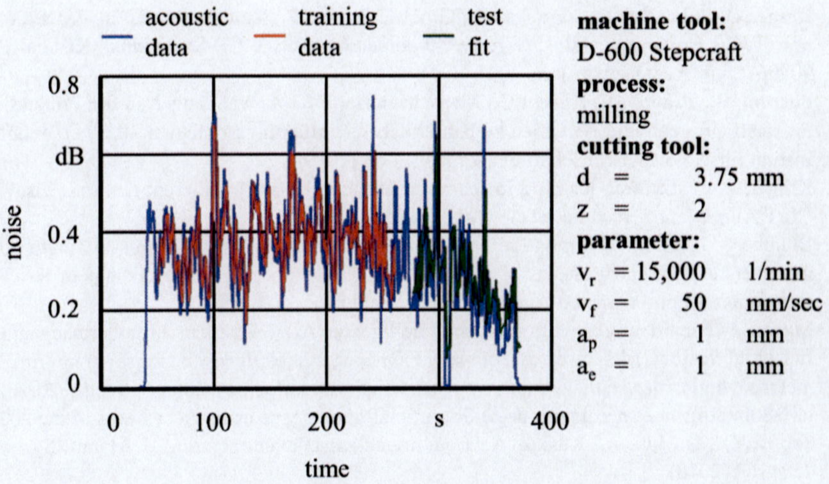

Fig. 5. Training and test results of the LSTM model on acquired acoustic data.

parameters acquired during the machining process. When an anomaly occurs, the trained model detects it through higher reconstruction errors with the confidency depending on a set threshold value. To enhance the model generalization, it can be further trained with sensor data using varied process parameters and diverse materials.

4 Results and Outlook

The presented approach focuses on linking AI methodological knowledge, work sciences and domain knowledge in the area of production technology. Due to a holistic approach regarding hard- and software interaction, data management and analysis, the potential for an increase of automation within production processes is shown while simultaneously relevant quality specifications regarding workpiece quality, process stability and cutting tool condition are met. Thereby, the potential for industrial implementation of the established approach within a variety of machining processes and production applications is given.

References

1. Uhlmann, E., Polte, J., Koutrakis, N.-S.: Holistic Concept Towards a Reference Architecture Model for Predictive Maintenance. In: Mourtzis, D. (ed.) Procedia CIRP 2021, vol. 104, pp. 1430–1433. Elsevier, Amsterdam (2021)
2. Uhlmann, E., Holznagel, T., Schehl, P., Bode, Y.: Machine learning of surface layer property prediction for milling operations. J. Manuf. Mater. Process. **5**(4), 104–115 (2021)
3. Uhlmann, E., Polte, M., Blumberg, J., Li, Z., Kraft, A.: Hyperparameter optimization of artificial neural networks to improve the positional accuracy of industrial robots. J. Mach. Eng. **21**, 47–59 (2021)

4. Krüger, J., Fleischer, J., Franke, J., Groche, P.: WGP Standpunkt: KI in der Produktion, WGP Homepage: https://wgp.de/wp-content/uploads/WGP-Standpunkt_KI-final_20190906-2.pdf. Accessed 06 June 2023
5. Pokorni, B., Braun, M., Knecht C.: Menschzentrierte KI-Anwendungen in der Produktion: Praxiserfahrungen und Leitfaden zu betrieblichen Einführungsstrategien 2021, Fraunhofer-Institut für Arbeitswissenschaft und Organisation (2019)
6. Krüger, J., et al.: Deep learning for part identification based on inherent features. Elsevier: CIRP Ann. Manuf. Technol. **68**(1), 9–12 (2019)
7. Blumberg, J., et al.: Deformation error compensation of industrial robots in single point incremental forming by means of data-driven stiffness model. In: Proceedings of the 26th International Conference on Automation & Computing, pp. 1–6
8. McKinsey, Smartening up with Artificial Intelligence (AI) - What's in it for Germany and its Industrial Sector?. McKinsey & Company Homepage: https://www.mckinsey.com/~/media/mckinsey/industries/semiconductors/our%20insights/smartening%20up%20with%20artificial%20intelligence/smartening-up-with-artificial-intelligence.ashx. Accessed 14 June 2023
9. Tao, F., Qi, Q., Liu, A., Kusiak, A.: Data-driven smart manufacturing. J. Manuf. Syst. **48**, 157–169 (2018)
10. Hunkeler, U., et al.: MQTT-S—A publish/subscribe protocol for Wireless Sensor Networks. In: 2008 3rd International Conference on Communication Systems Software and Middleware and Workshops (COMSWARE 2008), pp. 791–798, IEEE, Bangalore, India (2008)
11. Leitner, S.H., Mahnke, W.: OPC UA–service-oriented architecture for industrial applications. ABB Corporate Research Center, pp. 61–66 (2006)
12. Bormann, C., Castellani, A.P., Shelby, Z.: CoAP: an application protocol for billions of tiny internet nodes. IEEE Internet Comput. **16**(2), 62–67 (2012)
13. Silva, D., et al.: A performance analysis of internet of things networking protocols: evaluating MQTT, CoAP OPC UA. Appl. Sci. **11**, 4879 (2021)
14. Profanter, S., Tekat, A., Dorofeev, K., Rickert, M., Knoll, A.: OPC UA versus ROS, DDS, and MQTT: performance evaluation of industry 4.0 protocols. In: IEEE International Conference on Industrial Technology (ICIT), pp. 955–962. Melbourne, Australia (2019)
15. Trabesinger, S., et al.: Analysis of high frequency data of a machine tool via edge computing. Procedia Manuf. **45**, 343–348 (2020)
16. Plotly: Low-Code Data Apps. https://plotly.com/. Accessed 19 June 2023
17. Grafana: The open observability platform. https://grafana.com/. Accessed 19 June 2023
18. Zeuch, S., et al.: The NebulaStream platform: data and application management for the internet of things. In: Conference on Innovative Data Systems Research (CIDR), Amsterdam, Netherlands (2020)
19. Gaffney, K.P., et al.: SQLite: past, present, and future. Proc. VLDB Endowment **15**(12), 3535–3547 (2022)
20. PostgreSQL: The world's most advanced open source database. https://www.postgresql.org/. Accessed 19 June 2023
21. Raasveldt, M., Mühleisen, H.: Data management for data science-towards embedded analytics. In: Conference on Innovative Data Systems Research (CIDR), Amsterdam, Netherlands (2020)
22. Kùzu: an in-process property graph database management system built for query speed and scalability. https://kuzudb.com/. Accessed 19 June 2023
23. Baller, S.P., Jindal, A., Chadha, M., Gerndt, M.: DeepEdgeBench: benchmarking deep neural networks on edge devices. In: International Conference on Cloud Engineering (IC2E), pp. 20–30. IEEE, San Francisco, CA, USA (2021)
24. TensorFlow Serving: Serving Models. https://www.tensorflow.org/tfx/guide/serving. Accessed 19 June 2023

25. TORCHSERVE: a performant, flexible and easy to use tool for serving PyTorch models in production. https://pytorch.org/serve/. Accessed 19 June 2023
26. Fu, Y., et al.: Distributed scheduling problems in intelligent manufacturing systems. Tsinghua Sci. Technol. **26**, 625–645 (2021)
27. Wang, G., et al.: Building a replicated logging system with Apache Kafka. Proc. VLDB Endowment **8**(12), 1654–1655 (2015)
28. Narkhede, N, et al.: Kafka: The Definitive Guide: Real-Time Data and Stream Processing at Scale. O'Reilly Media, Inc., Sebastopol (2017)
29. Timescale: Time-series data simplified. https://www.timescale.com/. Accessed 19 June 2023
30. InfluxData: InfluxDB Times Series Data Platform. https://www.influxdata.com/. Accessed 19 June 2023
31. Apache IoTDB: database for internet of things. https://iotdb.apache.org/. Accessed 19 June 2023
32. Yang, F., Tschetter, E., Léauté, X., Ray, N., Merlino, G., Ganguli, D.: Druid: a real-time analytical data store. In: Proceedings of the ACM SIGMOD International Conference on Management of Data, pp. 157–168, Snowbird, UT, USA (2014)
33. MySQL Server. https://github.com/mysql/mysql-server. Accessed 30 July 2023
34. Oracle Database 23c. https://docs.oracle.com/en/database/oracle/oracle-database/23/index.html. Accessed 30 July 2023
35. Zaharia, M., et al.: Apache spark: a unified engine for big data processing. In: Commun. ACM **59**(11), 56–65 (2016)
36. Apache Hadoop. https://hadoop.apache.org/. Accessed 19 June 2023
37. Carbone, P., Katsifodimos, A., Ewen, S., Markl, V., Haridi, S., Tzoumas, K.: Apache Flink: stream and batch processing in a single engine. Bull. Tech. Committee Data Eng. **38**(4) (2015)
38. Apache Storm. https://storm.apache.org/. Accessed 19 June 2023
39. Paszke, A., et al.: PyTorch: an imperative style, high-performance deep learning library. In: Advances in Neural Information Processing Systems (2019)
40. Abadi, M., et al.: TensorFlow: a system for large-scale machine learning. In 12th USENIX Symposium on Operating Systems Design and Implementation, OSDI 2016, pp. 265–283 (2016)
41. UCCNC. https://cncdrive.com/UCCNC.html. Accessed 19 June 2023
42. Pylon. https://www.baslerweb.com/de/produkte/basler-pylon-camera-software-suite/. Accessed 19 June 2023
43. Yolov8n. https://github.com/ultralytics/ultralytics. Accessed 19 June 2023
44. CVAT. https://www.cvat.ai/. Accessed 19 June 2023
45. Aggogeri, F., Pellegrini, N., Tagliani, F.L.: Recent advances on machine learning applications in machining processes, 3 (2021)
46. Ahmad, S., Styp-Rekowski, K., Nedelkoski, S., Kao, O.: Distributed and operating systems autoencoder-based condition monitoring and anomaly detection method for rotating machines, pp. 4095–4096 (2020)

Efficient Process Monitoring in Clinching: Development of a Fast Predictor-Corrector Model for Clinch Analysis

U. Ahmed(✉) and A. Brosius

Chair of Forming and Machining Processes (FF), Technische Universität Dresden (TUD), 01069 Dresden, Germany

usama.ahmed@tu-dresden.de

Abstract. The clinching process in sheet metal production is affected by variations caused by material and production factors such as batch inconsistencies, non-uniform preconsolidation of sheet materials, and tool wear. These variations lead to fluctuations in important quality parameters such as undercut, neck, and bottom thickness, resulting in inconsistent joint quality. These inconsistencies can negatively impact the load-bearing capacity of a clinched joint. The research aims to enhance process monitoring by incorporating the recording of geometric joint characteristics. This will be achieved by using a rapid in-situ simulation during the joining process to determine the resulting quality parameters of the joint. A quick calculation model based on linear-elastic FEA will be developed and coupled with a semi-analytical approach to correct nodal displacements, thereby minimizing deformation energy in the system. This approach significantly reduces calculation time, as only a linear-elastic "forward" calculation is performed, without iterative loops. This novel process monitoring will enable medium-term control of clinching processes, laying the foundation for self-learning processes and intelligent process control.

Keywords: Clinching · Process Monitoring · Prediction-Correction Model

1 Introduction

The utilization of mechanical joining techniques such as clinching offers a viable approach to achieving lightweight construction with cost-effectiveness for joint manufacturing [1]. Clinching can effectively join different types of sheets, including those with coatings. Non-cutting clinching processes yield gas-tight and corrosion-resistant joints. The research under consideration focuses on clinching without cutting using a rigid die and a moving tool. Initially, the parts to be joined are positioned on the die and secured by the spring force exerted by the blank holder. In the subsequent step, the punch penetrates the parts, causing localized material displacement into the die. Compression occurs only when the punch advances further, resulting in necking of the material and a reduction in the base thickness. The joining material undergoes radial extrusion into the annular channel (rigid die), creating an undercut. Once the punch reaches its maximum force in

T. Bauernhansl et al. (Eds.): WGP 2023, LNPE, pp. 64–73, 2024.
https://doi.org/10.1007/978-3-031-47394-4_7

a force-controlled machine guide, the return stroke takes place. At this point, the clinch connection is formed, eliminating the need for post-treatment.

1.1 Process Monitoring

Efforts are underway in various areas of manufacturing technology to implement real-time process monitoring systems [2] and digital twins for intelligent monitoring and analysis of process behavior in tool manufacturing [3] and additive manufacturing [4]. Process monitoring systems play a crucial role in ensuring the required quality and documentation, particularly for safety-critical components. However, current process monitoring systems used in clinching processes do not provide insights into quality-relevant outcome variables. Assessing these variables, such as neck thickness, undercut, and bottom thickness, requires destructive post-process analysis. Presently, process monitoring cannot provide information about quality-relevant outcome variables, limiting the ability to detect significant changes in joint quality.

Process monitoring in clinching relies on evaluating the force-displacement curves in relation to predefined target data. Data necessary for this analysis is internally recorded during the process. Ideally, for a given tool set and sheet pairing, the resulting force-displacement curves remain constant [5]. However, this approach carries the risk of detecting pseudo faults, where the joint exhibits full load despite deviations in the comparison curve. Common evaluation methods for monitoring process curves include limit value, tolerance window, envelope curve methods, and the use of neural networks.

The limit value method is a simple and widely used approach in clinching process monitoring. It establishes a threshold for force and displacement measurements, flagging values that exceed this threshold as indications of faults. However, this method only detects a limited number of possible errors [5]. The tolerance window method monitors selected sections (tolerance windows) of the force-displacement curve distributed along a reference curve, considering a joint defective if the measured curve does not pass through any of these windows [5]. Another method, envelope analysis, involves generating a reference curve extended by a tolerance band. If a force-displacement curve falls outside this tolerance band, a faulty clinching process is assumed [5]. The accuracy of envelope curve evaluation may vary due to the wide tolerance, leading to undetected faulty connections, or a narrow tolerance, resulting in frequent process interruptions. Both extremes are undesirable in practical operations [6]. Alternatively, artificial intelligence (AI) methods, specifically neural networks, can be employed to evaluate and monitor the recorded force-displacement curve of a clinching process. Neural networks allow classification of process states even when correlations between process data are unknown. The learning process involves analyzing multiple datasets and assigning them to known process states, although it does not allow for clear parameter adjustments [5]. Overall, process monitoring enables complete control of the process, reduces the reject rate by halting production when necessary, and facilitates error analysis. However, it does not directly provide insights into characteristics of the clinching joint, such as neck thickness and undercut. Furthermore, this method is susceptible to pseudo defects, erroneously identifying defects that do not exist [7].

The objective of the ongoing research project is to expand process monitoring to include the recording of quality-relevant data related to geometric joint characteristics.

The approach involves a fast in situ simulation of the joining process, providing additional result data (neck thickness, undercut, and bottom thickness) based on the recorded force-displacement curves. The assessment of joint quality becomes more reliable, allowing for subsequent corrections from one joint to another by adjusting process parameters such as speed and immersion depth. In the future, adjustable adaptive tools could be developed to accommodate a wide range of variants using suitable models.

2 Methodology

A Python code is developed to perform a linear elastic finite element analysis which is then verified both analytically using the FE equations and numerically with a LS-Dyna simulation. The next step concerns the development of the contact between the tool and the joining sheets. An unconventional approach is used to develop the contact in order to save computational time and avoid nonlinearity. This is followed by updating the discrete mesh by means of numerical methods to minimize the deformation energy. The objective is that the numerical method such as steepest descent method gives the nodal position where the deformation energy is minimum until the nodal displacements are almost identical to those of an incremental elastic-plastic FEA.

2.1 Prediction-Correction Model

A prediction-correction model is developed for the fast analysis of the clinching process with the aim of process monitoring.

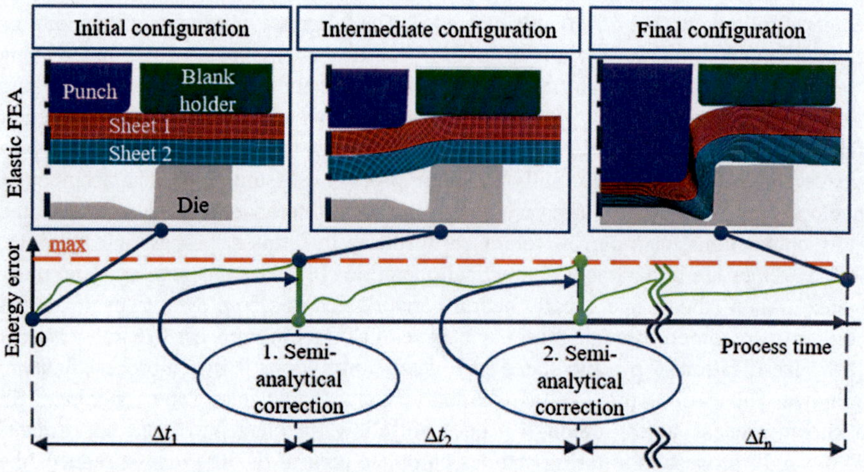

Fig. 1. Schematic illustration of the prediction-correction model

It combines linear elastic finite element analysis (FEA) with a semi-analytical approach to correct nodal displacements. Initially, a linear-elastic FEA is used for quick calculation of nodal displacements. However, as computational time progresses, an error arises due to the elastic-plastic behavior of the material. To quantify this error, the elastic deformation energy ($E_{elastic}$) is compared with the real energy ($E_{plastic}$) obtained by integrating the force-displacement curve. If the energy difference (E) exceeds a threshold, the linear-elastic calculation is aborted and $E_{elastic}$ is corrected. For this purpose, the element deformations calculated at this point are used and a comparative strain is determined, with which an initial plastic deformation energy $E_{plastic}$ is calculated via the known flow curve of the material. An optimization algorithm minimizes the difference between $E_{elastic}$ and the experimental $E_{plastic}$, making the nodal displacements similar to those obtained from an incremental elastic-plastic FEA. This correction step is repeated as needed. By avoiding iterative loops, this approach reduces FEA calculation time to a few seconds. The process monitoring model is shown in Fig. 1.

2.2 FEA Python Code

The finite element method (FEM) leads to an approximate solution of continuous problems. The behavior of these problems is described by partial, spatial and time dependent differential equations. Thereby, each state variable of a continuum has infinitely many values and thus each problem has infinitely many unknowns [8]. Nevertheless, in order to solve the problem, the basic principle of FEM is to decompose the problem into finite elements, i.e., a finite number of parts to be calculated. These elements are connected to each other at nodes. Thus, a complex problem becomes a solvable one [9].

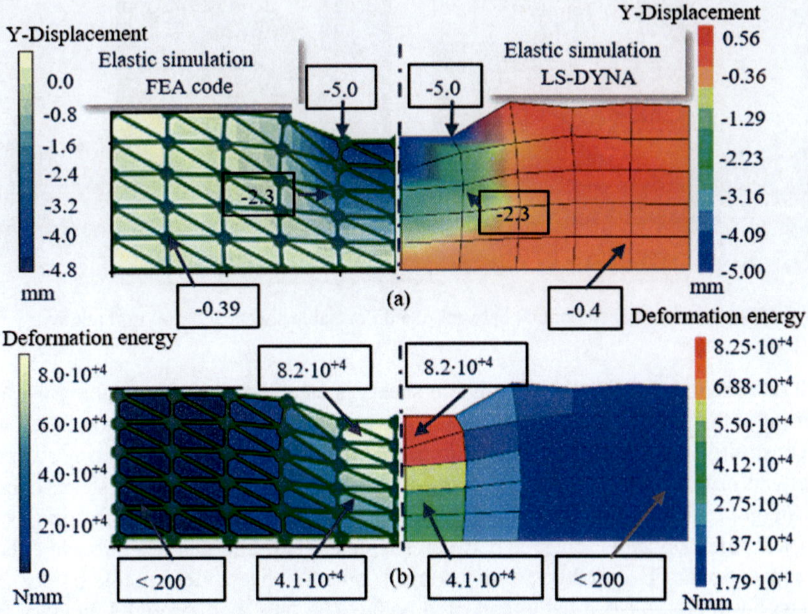

Fig. 2. Verification of the FE code with LS-DYNA simulation

At the beginning of the model development, a simple immersion process in a sheet metal is considered for model development. To validate the accuracy of the results, the FEA code is compared using an elastic simulation in LS-DYNA as shown in Fig. 2. The results of the FE code align closely with the LS-DYNA simulation. Both simulations capture the vertical displacement of -5 mm at the topmost central nodes, while accurately representing negligible deformation at the outer nodes. The distribution of stored deformation energy in the elements, shown in Fig. 2(b), is also consistent between the two simulations. The central elements exhibit higher deformation energy, whereas regions outside the center have lower energy due to limited deformation.

2.3 Development of the Contact

The clinch joint involves two types of contacts: one between the deformable sheets and another between the rigid and deformable bodies (top sheet/punch and base sheet/die). Conventional contact algorithms based on penalty factors and nonlinear equations used in commercial FEA software like LS-DYNA are not suitable for inline simulation due to time-consuming iterations associated with nonlinearity. Hence, a simplified and innovative approach is developed to establish contact between the sheets.

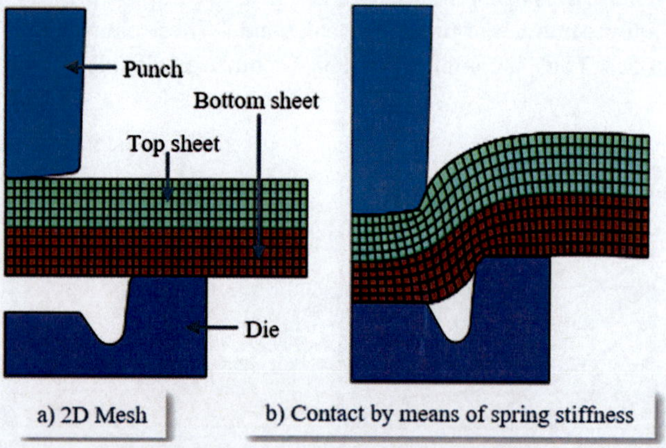

Fig. 3. Established contact between the deformable sheets and the rigid bodies

The contact between the deformable sheets is established by introducing a spring stiffness at the interface of the two sheets. The spring stiffness matrix is combined with the constitutive matrix of the entire system. By gradually increasing the spring stiffness, the effects on the contact are observed. If the spring stiffness is too low, the sheets overlap and merge, indicating improper contact. However, with a sufficiently high spring stiffness, a normal contact is achieved between the plates. This alternative approach replaces the conventional FEA method for developing contact between deformable bodies. The contact between the deformable and rigid bodies is established through boundary conditions, which secure the model against displacement and rotation along its axes. An "Active Set Strategy" is employed for defining rigid contact, utilizing the local coordinate system instead of the global one. For the punch and top sheet, nodes directly beneath the punch surface are selected and moved by the same value as the punch. When the nodes of the base sheet make contact with the die, the second contact algorithm is activated. The bottom nodes of the base sheet are constrained perpendicularly to the die to prevent overlap, while allowing movement tangentially to the die. The resulting contact, as depicted in Fig. 3(b), demonstrates the outcome of these established conditions.

2.4 Mesh Update and Energy Minimization

The linear-elastic simulation exhibits higher deformation energy compared to the elastic-plastic simulation. To address this, the deformation energy error is corrected by minimizing it as a function of displacement. By adjusting nodal positions, the equivalent strain in the element is modified, affecting the equivalent stress and total energy of the system. The equivalent strain is determined by mesh distortion, and the equivalent stress is derived from the yield curve. Convergence is attained using the conjugate gradient method with node displacement as a geometry variable.

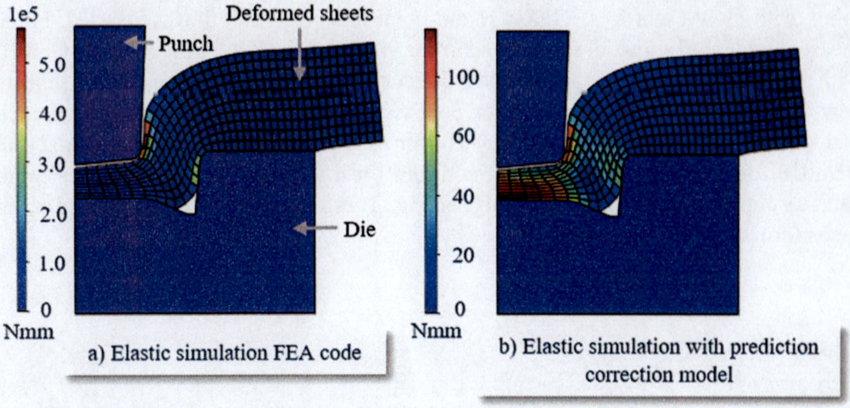

Fig. 4. Optimization of deformation energy as a result of the prediction-correction approach

In Fig. 4(a), the original elastic simulation is displayed, while Fig. 4(b) shows the elastic simulation with the prediction-correction approach. The joint combination includes two 1.5 mm thick AA5182 sheets, each discretized into 40 elements lengthwise and 5 elements thickness-wise for efficient simulation without compromising results. Punch and Matrix designations are A54100 and BD8014 respectively. The total deformation energy in the original simulation is 1.39e+7 Nmm, while in the corrected simulation, it reduces to 7.57e+3 Nmm, representing a 99.7% energy reduction due to mesh update and the correction step. The optimization algorithm minimizes energy after each deformation step, using the last mesh configuration for subsequent steps and iterations. The runtime for the elastic simulation is 8.53 s, whereas the updated elastic simulation computational time is just 25.26 s. Further reduction in computation time can be achieved through increased computational power or by switching to the C#.

3 Model Implementation for Process Monitoring

The prediction-correction model, previously implemented in a linear elastic finite element simulation, is utilized for inline simulation in the clinching process to monitor the process. Experimental force-displacement curves from the clinch joint are recorded and compared with calculated curves in the updated elastic simulation. The joining quality of the clinched joint is evaluated based on geometric quantities calculated by the model. In the conventional approach, the reaction force is calculated using the direct stiffness method, which is not suitable for the optimized elastic simulation. The global stiffness matrix in the updated elastic simulation is calculated from purely linear elastic equations, while in reality it must be calculated from a nonlinear elastic plastic material model. An alternative and simplified approach is used to calculate the reaction force, taking the derivative of the updated deformation energy with respect to displacement. This method calculates the reaction force for discrete displacements rather than a continuous force-displacement curve derived from experiments.

This study used the TZ-VSN machine by TOX® PRESSOTECHNIK GmbH & Co. KG, a servo-electric ironing machine with a 90 kN force capacity for producing clinch joints with closed matrices. TOX® round-point tools (8.0 mm die) were used. Force measurement used a triaxial force sensor (type 9067C) from Kistler Instrumente GmbH, with a 0–200 kN measuring range in the z-direction. Furthermore, high resolution sensors near the joining point were added for precise force-distance curve recording. Punch and matrix designations (A54100, BD8014 for AA5182/AA5182 clinch joint; A52100, BD8016 for HC340LA/HC340LA clinch joint) were utilized. A 30 mm edge-length square sample geometry minimized edge effects. These experiments were repeated seven times for robust data collection.

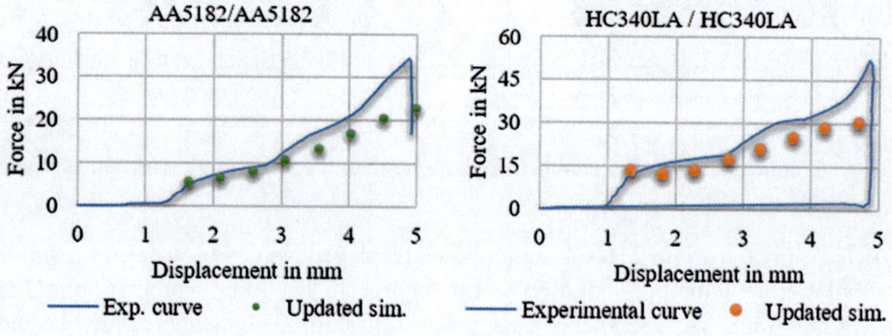

Fig. 5. Comparison of the force-displacement curve for two material combinations

Figure 5(a) presents the comparison between the continuous force-displacement curve obtained from experiments for the AA5182/AA5182 clinch combination and the calculated reaction force at discrete displacement intervals from the updated linear elastic simulation. Up to a displacement of 3.0 mm, the reaction force from the updated simulation aligns well with the experimental curve, with a percentage error ranging from 11.47% to 17.52%. However, the percentage error increases as the punch displacement increases, reaching 32.65% at the end of the clinch joint. Similar trends are observed for the HC340LA/HC340LA joint combination, with a percentage error ranging from 10.66% to 23.36% in the earlier stages and a higher deviation of 31.67% in the final stage of the simulation.

The percentage error between the experimental force-displacement curve and the updated linear elastic simulation arises due to the simplified and unconventional approaches employed in the FE code for inline simulation. These simplifications, including the use of linear elastic equations to save simulation time by avoiding nonlinearity and iterations, non-penalty-based contact modeling between deformable and rigid bodies, and the derivative-based force extraction, directly impact result accuracy. The updated elastic simulation with the prediction-correction approach offers reduced simulation time but sacrifices result accuracy. Therefore, a compromise is necessary to achieve acceptable results while minimizing simulation time.

3.1 Determining the Quality Relevant Parameters

The prediction quality of the inline process monitoring model is assessed based on key parameters such as neck thickness, undercut, and bottom thickness. A comparison is made between the results of the inline simulation and the values determined through experimentation. However, the calculation of these geometric parameters did not meet initial expectations in the updated linear elastic simulation. This discrepancy is attributed to two main factors. The first factor is related to the contact conditions or methods employed in the linear elastic FE simulation. To circumvent nonlinearity issues in contact modeling, a spring stiffness approach is utilized. While higher spring stiffness values ensure perfect normal contact, they also have a drawback. The high stiffness causes the nodes in contact to merge, preventing shear and hindering material flow in the tangential direction. Shearing of nodes is crucial for achieving desired geometric properties like neck thickness and undercut in a successful clinch joint.

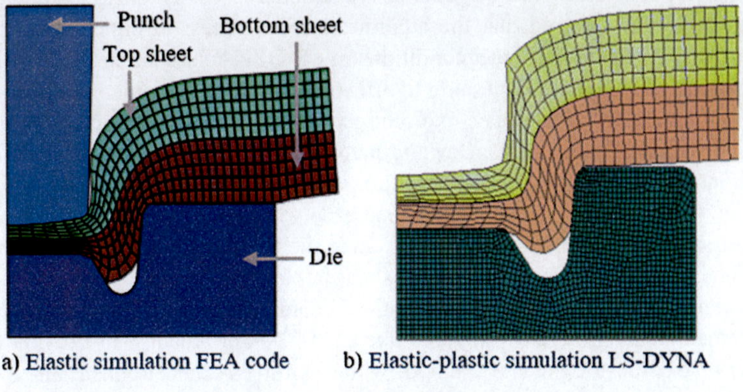

Fig. 6. Comparison of a linear-elastic FE simulation with an elastic-plastic model in LS-DYNA

To investigate the significance of node shearing resulting from material flow, a comparison is conducted between a linear-elastic simulation using FE code and an elastic-plastic simulation in LS-DYNA, as illustrated in Fig. 6. The LS-DYNA elastic-plastic simulation maintains the same geometric parameters, mesh size, element count, and boundary conditions as the linear-elastic simulation in the FE code. In the LS-DYNA simulation, the presence of shear or sliding of nodes at the contact surface is clearly evident, as depicted in Fig. 6(b). This tangential shear, arising from material flow, enables a profile that closely approximates the clinch joint. However, it is worth noting that even in the LS-DYNA elastic-plastic simulation, a fully successful clinch connection is not achieved.

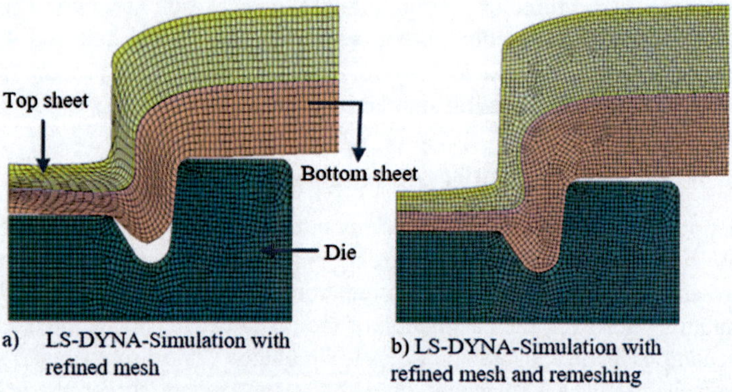

Fig. 7. Comparison of LS-DYNA simulations with (a) refined mesh and (b) remeshing

The LS-DYNA simulation is rerun with a refined mesh, resulting in a significantly enhanced profile that closely approaches a clinch connection as shown in Fig. 7(a). Despite this progress, the undercut is still not accurately captured. Notably, the implementation of adaptive remeshing in LS-DYNA brings about a flawless clinch joint that successfully encompasses the neck thickness and undercut as demonstrated in Fig. 7(b). Consequently, it is inferred that the attainment of a successful clinch connection is attributed to the nonlinear contact conditions in LS-DYNA simulation, which facilitate node shearing and adaptive remeshing to effectively represent the material flow.

Several approaches have been explored to achieve successful results while maintaining linear contact conditions. One alternative approach involves dividing the spring stiffness into local tangential and normal components, with the tangential stiffness close to zero ($k_x \approx 0$) and the normal stiffness extremely high ($k_y \approx \infty$). In a different approach, the stiffness matrix is modified to exclude shear components while retaining only normal ones, aiming to account for nodal shear but leading to sheet overlap.

Another option is to utilize the conventional nonlinear contact method employed in finite element analysis (FEA) programs like LS-DYNA or Abaqus. In FEA, nonlinear contact is based on the penalty method, which enforces contact constraints between surfaces. By introducing a penalty stiffness or contact force, the surfaces are prevented from penetrating or sliding apart. When the surfaces touch, the penalty stiffness or force increases, creating a restoring force that pushes them apart. However, implementing such a contact model introduces nonlinearity and significantly increases computation time, which contradicts the purpose of this research.

4 Conclusion

In conclusion, this research involved the development and validation of a linear elastic FEA code, which was successfully validated analytically and numerically. A prediction-correction approach was implemented to minimize deformation energy, resulting in a reduction of over 99%. Additionally, an unconventional approach established linear contact between deformable sheets and rigid tools. The force-displacement curves obtained from the updated linear-elastic simulation were compared to experimental curves, although with some accuracy limitations in favor of faster simulation time. The clinch connection did not yield the desired results due to non-conventional linear contact in FE simulation and the absence of remeshing elements to capture shear of nodes caused by material flow. Introducing non-linear contact models would require iterations and significantly increase computation time, which conflicts with the research goals.

Acknowledgement. The authors express their gratitude to the Laboratorium für Werkstoff- und Fügetechnik (LWF) at Paderborn University for providing the experimental force-displacement curves utilized in Fig. 5.

References

1. Eshtayeh, M.M., Hrairi, M., Mohiuddin, A.K.M.: Clinching process for joining dissimilar materials: state of the art. Int. J. Adv. Manuf. Technol. **82**, 179–195 (2016)
2. Kunze, R., Mallmann, G., Schmitt, R.: Inline plasma analysis as tool for process monitoring in laser micro machining for multilayer materials. Phys. Procedia **83**, 1329–1338 (2016)
3. Botkina, D., Hedlind, M., Olsson, B., Henser, J., Lundholm, T.: Digital twin of a cutting tool. Procedia CIRP **72**, 215–218 (2018)
4. Mukherjee, T., DebRoy, T.: A digital twin for rapid qualification of 3D printed metallic components. Appl. Mater. Today **14**, 59–65 (2019)
5. Khrebtov, P.: Neuartiges Verfahren zur Online-Prozessüberwachung und - Fehlerklassifizierung beim Durchsetzfügeverbinden von Blechen. Dissertation. TU Clausthal (2011)
6. Breckweg, A.: Automatisiertes und prozessüberwachtes Radialclinchen höherfester Blechwerkstoffe, Dissertation, Universität Stuttgart (2007)
7. Feldmann, K.: Qualitätssicherung in der Elektronikproduktion. tm – technisches Messen, München: Oldenbourg Verlag (1995)
8. Klocke, F.: Fertigungsverfahren 4. Umformen. Springer, Heidelberg (2017). https://doi.org/10.1007/978-3-662-54714-4
9. Schmid, E.: Nummerische Analyse des Clinchprozesses mit dreigeteilter Matrize. Dissertation. Universität Paderborn (2004)

Investigations on Continuous Transient Dynamic Analysis of Clinched Aluminum Sheets

G. Reschke(✉) and A. Brosius

Chair of Forming and Machining Processes, Technical University of Dresden, Dresden, Germany
gregor.reschke@tu-dresden.de

Abstract. Clinching is a representative of the mechanical joining processes defined by form and force closure. Traditionally, clinched joints are tested destructively, for example using shear tensile tests and metallographic sections. Transient dynamic analysis (TDA) offers the possibility of rapid non-destructive testing in addition to the established methods. TDA can be used during and after joining as well as during operational loading. In previously performed investigations TDA was carried out at discrete points and allowed the detection of defects in clinched joints. Utilizing a new approach, continuous measurement is made possible. First, the characteristic frequencies of the specimens are determined. The specimen is then excited harmonically with one specific determined characteristic frequency and the response is evaluated. The investigations were carried out on clinched shear tensile specimens made of aluminum. This allows the detection of emerging defects my means of TDA during the ongoing experiment.

Keywords: Active acoustic testing · Clinching · Non-destructive testing · Aluminum

1 Introduction

Clinching is a mechanical joining process where two or more components are joined by cold forming. The predominant binding mechanisms are force and form closure [1]. The load-bearing capacity of the clinch connection is decisively characterized by the undercut and the neck thickness [2]. These variables can only be measured destructively using metallographic micrographs or non-destructively using computer tomography [3]. Computer tomography in particular is a cost- and time-intensive testing method. The load-bearing capacity of clinched joints is determined primarily by destructive testing methods, such as shear or head tensile tests. No destructive test method can be used to evaluate in-service joints. There is a deficit of a rapid, non-destructive test method for the assessment of clinch connections.

T. Bauernhansl et al. (Eds.): WGP 2023, LNPE, pp. 74–83, 2024.
https://doi.org/10.1007/978-3-031-47394-4_8

Transient dynamic analysis (TDA) could close this gap. It is based on the targeted introduction of sound waves into a structure and evaluation of the signal damped by the joint. It is shown that this method can detect irregularities in clinch joints during joining [4] and under load [5]. Sadeghian et al. show additionally that geometric changes in the clinch point, such as a tool offset, result in a change in response behavior due to the changed formation of the joint [6]. This method can be applied to different joining processes as Wolf et al. [7] or Wang et al. [8] show. Since TDA is based on the propagation of sound waves, the method is not confined to specific materials. The damping, however, depends on the material as shown by the investigations of Köhler et al. [5]. The utilization of harmonic oscillations is particularly efficient with regard to experimental and numerical investigations.

In previous studies by Stephan et al. [9], TDA was always carried out at discrete points in time. Within a measurement cycle, the specimen was excited with a defined frequency, the response behavior was evaluated and consecutively the frequency was increased gradually. These measurements allow the assessment of a connection at a defined frequency and point in time. Due to the limited temporal resolution, emerging defects leading up to failure can only be detected to a limited extent. Therefore, a measurement setup was developed that enables continuous TDA.

In order to increase the temporal resolution using the continuous TDA, the measurement is fixed to one excitation frequency. Therefore, a suitable excitation frequency has to be determined in a first step. The hypothesis is tested whether the use of one of the natural frequencies of the structure for excitation is appropriate. After selecting a suitable frequency, the continuous TDA is carried out with simultaneous shear tensile testing of the specimen. The aim is the rapid detection of an emerging failure.

2 Materials and Methods

The investigations are carried out on clinched aluminum joints made of EN AW-6014 (Advanz™ 6F-e170, Novelis Inc., Atlanta, USA) with a sheet thickness of $t_s = 2$ mm. All samples were taken from one material batch to exclude batch influences such as sheet thickness deviations or mechanical property differences. The samples were joined in the T4 condition and subsequently transferred to T6. For this purpose, the specimens were aged for 20 min at $\theta = 185$ °C. The geometry of the single overlapping shear tensile specimen was defined according to ISO 12996. Deviating from this, an additional through-hole Ø 3.3 mm per plate was provided for mounting the piezoceramic stack actuators, see Fig. 1. Punches (A52100) and dies (BE8012) from TOX PRESSOTECHNIK GmbH & Co. KG (Weingarten, Germany) were used for clinching. The residual bottom thickness was set to $x = 0.7$ mm. For the shear tensile test, shim plates were used to ensure force concentricity.

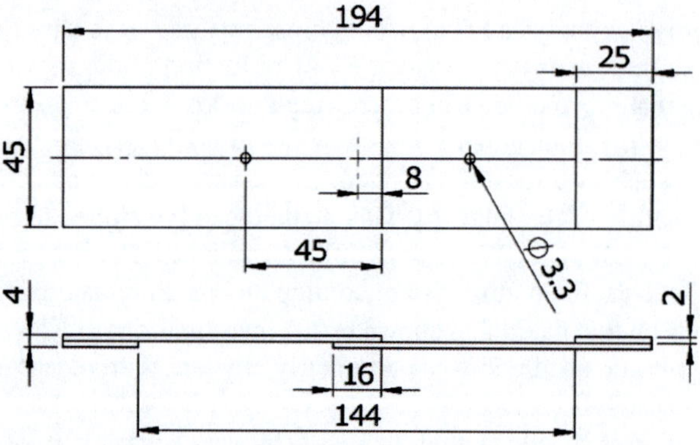

Fig. 1. Shear tensile specimen

To determine the natural frequencies, the specimens were clamped in a test bench, see Fig. 2 (left). The piezoelectric stack actuators (type P-016.00h, PI Ceramic GmbH, Lederhose, Germany) including the seismic mass were axially preloaded by a bolted connection. The bolted joint of size M3 was assembled with a tightening torque of $M = 2.2$ Nm. This defines the position of the stack actuators and sets up the compressive stress required for dynamic use. The actuator was mounted on the die side plate, the sensor is located on the punch side plate.

A universal testing machine type inspekt250 (Hegewald & Peschke Meß- und Prüftechnik GmbH, Nossen, Germany) was used for the shear tensile tests. The test speed during the shear tensile test is $v = 2$ mm/min.

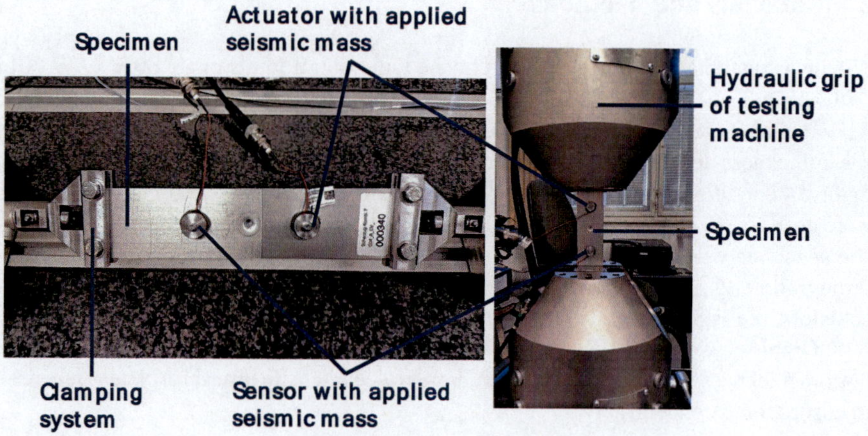

Fig. 2. Experimental setup

The actuator and the sensor are both equipped with an effective seismic mass of $m = 9.5$ g. This intensifies the initiated vibration at the actuator and, at the same time, increases the sensitivity of the sensor. The seismic masses used reduce the resonant frequency of the piezoelectric stack actuators from $f = 144$ kHz to $f \approx 76$ kHz.

LabView was used to control the measurement process. The sinusoidal excitation with an amplitude of 4 V is emitted via the analog output of the NI6356 data acquisition card (National Instruments Corp., Austin, Texas, United States) and amplified to $U = 67$ V by means of an analog signal amplifier type HV-LE150–100-EBW (piezosystem jena GmbH, Jena, Germany). The measured sensor signal passes through a $f = 1$ kHz high-pass filter (type EF111, Thorlabs Inc., Newton, New Jersey, United States) and a $f = 30$ kHz low-pass filter integrated in the Kistler 5018A charge amplifier used (Kistler Instrumente AG, Winterthur, Switzerland). The measurement data obtained are recorded using the NI6356. The sampling rate used is equivalent to 8 times the excitation frequency.

For the determination of the natural frequencies, the frequency band up to is analyzed gradually. In each measurement step, the specimen is excited for 2 s utilizing a defined excitation frequency. With the aim of evaluating the harmonic oscillation of the system, only the last 0.5 s are used for data acquisition and evaluation. The excitation frequency is then increased by 100 Hz and the measurement sequence repeats until the frequency $f = 20$ kHz is reached.

This method was adapted for the continuous TDA. The hardware described is used unchanged for continuous measurement. Deviating from the measurement procedure described above, the sample is continuously excited at one specific defined frequency and the response is recorded. To be able to evaluate signal changes during the test run in the frequency-discrete visualization, the response behavior is evaluated individually for defined data blocks.

Additionally, specimens joined with a lateral tool offset of $\Delta = 0.2$ mm are examined, see microsections in Fig. 3. The tool offset results in a changed clinch point formation, which can be shown by the different neck thicknesses.

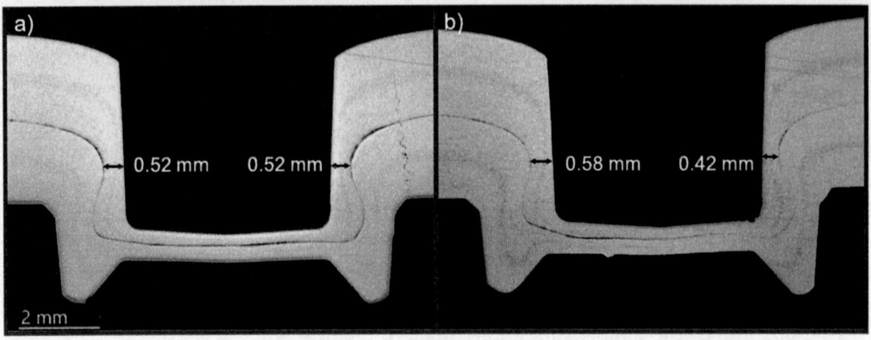

Fig. 3. Microsections of clinch joints with tool offsets of $\Delta = 0.0$ mm (a) and $\Delta = 0.2$ mm (b)

The tool offset was adjusted to reduce the neck thickness in the highly loaded area of the clinch joint. This shall result in a varying failure behavior with a crack initiation at lower displacements and force levels in the shear tensile test. The evaluation of these tests will be used to assess the performance of the continuous TDA.

3 Results and Discussion

For the determination of the natural frequencies, $n = 5$ specimens were tested without tool offset. These measurements were performed without additional load. The measured time-discrete response can be represented frequency-discrete by means of fast Fourier transformation, see Fig. 4. In this figure, the measured sensor signal has been plotted for each excitation frequency and linearly interpolated between the data points.

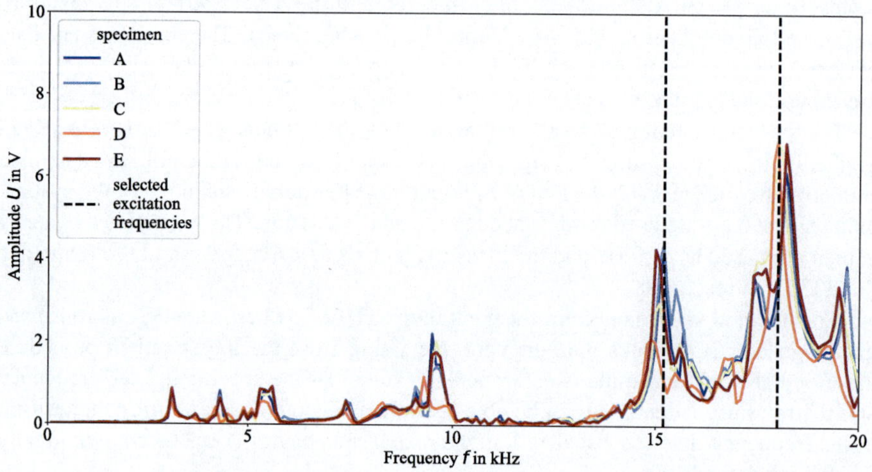

Fig. 4. Spectrogram of the TDA measurement with selected excitation frequencies

The curve shows amplitude heights of varying magnitude over the frequency band with clearly distinguishable peaks, especially in the frequency range above $f = 15$ kHz. The signal strength stands out significantly from the background noise, so that the measurement setup can be evaluated as functional. However, differences are noticeable when comparing the various curves. These can be attributed to the measurement setup on the one hand and to the investigated specimens on the other. For the measurement setup, variations in the clamping of the specimen and the position of the piezoelectric stack actuators are to be mentioned. Köhler et al. [10] shows numerically a strong spatial dependence of the amplitudes occurring during the TDA, which is confirmed experimentally by Stephan et al. [11]. Regarding the investigated samples, series fluctuations have to be taken into account, which mean a variance of the sample geometry and the clinch point formation. Concerning the sample geometry especially the overlap length and the position of the clinch joint are to be mentioned.

Summarizing, it is found that the response characteristics and thus the natural frequencies of each specimen are individual. Nevertheless, suitable excitation frequencies for continuous TDA are to be concluded from these measurements. Two different approaches are available for this purpose. On the one hand, each specimen can be excited with an individually selected natural frequency or, on the other hand, a frequency can be selected at which the amplitude of all specimens is as large as possible. The latter variant was chosen for the following investigations. This allows testing of the specimens with tool offset without prior determination of the natural frequencies.

As Köhler et al. [10] show, the lower frequencies up to $f = 2.5$ kHz are characterized by the oscillation of the entire structure for this sample geometry. This frequency range is therefore not suitable for the detailed analysis of the clinch joint. At the frequency ranges $f \approx 9.5$ kHz, $f \approx 15.2$ kHz and $f \approx 18$ kHz the highest amplitudes are measurable. The frequencies $f = 15.2$ kHz and $f = 18$ kHz are selected for further work because the large amplitudes mean a good signal to noise ratio and the significance of the measurement results increases.

For each of the selected excitation frequencies, $n = 3$ shear tensile tests were performed with continuous TDA. During the development of the measurement regime, the data basis and the temporal resolution of the continuous measurement were in focus. In this first step, completed data blocks, recorded for a measuring interval of 1 s, are analyzed individually. With regard to the test duration and speed of the shear tensile test, a good compromise between data basis and temporal resolution can thus be achieved.

Figure 5 shows the superposition of the TDA and the force-time curve of the shear tensile test. The measurement shown was recorded at an excitation frequency of $f = 15.2$ kHz. The synchronization of the data took place within the context of the data evaluation. For the representation of the TDA signal, the maximum of each data block was determined and linearly interpolated between the data points.

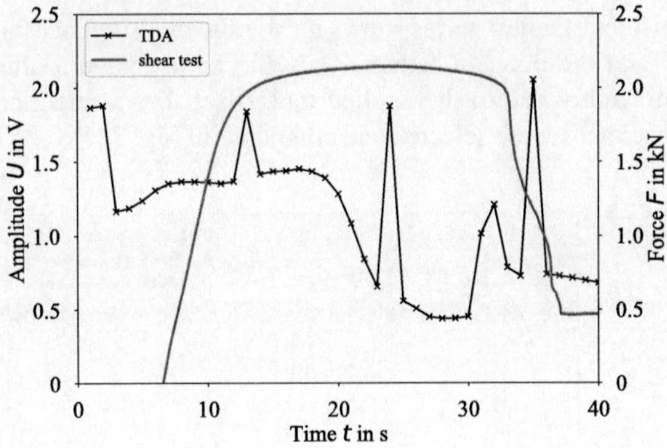

Fig. 5. Result of continuous TDA superimposed on force-time curve of shear tensile test

The TDA signal describes a monotonously decreasing curve over the shear tensile test, which is interrupted by several peaks. These peaks correlate with characteristic points of a shear tensile test and can be assigned to the transition from elastic to plastic deformation (flattening of the force-time curve), the exceeding of the uniform strain (exceeding of force maximum) and the crack formation and propagation (rapid force reduction). The superordinate monotonically decreasing curve can be correlated with the increasing damping due to increasing plastic deformation. The measurements show that the curve characteristics of the shear tensile test can be reproduced well using continuous TDA.

To better assess the sensitivity of continuous TDA, specimens with a tool offset were examined. Figure 6 shows a comparison of the shear tensile curves obtained. A total of $n = 19$ specimens without tool offset were tested in these investigations and are shown here in blue. The specimens with tool offset are shown in red.

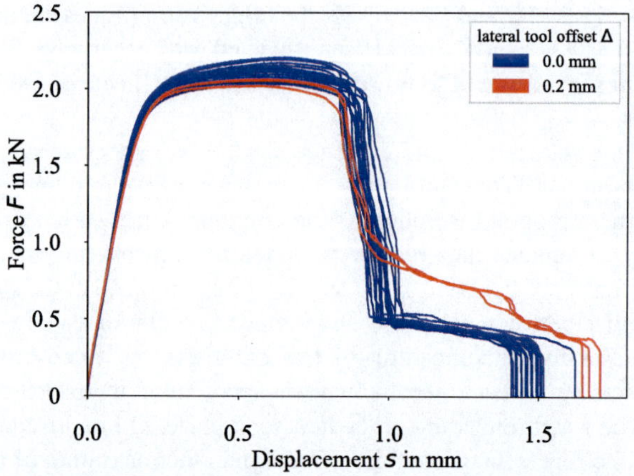

Fig. 6. Force-displacement curves of the shear tensile tests

The comparison of the test series shows no significant differences in the maximum forces achieved and the point of failure. Only the failure pattern after crack formation differs significantly due to the applied tool offset. For a direct comparison, two characteristic specimens were selected and compared in Fig. 7.

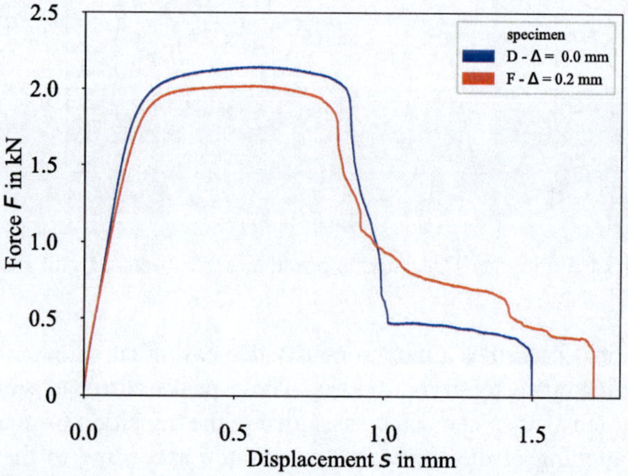

Fig. 7. Influence of the tool offset on force-displacement curves

Significant differences are shown with respect to the maximum shear tensile force, the initiation of failure and the failure behavior. It is necessary to analyze whether these differences are detectable in the TDA signal. For this purpose, the TDA measurements of these two specimens are compared in Fig. 8.

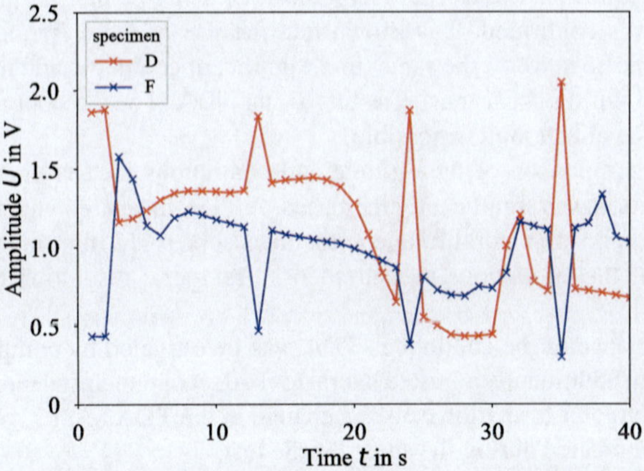

Fig. 8. Comparison of continuous TDA at different tool offsets

The TDA signal of sample "D" has already been discussed above. The curve of the sample with tool offset ("F") also shows a basically decreasing monotonicity with some strong deviations from this curve. It can be seen that the signal shows a significantly reduced amplitude at characteristic points. The measurements of the two specimens show peaks in opposite directions. The cause of this is currently not yet known and is the subject of further investigations. For a possible process control based on the TDA, the gradient of the TDA signal should therefore be considered. A single evaluation of the maximum amplitudes is not purposeful.

Comparing the force-displacement curves of the two specimens (cf. Fig. 7) reveals differences with respect to the initiation of failure. Specimen "F" with tool offset fails at a smaller displacement. It is expected that these differences can also be seen in the TDA. However, by comparing the TDA signals, it can be seen that the peaks occur at identical times. The different points of failure can therefore not be timely resolved. In further investigations, the temporal resolution will be further increased. For this purpose, a floating data evaluation should be used. This enables the evaluation of a sufficient amount of data with high temporal resolution at the same time. It needs to be investigated whether significant gradient changes can also be determined, which can be used for process evaluation or control.

The introduced tool offset means a change in the clinch joint formation and the interaction of the effective bonding mechanisms. Stephan et al. [11] already showed that this influences the damping behavior. The correlation of the TDA signal with the actual or maximum shear tensile force is still pending and cannot be clearly clarified due to the limited scope of the experiment. Using TDA, differences between samples can be detected very well. These signal differences cannot yet be assigned to specific properties of the clinch joint during shear tensile loading. Further research is required to create the linkage.

4 Conclusion

In these investigations, a continuous TDA was realized for the evaluation of clinched shear tensile specimens. The research hypothesis of using natural frequencies to excite the specimen was confirmed. It is shown that the use of these frequencies lead to a good signal-to-noise ratio. At the same time, significant geometric and material changes can be detected during shear tensile testing using TDA. Thus, a continuous condition assessment of the clinch joint is possible.

In a direct comparison of the discrete and continuous measurement methods, the temporal resolution was significantly increased. A TDA measurement at discrete times, as used to determine the natural frequencies, takes approx. 12 min. In the implemented configuration of the continuous measurement, a temporal resolution of 1 s could be achieved.

The performance of the continuous TDA was investigated by comparative tests on clinch joints with additionally adjusted lateral tool offset of punch and die. It is shown that the altered clinch joint formation causes a change in the TDA results, yet characteristic points are still indicated during the shear tensile test.

For further work, the evaluation of the gradient of the TDA signal will be more focused. The comparative measurements also show that the temporal resolution is still insufficient to resolve the minor variations of the shear tensile tests.

Acknowledgements. This research was funded by the Deutsche Forschungsgemeinschaft (DFG, German Research Foundation) – TRR 285/2 – 418701707.

References

1. Groche, P., Wohletz, S., Brenneis, M., Pabst, C., Resch, F.: Joining by forming - a review on joint mechanisms, applications and future trends. J. Mater. Process. Technol. **214**(10), 1972–1994 (2014). https://doi.org/10.1016/j.jmatprotec.2013.12.022
2. DVS EFB: Merkblatt 3420: Clinchen – Überblick. DVS Media GmbH, Düsseldorf (2021)
3. Gröger, B., et al.: Clinching of thermoplastic composites and metals—a comparison of three novel joining technologies. Materials **14**(9), 1–18 (2021). https://doi.org/10.3390/ma14092286
4. Köhler, D., Stephan, R., Kupfer, R., Troschitz, J., Brosius, A., Gude, M.: Investigations on combined in-situ CT and acoustic analysis during clinching. Key Eng. Mater. **926**, 1489–1497 (2022). https://doi.org/10.4028/p-32330d
5. Köhler, D., Stephan, R., Kupfer, R., Troschitz, J., Brosius, A., Gude, M.: In-situ computed tomography and transient dynamic analysis of a single-lap shear test with a composite-metal clinch point. In: Liewald, M., Verl, A., Bauernhansl, T., Möhring, H.C. (eds.) Production at the Leading Edge of Technology. WGP 2022. Lecture Notes in Production Engineering. Springer, Cham (2023). https://doi.org/10.1007/978-3-031-18318-8_28
6. Sadeghian, B., Guilleaume, C., Lafarge, R., Brosius, A.: Investigation of Clinched Joints – A Finite Element Simulation of a Non-destructive Approach. In: Behrens, B.-A., Brosius, A., Hintze, W., Ihlenfeldt, S., Wulfsberg, J.J. (eds.) WGP 2020. LNPE, pp. 116–124. Springer, Heidelberg (2021). https://doi.org/10.1007/978-3-662-62138-7_12

7. Wolf, A., Lafarge, R., Brosius, A.: A non-destructive testing method for joints by the measurement of the energy dissipation. Prod. Eng. Res. Devel. **13**(1), 99–106 (2019). https://doi.org/10.1007/s11740-018-0860-x
8. Wang, F., Huo, L., Song, G.: A piezoelectric active sensing method for quantitative monitoring of bolt loosening using energy dissipation caused by tangential damping based on the fractal contact theory. Smart Mater. Struct. **27**(1), 15–23 (2017). https://doi.org/10.1088/1361-665X/aa9a65
9. Stephan, R., Brosius, A.: Experimental measurement method and evaluation of an analytical approach for sound conduction through multiple clinched sheets. In: Dix, M., Kräusel, V. (eds.) SFU/ICAFT 2022, vol. 26 (2022). https://doi.org/10.3390/engproc2022026025
10. Köhler, D., Sadeghian, B., Troschitz, J., Kupfer, R., Gude, M., Brosius, A.: Characterisation of lateral offsets in clinch points with computed tomography and transient dynamic analysis. J. Adv. Joining Process. **5**, 100089 (2022). https://doi.org/10.1016/j.jajp.2021.100089
11. Stephan, R., Kalich, J., Füssel, U., Brosius, A.: Untersuchung des Einflusses des vorliegenden Bindemechanismus auf die Körperschalldämpfung in Clinchverbindungen. In: Zimmermann, M. (eds.) Werkstoffprüfung 2022, vol. 40 (2022)

Detection of Defective Deep Drawn Sheet Metal Parts by Using Machine Learning Methods for Image Classification

P. Tchasse(✉), A. Schenek, K. R. Riedmüller, and M. Liewald

Institute for Metal Forming Technology, University of Stuttgart, Holzgartenstraße 17, 70174 Stuttgart, Germany
papdo.tchasse@ifu.uni-stuttgart.de

Abstract. The progressing digitization in the manufacturing industry and the enhancements in the field of machine learning (ML) offer new approaches for monitoring manufacturing processes. In this context, the image classification using ML methods represents a well-solved task that can be leveraged for evaluating processes and products. From this perspective, this paper presents an application of ML methods in order to classify defective deep drawn parts. It proposes a methodical approach which includes not only the data acquisition using a low-cost equipment, but also the data pre-processing as well as the setup and training of the ML model. The described method is applied on an academic part of rectangular shape. After collecting a dataset of nearly 1000 images, a convolutional neural network was trained to identify three different quality classes. The trained model was then evaluated on a test dataset containing 30% of the samples and shows an accuracy of 96,77%.

Keywords: Machine Learning · Image Classification · Deep Drawing

1 Introduction

During deep drawing, a sheet metal blank is formed into a hollow part or a hollow part is formed into a similar part of smaller dimensions under tensile and compressive loads without changing its sheet thickness. Deep drawing is one of the most important processes in sheet metal forming and is used for the production of, for example, car body parts and white goods. However, manufacturing-related defects like cracks, wrinkles and partly waviness can occur during the production of such deep drawn parts, resulting in scrap or rejected products [1].

In this context, the expanding use of machine learning (ML) methods in production engineering and the increasing computer performances have provided a variety of new ways to identify defective sheet metal parts. ML is a domain of artificial intelligence, in which the algorithms are used to recognize patterns based on a previous optimization with example data. One of the subfields of ML, which focusses on learning patterns from image data and has already shown promising results in different use-cases, is deep learning with convolutional neural networks (CNN) [2, 3].

T. Bauernhansl et al. (Eds.): WGP 2023, LNPE, pp. 84–93, 2024.
https://doi.org/10.1007/978-3-031-47394-4_9

CNNs were introduced in [4] being derived from the study of the brain's visual cortex. The first network architecture that established the structure of CNNs as we know them today was introduced in [5]. Since their first application in the 1980s, CNNs have been used to perform many tasks as image recognition, voice recognition and natural language processing. Furthermore, they have been integrated in many systems like image search services, self-driving cars and automatic video classification [6].

In this paper, a state of art related to the image classification with CNNs in the field of quality assurance of sheet metal components is presented and methods are described, how a model can be built and optimized in order to successfully classify defective deep drawn parts.

2 Related Works

In the recent years, CNNs have been successfully used to address various challenges at different stages of product developments. In [7], for example, the authors combined finite element analysis and CNN to predict the buckling risk of automotive sheet metal panels during part design. The model was used to correlate part shapes with indentation points. In the study, the risk of buckling could be predicted with an accuracy of 90.1%. Another application of CNN in the process design and development is described in [8]. In this study, the authors attempted to deduce the appropriate manufacturing process for different parts based on their CAD representations. Here, five processes were considered: spinning, deep drawing, stretch forming, air bending and roll bending. For each process 300 parts were generated and a test accuracy of 89% was achieved.

Furthermore, the authors of [9] proposed a method to classify hot rolling strip surface defects based on the Wasserstein generative adversarial network. The model used was the SE-ResNet34, which was coupled with attention mechanisms in order to improve the classification accuracy. In this study, a test accuracy of 99.20% was achieved. Another study dealing with hot rolling strips addressed the recognition of surface defects based on CNN [10]. The study focused on five types of defects, specifically upwarps, black lines, cracks, slag inclusion and gas holes. As a result, it was shown that a recognition accuracy of 93.5% can be obtained in a short period of time depending on the CNN architecture used.

CNNs have also been used to derive the tool wear from workpiece images as described in [11]. For this purpose, blank workpieces were manufactured using 16 different wear states. By using a CNN, a classification accuracy of 99% could be obtained on a test dataset containing more than 1000 images.

3 Methodical Procedure

The study presented in this paper addresses the inspection of deep drawn parts using rectangular cups as an example. For this purpose, a corresponding deep drawing tool was mounted in the servo mechanical press (Aida NST-S2-6300(D)-305-150-SERVO) of the Institute for Metal Forming Technology in Stuttgart and 30 cup parts were manufactured. During the manufacturing of the rectangular cups, process parameters like lubricant quantity and blank holder force were varied in such a way that both defective and good parts were produced. Thereby, the defective parts showed two types of defects, namely cracks and dents. Afterwards, a multilabel binary classification task was carried out with the objective of classifying images of the produced parts into the three labels "good", "cracked" and "dented" by using CNNs. The pipeline for performing the multilabel classification task consisted of the four steps data collection, data preprocessing, model building and model training, described in the following subchapters.

For the model building, two approaches were considered. The first approach was to use an algorithm for hyperparameter optimization to automatically build a neural network from the bottom up by combining the different components of a CNN and applying some regularizations techniques. The second approach, which was compared with the first, comprised the use of a pre-trained model. The chosen pre-trained model was ResNet50 [12], which is considered in literature to be one of the best performing models in the ImageNet challenge [13].

3.1 Data Collection

For the purpose of data collection, a total of 30 rectangular cups of length $\times$ width $\times$ height $= 225 \times 180 \times 30$ mm was deep drawn with varying process parameters (e.g. lubrication, blank holder force). Afterwards, the manufactured parts were photographed with an 8MP color camera.

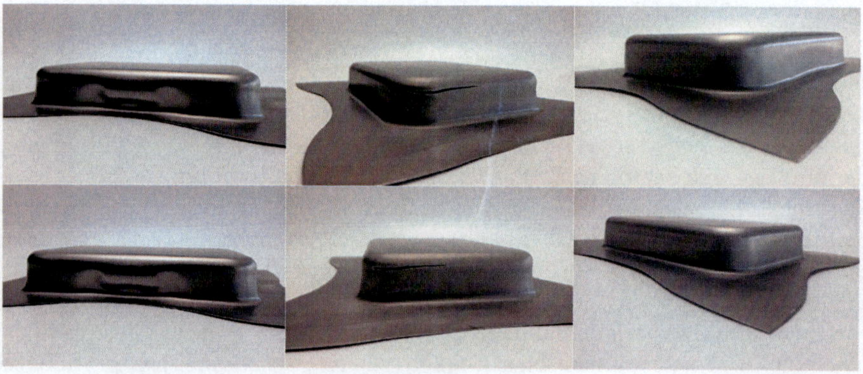

Fig. 1. Dataset of deep drawn samples and labels

Here, in order to enrich the training set in instances and variety and thus increase the data quality, each cup was captured in different positions by slightly shifting or rotating them in some directions. Finally, a data set of 940 images was collected, of which 70% were used for the training and the remaining 30% were used as test data to evaluate the generalization ability of the models. Figure 1 shows, for example, three labelled parts in different positions.

3.2 Data Preprocessing

The preprocessing operations were identical for the two compared models, i.e., the bottom-up CNN and the pretrained ResNet50, and comprised the three steps image resizing, data augmentation and normalization:

Image Resizing. Initially, the acquired images of the rectangular cups were reduced from 3264 × 2448 Pixels to a fixed size of 180 × 180 Pixels. On the one hand, this was done because the backpropagation algorithm used to compute the gradients of the loss function during the training cannot handle different sized inputs [14]. And on the other hand, large images would have required a huge amount of computational resources. The image size was chosen on the premise that the training patterns retain their aspect ratio and are still identifiable [5]. With the new image size, the smallest defects were of dimensions 15.7 × 10.2 Pixels.

Data Augmentation. After resizing the images, data augmentation was applied to artificially extent the training set for the CNNs by editing existing instances. For each training instance, additional variants were generated by using or combining operations such as zooming, rotation and shifting [15]. The advantage of data augmentation is that the models become more tolerant to variations in size, orientation, position and luminosity of the learned patterns [2]. Applying data augmentation helped to generate three times more data for the training, that means the training set increased from 658 to 2632 images.

Normalization. Generally, the gradient computation with the gradient descent algorithm converges faster if the input values are closed to zero [16]. Therefore, the value of the colours of each channel of a pixel (RGB) were divided by 255 in the course of the image data normalization. Thus, the colours of the input images were scaled to fit within the range [0, 1]. This approach regularizes the training by preventing the computed gradients of the loss function from growing exponentially, which is known as exploding gradients [2].

3.3 Model Building

The CNNs used for image classification consist of two different parts. The first and most important part is the convolutional basis, consisting of two fundamental building blocks: the convolutional layers and the pooling layers (Fig. 2). The second part, which is built on the top of the convolutional basis, is the classifier, which often represents a multilayer perceptron (MLP) [2]. In the following, the individual model components will be explained in more detail. To build the bottom-up CNN used in the present study, the different model components were combined by using hyperparameter optimization. For the model building, regularization methods were also considered for the purpose of solving two particular issues in ML, underfitting and overfitting.

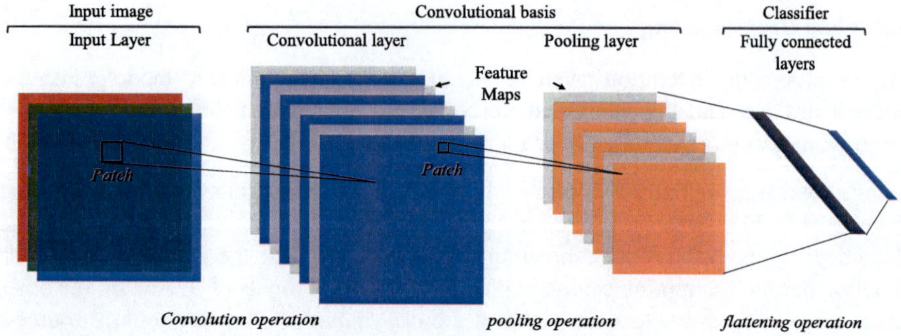

Fig. 2. Structure of a CNN for image classification

Convolutional Layers. Convolutional layers are the result of the convolution operation, which is performed by each neuron on its processed patch located on the previous layer in the network. The components of the convolutional layer are the feature maps. For each feature map, there is one filter. Given two consecutive layers A and B and considering two side by side neurons in the layer B, the degree of shift or overlapping of their neuron's neighbourhoods can be controlled by a parameter referred to as stride. Depending on the stride and the size of each neighbourhood, the neighbourhood of a neuron may be smaller than its receptive field at the border of the image. In this case, zeros can be automatically added in order to fit the patch in the receptive field. This approach is called padding [17]. For each feature map, the convolution operation is performed as:

$$z_{i,j} = f\left(b + \sum_{u=1}^{f_h} \sum_{v=1}^{f_w} x_{i',j'} . w_{u,v}\right) with \begin{cases} i' = u.s_h + f_h - 1 \\ j' = v.s_w + f_w - 1 \end{cases} \quad (1)$$

$z_{i,j}$ is the output on the current feature map at position (i, j). (f_h, f_w) is the dimension or size of the filter whose weights are $w_{u,v}$. $x_{i',j'}$ represents the pixel values from the previous layer, considering the different vertical and horizontal strides s_h and s_w. The activation function f used in this paper is the rectified linear unit (ReLU), as it is used in well-known ML architectures and is frequently recommended for image classification tasks [3, 5].

Pooling Layers. Inserting pooling layers between convolutional layers allowed to reduce the size of the processed instance and thus the computational load and the memory usage. Furthermore, pooling limits the risk of overfitting and favours the location invariance of the image processing, as the network is less susceptible to small shifts of the input image [5, 14]. The pooling layer executes an aggregation function, which can be the *max* and the *mean*. They respectively output the maximum and the mean value on every patch.

Multilayer Perceptron. The classification task was carried out by an MLP, which is a fully connected feedforward network. The activation function of the output layer of an MLP intended for the multilabel binary classification described in this paper is the logistic function [2]. The transition between the convolutional basis and the MLP is done by flattening the last output of the convolutional basis. Flattening is the representation of the three-dimensional information from the convolutional basis in a one-dimensional form that can be directly processed by the input layer of the MLP [5].

Model Regularization. Two particular methods have shown good results with regard to increasing the performances of CNNs. Those methods are dropout and batch normalization.

Dropout. Dropout is a popular approach to prevent or at least reduce the overfitting of neural networks [18]. The principle of dropout is to randomly drop network units and their connections to other units during the training. Randomly omitting hidden units help to prevent complex co-adaptation on the training data. Applying the dropout procedure can also be seen as training many different networks and then determining the test results as an average of the results of each different network. The Dropout applied in this study were controlled with a parameter called dropout rate. A very low dropout rate may not affect overfitting while a high rate may lead to underfitting [18, 19].

Batch Normalization. During the training of a Neural network, it can happen that the distribution of the inputs (*mean* and *variance*) of the network layers changes, extending its training duration. This phenomenon is known as internal covariate shift. For this reason a batch normalization was used in this study and consists in computing the mean and the variance of the training data for each batch and using these values to normalize the layers inputs and thus prevent the change of distribution inside of each batch [20].

Hyperparameter Optimization. Hyperparameters are parameters, that control the structure and the training of a model. Those parameters are not learnt during the training and should be set before any learning iteration [19]. In this study, the values of the hyperparameters of the models were set using a search algorithm.

The first step to optimize hyperparameters was to define a range or space of values for each hyperparameter to be set. This space constrains the search algorithms in a given scope. Afterwards, the algorithm inferred a combination of values out of the given spaces and this combination was used to build up a model that was evaluated on its accuracy. The whole process was repeated many times for different combinations of parameter values until the best combination of hyperparameters values was achieved [21].

The hyperparameters considered for the bottom-up CNN used in this study are summarized in Table 1. For each hyperparameter a space was defined based on the available computational resources and the recommended best practices [2, 17]. The search algorithm used was the hyperband often cited in literature [22]. The results of the hyperparameter optimization were the accuracies of different trained models and out of those accuracies, the best one was picked, whose corresponding model was the bottom up CNN shown in the Fig. 3. The bottom up CNN consists of the preprocessing steps, the succession of convolutional layers and pooling layers, which constitutes the convolutional basis and the classifier, comprising the flatten layer subjected to dropout, a batch normalization layer, a dense layer and at least the output layer.

Table 1. Values spaces for hyperparameters of the bottom up CNN.

Hyperparameter	Values space
Number of convolutional layers	[1, 5]
Number of pooling layers	[1, 5]
Pooling operation	(max, mean)
Number of filters per convolutional layer	(16, 32, 64)
Kernel size per convolutional layer	[3, 5]
Batch normalization layer before a convolutional layer	(yes, no)
Number of layers	[1, 5]
Number of neurons per layer	[10, 200]
Dropout rate	(0.1, 0.2, 0.3, 0.4, 0.5)
Batch normalization before layer	(yes, no)

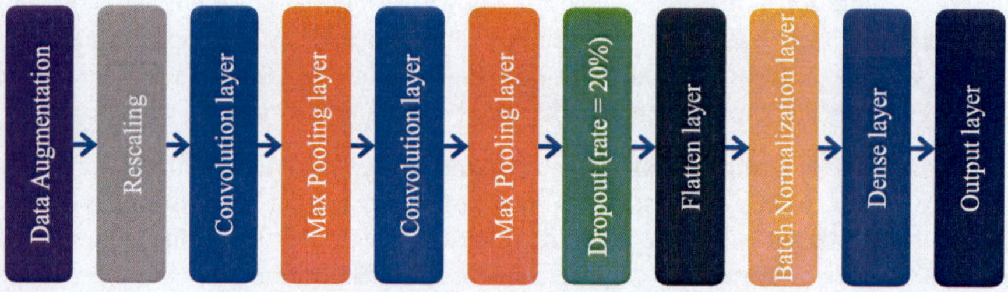

Fig. 3. Layer Sequence of the Bottom up CNN

3.4 Transfer Learning

For some applications, the training set may be particularly small, which would inevitably lead to an underfitting. Such cases can be handled using transfer learning. Transfer learning is the fine tuning of pretrained models in order to solve specific tasks. Pre-trained models have learned from datasets of different contents and sizes and can then be applied to different use-cases [2]. The pre-trained model considered in this study was the ResNet50. After loading a pre-trained model, the first step was to freeze the trained layers in order to preserve the information it contained. Then new layers were added on the top of the frozen ones. These new layers were then trained to learn the new features of the specific use-case. The Fig. 4 shows the layer sequence resulting from the transfer learning with the ResNet50. It shows the preprocessing steps, the convolutional basis of the ResNet50 and the classifier used in the Fig. 3.

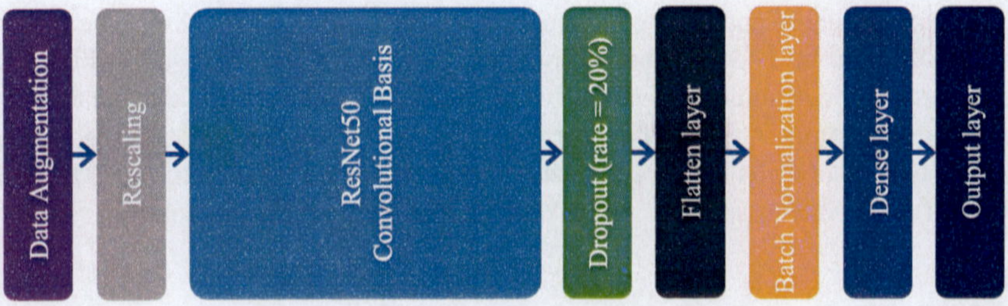

Fig. 4. Layer Sequence of the ResNet50

4 Results

The usual metric to evaluate the performances of a CNN on a classification task is the prediction accuracy, which is the ratio of the model predicted labels to the true labels. The performances of the two models trained in the present study were evaluated on a test dataset containing 282 images. The results showed that the bottom-up CNN and the pretrained ResNet50 could achieve accuracies of 96.77% and 95.74% respectively. Compared to their training accuracies of 96.78% and 96.12%, it could be stated that no overfitting occurred during the training. Figure 5 shows the confusion matrices of the two models. For every labelled image the confusion matrices show the repartition of the predictions between the different classes.

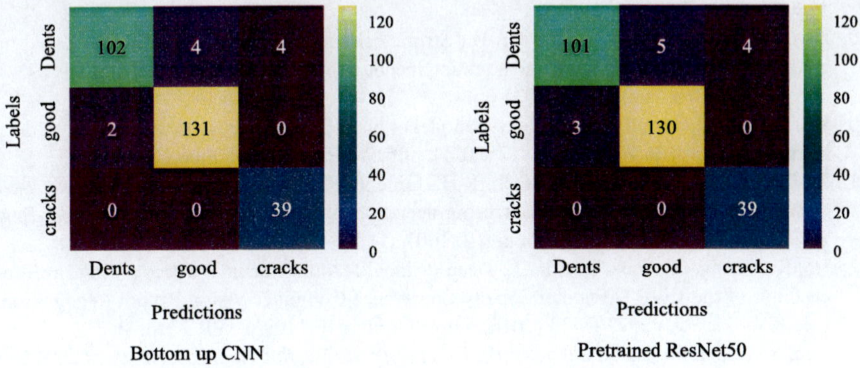

Fig. 5. Achieved accuracies on a test dataset containing 282 images

5 Conclusion

In this paper, the state of the art in applying machine learning methods for image classification in the field of quality assurance in sheet metal forming applications have been addressed. Approaches and methods were presented to build and optimize the training as well as to enhance the generalization ability of a model. These methods were applied on an academic use-case and showed a maximum accuracy of 96.77%. From the obtained insights, it could be deduced that CNN powered systems may be used in high clock rate automated environments, such as in the metal forming industry, despite their high computational complexity and if the correct infrastructure is given. A particular use-case could be the use of CNN to detect defects in progressive die manufacturing, where high stroke rates pose a major challenge for data processing and process control.

References

1. Lange, K.: Umformtechnik Handbuch für Industrie und Wissenschaft Band 3: Blechbearbeitung. Springer, Heidelberg (1990). https://doi.org/10.1007/978-3-662-10686-0
2. Chollet, F.: Deep Learning with Python, 2nd ed. (2021)
3. Lecun, Y., Bengio, Y., Hinton, G.: Deep learning. Nature **521**(7553), 436–444 (2015). https://doi.org/10.1038/nature14539
4. LeCun, Y.: Generalization and Network Design Strategies (1989)
5. LeCun, Y., Bottou, L., Bengio, Y., Haffner, P.: Gradient-based learning applied to document recognition. Proc. IEEE **86**(11), 2278–2323 (1998). https://doi.org/10.1109/5.726791
6. Lecun, Y., Kavukcuoglu, K.: Convolutional networks and applications in vision. http://ieeexplore.ieee.org/xpls/abs_all.jsp?arnumber=5537907
7. Lee, S., Quagliato, L., Park, D., Berti, G.A., Kim, N.: A buckling instability prediction model for the reliable design of sheet metal panels based on an artificial intelligent self-learning algorithm. Metals **11**(10), 1533 (2021). https://doi.org/10.3390/met11101533
8. Hamouche, E., Loukaides, E.G.: Classification and selection of sheet forming processes with machine learning. Int. J. Comput. Integr. Manuf. **31**(9), 921–932 (2018). https://doi.org/10.1080/0951192X.2018.1429668
9. Hao, Z., Li, Z., Ren, F., Lv, S., Ni, H.: Strip steel surface defects classification based on generative adversarial network and attention mechanism. Metals (Basel) **12**(2), 1–15 (2022). https://doi.org/10.3390/met12020311
10. Wang, D., et al.: Intelligent recognition model of hot rolling strip edge defects based on deep learning. Metals (Basel) **11**(2), 1–17 (2021). https://doi.org/10.3390/met11020223
11. Molitor, D.A., Kubik, C., Hetfleisch, R.H., Groche, P.: Workpiece image-based tool wear classification in blanking processes using deep convolutional neural networks. Prod. Eng. **16**(4), 481–492 (2022). https://doi.org/10.1007/s11740-022-01113-2
12. He, K., Zhang, X., Ren, S., Sun, J.: Deep residual learning for image recognition. In: Proceedings of the IEEE Computer Society Conference Computer Vision Pattern Recognition, vol. 2016-Decem, pp. 770–778 (2016). https://doi.org/10.1109/CVPR.2016.90
13. Hao, Y.: Convolutional neural networks for image classification. In: Proceedings of 2021 2nd International Conference on Artificial Intelligence on Computer Engineering ICAICE 2021, pp. 342–345 (2021). https://doi.org/10.1109/ICAICE54393.2021.00073
14. Lecun, Y., et al.: Handwritten digit recognition with a back-propagation network. In: Advances in Neural Information Processing Systems (NIPS 1989), Denver, vol. 2 (1990)
15. Hataya, R., Zdenek, J., Yoshizoe, K., Nakayama, H.: Meta approach to data augmentation optimization. In: Proceedings of 2022 IEEE/CVF Winter Confernce Applied Computer Vision, WACV 2022, pp. 3535–3544 (2022). https://doi.org/10.1109/WACV51458.2022.00359
16. LeCun, Y.A., Bottou, L., Orr, G.B., Müller, K.-R.: Efficient BackProp BT - neural networks: tricks of the trade, neural networks: tricks of the trade, vol. 7700, no. Chapter 3, pp. 9–48 (2012). http://link.springer.com/chapter/10.1007/978-3-642-35289-8_3/fulltext.html%5Cnpapers3://publication/doi/10.1007/978-3-642-35289-8_3

17. Géron, A.: Hands-on Machine Learning with Scikit-Learn, Keras, and TensorFlow. O'Reilly Media (2019)
18. Hinton, G.E., Srivastava, N., Krizhevsky, A., Sutskever, I., Salakhutdinov, R.R.: Improving neural networks by preventing co-adaptation of feature detectors, pp. 1–18 (2012). http://arxiv.org/abs/1207.0580
19. Goodfellow, I., Bengio, Y., Courville, A.: Deep Learning Ian, Foreign Aff., vol. 91, no. 5, pp. 1689–1699 (2016)
20. Glorot, X., Bengio, Y.: Understanding the difficulty of training deep feedforward neural networks. J. Mach. Learn. Res. **9**, 249–256 (2010)
21. Bergstra, J., Bengio, Y.: Random search for hyper-parameter optimization. J. Mach. Learn. Res. **13**, 281–305 (2012)
22. Li, L., Jamieson, K., DeSalvo, G., Rostamizadeh, A., Talwalkar, A.: Hyperband: a novel bandit-based approach to hyperparameter optimization. J. Mach. Learn. Res. **18**, 1–52 (2018)

Clustering of Learning Sub-models for Quality Prediction in a Resource-Efficient Tool Grinding Process

Berend Denkena, Marcel Wichmann, and Michael Wulf(✉)

Institute for Production Engineering and Machine Tools Hannover,
An der Universität 2, 30823 Garbsen, Germany
wulf_m@ifw.uni-hannover.de

Abstract. The prediction of workpiece quality in process planning, using machine learning models, is a common-researched topic. Until now, trained models were static and could not update themselves with new data. However, this aspect is crucial when considering the continuously changing manufacturing circumstances in regards to new process parameters, materials, and workpiece geometries. In addition, repeatedly training process models with an extended mixed dataset decreases the prediction quality due to the increased data divergence. This paper presents an approach to automatically generate sub-models, which maintain the prediction quality even if novel data is considered. The challenge is to define the amount and content of these sub-models through clustering. Tool grinding experiments will be conducted with different process parameters, materials, and workpiece geometries in order to obtain a divergent dataset. Subsequently, cluster approaches are compared to obtain dynamic growing models, which enable optimized planning for a more resource efficient process. Finally, the method will be generalized in order to ensure a process-independent usage.

Keywords: Machine Learning · Process Models · Tool Grinding

1 Introduction

Besides empirical and numerical modelling approaches, machine learning methods are being used more and more to describe relationships in a machining context [BRI06, DEN19, KÖN17]. The future vision of an autonomous machine tool needs to include this knowledge in order to perform a successful production [DIT21]. Because of their ability to map even complex and comprehensive processes, the automatic generation and evaluation of machine learning models is frequently researched [DEN20, KRÜ19]. In this research, the limits of modelling are specifically focussed on due to the limited complexity a model can include. To visualize this limit, Fig. 1 depicts three synthetically generated datasets and the support vector machines (SVM) of various dataset combinations. Their quality is evaluated using the mean absolute percentage error (MAPE) which is defined as the absolute model error per mean value of the modelled target parameter. A perfect model has a MAPE of 0%, while a larger MAPE indicates lower

T. Bauernhansl et al. (Eds.): WGP 2023, LNPE, pp. 94–103, 2024.
https://doi.org/10.1007/978-3-031-47394-4_10

model qualities [MYT16]. The combination of Dataset 1 and 2 in Fig. 1 results in a model quality between them. The 5.8%-MAPE of Dataset 2 is nearly average with the higher MAPE of Dataset 1 because all datasets have the same size. This effect can't be observed for the combination of all three datasets. Due to the inauspicious location of Dataset 3, the MAPE is even greater when compared to the single datasets' MAPEs. A possible solution could be clustering the datasets into two data clusters with one model each, instead of one model including all of the datasets. One cluster with Dataset 3 and another including Dataset 1 and 2 yields a compromise with fewer models and higher model quality, despite the included data complexity.

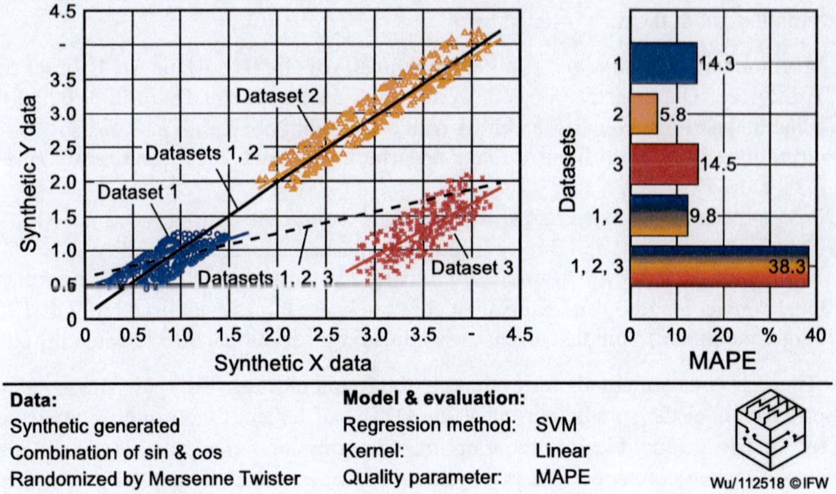

Fig. 1. Limited complexity of linear support vector machines on synthetic data

Currently, it is common practice to generate a separate model for each dataset (e.g. for each material-tool combination), resulting in a large amount of different models [DIT20, UHL21]. This process means that any new data, e. g. of a slightly different material, cannot be processed because the models have only a small definition range. Thus, an approach by which models can adapt to the database, and thus process new data points, is missing. This capability is essential for an autonomous machine tool with self-optimizing process planning and is, therefore, the subject of this paper.

In contrast to machine learning approaches, there is only limited research on clustering approaches in a machining context. Ochel et al. cluster continuous milling data as current signals, as well as axis position signals, to differentiate between geometrical workpiece features [OCH22]. Before the similarity to the predefined signal patterns per feature is calculated, the signals are roughly analysed by the spindle speed to omit tool changes. This approach is used by Brecher et al. to create a tool wear model for milling processes [BRE22]. It becomes clear that not just for the previously described data processing, but even for machine learning and clustering purposes, process knowledge is crucial [BRE20]. Those self-optimizing systems need to be able to automatically adapt to changing inputs [MÖH20]. For this aim, manually clustering using expert knowledge must be structured and automated by algorithm-based methods.

2 Approach

Based on the literature, before am automated hierarchical cluster algorithm is discussed as step towards an autonomous machine tool, manually clustering according to the most dominant process input parameters is presented. Analyzing both strategies successively, makes it possible to develop a general method of clustering for incorporating new data while maintaining prediction quality. The used dataset originates from helical flute grinding experiments conducted during the manufacturing of shank tools. Detailed information about the data is listed below.

- Machine tool: Vision 400 L (Walter Maschinenbau GmbH), Fanuc 31i-B5 control.
- Workpiece: Diameter D = 6, 10, 12 mm and core diameter D_c = (0.3, 0.5) x D; cemented carbide from Tigra GmbH with amount of cobalt = 3, 6, 10%.
- Grinding tool: Hybrid bonded diamond grinding wheels, shape 1A1, grain size = D33, D46, D54 μm.
- Process parameters: Feed rate v_f = 50, 100, 200 mm/min; cutting speed v_c = 15, 18, 22 m/s. Between two and three executions to ensure statistical reliability.
- Simulation data: Local cutting conditions by the software IFW CutS® in terms of local contact length l_g and equivalent chip thickness h_{eq} as established in [DEI10].
- Target parameter: Spindle current in % adjusted for frictional effects named DTRQ.

There are many methods for evaluating the found clusters [FRA10]. The given use case is to model the spindle current value DTRQ of helical flute grinding processes. Here, clustering should lead to more optimized regression model qualities within a set of clusters when compared to a comprehensive single model. The quality of a single cluster is evaluated by generating a regression model. The average quality of a cluster set can then be used for comparison with other sets. The quality of a single model can be expressed as a combination of the MAPE and the determination coefficient R^2. The latter describes the variance of explanation for due to the model in a range from 0 (not explained) to 1 (fully explained) [BAC16]. The equal modelling method is used for the regression models of all clusters in every cluster set, to ensure comparability. A linear SVM was chosen as the modelling method because of its short calculation times and successful performance in preliminary investigations [UHL22]. The modelling is done by Mathworks® Matlab using standard hyper parameters since an optimization of them wouldn't improve the model quality. Reliable regression models have similar prediction qualities in regards to both the validation data and the test data. Otherwise, the model is said to overfit or underfit. To omit models with an indication of under- or overfit, the MAPE and the R^2 of validation and testing were compared using a fitting indicator *Fit*, as established in Eq. 1. Preliminary work urge the use of models with $Fit \leq 10$. To ensure a statistical certainty, the modelling was done at least 20 times, for every model, with a 5-fold-cross validation and 20% unknown test data.

$$Fit = 0.5 * \left(\frac{MAPE(Testing)}{MAPE(Validation)} + \frac{R^2(Validation)}{R^2(Testing)} \right) \tag{1}$$

3 Manually Clustering Using Expert Knowledge

Defining clusters based on expert knowledge, strongly depends on the given use case. In terms of tool grinding, the available process input parameters, presented in the previous section, are possible indicators for clustering. Figure 2 presents the model quality for DTRQ of the grinding wheel D46 engaging a workpiece with 10% cobalt (D46T10). In the upper graph, with an increasing amount of D46T10 data points, a mostly constant model quality is already achieved using 30 data points. Including 60 data points, the final modelling quality is reached (green line in Fig. 2). This result shows that at least 30 observations are necessary in order to obtain a SVM regression model for helical flute grinding processes with an acceptable model quality as also found in [DIT20].

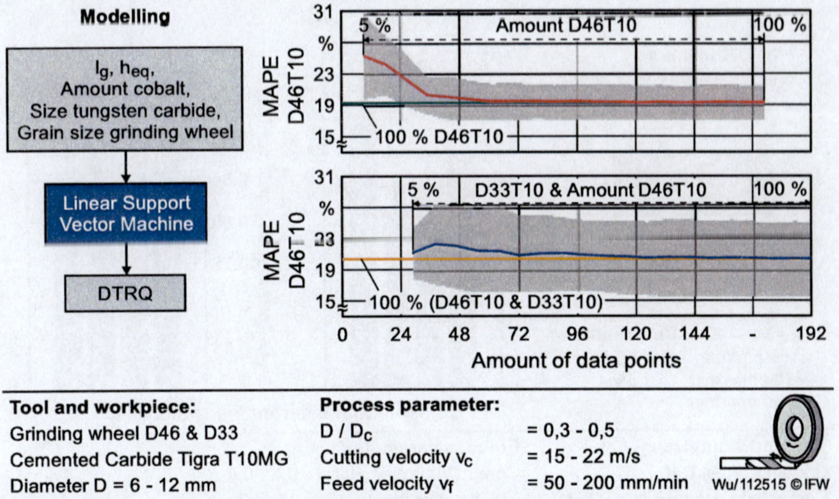

Fig. 2. Differing data influencing the quality of a common regression model (Color figure online)

For the lower graph of Fig. 2, additional data in terms of D33T10, is used to train the model. In comparison to D46T10 the datasets differ in the grinding wheel's grain size, which is reduced from 46 μm to 33 μm. Due to this variance, there is a base of 30 data points, which is consecutively increased by the D46T10 data. Even with this additional 30 data points of D33T10, when compared to the upper graph, the final model quality of D46T10 values is first reached with about 110 data points (30 out of D33T10 and 80 out of D46T10) in terms of the mean value. The modelling complexity in the lower graph was too high, thereby preventing the final model quality from being achieved before the 110th data point. In addition to the mean value, model variance is also affected by additional D33T10 data. This additional data allows for an even earlier attainment of the final model quality with 5% of D46T10 (lower graph) as compared to 9.5% of D46T10 (upper graph). Furthermore, the resulting range of variation for the MAPE doubles independently of the ratio of the data points from D33T10 to D46T10 included in the training data. This shows a crucial influence of the training datasets' composition on the model quality.

Besides the grinding wheel specification, other process variables are also possible for clustering based on expert knowledge. Surprisingly, models created with data that was clustered after the workpiece geometry parameters D or D_c fail immediately. In contrast, clustering by v_f or using local cutting conditions as a cluster indicator yields reliable performance as presented in Fig. 3. It becomes clear, that a combined dataset, including all three stages of feed rate, performs up to 7.7% (MAPE) worse than the individual stages' models. This difference is visible in both the MAPE and the R^2 model quality criterion. Remarkably the MAPE for the $v_f = 200$ mm/min dataset is clearly lower when compared with the $v_f = 100$ mm/min dataset, while their R^2 is mostly the same. This result confirms the differences in model quality, at least in regards to the used quality measurement method, are thereby supporting the use a combined MAPE and R^2 for objective model evaluation.

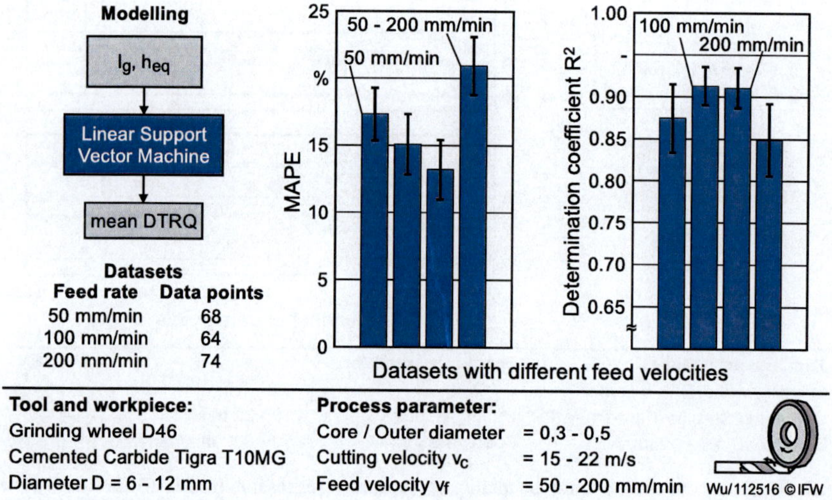

Fig. 3. Model quality after clustering D46T10 by feed rate v_f

The presented results lead to the following boundary conditions for generating and evaluating clusters for regression models:

1. The approach shall generate clusters with at least 30 data points each to ensure a proper model quality (in terms of helical flute grinding).
2. Additional data, which differs from the existing data, can drastically decrease model quality. The approach shall be able to identify and split those datasets into clusters.
3. Even a small number of additional data added to a huge dataset are able to reduce the model quality. The approach shall be sensible for those data points and allocate them into a new cluster.
4. Considering the previous boundary conditions, the approach shall minimize the number of clusters (and with this, the number of models).
5. The generated models should be evaluated based on several quality metrics.

The major disadvantage of this knowledge-based clustering is that well-educated and experienced employees are necessary to cluster the data. They have to decide which process parameter is sufficient as a cluster indicator, or even, if a combination of several parameters has to be taken into account. In addition, they have to define which data could decrease the model quality when combined with other data, although there is no established approach for measuring this data divergence. To conclude, besides the process knowledge, brief insights into data preprocessing and model creation are required for knowledge-based clustering. Another approach is to use hierarchical cluster algorithms which can handle the confusing number of data points systematically and within shorter durations.

4 Clustering Automated by a Hierarchical Algorithm

Hierarchical methods are well researched and often used in combination with a euclidean distance criterion to split data into clusters [FRA10]. The hierarchical method starts with a cluster for each point of the dataset and calculates a matrix of differences between every possible pair of clusters. Subsequently, the two clusters with the shortest distance are merged into a common cluster, whereby the number of clusters for the whole dataset decreases. The repetition of this procedure with new distance matrices and merged clusters leads, in the end, to one single cluster containing every point of the dataset. The generated structure of clusters is called "dendrogram" and enables the classification of data points into a given number of clusters [FRA10].

Differences in this hierarchical cluster method can be seen in the cluster linkage for euclidean distance measurements. For example, the distance between two clusters can be calculated by the shortest distance between two points of the clusters or by their farthest distance. There are many different linkage criteria, but to obtain clusters with low variance and comparable sizes (see Sect. 2), the ward linkage method is more suitable [FRA10, WAR63]. The method aims to minimize the overall error sum of squares (ESS) of a dataset with c points $P_{ij} \in \mathbb{R}^n$ in a cluster j by calculating ESS for every possible merge of two clusters, as presented in Eq. 2.

$$ESS = \min \sum\nolimits_{j}^{d} \sum\nolimits_{i}^{c} \left(P_{i,j} - o\left(\mathbb{R}^2\right)\right)^2 \tag{2}$$

The model quality of hierarchical clustered datasets with the ward linkage is depicted in Fig. 4 for different clustering input parameters and numbers of clusters. The grey bars going in the vertical direction present the model quality in regards to the whole D46T10 dataset as one single cluster. The horizontal bars in blue and yellow depict two, or rather, three clusters out of the D46T10 dataset. These clusters were generated by the cluster algorithm using different input parameters. As a benchmark, the manual clustering that utilized the v_f presented in Fig. 3 is done with this approach too.

Figure 4 points out that a clustering by the local cutting conditions l_g and h_{eq} is also suitable for achieving an improved model quality in comparison to the overall model with only one cluster (grey bar). In contrast to manual clustering, this method combines at least two input parameters to create the set of clusters. Because of the numeric optimization seen in Eq. 2, a manual approach can't achieve this same clustering in manageable time periods. Across all input parameters, Fig. 4 shows that three clusters (yellow) tend to result in in an average 2.7% lower MAPE than the two clusters (blue). However, with an increasing number of clusters, the Fit indicator increases with a mean of 8.2%, thereby decreasing model quality. This result supports the boundary conditions of Sect. 2, in regards to prefer a minimal amount of models.

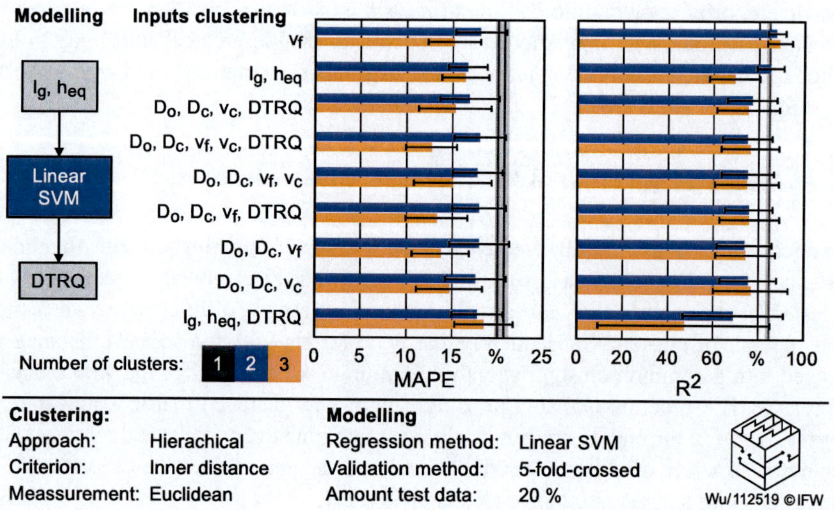

Fig. 4. Model quality of hierarchical clustered datasets by different clustering inputs (Color figure online)

The benefit of this automated approach, in comparison to the expert knowledge needed for manual clustering, is the structured proceeding that leads to an optimized clustering. The user can focus more on the evaluation and comparison of clusters rather than generating and testing them. Even combinations of process input parameters can be used as cluster inputs in this approach, which is more challenging and time consuming in the manual way. As presented in Fig. 4, an individual cluster isn't necessary for every stage of a process parameter. The algorithm-based approach can test every combination of process parameters for every number of desired clusters. This leads to reliable process models that enable more resource efficient production. The following section sumarizes the approach and gives guidance on how to modify the inputs for clustering purposes.

5 Generalization of the Method

A general description, as presented in Fig. 5, illustrates the usage of the automated clustering approach with a hierarchical algorithm independently of the machining process. On the left side, the regular use of the process model is depicted with unlabelled data and the predictions are calculated with the current model. The latter can be used, for example, in process planning or quality prognosis. When newly labelled data becomes available, the current model's quality can be evaluated by predicting the new input data.

Therefore, the current model is used to generate the labels for the data. A comparison of the predicted labels with the actual ones gives the prediction quality of the current model for the new, labelled data. MAPE, R^2, and Fit are used to ensure a sustainable assessment of model quality. If this new prediction quality equals or exceeds the current prediction quality, no action is needed. Conversely, new clusters and corresponding new models would have to be generated. In order to account for influences from measurement uncertainties and model building on the forecast quality, a threshold can be introduced. For example, clustering would only be triggered at a 5% reduction in model quality. To account for any possible influence on model quality, a new clustering is also conceivable for the first variation of a previously constant process input variable.

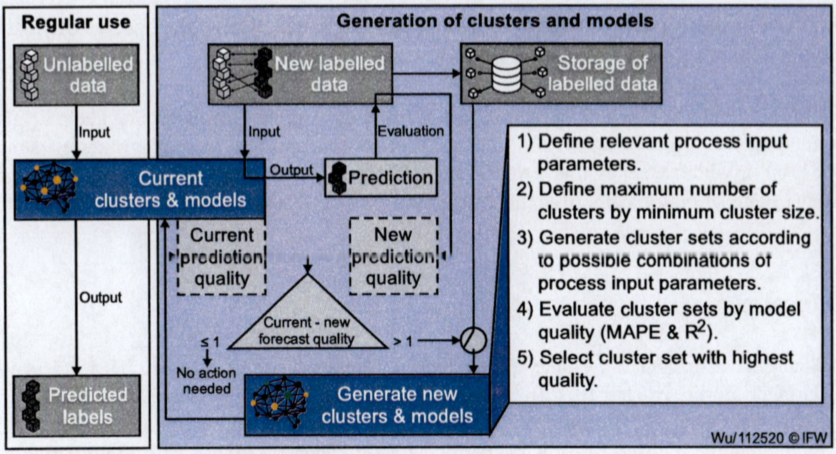

Fig. 5. Method of clustering for incorporating new data while maintaining prediction quality.

Generating new clusters requires the definition of relevant process input parameters included in the dataset. For example, helical flute grinding would include feed rate, cutting speed, local cutting conditions, workpiece material, and grinding wheel specifications. The maximum number of clusters is determined by the minimum number of data points per cluster. For every combination of the defined process input parameters a set of clusters is generated with the automated hierarchical cluster algorithm. This process enables the user to evaluate each cluster set. Therefore, a regression model is built for every cluster to compare the model quality. The most suitable set of clusters and models is then chosen to become the new current set for regular usage.

6 Conclusion and Outlook

The paper presented the limits of machine learning models in the case of complex tool grinding datasets concerning their dynamic growth during production. To overcome this challenge, datasets can be clustered to generate a set of models, instead of just a single one, for modelling the process relations. This clustering requires knowledge about the number and subdivision of the clusters. Both a manual and an algorithm-based approach for splitting datasets into suitable clusters were analysed to develop a general method of automated clustering. This method allows more than one process parameter to be considered in clustering, resulting in a reduced number of clusters, creating high quality process models for even complex datasets. Reliable process models are required in process planning to ensure a resource-efficient production. It can be seen as a step towards self-optimizing process models that automatically adapt to new processes. Such systems are necessary in order to achieve the vision of an autonomous and energy-efficient machine tool. Further work is necessary to explore this developed approach, including different linkage criteria and algorithms, like a density-based method. The presented approach uses a linear SVM to model the process relations and evaluate the clusters. Other methods, such as neuronal networks or gaussian process regressions, should be investigated too. Additionally an automated selection of the best set of clusters as well as the required calculation effort should be explored too.

Acknowledgement. The authors would like to thank the German Research Foundation (DFG) for funding the project LearnWZS - Learning process adaptation for tool grinding (number 445811009), which enables this investigation. Furthermore, the authors thank the Sieglinde-Vollmer Foundation for supporting this research.

References

Backhaus, K., Erichson, B., Plinke, W., Weiber, R.: Multivariate Analysemethoden (14. Aufl.). Springer, Berlin, Heidelberg (2016). https://doi.org/10.1007/978-3-662-46076-4

Brecher, C., Ochel, J., Lohrmann, V., Fey, M.: Maschinelles Lernen zur Prädiktion der Bauteilqualität. Zeitschrift für wirtschaftlichen Fabrikbetrieb **115**(11), 834–837 (2020)

Brecher, C., Lohrmann, V., Wiesch, M., Fey, M.: Clustering zur Bestimmung von Werkzeugverschleiß. Zeitschrift für wirtschaftlichen Fabrikbetrieb **117**(4), 218–223 (2022)

Brinksmeier, E., et al.: Advances in modeling and simulation of grinding processes. CIRP Ann. **55**(2), 667–696 (2006)

Deichmueller, M., et al.: Determination of static and dynamic deflections in tool grinding using a dexel-based material removal simulation. In: CIRP 2nd International Conference Process Machine Interactions 2010. Vancouver, Canada (2010)

Denkena, B., Dittrich, M.-A., Böß, V., Wichmann, M., Friebe, S.: Self-optimizing process planning for helical flute grinding. Prod. Eng. Res. Devel. **13**(5), 599–606 (2019)

Denkena, B., Dittrich, M.-A., Lindauer, M., Mainka, J., Stürenburg, L.: Using AutoML to optimize shape error prediction in milling processes. MIC Procedia **20**(1), 160–165 (2020)

Dittrich, M.-A., Uhlich, F.: Self-optimizing compensation of surface deviations in 5-axis ball-end milling based on an enhanced description of cutting conditions. CIRP J. Manuf. Sci. Technol. **31**(1), 224–232 (2020)

Dittrich, M.-A.: Autonome Werkzeugmaschinen - Definition, Elemente und technische Integration. Habilitation, Gottfried Wilhelm Leibniz Universität Hannover (2021)

Frades, I., Matthiesen, R.: Overview on techniques in cluster analysis. In: Bioinformatics Methods in Clinical Research, pp. 81–107. Humana Press, Totowa, USA (2010)

Königs, M., Wellmann, F., Wiesch, M., Epple, A., Brecher, C.: A scalable, hybrid learning approach to process-parallel estimation of cutting forces in milling applications. In: WGP-Jahreskongress Aachen 2017, vol. 7, pp. 425–432. Apprimus, Aachen (2017)

Krüger, J., Fleischer, J., Franke, J., Groche, P.: WGP-Standpunkt KI in der Produktion. Wissenschaftliche Gesellschaft für Produktionstechnik WGP e.V. (2019)

Möhring, H.-C., Wiederkehr, P., Erkorkmaz, K., Kakinuma, Y.: Self-optimizing machining systems. CIRP Ann. **69**(2), 740–763 (2020)

Myttenaere, A.D., Golden, B., Le Grand, B., Rossi, F.: Mean Absolute Percentage Error for regression models. Neurocomputing **192**, 38–48 (2016)

Ochel, J., Fey, M., Brecher, C.: Semantically meaningful segmentation of milling process data. In: Behrens, B.A., Brosius, A., Drossel, W.G., Hintze, W., Ihlenfeldt, S., Nyhuis, P. (eds.) Production at the Leading Edge of Technology, pp. 319–327. WGP 2021. LNPE. Springer, Cham (2022). https://doi.org/10.1007/978-3-030-78424-9_36

Uhlich, F.: Lernende Prozesssimulation für die Prognose und Kompensaiton von Formabweichungen in der Einzelteilfertigung. Dr.-Ing. Diss., Gottfried Wilhelm Leibniz Universität Hannover (2022)

Ward, J.H.: Hierarchical grouping to optimize an objective function. J. Am. Stat. Assoc. **58**(301), 236–244 (1963)

Check for updates

Improving Efficiency and Sustainability Through Predictive Tool Wear Monitoring During Manual Drilling of CFRP

Malte Flehmke(✉), Sebastian Junghans, Andreas Jessen, Ganna Shchegel, Carsten Möller, and Jan Hendrik Dege

Institute of Production Management and Technology (IPMT), Hamburg University of Technology (TUHH), 21073 Hamburg, Germany
malte.flehmke@tuhh.de

Abstract. Aircraft assembly involves many high-precision rivet holes, with around one-third drilled manually by experienced workers. Wireless electrical machines are increasingly replacing pneumatic drilling machines to reduce energy consumption. Nevertheless, the lifespan of the cutting tools varies between 20 and 70 drill holes in carbon fiber reinforced polymer (CFRP) depending on the individual experience and constitution of the worker, resulting in a large waste in cemented carbide. This article proposes a method that uses the internal sensors of the drilling machine to predict tool wear during manual drilling. The main challenge is the rapid wear of the tools and the unknown feed rate, which depends on the individual workers' constitution and experience. By predicting tool wear, this method can help reduce waste and improve the efficiency, sustainability, and precision of manual drilling operations in aircraft structural assembly.

Keywords: manual drilling · tool wear · tool monitoring

1 Introduction

The aircraft industry has mostly recovered from the COVID-19 crisis. With global air traffic growing at 3.6% per annum, the Airbus Global Market Forecast 2022 predicts a need for 39,490 new passenger and freighter aircrafts over the next 20 years [1]. Due to the predominant use of rivets in the construction of aircraft structures, there is a significant demand for rivet holes, such as approximately 250,000 for a medium-sized aircraft [2], of which approximately 30% are produced manually using hand-guided automatic drilling units (ADUs) [3]. With the added demands of operational sustainability and manufacturing efficiency, reducing carbon footprint and material waste is a primary goal for the foreseeable future. Moreover, there is a projected rise in the adoption of lightweight, high-performance materials such as carbon fiber reinforced polymer (CFRP) to decrease the weight of the aircraft and consequently enhance fuel efficiency [4]. However, the increasing use of these materials presents novel manufacturing challenges. Due to the highly abrasive nature of CFRP fibers, the machining of CFRP necessitates the

T. Bauernhansl et al. (Eds.): WGP 2023, LNPE, pp. 104–113, 2024.
https://doi.org/10.1007/978-3-031-47394-4_11

use of specialized cemented carbide tools [5]. These tools consist of tungsten carbide and cobalt [6]. Notably, cobalt is frequently mined under unfavorable working conditions [7]. Furthermore, the manufacturing of cemented carbide is a highly energy-intensive process [8], involving sintering and coating procedures. The substantial energy consumption during production underscores the need to maximize the optimal utilization of each tool.

Today, the decision to replace a tool is generally based on the experience of the individual operator. This results in vastly different service life for different tools depending on the operator. Tool life can vary from 20 to 80 boreholes in CFRP. The continuous transition from pneumatic to electric and sensor-integrated ADUs offers the potential for automated process monitoring through the recording and subsequent analysis of process data. A new method is therefore being developed to determine the state of wear of the drilling tool based on internal sensor data from the machine, rather than on the individual feel of the operator. This advancement allows for a reduction in manual control and rework. The method is designed to ensure that a tool is not changed too early, where cemented carbide is wasted, or too late, where the risk of manufacturing errors increases significantly [9].

Various approaches to monitoring tool wear during drilling have been documented in the literature. The first attempts were made by Thangaraj et al. in the 1980s, where the authors used thrust force measurements to track the wear of the drilling tool [10]. In [11], a vibration signature analysis was carried out during the drilling of cast iron It was observed that an increase in tool wear correlated with an increase in vibration amplitude. In [12], a model was developed to relate thrust force, drilling parameters, and tool wear, with the aim of using the model to monitor tool wear. According to the authors, feed rate, cutting speed, and tool wear are the most important factors affecting the thrust force. In [13], the influence of tool wear on different sensor signals such as cutting temperature, thrust, and cutting torque was investigated. In [14] and [15], time series data from several sensors such as vibration, acoustic emission, thrust, and cutting torque were analyzed and transformed into features to train an artificial neural network (ANN) for tool wear prediction. However, a method that is completely independent of the cutting parameters is preferred. For CFRP drilling, [16] presented a method for the direct measurement of cutting-edge roundness (CER) of uncoated tools. In [17] a similar approach was developed for carbide inserts in turning CFRP. However, measuring CER during the normal production process is not a feasible method in a real production environment. As an alternative, the authors in [18] developed a method to monitor tool wear during milling, independent of the cutting condition. This method has not yet been successfully adapted for CFRP drilling. This means that there is a gap in the literature for a cutting condition-independent method for monitoring tool wear in CFRP. This study explores the potential for predicting tool wear in manual drilling by collecting process data and analyzing them regarding model correlations and relationships.

Section 2 describes the new approach and the relevant theoretical background. In Sect. 3 the methodologies for acquiring testing data are described. The modeling process is presented in Sect. 4 and the model is evaluated in Sect. 5. Finally, in Sect. 6 the obtained results are summarized and interpreted within their relevant context.

2 Theoretical Background

Cutting parameters, tool geometry, materials, and cutting environment play a significant role in the wear progression of drills used for CFRP laminates [19]. Particularly, cutting speed and feed rate are crucial parameters that govern the drilling kinematics of CFRP laminates, as they directly influence the tribo-mechanical interaction between the cutting tool and the workpiece specimen. For instance, when the cutting speed is increased, it leads to an increased wear progression of the drilling tool [20]. The primary wear mechanism involved is the rounding of the cutting edge caused by abrasion of the cemented carbide against the highly abrasive carbon fibers. This increase in cutting-edge rounding subsequently raises the drilling thrust and cutting torque for equal process parameters, which leads to a rise in delamination [16]. However, as the drilling thrust force and cutting torque are also highly correlated with feed rate and cutting speed [12], they are unsuitable as a tool wear criterion for manual drilling. In manual drilling, the feed rate is controlled by the individual operator and can vary depending on the person, mood, or time of day. As thrust force and cutting torque are highly dependent on feed rate, there is no guarantee that an increase in cutting torque is due to increased tool wear and not to a temporary increase in feed rate by the operator.

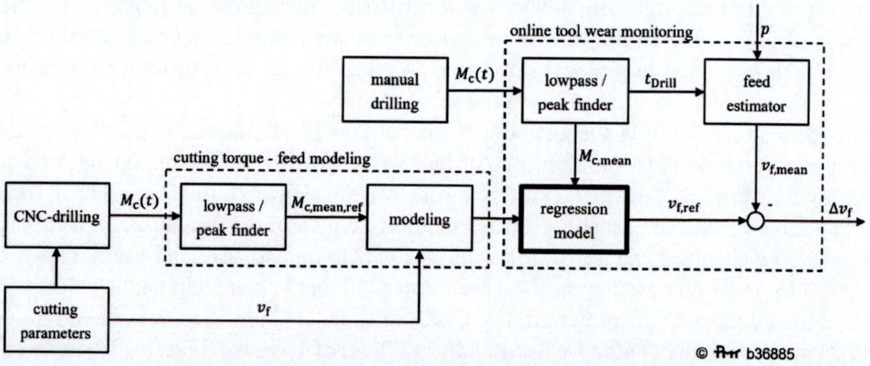

Fig. 1. Approach for tool wear monitoring in manual drilling

To circumvent this problem, an empirical approach to determine a variable that is feed-independent can be applied. The approach was developed in [21]. Feed force and cutting torque are correlated with feed and cutting speed. For example, when drilling epoxy CFRP fabric, both force components increase significantly with the feed rate v_f and decrease slightly with the cutting speed v_c [22]. This relation holds for a wide variety of tool geometries under dry-cutting conditions. A polynomial regression model of order 2 proved to capture this correlation well [21]. To determine the model for the manual drilling process, the manual drilling machine is fixed to a CNC machine where feed and cutting speed can be set exactly by numerical control. Under varying feeds, the cutting torque is measured. The reference cutting torques $M_{c,mean,ref}$ and the known feed can then be used to determine the empirical model. The resulting empirical model can then be applied to a manual drilling process. To do this, a theoretical feed $v_{f,ref}$ is calculated from the inverse model using the torque input data from the sensors $M_{c,mean}$. With an increase in tool wear, the actual mean feed $v_{f,mean}$, which is calculated from the thickness of the laminate p and the actual drilling time t_{Drill}, is going to get larger [12]. Thus, the resulting feed difference Δv_f between $v_{f,mean}$ and $v_{f,ref}$ increases with tool wear and can be monitored during the drilling process or set as a tool change criterion. It must be noted that this empirical model is only viable for the given tool-material combination. For different process parameters, the empirical model must be adjusted. The full approach is presented in Fig. 1 as exemplary for the cutting torque. The approach can similarly be applied to the thrust force.

3 Methodology

To eliminate the unknown feed rate and create a model in the initial investigations, the manual drilling machine is mounted in a CNC machine. The process data is saved and analyzed for modeling. The feed rate is varied in the machine. This is done to quantify the influence of different feed rates on the drilling process. Finally, boreholes are produced manually, while process data is measured in the same way as before. After a set number of drilling operations, the CER of the given tool is evaluated using a Keyence VHX-7000N digital microscope to objectively document tool wear. The test series are recorded in Table 1. They are not carried out sequentially but in an equally distributed random order. The randomization minimizes the effect of tool wear on the resulting model. The cutting speed is set to $v_c = 55$ m/min for all investigations. Furthermore, all tests are performed under dry conditions.

Table 1. Experimental design of the test series

	Name	Tool	Process parameter		Feed	No. of drillings
			v_c [m/min]	v_f [m/min]		
modeling	S1	T1	55	0.05	Auto	10
	S2	T1	55	0.1	Auto	10
	S3	T1	55	0.2	Auto	10
	S4	T2	55	0.05	Auto	10
	S5	T2	55	0.1	Auto	10
	S6	T2	55	0.2	Auto	10
validation	S7	T3	55	-	Manual	-

3.1 Experimental Setup

Figure 2 shows the experimental setup. The manual drilling machine is a SlimADU prototype from Johannes Lübbering GmbH. The machine is mounted in a Deckel FP3A CNC machining center using a 90° angle bracket on the spindle gearbox. The cutting torque is measured directly on the spindle using a force-sensitive adapter. Strain gauges inside the force-sensitive drill adapter allow for strain measurements that can be converted to feed force and torque. To safeguard the electronics from cooling lubricant, chips and dust, they are enveloped in resin. The sensor data from the adapter is transmitted wirelessly via a Wi-Fi connection to a base station. The drilling tool is clamped in a three-jaw chuck. In order to prevent displacement in the feed direction, the CFRP specimen is clamped onto a perforated steel plate.

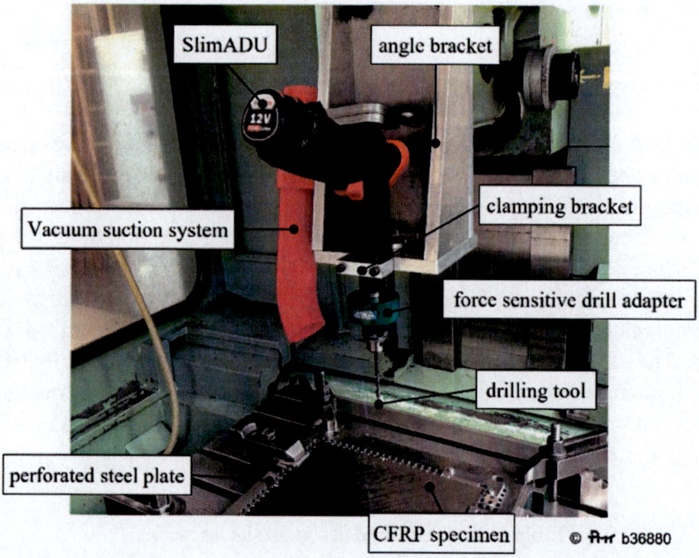

Fig. 2. Test setup in the work envelope of the Deckel FP3A machining center

3.2 Tools

The drilling tool is an uncoated cemented carbide twist drill, designed for manual drilling operations in CFRP. Tool diameter is $d = 4.9$ mm with two flutes. The point angle is $\sigma = 85°$, it has 4 facets, the rake angle is $\gamma_f = 30°$, the clearance angle is $\alpha_f = 22.5°$ and the cutting edge rounding is CER = 21 μm for Tool 1 and CER = 20 μm for Tool 2.

3.3 Materials

The CFRP material used for testing is the prepreg material Hexply6376 C-HTS(12K)-10–35-ATL300 with a thickness of $p = 4$ mm and 16 layers. The fiber orientation is laid according to $[45°, 90°, 0°, -45°]_S$. A single sheet of size 300 × 300 mm^2 was used. The boreholes are spaced 10 mm apart.

4 Modeling

The measured data allows the derivation of a model that establishes a relationship between feed and cutting torque. To determine the mean torque, the time series signal is filtered to remove measurement noise. After that, the mean value is calculated. A 1st- and a 2nd-order polynomial regression model are derived by the least-squares method and compared by calculating the root mean squared error (RMSE) which denotes the measure between the differences of the values predicted by the respective model and the values observed.

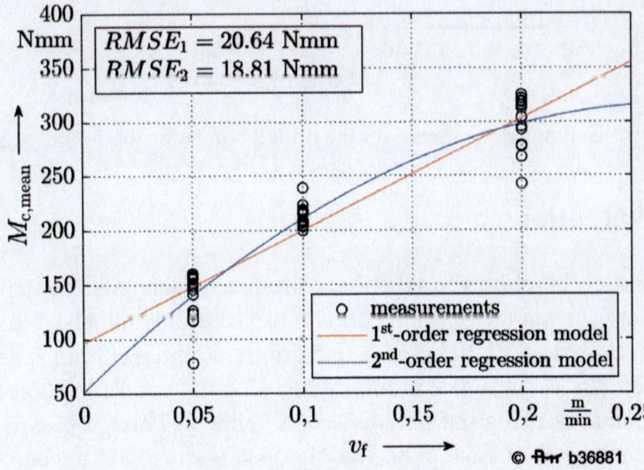

Fig. 3. Measurements, 1st- and 2nd-order regression model of the mean cutting torque

For the given cutting conditions the 1st-order polynomial equation is

$$M_c = 1034.4\text{Nmin}v_f + 96.549\text{Nmm}. \tag{1}$$

The respective RMSE is $RMSE_1 = 20.64$ Nmm. For the 2nd-order polynomial the equation is

$$M_c = -3583.3\frac{\text{Nmin}^2}{\text{mm}}v_f^2 + 1950.1\text{Nmin}v_f + 52.162\text{Nmm} \tag{2}$$

with $RMSE_2 = 18.81$ Nmm. Figure 3 shows the calculated regression models and the measurement data. Compared to the 1st-order model, the 2nd-order polynomial models show a better fit of the data as indicated by a lower RMSE. Hence in further investigations, the 2nd-order model is used.

Figure 4 shows the CER over the number of drilling operations performed during the process data acquisition process. The CER does not change significantly during the reference drilling operations. The increase of the CER due to abrasion is approximately 1–5 μm. The low raise in CER means that the model did not capture the influence of tool wear on cutting torque.

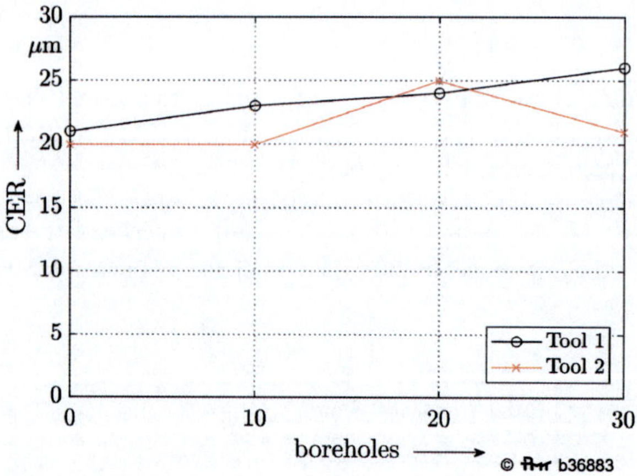

Fig. 4. Cutting edge radius over the drilled boreholes for Tools 1 and 2

5 Model Evaluation

To assess the model, boreholes in CFRP are produced manually. To expedite the tool wear progress between manually drilled boreholes, the tool is subjected to artificial wear using a CNC machine to drill 30 holes in one program with a cutting speed $v_{c,cnc} = 60$ m/min and a feed rate $v_{f,cnc} = 0.2$ m/min. After 30 CNC controlled holes, 5 boreholes are produced manually using the parameters in Table 1. This process is repeated. To do the manual drilling, the CFRP plate was clamped vertically in a vice. All evaluation holes were performed on one day by two different workers alternating for each hole. The CER is measured only once during the manual drilling operations and is assumed to stay constant for all five drillings. This assumption can be made since during the reference tests, the CER did not change significantly over 30 drilling operations, see Fig. 4. The data acquired from the manual drilling operations is then fed into the pre-determined 2nd-order model for the cutting torque to calculate the reference feed $v_{f,ref}$. To accomplish this, an inverse cutting torque model must be calculated. However, due to the lack of an inverse function in Eq. 2, the solution for this issue necessitates restructuring Eq. 2 to.

$$0 = -3583.3v_f^2 + 1950.1v_f + 52.1625 - M_{c,mean}, \tag{3}$$

and finding the roots of this new equation. The roots can be determined as

$$v_{f,ref1,2} = 1.0884 \pm \sqrt{1.1847 + \frac{52.1625 - M_{c,mean}}{3583.3}}. \tag{4}$$

Plugging a measurement of $M_{c,mean}$ into Eq. 4 yields two solutions of which only one is a feasible solution. In the final step, the actual feed is approximated by

$$v_{f,mean} = \frac{p}{t_{Drill}}. \tag{5}$$

The actual drilling time, t_{Drill}, is automatically determined by identifying change-points within the cutting moment timeseries data. The mean and the reference feed can then be compared and plotted over the CER, using

$$\Delta v_f = v_{f,ref} - v_{f,mean}. \tag{6}$$

With an increasing CER, which defines additional wear, the difference between the actual and the theoretical feed increases, see Fig. 5. This is due to the CER influencing the cutting dynamics that is not captured in the model. From a CER of 25 μm to a CER of 40 μm the mean of the feed difference Δv_f increases by over 100%. Furthermore, the calculated values for the feed difference can vary between individual boreholes. This is due to the high variability in manual drilling. For an online implementation, a moving average of at least 5 samples should therefore be used to smooth out irregularities.

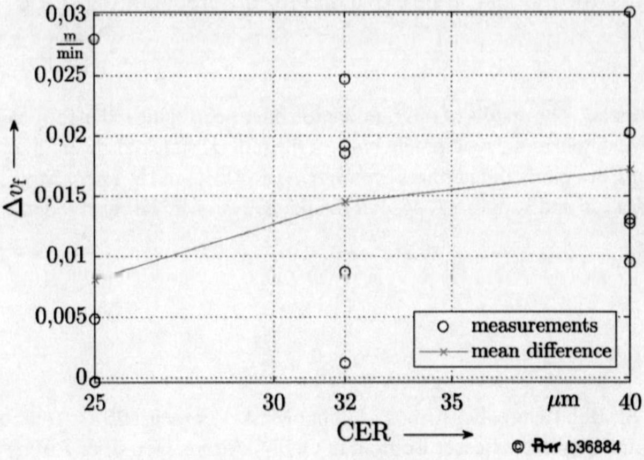

Fig. 5. Resulting feed difference Δv_f between actual mean feed $v_{f,mean}$ and reference feed $v_{f,ref}$ for increasing CER

6 Conclusion and Outlook

In conclusion, this study aimed to address the need for a feed independent method to monitor tool wear in manual drilling of CFRP. The use of specialized cemented carbide tools in CFRP drilling presents manufacturing challenges, and optimizing tool usage is crucial due to the energy-intensive production process associated with the extraction of tungsten and cobalt, that are key components of these tools.

- By means of an empirical model which is determined in a controlled environment, the wear progression of the drilling tool can be monitored. This enables the implementation of a system for monitoring the tool health and supporting the operators in their decision to change the tool leading to less waste and reduced risk of errors, which can lead to decrease the cost per rivet.
- The comparison between the theoretical and actual mean feed rates revealed that as the CER increased (indicating tool wear), the difference between the two feed rates also increased. This demonstrates the effect of tool wear on the cutting dynamics that were not accounted for in the model.

- With the given sensor setup, the measurement for the cutting torque showed promising results.

Further research is needed to refine the model and validate its effectiveness in real production environments.

- Development and construction of a thrust force-sensitive manual drilling machine to validate the method for the thrust force is in progress.
- Since the cutting torque is directly proportional to the current of the spindle motor, this might also be feasible as a criterion for model creation. The SlimADU used in this work is already equipped with current sensors for the spindle motor.
- Lastly, the developed approach can be implemented in an online real-time environment. For this, the model can be expanded for different cutting speeds as was shown in [21].

Acknowledgements. The results of this publication have been achieved in the project VerticalE2E funded by the investment and development bank of the federal state of Lower Saxony (NBank). We would like to thank our industrial partners Premium AEROTEC and Marposs Monitoring Solutions for their collaboration and support. A special thanks goes to Mr. Till Kamlade for supporting the practical testing.

References

1. Shparberg, S., Bob Lange, B.: Airbus - Global Market Forecast 2022. Toulouse, (2022)
2. Airbus Group: Qualität – Besser Bohren, in ONE – Airbus News For Airbus People (2015)
3. Stuhrmann, J., Ramme, L.: Herausforderungen und Entwicklungen bei der zerspanenden Bearbeitung von Verbundwerkstoffen in der Luftfahrtindustrie: Exkursion TUHH (2019)
4. Zhang, J., Lin, G., Vaidya, U., Wang, H.: Past, present and future prospective of global carbon fibre composite developments and applications. Compos. Part B: Eng. 250, S. 110463 (2023). https://doi.org/10.1016/j.compositesb.2022.110463
5. Wolfgang, H.: CFK-Bearbeitung. Springer Heidelberg (2021)
6. Schatt, W., Wieters, K.-P., Kieback, B.: Pulvermetallurgie. Springer, Heidelberg (2007)
7. Slack, J.F., Kimball, B.E., Shedd, K.B.: Cobalt. U.S. Geological Survey, Reston, VA, (2017)
8. Karpuschewski, B., Kalhöfer, E., Joswig, D., Rief, M.: Energiebedarf für die Hartmetallherstellung. Zeitschrift für wirtschaftlichen Fabrikbetrieb, pp. 496–501 (2011)
9. Xu, J.: A review on tool wear issues in drilling CFRP laminates. Front. Mater. 9, Artikel 990773 (2022). https://doi.org/10.3389/fmats.2022.990773
10. Thangaraj, A., Wright, P.K.: Drill wear sensing and failure prediction for untended machining. Robot. Comput. Integr. Manuf. **4**(3–4), 429–435 (1988). https://doi.org/10.1016/0736-5845(88)90014-2
11. El-Wardany, T.I., Gao, D., Elbestawi, M.A.: Tool condition monitoring in drilling using vibration signature analysis. Int. J. Mach. Tools Manuf. **36**(6), 687–711 (1996). https://doi.org/10.1016/0890-6955(95)00058-5
12. Iliescu, D., Gehin, D., Gutierrez, M.E., Girot, F.: Modeling and tool wear in drilling of CFRP. Int. J. Mach. Tools Manuf. **50**(2), 204–213 (2010). https://doi.org/10.1016/j.ijmachtools.2009.10.004

13. Ramirez, C., Poulachon, G., Rossi, F., M'Saoubi, R.: Tool wear monitoring and hole surface quality during CFRP drilling. In: Procedia CIRP, vol. 13, pp. 163–168 (2014). https://doi.org/10.1016/j.procir.2014.04.028
14. Caggiano, A., Nele, L.: Artificial neural networks for tool wear prediction based on sensor fusion monitoring of CFRP/CFRP stack drilling. IJAT **12**(3), 275–281 (2018). https://doi.org/10.20965/ijat.2018.p0275
15. Zhu, G., Hu, S., Tang, H.: Prediction of tool wear in CFRP drilling based on neural network with multicharacteristics and multisignal sources. Compos. Adv. Mater. 30(2021). https://doi.org/10.1177/2633366X20987234
16. Faraz, A., Biermann, D., Weinert, K.: Cutting edge rounding: an innovative tool wear criterion in drilling CFRP composite laminates. Int. J. Mach. Tools Manuf. **49**(15), 1185–1196 (2009). https://doi.org/10.1016/j.ijmachtools.2009.08.002
17. Denkena, B., Lucas, A., Bassett, E.: Effects of the cutting edge microgeometry on tool wear and its thermo-mechanical load. CIRP Ann. **60**(1), 73–76 (2011). https://doi.org/10.1016/j.cirp.2011.03.098
18. Nouri, M.; Fussell, B.K.; Ziniti, B.L.; Linder, E.: Real-time tool wear monitoring in milling using a cutting condition independent method. In: International Journal of Machine Tools and Manufacture 89, S. 1–13 (2015). DOI: https://doi.org/10.1016/j.ijmachtools.2014.10.011
19. Hintze, W., et al.: Digital assistance systems for smart drilling units in aircraft structural assembly. In: 7th International Workshop on Aircraft Systems Technologies, pp. 255–266 (2019)
20. Rawat, S., Attia, H.: Wear mechanisms and tool life management of WC–Co drills during dry high speed drilling of woven carbon fibre composites. Wear **267**(5–8), 1022–1030 (2009). https://doi.org/10.1016/j.wear.2009.01.031
21. Faraz, A.: Experimental study on delamination, mechanical loads and tool wear in drilling of woven composite laminates. Dissertation. Technische Universität Dortmund, Dortmund (2011)
22. Attia, H., Sadek, A., Meshreki, M.: High speed machining processes for fiber-reinforced composites. Mach. Technol. Compos. Mater. 333–364 (2012)

Feature-Based Analysis of Acoustic Emission Signals for Wear Monitoring in Centerless Through-Feed Grinding

Daria Gelbich[1(✉)], Peter Breuer[1], Hou Zeng Man[1], Marco Becker[1], Sebastian Barth[1], Philipp Niemietz[1], and Thomas Bergs[1,2]

[1] Laboratory for Machine Tools and Production Engineering (WZL) of RWTH Aachen University, Aachen, Germany
d.gelbich@wzl-tf.rwth-aachen.de
[2] Fraunhofer Institute for Production Technology IPT, Aachen, Germany

Abstract. Wear monitoring is a critical aspect of maintaining the health and performance of machinery. During centerless grinding, grinding and regulating wheels as well as work rest blades wear out over time due to contact with workpieces, which impairs the geometrical accuracy and surface quality of the processed parts, and ultimately the productivity of the process. Acoustic emission sensors are a promising source of information for wear monitoring which in turn allows to optimize dressing intervals and prevent rejects. During an experimental series, acoustic emission signals were collected from a centerless through-feed grinding process to identify changes in the signal that could be indicative of wear. The collected acoustic emission signals were preprocessed by digital filtering and extracting a comprehensive set of features. A one-class support vector machine was used to quantify the signal evolution over the span of the experimental series. The resulting resemblance of the signal evolution to the grinding wheel wear effects observed in the geometric properties of the workpieces from the series suggests the validity of this approach.

Keywords: Wear Monitoring · Acoustic Emission Signals · Centerless Grinding

1 Introduction

Wear monitoring plays a critical role in maintaining machinery as it enables timely detection and intervention to prevent expensive breakdowns and ensure optimal performance. To effectively address these challenges, it is essential to carefully choose parameters that can offer real-time insights into internal conditions and mechanical faults of the device [1].

In the context of centerless grinding processes, traditional methods of wear monitoring, such as visual inspection, manual measurements, or fixed grinding wheel dressing intervals, have been commonly used. However, these methods have limitations that hinder their effectiveness in real-time wear monitoring. Visual inspection relies on human judgment and may suffer from subjectivity and inconsistency. Manual measurements,

T. Bauernhansl et al. (Eds.): WGP 2023, LNPE, pp. 114–123, 2024.
https://doi.org/10.1007/978-3-031-47394-4_12

although more quantitative, can be time-consuming and impractical for continuous monitoring. Similarly, fixed grinding wheel dressing intervals may not capture wear-related issues in a timely manner, leading to potential quality issues and production delays.

The complex nature of centerless grinding processes, influenced by factors like the grinding wheel, grinding and dressing conditions, and machine setup, further underscores the importance of tool wear monitoring in this field [2]. The condition of machinery can be also monitored using various techniques, including vibration analysis, thermography, and ultrasonic inspection. These techniques can help detect early signs of degradation, thereby allowing for timely intervention. Vibration has been used to pinpoint the potential sources of vibrations and identify the component with faulty behavior [3]. Thermography was also used to locate potential issues using hot or cold spots in a machine [4], while the ultrasonic inspection is a reliable method that gives valuable information about the internal state of the machine through the analysis of sound waves [5].

In the context of grinding processes, acoustic emission (AE) signals have been found to be particularly useful for monitoring wear, which implies a high probability of successful applications in centerless grinding. Especially, detecting grinding wheel wear, loading, chatter analysis, collision detection, grinding burn and cracks, gap elimination, process control, and dressing/truing verification have been deemed feasible [6]. Signal processing methods in both time and frequency domains, such as time direct analysis (TDA), singular spectrum analysis (SSA), Fourier transform, and wavelet transform, have been utilized for online process monitoring. Moreover, different predictive techniques including multivariate regression, artificial neural networks, and support vector machines (SVM) have been applied considering the correlating features of the parameters under study [7].

Previous research has provided evidence of the feasibility of utilizing vibration and acoustic emission signals for monitoring grinding processes. However, the effectiveness and practicality of implementing these approaches for efficient online monitoring in centerless through-feed grinding processes have not yet been thoroughly investigated and validated. Although the feature-based analysis of AE signals appears to be applicable for wear monitoring in centerless through-feed grinding, no thorough investigations and validations have been performed yet. Further research is required to establish the consistency of tool wear detection by these monitoring methods with the measurement of corresponding process conditions and result variables. Therefore, the effectiveness of the method, specifically in the context of centerless through-feed grinding operations, is still to be investigated.

By enhancing wear monitoring in centerless grinding processes, this research makes a distinct contribution by focusing on the development of a one-class support vector machine (SVM) [8] model for evaluating the process state in centerless through-feed grinding using acoustic emission (AE) signals. The one-class SVM is a machine learning algorithm specifically designed for anomaly detection, capable of identifying instances that deviate from the norm. In the context of wear monitoring, the one-class SVM can effectively detect patterns in AE signals, indicating changes in machine behavior possibly relating to wear. To validate the proposed approach, the research will conduct experimental studies in real-world centerless through-feed grinding operations, specifically in an industry-like process setup. This validation will demonstrate the practical applicability, reliability, and effectiveness of the one-class SVM model for wear monitoring.

The paper is structured as follows: Section 2 presents the methodology, including the experimental setup, data acquisition, feature-based process analysis, and the use of the one-class Support Vector Machine (SVM) algorithm for grinding tool condition monitoring. Section 3 provides a summary of the results and includes a discussion of their implications. Section 4 concludes and provides an overview of the entire research, highlighting its contributions and potential directions for future research.

2 Methodology

2.1 Experimental Setup

A partially hardened C53 steel with an initial workpiece diameter of $d_{w,0}$= 15.1 mm was ground with a depth of cut of $a_e = 0.3$ mm. Workpieces with two different lengths were examined in order to investigate any correlation between the grinding wheel width to workpiece length ratio on the process results. The workpiece length for the short workpiece was $l_{w,1}$= 284 mm and for the long workpiece was $l_{w,2}$= 425 mm. The centerless through-feed grinding experiments were carried out on a Mikrosa Kronos S250 grinding machine. The cooling lubricant used was a Hocut 4570 emulsion from Quaker Houghton at a concentration of $c = 6\%$. A vitrified bonded grinding wheel of grit size F90 from Effgen Lapport was used for the tests. The regulating wheel was rubber-bonded with the corundum grit size F120 and from the company Effgen Lapport.

The grinding wheel has been profiled with a run-in zone, grinding zone and spark-out zone. At a total grinding wheel width of $b_{gw,tot} = 250$ mm, the grinding zone took a width of $b_{gw,2}$= 150 mm. Grinding was performed with an elevation of $h = 3.7$ mm over the center and a regulating wheel inclination of $\alpha_r = 3.0°$.

In the grinding investigations, the circular shape deviation f_k, the straightness deviation f_g, the cylindrical shape deviation f_{cyl}, the arithmetic mean roughness R_a and the maximum height of profile R_z were measured. For the form deviations f_k, f_g, f_{cyl}, a form measurement machine of the type MMQ-400 from the company Mahr was used. The roughness values R_a, R_z were measured using an Etamic Nanoscan roughness and countour measurement machine from the manufacturer Hommel. The grinding wheel was dressed in an identical manner before each test point. After the dressing process, the grinding wheel wear behavior was investigated. V_w was continued until a material removal of V_w= 800,000 mm^3 was achieved. At this point, the grinding wheel wear was considered critical due to a diameter deviation of the ground parts in all test points considered. Acoustic emission data were recorded during the process using a measuring chain for acceleration sensors of the type Piezotron/8852A from the company Kistler was used and set to a sampling rate of 1 MHz. During the tests in centerless through-feed grinding, the AE sensor was placed a few millimeters below the workpiece in the run-in zone of the work rest, which lead to well analyzable results. The test setup and the sensor are shown in Fig. 1

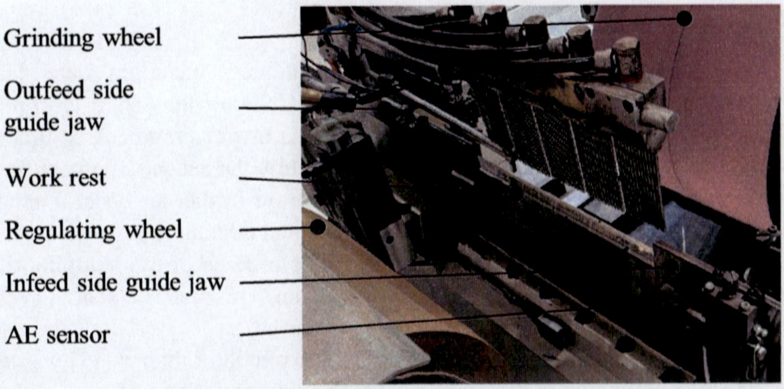

Fig. 1. Experimental configuration and sensor placement.

2.2 Data Acquisition

For the grinding experiments carried out, the grinding wheel radial wear was measured on the basis of the dimensional deviations of the workpieces as a function of the material removal V_w. It was found that increasing dimensional deviations are present starting from a material removal of $V_w = 500{,}000$ mm^3 after the previous grinding wheel dressing process. It can be assumed that this is due to a macroscopic grinding wheel wear state in which the entire spark-out zone width has been involved in the material removal process, following a diameter loss of the grinding zone. For the given experimental setup, an increase in the circular shape deviations f_k as well as straightness deviations f_g occurred starting from a material removal of $V_w = 300{,}000$ to $500{,}000$ mm^3.

The AE dataset utilized in this research includes four distinct experiments involving two different work rest geometries and two different workpiece lengths. In the first experiment, focusing on the standard support rail geometry, data were recorded for the long workpieces spanning from the 55th to the 270th workpiece (164,000 mm^3 $< V_w$ $<$ 803,000 mm^3). This range was obtained due to a measurement issue encountered during the experiment. For the short workpieces, the AE signal was captured from the 1st workpiece up to the 270th workpiece (0 mm^3 $< V_w$ $<$ 803,000 mm^3). The second experiment involved a non-standard work-rest geometry which was characterized by different workpiece heights on the infeed and outfeed sides of the grinding gap. The exact design of the work rest was not of further importance for the investigations in this study. This experiment merely served to obtain further test data in a divergent process. Herein, the AE signals were recorded for all repetitions of both long and short workpieces (in every case 0 mm^3 $< V_w$ $<$ 803,000 mm^3). For the long workpieces, data was collected continuously for 270 consecutive workpieces. Similarly, for the short workpieces, the AE signals were recorded for a total of 390 consecutive workpieces. This experimental design allowed for comprehensive data collection across various workpiece ranges, enabling a comprehensive analysis of the AE signals in relation to the different geometries and workpiece sizes.

2.3 Feature-Based Process Analysis

Time series analysis encompasses various techniques, including deep learning approaches such as convolutional neural networks (CNN) or long-short term memory neural networks (LSTM), which can directly process raw or preprocessed time series data without prior feature extraction [9]. Alternatively, the second approach involves transforming the time series into a feature space before further analysis. It entails for instance identifying suitable features from the statistical domain (e.g., mean or median) or the spectral domain (e.g., FFT coefficients). The process of transforming the raw data into a feature space is known as feature extraction. This step is essential to capture relevant patterns and characteristics of the time series [10].

In this paper, the TSFEL (Time Series Feature Extraction Library) [11] was used as a tool for feature extraction in the context of the acoustic emission (AE) signal analysis. The utilization of the TSFEL library allowed for the extraction of a diverse set of features from the AE signal. These features cover various domains, including statistical, spectral, and entropy-based measures [12]. A total of 260 features were extracted for analysis, encompassing various signal characteristics such as FFT mean, Wavelet variance, and absolute energy. These features were selected to provide a comprehensive representation of the data and capture relevant information related to the studied phenomenon. The FFT mean feature quantifies the average frequency content of the signals, while the Wavelet variance feature captures the variability of the signal across different scales. Additionally, the absolute energy feature reflects the overall magnitude of the signals.

2.4 Grinding Tool Condition Monitoring Using the One-Class Support Vector Machine (SVM) Algorithm

Several researchers have explored the application of machine learning algorithms for tool condition monitoring. Sick [13] proposed tool wear monitoring using Artificial Neural Networks (ANN), while Shi and Gindy [14] developed a tool wear prediction model using Least Square Support Vector Machines (SVM). Elangovan et al. [15–17] utilized Bayesian classifiers, decision trees, and SVM for single point tool condition monitoring in turning processes based on vibration signals. Wang et al. [18] studied machine tool conditions using the SVM algorithm. Krishnakumar et al. [19] applied decision trees and ANN for multipoint tool condition monitoring in high-speed machining. Zhang [20] implemented neuro-fuzzy models for tool wear studies using vibration signals. Arun et al. [21] experimentally evaluated grinding wear in cylindrical grinding processes using vibration-based techniques, comparing the performance of various machine learning classifiers [22]. This paper introduces a novel approach for tool wear monitoring based on a one-class support vector machine (SVM) algorithm. Unlike traditional supervised learning methods, our proposed framework operates in an unsupervised manner, focusing on anomaly detection rather than classification. By leveraging the one-class SVM, we aim to address the limitations of existing tool wear monitoring techniques in centerless grinding processes. Our approach utilizes acoustic emission signals as input data, capturing the unique patterns associated with tool wear.

One-class SVM works on the basic idea of minimizing the hypersphere of the single class of examples in training data and considers all the other samples outside the hypersphere to be outliers or out of training data distribution. Figure 2 illustrates the concept of using a one-class SVM to create a hypersphere for classifying out-of-training-distribution data. The process of obtaining the hypersphere, defined by its center c and radius r, involves solving the following constrained optimization problem:

$$\min_{r,c} r^2 \textit{ subject to } ||\Phi(x_i) - c||^2 \leq r^2 \forall i = 1, 2, \ldots, n \tag{1}$$

However, a more flexible formulation that allows for a certain degree of tolerance towards outliers is given by:

$$\min_{r,c,\zeta} r^2 + \frac{1}{\nu n} \sum_{i=1}^{n} \zeta_i \tag{2}$$

$$\textit{subject to}, ||\Phi(x_i) - c||^2 \leq r^2 + \zeta_i \forall i = 1, 2, \ldots, n \tag{3}$$

In the given formulation, the function Φ represents a transformation function that maps the original data points x_i into a different space, the mathematical variable ζ_i, represents scalars that describe distances from the hypersphere around c to datapoints outside of the sphere and ν is a positive parameter that determines the compromise between the volume of the sphere and the number of outliers [23].

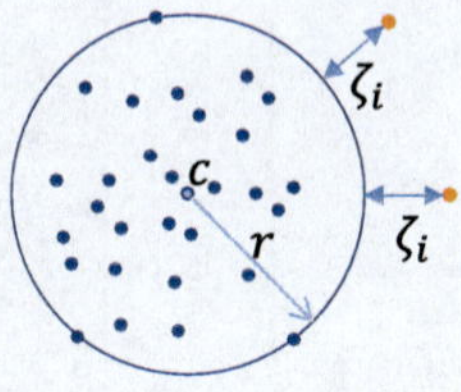

Fig. 2. Hypersphere representation of target data with center c and radius r.

Following the feature extraction process, the extracted features serve as inputs for the algorithm. During the training phase, the algorithm learns to differentiate normal operating conditions from abnormal or worn-out conditions by establishing a boundary in the feature space. Subsequently, the trained one-class SVM is employed to classify new acoustic emission signals based on their similarity to the learned normal behavior. By assessing the proximity of the extracted features from these signals to the established boundary, the algorithm identifies instances that deviate significantly from the norm, indicative of potential wear in the grinding tool.

3 Results and Discussion

To establish a comparison between data analysis and real-world experiments, it is necessary to consider the computational factors relating to the number of short and long workpieces, as well as the value of cutting volume (V_w). The conversion between the number of parts and the V_w value is determined using specific factors, as outlined in the provided Table 1.

Table 1. Conversion factors for establishing comparison between number of workpieces and cutting volume V_w.

Workpiece length l_w [mm]	Cutting volume $= \pi \frac{d_{wA}^2 - d_{wE}^2}{4} \cdot l_w$ [mm^3]
Short (284)	1987.4
Long (425)	2974.1

Based on the measurements obtained from the experiment, it is observed that an increase in the novelty score of the analysis occurs between V_w values of 300,000 mm^3 and 500,000 mm^3. During this range, there is also a significant rise in both the straightness deviation (f_g) and the cylindricity deviation (f_{cyl}) due to grinding wheel wear.

This finding suggests that as the V_w value increases within this specific range, the deviation in straightness and cylindricity becomes more pronounced. This can be attributed to the wear of the grinding wheel, which impacts the accuracy and precision of the workpiece geometry.

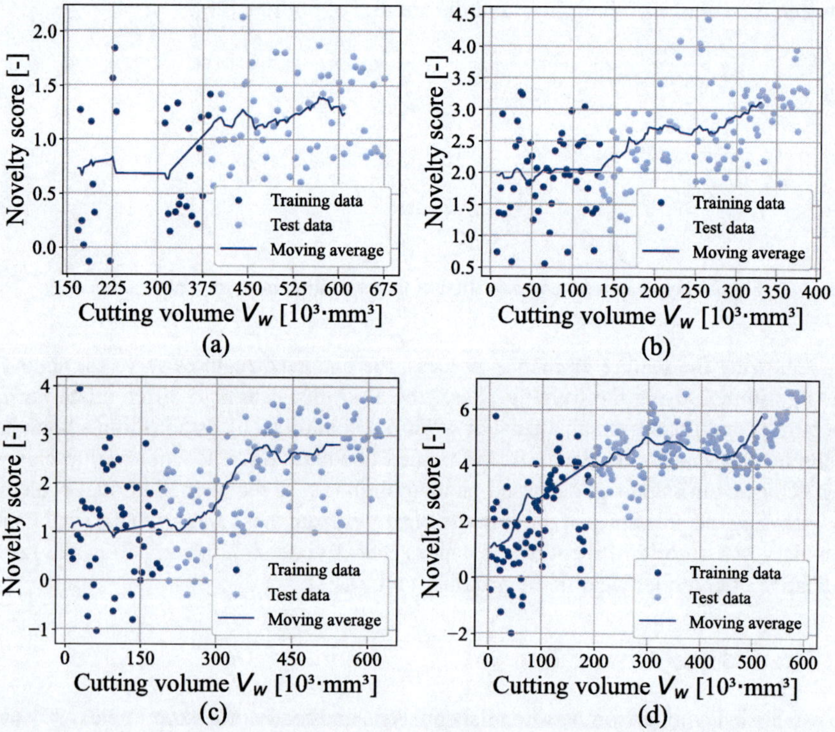

Fig. 3. Figures 3(a) and 3(b) showcase the application of novelty detection on long and short workpieces with a standard work rest geometry, while Figs. 3(c) and 3(d) demonstrate novelty detection on long and short workpieces with a non-standard work rest geometry.

The figures show that there is a comparatively large scatter in the calculated novelty scores. Nevertheless, there was good agreement between the moving average of the novelty scores and the wear-related rise in component shape deviations.

Figures 3(a) and 3(b) depict the application of novelty detection on both long 3(a) and short 3(b) workpieces with a standard work rest geometry. For the short workpieces, 50 training points were used for subsequent prediction, while the long workpieces employed 30 training points. Similarly, Figs. 3(c) and 3(d) demonstrate novelty detection on both long 3(c) and short 3(d) workpieces with a non-standard work rest geometry. In this case, 75 training points were utilized for the short workpieces, and 50 training points were employed for the long workpieces. The variation in the number of training points allows for tailored prediction and analysis based on the specific characteristics of each workpiece.

The presence of bends in the graphs reflects interesting observations regarding tool wear. Specifically, in Figs. 3(b), 3(c) and 3(d), there is a notable change in the trend, indicating a distinct behavior related to tool wear. The bend in the graph signifies a critical point where the tool wear effects become more pronounced, leading to a deviation from the previous trend. This bending point can serve as a reference to identify the threshold at which the tool's condition significantly impacts the machining process. The significance of this observation cannot be reliably confirmed in Fig. 3(a) due to failures in the recording of measurements, which necessitated the use of more than 300,000 mm^3 for training data. Nevertheless, the findings from the first and second experiment series remain compelling and warrant further investigation.

In summary, the key finding of this research is the consistent observation of a significant increase in the novelty score between cutting volume values of 300,000 mm^3 and 500,000 mm^3 in the analyzed data. This trend was observed regardless of the specific workpiece series or geometry under investigation. Consequently, this finding indicates the critical relevance of this cutting volume range in detecting novel or anomalous instances within the grinding process. Furthermore, the presence of bends in the graphs illustrates a distinct behavior related to tool wear, signifying a threshold at which the tool's condition significantly impacts the machining process. These findings highlight the importance of further investigation into the factors influencing tool wear and the relationship between anomalies and grinding process variables.

4 Conclusion and Future Work

This research aimed to improve wear monitoring in centerless grinding processes using acoustic emission (AE) signals by integrating a one-class Support Vector Machine (SVM) algorithm. The combination of feature extraction using TSFEL and SVM algorithm proved to be effective in detecting anomalies during centerless through-feed grinding operations. The analysis of various workpiece geometries and sizes successfully identified wear-related changes in the grinding process, specifically the impact of grinding wheel wear on workpiece geometry accuracy. The observed increase in novelty score and the presence of bending points in the graphs provided evidence of the threshold at which grinding wheel wear significantly affected the machining process, supported by geometry measurements of the workpieces.

The key finding of this research work is the successful detection of deviations associated with grinding wheel wear through AE signal analysis. This finding contributes to the understanding of wear monitoring in centerless grinding operations and highlights the importance of timely intervention and preventive maintenance measures to optimize machine performance and improve product quality.

While this study shows promising results, there are several limitations that need to be addressed in future research. The limited variation in geometry and parameters in the current study calls for further investigation to verify the applicability of the proposed approach in different settings. Additionally, exploring additional features such as vibration and investigating featureless methods that can extract information directly from raw data could provide a more comprehensive understanding of the grinding process and improve wear detection accuracy. Increasing the data sample size and refining the one-class SVM model by fine-tuning parameters, exploring different kernel functions, or considering alternative anomaly detection algorithms is essential to enhance the approach's generalizability and robustness. Future research should also focus on optimizing the model, validating it in larger-scale industrial environments, exploring different grinding processes, and expanding the range of parameters under monitoring. Advancements in these areas will significantly contribute to improving process control and ensuring high-quality results in grinding operations.

References

1. Nithin, S.K., Hemanth, K., Shamanth, V., Mahale, R.S., Sharath, P.C., Patil, A.: Importance of condition monitoring in mechanical domain. Mater. Today: Proc. **54**, 234–239 (2022)
2. Kim, H.Y., Kim, S.R., Ahn, J.H., Kim, S.H.: Process monitoring of centerless grinding using acoustic emission. J. Mater. Process. Technol. **111**(1–3), 273–278 (2001)
3. What can vibration analysis detect. https://lightningchart.com/blog/what-can-vibration-analysis-detect/
4. Passive and Active Thermography. https://movitherm.com/knowledgebase/passive-and-active-thermography/
5. Ultrasonic Testing, https://www.flyability.com/ultrasonic-testing
6. Webster, J., Dong, W.P., Lindsay, R.: Raw acoustic emission signal analysis of grinding process. CIRP Ann. **45**(1), 335–340 (1996)
7. Plaza, E.G., Chen, X., Ouarab, L.A.: Abrasive feature related acoustic emission in grinding. In: 25th International Conference on Automation and Computing (ICAC), Lancaster, UK, pp. 1–6 (2019)
8. Schölkopf, B., Williamson, R.C., Smol, A., Shawe-Taylor, J., Platt, J.: Support vector method for novelty detection. In: Advances in Neural Information Processing Systems, vol.12. no.3 (2000)
9. Fawaz, I.H., Forestier, G., Weber, J., Idoumghar, L., Muller, P.A.: Deep learning for time series classification: a review. Data Min. Knowl. Disc. **33**, 917–963 (2019)
10. Unterberg, M., Voigts, H., Weiser, I.F., Feuerhack, A., Trauth, D., Bergs, T.: Wear monitoring in fine blanking processes using feature based analysis of acoustic emission signals. In 54th CIRP Conference on Manufacturing Systems, vol. 104. no. 2021, pp. 164–169 (2021)
11. Barandas, M., et al.: TSFEL: time series feature extraction library. SoftwareX **11**, 100456 (2020)
12. Time Series Feature Extraction Library Documentation. https://tsfel.readthedocs.io/en/latest/

13. Sick, B.: Online and indirect tool wear monitoring in turning with artificial neural networks: a review of more than a decade of research. Mech. Syst. Signal Process. **16**(4), 487–546 (2002)
14. Shi, D., Gindy, N.N.: Tool wear predictive model based on least squares support vector machines. Mech. Syst. Signal Process. **21**(4), 1799–1814 (2007)
15. Elangovan, M., Devasenapati, S.B., Sakthivel, N.R., Ramachandran, K.I.: Evaluation of expert system for condition monitoring of a single point cutting tool using principle component analysis and decision tree algorithm. Expert Syst. Appl. **38**(4), 4450–4459 (2011)
16. Elangovan, M., Ramachandran, K.I., Sugumaran, V.: Studies on Bayes classifier for condition monitoring of single point carbide tipped tool based on statistical and histogram features. Expert Syst. Appl. **37**(3), 2059–2065 (2010)
17. Elangovan, M., Sugumaran, V., Ramachandran, K.I., Ravikumar, S.: Effect of SVM kernel functions on classification of vibration signals of a single point cutting tool. Expert Syst. Appl. **38**(12), 15202–15207 (2011)
18. Wang, P., Meng, Q., Zhao, J., Li, J., Wang, X.: Prediction of machine tool condition using support vector machine. J. Phys. Conf. Ser. **305**(1), 012113 (2011)
19. Krishnakumar, P., Rameshkumar, K., Ramachandran, K.I.: Tool wear condition prediction using vibration signals in high speed machining (HSM) of titanium (Ti-6Al-4V) alloy. Procedia Comput. Sci. **50**, 270–275 (2015)
20. Zhang, C., Yao, X., Zhang, J., Jin, H.: Tool Condition monitoring and remaining useful life prognostic based on a wireless sensor in dry milling operations. Sensors **16**(6), 795 (2016)
21. Arun, A., Rameshkumar, K., Saimurugan, M.M., Sumesh, A.: Experimental evaluation of grinding wheel wear using vibration based technique, In Proceedings of International Conference IRAM, IIT Indore (2013)
22. Arun, A., Rameshkumar, K., Unnikrishnan, D., Sumesh, A.: Tool condition monitoring of cylindrical grinding process using acoustic emission sensor. Mater. Today: Proceedings **5**(5), 11888–11899 (2018)
23. Noumir, Z., Honeine, P., Richard, C.: On simple one-class classification methods. In: 2012 IEEE International Symposium on Information Theory Proceedings, pp. 2022–2026. IEEE (2012)

Cross-Machine Comparison of the Usability of Internal Machine Control Data for Process Monitoring in Machining Applications

D. Gutsche(✉), T. Reeber, P. Georgi, and H.-C. Möhring

University of Stuttgart - Institute for Machine Tools, Holzgartenstraße 17, 70174 Stuttgart, Germany

daniel.gutsche@ifw.uni-stuttgart.de

Abstract. The use of machine control data for process monitoring has many advantages. It represents a cost-effective alternative to expensive external measurement devices and can be retrofitted to existing CNC machines. External sensors can be affected by the machining environment, require setup, and are costly to procure. Furthermore, they influence the process since they are part of the machining process because they replace either the workpiece clamping technology or the toolholder. Machine control data such as spindle currents are in proximity to the process without directly affecting it. Since these data are already known to the NC, they can be used for adaptive control applications. This paper focuses on investigating the differences in position control data on three different machine tools. Using the same reference process. Correlations between feed forces and motor currents are calculated and compared. Correlations of motor currents between different machine tools performing the same process are analysed.

Keywords: Machining · Monitoring · Computer numerical control (CNC)

1 Introduction and Motivation

As the field of manufacturing continues to evolve, there is an increasing need for efficient, economical, adaptable, and autonomous machining solutions. One area that has received significant attention is the monitoring and control of machining processes. The digitalization laboratory (DigiLab) is installed for this purpose at the Institute for Machine Tools (IfW) at the University of Stuttgart, as can be seen in Fig. 1. It features a connected test field with 7 fully connected machines of various type and age. The IfW has been working for years on the design, layout and realization of self-optimizing manufacturing systems in the field of mechanical production [4]. Traditionally, this has been achieved using external measurement devices, which, although practical, have several notable disadvantages. They are often expensive to procure, require experts to set up, and can influence the machining process. This can be circumvented by leveraging position control data directly from the computerized numerical control (CNC). This approach is desirable as it can be retrofitted to existing machinery, thus eliminating the

T. Bauernhansl et al. (Eds.): WGP 2023, LNPE, pp. 124–132, 2024.
https://doi.org/10.1007/978-3-031-47394-4_13

need for costly upgrades or replacements. Utilizing control variables such as spindle currents offer an opportunity to develope adaptive control applications (AC) across multiple machines. The comparative examination of internal control data across machines of varying kinematics, age, and size using a consistent reference process will provide valuable insights into the versatility and applicability of this approach as well as how to translate insights and control strategies from one machine to antoher. The goal is to develop generalized machine data-based process control strategies that can be translated to any machine configuration on the shop floor.

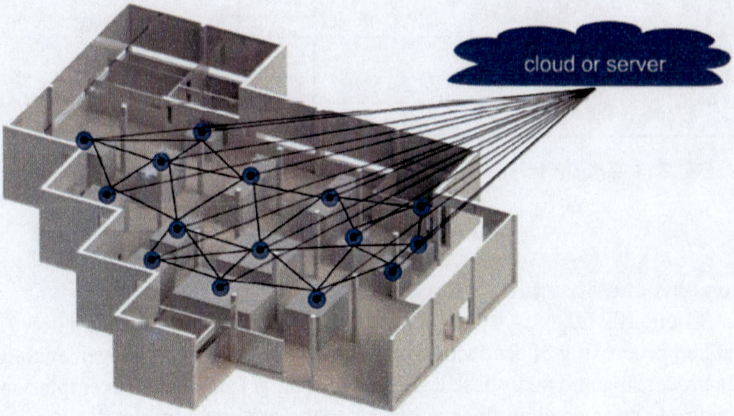

Fig. 1. DigiLab at the IfW

2 State of the Art on the Use of Machine Control Data for Process Monitoring

Capabilities for Data Acquisition

Mutliple strategies are known to extract control internal data from machine tools [1, 5, 8]. Internal control data as a business model rose to prominence in the latter years, where special combinations of hard- and software – so called edge devices – facilitate the extraction of control internal data without increasing the load on the control device itself. Furthermore, data aggregation and preprocessing before forwarding the data to higher automation levels decrease the load in the company network while giving highly valuable insights to controlling and production planning. Early approaches to the use of internal control data, and more specifically data relating to the position control loops, are shown by Kaever [3]. The author presents a prototype system for detecting both cutter breakage and overload. A good transferability of the basic principle is shown between different machines, but not so in terms of different processes.

Extracting Data from Machine Control Systems

Different possibilities on the machine control level to acquire process data are illustrated in Fig. 2. The interfaces to the programmable logic controller PLC and the machine tool's NC are shown.

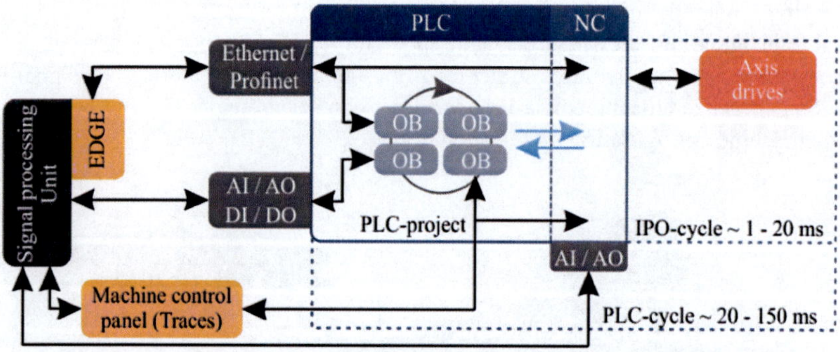

Fig. 2. Interfaces for the data communication in machine tools based on [6]

Data Domains and Structure of the Data

Siemens Sinumerik Edge as a means to extract internal control data allows for the decentralized processing of production data directly at the machine level, enabling real-time data processing and analytics. Trabesinger et al. [9] showed an exemplary network infrastructure to connect shop floor machines to the,as shown in Fig. 3.

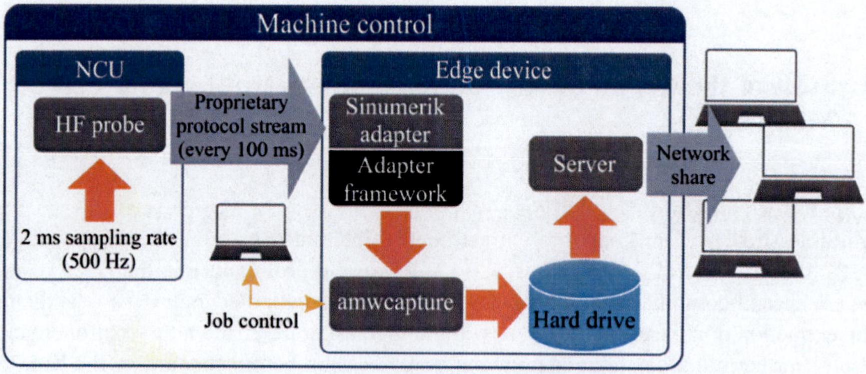

Fig. 3. Data capturing architecture with Sinumerik Edge based on [9]

They describe the data capturing of high-frequency and low-frequency drive signals at sampling rates of 2 ms and 100 ms, respectively, and stored the data using the edge app analyze my workpiece capture (AMW) [7]. The capturing of High-Frequency (HF) data was implemented using the HF probe inside the SINUMERIK 840d sl NCU, and the capturing of Low-Frequency (LF) utilizing an encrypted OPC UA server-client connection. Table 1 shows an excerpt of the available data that can be recorded via AMW. Depending on the parameterization and the machine, these data are available for all axes or drives of the machine [7]. This method of internal control data acquisition was also chosen for this paper.

Table 1. Excerpt of the available HF-data

Signal	Signal Description	unit
DES_POS	Position setpoint after fine interpolator	mm or °
CTRL_DIFF	Controller difference, compensation included	mm or °
VEL_FFW	Feed rate (forward), at the output of the position controller	mm/s or °/s
CMD_SPEED	Speed setpoint	mm/s
ENC1_POS	The actual position value of encoder 1	mm or °
TORQUE	Drive torque setpoint	Nm or N
POWER	Smoothed effective drive power	W
CURRENT	The actual current value of the axis/spindle	A
LOAD	Drive load	%

Comparison Methods
Their linear dependence can represent the correlation coefficient of two matrices. If each variable has N scalar observations, the correlation coefficient is defined as in Eq. (1) [2].

$$\rho_{(x,y)} = \frac{1}{N-1}\sum_{i_1}^{N}\left(\frac{x_i - \mu_x}{\sigma_x}\right)\left(\frac{y_i - \mu_y}{\sigma_y}\right) \quad (1)$$

$\mu_{x,y}$ and $\sigma_{x,y}$ are the mean and standard deviation of x and y

3 Methods

General insight into the process as well as supervision, prediction and adaptive control strategies are a major focus of research efforts in the machining sector. In most cases, these aforementioned research findings are based on an extensive list of experiments conducted on a single machine. Ideally, these findings are described as universally applicable empirical relationships based on mechanicalistic and materialistic principles.

But in most cases, the description of the systems with their various input factors and the subsequent formulation of such models is difficult.

Machines and Experimental Setup

To translate these research findings, the individual behavior of each machine has to be examined and efforts have to be made to describe their differences mathematically.

For this, multiple drilling processes have been conducted on different machines. Three connected machine tools were used for the experiments. Table 2 shows the different CNC-machine types in use. The general setup of the experiments can be seen in Fig. 4. Parallel measurements of forces and relevant control data were conducted. The respective data sources are synchronized in time. This was achieved through linking both data sources to the same trigger signal source, which is activated via G-Code. The sampling rate of the external force measurement was choosen to be 10 Ks/sec, while the internal control data measurement is linked to the position control cycle of the individual machine, which was 500 s/sec for all three machines. Drilling experiments were conducted using a spiral drill with a diameter of 8 mm and no cooling. The workpiece material consisted of 42CrMo4 (Table 3).

Table 2. Machine tools used for the experiments

Machine ID	Name	Spindle Power	Machine type	Internal Control Data Measurement	External Force Measurement
M1	Grob G552	65 kW	Horizontal 5-axis milling centre	X	X
M2.1	Index R200 – Live Tool	11 kW	Mill-Turn centre	X	
M2.2	Index R200 – Static Tool	15 kW			
M3	DMG Lasertec 65 3D Hybrid Index R200	25 kW	Vertical 5-axis milling centre	X	X

Data Analysis Strategy

In the context of this paper, the main indicator to compare the drilling processes on the examined machine tools is the moment-forming current I(q) of each axis. The index "s" hereby refers to the spindle used to perform the cutting movement, whereas the index "z" indicates the axis performing the feed movement. Furthermore, the position control lag along the feed axis of the drilling process will also be examined, hereby referred to as "ΔS". To determine the feasibility of a cross machine comparison of the moment-forming currents additional force measurements are performed alongside the recording of control parameters.

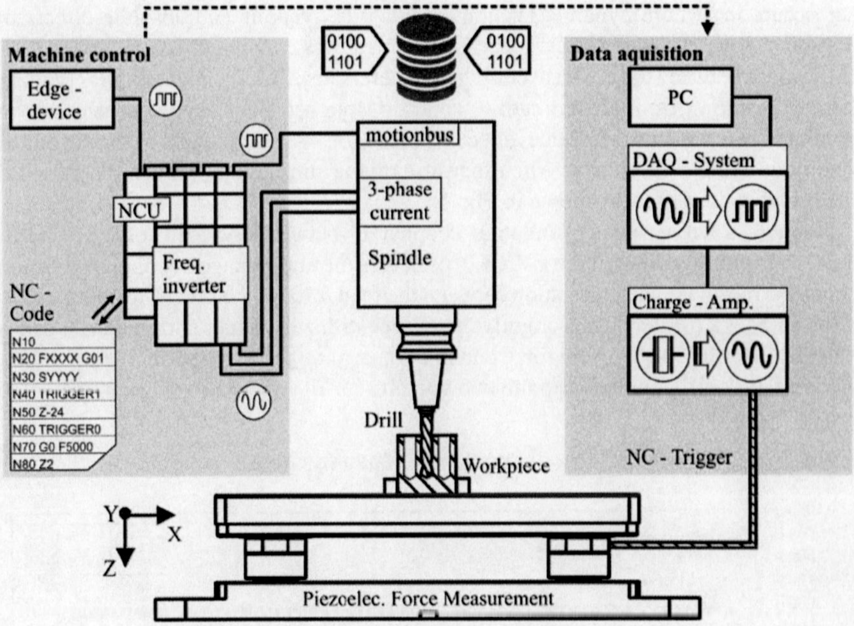

Fig. 4. Structure of the measuring chain for process data acquisition

Table 3. Experimental parameters and variations

Parameter			Variations
Feed rate per revolution	f	mm/rev	0.10 and 0.20
Feed rate	v_f	mm/min	636.62 and 1,591.55
Rotational speed	n	min^{-1}	3,183.10 and 3,978.87
Cutting speed	v_c	m/min	80 and 100
Repetitions per Parameter			7
Sum of experiments per machine			56

4 Results and Discussion

Force measurements are recognized as reliable indicators of various effects in machining processes, including wear or alterations in machining parameters and conditions. To validate the effectivness of internal control data as a means of describing a drilling process and translating AC-strategies, correlation coefficients were computed across the experimental design between the external force data and internal control data. A sample of raw data is shown in Fig. 5. Visually the correlation between externally measured and control internal data is apparent. As the external measurements are sampled with a 20× higher sampling rate than the control data, they can be used to check for events that occure in a more dynamic fashion. To include dynamic and absolute effects the root-mean-square-value (RMS) for each experiment was computed. A comparison of the RMS values between the different cutting parametersclearly demonstrates the anticipated changes resulting from altered cutting conditions across all machines. Therefore, the calculated correlation coefficients reflect the precision of detection of a change in cutting conditions using control data. When focusing on the correlation between the measured forces and control data, as shown in Fig. 5

Table 4, a strong mean correlation is apparent between the feed force F_z and the moment-forming current of the z-Axis $I(q)_z$ across the entire range of conducted experiments. Furthermore, the correlation between the feed force and moment-forming current of the spindle $I(q)_s$ displays a slightly lower, yet still substantial, correlation. It can be seen that the relative change in force and in current in response to the alternation of feed rate is slightly different between the two domains.

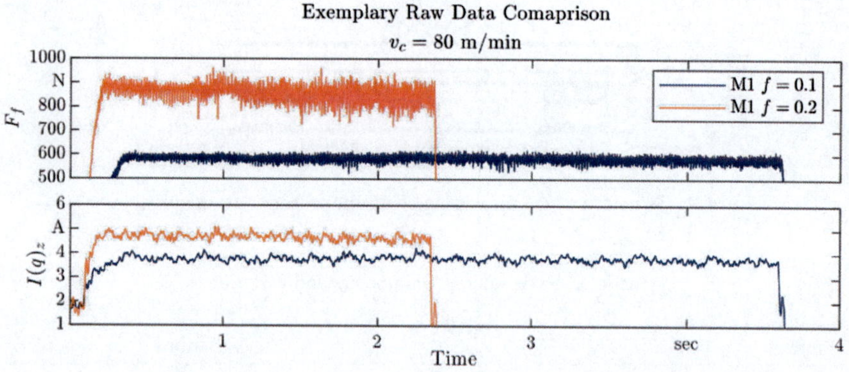

Fig. 5. Exemplary plot of raw data obtained at two different feed rates

Table 4. Correlation coefficients between Machines 1 and 3

Machine	Machine 1	Machine 3
R between Fz and I(q)z	0.9841	0.996
R between Fz and I(q)s	0.9682	0.8890

In Fig. 6, the root mean square (RMS) was calculated for the configuration types M1, M2.1 and M3 for each experimental parameter, which lead to 56 RMS – Values for each variable and each machine. An example can be seen in Fig. 6b) for the moment forming current of the spindle, Iq(s). Four stair-effects can be seen, which denote the change in cutting parameters every seven drillings, the first change being a change in feed rate (0.1 to 0.2 mm/rev) and then a change in cutting velocity (80 to 100 m/min).The RMS was computed over the whole active cutting interval. As can be seen exemplary in the RMS of Iq(S), comparable shifts can be seen on all three machines when a change in parameters occurs. In detail, it can be seen that the relative change for M1 and M3 (the 5-axis-machining centres) is the same. A more pronounced relative change between feed rates can be seen for M2.1, which is the mill-turn-centre, which could be explained through a smaller spindle, which in turn needs to pull more current to cut the increased chip cross section. The plot shows that relative differences in machine behavior exist which need to be accounted for. Figure 6 depicts the general Pearsons correlation coefficient matrix, which describes the correlation of the respective RMS along the design of experiments for each variable and with different machine pairings. The in-machine correlations of each machine pairing for both Iq(z) and Iq(s) can be seen as very pronounced, with a slight deviation in the comparison between M3 (Heavy machining centre) and M2.1 (mill-turn centre) with regard to the spindle. This difference could be explained through the gap in spindle motor power, as described in Table 2, because higher chip removal rates lead to a different relative behavior of the lighter and heavier spindle.

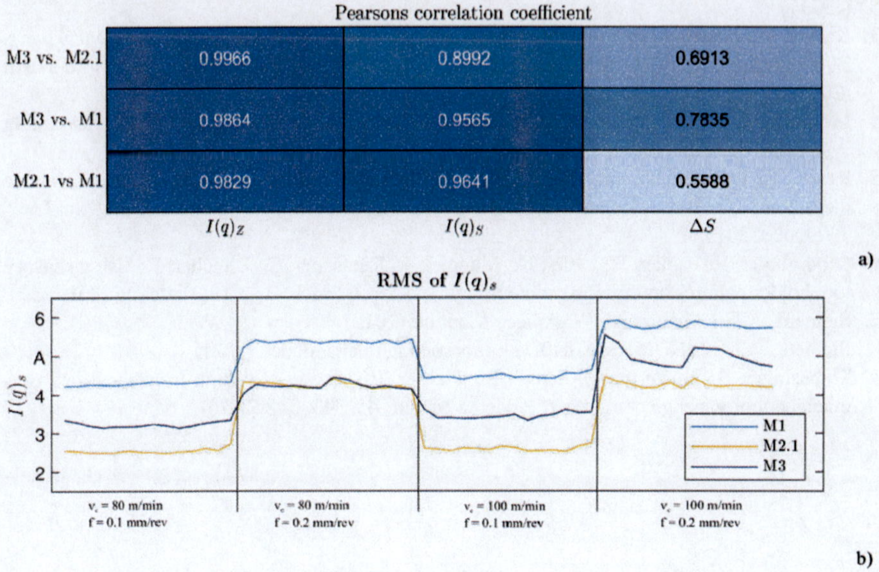

Fig. 6. a) Pearsons correlation coefficient matrix for in – machine relationships between Iq(s), Iq(z) and ΔS (above) of the RMS for all experiments – b) exemplary comparison of the RMS of Iq(s) for all machines along the design of experiments (bottom).

5 Conclusion and Outlook

The results presented in this paper show that different machine concepts differ in the response of internal control data to changes in cutting parameters. The relative response in the time domain is more or less the same for different machine concepts and sizes. However, if absolute values in terms of amplitude and dynamic behaviour are required for certain applications, then the differences between individual machine tools need to be considered. Different approaches could be considered: One would be to find a base value to describe different machine concepts, such as drive type, size and general drive and spindle concept. AI models offer the possibility of describing non-linear relationships, but generally require large amounts of training data before being able to perform adequate calculations. Combinations of the two approaches are also possible. In the future, further research will be conducted at the IfW to build such models and to further exploit the possibilities of control data to assess and improve the performance of machining operations on an industrial scale.

References

1. Bergmann, B., Reimer, S.: Online adaption of milling parameters for a stable and productive process. CIRP Ann. **70**(1), 341–344 (2021)
2. Fisher, R.A.: Statistical Methods for Research Workers. Oliver & Boyd, Edinburgh, London (1971)
3. Kaever, M.: *Steuerungsintegrierte Fertigungsprozeßüberwachung bei spanender Bearbeitung*. Zugl.: Aachen, Rhein.-Westf. Techn. Hochsch., Diss., 2004. Books on Demand, Norderstedt (2005)
4. Möhring, H.-C., Wiederkehr, P., Erkorkmaz, K., Kakinuma, Y.: Self-optimizing machining systems. CIRP Ann. **69**(2), 740–763 (2020)
5. Resch, T.: *Edge Computing: Digitalization in machine tool manufacturing*. https://blog.siemens.com/2020/04/edge-computing-digitalization-in-machine-tool-manufacturing/?linkId=300000000381534
6. Schörghofer, P., Pauker, F., Leder, N., Mangler, J., Ramsauer, C., Bleicher, F.: Using sensory tool holder data for optimizing production processes. J. Mach. Eng. **19**(3), 44–56 (2019)
7. Siemens, A.G.: Analyze MyWorkpiece /Capture (2019)
8. Siemens, A.G.: SINUMERIK 840D sl: Inbetriebnahmehandbuch (2021)
9. Trabesinger, S., Butzerin, A., Schall, D., Pichler, R.: Analysis of high frequency data of a machine tool via edge computing. Procedia Manuf. **45**, 343–348 (2020)

Data-Based Model Parametrization of Quality Predictive Material Removal Simulations

Matthäus Loba(✉), Christian Brecher, and Marcel Fey

Laboratory for Machine Tools and Production Engineering (WZL), RWTH Aachen University, Steinbachstraße 19, 52074 Aachen, Germany
m.loba@wzl.rwth-aachen.de

Abstract. The ability to control product quality in milling processes is an essential target variable. Quality deviations can have different causes, which can be both systematic and stochastic. Tracing data from numerous sensors of modern machine controls and combining them with a material removal simulation (MRS) allows to predict the workpiece quality in a process-parallel manner. The accuracy of the prediction depends on the underlying model and its parametrization. Model parametrization usually requires measurements during which the machine tool cannot be used. Additionally, the necessary measuring equipment is often not available in companies. All this makes the implementation of MRS difficult in the manufacturing industry. This article presents an approach that circumvents the abovementioned disadvantages. With the help of a MRS, machine data is refined and contextualized with quality data. The refined and contextualized data is then used to fit various models, such as stiffnesses and process force. In future the presented approach should enable a cost-effective parametrization without machine downtime. With the parameterized models, MRS that predict quality can be used both in advance during CAM planning and in a process-parallel manner for quality monitoring.

Keywords: Machine Data Analysis · Process Modelling · Model Parametrization

1 Introduction

During machining, different effects occur which influence workpiece quality, productivity and availability of the tools and the machine (e.g. due to wear). During process planning, these three parameters must be optimized to enable an economical production [1]. With regard to productivity and tool wear, tool manufacturers already offer a preselection of optimized parameters. However, the tool manufacturer cannot predict the workpiece quality in its full complexity, since in-depth knowledge, e.g. about the stiffness behavior of the machine tool, is usually necessary. In previous work, the prediction of workpiece quality has been addressed in CAM planning [2] as well as in process monitoring [3]. The prediction accuracy of these simulations depends on the one hand on the discretization and on the other hand on the models behind it. Especially the model parametrization is still very time-consuming and costly, due to the necessary

T. Bauernhansl et al. (Eds.): WGP 2023, LNPE, pp. 133–142, 2024.
https://doi.org/10.1007/978-3-031-47394-4_14

measurements which have to be performed out of production mode. At the same time, the measurements require special equipment and qualified personell. All these factors lead to the fact that the solutions are currently not used on an industrial level. An approach that uses existing resources and does not disrupt production is therefore necessary.

In this paper, a concept is presented that addresses this challenge. In a first step, the state of the art of modeling certain effects and parametrizing their model is discussed. Subsequently, a concept is presented that enables a model parametrization based on machine and quality data. The procedure of the concept is conducted for separate models. Finally, a summary and an outlook are given.

2 Machining Errors

In recent years, numerous models and approaches have been developed along the digital process chain that enable quality prediction in machining. The influences on the workpiece quality can be classified according to different criteria. Apart from the classification according to the degree of order of the defect on the workpiece (cf. [4]), the influences can also be classified according to their source (machine-, workpiece-, tool- and process-related) [5] or the type of generation [6]. When classifying according to the type of generation, it should be noted that some of the influences interact with each other. For example, the cutting edge geometry and its position (tool runout) have an influence on mechanical and thermal factors such as the process force or the heat generation in the process. Altogether, four types of influence are defined: 1) numerical (control) errors, 2) thermal-induced errors, 3) load errors due to a static or dynamic forces and 4) geometric-kinematic errors. In this article, this type of classification is used, since it is considered to be the most useful for the systematic approach of the concept. In addition to [6], the influence of post-processing, the form and location deviation of the raw part and the influence of the clamping force deflection are added.

Mayr et al. state that up to 75% of all geometric defects on the workpiece can be related to thermal influences [7]. In a case study [8], Schmitz investigated the individual influences with regard to effects on the workpiece based on the circle-rhombus-square sample component (NAS979). In the worst case scenario, 76% are due to cutting forces, 18% to geometric and numerical errors, and 6% to thermal expansion of the spindle [8]. Depending on the process parameter setting and running time of the machine, the dimensions can vary [8]. At the same time, the production of the workpiece is a three-axis machining process, so that errors due to post-processing or path errors of the control are rather negligible. Furthermore, the clamping situation is selected in such a way that distortion due to clamping forces tends to be negligible. The variety of workpieces, tools, machine tools, process parameters and environmental influences is so high that the extent of each influence cannot be quantified generally. Rather, this challenge must always be addressed in the context of the entire manufacturing process. In the following, the modeling of the individual errors will therefore be discussed. The focus here is primarily on the model parametrization.

Numerical errors can occur due to discretization or miscalculation. They occur mainly in five-axis machining and arise during postprocessing after CAM programming or to the path interpolation of the NC. Both a too high and a too low point density can lead to machining errors [9]. If the point density is too low, faceting of the surface can occur. If the point density is too high, the feed rate may decrease. The resulting decrease in process force can also lead to machining errors. This can be avoided by continuous-path mode. A disadvantage of this mode is that the programmed contour won't be traversed precisely [10]. As a further numerical error, the interpolation error of the control, which becomes apparent as chord error, should be mentioned [11]. Numerical Errors can be reproduced by using machine-internal data or data from a virtual control (VNCK) [12].

Thermal-induced errors are split into two categories. On the one hand, different thermo-elastic expansions occur on the machine side due to heat sources (e.g. power losses of the drives), which influence the machine accuracy. On the other hand, thermal expansions occur in the cutting zone as a result of the cutting energy induced. Most current work in this area uses finite element method (FEM) [5, 8, 13, 14]. Increasingly, however, hybrid approaches are finding an application as they allow faster calculation of the displacements [15]. One way to measure the thermo-elastic machine behavior is presented in [16]. Using position sensing detector sensors and a thermally stable laser frame, 13 out of 21 errors for three-axis machines are detected. In [17], Denkena et al. present a method to model the thermo-mechanical behavior in the cutting zone. By means of a dexel-based material removal simulation (MRS), they calculate the chip cross-section and the resulting forces as well as the generated heat flux. An FE simulation is used to get the actual workpiece expansion, which is stored in the respective dexels. Subsequently, an inverse displacement of the triangulated tool surface is performed.

Load errors always dependent on both the forces and torques that occur as well as the interacting stiffnesses.

Existing approaches already allow to predict process forces by means of machine data. In most cases, these are often current based [18] or rely on a prediction based on the difference of the encoders [19]. To parameterize the models, load-deformation curves are used. At the same time, air cuts are run to determine the influence of motion on the correlating variables. At low or reversing velocities these approaches often reach their limits (e.g. due to static friction). In addition, the model parameters are affected by further influences, e.g. due to the running time of the machine and past machining operations, which cannot always be reproduced due to their variance. Hybrid approaches provide an improvement of the prediction accuracy. In [20], Königs et al. use an artificial neural network (ANN) for process-parallel process force estimation. Here, in addition to the machine-internal data, the engagement conditions from the process-parallel MRS are used as input parameters. In [21], Denkena et al. use an ANN with long-short-term memory (LSTM) with motor currents and position data, as well as the variables derived from them, such as axis velocity, acceleration and jerk, as input data for predicting the process force. A disadvantage of approaches for process force prediction using machine data is that dynamic effects cannot be modelled due to the low sampling rate (~500 Hz). Additional sensor technology is required for this purpose. Brecher et al. use a spindle-integrated force sensor (SIFS), which determines the process force via the displacement of the bearing rings [22]. Postel et al. use accelerometers on the main spindle housing to predict the process forces [23]. Furthermore, Denkena et al. use strain gauges on the spindle slide to determine the process force as well as the tool stiffness [24]. All approaches have in common that in each case the transfer functions between tool center point (TCP) and the sensor locations must be known.

For the prediction of the stiffness at the TCP (tool + machine), FE-based [25] or analytical [26] approaches are mostly used. By means of dynamic substructure coupling, the machine stiffness can also be modeled [25]. In [27], an approach is presented that uses machine and quality data to determine the resulting stiffness. Using machine internal data, a digital shadow is created that stores force information on the part. Based on the shape deviations between the digital and real part, a calculation of the TCP stiffness is performed.

Approaches for modeling the stiffness on the workpiece side deal with both the deformation of thin-walled components [26] due to the occurring process forces and the deformation due to the clamping situation [28]. These approaches are also mostly FE-based.

Geometric-Kinematic Errors are all those errors that occur due to the interacting geometries in the load-free state. These errors must be taken into account when modeling the kinematics and the individual geometries. On the workpiece side, these are errors in the raw geometry (e.g. casting allowances) and positioning errors. Machine data can be used to detect tool engagement [29]. This theoretically offers the possibility to predict casting allowances, provided that all other geometries are known. In addition, positioning error can be reduced by continuous calibration of the blanks.

On the machine side, the errors of the individual axes and their position relative to each other should be mentioned. These can be measured both directly and indirectly. A direct measurement can be achieved, e.g. by means of laser measurement systems or double-ball bars. An indirect measurement is possible by the use of specimen workpieces. The individual defects and their measurements are summarized in ISO230 [30].

From tool side, geometric errors are caused by tool runout and tool geometry. Depending on the level of detail of the tool (cf. [31]), different orders of errors can be reproduced. When representing the tool as a hull body, form deviations and the waviness can be modeled. By an additional consideration of the cutting edge geometry a mapping of the roughness of the third order is possible (grooves). The tool geometry is measured in the production scheduling. The individual cutting edge diameters at the TCP as well as the tool runout in the holder can be measured. Clamping in the main spindle causes an additional displacement of the tool rotation axis. At the same time, wear occurs on the tool cutting edges as a result of machining. Schmidt investigated both the calculation of tool runout and cutting edge wear with the help of a force measuring platform [5]. At the same time, there are approaches that use machine data to monitor tool wear [32].

Conclusion of the State of the Art. The current state of the art already allows the modeling of many effects. The numerical effects can already be mapped well with the help of VNCK [12]. When considering the errors according to their source, it becomes apparent that geometric kinematic errors on the machine side can be partially recorded by means of specimen workpieces [30]. The modeling of thermal influences on both the machine and the process side is mostly done by means of FE simulations [5, 8, 13, 14, 17]. Variables that influence the process force, such as tool wear or runout, can be predicted indirectly [5]. However, high-frequency force data are often required here.

Controllers, however, offer only low sampling rates, so that the prediction possibilities are limited. Furthermore, a concept to predict tool and machine tool stiffness based on machine and quality data is already available [27].

3 Concept

The previous framework for quality prediction can be divided into a real and a virtual part. In real world, the product is first designed in its target geometry and the manufacturing process is planned and prepared. These data are virtually pre-processed to enable process simulation. In practice, this is followed by the production of the product and its quality control. During CNC-Process, data from the machine's internal sensors can be recorded. This data is transferred to a process simulation. Here, together with the pre-processed data, the data is processed within the model chain. In addition to the MRS, typical models describe the machine kinematic or stiffness models of the interacting components (machine, workpiece, tool). At the end of the process simulation, a digital shadow is created. By comparing the real and virtual component, deviations can be determined. These are dependent on the individual models as well as their parametrization. The parameterization takes place only once. By adding an additional parameterization layer, the individual information from the virtual models, the machine data as well as the real and virtual quality data are now connected with each other and a parameter fitting is to be made possible. The entire concept is shown in a simplified form in Fig. 1. Based on this, the procedure will be explained exemplarily in the following for two models.

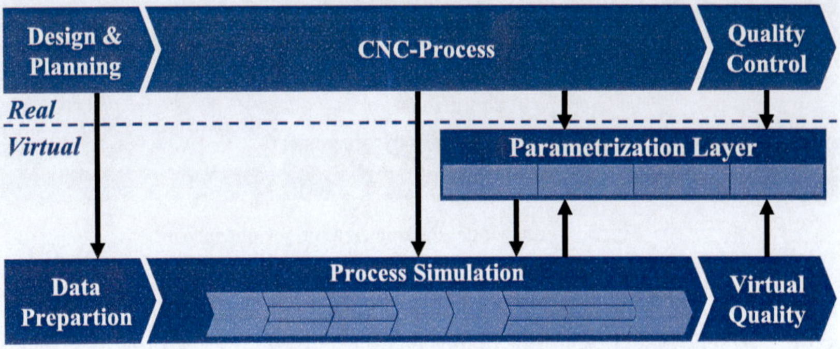

Fig. 1. Data-Based Model Parametrization – Concept

4 Data-Based Model Parametrization

For a data-based model parametrization, a deeper analysis of the models and their relationships is necessary. Due to the large number of machining errors, as mentioned in Sect. 2, the procedure will be explained using two examples. One is a process force model and the other is a stiffness model.

In a first step, it shall be assumed that only models describing the ideal state are available. Considering the force model, the resulting current at the main spindle I_{Spindle} consists of a cutting force component I_{cut}, and several components related to the spindle position s, speed n, acceleration a and jerk j. By means of an air cut, these process force-independent components can be recorded so that the linear relationship between cutting force $\boldsymbol{F}_{\text{cut}}$ and the effective cutting edge radius $\boldsymbol{r}$ of each tool edge and current remains.

$$I_{\text{Spindle}} = \underbrace{K_{I,M}\boldsymbol{F}_{\text{cut}} \cdot \boldsymbol{r}}_{I_{\text{cut}}} + \underbrace{I_{\text{j}}(\text{j}) + I_{\text{a}}(\text{a}) + I_{\text{n}}(\text{n}) + I_{\text{s}}(\text{s})}_{\text{Parameterization based on air cuts}} \quad (1)$$

Kienzle [33] calculate the cutting force for each cutting edge in dependency to the uncut chip thickness h and width b:

$$F_{\text{cut,Kienzle},i} = k_{c,i} \cdot b \cdot h^{1-m_{c,i}} \quad (2)$$

The cutting conditions can be calculated with a MRS. Assuming that the force coefficients $k_{c,i}$ and m_c are known and constant for a sufficiently long time, the constant $K_{I,M}$ can be determined by changing the chip thickness or width:

$$K_{I,M} = \frac{\Delta I_{\text{cut}}}{\Delta \boldsymbol{F}_{\text{cut}} \cdot \boldsymbol{r}} = \frac{\Delta I_{\text{cut}}}{\Delta M_{\text{cut}}} \quad (3)$$

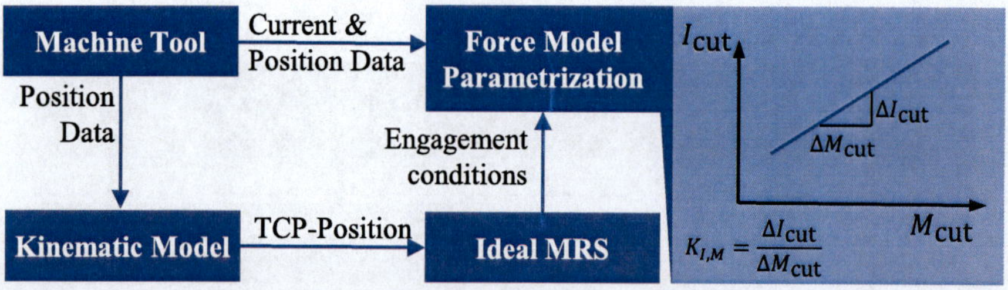

Fig. 2. Force Model Parametrization – Concept

For the determination of $K_{I,M}$, it has to be ensured that factors which have an influence on the engagement, e.g. tool runout or tool deflection, are so low that they are negligible. Figure 3 shows the force induced spindle current I_{cut} during slot milling of C45 with multiple Ø16 mm milling tools (Fraisa Favora P45317610) in different holder types. Each color shows one slot of in different parts. The selected cutting parameters consist of a feed velocity of 525 mm/min, a spindle speed of 2585 rpm, and a cutting depth of 5 mm. For evaluation purposes, a distance of 10 mm was used. It becomes evident that while as I_{cut} increases over time due to wear, the peaks and valleys remain at a constant spindle position. Based on the extrema, therefore a prediction of the cutting edge location is possible. Afterwards, the chip thickness can be calculated.

Using Eq. 2, $\boldsymbol{F}_{\text{cut}}$ is predicted. Here, the values $k_{c,i}$ and m_c are taken from [34]. Based on $\boldsymbol{F}_{\text{cut}}$ M_{cut} is calculated and $K_{I,M}$ is fitted. The result for one slot is shown in

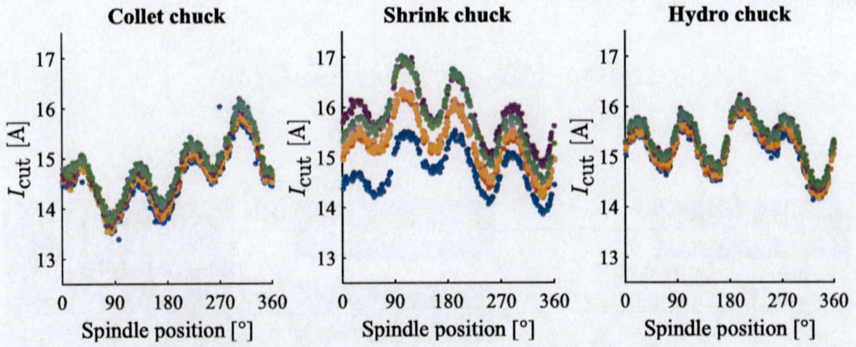

Fig. 3. Force induced spindle current I_{cut} during slot milling

Fig. 4. Besides, all results are in Table 1 summarized. The reciprocals of $K_{I,M}$ can now be used to calculate $\boldsymbol{F}_{cut}$ during machining. Besides, further factors, which are related to the force signals, can be calculated. Tool runout can be determined by observing force oscillations, and tool wear by the increase of cutting force coefficients [5]. For the determination of further geometric-kinematic errors, ISO230 can be used [30].

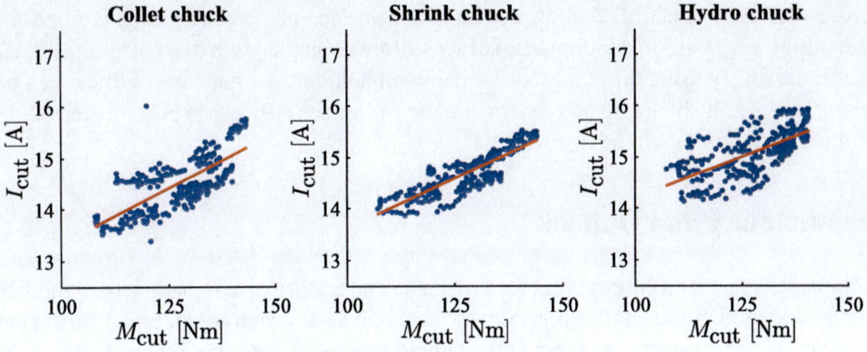

Fig. 4. Force Model Parametrization – Results

Table 1. Force Model Parametrization – Results

	• Part 1	• Part 2	• Part 3	• Part 4	• Part 5	Mean
Collet Chuck	0.0427 A/Nm	0.0430 A/Nm	0.0431 A/Nm	0.0437 A/Nm	0.0430 A/Nm	0.0431 A/Nm
Shrink Chuck	0.0387 A/Nm	0.0476 A/Nm	0.0463 A/Nm	0.0502 A/Nm	0.0549 A/Nm	0.0476 A/Nm
Hydro Chuck	0.0320 A/Nm	0.0322 A/Nm	0.0326 A/Nm	0.0324 A/Nm	0.0323 A/Nm	0.0323 A/Nm

To calculate TCP stiffness k_{TCP}, a virtual workpiece as described in [27], is generated where all stiffnesses are assumed as infinite. In total, three single stiffnesses are connected in series: 1) cutting part k_{cutting} (cantilever beam), 2) holder part k_{holder} (static) and 3) machine part k_{machine} (static):

$$k_{\text{TCP}} = \left(k_{\text{cutting}}^{-1} + k_{\text{holder}}^{-1} + k_{\text{machine}}^{-1}\right)^{-1} \tag{4}$$

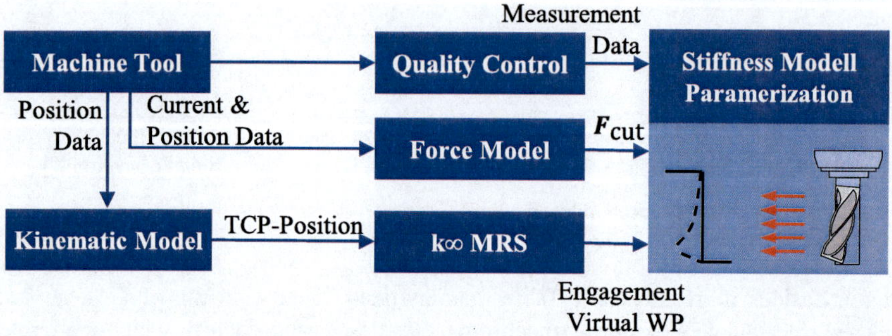

Fig. 5. Stiffness Model Parametrization – Concept

The cutting part is modeled as cantilever beam with a uniform load over the cutting depth. By superposing the real and virtual workpiece, the deviation is calculated. Subsequently, the product of Young's modulus and surface moment of inertia is fitted for the cutting edge part. In addition, the holder stiffness together with the machine stiffness is also fitted. By using different tool-holder combinations, the machine stiffness can be extracted later. It should be noted here that the workpiece stiffness is to be neglected. In further work this should be investigated.

5 Summary and Outlook

This paper presents a concept that uses machine and quality data to parametrize models for quality prediction. For this purpose, the machine data, which are recorded during the production, are evaluated. Subsequently, individual models are parameterized separately by means of a parameter fitting. The procedure of the concept is described concretely for two models. One is a process force model and the other is a stiffness model. By means of a material removal simulation, the data are brought together in a common context. Here, the engagement variables such as cutting depth and cutting width are determined. By means of a relative analysis, force models can then be fitted. At the same time, the fitted force models can be used to determine other variables, such as tool runout. Furthermore, stiffnesses at the TCP can be determined by superposing digital and real workpiece. In the future, the presented concept can be used to parametrize quality-predictive material removal simulations cost-efficiently during regular production. Therefore this concept can be an enabler for the model parametrization of self-optimizing machining systems as described in [35].

Acknowledgments. Funded by the Deutsche Forschungsgemeinschaft (DFG, German Research Foundation) under Germany´s Excellence Strategy – EXC-2023 Internet of Production – 390621612.

References

1. Brecher, C., Epple, A., Fey, M., Königs, M., Neus, S., Wellmann F.: Lernende Produktionssysteme. In: Brecher C, Klocke F, Schmitt R et al. (eds) Internet of Production für agile Unternehmen: AWK Aachener Werkzeugmaschinen-Kolloquium 2017, 18. bis 19. Mai, pp 135–195. Apprimus Verlag, Aachen (2017)
2. Brecher, C., Wellmann, F., Epple, A.: Quality-predictive CAM simulation for NC milling. Procedia Manuf. **11**, 1519–1527 (2017). https://doi.org/10.1016/j.promfg.2017.07.284
3. Königs, M., Brecher, C.: Process-parallel virtual quality evaluation for metal cutting in series production. Procedia Manuf. **26**, 1087–1093 (2018)https://doi.org/10.1016/j.promfg.2018.07.145
4. DIN Deutsches Institut für Normung e. V.: DIN 4760:1982–06, Gestaltabweichungen; Begriffe, Ordnungssystem 01.040.17, 17.040.20(4760) (1982)
5. Schmidt, C.: Einflussgrößensensitive Simulation und Überwachung von Fräsprozessen. Berichte aus dem IFW. PZH Produktionstechn. Zentrum, Garbsen (2011)
6. Hocken, R.J.: Technology of machine tools. Volume 5. Machine tool accuracy, United States (1980)
7. Schmitz, T.L., Ziegert, J.C., Canning, J.S., Zapata, R.: Case study: a comparison of error sources in high-speed milling. Precis. Eng. **32**, 126–133 (2008). https://doi.org/10.1016/j.precisioneng.2007.06.001
8. Mayr, J., Jedrzejewski, J., Uhlmann, E., et al.: Thermal issues in machine tools. CIRP Ann. **61**, 771–791 (2012). https://doi.org/10.1016/j.cirp.2012.05.008
9. Wellmann, F.: Datengetriebene, kontextadaptive Produktivitätssteigerung von NC-Zerspanprozessen, 1st edn. Apprimus Wissenschaftsverlag, Aachen (2019)
10. Siemens, A.G: Programmierhandbuch Grundlagen (2018)
11. Brecher, C., Weck, M. (eds.): Werkzeugmaschinen Fertigungssysteme 3. Springer, Heidelberg (2021)
12. Erkorkmaz, K., Altintas, Y., Yeung, C.-H.: Virtual computer numerical control system. CIRP Ann. **55**, 399–402 (2006). https://doi.org/10.1016/S0007-8506(07)60444-2
13. Mayr, J.: Beurteilung und Kompensation des Temperaturganges von Werkzeugmaschinen, ETH Zurich (2009)
14. Vettermann, J., Steinert, A., Brecher, C., Benner, P., Saak, J.: Compact thermo-mechanical models for the fast simulation of machine tools with nonlinear component behavior. at - Automatisierungstechnik **70**, 692–704 (2022). https://doi.org/10.1515/auto-2022-0029
15. Brecher, C., Fey, M., Loba, M., et al.: Sustainability in Production Lines. In: Schmitt RH, Bergs T, Brecher C et al. (eds.) Empower Green Production. Conference proceedings, pp 202–232. Fraunhofer-Gesellschaft (2023)
16. Brecher, C., Spierling, R., Fey, M., Neus, S.: Direct measurement of thermo-elastic errors of a machine tool. CIRP Ann. **70**, 333–336 (2021). https://doi.org/10.1016/j.cirp.2021.04.084
17. Denkena, B., Schmidt, A., Henjes, J., Niederwestberg, D., Niebuhr, C.: Modeling a thermo-mechanical NC-simulation. Procedia CIRP **8**, 69–74 (2013). https://doi.org/10.1016/j.procir.2013.06.067
18. Aslan, D., Altintas, Y.: Prediction of cutting forces in five-axis milling using feed drive current measurements. IEEE/ASME Trans. Mechatron. **23**, 833–844 (2018). https://doi.org/10.1109/TMECH.2018.2804859

19. Fey, M., Epple, A., Kehne, S., et al.: Verfahren zur Bestimmung der Achslast auf Linear- und Rundachsen G01L 1/04 (2016)
20. Königs, M., et al.: A scalable, hybrid learning approach to process-parallel estimation of cutting forces in milling applications. Robert Schmitt Günther Schuh (Publ.) **7**, 425–432 (2017)
21. Denkena, B., Bergmann, B., Stoppel, D.: Reconstruction of process forces in a five-axis milling center with a LSTM neural network in comparison to a model-based approach. JMMP **4**, 62 (2020). https://doi.org/10.3390/jmmp4030062
22. Brecher, C., Eckel, H.-M., Motschke, T., Fey, M., Epple, A.: Estimation of the virtual workpiece quality by the use of a spindle-integrated process force measurement. CIRP Ann. **68**, 381–384 (2019). https://doi.org/10.1016/j.cirp.2019.04.020
23. Postel, M., Aslan, D., Wegener, K., Altintas, Y.: Monitoring of vibrations and cutting forces with spindle mounted vibration sensors. CIRP Ann. **68**, 413–416 (2019). https://doi.org/10.1016/j.cirp.2019.03.019
24. Denkena, B., Litwinski, K.M., Boujnah, H.: Detection of tool deflection in milling by a sensory axis slide for machine tools. Mechatronics **34**, 95–99 (2016). https://doi.org/10.1016/j.mechatronics.2015.09.008
25. Albertelli, P., Goletti, M., Monno, M.: An improved receptance coupling substructure analysis to predict chatter free high speed cutting conditions. Procedia CIRP **12**, 19–24 (2013). https://doi.org/10.1016/j.procir.2013.09.005
26. Agarwal, A., Desai, K.A.: Tool and workpiece deflection induced flatness errors in milling of thin-walled components. Procedia CIRP **93**, 1411–1416 (2020). https://doi.org/10.1016/j.procir.2020.04.101
27. Loba, M., Brecher, C., Fey, M., Roenneke, F., Yeh, D.-F.: Determination of tool and machine stiffness based on machine internal and quality data. In: Liewald, M., Verl, A., Bauernhansl, T., et al. (eds.) Production at the Leading Edge of Technology WGP 2022. LNCS, pp 335–345. Springer, Cham (2023). https://doi.org/10.1007/978-3-031-18318-8_35
28. Knape, S., Königs, M., Epple, A., Brecher, C.: Increasing accuracy of material removal simulations by modeling workpiece deformation due to clamping forces. In: Schmitt, R., Schuh, G. (eds.) WGP 2018, pp. 72–80. Springer, Cham (2019). https://doi.org/10.1007/978-3-030-03451-1_8
29. Brecher, C., Ochel, J., Fey, M.: Datengetriebene Werkzeugeingriffsdetektion für Fräsprozesse. Zeitschrift für wirtschaftlichen Fabrikbetrieb **117**, 784–789 (2022). https://doi.org/10.1515/zwf-2022-1146
30. ISO International Organization for Standardization: Test code for machine tools - Part 1: Geometric accuracy of machines operating under no-load or quasi-static conditions 25.080.01(230–1:2012(E)) (2012)
31. DIN Deutsches Institut für Normung e. V.: DIN 4003–1:2017–10, Konzept für den Aufbau von 3D-Modellen auf Grundlage von Merkmalen nach DIN_4000_- Teil_1: Übersicht und Grundlagen 21.020(4003–1:2017–10) (2017)
32. Xi, T., Benincá, I.M., Kehne, S., Fey, M., Brecher, C.: Tool wear monitoring in roughing and finishing processes based on machine internal data. Int. J. Adv. Manuf. Technol. **113**, 3543–3554 (2021). https://doi.org/10.1007/s00170-021-06748-6
33. Kienzle, O.: Die Bestimmung von Kräften und Leistungen an spanenden Werkzeugen und Werkzeugmaschinen. VDI-Z, **94**, 299–305 (1952)
34. Apprich, T., Brenner, J., Dambacher, M., Dreher, F., Fischer, G., et al.: Tabellenbuch für Zerspantechnik. 1st edn. Verl. Europa-Lehrmittel Nourney Vollmer, Haan-Gruiten (2015)
35. Möhring, H.-C., Wiederkehr, P., Erkorkmaz, K., Kakinuma, Y.: Self-optimizing machining systems. CIRP Ann. **69**, 740–763 (2020). https://doi.org/10.1016/j.cirp.2020.05.007

Automated Identification of Components of Feed Axes

Alexander Puchta(✉), Marvin Frisch, and Jürgen Fleischer

Wbk Institute of Production Science, Kaiserstrasse 12, 76131 Karlsruhe, Germany
alexander.puchta@kit.edu

Abstract. Monitoring solutions like simulation models and digital twins enable methods like predictive maintenance, cutting down on idle time and improving the effectiveness and lifetime of machine tools. Creating these models for brownfield equipment is therefore sensible from both an ecological and economical view. Designing them, however, is a time-consuming process that requires a lot of expertise. Automating this has the potential to increase both the number and the quality of models. This paper presents an approach to automatically identify the parts of the feed axes of machine tools based on reference runs. For this, the control signals of exemplary machines are analyzed in order to develop a rule-based system to differentiate variations of parts. Furthermore, approximations of certain system parameters like gear ratios are determined. This constitutes a first step towards fully automated generation of functional models.

Keywords: Manufacturing system · Identification · Machine Tool

1 Introduction

Maintenance is an important part of modern manufacturing. In order to increase the Overall Equipment Effectiveness, short and predictable maintenance times are essential. To achieve this, a simulation model or digital twin of the machine in question is required [1–3]. Creating these models, especially for brownfield equipment, is a costly and time-consuming process that requires a lot of expertise and knowledge about machine tools. Automating this process has the potential to significantly increase the use and quality of such models, especially for small and medium enterprises. This is particularly important for machine tools, where long lifetimes and high replacement costs make retrofitting a viable option from both an environmental and economic perspective [4].

An approach that does not require additional sensory equipment further lowers the barrier of entry into the methods of industry 4.0. Therefore, only existing signal data such as motor current, torque, power and position signals should be used. The first step in automatically generating a model of a feed axis is to identify its components. This requires a general approach that builds on the provided signal data, generalizes to different types of components and expands the knowledge base. Based on this, a first-generation model can be created by combining component models into a modular representation of the feed axis. Its parameters can then be fine-tuned. To make this

T. Bauernhansl et al. (Eds.): WGP 2023, LNPE, pp. 143–151, 2024.
https://doi.org/10.1007/978-3-031-47394-4_15

process faster and more reliable in its output, important system parameters like gear ratios need to be roughly estimated to work as a starting point.

A well generalized identification of components serves both as a first step of model creation and as an information gain about the behavior of the components and their influence on the signals. As a result, models can be created that better represent real-world conditions. For a thorough automation of the modeling and scalability of the approach, an additional automated parameter identification [5] can be used to provide the required signals.

2 State of the Art

Huang et al. demonstrate an approach to generate component-based simulation models by the example of a railway system. Different types of crossings are sourced from a database and combined based on a graph of the railway network. Further data sources like timetables and network graphs are then used to expand the model. It can then be validated and adapted to real life data. Even though the subject data is completely different, the general approach of identifying components using data and combining component models from a database is universally applicable and is used in this work [6].

Martinez et al. develop an approach to automate the generation of digital twins of process plants. A static model or an alternative representation of the process is used to obtain a first model. Its parameters are improved using a 3D CAD representation of the plant. It is then introduced into the real system loop to synchronize operations and reflect changes in the plant. It shows good results, but still requires knowledge about the inner workings of the plant to create the 3D-model [7].

Lugaresi and Matta use a data-driven approach to create a digital twin for an assembly line. Automatically generated logs are used to determine the steps of the production and the process times. The human handler only needs to define the grade of the model, the degree of automation is very high. However, the complexity of the model is low, it is limited to the system level and does not include details within the assembly [8].

Sommer et al. use a 3D scan to create a CAD model of a production facility. Machines are automatically identified within the resulting pointcloud, loaded from a library, and placed in a model of plant. It can then be expanded into a full simulation model by specifying parameters like process times and batch sizes. These are not automatically derived but must be implemented manually. Like the last approach, it results in a system model that cannot be used to analyze the machines in detail [9].

In conclusion, no method to automate the generation of simulation models with enough detail within in the machine to use it for predictive maintenance currently exists. An approach focused on machine tools that can accurately represent their behavior and inner workings, using as few additional sensors as possible, is needed.

After the component identification, [10] describes the continuation of the approach by a parameterization of the model. For this, a genetic algorithm is used. The signals of the machine during a simple reference run are compared to the output of an existing simulation model and parameters such as stiffness, masses and damping constants are estimated. The estimation works well for the latter, but needs to be improved for the stiffness and mass. Furthermore, a modular simulation model is developed and shown to be able to replicate the behavior of the axis, validating the general approach [10].

3 Component Identification

To identify the components, the signal data during a reference run needs to be analyzed. Sampling these is done several hundred times a second, resulting in a large number of data points. It is not feasible to use this as input to a machine learning method, as the amount of training data needed to achieve good performance would be too large. Therefore, the signals are manually analyzed in order to find general rules by which to differentiate different components. This has the added benefit of increasing the knowledge about the signals.

In general, the driving part of a feed axis consists of four parts: A motor, a gearbox, a rotational-translatory transmission (RTT), and a guide. For each of these components, several different variations are possible. Due to the high number of possible combinations, identifying the components separately from each other simplifies the decisions and reduces amount of data required. Cascade classification is used within the resulting individual classifiers to further reduce complexity. The position signal during the reference run used to generate the signal data for the identification is displayed in Fig. 1. The run starts with two drives of 3 mm with a speed of 5 mm/min in positive and negative direction respectively. From there, the distance is increased to 200 mm. Positive and negative directions alternate, with the speed increasing to 150, 300 and 2000 mm/min before increasing steadily by 2000 mm/min per section until the maximum speed of 40000 mm/min is reached. The axis pauses for 1 s between sections. The sampling frequency for all signals is 500 Hz.

3.1 Motor Classification

For the motor, four different types are considered: A DC-motor, stepper motor, synchronous AC-motor and asynchronous AC-motor.

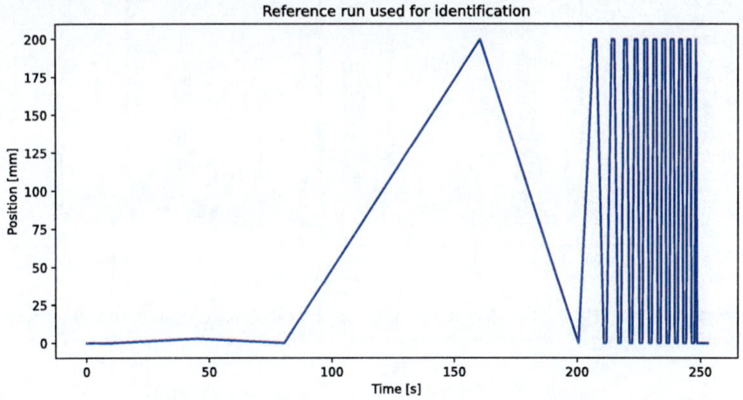

Fig. 1. Position of the axis during the reference run.

Stepper Motor Identification. This is the first level of the cascade. It uses the current signal. Due to the operating principle of a stepper motor, creating a rotating electromagnetic field by cyclically energizing several separate coils, the direction of the current on the common wire does not depend on the rotational direction. A DC-Motor inverts the current to rotate in the opposite direction, and the sign of a three-phase current indicates the direction of rotation. If the current signal has a constant sign, the motor can be identified as a stepper motor.

DC-Motor Identification. To separate DC-motors from three-phase current motors, the Fourier transform of the current signal for a section of the reference run is calculated (shown in Fig. 2). Three-phase current motors show strong frequencies $f_{Current}$ based on their number of coils n_{Coils} and the rotational frequency $f_{Rotation}$:

$$f_{Current} = n_{Coils} * 3 * f_{Rotation} \tag{1}$$

DC-Motors do not show any dominant frequencies and can be identified by this fact. The exact frequencies for a known rotational speed can be used to determine the number of coils, should they be needed for the model.

Three-Phase Current Motor Differentiation. As the last level of the motor cascade, synchronous and asynchronous motors are to be differentiated. For this, the torque/current ratio is calculated for every sampling point. For synchronous motors, this ratio is constant except for measurement noise, for asynchronous motors, the ratio shows different sections especially for low currents (see Fig. 3). This difference can be detected by k-means-clustering the ratios for k = 1 and k = 2 and comparing the sum of squared distances: For asynchronous motors, the difference in summed distance is larger than for synchronous motors. In the case of the latter, the determined torque/current ratio can be implemented into the simulation model.

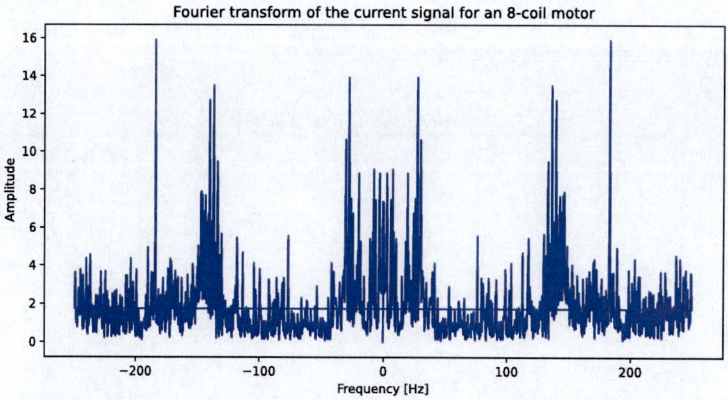

Fig. 2. Fourier transform of the current signal for an 8-coil synchronous motor with a rotational frequency of 7.59 Hertz.

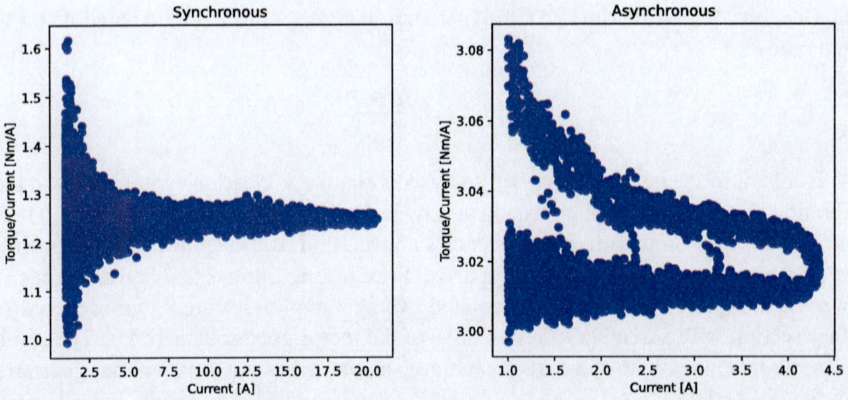

Fig. 3. Torque/Current ratio over current for a synchronous and an asynchronous motor

3.2 Rotational-Translatory Transmission Classification

The two main types of RTT are ball screws and gear racks. They differ in their feed ratios, where ball screws are usually under 40 mm/revolution and gear racks over 100 mm/revolution of the driving gear. Therefore, the feed ratio is determined both for differentiation and as a starting point for the parameter estimation.

The first step is to calculate the feed ratio of the entire axis. This is done by averaging the power, speed and torque signals for one section of the test drive. A section with high speeds is used for this because the power signal becomes more reliable the higher its value is. Using the average speed $\overline{v}$, the average torque $\overline{M}$, and the average power $\overline{P}$, the feed ratio j, that is the feed per revolution of the motor, is calculated:

$$j_{total} = \overline{v} * \overline{M} * 2 * \pi * \frac{1}{\overline{P}} \tag{2}$$

Using this, the average number of revolutions of the motor for a section can be determined from the average speed.

To determine the RTT feed ratio, the difference between the target and actual position is calculated. A Fourier transform is then performed on a single section of this difference. The number of revolutions of the screw drive or driving gear respectively show as a strong frequency in the position difference, as does the number of revolutions of the motor. It is generally expected that the RTT rotates more slowly than the motor. Therefore, only frequencies smaller than the number of revolutions of the motor are considered. From those, the one with the highest amplitude is chosen as the estimate of the number of revolutions of the RTT. The feed ratio can then be calculated by using it as the divisor of the average speed. Using the feed ratio, the RTT is classified.

3.3 Gearbox Classification

The types of gearboxes considered are planetary gears, timing belts and spur gears. Additionally, it is possible that the axis does not include a gearbox at all.

No Gearbox Classification. As the first step, the gear ratio i is calculated using the feed ratios:

$$i_{Gear} = \frac{j_{Linear}}{j_{total}} \quad (3)$$

If this ratio is equal to 1, further analysis is required. Gearboxes in general reduce vibration, especially at higher frequencies, by increasing the stiffness of the axis. Therefore, the average amplitude for frequencies above 100 Hz is calculated for the position difference for every section of the test drive. Prior testing implies that higher speeds are more meaningful, therefore the calculated average amplitudes are again averaged for every section with a number of revolutions of the motor greater than 10 Hz. The results for three machines with three axes each are shown in Fig. 4, axes without gearboxes are represented with crosses, axes with gearboxes as circles. The results indicate that a differentiation is possible, especially when combined with the additional factor of the gear ratio.

Planetary Gear Classification. The advantage of planetary gears is their allowance for higher gear ratios; therefore, the gearbox is classified as that if its gear ratio is higher than 12.

Timing Belt Classification. The last step is to distinguish between timing belts and spur gears. In the case that $i_{Gear} = 1$, a timing belt is assumed due to their better suitability as offset gears. If $i_{Gear} > 1$, the reversal of the rotation that spur gears yield can be used. Given the motor is a three-phase current one, the sign of the current signal shows the rotational direction of the motor. If the motor rotates positively and the axis moves negatively, the rotational direction is reversed at some point. If the position of the motor relative to the axis and the positive direction of the axis is known, it can be determined whether the reversal stems from the gearbox. However, this requires manual input and additional knowledge not gained from the control signals.

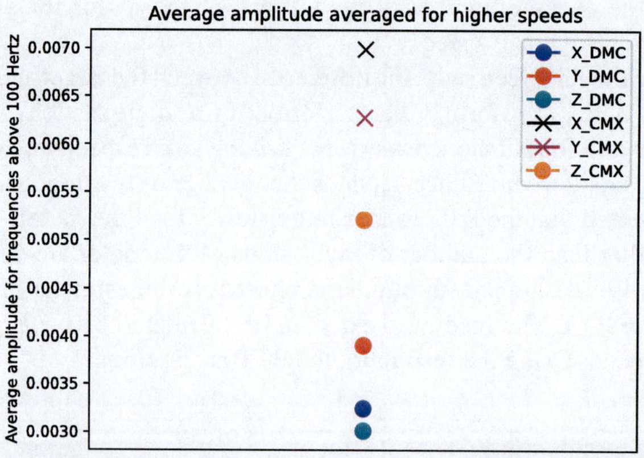

Fig. 4. Average amplitude for frequencies over 100 Hertz for the position difference averaged over multiple sections, axes without gearboxes as crosses, axes with gearboxes as circles.

3.4 Guide Classification

The last component to be identified is the axis guide. The possibilities are a hydrodynamic, hydrostatic or roller guide. They can be differentiated by the friction curve, which is the friction force versus the relative speed: Hydrodynamic guides show a stick-slip phenomenon where the friction force decreases and then increases again, roller guides show a logarithmic-like curve and hydrostatic guides show an almost linear curve. Since the friction force cannot be measured directly without additional sensors, it is estimated from the motor torque. Only the last two sections of the reference run are used for this, because continuous travel from high positive speed to high negative speed best reflects the friction conditions around $v = 0$. The speed range is divided into ten equally great parts and the maximum torque is calculated for all sample points within this range. The resulting ten torque samples are normalized and classified using a decision tree classifier. The decision tree is trained using synthetic friction curves with additional noise and on those has a training and test accuracy of 97 and 100%, respectively.

4 Results and Discussion

To test the identification, it is run on the three machine tools already shown in Fig. 4. They are a DMG MORI DMC 60H (DMC) and a DMG MORI CMX 600V (CMX).

The motors of the axes are correctly identified as either synchronous or asynchronous for all available axes. None of the axes have a DC- or stepper motor.

The gearboxes are at least partially correctly identified for all axes of all machines. Because the differentiation of timing belts and spur gears in some cases requires manual input it only gets partial credit for those.

The gear ratios are estimated within 1% error for most of the axes. The rotational B axis of the DMC has a planetary gearbox with $i_{Gear} = 36$ which is estimated as $i_{Gear,estimate} = 40.42$.

The RTTs are correctly identified as ball screws for all translatory axes. None of the axes have a gear rack, however, the rotational B axis of the DMC has a rotary equivalent which would be classified as a gear rack.

The feed ratios are correctly determined for all translatory axes except the z axis of the CMX, where the estimated ratio is three times the actual ratio. The reason for this significant discrepancy in this particular axis requires further analysis. For the other axes, original estimates are inaccurate by up to 5%. However, since screw drives are standard parts, an adjustment towards a common ratio is possible. For the rotational B axis of the DMC, $j_{Rotation} = 71.01°$ is estimated as $j_{Rotation,estimate} = 81.63°$. Possible reasons for this include the rotary nature of the axis not being fit for the classifier or difficulties arising from larger feed ratios.

The guides are correctly identified as roller guides for all axes of the DMC. For the CMX, two of the three translatory axes are wrongly classified as hydrodynamic instead of roller guides. This implies that either training on synthetic data is not enough and more training data is needed or that using torque as a replacement for friction force is not suitable.

The results of the identification for the axes of the DMC excluding the spindle are shown in Table 1 the results for the axes of the CMX are shown in Table 2. For numerical values, the deviation from the ground truth is displayed in angular brackets.

Table 1. Identification Results for the DMC

Axis	X	Y	Z	B
Motor	Synchronous	Synchronous	Synchronous	Synchronous
Gearbox	T.B./S.G	T.B./S.G	T.B./S.G	Planetary
RTT	Ball Screw	Ball Screw	Ball Screw	-
Guide	Roller	Roller	Roller	Roller
i_{Gear}	2.280 [0.5%]	2.288 [0.2%]	2.276 [0.6%]	40.24 [12.3%]
$j_{Linear/Rotation}$	20 [0%]	20 [0%]	20 [0%]	81.63 [14.9%]

Table 2. Identification Results for the CMX

Axis	X	Y	Z
Motor	Synchronous	Synchronous	Synchronous
Gearbox	None	None	T.B./S.G
RTT	Ball Screw	Ball Screw	Ball Screw
Guide	Hydrodynamic	Roller	Hydrodynamic
i_{Gear}	1 [0%]	1 [0%]	3.5974 [199.8%]
$j_{Linear/Rotation}$	12 [0%]	12 [0%]	36 [200%]

5 Conclusion and Outlook

The approach of identifying feed axis components by analyzing the control signals using a rule-based classification shows mixed results. The motor and RTT classification as well as the gear and feed ratio estimation are promising. The gearbox identification works partially, but does not distinguish between timing belts and spur gears. The guide identification approach is either flawed or needs real training data for its decision tree classifier.

Future work needs to expand the test domain to ensure generalization for different machine tools. In addition, alternative classifications for the gearbox and guide need to be investigated and the simulation modules for all possible components need to be designed in order to test the combination of the component identification and parameter estimation.

Acknowledgement. We extend our sincere thanks to the German Federal Ministry for Economic Affairs and Climate Action (BMWK) for supporting this research project 12IK001ZF "Software-Defined Manufacturing for the automotive and supplying industry https://www.sdm4fzi.de/".

References

1. Aivatolis, P., Georgoulias, K., Chryssolouris, G.: The use of digital twin for predictive maintenance. Int. J. Comput. Integr. **32**, 1067–1080 (2019)
2. Armendia, M., Ghassempouri, M., Ozturk, E., Peysson, F.: Twin-Control: A Digital Twin Approach to Improve Machine Tools Lifecycle. Springer, Cham (2019)
3. Him, L.C., Poh, Y.Y., Pheng, L.W.: Improvement of overall equipment effectiveness from predictive maintenance. In: International Conference on Digital Transformation and Applications (ICDXA) (2020)
4. Ilari, S., Carlo, F.D., Ciarapica, F.E., Bevilacqua, M.: Machine tool transition from Industry 3.0 to 4.0: a comparison between old machine retrofitting and the purchase of new machines from a triple bottom line perspective. Sustainability **13**(18), 10441 (2021)
5. Gönnheimer, P., Ströbel, R., Fleischer, J.: Analytical approach for parameter identification in machine tools based on identifiable CNC reference runs. In: Liewald, M., Verl, A., Bauernhansl, T., Möhring, HC. (eds.) Production at the Leading Edge of Technology. WGP 2022. LNPE, pp. 494–503 (2022). https://doi.org/10.1007/978-3-031-18318-8_50
6. Huang, Y., Seck, M., Verbraeck, A.: From data to simulation models: component-based model generation with a data-driven approach. In: Proceedings of the 2011 Winter Simulation Conference (WSC) (2011)
7. Martinez, G., Sierla, S., Karhela, T., Vyatkin, V.: Automatic generation of a simulation-based digital twin of an industrial process plant. In: IECON 2018 - 44th Annual Conference of the IEEE Industrial Electronics Society (2018)
8. Lugaresi, G., Matta, A.: Automated manufacturing system discovery and digital twin generation. J. Manuf. Syst. **59**, 51–66 (2021)
9. Sommer, M., Stjepandic, J., Stobrawa, S., von Soden, M.: Automatic generation of digital twin based on scanning and object recognition. In: Transdisciplinary Engineering for Complex Socio-technical Systems, pp. 645–654. IOS Press (2019)
10. Puchta, A., Riegel, V., Barton, D., Fleischer, J.: Auto-identification of dynamic axis models in machine tools. In: 16th CIRP Conference on Intelligent Computation in Manufacturing Engineering (2022)

Development of a Tool System for Internal Contouring of Small Diameters

Timo Rinschede[1](✉), Moritz Fuß[2], Rainer Brockmann[3], and Dirk Biermann[1]

[1] Institute of Machining Technology, TU Dortmund, 44227 Dortmund, Germany
timo.rinschede@tu-dortmund.de
[2] BGTB GmbH, 44149 Dortmund, Germany
[3] Lühmann & Brockmann CNC-Technik GmbH, 29303 Bergen, Germany

Abstract. Both conventional processes such as internal turning and special processes such as bottle boring do not allow complex internal machining of small bores at large depths. For this reason, a new type of tool system is being developed for internal contouring of small diameters even at large machining depths. In a first development step, preliminary investigations of internal machining by means of internal turning were carried out to characterize the existing limits and to select suitable cutting inserts and cutting parameters. This was followed by the development, design and manufacture of the tool system, which was initially designed for guide bores with a diameter of $D = 14$ mm and a length-to-diameter ratio > 20. The development of the tool head was implemented using finite element analyses.

Keywords: Inner contour · Internal machining · Deep holes

1 Introduction

Increasing demands are being placed on components in terms of both the material used and the design. With regard to the design, more and more complex shapes are required in order to meet increasing functionality, higher load-bearing capacity and a consistent striving for lightweight construction. In this context, the available processing methods and tools focus primarily on external machining, while the possibilities for internal machining are severely limited. Other processes are also limited to internal operations with limited depths (internal turning) or are only available for larger internal diameters (bottle boring). However, in areas such as medical technology, aerospace and mobility in particular, there is a need to produce internal contours at large component depths even with small internal diameters. For example, common rails could be internally contoured to increase their load-bearing capacity, the weight of drive shafts could be reduced in a targeted manner, and functional surfaces could still be introduced into waveguides at large depths.

Due to the local machining in a component, internal turning is subject to some special features and requirements. If the area to be machined is deep in the workpiece, the tool and boring bar must have the appropriate length to reach this area. The smallest diameter

T. Bauernhansl et al. (Eds.): WGP 2023, LNPE, pp. 152–162, 2024.
https://doi.org/10.1007/978-3-031-47394-4_16

of the part defines the maximum geometry that the tool and boring bar may assume, so for small diameters and large depths, the l/D-ratio of the boring bar is significantly large. This results in a susceptibility to vibrations and deflections and consequently low surface qualities and dimensional accuracies. Chatter vibrations in particular can also lead to a shortened tool life due to the breakout of cutting edges. The shank of the tool must be designed to be stable in order to prevent deflection due to lateral forces. Carbide-reinforced boring bars improve the process conditions, and the use of damping elements in the boring bar reduces vibrations and thus also optimizes the process [1–3].

A key issue in internal turning is the buildup of forces, which can be actively influenced by the cutting geometry and the tool. Compared to external turning, larger forces occur in internal turning due to a concave internal surface, which leads to a higher chip upset. The shear angle Φ increases, which increases the cutting force and with it the cutting force. In general, compared to external turning, there is a 20% increase in the cutting force, but this value can still be corrected upward as well as downward by the dependence on the setting angle and diameter [2, 4].

Chip evacuation is also a challenge in internal turning, since the chips cannot move unhindered from the cutting edge to the outside. This is often solved by supplying cooling lubricant under high pressure, whereby there are currently two possibilities for this. On the one hand, the cooling lubricant can be fed to the cutting edge from the outside; on the other hand, in some cases it makes more sense to use tools with an internal coolant supply. The chip formation also limits the drill tube diameter, since this must be so small that there is still sufficient space between the workpiece and the drill tube for the chip removal [2, 3]. Furthermore, the combination of internal turning and a joining process is possible. In this case, the component consists of several parts and is first internally turned and then joined by material, form or friction. In this context, the respective joint represents the weak point of the system due to micro- and macroscopic defects. In addition, the process chain and the number of machines to be used are extended.

Another process used is bottle boring, which is a combination of deep hole drilling with the use of NC-controlled machines with two independent feed axes. The technology is able to create contours similar to external machining in a bore. For this purpose, a cutting edge moves radially in the bore, while guide pads support the tool against the bore wall. The basic process-independent design consists of an bottle boring tool located in the workpiece, a special drill pipe and a drive unit outside or inside the workpiece. The tool carries a cutting edge which can be controlled and radially adjusted by the drive unit via a push rod located in the drill pipe and actuating a cutting edge slide. This represents the first of the two axes, the second axis with the feed in longitudinal direction is realized by moving the boring bar by means of a bearing block or headstock. Almost any internal contours can be machined within the limits of the maximum radial travel of the cutting edge. NC control can increase process accuracy by taking into account cutter wear and other effects. Many geometries can be produced for the first time and other machining operations that were previously produced by form drilling and turning are significantly simplified and optimized both economically and in terms of process technology [5].

This technology is mostly used on deep hole drilling machines, since here the necessary guide bore can also be produced and a large machine set-up is available for machining long components. Due to the complex mechanical structure, bottle boring tools are only used from a diameter of D = 40 mm [6]. In addition, there are other processes for internal machining, such as internal milling, form drilling, electrochemical ablation and forming, which have further significant limitations with regard to the productive as well as precise manufacture of internal contours.

The increasing demand for miniaturized products and the necessity to produce complex contours independent of depth and diameter show the need for a process that closes this existing gap in internal machining. For this reason, the development of an innovative tooling system that can overcome the previous challenges in machining small components at large machining depths with respect to internal contours is carried out. In this paper, the topic of internal turning is first examined in more detail in order to analyze the limitations of this process. This will also include a suitable pre-selection of cutting inserts and cutting parameters for subsequent internal contouring by means of the tool system to be developed. The concept of this tooling system will be explained in the following and the associated individual components and development steps will be discussed.

2 Experimental Setup

2.1 Workpiece Material and Tool

Two materials were used for the fundamental trials on internal turning. Firstly, the quenched and tempered steel 42CrMo4+QT (AISI4140) was used, which has high strength as well as ductility. This material is widely used in the automotive and mechanical engineering industries for highly stressed parts. In many cases, the machining of the material also includes deep hole drilling. On the other hand, the superalloy Inconel 718 is used in the aerospace, automotive and chemical industries [7]. Its applications are based on high temperature resistance, high creep resistance, and resistance to corrosion and chemical attack [8]. In addition to the above-mentioned properties, the material has a low thermal conductivity, which leads to a high thermal tool load during machining [9, 10]. Moreover, carbides and strength-increasing precipitates cause abrasive wear on the cutting edge of the tool. Consequently, tools for machining Inconel 718 are subject to pronounced wear behavior. The specimens made of both materials have an outer diameter of d = 30 mm and a through hole of D = 14 mm.

Cutting inserts for internal machining of small diameters (nominal diameter d = 11 mm) were used as tools. Four different insert types were implemented, each with two different cutting material grades (X800 and X510). This includes the insert types reverse turning, grooving, full radius and profile turning, which are shown in Fig. 1. The cutting inserts are screwed to a carbide clamping holder, which allows a maximum machining depth of l = 64 mm. The setting angle is $\kappa = 90°$ for all cutting inserts.

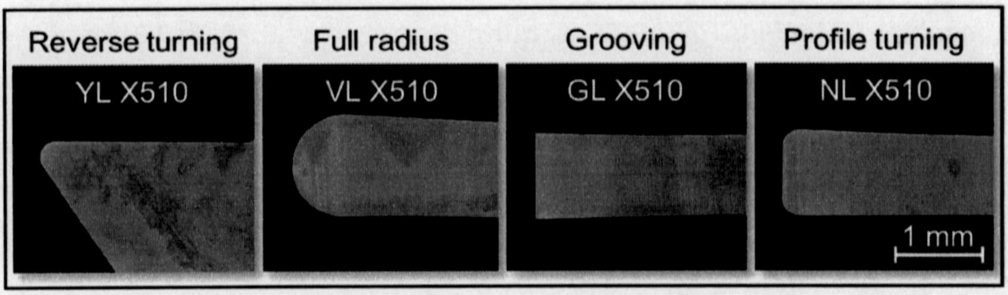

Fig. 1. Overview of the inserts used for internal turning

2.2 Machine Tool

A Monforts RNC200A CNC lathe is used for the tests. The lubricant used is Blasomill 15 oil from Blaser Swisslube, which is applied in the form of flood lubrication. The workpiece samples are clamped in a three-jaw chuck.

2.3 Methods of Analysis

Different analysis methods are used to evaluate the internal turning process. First, the mechanical tool loads are recorded during the process using a Kistler 9121 force plate and the Kistler 5019 charge amplifier connected to it. In addition, the surface quality produced is evaluated using the Confovis Duo Vario confocal microscope on cut-up workpieces. Likewise, the chip shape was documented and studied for all test series.

3 Fundamental Studies on Internal Turning

In the fundamental tests, an internal longitudinal turning process was carried out in the pulling variant using the cutting inserts and materials shown. The cutting parameter variation in the test series included the adjustment of cutting depth, feed rate and cutting speed. For the material 42CrMo4+QT the cutting parameters were $f = 0.01 \ldots 0.12$ mm ($\Delta f = 0.01$ mm), $a_p = 0.25 \ldots 1.50$ mm ($\Delta a_p = 0.\ 25$ mm) and $v_c = 20 \ldots 120$ m/min ($\Delta v_c = 20$ m/min), for Inconel 718 the following values were used due to the higher tool load: $f = 0.01 \ldots 0.06$ mm ($\Delta f = 0.01$ mm), $a_p = 0.1 \ldots 0.6$ mm ($\Delta a_p = 0.1$ mm) and $v_c = 15 \ldots 35$ m/min ($\Delta v_c = 5$ m/min). Variation of one parameter was performed while keeping the other two parameters constant. The tests were repeated in each case, and in the event of deviation between the two series of tests, a third run was implemented.

In addition, supplementary investigations were carried out on tool life over an operation time of $t_h = 30$ min. Cutting inserts with cutting material grade X800 were used for machining the material 42CrMo4+QT, while Inconel 718 was machined with cutting material grade X510. The evaluation of the tests was carried out for all test series; Fig. 2 first shows the mechanical tool loads of the basic parameters for the four cutting inserts in order to build up a general understanding. Characteristic aspects of the other parameter combinations are discussed in the following.

Despite a chip cross-section of the same nominal value, varying cutting forces result, which are due to the shape of the insert. On the one hand, differences in the mean value

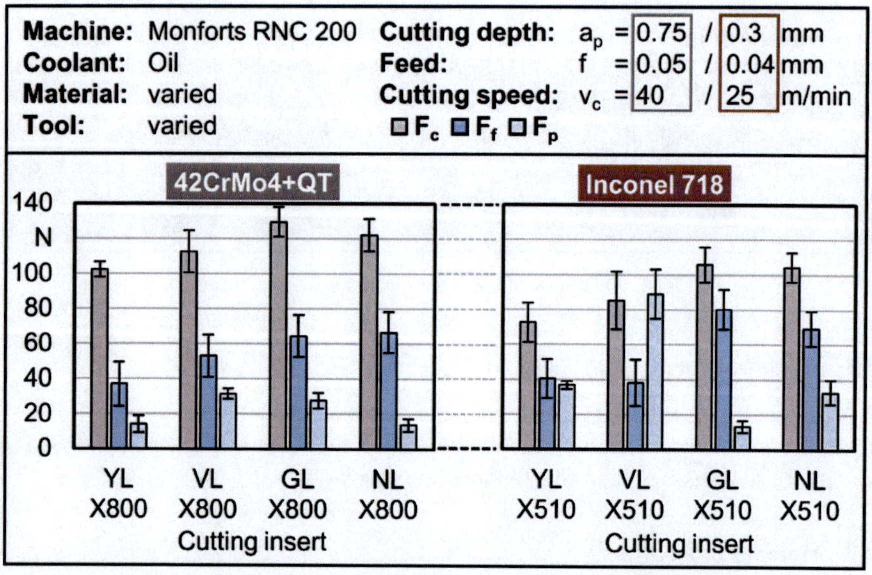

Fig. 2. Mechanical tool loads when using different insert shapes

of the loads occur, and on the other hand, differences in the recorded standard deviation of the force components are obtained, which can be attributed to the appearance of vibrations. The lowest mechanical tool loads as well as the most stable process can be observed when using the reverse turning plate (YL800, YL510). The use of the full radius insert causes more vibrations, and high passive forces are also observed. The shape of the full radius insert causes a pronounced force flow to the tool shank. This can cause the tool to be displaced, which can lead to dimensional errors and vibrations, especially in the case of small, long-projecting tools. This characteristic is particularly evident with the harder material Inconel 718, which is more challenging to machine. On the other hand, the use of this insert can be considered useful because complex contours can be produced with the same setting angle. The other two cutting inserts, GL and NL, mainly produce high feed forces. The fully comprehensive cutting parameter variation with adaptation of cutting depth, cutting speed and feed rate was implemented for all inserts and materials. The surface finish produced can be used to evaluate the turning process. Figure 3 shows the roughness values for machining the material 42CrMo4+QT using the YL and VL inserts.

The feed rate variation causes different effects for the tools used. Higher feed rates produce workpiece surfaces with increasing roughness parameters for the YL insert. On the other hand, the surface finish of the full radius insert increases with higher feed rates. The differences here can be partly attributed to the two corner radii (YL: $r_E = 0.2$ mm, VL: $r_E = 0.6$ mm), so that the full radius insert must undergo a greater load in order to stabilize the cutting process. The selection of the depth of cut also has an influence on the process. A lower roughness of the workpiece surface can be produced at higher cutting depths using the reverse turning insert. When using the full radius insert, there

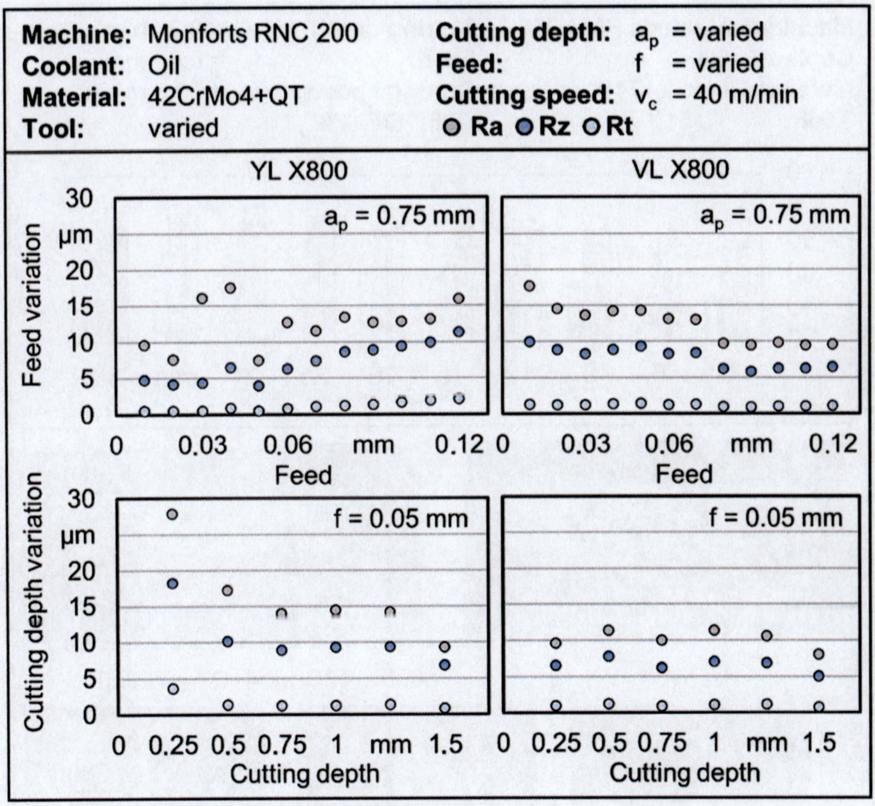

Fig. 3. Roughness characteristics under variation of feed rate and depth of cut using the YL and the VL cutting insert

is no correlation between cutting depth and roughness characteristics. It is therefore recommended to use medium feeds (f = 0.06 mm) with increased depth of cut (a_p = 0.1 mm) for the YL insert. The VL insert should be used at medium depths of cut (a_p = 0.75 mm) and higher feeds (f = 0.1 mm). The evaluation of all test series could also show that the cutting speed has no significant influence on the mechanical tool loads and the surface quality, although it can be assumed that the higher thermal load at increased cutting speeds will result in greater tool wear. Based on these tests, the tool life tests for the reverse turning and the full radius insert were performed for both materials, since these inserts are more suitable for longer bottle boring operations based on their shape and the tool loads previously analysed. Figure 4 represents the mechanical tool loads as well as the tool wear on rake and flank face at the end of the cutting time for the YL insert. In addition, the wear states for the other investigated tool life tests are shown after an operation time of t_h = 30 min.

During the tests, increasing tool loads can be observed due to the progressive wear. In the process, a vibration-prone process also sets in at the end of the operating time. The cutting insert is subject to pronounced abrasive wear, which erodes the coating and

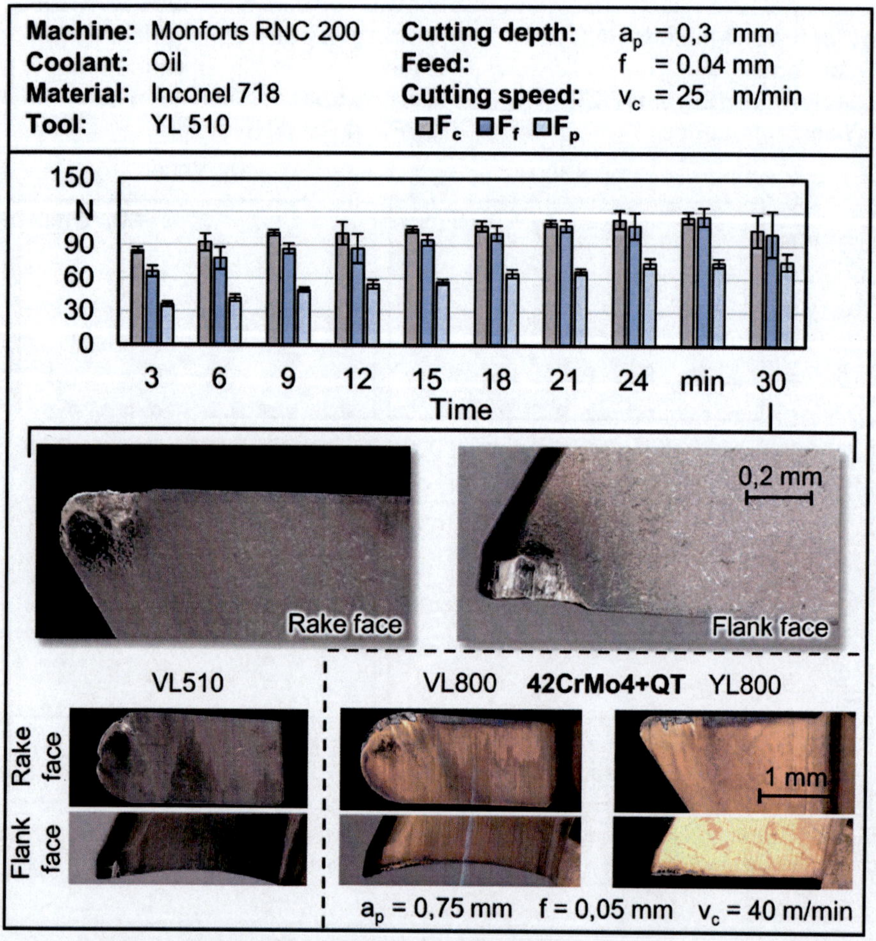

Fig. 4. Tool life tests for internal turning when machining Inconel 718

the substrate on the flank and rake face. Consequently, the chipping of the cutting edge decreases considerably and in this way causes the higher cutting force components. Similar behaviour of the mechanical tool loads can be seen with VL inserts, but with a stronger characteristic of the passive force due to the shape-related deflection. The wear behaviour is also characterised by a strong notch effect at the level of the cutting depth. Despite the higher cutting parameters, the wear phenomena are lower when machining the material 42CrMo4+QT. Abrasive wear mainly occurs on the rake face and also a notch effect on the flank face in the case of the YL turning insert. However, the investigations show that the tool can also be used for machining difficult-to-machine materials over a longer cutting time and is consequently suitable for the internal contouring of longer and larger chambers. In addition, the basic investigations were able to show which cutting parameters have to be used with which cutting insert in order to achieve a high level of process reliability and the required surface quality.

4 Tooling System for Internal Contouring

For the development of the tooling system, the aim is to combine the advantages of internal turning with those of bottle boring technology. For this purpose, a system was designed which consists of three essential subassemblies: Drive and connection unit, special boring bar and tool head. The first development was implemented for workpieces with an internal diameter of D = 14 mm; the model is shown in Fig. 5.

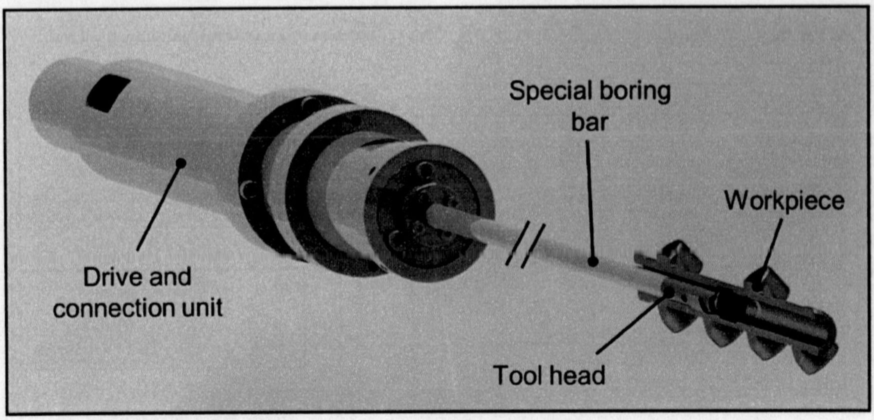

Fig. 5. Illustration of the novel tool system for internal contouring

The drive and connection unit ensures the connection of the tool system to the machine tool and consequently has to be adapted to different tool holders. In addition, the unit includes a linear actuator, which carries out the traverse movement of an inner push rod and thereby the radial extension of the tool cutting edge. The actuator is integrated into the control system of the machine and applied as an axis. It is also possible to feed coolant into the system via a transfer at the unit. The coolant is then transferred to the tool head via the special boring bar. This system connects the tool head and the drive unit and consists of an outer BTA tube and an inner push rod. The tool head is a guided element, which has a bolted-on cutting insert. The radial extension of the cutting edge is converted against a contour bevel via the elastic behavior of the base body when the push rod is moved. The guide bar is made of carbide and follows the guide principle of a single-lip drill. It has the same diameter as the pilot bore and is supported in the borehole without affecting its dimensions or the marginal zone structure of it. For precise internal contouring, the exact knowledge of the setting behavior is essential. Therefore, during tool development, this behavior of the tool head was simulated using a finite element model in Ansys software. Figure 6 shows the FEM-based simulation model taking into account elastic deformation analysis as well as stress analysis based on the shape modification hypothesis.

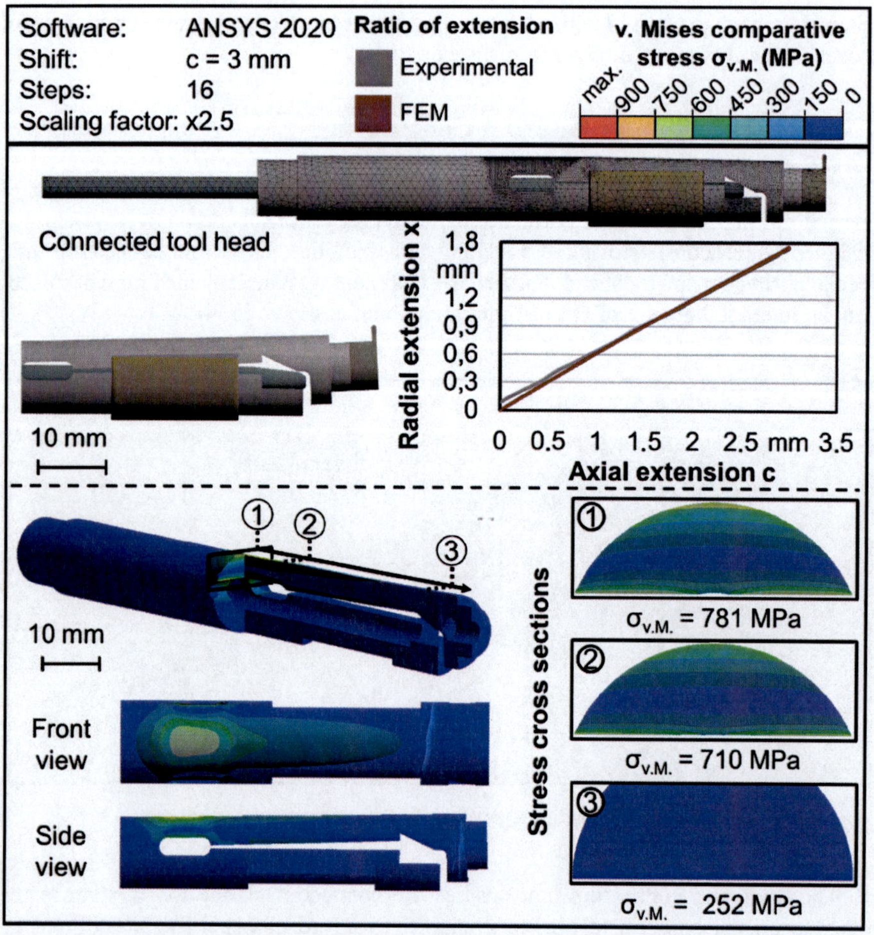

Fig. 6. FEM analysis of the tooling system's setting behavior and the comparative stresses on the tool system

For the model, an adapted modeling of the boundary conditions and the meshing of the tool head and push rod was performed. For the contact conditions, a bonded contact between the guide bar and the tool head was chosen due to the existing fixed clamping. In addition, there is a friction contact between the push rod and the main body. With regard to the clamping conditions, there is a fixed position of the base body due to the connection to the BTA tube. The guide bar is subject to a cylindrical bearing in the guide bore. Tetrahedrons represent the underlying model elements; the special feature here is that finer meshing was carried out in areas with probable deformation.

The infeed of the push rod is performed up to a maximum extension in steps of 0.2 mm. In doing so, the simulation can approximately reproduce the setting values of the real system, which were determined experimentally. The values can finally be used to enable targeted CNC programming of complex internal contours. Furthermore, this also allows the contour to be adjusted and simulated again. The stress analysis in the maximally extended state shows that the elastic range of the base material is not exceeded and thus no plastic deformation of the tool head occurs. Consequently, the selected inner contour is well suited and can contribute to a permanently process-safe use of the tool system. In the future, the consideration of the process forces could also be integrated into the model. Consequently, the cutting forces from the basic tests for internal turning can be aligned with the cutting insert. Presumably, however, this will not lead to the elastic range of the materials used being exceeded, since the cutting force in particular is subject to a different orientation and thus does not further increase the existing stresses.

5 Conclusion and Outlook

There is a wide-ranging need for internally contoured components in various areas of application in order to increase their functionality and load-bearing capacity as well as to reduce their weight. With the existing state of the art, there is an area that is not reasonably covered by any machining process. Specifically, this is the machining of small bores at great depths. For this reason, the development of a tool system is aimed at, which can process exactly this area. For this purpose, fundamental trials on internal turning with a bore diameter of $D = 14$ mm were first carried out in order to highlight the essential characteristics of this conventional machining process. It was found that there is a pronounced susceptibility to vibration in a wide range of cutting parameters, which is reflected in fluctuating mechanical tool loads and negative marks on the workpiece surface. In particular, the machining of demanding materials such as Inconel 718 clearly limits the cutting parameters that can be used, so that cutting splitting is often necessary. A comparison between different inserts has shown that for pulling internal contouring of complex contours, full radius and reverse turning inserts are also suitable over a longer cutting time.

In the following, the development and design of the novel tool system was implemented, which uses a guided tool head for a vibration-avoiding design. The radial extension of the cutting edge is based on material elasticity and is achieved by an axial infeed of an integrated linear actuator. In this context, FEM simulations of the setting behavior were implemented in order to be able to adapt the inner contour of the tool head. In the future, initial trials with the tool system will be carried out on a deep hole drilling machine to test its functionality. In a further step, a second tool system is to be developed which has a rotation capability to enable use on a larger number of machine tools. Overall, the development results in a large production technology potential by closing a gap in the status quo of internal machining processes.

This research and development project is funded by the German Federal Ministry of Education and Research (BMBF) within the “SME-innovative: Production Research” funding measure and managed by the Project Management Ageny Karlsruhe (PTKA). The author is responsible for the content of this.

References

1. Pauksch, E., Holsten, S., Linß, M., Tikal, F.: Zerspantechnik – Prozesse, Werkzeuge, Technologien, 12th edn. Vieweg+Teubner Verlag, Wiesbaden (2008)
2. Schönherr, H.: Spanende Fertigung, 1st edn. Oldenbourg Wissenschaftsverlag, Berlin (2002)
3. Sandvik Coromant Homepage. https://www.sandvik.coromant.com/. Accessed 06 June 2023
4. Hennermann, H., Dix, N.: Kleine Zerspanungslehre, 1st edn. Carl Hanser Verlag, München (1967)
5. Buse, B., Bergmann, K., Fuß, H.: Flexible innenbearbeitung mit rechnergesteuerten Bohrwerkzeugen. In: Weinert, K.: Spanende Fertigung, 1st edn. Vulkan Verlag, Essen (1994)
6. Metzger, M.: Innenkonturieren und Walzen tiefgebohrter Grundbohrungen. Vulkan Verlag, Essen (2019)
7. Veselovac, D.: Process and product monitoring in the drilling of critical aero engine components. Apprimus Verlag, Aachen (2013)
8. Renhof, L.: Mikrostruktur und mechanische Eigenschaften der Nickellegierung IN 718. München (2007)
9. Fu, S., Dong, J., Zhang, M., Wang, N., Xie, X.: Research on Inconel 718 type alloys with improvement of temperature capability. In: 7th International Symposium on Superalloy 718 and Derivates, pp. 281–296. Wiley, Pittsburgh (2010)
10. Özkaya, E., Bücker, M., Strodick, S., Biermann, D.: A thermomechanical analysis leading to a novel flank face design providing longer tool lives for tools used in the drilling of Inconel 718. Int. J. Adv. Manuf. Technol. **102**, 2977–2992 (2019)

Sensorised Tool Holders for Monitoring the Process Forces During High-Speed Internal Grinding

Tountzer Tsagkir Dereli[1](✉), Monika Kipp[1], Dirk Biermann[1], Rainer Wunderlich[2], and Robin Marxer[2]

[1] Institute of Machining Technology, TU Dortmund University, Baroper Straße 303, 44227 Dortmund, Germany
tountzer.tsagkirdereli@tu-dortmund.de

[2] pro-micron GmbH, Gottlieb-Daimler-Straße 6, 87600 Kaufbeuren, Germany

Abstract. The application of sensor technology for monitoring and analysing manufacturing processes is becoming increasingly important and is well established for manufacturing processes with geometrically defined cutting edges. However, in grinding processes such as internal grinding, the use of sensor-based tool holders reaches its limits. Especially in these demanding processes, the use of sensor technology offers a high potential for increasing manufacturing accuracy, performance and process reliability. In this context, a demonstrator of a tool holder for high-speed applications with integrated measurement technology has been developed. In experimental investigations, the tool holder has been qualified for internal grinding processes. It has been possible to perform reliable high-speed internal grinding processes even at high tool speeds of up to 30,000 1/min.

Keywords: Internal grinding · sensorised tool holder · high-speed grinding

1 Introduction

The monitoring of process forces is an efficient solution for reducing the process times, for enhancing the product quality and to avoid reworking of parts or to gain further information concerning the wear stadium of tools. Process monitoring is already established in several applications. Especially in grinding operations, the monitoring of the process and the grinding tool is challenging. This is due to the high circumferential speeds, which is of high relevance especially for high-speed and high-performance grinding processes. Furthermore, there is a requirement regarding high sensitivity of the sensors in order to gain detailed information of processes characterized by low loads and low changes. As a result, there is a need for adapted sensors and equipment to fulfill the specific requirements.

T. Bauernhansl et al. (Eds.): WGP 2023, LNPE, pp. 163–173, 2024.
https://doi.org/10.1007/978-3-031-47394-4_17

2 Tool Holder Development for Internal Grinding

In industrial applications, there are different possibilities for measuring the mechanical load in manufacturing processes [1–3]. One possibility is to measure the power of the spindle motor. This allows conclusions to the torsional moments or tangential forces [2]. Nonetheless, it is not possible to infer bending moments or normal forces from spindle power. But the normal forces play an important role in internal grinding processes. In these applications, a further challenge is the compliance of the used spindle systems [4, 5]. As a result of the normal forces induced by grinding process, the tool and workpiece deflect, which leads to shape deviations [4]. Therefore, it is important to gain information with regard to these measurement values.

Other systems for measuring process loads can include accelerometers, piezoelectric measuring systems and strain gauges that can be integrated into a tool holder [1–3, 6]. Measuring process forces by a rotational dynamometer used as a tool holder based on piezoelectric force measurement is very accurate in terms of measurement resolution and widely used in the research environment. New developments also show the possibility of working without stators [7]. Other systems introduced for processes with geometrically defined cutting edges are based on accelerometers [8, 9]. Sensorised tool holders based on strain gauges offer the possibility to work without stators [10, 11]. These systems are often used in industrial applications [10]. They have the potential to work at high rotational speeds, which is of interest in grinding processes [11, 12]. Based of deformations, that have to occur at the tool holder, the loads can be measured. This has to be taken into account concerning the process performance in case of compliant systems. Solutions that do not reduce the stiffness are desirable. Strain gauges based on metal and semiconductors are frequently used [3]. The latter solution has higher sensitivity, so that smaller deformations on the tool holder can be sufficient to measure the process forces with higher accuracy. But this system is more temperature sensitive [3], which complicates use at high rotational speeds. In order to expand the use for high-speed internal grinding processes, the limits of the maximum rotational speeds for the measuring systems have to be increased even further. Rotational speeds up to $n = 40{,}000\ \text{min}^{-1}$ are common in high speed internal grinding and offer a wide range of benefits regarding process performance. Reduced process forces and improved surface qualities are advantages in this context [13]. Due to the circumstances in internal grinding processes there is still a need for research and further development of sensor technology for process monitoring of the mechanical load in this application [3, 6, 14–16]. The following requirements can be derived for the development of a sensorised tool holder.

First of all, the holder geometry has to be designed in such way, that there is no interfering contour, which could limit bore grinding processes, especially of long protuding chuck components. Furthermore, the tool holder has to be suitable for high rotational speeds. In addition, an increased sensitivity of the holder is required in order to be able to identify low process loads. These can occur in certain applications, such as dressing of the small grinding wheels that are exposed to increased wear. An enhanced sensitivity opens up the possibility to identify even minor variations in the process loads during a grinding operation. Thus, the need to adjust the process strategy or to condition the tool can be detected.

Taking these points into account, a demonstrator has been designed in comparison to a commercially available tool holder for grinding processes, which are both shown in comparison in Fig. 1. The shape of the demonstrator has no interfering contours, so that the common internal cylindrical grinding operations can be carried out without geometric limitations.

a) spike_mobile b) demonstrator

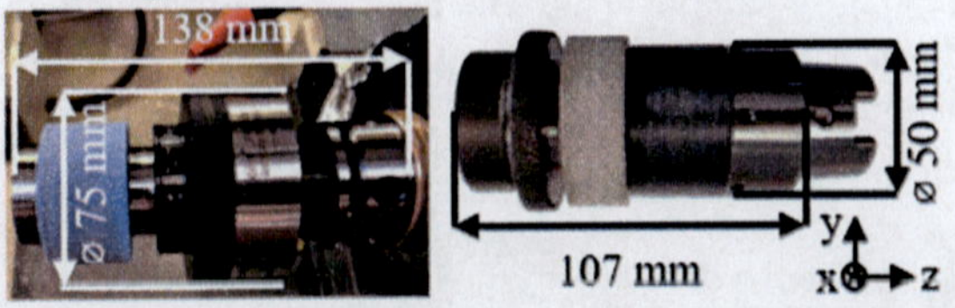

c) Compliance under bending

X-axis:
0,397 μm/Nm
Y-axis:
0,380 μm/Nm

d) Noise, minimum resolution and max. rotational speed

		spike_mobile	**demonstrator**
Measurement Noise RMS	Bending moment M_b in Nm	0.17	0.055
	Torsional moment M_t in Nm	0.30	0.12
Min. Resolution	Bending moment M_b in Nm	0.00748	0.000955
	Torsional moment M_t in Nm	0.0129	0.00149
Max. rotation speed in min^{-1}		15.000	> 30.000

Fig. 1. a) Sensorised tool holder spike mobile by pro-micron for grinding, b) Demonstrator of a sensorised tool holder for high-speed internal grinding; c) Compliance of the demonstrator under bending; d) Sensitivity of the demonstrator

The accumulators, the charging interface and the electronics, which are located on the outer ring in Fig. 1a) have been placed closer to the rotational axis of the tool holder in Fig. 1b). Therefore, it is possible to keep the centrifugal forces acting on the electronic components and accumulators as low as possible. To ensure both, high stiffness and high measurement sensitivity of the holder, the tool holder geometry has been optimised using the finite element method. The aim was to concentrate the deformation of the holder under load in the area, where the strain gauges are located. In addition, the holder length has been kept as short as possible in order to minimise deflection at the grinding tool under bending. Before using the sensorised tool holder, the noise and sensitivities of all axes and the compliance under bending load of the demonstrator, were empirically determined at pro-micron. The values determined are listed in Fig. 1c) and 1d). The resolution of the demonstrator has been enhanced and the measurement noise reduced. This increases the measurement accuracy and sampling rate, which is advantageous at high speeds.

3 Experimental Conditions for Internal Cylindrical Grinding with Sensorised Tool Holder

In order to qualify the sensorised tool holder grinding experiments have been performed. Selected scenarios for internal plunge and internal traverse grinding have been considered in order to validate the sensorised tool holder in terms of sensitivity and stiffness. An additional focus is set on different boundary conditions to identify the effects on the measuring results and to analyse the wear on the grinding tool as for example the change of the topography. The last point is important with respect to process stability and the achievement of a constant manufacturing quality.

In the grinding investigations, bearing components out of the workpiece material 100Cr6, hardened to 60 ± 2 HRC, have been machined on the internal grinding machine *Overbeck Danobat IC 400*. The material 42CrMo4 + QT has also been processed, which has been used to promote clogging in sample tests. The used process parameters and additional test conditions are given in the figures concerning the results. The experimental setup for the grinding tests is shown in Fig. 2.

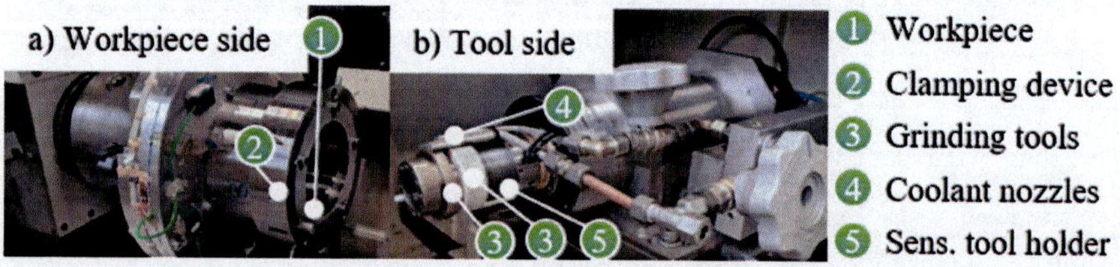

Fig. 2. Experimental setup for grinding tests; a) workpiece side; b) tool side

For internal plunge and traverse grinding, two different grinding wheels have been mounted on the tool holder and have been separated from each other by a distance ring. For conventional internal plunge grinding operations a cylindrical vitrified bonded aluminium oxide grinding wheel of the type *35A 80 K10 V170* from the manufacturer *Krebs & Riedel Schleifscheibenfabrik GmbH & Co. KG* has been used. These tools had an outer diameter of $d_{t,c} < 60$ mm and a grinding wheel width of 20 mm. In addition, electroplated cBN grinding tools with an outer diameter of $d_{t,e} = 60$ mm and grit sizes of $d_g = \text{B126}$ and $d_g = \text{B181}$ from the manufacturer *August Rüggeberg GmbH & Co. KG* have been used. The tools have a bifunctional design, consisting of a conical roughing zone and a cylindrical finishing zone and are especially appropriate for high-performance internal traverse grinding processes. In the case of an internal traverse grinding process, the major material removal takes place in the conical roughing zone. This area covers a width of 4 mm with a roughing zone angle of $\chi = 3°$. Using the cylindrical finishing zone, which has a width of $w_f = 4$ mm, the surface has been smoothed.

4 Bore Grinding with Sensorised Tool Holder

In the first step, an internal plunge grinding process has been considered. These processes are used widely in industrial applications. Internal plunge grinding is a manufacturing process, where the used tool is usually in contact over the entire bore length of the workpiece. A continuous radial infeed achieves the material removal. Bores of functional parts can be manufactured productively in this process. Setting a high coverage ratio between grinding wheel and workpiece allows to achieve a high surface quality. Due to the large contact width between tool and workpiece, high global process normal forces occur during internal plunge grinding. The result is that the slim toolholders used for internal grinding are displaced. This causes shape deviations and limits process performance. Therefore, in practice, spark out phases are used to reduce the wheel displacement and minimize the shape deviation.

The conducted grinding process is divided in three phases. In the first phase, a roughing process is carried out with a specific material removal rate of $Q'_w = 1$ mm^3/(mm*s). This is followed by a finishing process, in which the specific material removal rate is halved and the coverage ratio is increased in order to produce a targeted surface topography. In the last step, the spark out phase takes place, in which no more infeed occurs and the previously deflected grinding tool can relieve itself. The demonstrator presented in Sect. 2 is intended to reliably record the bending and torsional moments. Using the sensorised tool holder, it is the aim to identify the different process phases. This offers the opportunity to optimize the process efficiency and spark out times. The bending and torsional moments recorded during the grinding test are shown in Fig. 3.

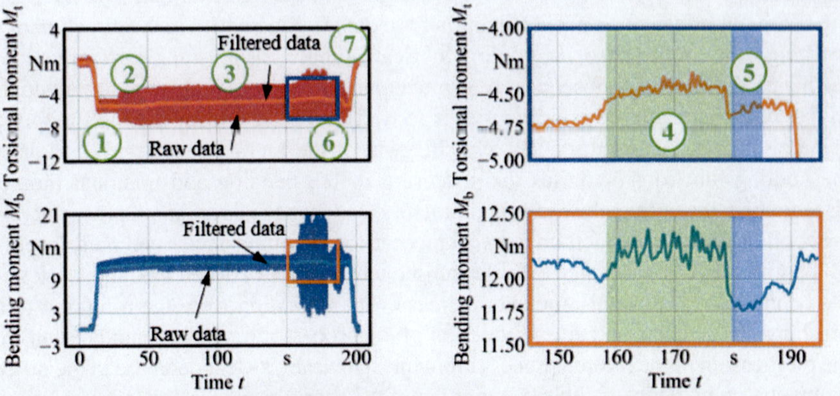

1. Rotational speed increase
2. Start of cooling lubricant supply
3. Positioning of grinding tool
4. Grinding process
5. Spark out phase
6. End of cooling lubricant supply and return of spindle
7. Spindle stop

System parameters	
Process:	Internal plunge grinding
Workpiece:	100Cr6, 60 ± 2 HRC
Cooling:	Grinding oil
Grinding tool:	35A 80 K10 V170, $d < 59$ mm

Process parameters	
Cutting speed:	$v_c = 80$ m/s
Radial stock removal (total, roughing, finishing):	$a_{e,tot} = 0.1$ mm; $a_{e,r} = 0.075$ mm; $a_{e,f} = 0.025$ mm
Spark out time:	$t = 5$ s
Speed ratio:	$q = 70/120/120$
Spec. material removal rate:	$Q'_w = 1...0,5$ mm³/(mm·s)

Fig. 3. Internal plunge grinding process and detailed view during grinding process

The measurement data recorded with the sensorised tool holder has been analyzed using a program implemented in Python. The bending moment and the torsional moment have been processed as followed: First, the amount of the resulting bending moment has been determined from the individual components measured in the x- and y-directions. This is followed by a correction of the zero level of bending and torsional moment as well as a linear drift correction. In order to detect the constant parts of the process moments during the engagement, a low-pass filter with a frequency of 1 Hz and a transition range with a width of 2 Hz has been used.

The grinding process is shown from the start of tool rotation until the spindle stops after the grinding process. In the first phase, the grinding spindle accelerates and the set rotation speed of the grinding tool is reached after a few seconds. During this phase, there is an increase in the torsional and bending moments acting on the tool holder, leading to a constant moment offset. A prior operational balancing of the tool holder at nominal speed has been performed in order to keep possible imbalances as low as possible. In the second phase, the cooling lubricant supply is started at low pressure. This leads to an increase in the spread values of the raw signals in the torsional moment, as most of the cooling lubricant is directed tangentially to the grinding wheel in the direction of the working zone. In this case, the cooling lubricant supply has no significant influence on the bending moments. Subsequent feed movements take place in the air without contact to the workpiece and the grinding tool is moved to the set starting position before the grinding process. The subsequent switching-on of the high-pressure pump, to deliver the cooling lubricant increases the scattering of the bending and torsional moment. Especially in the case of the bending moment, these scatterings are more pronounced, as the cooling lubricant impacts on the workpiece in the clamping device and a displacement of the grinding tool takes place. In the subsequent contact between tool and workpiece, the bending and torsional moments increase. The transition from the roughing to the finishing phase and an expected reduction of the process moments cannot be seen from the progression of the filtered process moments. However, a clear decrease in the process moments can be noticed with the beginning of the spark-out phase and lasts for a period of 5 s. In the last phase, the cooling lubricant supply is stopped immediately while at the same time the grinding tool moves in the opposite direction in fast motion, which leads to an increase of the bending moment. Subsequently, the rotational speed of the grinding tool stops and the process moments are reduced to the zero point.

In order to investigate the response of the sensorised tool holder in a further internal grinding process, the internal traverse grinding process initially is examined at a cutting speed. Internal traverse grinding is an efficient machining process that uses two-part grinding tools consisting of a conical roughing zone and a cylindrical finishing zone. Due to the tool geometry used, this process is characterised by variable engagement conditions. A high radial stock is set before contact between the tool and the workpiece. This is followed by contour-parallel machining along the bore length. The roughing zone of the grinding tool is used to remove the main material, while the finishing zone is applied to smooth the surface. Using high-speed conditions, the advantages of this grinding process are increased. The processing of the raw data has been done analogous to the previous investigation. The raw and filtered data are presented in the Fig. 4.

The data shown refers to the process state after the grinding wheel has accelerated to nominal speed, the positioning process takes place and the cooling lubricant is supplied with a low volume flow. A high increase in bending and torsional moments follows when the conical roughing zone comes in contact with the workpiece due to the high loads generated in the rough grinding process. The rise in process moments is reduced when the cylindrical finishing zone also engages. A quasi-stationary phase follows, with almost constant material removal. A decrease of the process moments occurs after the roughing zone gradually exits and a rebound of the grinding tool occurs. This can be recognised by the over-swinging of the bending moment. As a result, more material is removed in the run-out area of the workpiece with the finishing zone and a deviation in the shape of the component results. This case is known in internal traverse grinding [4]. The torsional moment indicates that at the point where the local minimum of the bending moment is recorded, material is removed by the finishing zone. After that, a further feed movement takes place without the tool and workpiece being in contact. At this stage, the cooling lubricant supply in the fixture is still active. As a result, turbulence is increasingly created by the cooling lubricant as the grinding tool approaches the bottom of the fixture bore. This turbulence leads to an increased scattering of the raw signals of the torsional moment.

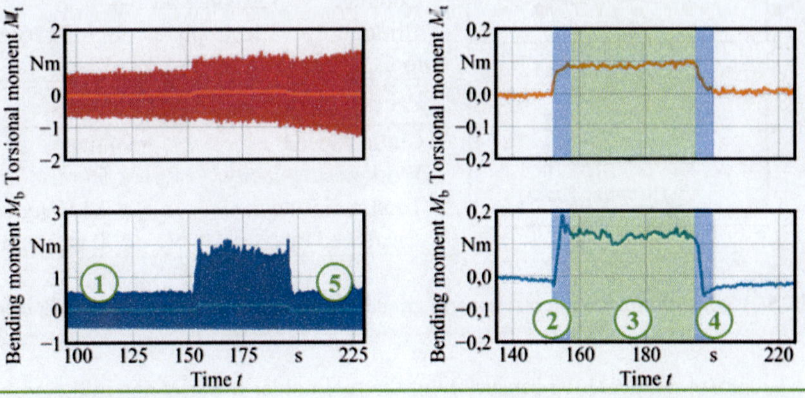

1. Positioning
2. Running-in phase
3. Constant tool contact
4. Grinding wheel exits the workpiece
5. Feed movement without contact to the workpiece

System parameters	
Process:	Internal traverse grinding
Workpiece:	100Cr6, 60 ± 2 HRC
Cooling:	Grinding oil
Grinding tool:	Electroplated cBN, B181, $d = 60$ mm, $\chi = 3°$

Process parameters	
Cutting speed:	$v_c = 25$ m/s
Workpiece velocity:	$v_w = 1.56$ m/s
Total radial stock:	$a_{e,tot} = 0.100$ mm
Axial feed rate:	$v_{fa} = 24$ mm/min

Fig. 4. Internal traverse grinding process and detailed view during grinding process

In order to qualify the process suitability of the sensorised tool holder for grinding processes with high loads, the use of the demonstrator was extended to the application of high-performance internal traverse grinding. The following process was carried out at a grinding tool velocity of 94 m/s. The bending and torsional moments during the process are shown in Fig. 5. At a nominal rotation speed of 30,000 min^{-1}, the demonstrator is still usable and measurement signals can be successfully received. In contrast to the grinding process with reduced grinding tool speed, a drift can be seen in the time interval between reaching the target speed and the first contact between tool and workpiece (t = 75 s). This is probably due to heating up the tool holder as a consequence of the high circumferential speed. After acceleration and switching on the cooling lubricant supply, a high-pressure pump is activated in order to improve the local thermal load dissipation during the internal traverse grinding process. This change is clearly visible in the development of the torsional moment from about 50 s onwards. This is followed by a high mechanical load during the actual grinding process.

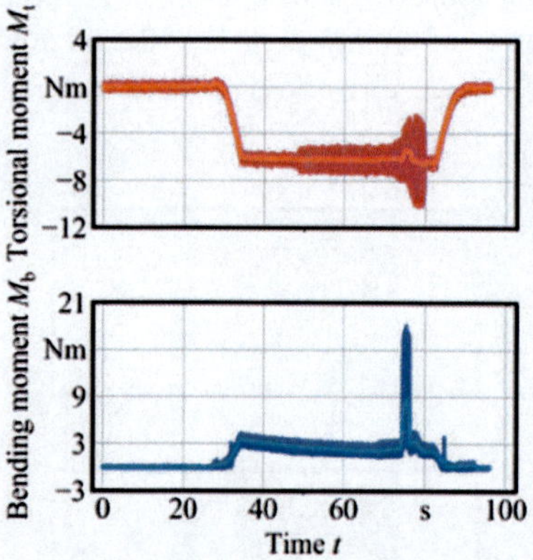

System parameters	
Process:	Internal traverse grinding
Workpiece:	100Cr6, 60 ± 2 HRC
Cooling:	Grinding oil
Grinding tool:	Electroplated cBN, B126, d = 60 mm, χ = 3°

Process parameters	
Cutting speed:	v_c = 94 m/s
Workpiece velocity:	v_w = 1.56 m/s
Total radial stock:	$a_{e,tot}$ = 0.150 mm
Axial feed rate:	v_{fa} = 600 mm/min

Fig. 5. High performance internal traverse grinding process using a sensorised tool holder

In order to investigate the sensitivity of the tool holder to variations in the process parameters, the feed rate in grinding processes is varied in three stages. To reduce additional influences, a cutting speed of 30 m/s is used and a reduced cooling lubricant supply is applied. In addition, the material 42CrMo4+QT is used to cause material clogging. Figure 6 shows the process moment characteristics shortly before and after tool engagement.

The rise in feed rate is accompanied by an increase in the maximum bending and torsional moment, which is consistent with the basis of the grinding technology. The scattering in the raw signals of the bending moment that occurs at a feed rate of 50 mm/min can be interpreted as indicators of material clogging. This causes a stronger pressing than chipping through the grinding wheel. This material clogging occurred to a small extent on the grinding tool during the grinding process. In a final test, the radial allowance was doubled and the feed rate was increased further to cause more material clogging and to better determine when it would occur, see Fig. 7.

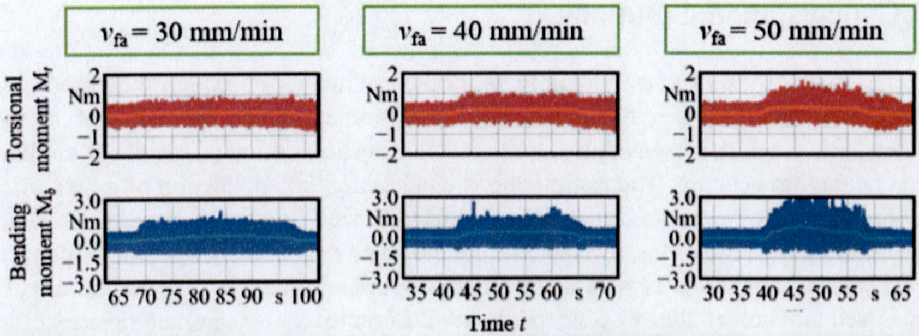

System parameters	
Process:	Internal traverse grinding
Workpiece:	42CrMo4 + QT
Cooling:	Grinding oil
Grinding tool:	Electroplated cBN, B181, $d = 60$ mm, $\chi = 3°$

Process parameters	
Cutting speed:	$v_c = 30$ m/s
Workpiece velocity:	$v_w = 0.5$ m/s
Total radial stock:	$a_{e,tot} = 0.05$ mm
Axial feed rate:	$v_{fa} = 30...40...50$ mm/min

Fig. 6. Variation of axial feed rate during internal traverse grinding process

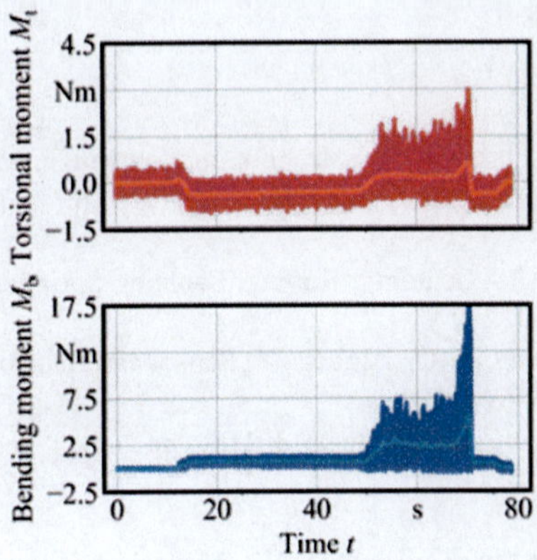

System parameters	
Process:	Internal traverse grinding
Workpiece:	42CrMo4 +QT
Cooling:	Grinding oil
Grinding tool:	Electroplated cBN, B181, $d = 60$ mm, $\chi = 3°$

Process parameters	
Cutting speed:	$v_c = 30$ m/s
Workpiece velocity:	$v_w = 0.5$ m/s
Total radial stock:	$a_{e,tot} = 0.1$ mm
Axial feed rate:	$v_{fa} = 60$ mm/min

Fig. 7. Idenfication of cloggings during internal traverse grinding

Shortly after the engagement of the grinding wheel, pronounced scattering of the bending moment is visible. With an extended process duration, a high increase in the bending and torsion moments has been observed, so that the grinding process has been interrupted as a consequence. As a result, extensive material clogging has been identified on the grinding tool. Due to the application of the sensorised tool holder, there is the possibility to identify change on the grinding wheel.

5 Conclusion and Outlook

In order to qualify sensorised tool holders for the use in high-speed grinding processes, there is the need for a specific adjustment of the geometry and components of the tool holder. In this regard, design adjustments as wells as process oriented specifications have to be taken into account. The examinations concerning the application of a sensorised tool holder in different grinding processes, characterized by varying process strategies and therefore load situations, revealed the potentials as well as challenges in this context. The designed tool holder is suitable for internal grinding processes. Due to the use of a sensorised tool holder, it is possible to identify different stages in internal traverse grinding processes. In internal plunge grinding with smaller differences between the roughing and finishing phases, it was not possible to identify the individual phases explicitly, so that there is the desire for a further increase of the sensitivity. This extends the possibilities of process control with regard to different process phases especially concerning roughing and finishing. Moreover, a further increase in sensitivity offers potentials concerning the process monitoring of dressing processes, which will be investigated in ongoing studies.

Acknowledgements. Funded by Federal Ministry for Economic Affairs and Climate Action under the program Zentrales Innovationsprogramm Mittelstand (ZIM), KK5072208.

References

1. Kuttner, T., Rohnen, A.: Praxis der Schwingungsmessung, Messtechnik und Schwingungsanalyse mit MATLAB, 2nd edn. Springer, Wiesbaden (2019). https://doi.org/10.1007/978-3-658-25048-5
2. Klocke, F-, König, W.: Manufacturing Processes 2 – Grinding, Honing, Lapping. Springer, Berlin (2019). https://doi.org/10.1007/978-3-540-92259-9
3. Bleicher, F., Biermann, D., Drossel, W.-G., Moehring, H.-C., Altintas, Y.: Sensor and actuator integrated tooling systems. CIRP Ann. **72**(2), 673–696 (2023)
4. Brecker, M., Weck, M.: Werkzeugmaschinen Fertigungssysteme 1 Maschinenarten und Anwendungsbereiche, 9th edn. Springer, Berlin (2019). https://doi.org/10.1007/978-3-662-46565-3
5. Tsagkir Dereli, T., Schmidt, S., Furlan, T., Holermann, R., Biermann, D., Menzel, A.: Simulation based prediction of compliance induced shape deviations in internal traverse grinding. J. Manuf. Mater. Process. **5**(2), 60 (2021)
6. Ayman, M, Hassan, M, M'Saoubi, R, Attia, H.: Tool condition monitoring for high-performance machining systems - a review. Sensors **22**(6) (2022)
7. Kistler Homepage. https://www.kistler.com/DE/de/cp/kabellose-rotierende-zerspankraft-dynamometer-9170b/P0001250. Accessed 15 June 2023
8. Schunk GmbH, iTendo^2. https://schunk.com/de/de/tools/werkzeughalter-quickfinder/itendo2/c/PGR_6468. Accessed 15 June 2023
9. Bleicher, F., Schörghofer, P., Habersohn, C.: In-process control with a sensory tool holder to avoid chatter. J. Mach. Eng. **18**(3), 16–27 (2018)
10. Schuster, A., Rentzsch, H., Ihlenfeldt, S.: Energy self-sufficient, multi-sensory tool holder for sensitive monitoring of milling processes. Procedia CIRP **117**, 80–85 (2023)
11. Promicron. https://www.pro-micron.de/spike/. Accessed 15 June 2023
12. Keil, S.: Drehnungsmessstreifen, 2nd edn. Springer, Wiesbaden (2017)

13. Rowe, W.B.: Principles of Modern Grinding Technology, 2nd edn. William Andrew Publishing (2014)
14. Schörghofer, P., Pauker, F., Leder, N., Mangler, J., Ramsauer, C., Bleicher, F.: Using sensory tool holder data for optimizing production processes. J. Mach. Eng. **19**(3), 43–55 (2019)
15. Mauthner, G., Votruba, W., Ramsauer, C., Plessing, L., Trautner, T., Bleicher, F.: Development of a CAM-in-the-loop system for cutting parameter optimization using an instrumented tool holder. Procedia CIRP **107**, 326–331 (2022)
16. Schmidt, N., Tsagkir Dereli, T., Furlan, T., Holtermann, R., Biermann, D., Menzel, A.: Towards the prediction of compliance influences on shape deviations in internal traverse grinding in production at the leading edge of technology. In: Behrens, B.-A., Brosius, A., Hintze, W., Ihlenfeldt, S., Wulfsberg, J.P. (eds.) WGP 2020. Lecture Notes in Production Engineering, pp. 304–314. Springer, Heidelberg (2021). https://doi.org/10.1007/978-3-662-62138-7_31

Cross-Process Chain Analysis on Gear Quality and Sustainability

Hanwen Zhang[1](✉), Gonsalves Grünert[1], Mareike Solf[1], Jens Brimmers[1], Sebastian Barth[1], and Thomas Bergs[1,2]

[1] Laboratory for Machine Tools and Production Engineering (WZL) of RWTH Aachen University, Campus Boulevard 30, 52074 Aachen, Germany
h.zhang@wzl.rwth-aachen.de

[2] Fraunhofer Institute for Production Technology (IPT), Steinbachstraße 17, 52074 Aachen, Germany

Abstract. The aim of this study is to use a Shapley value-based machine learning algorithm to investigate the cross-process chain impacts of the processes involved in a pinion shafts manufacturing chain on final gear quality with sustainability considerations. Process data is collected during gear hobbing and profile grinding and analyzed using a neural network model to calculate Shapley values. The results quantify the influence of process parameters and trace data of each process on the gear quality, thereby improving understanding and leading to better control and sustainability of processes within the manufacturing chain. By achieving better process control and evaluating sustainability, this study not only helps to save time, energy, and material resources but also promotes sustainable manufacturing practices that minimize waste and the environmental impact of gear production. Overall, this study provides valuable insights into the importance of processes in the gear manufacturing chain, aiding in its optimization.

Keywords: Shapley Value Analysis · Gear Quality · Sustainability

1 Introduction

The manufacturing industry aims for improved efficiency, quality, and sustainability in its production processes. It operates through a multistage system using a series of interconnected processes, involving diverse components and stations working together towards the completion of the final product [1]. Within a manufacturing chain, each individual process contributes to the overall quality of the final product. Understanding the cross-process chain impacts and their influence on the product quality is essential for achieving optimal performance and sustainability, which is especially relevant for highly loaded and efficiency-relevant components such as gears.

Traditionally, process optimization relies on empirical methods for decision-making and quality control. However, these empirical methods face limitations, particularly when dealing with complex systems where underlying models may become intractable,

T. Bauernhansl et al. (Eds.): WGP 2023, LNPE, pp. 174–184, 2024.
https://doi.org/10.1007/978-3-031-47394-4_18

and individual processes are often analyzed and optimized independently, without considering the process chain as a whole. Advancements in machine learning algorithms have opened up new possibilities for gaining deeper insights into processes, enabling us to investigate cross-process relationships, their impact on product quality, and improve sustainability in the manufacturing chain.

2 State of the Art

A manufacturing process chain refers to a series of manufacturing processes that directly contribute to the final component characteristics. Understanding how the individual processes and their relation influence the final component becomes crucial for optimizing process chains and components. As a previous approach to investigate relationships and dependencies between variables representing different process steps, Beckers et al. developed a stochastic model that represents each process and captures the aspects of their dependencies. To achieve this, a graphical model was derived to describe the manufacturing process chain, utilizing the concept of graphical models known for their ability to represent complex distributions and analyze structural relationships [2]. However, the graphical model is not able to capture unknown dependencies that still needs to be explored. The model also does not provide the possibility to be adapted to dynamically changing manufacturing environment.

Gear manufacturing commonly includes a manufacturing process chain, where a series of specific processes are performed to produce gears with desired characteristics and meet specific requirements. As a crucial component in mechanical systems, gears are responsible for mechanical power transmission and offer advantages through their simple construction, compact size, high operational reliability, and significant impact on the overall efficiency of the mechanical system [3, 4]. Examining the process relationships of gear manufacturing offers potential to predict and improve part quality leading to better load-carrying capacity and transmission efficiency.

Machine learning algorithms can analyze vast amounts of data, detecting hidden relationships, and adapt on environmental changes over time [5]. Consequently, these algorithms are broadly adapted in the field of quality assurance [6]. However, the issue arises when we lack an understanding of how the algorithm arrived at its suggested solution. This problem, known as the explanability problem, can create obstacles in our efforts to improve the underlying algorithm [7, 8]. An attempt to solve the explanability problem is SHAP (Shapley Additive Explanation). SHAP is an unified framework to interpret machine learning model predictions by generating unique Shapley values for features involved in the prediction. Unlike other approaches, SHAP guarantees unique Shapley values, indicating the importance for each feature. The Shapley value, derived from cooperative game theory, provides a fair and effective approach to quantify the influence of each in the prediction involved feature to the final prediction [7].

To achieve the goal of optimizing the sustainability e.g. of processes and process chains, Life Cycle Assessments (LCA) based on the ISO14040/44 standards are widely employed [8, 9]. These assessments provide a comprehensive understanding of the environmental impacts associated with a product's entire life cycle, from raw material extraction to end-of-life disposal. However, LCA offers many degrees of freedom, which limits the direct comparability of different calculations. Therefore, Bergs et. al have developed a method which allows a systematic evaluation of manufacturing processe [10]. Based on this, further models were developed which enable the specific evaluation of individual tools and manufacturing processes [11–13].

Overall, the lack of approaches investigating the interrelationships between manufacturing processes and their impact on part quality, coupled with the limitations of existing methods for process optimization, highlights the need for a comprehensive approach combining machine learning algorithms with process knowledge considering quality and sustainability simultaneously.

3 Objective and Procedure

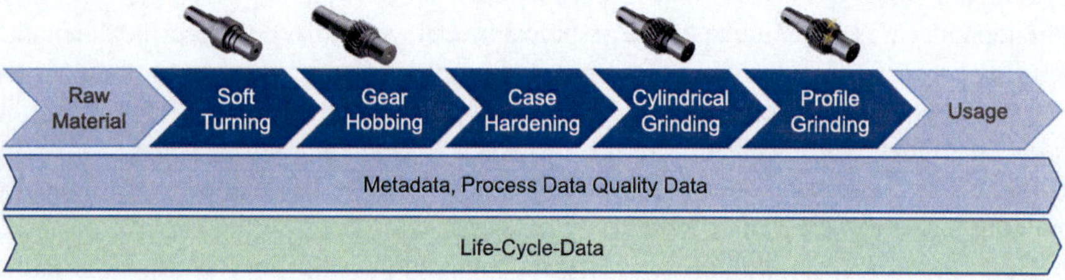

Fig. 1. Process chain of pinion shaft manufacturing

Based on the identified need for a methodology that integrates quality and sustainability factors to optimize the manufacturing chain, the objective of this study is to *develop an approach that investigates the cross-relationships between manufacturing processes and their impact on gear quality, while simultaneously considering the factors of sustainability, in order to optimize the manufacturing process in gear manufacturing.* In pursuit of this objective, the study is carried out on a use case of the pinion shaft process chain for gearboxes in electric cars, see Fig. 1.

To achieve our objective, we will utilize the Shapley value-based machine learning algorithm to analyze the process parameters and trace data specifically within the process chain. The Shapley value analysis aims to quantify the individual contributions of each process step to the overall gear quality. By understanding the relative importance of each process parameter, we can enhance process control and optimization, leading to improved gear quality and sustainability within the given process chain.

Additionally, we will conduct an analysis using LCA methodologies based on ISO14040/44 standards, providing an evaluation of the environmental impacts associated with different manufacturing processes and improve the decision-making progress for a sustainable process chain. This makes it feasible to use LCA to examine how altering process parameters affect the environment. By examining the impact of process parameters on the environment, LCA seeks opportunities to decrease environmental impact so that an ecological-economic evaluation is possible [10, 14].

4 Methodology

In this section, we employ two key approaches to investigate the impact of the manufacturing chain on gear quality and sustainability. Firstly, we utilize Shapley value analysis, using a machine learning algorithm, to quantify the influence of process parameters and trace data on gear quality. Secondly, we conduct LCA to assess the environmental impacts of the manufacturing process. By combining these approaches, we gain insights into the factors affecting gear quality and environmental sustainability. The use case manufacturing chain involved the production of eight groups of pinion shafts, each group having different settings for process parameters at various stages of the process chain. The variations between each experimental group are outlined in Table 1. These different groups ensure the diversity of the data and provide the foundation for analyzing the impact of each process on the quality of the pinion shafts.

Table 1. Overview of the different experimental groups

experimental groups	soft cutting	gear hobbing	cylindrical grinding	profile gear grinding
U1	reference	reference	reference	reference
U2	no finishing	concentricity error 1	reference	reference
U3	no finishing	concentricity error 1	reference	no finishing
U4	worn tool	concentricity error 2	reference	reference
U5	no finishing	no finishing	productivity	reference
U6	reference	reference	productivity	productivity
U7	productivity	productivity	reference	reference
U8	productivity	productivity	productivity	productivity

4.1 Shapley Value Analysis

During the research, we focused on processes related to gear manufacturing and validated the results on the given process chain. The Shapley value analysis consists of three steps including data preparation, building a neural network to describe the relationship between gathered data, and utilizing the Shapley value to explain the relationship between input and output data. The results demonstrate the influence of process parameters and trace data from various processes on the final quality of the gears. By combining process data and trace data, which are detailed logs that records the sequence of events or actions within a system, the machine learning model get more input features, which usually leads to a better training accuracy [15].

Data Preparation. The first step is to gather data from the process chain. The collected data includes process parameters, trace data, and quality measurements. The data from various stages of the process chain related to gear manufacturing was gathered. Soft turning and cylindrical grinding were excluded from consideration as they do not directly contribute to the final gear quality. The collected data was then prepared and divided into input and output datasets categorized by features (see Fig. 2). To analyze the cross-process chain impact on gear quality, the input data consisted of process parameters and trace data from both gear hobbing and profile grinding, while the output data represented the quality measurements of the final gear.

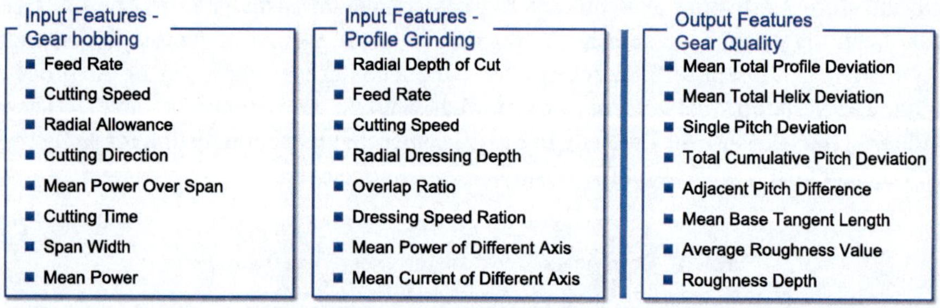

Fig. 2. Input and Output Features

Neural Network Model for Process Chain Analysis. The next step in the analysis is to construct a neural network model using the input and output data to represent the relationship between the data. A neural network model is created using the data collected from gear hobbing and profile grinding, along with the final gear measurements. The dataset is splitted into training data, validation data, and test data for cross-validatory purposes [16]. The model was trained with a learning rate of 0.001, using 80% of the data for training, 10% for validation, and 10% for testing. To prevent overfitting, a dropout layer of 0.25 was implemented in the network structure. It is important to note that the dataset consisted of a relatively small number of data points, totaling less than 250 observations.

Derivation of Shapley Value. The last step is to calculate the Shapley value for the input features in Table 1. The Shapley value is a concept derived from cooperative game theory that aims to fairly distribute the total value generated among the participants. It provides a quantitative measure of the contribution made by each individual in a cooperative setting, considering all possible combinations and permutations of players. By accounting for the contributions, the Shapley value offers insights into the relative importance and influence of each player within the cooperative framework. [17].

A game is a tuple (N, v) with a set of players $N = \{1, 2, \ldots, n\}$ and $n \epsilon \mathbb{N}$. The coalition function v assigns a real number $v(S)$ to each subset $S \subseteq N$ with $v(\varnothing) = 0$. The Shapley value $\Phi_i(v)$ of the player $i \in N$ is defined as

$$\Phi_i(v) = \sum\nolimits_{S:i \notin S} \frac{|S|!(n-1-|S|)!}{n!}(v(S \cup \{i\}) - v(S)) \tag{1}$$

For a given subset of players $S \subseteq N$ with $i \in N$, the contribution of player i $(v(S \cup \{i\}) - v(S))$ is defined as the difference of the contribution of the coalition S without and with the player i. The term $|S|!(n-1-|S|)!$ relates to the order of player i. In the subset $S \subseteq N\backslash\{i\}$ there are $|S|!$ possible orders for the players before i and $(n-1-|S|)!$ possible orders for the players behind player i. By considering all possible coalitions and their corresponding contributions, the Shapley value provides a measure of the individual's importance within a cooperative framework [17].

In our context, we can draw an analogy between the neural network and a game, where the input features act as players, and the output features represent the outcome of the game. Each player (input feature) contributes to the overall outcome. The neural network, as the game, processes the inputs and generates the corresponding outputs based on the interactions between these features. By utilizing this analogy, we can determine the impact of process parameters and trace data on final gear quality.

4.2 Life Cycle Assessment

This paragraph sets out the method for the ecological assessment via LCA. The LCA is divided in four different phases. The goal of the conducted LCA is to evaluate the environmental impact of the manufacturing process chain of the pinion shaft. In this gate-to-gate approach only the value adding processes are considered, which forming the product system. The hardening process was excluded because it was carried out by a supplier and no specific data was available. Table 1 shows the design of experiments. Different process parameters were varied to investigate their influence on the environmental effects. The reference flow or functional unit is the manufacturing of one pinion shaft. The IPCC method was used to calculate the possible global warming potential (GWP), representing the environmental impact. [18].

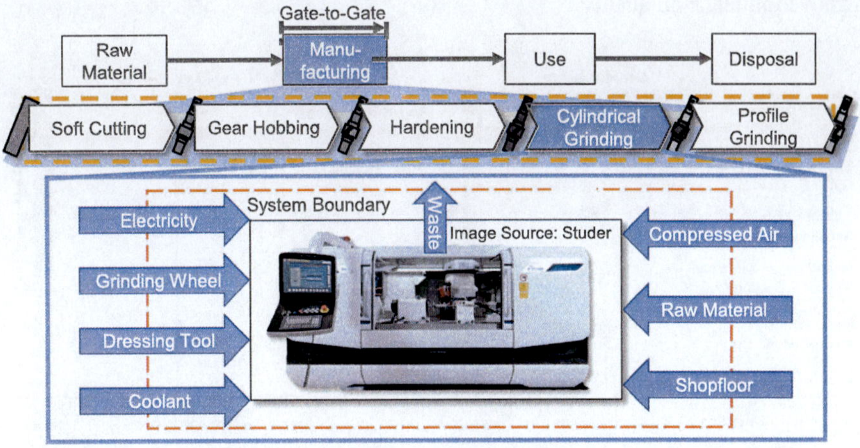

Fig. 3. Superordinate product system using the example of the grinding process.

In the Life Cycle Inventory (LCI) all necessary data is collected. The foreground data was measured on the shopfloor or gathered manually. For the secondary data the LCI-database ECOINVENT v3.9 (Allocation, cut-off by classification) was used. Figure 3 shows the considered energy and material flows using the example of the grinding process. This includes energy, raw materials, supplies and waste [19].

For the Life cycle impact assessment, the Software OpenLCA v.1.11 was used [20]. A parameterized model was created, which allowed the calculation of all test points. In addition to the GWP, the cost of pinion shaft manufacturing was calculated. The following manufacturing costs of time, workpiece, machine hour, operating resources like coolant and tools are assumed [21]. The results are shown and discussed in Sect. 5.2.

5 Experimental Results

This chapter presents the experimental results of the research, focusing on two key aspects: the Shapley value analysis and the LCA. In the first part, we show the results of the Shapley value analysis. This analysis provides insights into the influence and contribution of each process parameter and trace data on final gear quality, meaning the relationships between the individual process and the gear quality after profile grinding are captured using neural network. The second part of this chapter presents the results of the LCA, which assesses the environmental impacts of the manufacturing process.

5.1 Results of Shapley Value Analysis

Figure 4 shows the results of the Shapley value analysis for gear hobbing and profile grinding, focusing on the features that have the most impact on gear quality. The y-axis represents these influential features, while the x-axis represents the calculated Shapley values, which are directly correlated to the quality itself. Negative Shapley values indicate an quality improvement with increasing feature values, while positive values suggest a negative influence on quality.

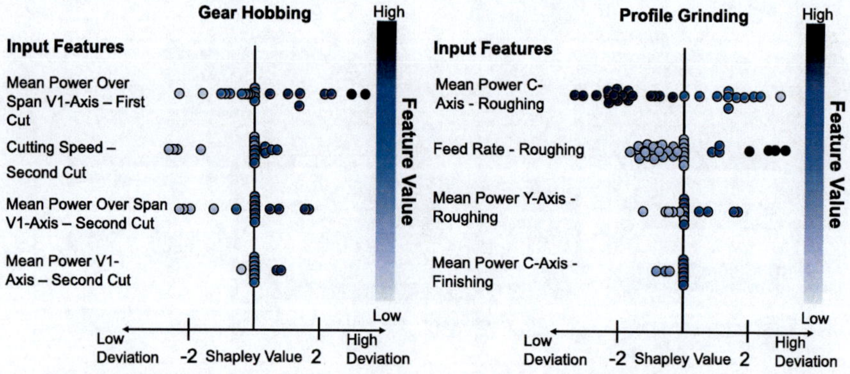

Fig. 4. Features with most impact on final gear quality after grinding

In this figure, each point represents the data of a manufactured pinion shaft. The color of each point reflects the corresponding feature value associated with that data point. By observing the distribution of points, we can identify patterns regarding the impact of different feature values on the final gear quality. To avoid overlapping points, the use of parallel points in the figure is employed.

For gear hobbing, the input features "Mean Power Over Span V1-Axis - First Cut," "Cutting Speed - Second Cut", "Mean Power Over Span V1-Axis - Second Cut", and "Mean Power V1-Axis - Second Cut" stand out as having the greatest impact on final gear quality. The axis V1 refers to the spindle axis of the gear hobbing process. Notably, it is to observe that the first cut also has a significant impact. This phenomenon can possibly be attributed to the greater deviations of the tooth quality because of the first cut, which may be hard to compensate by subsequent processes.

Regarding profile grinding, the input features with the highest impact on final gear quality are "Mean Power of Axis C in Roughing Mode", "Feed Rate in Roughing Mode", "Mean Power of Axis Y in Roughing Mode", and "Mean Power of Axis C in Finishing Mode". In this context, the C-axis represents the spindle axis, and the Y-axis represents the feed axis of profile grinding. Noteworthy is that the roughing factors generally exhibit a more substantial influence than the finishing factors. This could be due to the experimental focus on roughing in a particular series, influencing the behavior of the neural network. Another finding is the observation that higher spindle power tends to result in improved quality, while the relationship between spindle power and feed rate shows a counterintuitive pattern. Commonly, increased spindle power means higher cutting forces and more aggressive material removal, causing more tool wear and rougher surfaces [22]. However, this unusual relationship between spindle power and gear quality is also evident in the raw data, which matches the trained behavior of the neural network. This can be related to the averaging technique used for the discontinuous grinding traces. A method to accurately represent and evaluate the signals generated by the discontinuous grinding process will enable a more precise analysis and understanding of the relationship between process parameters and gear quality.

In summary, the Shapley value analysis revealed that the first cut during gear hobbing leads to deviations in tooth quality that cannot be easily fixed with subsequent processes. Therefore, our focus should not only be the second cut. It is also crucial to find an appropriate technique to accurately represent data in discontinuous grinding processes that involve interruptions, in order to avoid misleading data.

5.2 Results of LCA

In this chapter, the results of the LCA and economic evaluation are presented and discussed. Figure 5 shows the LCA results for the manufacturing of the pinion shaft. For this purpose, the GWP is plotted over the different experimental groups.

The left plot shows the gate-to-gate approach, including only the value adding manufacturing processes which were carried out inhouse. The right plot shows the GWP from the upstream supply chain of the raw material. Due to the changes in the process parameters, a change in GWP is visible across the experimental groups. Between group 4 and 6 a reduction of the GWP by 6.6% (260 g CO_2-eq) were realized. If the scope is extended to include the raw material production the reduction of the GWP between group 4 and 6 is 2.7%.

Figure 6 shows the GWP of the pinion shaft manufacturing without the raw material plotted against manufacturing costs. The optimization direction is towards the lower left

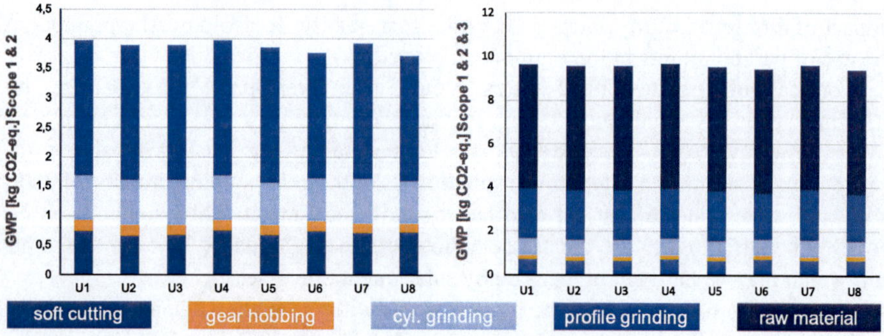

Fig. 5. LCA results for pinion shaft manufacturing

corner. In this experimental setup, the reduction of GWP correlates with the manufacturing costs. This is because process parameters like cutting speed shorten the manufacturing main time, leading to less energy consumption of the machine tool. The reduction in energy consumption in turn leads to a reduction in GWP. Although tool wear is considered by estimations of tool life, its influence on GWP is very low. As a result, manufacturing time and energy consumption are crucial in the ecological and economic analysis. Thus, increasing the feed rate is desirable as long as quality targets can be met.

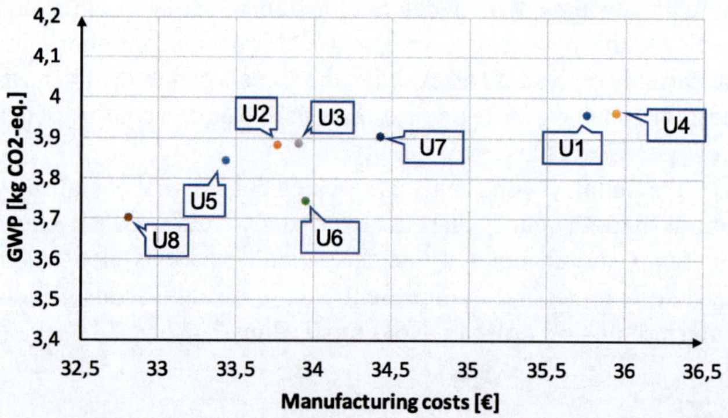

Fig. 6. GWP plotted against manufacturing costs in pinion shaft manufacturing.

6 Conclusion and Outlook

In this study, we employed a neural network to analyze the cross-process relationship between process parameters and gear quality and implemented an ecological assessment to evaluate the environmental impact on the manufacturing process chain of a pinion shaft. By utilizing this approach, we identified the key variables that significantly influence the gear manufacturing process chain with the potential of identifying hidden relationships between process parameter, trace data and quality measurements, and gained insights to reduce CO_2 emissions in manufacturing.

Our trial demonstrates a significant opportunity to reduce CO_2 emissions in manufacturing by selecting the optimal process parameters. Achieving a reduction of 200 g CO_2-equivalent per pinion shaft can have a massive impact, especially in large-scale production with many units. Based on the findings, several future directions can be pursued. Firstly, there is a need to develop a dedicated method for effectively analyzing signals from discontinuous processes, enabling a comprehensive understanding of their impact on workpiece quality. Secondly, the development of predictive models for assessing workpiece quality prior to specific processes can help to improve part quality and avoid scrap. Although the study showed that raw material has the largest impact on CO_2 emissions, improving part quality and therefore gear performance as well as avoiding scrap can have a significant contribution to CO_2 emission reduction.

References

1. Shi, J., Zhou, S.: Quality control and improvement for multistage systems: a survey. IIE Trans. **41**, 9 (2009)
2. Beckers, A., et al.: Digitalized manufacturing process sequences – foundations and analysis of the economic and ecological potential. CIRP J. Manuf. Sci. Technol. **39** (2022)
3. Heider, M.K.: Schwingungsverhalten von Zahnradgetrieben. TU München (2012)
4. Wittel, H., Spura, C., Jannasch, D.: Roloff/Matek Maschinenelemen-te (Normung, Berech nung, Gestaltung. Springer Fachmedien Wiesbaden GmbH (2021)
5. Nilsson, N.J.: Introduction to Machine Learning 2005 (1998)
6. Tercan, H., Meisen, T.: Machine learning and deep learning based predictive quality in manufacturing: a systematic review. J. Intell. Manuf. **33** (2022)
7. Lundberg, S., Lee, S.-I.: A unified approach to interpreting model predictions (2017)
8. DIN EN ISO 14040:2021-02, Umweltmanagement_- Ökobilanz_- Grundsätze und Rahmenbedingungen; Deutsche Fassung EN_ISO_14040:2006_+ A1:2020. Beuth, Berlin
9. DIN EN ISO 14044:2018-05, Umweltmanagement_- Ökobilanz_- Anforderungen und Anleitungen; Deutsche Fassung EN_ISO_14044:2006_+ A1:2018. Beuth, Berlin
10. Bergs, T., Grünebaum, T., Rey, J., Barth, S., Goldmanns, M.: A methodology for the ecological and economic assessment of manufacturing process sequences. Procedia CIRP **90** (2020)
11. Bergs, T., Grünebaum, T., Fricke, K., Barth, S., Ganser, P.: Life cycle assessment for milling of Ti- and Ni-based alloy aero engine components. Procedia CIRP **98** (2021)
12. Fricke, K., Gierlings, S., Ganser, P., Venek, T., Bergs, T.: Geometry model and approach for future Blisk LCA. IOP Conf. Ser.: Mater. Sci. Eng. **1024**(1), 12067 (2021)
13. Grünebaum, T., Hermann, L., Trauth, D., Bergs, T.: Towards sustainable production: a methodology to assess influences between life cycle phases in tool manufacturing. Procedia CIRP **80** (2019)
14. Grünert, G., Grünebaum, T., Beckers, A., Stauder, L., Barth, S., Bergs, T.: Methodology for the selection of manufacturing technology chains based on ecologic and economic performance indicators. J. Manuf. Syst. **66** (2023)
15. Müller, A.C., Guido, S.: Introduction to Machine Learning with Python. A Guide for Data Scientists. O'Reilly Media, Sebastopol (2017)
16. Picard, R.R., Berk, K.N.: Data splitting. Am. Stat. **44**(2), 140 (1990)
17. Peters, H.: Game Theory: A Multi-leveled Approach. Scholars Portal, Berlin (2008)
18. Change, I. P. o. C.: Climate Change 2021 – The Physical Science Basis. Cambridge University Press (2023)

19. Wernet, G., Bauer, C., Steubing, B., Reinhard, J., Moreno-Ruiz, E., Weidema, B.: The ecoinvent database version 3 (part I): overview and methodology. Int. J. Life Cycle Assess **21** (2016)
20. Ciroth, A.: ICT for environment in life cycle applications openLCA—a new open source software for life cycle assessment. Int. J. Life Cycle Assess. **12**, 4 (2007)
21. Beckers, A., Stauder, L., Grünebaum, T., Barth, S., Bergs, T.: Design of economically-optimized manufacturing process sequences using cross-process models. CIRP J. Manuf. Sci. Technol. **33** (2021)
22. Klocke, F.: Fertigungsverfahren. Zerspanung mit geometrisch unbes-timmter Schneide. VDI-Buch. Springer, Berlin (2018)

Modular Digital Production Twin as Enabler for Sustainable Value Creation – The Case of Urban Environments

M. Hertwig[1](✉), A. Werner[1], N. Zimmermann[1], and K. Hölzle[1,2]

[1] Fraunhofer IAO, Fraunhofer Institute for Industrial Engineering IAO, Nobelstrasse 12, 70569 Stuttgart, Germany
michael.hertwig@iao.fraunhofer.de

[2] Institute of Human Factors and Technology Management, University of Stuttgart IAT, Nobelstrasse 12, 70569 Stuttgart, Germany

Abstract. Climate change and progressing urbanization urge to increase sustainability in all areas – not only, but especially in manufacturing. Digitization and particularly digital twins enable a more efficient, decentralized, and small-scale production. This helps companies to reduce the usage of natural resources and the overall environmental impact. Corresponding modular digital twins represent one configuration and must be configured specifically according to individual production orders. The determination of required type, amount and linkage of integrated data and partial models of the production processes and plants is complex. However, this determination is required to choose suitable tools for the digital factory and to adapt processes accordingly.

The objective of this paper is to provide a concept for a modular digital production twin with integrated partial models as enabler for sustainable value creation. The concept will be concretized by means of requirements from an urban environment.

Keywords: Digital Twin · Sustainability · Urban Production

1 Introduction and Initial Situation

Companies have always been confronted with an increasing number and complexity of production-related requirements due to different production orders. However, the division of labor with many specialist departments, work packages and components does not lead to a reduction in complexity, but rather to a shift to the interfaces. In addition, the number and complexity of requirements is continuously increasing due to sustainability requirements from customers, investors, regulatory requirements and public legitimation pressure. Hence, value creation and primarily production has been undergoing continuous transformation, shaped by external dynamics and forces as well as new types of technological capability. Production can be grouped into paradigms [1] based on the respective concise driving forces. Since around 2020, sustainable production has been referred to as the current production paradigm [2]. In order to realize sustainable production, four core elements have been identified as levers [3]:

T. Bauernhansl et al. (Eds.): WGP 2023, LNPE, pp. 185–194, 2024.
https://doi.org/10.1007/978-3-031-47394-4_19

- Significant increase in resource efficiency.
  To ensure a decoupling of growth and resource consumption, innovation in technology and process must be exploited to raise efficiency potential [4].
- Implement circular material flows and supporting business models.
  The further development of material cycles towards reuse and recycling approaches must be strengthened. It is also important to develop approaches to dematerialization and to create the accompanying incentives to realize massive savings in greenhouse gases [5].
- Introduction of eco-design principles in all products and processes.
  Through the re-design of products, their footprint must be reduced or even neutralized to stop the degradation of the environment [6].
- Change the prevailing economic paradigm and fiscal frameworks.
  There needs to be a shift in focus from labor productivity to maximizing productivity gains through resource management. This requires a redesign of the tax system, as labor taxation is a major part of government revenue, with environmental taxation being almost non-existent [7].

Digitization and, in particular, digital twins act as enablers to achieve these core elements by, on the one hand, creating transparency within defined system boundaries [8] and, on the other hand, being able to exploit optimization potential even before actual production by generating model-based or simulative sustainability-related production information [9]. The use of digital twins and the simulative generation of sustainability-related production information offer companies the opportunity to optimize their production processes, use resources more efficiently and integrate sustainable practices into their production processes and thus to strengthen their competitiveness. However, completely new challenges arise for companies through the introduction of digital twins. On the one hand, the requirements for the digital twin must be integrated into the development process of the digital twin at an early stage [10]. On the other hand, digital twins represent an order-specific configuration [11] and must be structured differently according to requirements and therefore as modularly as possible. Consequently, the type and number of integrated data and partial models and their model complexity of the respective digital twin are difficult to determine.

Hence, the objective of this paper is to provide a concept for a modular digital production twin with integrated partial models as enabler for sustainable value creation. The concept will be concretized by means of requirements from an urban environment.

2 Modular Digital Production Twin

Modularization of a digital twin improves its flexibility, scalability, and reusability. It enables a more efficient and effective representation of the physical system and facilitates easier integration with other systems or models. The interconnectedness is ensured by designing modules, i.e., partial models and data sets, to communicate and interact with each other. This allows for a holistic representation of the entire system and facilitates the exchange of data and information between different modules. Scalability and flexibility are enabled, since the digital twin can easily be scaled up or down by adding or removing modules as needed. This provides flexibility to adapt the digital twin to different levels of detail or complexity and allows for an easy integration with other systems or models.

The approach of this paper includes the conception of a modular digital twin with a type and number of integrated data and partial models to be defined, which can be configured depending on the use case – with resulting requirements. Figure 1 visualizes the overall concept for the use of a modular digital production twin as enabler for sustainable value creation and will be described in the following subsections.

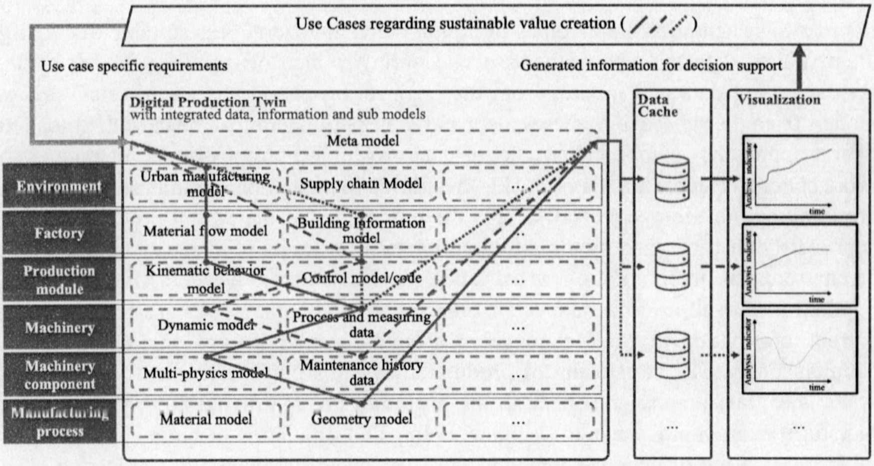

Fig. 1. Holistic concept for the use of the modular digital production twin

2.1 Linking Data and Partial Models

The modular digital twin represents a decentralized data model of the assets under consideration, which links partial models and data relevant to the production order. These include, for example, the material flow model, attributes, kinematic behavior model or process and measurement data. These data and partial models are often already available in (production-related) IT systems in companies, but their potential is not fully exploited. A superordinate metamodel and the requirements it manages are used to select the partial models and data required for the use case regarding sustainable value creation and to link them via a digital thread. This digital thread provides access to distributed data, aggregates relevant partial models, and thus enables the demand-based and context-sensitive provision of information for various actors along the value network. Applying this requirements-driven model configuration, the targeted sustainability-related output information can be generated on a model basis using suitable IT systems and by performing simulations. Prior to this, the IT systems engineering process must be planned and executed in a use case-specific manner. It must be specified exactly which input variables are required in the respective IT system and which output is to be generated. For the selection of suitable production-related IT systems, a methodical procedure including a multi-criteria system evaluation is recommended, considering the prevailing framework conditions in the respective company.

Full potential can be achieved if all data over the entire life cycle of all production and factory assets is collected, processed, and analyzed in the modular digital production twin in order to provide the resulting information in a context-sensitive manner according to defined user roles.

2.2 Systematization of Modules of the Digital Twin

In order to systematize the modules of digital twins in general, it is suitable to classify the partial models and data of the assets according to their origin along the life cycle. By combining data and models from the engineering phase with mostly data-driven models from the subsequent phases, data and information gaps can be closed. So-called hybrid approaches combine strengths of data-driven and physics-based techniques to avoid concept-specific limitations [12]. In addition to the temporal dimension, i.e., data origin along the life cycle of all assets, the modularity of the digital production twin is primarily characterized by the different hierarchical levels of a factory embedded in an environment. Integration of partial models and data across all hierarchical levels is required so that all user roles, specialist disciplines and departments are supplied with current information via the metamodel as a "single source of truth" and can update it simultaneously. The structuring of production systems into hierarchical levels is not a new idea, but already goes back to the beginnings of the digital factory. In various literature, an attempt is made to assign modeling and simulation techniques to the identified levels, each of which is intended to solve specific problems in production. The subdivision of a factory embedded in the environment into different hierarchical levels enables a systematization of the necessary models and associated with this, the respective modeling or simulation technique. The individual models serve as components of the digital twin and represent the partial models of the modular digital twin. Taking into account relevant literature on the subdivision of the factory into different hierarchy levels [13–17], the systematics shown in Fig. 2 was derived for this work in order to modularize the digital twin. Figure 2 presents a qualitative assessment for the hierarchy levels derived for this paper with respect to the degree of abstraction and granularity of the models. Furthermore, dominating models for the respective level are named and possible simulation techniques are described. In addition, exemplary sustainability-related key performance indicators (KPI) in the context of production in the broader sense are listed.

Hierarchy Level	Evaluation	Dominating model	Simulation technique	Description	Exemplary sustainability KPI
Environment		Socio-economic model	System Dynamics	▪ Modeling, simulating, analyzing, and designing dynamically complex issues ("dynamic complexity") ▪ Application mostly for socioeconomic systems	Water consumption
Factory	Degree of abstraction	Material flow model	Event Discrete simulation	▪ Simulation progress by processing a list of events ▪ Dynamic monitoring of the entire factory process, e.g. throughput, malfunctions, etc.	Material waste
Production module		Kinematic behavior model	3D motion simulation	▪ Pure observation of the motion sequence without mapping of the dynamic behavior ▪ Investigation of reachability, trajectories & collision	Energy consumption
Machinery		Dynamic model	Multi-body simulation (MBS)	▪ Basis: Kinematic behavior model ▪ Extension of the 3D motion simulation with dynamic behavior (frequencies, vibrations, natural oscillations, etc.)	Product lifetime
Machinery component	Granularity	Multi-physics model	Function simulation	▪ Dynamic behavior of individual components (hierarchy possible) ▪ Based on signal flow and object-oriented modeling ▪ Integration of virtual sensors possible (comparison to real sensor values)	Remaining Useful Life
Manufacturing process		Material model	Process simulation	▪ Particle formation simulation ▪ Material removal: geometric-physical simulation (FEM, CFD etc.)	Material fatigue

Fig. 2. Advanced hierarchy levels to systematize models regarding digital factory

3 The Case of Urban Environments

Urban environment can be understood as the close connection of living, working, leisure and traffic [18]. The consideration of these ecotones with an increased mix of uses holds great potential for the positive development of businesses and the surrounding area, as the coexistence of different types offers a wider range of solutions and opportunities. Urban and industrial structures have always been adapted to social needs and have therefore grown over time. The increasing separation of functions is currently presenting urban spaces with massive challenges, which are evident in traffic congestion, excessive emissions, and the segregation of social milieus [19]. The transformation and growth of existing structures lead to conflicts of use.

In these conflict-ridden spaces, propagating compatible production and life is a challenge for everyone involved [20]. However, the preservation of production work in urban areas is an approach for addressing challenges by solving or mitigating root causes [21]. Producing companies are important for cities, as they provide jobs and necessary income. Accordingly, it must be possible to reconcile the requirements of industry and the middle classes, or to create a balance of interests and new attractive offers.

Urban production can be divided into three clusters according to their characteristics [22]:

- Urban manufactory: Manufactories are small enterprises that produce goods with a high value proposition. These enterprises are rarely highly integrated but have a small specific organizational structure for rapid action. To manufacture the products, they often fall back on a far-reaching network in which they act as a hub for the specific order.
- Urban industry: In urban industry, automated processes based on the division of labor are adapted to the particularities of urban space by stacking the processes and accommodating different uses near each other.

- Urban agriculture: Urban agriculture describes the professional cultivation of land to produce food or horticultural crops in urban space or in the direct vicinity of housing and industry.

To realize industrial, sustainable production in an urban context, a change of mind is required, because the design of the factory, production processes and material flows must be compatible with the city [23]. This creates a complex system of interdependencies, whereby the components of the urban system must be interconnected, and symbioses must emerge [24]. To create added value for all stakeholders, the perspectives and self-image of companies and the environment must change in similar ways. The development of certain design approaches for urban production is intended to promote acceptance and increase compatibility in urban areas. [25].

3.1 Modeling of Urban Interdependencies

The interdependencies between a factory and the units of the urban environment are highly complex and sometimes generate repercussions. Due to this aspect, a model with a long-term perspective as well as an overarching systematic approach is required. Since changes in the urban ecosystem have slower cycles than in industry and technology, the modeling approach must also take this aspect into account.

In general, the modeling of the relevant relationships requires a concrete definition of the field of analysis [26]. In the context of production in an urban environment, the factory is considered as a black box that interacts with input and output flows and effects on elements of the environment [27]. The interactions are classified and characterized according to the three sustainability perspectives, always with reference to the production perspective. The information provided by the partial models of the production will be interlinked with the sustainability perspective of urban surroundings. Therefore, the model consists of four perspectives.

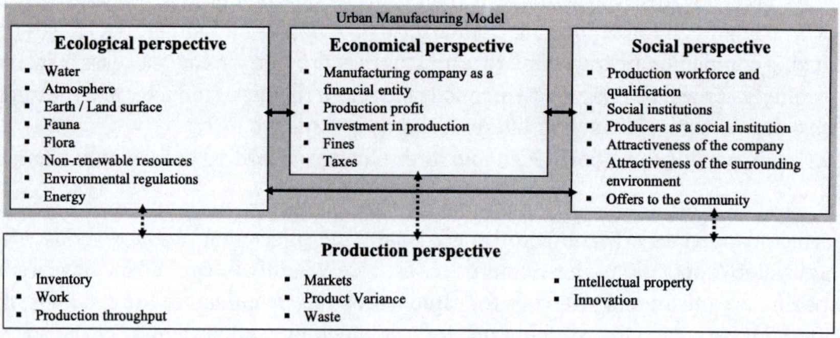

Fig. 3. Interdependencies of the urban manufacturing model

As a method for describing higher level relationships with corresponding feedbacks or reciprocal relationships, system dynamics has proven to be effective [28]. Causal loop diagrams can be used to describe complex interdependent structures [29]. Through a detailed analysis of the overall model, self-reinforcing relationships can thus be identified. Causal loop diagrams are the basis for the development of quantitative models, which can be used to perform numerical simulations [30].

3.2 Sustainability KPIs to Reflect the Current State of a Company

First approaches on KPI development can be found in research projects like Ultra-Efficiency Factory [31] and Positive Impact Factory [32]. They have in common that they develop a vision of the production of the future that is in harmony with the ecological and social environment. Some of the concepts developed are very detailed, which can help companies to start the transformation towards sustainable production. In many cases, however, the current situation of the companies is not sufficiently put into context. This is because the visions are not universal and different measures and characteristics are required for each company. In this context, maturity models can serve both as an overview of the current state and to determine a target state to strive for [33].

To precisely consider the current state of the business and to support the sustainable transformation from there, data-driven projections regarding future development are needed. With the modular digital production twin regarding the hierarchy level for environment, it is possible to consider potential future features and effects. This is because the measures and concepts described can be validated by means of the urban manufacturing model. The suitability and contribution of individual or associated measures to sustainability can be predicted for the future with the support of simulation. This provides decision support for the company management, that supports a targeted investment in transformation.

Since current company data is available from various partial models of the digital twin, the Urban Manufacturing Model can be adapted with up-to-date company-specific data and a prediction can be derived based on the current state. The company management selects various measures and uses them in a simulated scenario. In this way, the characteristics of sustainability KPI are propagated with the aim of determining the future characteristics of sustainable production.

The expression of selected analysis parameters depends on the specific scenario. Figure 4 shows an example of four indicators, each of which is compared with the base scenario in three different cases. By entering the time when the measures describing the scenario become effective (dotted line in the figure), it is possible to evaluate their impact on the selected metrics.

The visualization of the relevant sustainability indicators as well as characteristics regarding the relationships to the urban environment serve as a recommendation for management decisions. After all, to continue the change, management must record the selection of measures and release investment funds or initiate interaction with the relevant stakeholders to ensure that the measures are implemented.

Currently, the adaptation is done with static data that was manually collected in the company. The currently existing IT landscape stores most of the data, which only had to be exported and imported into the partial model for the simulation. Since not all processes and information are captured and stored on the IT side, the data that was not available and had to be captured manually. With the increasing use of the modular digital twin, all data can be extracted and used from existing systems in the future. This

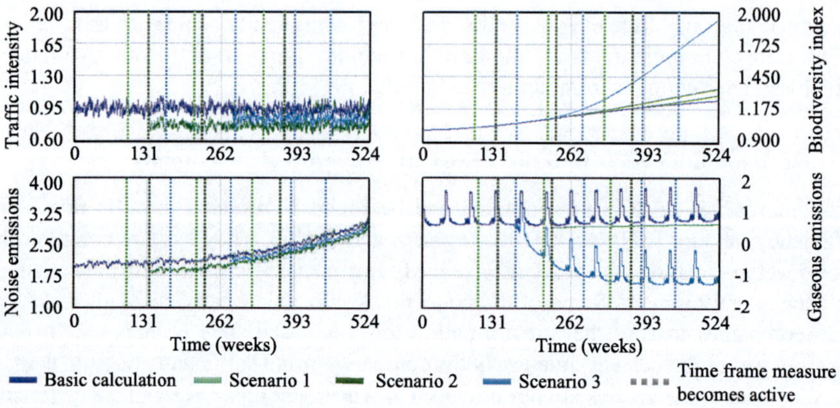

Fig. 4. Exemplary indicators (traffic intensity, noise emissions, biodiversity index, gaseous emissions) for a medium-sized company for three selected analysis scenarios

requires appropriate interfaces, which must be designed differently in each company, as the initial situation is individual for each organization. In addition, data identification is also a company-specific task, as each organization has significant differences in how it stores and holds its data.

4 Summary and Further Research Work

In summary, data-driven projections and simulation are needed to support sustainable business decisions and thus sustainable transformation. By utilizing a modular digital production twin including the urban manufacturing model as a partial model, companies can validate and predict the future impact of sustainability measures, providing decision support for management and facilitating targeted investment in transformation. The visualization of sustainability KPI and their relationship to the urban environment guides management decisions and enables the implementation of selected measures for continued change.

The modular structure of a digital twin increases flexibility, scalability, and reusability. The interconnected modules enable a holistic representation of a company and facilitate data exchange. Further research work must address integration and interoperability to facilitate seamless data exchange across hierarchical levels and organizational departments. Additionally, optimal module configuration methods should be explored to identify the most suitable combination of modules for specific use cases, considering sustainability requirements. Research should also emphasize lifecycle-wide data collection and analysis techniques, hybrid modeling approaches, the standardization of sustainability-related KPI, as well as validation and verification methods to enhance the accuracy and effectiveness of modular digital twins.

References

1. Boër, C. R., Dulio, S.: Mass Customization and Footwear: Myth, Salvation or Reality? A Comprehensive Analysis of the Adoption of the Mass Customization Paradigm in Footwear, from the Perspective of the EUROShoE (Extended User Oriented Shoe Enterprise). Springer, New York (2007). https://doi.org/10.1007/978-1-84628-865-4
2. Jovane, F., Koren, Y., Boër, C.R.: Present and future of flexible automation: towards new paradigms. CIRP Ann. **52**(2), 543–560 (2003)
3. Leuenberger, H., Mehdi, H.: Sustainable Production: can industry go truly green? Development **58**(4), 492–499 (2015)
4. von Weizsacker, E., Hargroves, K., Smith, M.H., Desh, C., Stasinopoulos, P.: Factor five. Transforming the Global Economy through 80% Improvements in Resource Productivity. Earthscan, London (2009)
5. Dittrich, M., Giljum, S., Lutter, S., Polzin, C.: Green economies around the world: implications of resource use for development and the environment. Sustainable Europe Research Institute (SERI), Vienna (2012)
6. European Union, Ecodesign Your Future: How Ecodesign can help the environment by making products smarter. European Union, Brussels (2012)
7. Groothuis, F.: New era new plan fiscal reforms for an inclusive circular economy. Austerlitz, The Netherlands (2014)
8. Miehe, R., Waltersmann, L., Sauer A., Bauernhansl, T.: Sustainable production and the role of digital twins Basic reflections and perspectives. J. Adv. Manuf. Process. (2021)
9. Barni, A., Fontana, A, Menato, S, Sorlini, M, Canetta, L., Exploiting the digital twin in the assessment and optimization of sustainability performances. In: International Conference on Intelligent Systems (IS), pp. 706–713 (2018)
10. Werner A., Lentes J., Zimmermann N.: Digitaler Zwilling zur vorausschauenden Instandhaltung in der Produktion - Physikbasierte Modellierung und Simulation zur Optimierung datengetriebener Modelle. In: Stuttgarter Symposium für Produktentwicklung, May 2021, pp. 139–150 (2021)
11. Eigner, M., Detzner, A., Schmidt, P.H., Tharma, R.: Definition des digital twin im produktlebenszyklus. ZWF Zeitschrift wirtschaftlichen Fabrikbetrieb **114**(6), 345–450 (2019)
12. Anger, C., Schrauder, R., Klingauf, U.: Unscented Kalman filter with gaussian process degradation model for bearing fault prognosis. In: Proceedings of First European Conference of the Prognostics and Health Management Society, Dresden, Germany, 3–5 July 2012, pp. 202–213 (2012)
13. Bracht, U., Geckler, D., Wenzel, S.: Digitale Fabrik - Methoden und Praxisbeispiele - Basis für Industrie 4.0. 2nd edn. Springer, Berlin (2018)
14. Dießen, S.: Systemkopplung zur komponentenorientierten simulation digitaler produkte. Zurich, ETH, Dissertation (2002)
15. Kiefer, J.: Mechatronikorientierte Planung automatisierter Fertigungszellen im Bereich Karosserierohbau. University of the Saarland, Dissertation, Saarbrücken (2007)
16. Westkämper, E., Niemann, J., Warschat, J., Scheer, AW., Thomas, O.: Methoden der digitalen Planung. In: Bullinger, H.J., Spath, D., Warnecke, HJ., Westkämper, E. (eds.) Handbuch Unternehmensorganisation. VDI-Buch, pp. 515–568. Springer, Berlin (2009). https://doi.org/10.1007/978-3-540-87595-6_8
17. Lee, J., Jin, C., Liu, Z.: Predictive big data analytics and cyber physical systems for TES systems. In: Redding, L., Roy, R., Shaw, A. (eds.) Advances in Through-life Engineering Services, pp. 97–112. Springer, Cham (2017). https://doi.org/10.1007/978-3-319-49938-3_7
18. Juraschek, M., Vossen, B., Hoffschröer, H., Reicher, C., Herrmann, C.: Urbane produktion: ökotone als analogie für eine nachhaltige wertschöpfung in städten. In: Redlich, T., Moritz, M.,

Wulfsberg, J. (eds.) Interdisziplinäre Perspektiven zur Zukunft der Wertschöpfung, pp. 195–207. Springer, Wiesbaden (2018). https://doi.org/10.1007/978-3-658-20265-1_15
19. Fricke, A., Siedentop, S., Zakrzewski, P., Reurbanisierung in baden-württembergischen Stadtregionen. In: Arbeitsberichte der ARL, vol. 14. Hannover (2015)
20. Förster, A., Schmitt, G.: New urban quarters, planning the city on a large scale again, pNd – rethinking planning, vol.1. RWTH Aachen University (2021)
21. Lange, B.: Offene Werkstätten und postwachstumsökonomien. Kollaborative orte als wegbereiter transformativer wirtschaftsentwicklungen? Zeitschrift Wirtschaftsgeographie **61**(1), 38–55 (2017)
22. Brandt, M., Gärtner, S., Meyer, K.: Urbane Produktion – ein Versuch einer Begriffsdefinition. In: Forschung Aktuell, vol. 08, pp. 1–14 (2017)
23. Spath, D., Lentes, J.: Urban Production to advance the competitiveness of Industrial Enterprises. In: 22nd International Conference on Production Research, Inguassu Falls, Brazil (2013)
24. Bauer, W.: Zukunftsfähige Produktion. Wirtschaftlich produzieren im städtischen Umfeld, Innovationswerkstatt "zukunftsfähige Produktion". Fraunhofer IAO, Stuttgart (2015)
25. Stiehm, S.: Gestaltungsparameter für die (re-) integration von produktion in den urbanen raum im kontext von industrie 4.0. IIF - Institut für Industriekommunikation und Fachmedien GmbH (2017)
26. Barlas, Y.: Formal aspects of model validity and validation in system dynamics. Syst. Dyn. Rev. **12**(3), 183–210 (1996)
27. Kibira, D., Jain, S., McLean, C.R.: A system dynamics modeling framework for sustainable manufacturing. In: 27th System Dynamics Conference (2009)
28. Abdolia, S., Kara, S., Hauschild, M.Z.: An investigation into environmental impact of system of systems. Procedia CIRP **80**, 584–589 (2017)
29. Haraldsson, H.V., Sverdrup, H.: Finding simplicity in complexity in biogeochemical modelling. In: Environmental Modelling: Finding Simplicity in Complexity, pp. 211–213. Wiley, New York (2003)
30. Binder, T., Vox, A., Belyazid, S., Haraldsson, H.V., Svensson, M.: Developing system dynamics models from causal loop diagrams. In: Proceedings of the 22nd International Conference of the System Dynamic Society, pp. 1–21 (2004)
31. Lentes, J., Mandel, J., Schließmann, U., Blach, R., Hertwig, M.: Competitive and sustainable manufacturing by means of ultra-efficient factories in urban surroundings. Int. J. Prod. Res. (2016)
32. Herrmann, C., Blume, S., Kurle, D., Schmidt, C., Thiede, S.: The positive impact factory-transition from eco-efficiency to eco–effectiveness strategies in manufacturing. Procedia CIRP **29**, 19–27 (2019)
33. Lentes, J., Hertwig, M., Zimmermann, N., Mahlau, L.-M.: Development path for industrial enterprises towards urban manufacturing. In: 24th International Conference on Production Research, Poznan, Poland, pp. 69–74 (2018)

A Quality Data Model Based on Asset Administration Shell Technology to Enable Autonomous Quality Control Loops

A. Bilen[1(✉)], F. Stamer[1], S. Behrendt[1], and G. Lanza[1,2]

[1] Karlsruhe Institute of Technology, Kaiserstraße 12, 76131 Karlsruhe, Germany
ali.bilen@kit.edu

[2] Global Advanced Manufacturing Institute (GAMI), KIT China Branch, Suzhou 215123, China

Abstract. In modern markets, increasing quality requirements necessitate high performance quality assurance processes to guarantee the fulfillment of these requirements sustainably. Thus, the quality assurance must be able to take targeted countermeasures which efficiently correct process deviations. In this context, quality control loops play a major role for quality monitoring and control. Currently, however, a great deal of manual effort goes into the design and implementation of mostly knowledge-based control logics. A heterogeneous data landscape and the resulting data preparation processes cause high effort. Autonomous quality control loops represent a new development and are intended to provide an efficient and data-based approach to setting up quality control loops. As an enabling step for autonomous quality control loops, this paper discusses the development of an Asset Administration Shell based, standardized quality data model.

Keywords: Autonomous Quality Control · Quality Data Model · Asset Administration Shell

1 Motivation

Today's manufacturing landscape presents numerous obstacles, including unpredictable demand patterns and growing demands for individualized products, which impact both the quantity and quality of units produced [1]. To remain competitive in this relentless market, businesses must adopt an agile approach to these shifting conditions, utilizing efficient, adaptable production systems that can accommodate undeveloped processes with speed and effectiveness. This adaptability can help reduce the time to reconfigure production systems [2, 3].

T. Bauernhansl et al. (Eds.): WGP 2023, LNPE, pp. 195–204, 2024.
https://doi.org/10.1007/978-3-031-47394-4_20

A comprehensive quality assurance system within an organization is essential for efficient ramp-ups, aiming for continuous and quick determination of manufacturing parameters maximizing product quality. To achieve this, automatic quality control loops (QCL) are used, which contain software-defined logics for regulating quality on the basis of quality and process data [4, 5]. Currently, with the recent developments in AI, there is a clear shift away from explicit, rule-based logics towards implicit, data-based models that represent the correlation between process parameters and the features of the manufacturing products. This surge concurrently demands an intensified focus on data preparation processes [6].

The absence of a unified data standard for quality and process data in production leads to laborious processes for data acquisition and analysis. Thus, a standardized data model is necessary - especially with the great advantage of being able to transfer programmed logics easily and to use synergy effects [8]. This paper examines existing approaches in literature for representing data in defined structures in the context of manufacturing. This paper builds upon previous work that has explored the fundamental approach for the development of data models in the context of ***Autonomous Quality Control Loops** (AQCL)* [7]. The development of a standardized data model that encapsulates quality and process data is a critical step in improving quality assurance mechanisms and enabling Autonomous Quality Control in contemporary manufacturing processes.

This research serves as a cornerstone for subsequent investigations and the development of a generic data model grounded in Asset Administration Shell (AAS) technology. It specifically focuses on measured quality data, while the standardization of process data storage will be investigated separately. This data model will serve as a basis for Autonomous Quality Control, aiming to empower manufacturing processes with greater precision, efficiency, and autonomy through more efficient deployment of quality control loops, direct bilateral adjustment of process parameters, and easy transfer of developed logic to other systems with minimal effort [7].

2 Fundamentals

2.1 Autonomous Quality Control Loops

Quality control loops are control systems that operate in a closed-loop fashion [9]. They utilize measured quality and process data to determine optimal process parameters for the specific process that generates the desired features. This optimization is achieved through the use of explicit (rule-based/heuristic) or implicit (data-based, learning) logic [9, 11]. In order to provide a consistent description of QCLs, an analogy is drawn to control loops in measurement and control technology.

The schematic structure of a generic control loop model is depicted in Fig. 1:

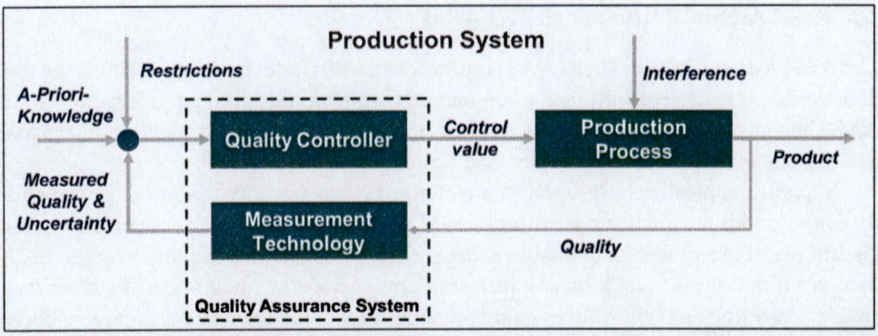

Fig. 1. Quality Control Loop (wbk – Institute for Production Science)

The center of the quality control loop is the quality controller. It contains implicit or explicit logic that describes the correlation between the control variables of the process and the quality result. With this correlation, the quality control is done by optimizing a quality function which can depend on various production parameters.

At the core of the quality control loop lies the quality controller, which incorporates logic defining the relationship between the process' control variables $P_1, P_2, \ldots, P_n$ and the resulting output of the process, i.e. the quality Q. This correlation enables the optimization of a quality function $Q(P_1, P_2, \ldots, P_n)$, which may depend on multiple parameters [9]. Equation (1) formally expresses this relationship:

$$P_{opt} = \begin{pmatrix} P_{1,max} \\ P_{2,max} \\ \ldots \\ P_{n,max} \end{pmatrix} = argmax(Q(P_1, P_2, \ldots, P_n)) \tag{1}$$

An increasing value for the function $Q(P_1, P_2, \ldots, P_n)$ goes hand in hand with a decreasing difference between target and actual quality.

The optimal combination of parameters for enhancing the quality outcome is represented by the vector P_{opt}. The design of the quality controller entails a significant amount of manual effort, depending on the complexity of the system being considered. Moreover, the input data used in most cases need to undergo pre-processing to ensure their suitability for use by the controller [12].

To address these challenges, the concept of AQCL comes into play. According to Wolfgang Wahlster and Frank Kirchner [13] autonomous systems act autonomously by learning and solving complex tasks, and can respond to unpredictable events. In the present context of QCL, autonomy is pursued through data-driven and preferably process-independent modeling for correlation representation, along with standardized provision of quality and process data. Thus, a unified description to comprehensively map quality data is needed.

2.2 Asset Administration Shell Technology

The Asset Administration Shell (AAS) is a key element in Industry 4.0 (I4.0). Serving as a digital twin, it represents physical components within an industrial environment. These AAS' encompass digitally all relevant information about their physical counterparts, detailing their properties, conditions, and functionalities [1, 14, 15].

A quality controller built upon this technology can significantly enhance the ability for autonomous quality control in several ways. Embedded with machine learning algorithms, these models can enable independent decisions making and execute quality control measures, like adjusting process parameters. The integration of sensor data enables detection and reporting of quality deviations, paving the way for rapid responses and enhancing overall process efficiency by preventing unnecessary waste or rework.

A primary function of the AAS is to offer a standard and efficient mechanism for gathering, organizing, and transmitting data within an I4.0 system. The AAS plays a pivotal role in the I4.0 landscape, and is regarded as the foundational element of interoperability within this context. Both the AAS' meta model and its interface are standardized, allowing for the homogenizing of all heterogeneous systems that exist in the industrial environment. The AAS provides a universal modeling language and a common consensus for exchanging information, facilitating interoperability among a production infrastructure composed of diverse technologies. The potential benefits of using AAS, particularly in terms of interoperability, scalability, and standardization potential across various production systems, will be explored in this scientific work [8, 15].

Components and Hierarchy of the AAS Meta Model

Subsequently, the components of the AAS are visualized with their hierarchical dependencies in Fig. 2:

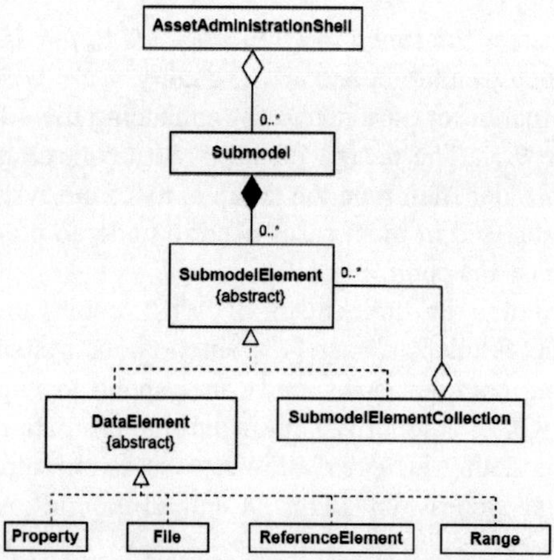

Fig. 2. Components and hierarchy of the AAS [15]

The structure of the AAS is defined by the Federal Ministry for Economic Affairs and Climate Action (BMWK) [15].

In general within each Submodel, Submodel Elements are used to represent individual data points or attributes. These can be properties, operations, or events related to the aspect of the asset that the Submodel represents. Submodel Elements can contain static information (like a machine's technical specifications) or dynamic data (like current temperature or pressure). They provide the fine-grained, detailed information that is needed to control and optimize production processes [15]. In this context, a quality data model, which is developed subsequently, can be perceived as an AAS in itself, comprised of several Submodels which are then filled with Submodel Elements.

3 Sate of the Art for Quality Data Models and Usage of AAS

Implementing universal data standards in today's technological environment is complex and challenging due to specific application requirements, the lack of universally accepted communication standards, economic considerations, business system requirements, and privacy concerns [8]. These complexities and use case specific requirements create considerable obstacles to the widespread adoption and implementation of such standards. In industry, there exist use case and technology-bound island solutions. This paper aims to develop an overarching standard for the orchestration of quality data. In the following, approaches will be presented that aim at standardized and interoperable data exchange, especially in the area of quality assurance.

3.1 Existing Data Models for Product Data

Quality assurance in manufacturing is a key challenge that has conditioned various approaches to model quality. Three prominent approaches are the Quality Information Framework (QIF) 3.0 [16] and ISO 10303 (STEP AP 242) [17].

White *et al.* [18] describes QIF 3.0 as well as STEP AP 242 in the context of model based product data definitions as following:

QIF 3.0

The Quality Information Framework (QIF) 3.0 is a Digital Metrology Standards Consortium (DMSC) standard that supports digital concepts in engineering applications. It is a collection of XML-schemas that ensure both data integrity and data interoperability in model-based enterprise implementations.

QIF 3.0 is particularly relevant to inspection process planning and focuses on the development of heuristic models, i.e. standards for representing rules used in this context. It provides support for a wide range of measurement device types and allows the specification of measurement resource rules. This allows OEMs, for example, to specify the type of coordinate measuring machine. QIF 3.0 also has enhanced features such as improved PMI (Procut Manufacturing Information) modeling and improved traceability in statistics that differentiate it from previous versions [18].

STEP AP 242

ISO 10303, known as "STEP" (Standard for the Exchange of Product model data), is an ISO standard for the computerized representation and exchange of PMIs. It is an ASCII-based format that can represent 3D objects in CAD and related information.

STEP is designed to describe product data independently of any specific system throughout the lifecycle. This makes it suitable for neutral file exchange, but also for implementing and sharing product databases and archives. STEP can be used in CAD, CAM, CAE, PDM/EDM and other CAx systems and handles product specific data such as mechanical and electrical design, geometric dimensioning and tolerancing as well as manufacturing and analysis information [18].

3.2 AAS Based Data Modelling in Production Environments

No specific approach to standardized mapping of quality data based on AAS exists in the research landscape. Accordingly, other use cases of the application of AAS in the manufacturing environments will be presented here in order to elaborate the key findings of these investigations.

In the world of digital twins and I4.0, there are various approaches to implement AAS. A number of organizations and research projects have focused on creating Submodel templates that facilitate communication and interoperability of systems by serving as domain standards with a specific modeling scope and goal.

The Industrial Digital Twin Association (IDTA) e.g. is an organization focused on the promotion and development of digital twins and asset administration shell technology. They provide Submodel templates to facilitate and standardize the implementation of administration shell technology. The restricted availability and lack of open-source access to Asset Administration Shell (AAS) technologies limit their applicability as a generic standard, as the requisite universality is undermined by paywalls and lack of public accessibility [19].

The "Interopera" research project has also created Submodel templates for various production applications. These include, for example, various use cases from laser processing or semiconductor production, for which various system-, processes- and product-related Submodels are specified by templates [20]. However, most of these templates are still in development and not publicly available.

Analysis of literature showed that there is currently no industrial standard for representing quality and process data designed for control purposes available. Such a model should automatically store data in a standardized manner, offering benefits such as enhanced efficiency, accuracy, real-time monitoring, analysis, interoperability, regulatory compliance, and quality reporting [8, 21].

4 A Quality Data Model Based on AAS-Technology

To close the identified gap, a novel data model has been designed to address the needs of quality control in manufacturing with the use of existing I4.0 technology to ensure interoperability and avoid proprietary solutions. Thus, this data model, as shown in Fig. 3 and Fig. 4, exploits the AAS framework and capitalizes on the hierarchical organization of Submodels and Submodel Element Collections (SMCs) to accurately represent the multifaceted nature of product information and quality data.

At the core of this structure is the "QualityData" Submodel. This Submodel serves as a dedicated repository for the entirety of data and information pertinent to product quality control. It is to be noted that the "QualityData" Submodel forms a component of a more comprehensive, currently not specified Product-Model, which contains further connections to other Submodels – e.g. production Resources, which are used to create the considered product.

An integral part of the "Quality Data" Submodel is the Submodel Element Collection "ProductionProcedures". This collection is designed to encompass the various production steps involved in the manufacturing process. Each production step is in turn linked to specific quality features generated during that particular step, thus establishing a clear correlation between the manufacturing process and the resulting quality features.

Nested within the "Production Procedures" collection is another Submodel Element Collection labeled "Features". This collection houses the quality features obtained during each production step by providing an organized structure for feature storage and retrieval.

Furthermore, each quality feature is extensively characterized through a set of descriptive properties and values. These include "Feature Type", which classifies the nature of the feature (e.g., geometric feature, surface feature), "Function", which describes the purpose served by the feature, and "Unit", which specifies the unit of measurement for the feature. Additionally, the expected value of the feature ("Target Value") is included, along with tolerance limits for process control including control limits and warning limits.

The "References" subcategory is also included to store necessary references for feature definition, such as planes, axes, etc. The data model also incorporates information

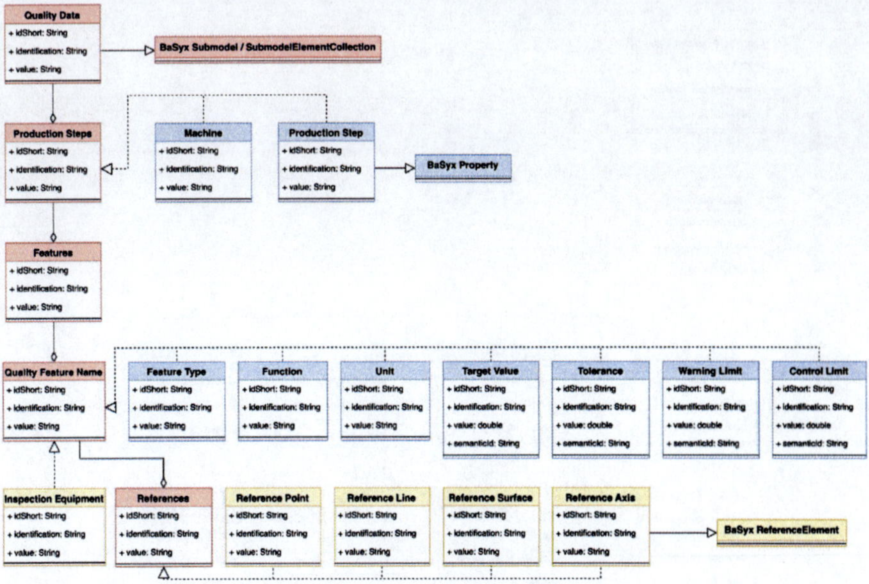

Fig. 3. Developed Quality Data Model (1)

about the inspection tool used, along with its specifications, enabling future tolerance analyses.

An important aspect of the model is the inclusion of a Submodel Element Collection for individual measurement results obtained from batch production samples. These samples, representative for the overall production quality, form a crucial part of statistical process control. This collection, named "Sample Data", houses data related to sample identification number, date, and the corresponding measurement values, among other things.

This data model provides a comprehensive and structured approach to quality data representation, aligning with the principles and technological developments of I4.0. By integrating all elements, it forms a robust foundation for the digital representation of quality control processes and data in the manufacturing sector. The modular modeling structure of the AAS framework is exploited to ensure future extensions of the quality model, providing backwards compatibility and ensuring long-lasting interoperability of quality data.

The foundation laid by this data model is a critical first step towards the development of a generic quality control and correlation analysis system in future research. The extensive standardization and characterization of quality data paves the way for automated detection of correlations and for the future construction of quality prediction models. This approach significantly enhances efficiency by reducing manual effort for data analysis and model creation, while fostering continuous learning from the data to improve prediction accuracy and process parameter optimization.

The model's practical application is achieved by hosting it on a server with orchestration capabilities. This server maintains high performance as data volumes increase.

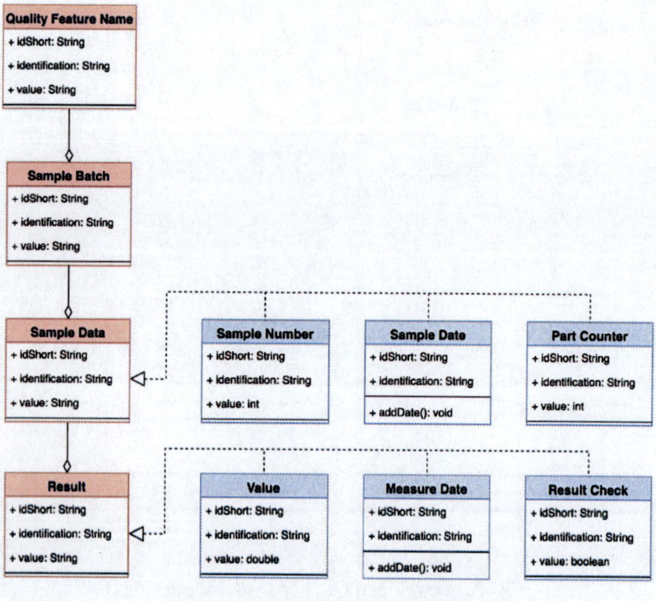

Fig. 4. Developed Quality Data Model (2)

Data exchange is facilitated through a REST API, allowing seamless interaction with the quality data and marking a step towards a more interconnected manufacturing sector.

5 Summary and Outlook

This work addresses the critical need for a standardized way to represent quality data in autonomous quality control loops. It introduces a novel data model, within the AAS framework, designed to enhance efficiency and interoperability in quality assurance processes. This model aids by serving as a basis for swiftly resolving production issues, providing a fundamental for significant boost to dynamic manufacturing systems. The data model was designed based on a manufacturing process view with generic consideration of all relevant process-, production- and resource-related information.

Future work involves broadening the model to include general process data, enabling automated correlation analysis between machine parameters and quality outcomes. The model's robustness will be tested and validated in a use case involving a correlation analysis tool, paving the way for automated quality control. Furthermore, the general applicability of the model will be examined by applying it to different domains and use cases. The ultimate vision is to enable holistic and efficient quality assurance services that contribute to the realization of I4.0.

References

1. Tantik, E., Anderl, R.: Integrated data model and structure for the asset administration shell in industrie 4.0. Procedia CIRP **60**, 86–91 (2017). https://doi.org/10.1016/j.procir.2017.01.048
2. Rösiö, C., Säfsten, K.: Reconfigurable production system design-theoretical and practical challenges. Jo. Manuf. Technol. Manage. (2013)
3. Beyerer, J., Bretthauer, G., Hofmann, C., Lanza, G.: Wandlungsfähige produktion für die kreislaufwirtschaft. Automatisierungstechnik **70**(6), 501–503 (2022). https://doi.org/10.1515/auto-2022-0063
4. Kim, H., Lin, Y., Tseng, T.-L.B.: A review on quality control in additive manufacturing. Rapid Prototyping J. (2018)
5. Gauder, D., Gölz, J., Jung, N., Lanza, G.: Development of an adaptive quality control loop in micro-production using machine learning, analytical gear simulation, and inline focus variation metrology for zero defect manufacturing. Comput. Ind. **144**, 103799 (2023)
6. Cioffi, R., Travaglioni, M., Piscitelli, G., Petrillo, A., de Felice, F.: Artificial intelligence and machine learning applications in smart production: progress, trends, and directions. Sustainability **12**(2), 492 (2020)
7. Bilen, A., Stamer, F., May, M., Lanza, G.: A development approach for a standardized quality data model using asset administration shell technology in the context of autonomous quality control loops for manufacturing processes. In: Euspen's 23rd International Conference & Exhibition, Copenhagen, DK, vol. 2023 (2023)
8. Inigo, M.A., Porto, A., Kremer, B., Perez, A., Larrinaga, F., Cuenca, J.: Towards an asset administration shell scenario: a use case for interoperability and standardization in industry 4.0. In: NOMS 2020–2020 IEEE/IFIP Network Operations and Management Symposium, pp. 1–6 (2020)
9. Lanza, G., et al.: In-line measurement technology and quality control. In: Gao, W. (eds.) Metrology. Precision Manufacturing, pp. 399–433. Springer, Singapore (2019). https://doi.org/10.1007/978-981-10-4938-5_14

10. Schick, M., Mubarak, H., Haueis, M.: Modellierung und Steigerung der Güte von Qualitätsregelkreisen. Zeitschrift wirtschaftlichen Fabrikbetrieb **104**(12), 1151–1157 (2009)
11. Rosenberger, M., Schellhorn, M., Linß, G.: New education strategy in quality measurement technique with image processing technologies-chances, applications and realisation. Acta IMEKO **2**(1), 56–60 (2013)
12. Teutsch, C., Schenk, M.: Qualitätsdatengesteuerte Produktion. Werkstatttechnik online, 11/12 (2017)
13. Wahlster, W., Kirchner, F.: Autonome Systeme: Technisch-wissenschaftliche Herausforderungen und Anwendungspotenziale (2015)
14. Ye, X., Hong, S.H.: Toward industry 4.0 components: insights into and implementation of asset administration shells. IEEE Ind. Electron. Mag. **13**(1), 13–25 (2019)
15. Federal Ministry for Economic Affairs and Climate Action (BMWK): Details of the Asset Administration Shell: Part 1 - the exchange of information between partners in the value chain of Industrie 4.0 (2022)
16. Digital metrology Standards Consortium (DMSC). QIF 3.0. qifstandards.org. Accessed 9 June 2023
17. International Organization for Standardization (ISO), ISO 10303-242:2020-04, Industrial automation systems and integration - Product data representation and exchange - Part 242: Application protocol: Managed model-based 3D engineering. https://www.beuth.de/de/norm/iso-10303-242/324323381
18. White, M., Holterman, E., Bakker, T., Maggiano, L.: Open standards for flexible discrete manufacturing in the model-based enterprise: US Department of Commerce, National Institute of Standards and Technology (2020)
19. Industrial Digital Twin Association. About IDTA. https://industrialdigitaltwin.org/. Accessed 19 June 2023
20. Standardization Council Industrie 4.0. InterOpera. https://interopera.de/. Accessed 19 June 2023
21. Behrendt, S., et al.: Software-Defined Manufacturing for the Entire Life Cycle at Different Levels of Production. In: Kiefl, N., Wulle, F., Ackermann, C., Holder, D. (eds.) SCAP 2022. ARENA2036, pp. 25–34. Springer, Cham (2023). https://doi.org/10.1007/978-3-031-27933-1_3

Systematic Investigation of Influences on the Repeatability of Vertically Articulated Robot Arms

Philip Gümbel(✉) and Klaus Dröder

Institute of Machine Tools and Production Technology, Technische Universität Braunschweig, 38106 Braunschweig, Germany
p.guembel@tu-braunschweig.de

Abstract. Vertically articulated robot arms (VAR) are highly flexible and cost-effective robots, representing the largest market share of industrial robots. Due to their kinematic structure, they are highly flexible but suffer from low repeatability. As a result, VARs are rarely used for precision assembly tasks, as expensive and complex compensation mechanisms are required. To enable these applications, a deeper understanding of the influences on their repeatability is required. Existing research largely analyzes the influencing factors in isolation, however, comprehensive systematic studies are lacking. Therefore, this work investigates a systematic analysis of the influences on the repeatability of VAR. To this end, a comprehensive experimental design was created and repeatability experiments were carried out using a KUKA-LWR4+. Influencing variables considered include, among others, the robot's pose, thermal effects, backlash and external moments. Finally, practical recommendations for the use of VAR in repeatability critical processes are given.

Keywords: Industrial Robot · Repeatability · Influences

1 Introduction

Industrial robots and in particular Vertically Articulated Robot Arms (VAR) are key enablers of an automated industrial production. Due to the kinematic structure of serially connected rotary joints, VARs are characterized by an excellent ratio of floor space to work area and high flexibility. As a result, VAR are used in particular for handling tasks that require a high degree of dexterity. For these technical reasons as well as the comparatively low purchase costs, VARs account for 67% of the industrial robot market [1]. However, due to their complex kinematic structure, VARs have comparatively low accuracy, stiffness and dynamics. In contrast to Cartesian systems such as machine tools or gantry robots, these properties are often not constant over the workspace, but strongly dependent on the robot's pose. In order to increase the accuracy, different strategies are used in industry and research to calibrate kinematic models, dynamic models or entire processes. What these methods have in common is that they compensate for systematic errors and thus increase absolute accuracy. Conventional industrial manufacturing processes, such as pick-and-place tasks, are typically programmed by teaching,

T. Bauernhansl et al. (Eds.): WGP 2023, LNPE, pp. 205–217, 2024.
https://doi.org/10.1007/978-3-031-47394-4_21

which effectively compensates for systematic errors. Repeatability (*RP*) is therefore the limiting factor for this class of processes. Due to their stochastic nature and limited observability, active compensation in the form of calibration to improve *RP* is not feasible. Possible approaches to increase *RP* are therefore mainly based on densifying the system against these types of errors. This densification requires a detailed understanding of the factors influencing *RP*. Existing scientific publications mostly examine possible influencing factors in isolation. This is where the present work comes in with a comprehensive experimental design for the systematic investigation of error influences on the *RP* of a VAR.

2 Related Work

The majority of existing publications in the field of improving positioning accuracy of VAR address the kinematic calibration of industrial robots. These approaches use coordinate measurement systems such as laser trackers [2–5], ballbars [6–8] or other measurement systems [9, 10] to approximate Denavit-Hartenberg (D-H) parameters [11] or parameters of extended kinematic models. They are thus able to reduce manufacturing inaccuracies of the kinematic chain, which improves systematic positioning accuracy and therefore the absolute accuracy.

Other research investigates the calibration of the compliance of the robot structure. Klimchik et al. [12] propose an approach for calibrating robotic manipulators by considering practical identifiability and the elastic properties of both links and joints. Their approach incorporates physical algebraic and statistical model reduction methods, leveraging sparseness in the stiffness matrix and heuristic selection of non-identifiable parameters. Zhang et al. [13] apply the Monte-Carlo method for considering uncertainties in D-H parameters. The Cartesian stiffness is then identified through an experiment using a laser tracker and a cable pulley system.

Others like Han et al. [14] apply methods for identifying a robots dynamic model. They integrate Weighted Least Squares (WLS) with linear matrix inequality constraints, nonlinear friction models and optimization of exciting trajectories considering measurement noise. Wu et al. [15] present a closed-loop identification procedure involving a static continuous friction model for joint friction compensation and utilizes Modified Fourier Series (MFS) as exciting trajectories.

In addition to the calibration of the VAR, various publications [16] address the calibration of complete processes / robot cells. Ge et al. present a procedure for electronics assembly applications robot cells, specifically targeting electronic assembly applications where trays and tools require frequent but slight changes. The method utilizes a displacement sensor, a ball, and three fixed intersected planes to establish the work object coordinate, and proposes a cyclic cross method for orthogonalization. Tool coordinate calibration is achieved by fitting the ball center and tool offset using the Levenberg-Marquardt fitting algorithm.

More recently, publications tend to apply machine learning methods such as artificial neural networks for process calibration also. Bobka et al. [17] explore the use of deep artificial neural networks, to learn and optimize process parameters of a pick-and-place process of thin sheets. Correlating large datasets of assembly errors and the pose of the picked components the authors were able to increase the accuracy significantly.

All approaches listed thus far are of calibrating nature, address systemic errors and have no influence on stochastic errors. Work addressing a reduction or characterization of stochastic influences is comparatively rare. Brethé et al. [18] model the *RP* stochastically by mapping errors resulting from joint deviations to Cartesian space using an experimentally determined encoder sensitivity. In [19] Brethé et al. apply this procedure to a SCARA robot and validate it experimentally. Further approaches using this methodology include the work of Brandstötter et al. [20] investigating the physical upper bound of pose *RP* as a function of the joint resolution, the work of Kluz et al. investigating the influence of the robots pose [21] on the *RP* and the influence of the end effectors (EEF) orientation during handling of cylindrical parts on the *RP* [22]. Furthermore in [23] the development and experimental demonstration of an approach based on the kinematic sensitivity and fundamental mechanical assumptions and without requirement of experimental parameter identification similar to [18] and [21] is presented.

Overall, existing publications tend to primarily focus on modeling the *RP* as a function of joint errors. Addressing more influences experimentally, Slamani et al. [24] investigate varying configurations, tool center point (TCP) speeds and payloads as well as their influence on backlash and *RP*. Modeling these using a polynomial function, the authors find a significant influence of TCP pose and speed and on the backlash, while payload seemed not to affect it.

The following chapters aim to extend these mostly isolated investigations by systematically analyzing a more vast number of potential influences on *RP* in a controlled setup using a vast amount of measurements.

3 Methods and Materials

The influences chosen for investigation in this work are the robot's pose, velocity, acceleration, EEF mass, temperature changes due to self-heating, external torques and externally coupled sinusoidal oscillations. The chosen robot is a KUKA LWR4+ due to the large amount of publically available data and scientific investigations [25–27]. In order to apply the listed influences to the robot, the experimental setup shown in Fig. 1 was used. All measurements utilize a capacitive 3-axis measuring system (Micro-Epsilon capaNCDT600 with S600-1 heads) with a measuring range of 1 mm with a theoretical dynamic resolution of 0.2 μm at 6 kHz (0.04 μm static). The capaNCDT600's analogue outputs (0–10 V) were digitized using a National Instruments NI6259 USB ADC. For the sensors' reference surfaces a ground reference cube was installed to the EEF. For additional tracking of the joint's positions and torques during the measurements the KUKA Fast Research Interface (FRI) [28] in conjunction with the Robot Operating System (ROS) [29] were used. For tracking robot temperatures, 10 kΩ NTC thermistors were installed in strategically chosen position on each robot link (see Sect. 4 for details). Unbalanced Wheatstone bridges and the NI 6259 system were used to digitize the temperatures in sync with the robots measurements.

To vary the EEF weight, sets of disks and spacers were attached between the reference cube and the TCP. Using these, the EEF mass could be increased by up to 4 kg in 1 kg steps without shifting the measuring cube relative to the TCP. For applying constant external torques to the robots first joint, exchangeable lever arms with 100 mm, 150 mm

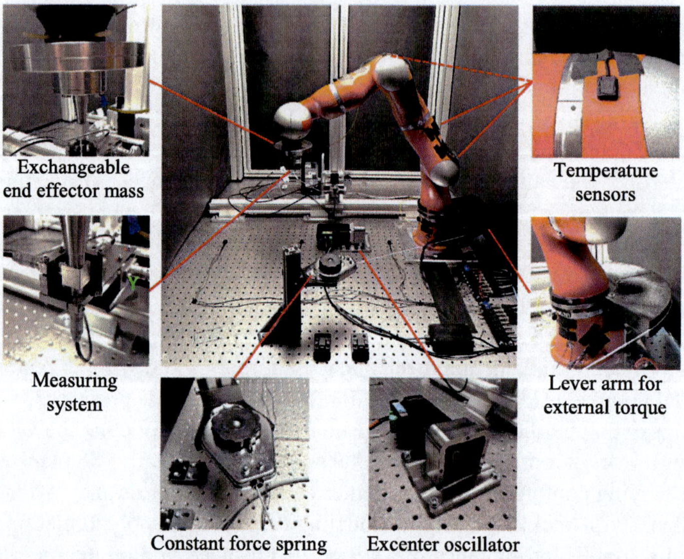

Fig. 1. Experimental setup based on KUKA LWR4+ robot measuring systems and devices for applying external torques and vibrations.

and 200 mm of effective leverage were fixed to the robots first movable link and loaded by a constant force spring with 20 N. To simulate externally coupled sinusoidal oscillations an excenter oscillator with an adjustable *RPM* from 1 and 2400 1/min (1 to 40 Hz) as well as interchangeable weights (with the same radius of the excentricity) of 57 g, 85.5 g and 114 g were utilized.

In order to enable a comparison, a reference trajectory with the measuring pose at $P(500\ \text{mm}, 0\ \text{mm}, 300\ \text{mm}, 0°, 0°, 0°)$ an EEF mass of 2 kg, a velocity of 0.11 m/s and an acceleration of 1.1 m/s^2 was chosen. In the following investigations the measuring pose was only varied in *X* direction (from now on TCP-Base distance). In each measuring cycle the robot moves along a path simulating a pick and place process by moving up linearly 100 mm in *Z*, then 300 mm in *Y* and again 100 mm down in *Z* before reversing the motion back to the starting pose (see Fig. 2). After reaching its initial pose it is maintained for 30 s to ensure complete settling of any residual control oscillation before moving on with the next cycle. The capacitive sensors were constantly sampled at 1000 Hz. For calculating *RP*, measurements in the span from second 25–29 after reaching the measuring pose were taken and averaged. For each *RP* measurement these cycles are repeated 100 times for statistical certainty.

Due to the large number of adjustable dimensions this experimental setup offers the design of experiments was split into factors that are investigated with regards to possible interactions as well as factors investigated in isolation. Due to their expected mechanical interaction the mass, pose, velocity and acceleration were chosen to be investigated together, while thermal effects, external torques and external vibrations were tested in isolation.

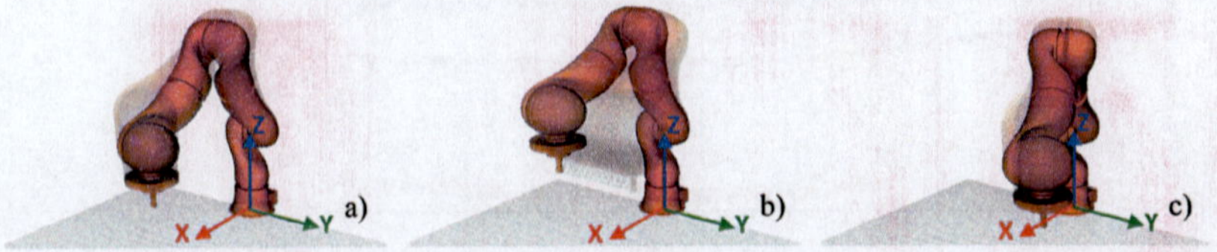

Fig. 2. Reference trajectory; a) 100 mm motion in z-direction, b) 300 mm in y-direction, c) – 100 mm in z-direction.

4 Self-heating as a Result of Waste Heat from Drives

As suggested by literature a robots temperature can cause significant thermal expansion and a resulting shift of the TCP. This thermal expansion is of systematic nature, and as such can be modeled and compensated for. If temperatures or variations thereof are not known they may be considered quasi-stochastic errors and contribute to *RP* of the robot. To determine the temperature's effect and to rule out its influence, a set of thermistors (see Fig. 1) was attached to the robot. The positions of these sensors were chosen at points, which exhibit the largest thermal gradient during heating up and cooling down. To identify these, thermal imaging (Fig. 3) was used during repeating cycles of the reference trajectory (see Fig. 2). The points of the largest thermal gradients were found to be near the joints motors. For each link, a temperature sensor was mounted with a thermal adhesive, thermal insulation and 3D printed covering caps near its base joint.

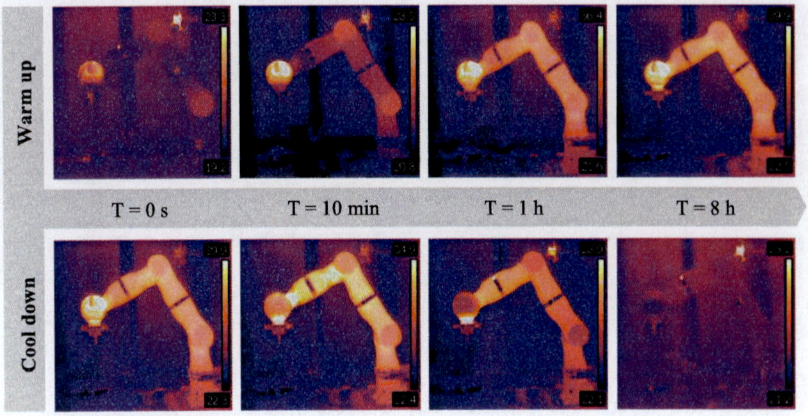

Fig. 3. Thermal images during warming and cooling down over 8 h.

For quantifying the thermal drift, the robot was started from a cold state and ran repeated cycles of the reference trajectory for an extended amount of time. Figure 4 shows the resulting shift of the TCP as well as the rolling repeatability RP_r over 100 measurements as well as the corresponding temperature changes of the links. RP_r converges after roughly three hours, even though further temperature changes of up to 2 degrees were measured until thermal convergence after around 7 h.

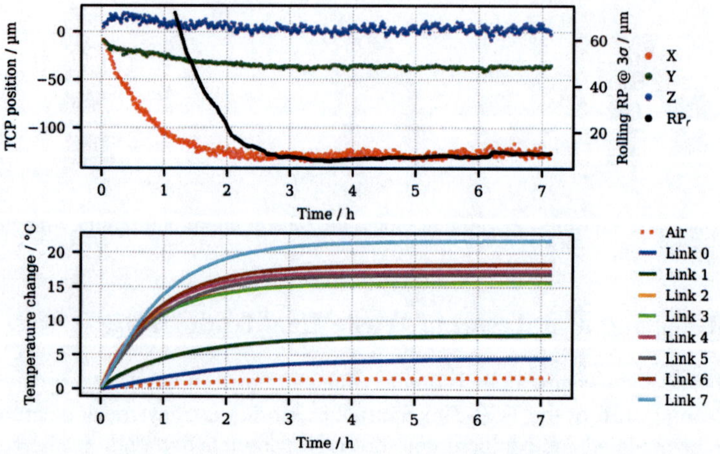

Fig. 4. Shift of TCP and rolling RP_r during warmup.

With the robot at temperature, the robot was then kept in the reference pose for an extended amount of time to simulate a larger idle time which could represent industrial idle times such as an omitted shift in series production.

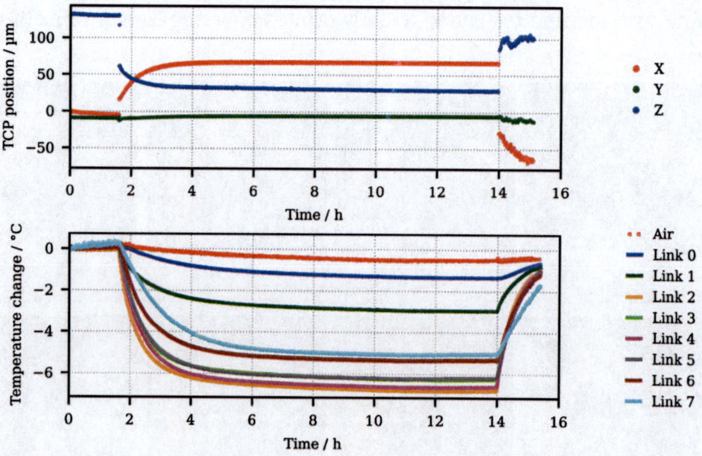

Fig. 5. Cooling due to engaged breaks and heating up during measurements.

As seen in Fig. 5, the temperatures stay mostly constant for roughly two hours before the robot's breaks automatically engage, taking load of the servomotors and decreasing the thermal flux. Furthermore, a significant initial sagging of the EEF while engaging the breaks is observable. Due to the symmetry of the chosen robot pose Y was largely unaffected. After approximately 12 h, the robot continues with a measurement cycle which results in significant heating up during the measurement and an accordingly worse RP of 28.18 μm.

5 Mass, Pose, Velocity and Acceleration

With the capability to quantify the self-heating's contribution to the RP and rule out its effect in further measurements (by use of warm up procedures), the design of experiments for the combined investigation of mass, pose, velocity and acceleration was carried out. The goal of this investigation is to identify the main effects and possible interactions between these dimensions without assuming them linear. To deal with the high dimensionality a response surface design, specifically the central composite design shown in Table 1 was chosen. In this design, the mass was varied from 0 to 4 kg of additional loading. The pose was varied by adjusting the TCP-Base distance from 300 mm to 700 mm. Upper limits for velocity and acceleration of 0.2 m/s and 2 m/s^2 were chosen in conjunction to ensure the robot can reach its target velocity over the 300 mm motion of the reference trajectory. The lower limits were chosen as a tenth of the upper limit (0.02 m/s and 0.2 m/s) to maintain a reasonable duration for 100 measurements of about 2:15 h in the worst case. While carrying out the experiments, all temperatures fluctuations stayed below 1 °C, therefore influences of self heating were minimized.

The measured RPs were analyzed by fitting a quadratic regression model resulting in Eq. (1) with the mass M, the TCP-Base Distance P, the velocity V and acceleration A:

$$\begin{aligned} RP = {} & 34,2 - 3,44\,M - 0,0749\,P + 20,0\,V - 9,85\,A + 0,022\,M * M \\ & + 0,000065\,P * P - 17\,V * V + 0,27\,A * A + 0,00733\,M * P - 7,90 * V \\ & + 0,959\,M * A - 0,0292\,P * V + 0,01376\,P * A + 11,4\,V * A \end{aligned} \tag{1}$$

While the resulting R^2 score of 79.00% suggests a good fit, a low predicted R^2 score of 1.22% and high prediction error sum of squares score (*PRESS*) of 134.8 suggest overfitting. The Pareto Chart of Standardized Effects (see Fig. 6) supports this interpretation by only showing three significant effects, namely the pose, squared pose and the mass-pose interaction, with the latter being on the very border of significance. The Main Effect Plot (see Fig. 6) further supports this with mass, velocity and accelerations showing a severely more flat curve compared to position.

The influence of TCP-Base distance can further be interpreted in conjunction with kinematic sensitivity – a reduction in TCP-Base distance results in a decrease of the Jacobians condition number. This can be understood geometrically as increasing the leverage of the joints (by decreasing the lever length). A pose optimization like suggested in [23] seems as such beneficial.

While the effects of mass, velocity and acceleration appear small, severe overshoots of the measuring pose were observable for high masses and accelerations. These effects are not investigated further in this work, as a result conservative parameters for acceleration, mass and velocity are suggested.

Table 1. Chosen central composite design.

Mass / kg	TCP-Base Distance / mm	Velocity / m s^{-1}	Acceleration / m s^{-2}	RP / μm
2	500	0.110	0.20	11.74
3	400	0.065	0.65	12.36
2	500	0.200	1.10	11.19
2	500	0.110	1.10	12.85
3	600	0.155	0.65	14.93
1	600	0.155	0.65	13.19
1	600	0.065	1.55	14.62
0	500	0.110	1.10	9.58
2	500	0.110	2.00	12.90
2	700	0.110	1.10	18.33
1	400	0.065	0.65	12.56
1	400	0.155	0.65	14.97
3	400	0.065	1.55	11.27
3	600	0.065	1.55	17.73
2	500	0.110	1.10	13.83
1	600	0.155	1.55	16.07
3	400	0.155	1.55	12.79
4	500	0.110	1.10	14.81
2	500	0.110	1.10	12.90
1	600	0.065	0.65	14.00
2	300	0.110	1.10	11.07
2	500	0.110	1.10	14.26
1	400	0.155	1.55	12.08
2	500	0.110	1.10	11.38
1	400	0.065	1.55	11.61
3	600	0.065	0.65	15.04
2	500	0.110	1.10	11.61
3	600	0.155	1.55	16.82
3	400	0.155	0.65	9.70
2	500	0.110	1.10	12.29
2	500	0.020	1.10	12.75

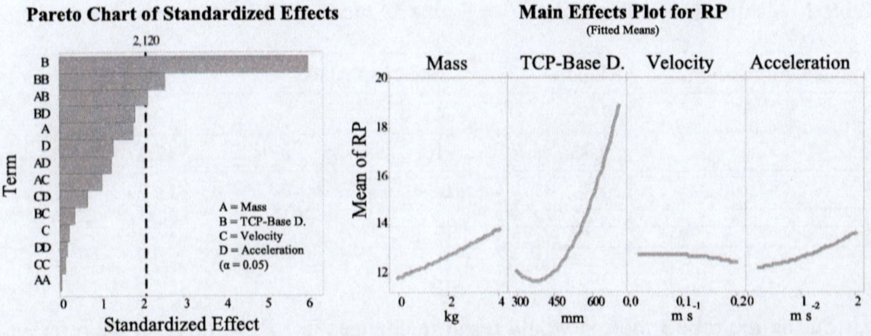

Fig. 6. Results of the Experiments: Standardized Effects and Main Effects.

6 External Torques and Torque Direction Changes

To investigate the effects of external toques on *RP*, the setup shown in Fig. 1 was utilized to preload the robots first axis with 2 Nm, 3 Nm and 4 Nm resulting in the torques along the reference trajectory as shown in Fig. 7.

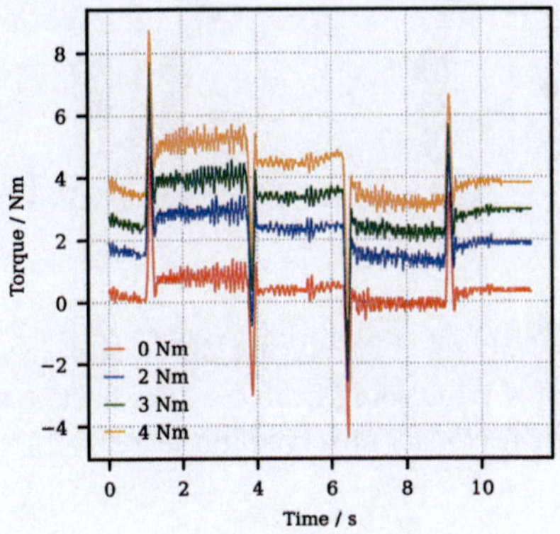

Fig. 7. Torques externally applied to axis 1 during the reference trajectory

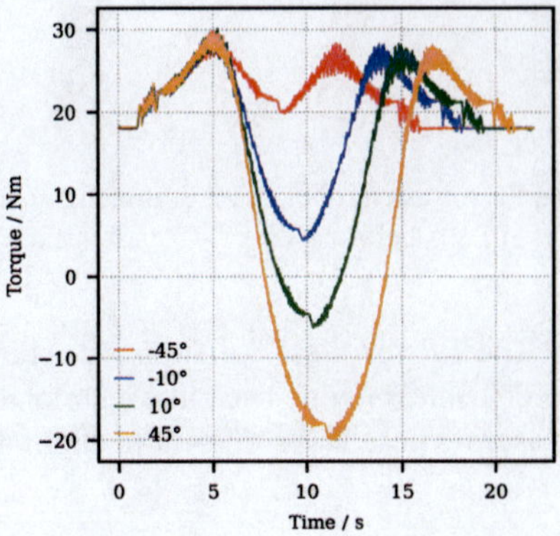

Fig. 8. Torques in axis 2 when approaching and passing the apex

For these experiments, the central point of the previously shown central composite design (2 kg, 500 mm, 0.11 m/s, 1.1 m/s^2) was chosen to provide reliable reference measurements. For all three cases *RP* (Table 2) was significantly lower when compared to the unloaded references. Statistical significance was verified using ANOVA.

The improvement in *RP* is likely a result of the preloaded axis usually being parallel with gravity and thus free from any torques resulting from the robots mass. As such preloading the axis may have reduced present backlash. The tested scenarios show no trend of further improvement by increasing the preload.

Considering the possible presence of backlash in the robots axis, another experiment was carried out to investigate whether changes in the direction of torques acting on the

Table 2. Measured RP – Torques/axis 0

Applied Torque/Nm	RP/μm
0	12.73
2	9.43
3	10.43
4	9.70

Table 3. Measured RP - Varied distance to apex

Distance to Apex/°	RP/μm
−45	12.22
−10	12.27
10	13.27
45	12.88

axis during the robots motion would result in changes in *RP*. To this end the reference trajectory was modified, so that the robot no longer moves 300 mm along *Y* but rotates joint 1 and 2 upwards, before moving down again for the measurement (see Fig. 9). The endpoint of this rotation was varied as such that the robot stopped at –45°, –10°, 10° and 45° relativ to the apex. With the link and EEF masses moving over the apex of this motion, the torques acting on axis 2 flipped as shown in Fig. 8.

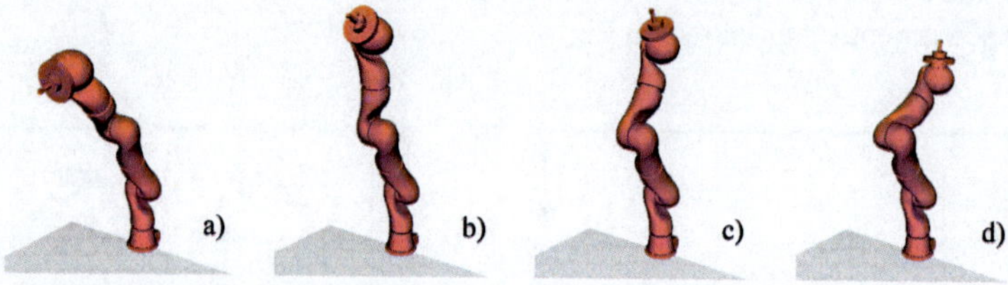

Fig. 9. Adjusted robot poses to investigate moving into and back out of backlash: a) 45° before apex, b) 10° before apex, c) 10° past apex, d) 45° past apex.

The *RP* for these variations are shown in Table 3. No relevant differences in *RP* were visible, moving back into a preloaded state by returning from the apex back to the measuring pose was sufficient to eliminate the previously observed influence.

7 Externally Coupled Sinusoidal Oscillations

Due to their serial structure VAR offer comparatively low stiffness when compared to other kinematic structures. As such VAR are typically more susceptible to vibration excitation from to external oscillations. To investigate the influence of such external oscillations, frequency sweeps from 1 to 40 Hz with the three varying oscillator masses of 57 g, 85.5 g and 114 g and at 400 mm, 500 mm and 600 mm of TCP-Base distance were carried out. During these sweeps, the frequencies were increased in 1 Hz steps and held for 10 s. Figure 10 shows an exemplary plot of the resulting response for X, Y and Z. Using the Euclidian vector norm, the overall frequency response for all experiments calculated (see Fig. 11).

Comparing the results, peaks at around 10 Hz (mainly resulting from oscillation in Y) and around 30 Hz resulting are observable. As expected increasing the oscillating mass

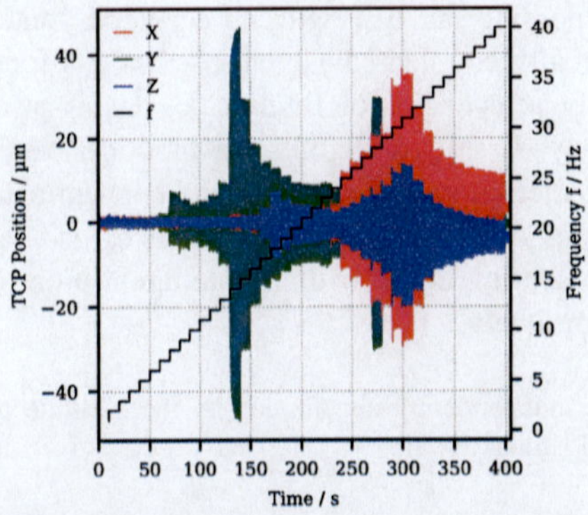

Fig. 10. Exemplary frequency sweep from 1 to 40 Hz for robot at 500 mm and with 57g oscillator

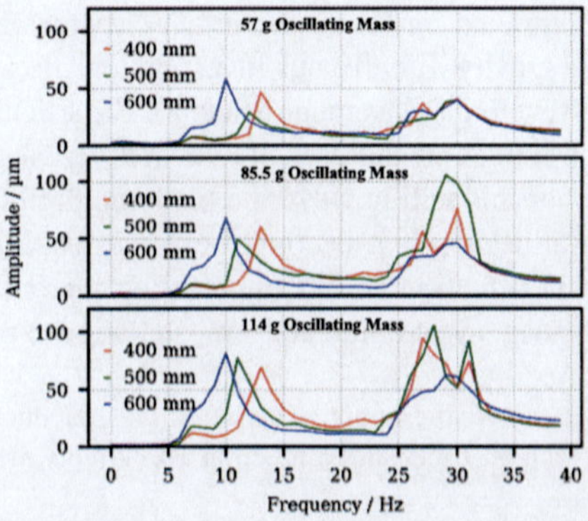

Fig. 11. Overall frequency responses for 57 g, 85.5 g and 114 g

increases the robots oscillation amplitude. Decreasing the TCP-base distance results in a upwards shift of the lower peak from 10 Hz (600 mm) to 13 Hz (400 mm), likely due to increased stiffness in this pose. To investigate the impact of the oscillation on the *RP* measurements, an oscillation using the lowest oscillator mass of 57 g and a frequency of 30 Hz was applied to the reference trajectory. The measured *RP* of 11.88 µm was unaffected. This is likely due to the averaging of the measured poses over 5 s, in which positive and negative oscillations cancel each other out. If the oscillation were to be present during a pick or place action, an additional error of up to the maximum oscillation amplitude may occur.

8 Summary and Recommendations

This work provides a comprehensive study of influences on the *RP* of a KUKA LWR4+. The investigation of thermal influences showed a significant shift of the TCP of up to 120 µm during warmup. Thermal fluctuations as a result of cycle times can mostly be ruled out for as long as breaks remain inactive. In practice, either a warmup procedure after each prolonged engagement of the breaks or monitoring of joint temperatures and thermal calibration is suggested. The surface response design for varying mass, pose, velocity and acceleration resulted in only pose showing a reliably significant influence. The position control algorithm utilized by the robots manufacturer handles varying dynamics well. As a practical suggestion we however encourage choosing these parameters conservatively, as we observed significant overshoots during the approach of the measuring pose. As we suggested in [23] an optimization of the robots pose for lowest sensitivity seems appropriate to increase *RP*. Our investigations further found a statistically significant improvement of *RP* by preloading otherwise unloaded axis with an external torque. In practical application, applying a constant torque can be achieved for the robots first axis with a simple setup as shown in this investigation. For other unloaded axis along the kinematic chain this may however require more mechanically complex setups. Alternative preloading these axis' could be achieved by choosing robot poses optimized based on the gravity vector and possibly an EEF with an eccentric center of gravity. Finally our investigation of externally coupled sinusoidal oscillations fond a severe displacement of the TCP, particularly at its excitation frequencies. Using pose optimization these excitation frequencies may be shifted slightly, the practical benefit of this will however be likely negligible. Alternatively we recommend conventional approaches that decouple external oscillations or possibly a combination of oscillation detection (for example using inertial measuring units) with an intelligent process management that reacts on oscillations appropriately.

Acknowledgements. This work was conducted independently and funded by the Institute of Machine Tools and Production Technology of TU Braunschweig.

References

1. Feldmann, K., Schöppner, V.: Handbuch Fügen, Handhaben und Montieren, vol. 5. Carl Hanser Verlag GmbH Co KG (2013)
2. Li, Z., Li, S., Luo, X.: An overview of calibration technology of industrial robots. IEEE/CAA J. Autom. Sin. **8**(1), 23–36 (2021)
3. Lee, J.-W., et al.: Industrial robot calibration method using denavit—Hatenberg parameters. In: 2017 17th International Conference on Control, Automation and Systems (ICCAS). IEEE (2017)
4. Nubiola, A., Bonev, I.A.: Absolute calibration of an ABB IRB 1600 robot using a laser tracker. Robot. Comput.-Integr. Manuf. **29**(1), 236–245 (2013)
5. Santolaria, J., Conte, J., Ginés, M.: Laser tracker-based kinematic parameter calibration of industrial robots by improved CPA method and active retroreflector. Int. J. Adv. Manuf. Technol. **66**, 2087–2106 (2013)
6. Wu, Z., et al.: Kinematic calibration for industrial robot using a telescoping ballbar. In: Chan, C.S., et al. (eds.) Intelligent Robotics and Applications. ICIRA 2020. LNCS, vol. 12595, pp. 285–295. Springer, Cham (2020). https://doi.org/10.1007/978-3-030-66645-3_24
7. Nubiola, A., Bonev, I.A.: Absolute robot calibration with a single telescoping ballbar. Precis. Eng. **38**(3), 472–480 (2014)
8. Yang, P., Guo, Z., Kong, Y.: Plane kinematic calibration method for industrial robot based on dynamic measurement of double ball bar. Precis. Eng. **62**, 265–272 (2020)
9. Nubiola, A., et al.: Comparison of two calibration methods for a small industrial robot based on an optical CMM and a laser tracker. Robotica **32**(3), 447–466 (2014)
10. Icli, C., Stepanenko, O., Bonev, I.: New method and portable measurement device for the calibration of industrial robots. Sensors **20**(20), 5919 (2020)
11. Denavit, J., Hartenberg, R.S.: A kinematic notation for lower-pair mechanisms based on matrices, pp. 215–221 (1955)
12. Klimchik, A., et al.: Identification of the manipulator stiffness model parameters in industrial environment. Mech. Mach. Theory **90**, 1–22 (2015)
13. Zhang, X., et al.: Stiffness identification for serial robot manipulator based on uncertainty approach. In: Jeschke, S., Liu, H., Schilberg, D. (eds.) Intelligent Robotics and Applications. ICIRA 2011. LNCS, vol. 7102, pp. 378–388. Springer, Berlin, Heidelberg (2011). https://doi.org/10.1007/978-3-642-25489-5_37
14. Han, Y., et al.: An iterative approach for accurate dynamic model identification of industrial robots. IEEE Trans. Robot. **36**(5), 1577–1594 (2020)

15. Wu, W., et al.: Closed-loop dynamic parameter identification of robot manipulators using modified Fourier series. Int. J. Adv. Robot. Syst. **9**(1), 29 (2012)
16. Ge, J., et al.: An automatic industrial robot cell calibration method. In: ISR/Robotik 2014; 41st International Symposium on Robotics. VDE (2014)
17. Bobka, P., et al.: Development of an automated assembly process supported with an artificial neural network. J. Mach. Eng. **18**, 28–41 (2018)
18. Brethé, J-F., Brayima, D.: A stochastic ellipsoid approach to repeatability modelisation of industrial manipulator robots. In: IEEE/RSJ International Conference on Intelligent Robots and Systems, vol. 2. IEEE (2002)
19. Brethé, J.-F., et al.: Modelling of repeatability phenomena using the stochastic ellipsoid approach. Robotica **24**(4), 477–490 (2006)
20. Brandstötter, M., Hofbaur, M.: The physical upper bound on pose repeatability of general serial manipulators based on joint resolution. Int. J. Mech. Robot. Syst. **3**(2–3), 193–210 (2016)
21. Kluz, R., Trzepieciński, T.: The repeatability positioning analysis of the industrial robot arm. Assem. Autom. **34**(3), 285–295 (2014)
22. Kluz, R., Trzepieciński, T.: Analysis of the optimal orientation of robot gripper for an improved capability assembly process. Robot. Auton. Syst. **74**, 253–266 (2015)
23. Gümbel, P., He, X., Dröder, K.: Precision optimized pose and trajectory planning for vertically articulated robot arms. Procedia CIRP **106**, 185–190 (2022)
24. Slamani, M., Nubiola, A., Bonev, I.A.: Modeling and assessment of the backlash error of an industrial robot. Robotica **30**(7), 1167–1175 (2012)
25. Gaz, C., Flacco, F., De Luca, A.: Identifying the dynamic model used by the KUKA LWR: a reverse engineering approach. In: 2014 IEEE International Conference on Robotics and Automation (ICRA). IEEE (2014)
26. Shareef, Z., Mohammadi, P., Steil, J.: Improving the inverse dynamics model of the KUKA LWR IV+ using independent joint learning. IFAC-PapersOnLine **49**(21), 507–512 (2016)
27. Bischoff, R., et al.: The KUKA-DLR lightweight robot arm-a new reference platform for robotics research and manufacturing. In: ISR 2010 (41st International Symposium on Robotics) and ROBOTIK 2010 (6th German Conference on Robotics). VDE (2010)
28. Schreiber, G., Stemmer, A., Bischoff, R.: The fast research interface for the kuka lightweight robot. In: IEEE Workshop on Innovative Robot Control Architectures for Demanding (Research) Applications How to Modify and Enhance Commercial Controllers (ICRA 2010). Citeseer (2010)
29. Quigley, M., et al.: ROS: an open-source robot operating system. In: ICRA Workshop on Open Source Software, vol. 3, no. 3.2 (2009)

Analysis of the Influence of the Exit Geometry and Nozzle Orientation on the Strand Quality When Depositing Curves and Layers in the FLM Process

Michael Krampe(✉) and Bernd Kuhlenkötter

Chair of Production Systems, Ruhr University Bochum, Universitätsstraße 150, 44801 Bochum, Germany
Krampe@lps.ruhr-uni-bochum.de

Abstract. The aim of this paper is to investigate the effects of different nozzle exit geometries in the FLM process on the deposition of curves and the interconnection of multiple lanes. To investigate the effects of the different nozzle exit geometries, curves were printed with different nozzles, as well as compounds of lanes next and on top of each other. Three unusual exit geometries, triangle, square and hexagon, were compared with the usual circular nozzle shape. Furthermore, the effects of a rotating nozzle were investigated when printing the curved lanes. The results show that the hexagonal exit geometry is the only exit geometry that produces an almost constant profile contour independent of orientation, which strongly resembles the profile contour produced with a circular exit geometry. Both, the triangular and the square exit geometry, exhibit strongly differing profile contours, due to the orientation during the curved deposition process. The evaluation of the multilayer specimens underlines the influence of the exit geometries on the contact areas, void sizes, and the distortion of the profiles.

Keywords: Additive manufacturing · fused layer modelling · nozzle exit geometry

1 Introduction

In the contemporary industrial sector, high-tech areas, and even in the environmental domain, additive manufacturing is gaining increasing importance [1, 2]. Fused Layer Modelling (FLM) is arguably the most well-known method in additive manufacturing, particularly suited for the production of prototypes and small batches due to its low financial barriers. In the FLM process, a thermoplastic wire is transported into a melting system using an extruder, heated until it reaches a pasty consistency, and subsequently extruded through a nozzle. The extruded material is deposited in predetermined paths on a build plate or previously created layer, where it melts the already deposited strands of material and forms a solid connection after cooling [3].

Standardly, the nozzles of an FLM system have a circular exit opening of constant size. If the component requires a high level of detail, a nozzle with a small diameter,

T. Bauernhansl et al. (Eds.): WGP 2023, LNPE, pp. 218–226, 2024.
https://doi.org/10.1007/978-3-031-47394-4_22

for example, between 0.1 mm and 0.4 mm, can be used, but this slows down production speed. Nozzles with a larger diameter, up to 2 mm or more, can reduce construction time but at the expense of possible detail accuracy. Hence, the user of the FLM system has the opportunity to decide whether the focus should be more on speed or detail by selecting the nozzle diameter. To counteract this conflict, there have long been approaches for nozzles where the nozzle opening can be variably adjusted, allowing distinction between fast or detailed printing within one construction process without a nozzle change. Some of these approaches come from research work, while others have been outlined directly in patent descriptions.

Switching between these pressure characteristics as variably as possible would contribute greatly to the sustainability of the FLM process. By fine-tuning the printing process, materials and energy costs can be saved without sacrificing quality.

Brooks et al. [4] present a concept for a two-stage nozzle that enables switching between two different nozzle diameters. However, for this work, approaches with continuously adjustable nozzles are of greater interest. In the Chinese patent by Ge et al. [5], a nozzle is presented that has a fine adjustment mechanism that allows the nozzle opening to be continuously adjusted. In this case, the nozzle opening consists of six components, resulting in a hexagonal opening area, and thus, the nozzle diameter is not circular. Other approaches also lead to cross-sectional areas of the nozzle diameter that are not circular. A patent by Desktop Metal [5] features a triangular geometry, while Xu et al. [6] have developed a concrete printing nozzle with a square opening geometry.

Although the FLM process typically uses a circular nozzle opening geometry, a mechanically continuously adjustable opening diameter seems to require a non-circular geometry. This is primarily due to the small form factor of the required mechanics. After initial work has already investigated the impacts of a non-circular nozzle opening [7], this paper now describes further investigations into the influence of nozzle shapes on curve printing and the connection between multiple deposited paths. For this, experimental studies with a robot-assisted FLM test stand are implemented. Semi-circular strands are created both with nozzle rotation, i.e., contour-following orientation change, and with a static orientation of the exit geometry. These are then examined and compared regarding their profile contour using a laser scanner and a microscope. In addition, straight strands are stacked on top of each other and laid flush next to each other. Using microscopy images, the influence of the exit geometries and orientations with regard to contact surface size, cavity sizes, and profile distortion due to overlap is investigated.

2 Related Work

This paper directly builds on the preliminary experiments done by the authors. Since there are no nozzles with non-circular exit geometries, some are manufactured for the experiments. Round, square, triangular, as well as hexagonal nozzles are produced. For each shape, three different sizes are made, in which the cross-sectional areas of the nozzle openings are equivalent to round nozzles with diameters of 0.6 mm, 0.8 mm, and 1.0 mm. The nozzles are manufactured with the aid of a milling cutter with a diameter of 0.3 mm, resulting in radii of 0.15 mm in the corners of the geometric shapes. Subsequently, paths are laid with these nozzles in a test setup. Table 1 shows the process parameters for using the 1.0 mm nozzles.

Table 1. Process parameters for the exit size 1 mm (straight lines)

Factor	Unit	Stages				
Geometry of the exit	–	Circle, Rectangle, Triangle, Hexagon				
Orientation of the nozzle	–	Corner, Edge				
Nozzle diameter (eq.)	mm	1				
Temperature heating bed	°C	60				
Temperature nozzle	°C	220				
Printing speed	mm/s	20				
Extruder speed	mm/min	44,5	64,86	83,96	101,8	118,37
Layer thickness	mm	0,25	0,375	0,5	0,625	0,75

The paths laid down are subsequently recorded with a laser line scanner, and the average profile progressions were calculated and depicted. Figure 1 shows the cross-sectional profiles for the 1.0 mm nozzles with round, triangular, and hexagonal exit geometries. The orientation of the nozzle during deposition is also examined by distinguishing, for example, whether the tip or the lower edge of the triangle is in the direction of movement in the case of the triangular exit opening. Since no undercuts could be detected with the laser line sensor, the strands are also cut with a scalpel and the cross-sections are examined microscopically. Figure 2 shows the microscopic images of the previously

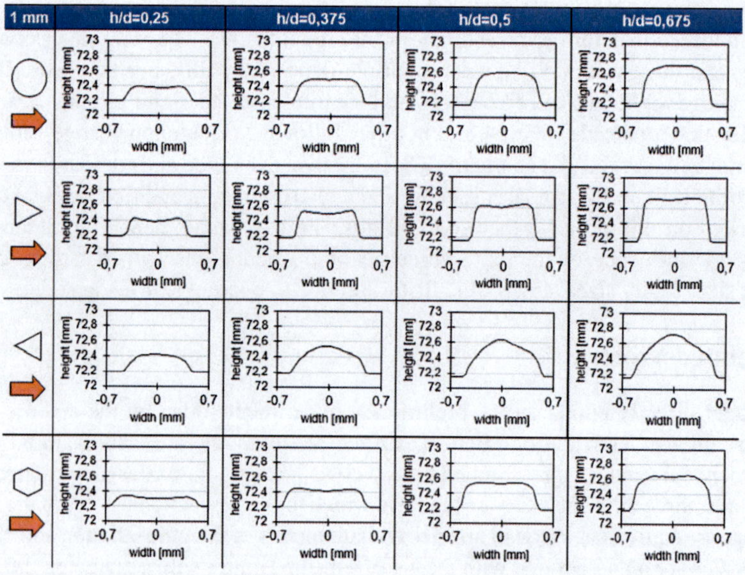

Fig. 1. The cross-sectional profils of the 1 mm nozzles

shown cross-sectional profiles. In summary, it can be seen that the nozzle exit geometries, and especially the orientation of the geometries to the process direction, have a strong influence on the appearance of the deposited strand.

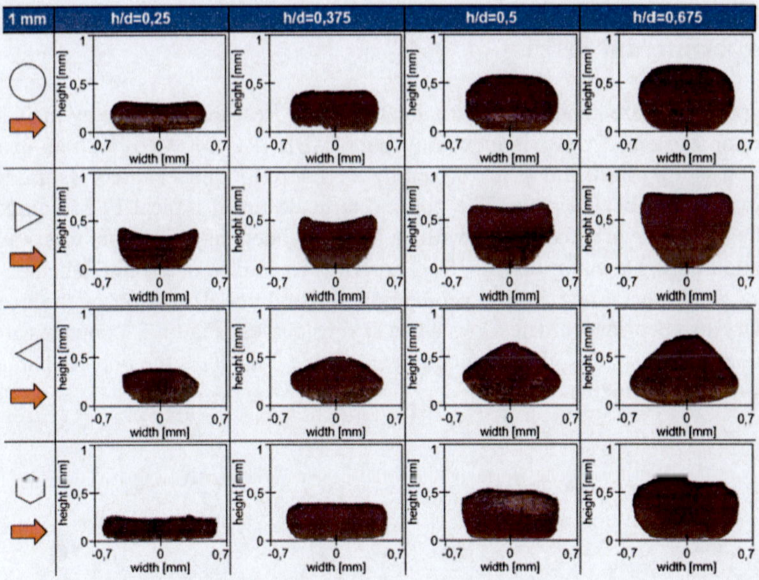

Fig. 2. The microscopy images of the cross-sections of the 1 mm nozzle

Apart from these experiments, there is little literature dealing with the influence of different nozzle exit geometries. However, the authors Papon et al. [8–10] have published three papers on this topic.

First, the authors investigate in a numerical simulation how a rectangular and star-shaped exit geometry affect the pressure loss and the distribution of the melt outflow velocity. Compared to the circular cross-section, the pressure drop for the star-shaped and rectangular opening is slightly increased. The speed distribution in the radial direction shows a drop from the centre to the outer edge for all shapes, with the star shape resulting in a more even drop [10].

Subsequently, Papon et al. compare the mechanical properties of FLM components produced with circular and rectangular exit geometries. The rectangular nozzle significantly reduces the air inclusions between the strands, leading to an increase in the mechanical strength of the component [8]. In a further publication, Papon et al. conduct a numerical simulation to investigate how different exit geometries influence the shape of the extruded strand. The predictions of the simulation are then compared with experimental results. The three exit geometries from the previous studies (circle, rectangle, and star) are examined, with the cross-sections being adjusted so that all shapes had the same diameter of 0.4 mm. Crucial here, however, is that the strands are not deposited by the nozzle, and thus they were not deformed between the nozzle and the build platform [9].

The preliminary experiments of the authors and the existing literature thus do not show the possible influence of the nozzle exit geometry on more complex applications such as printing curves or connecting strands. This research gap is now to be addressed in the following through experiments.

3 Experimental Setting

The experiments are conducted in the Learning and Research Factory by the Chair of Production Systems. A 6-axis industrial robot (ABB IRB 2400-16) with an extrusion head, consisting of a Bulldog XL driven by a NEMA 17 stepper motor, is used to lay down paths on a heated plate. The hot end is made up of typical FLM components [7]. The robot provides the nozzle with an additional degree of freedom to examine its orientation during a curve driving. The experiments aim to determine whether, when printing with special nozzle exit geometries, the additional rotation of the nozzle is necessary to keep inaccuracies low when laying curves. Figure 3 compares the two deposition strategies that are being compared. Table 2 displays the process parameters used for the experiments.

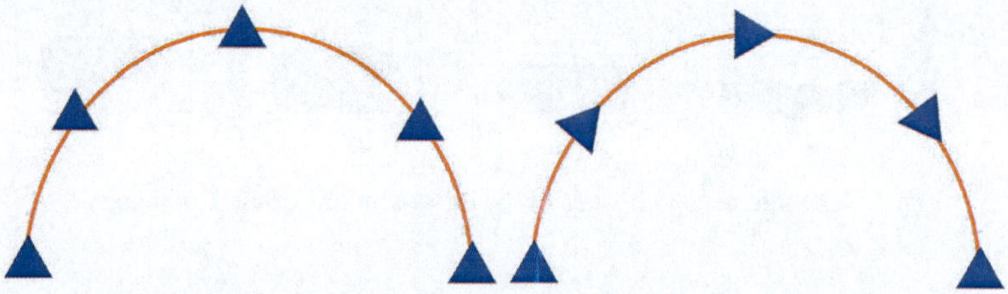

Fig. 3. Deposition strategies, left: without rotation, right: with rotation.

Table 2. Process parameters for the depositing of curved paths

Factor	Unit	Stages
Geometry of the exit	–	Circle, Rectangle, Triangle, Hexagon
Orientation of the nozzle	–	Corner, Edge
Rotation of the Nozzle	–	Yes, No
Nozzle diameter (eq.)	mm	1
Temperature heating bed	°C	60
Temperature nozzle	°C	220
Printing speed	mm/s	20
Extruder speed	mm/min	108,579
Layer thickness	mm	0,675

Additionally, it is intended to determine how multiple laid strands behave depending on the nozzle exit geometry. How do they connect, and what might the interstitial spaces look like.

4 Results

To investigate the influence of nozzle geometry on the deposition of curves, as well as the necessity of an additional rotational axis of the nozzle, several experiments are conducted with different nozzle exit geometries. Figure 4 exemplarily and simplistically shows the result with the triangular nozzle and the influence of the additional rotation. The upper experiment in the figure does not use additional rotation and it is clearly noticeable that the profile of the deposited strand changes several times along the curve as a result. In the lower picture, additional rotation is used, causing the tip of the triangle to continuously follow the path and thus the profile of the deposited path always remains the same. However, the triangle was already the geometry with the most extreme effects on the deposited profile in previous experiments.

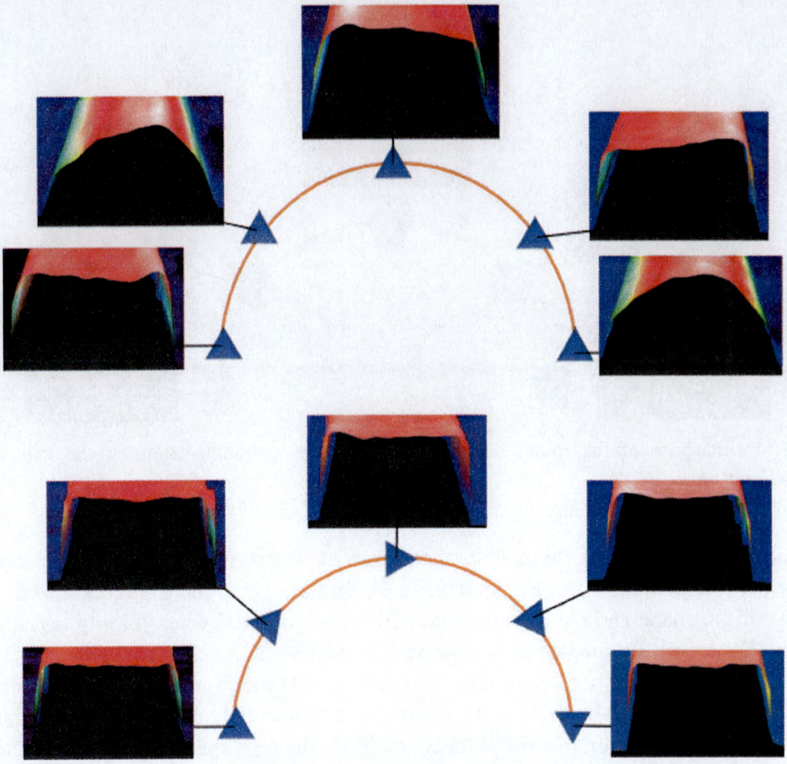

Fig. 4. Comparison of the triangular nozzle with (bottom) or without (top) rotation when depositing a curve.

The previous tests also show that the difference in the paths between the hexagonal nozzle and the round nozzle is much smaller. This is also evident in the deposition of the curved paths. While the round nozzle does not require a differently rotated additional axis due to its geometric shape, there is hardly any difference compared to the hexagonal nozzle when curves are deposited (see Fig. 5). This shows that when using a hexagonal nozzle, no additional rotatable axis will be necessary in the FLM process.

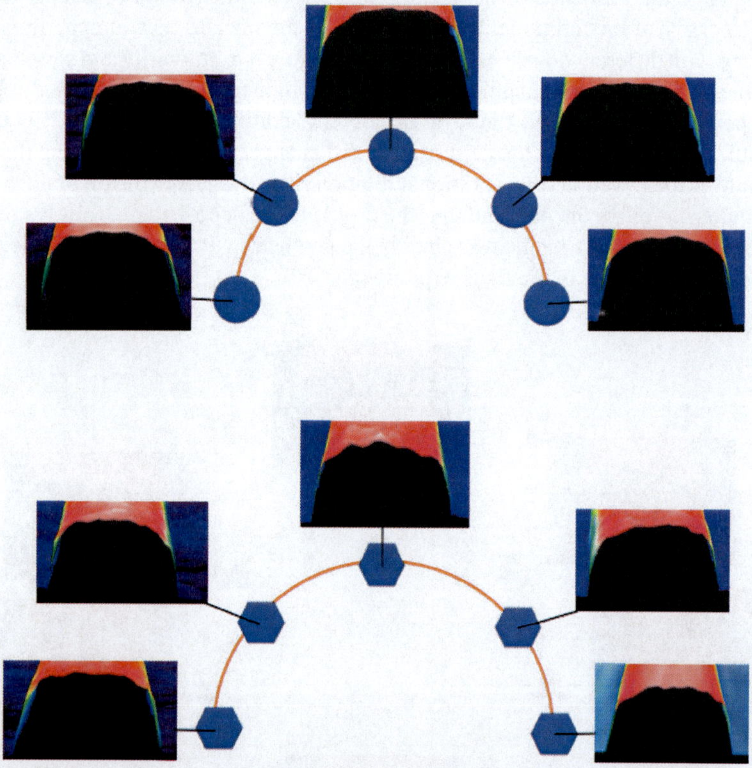

Fig. 5. Comparison of the round and hexagonal nozzle without rotation of the axis when depositing a curve.

Similar influences of the nozzle shapes are also shown when connecting several linear deposited strands (see Fig. 6 and Fig. 7). In the case of the triangular nozzle, the effects of the shape and the feed direction could be observed when joining two layers, but these were much smaller than expected. Even with a 2×2 matrix, the effects on the lower, more strongly pressed layer are very small, which means that the choice of feed direction seems to be irrelevant for this. Nevertheless, slight differences are visible. However, the summation of several inaccuracies in the tests may also play a role here.

When comparing the hexagonal and the round nozzle, only small differences in the shape of the deposited strands become visible again. The spaces created are also very similar, apart from a few inaccuracies.

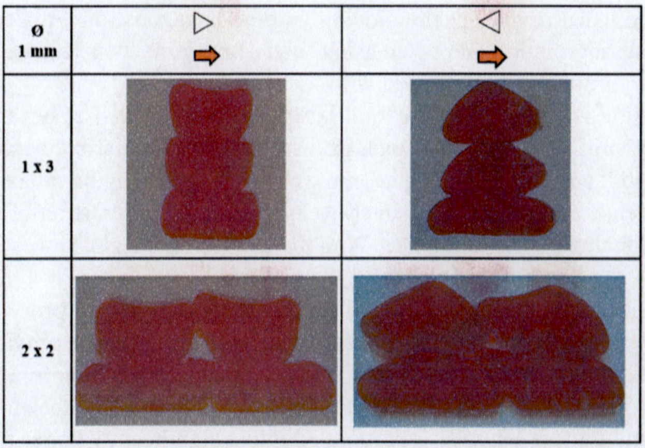

Fig. 6. Connection of several linear deposited strands with a triangular nozzle and different feed directions.

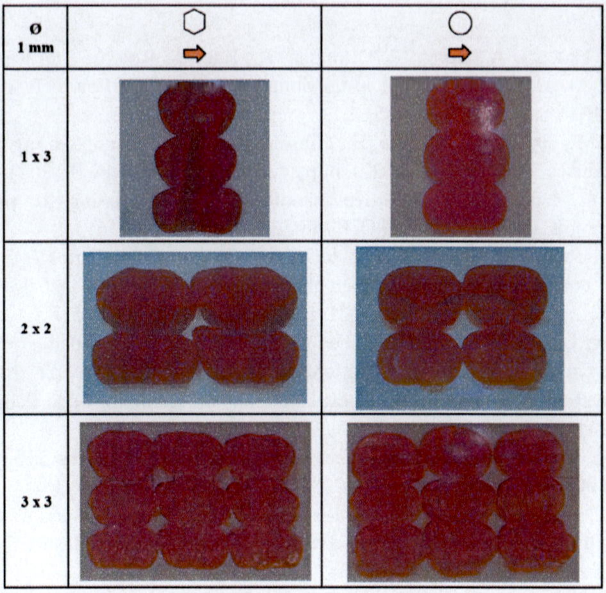

Fig. 7. Connection of several linear deposited strands with a hexagonal and a round nozzle

5 Conclusion

In summary, the results show that the nozzle shape not only influences the delivery of straight webs, but also plays a role in generating curves. The profile of the deposited curve can vary depending on the rotation of the nozzle. As with the straight paths, these effects are shown as a function of the number of corners of the geometry. Accordingly, the effects are more pronounced with triangular nozzles than with hexagonal ones. When comparing the usual round nozzle opening and the hexagonal one, differences are so small even without rotation that when using such a hexagonal nozzle, an additional axis of rotation can probably be dispensed with.

The situation is similar when several layers are connected. The hexagonal nozzle, which is very similar to the round one, also creates very similar connections between the strands and layers. On the opposite, the webs from the triangular nozzle partly have a strong influence on the already printed webs. Again, the orientation of the nozzle to the process direction has an influence. With the hexagonal nozzle, however, this can be neglected.

In terms of the previously mentioned sustainability of the FLM process, the results indicate that the use of a size-variable nozzle with a hexagonal shape provides an opportunity to make the process more flexible and sustainable. Furthermore, it appears that an additional axis could be omitted, which would make the new technology smaller and lighter, and thus, potentially incorporated into a significantly larger number of 3D printers in the future.

References

1. Javaid, M., Haleem, A., Singh, R.P., Suman, R., Rab, S.: Role of additive manufacturing applications towards environmental sustainability. Adv. Ind. Eng. Polym. Res. (2021). https://doi.org/10.1016/j.aiepr.2021.07.005
2. Khorasani, M., Ghasemi, A., Rolfe, B., Gibson, I.: Additive manufacturing a powerful tool for the aerospace industry. RPJ (2022). https://doi.org/10.1108/RPJ-01-2021-0009
3. Gebhardt, A., Hötter, J.-S.: Additive manufacturing. 3D printing for prototyping and manufacturing. Hanser, Cincinnati, OH, Munich, Cincinnati (2016)
4. Brooks, H., Rennie, A., Abram, T., McGovern, J., Caron, F.: Variable fused deposition modelling - concept design and tool path generation (2011)
5. Ge, M., Lu, J.: Nozzle bore adjustable 3D print Patent CN208774065U
6. Xu, J., Ding, L.: Volume - forming 3D concrete printing using a variable - diameter square nozzle. In: Creative Construction Conference 2018 - Proceedings. Creative Construction Conference 2018, pp. 104–113. Budapest University of Technology and Economics (2018). https://doi.org/10.3311/CCC2018-015
7. Krampe, M., Walter, F.: Experimental investigation of the influence of different nozzle exit geometries in fused layer modeling (2022)
8. Papon, E.A., Haque, A.: Tensile properties, void contents, dispersion and fracture behaviour of 3D printed carbon nanofiber reinforced composites. J. Reinforced Plastics Compos. (2018). https://doi.org/10.1177/0731684417750477
9. Papon, E.A., Haque, A., Sharif, M.A.R.: Numerical study for the improvement of bead spreading architecture with modified nozzle geometries in additive manufacturing of polymers. RPJ (2021). https://doi.org/10.1108/RPJ-05-2019-0142
10. Papon, M.E.A., Haque, A., Sharif, M.A.R.: Effect of nozzle geometry on melt flow simulation and structural property of thermoplastic nanocomposites in fused deposition modeling. In: American Society for Composites 2017. American Society for Composites 2017, 23 October 2017–25 October 2017. DEStech Publications, Inc., Lancaster, PA (2017 - 2017) (2017). https://doi.org/10.12783/asc2017/15339

Challenges in Establishing and Measuring Performance Criteria of Industrial Robots – Accuracy and Repeatability

M. Stucki[1,2](✉), L. Binnemann[1,2], N. Terei[1,2], R. Wiemann[1], and A. Raatz[1,2]

[1] Institute of Assembly Technology, Leibniz University Hannover, An der Universität 2, 30823 Garbsen, Germany
stucki@match.uni-hannover.de

[2] Cluster of Excellence PhoenixD (Photonics, Optics, and Engineering – Innovation Across Disciplines), Hannover, Germany

Abstract. Manufacturers of industrial robots rarely specify their products with reliable parameters for position accuracy and repeatability. Often the specifications only refer to individual axes or provide too little information about the distribution of the measured values. The ISO 9283:1998 "Manipulating industrial robots – Performance criteria and related test methods" provides instructions for the measurement of performance criteria, but is usually not applied. This paper introduces ISO 9283 with a focus on accuracy, repeatability and position stabilization properties. The standard is illustrated by measurements on a cartesian precision robot of the type Autoplace 411 (Unitechnologies SA). Based on the experience and the measurement, the authors discuss the advantages and disadvantages of the classification according to ISO 9283. Finally, we present proposals on how performance criteria can be defined more appropriately.

Keywords: Robotics · Precession Assembly · Accuracy

1 Introduction

Robots still play a major role in industrial automation. There are different types of robots with a wide range of properties and accuracy. Collaborative robots are used in interaction with humans and industrial robots for non-interactive automation such as in the automotive industry. Micromechanical applications often require high precision, so special SMT automats or precision robots perform these tasks. For process design, it is necessary to know the characteristics of these robots.

There were efforts to establish international standards to measure the characteristics of robots in the 1990s. The result are some ISO standards, but especially the standard ISO 9283 [1] brought specific performance criteria and related test methods. However, the industry did not use this standard for a long time [2]. In the commercial environment, there are widely differing specifications for industrial robots. Frequently, the manufacturer specifies the accuracy of the individual axes. Often there is no indication of the

T. Bauernhansl et al. (Eds.): WGP 2023, LNPE, pp. 227–237, 2024.
https://doi.org/10.1007/978-3-031-47394-4_23

variation to which the accuracy refers. Within the last decade, more and more manufacturers have started to reference some of the performance criteria. Nevertheless, these references in available data sheets are insufficient and lack specifications.

If precision is required, reliable accuracy and repeatability values are essential for the process development. A theoretical predictive handling accuracy can be calculated with normal distributed values. But the ISO 9283 performance criteria do not fulfil these requirements. Therefore, we discuss the ISO 9283 with a focus on accuracy and repeatability.

Firstly, we will introduce the standard and present measurements on a precision robot according to the ISO. Secondly, we will discuss the strengths and weaknesses of the standard. Finally, we will present proposals for how performance criteria can be defined more appropriately for process design.

1.1 Relevant Standards for Industrial Robots

There are several standards for the description of industrial robots. The most relevant for the description of the characteristics are:

- ISO 8373 – Vocabulary [3]
- ISO 9283 – Performance criteria and related test methods [1]
- ISO 9787 – Coordinate Systems [4]
- ISO 9946 – Presentation of characteristics [5]

There are also standards for safety requirements (ISO 10218 and ISO 15066) and electrical tests (ISO 11062). In addition, DIRECTIVE 2006/42/EC applies within the EU.

2 Performance Criteria According to ISO 9283

The ISO 9283 standard provides detailed instructions for the measurement of accuracy parameters. It refers to accuracy and dynamic behavior. In the following, the definition and measurement methods for pose parameters are explained, whereby simplifications are made in the reproduction of the ISO standard to increase understanding and to keep the scope of the paper.

According to ISO, measurements for the performance criteria are performed with 100% of the maximum load and at 100% of the speed. The measurements are done at 5 positions within a cube that is located in the working space with the "greatest anticipated use" [1]. The cube should have a maximum volume and must be oriented perpendicular to the base coordinate system.

Figure 1 shows the location of the measurement points within the cube. Measurements of pose parameters are usually done with 30 repetitions, dynamic measurements with 3 repetitions and path parameters with 10 repetitions (compare ISO 9283 Table 5 for details [1]).

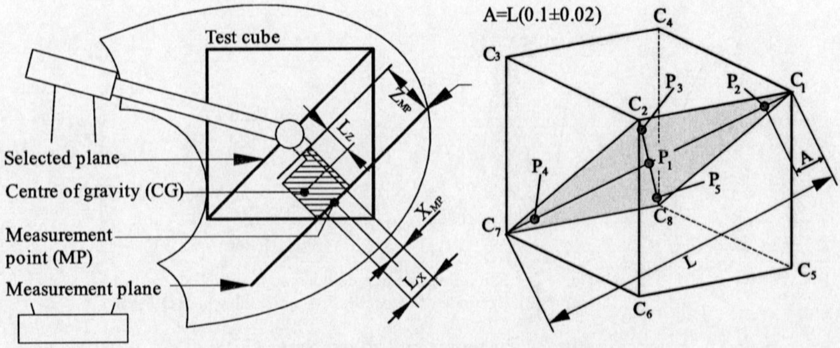

Fig. 1. Workspace with cube and planes (based on [1])

2.1 Pose Accuracy (Unidirectional) (AP)

ISO 9283 defines pose accuracy as the "deviation between a command pose and the mean of the attained poses when approaching the command pose from the same direction" [1]. It is divided into positioning accuracy (AP_P) and orientation accuracy ($AP_{a/b/c}$). The pose accuracy is especially important in fully automated processes when the pose is set by simulation or computer-aided programming. This is common in SMT manufacturing because poses are obtained from the design of the printed circuit board. Generally, the pose accuracy can be improved by calibration or mapping. With a taught pose, this pose accuracy is irrelevant.

2.2 Pose Repeatability (Unidirectional/Multidirectional) (RP)

"Pose repeatability expresses the closeness of agreement between the attained poses after n [n = 30] repeat visits to the same command pose in the same direction" [1]. The positioning repeatability RP_l is calculated as the sum of the mean deviation $\bar{l}$ and three times the standard deviation S_l of l (1).

RP indicates the general inaccuracy when approaching a pose. The deviation is caused by stochastic error. Reducing this misalignment is difficult to achieve because it usually requires the modification of mechanical components.

$$RP_l = \bar{l} + 3S_l \tag{1}$$

2.3 Position Stabilization Time and Position Overshoot

At high speed and high load, oscillations can occur when approaching a pose. The time needed to keep a pose stable is called position stabilization time. In dependence of this is the position overshoot, which indicates the maximum distance by which the attained pose is missed. The limit band for these measurements is defined as the pose repeatability RP_l or may be defined by the manufacturer. Figure 2 illustrates stabilization time and overshoot in a qualitative diagram.

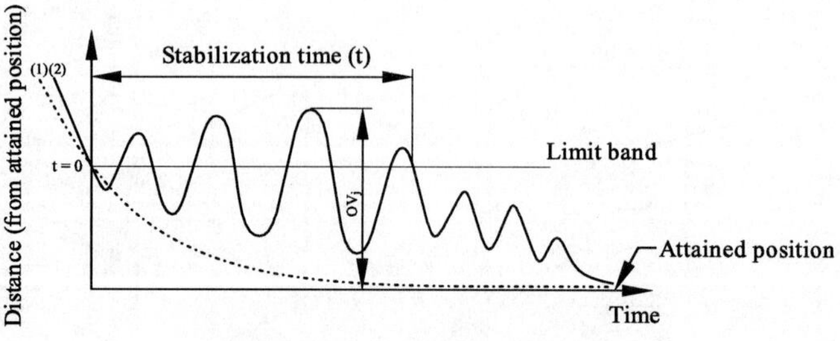

Curve (1): Example of an overdamped approach
Curve (2): Example of an oscillation approach where OV exists

Fig. 2. Position stabilization time [1]

3 Example Measurements on a Precision Robot

To illustrate the measurement of position accuracy, position repeatability, stabilization time and position overshoot, we present the results of a cartesian precision robot of the type Autoplace 411 (Unitechnologies SA). This robot has 4 axes (x, y, z, and c (= θz)). Therefore only the position repeatability was tested. This measurement was part of the site acceptance test (SAT) to verify the specification limit of $RP_l \leq 1.5\ \mu\text{m}$.

3.1 Metrology

The measurements were performed with a distance measuring system capaNCDT600 from MICRO-EPSILON with three capacitive sensors of type S-600-0.2 arranged at a right angle. The system has a static resolution of 0.008 μm and covers a measuring range of 200 μm. The small measuring range only allows the repeatability measurement, as the sensors must be repositioned in the working space for each measuring point. Each point was reached from a waypoint with a 50 mm distance in each axis. P1 should be in the center of the diagonal plane (see Fig. 1). The robot was already implemented and an assembly station blocked the center of the working space. Therefore the P1 position is approx. 20 mm off-centered in the z-axis to the pre-defined ISO 9283 position. Figure 3 shows the schematic drawing and a picture of the sensor heads.

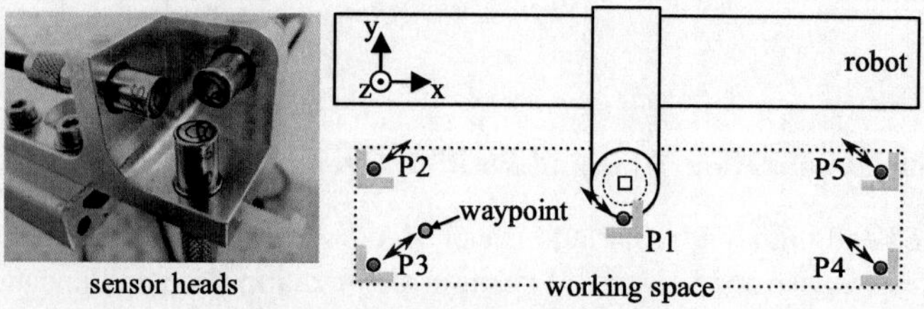

Fig. 3. Test setup and working space

The load during the tests does not comply with ISO, because of a missing specification for maximum load for the Robot. Instead, the usual load of approx. 200 g was applied to the robot head, which matches the weight of the components for the specified SAT process. Later, the maximum load was estimated at 1.000 g.

3.2 Position Repeatability

For each measuring point (P1-P5) the results show slightly different values (see Table 1). On the one hand, this difference is due to general fluctuations; on the other hand, robots also have preferred working areas. Usually, cartesian robots are very even in this regard. One advantage of measuring at several points is that the area with the best performance can be identified within the working space. Thus, critical processes can be planned within this zone.

Table 1. Repeatability and associated values

all values in μm	**P1**	**P2**	**P3**	**P4**	**P5**
$\bar{l}$	0.23	0.28	0.17	0.35	0.42
S_l	0.099	0.12	0.086	0.116	0.2
RP_l	0.52	0.64	0.43	0.78	1.03
$max(\Delta x_1, \ldots, \Delta x_n)$	0.34	0.38	0.24	0.66	0.77
$max(\Delta y_1, \ldots, \Delta y_n)$	0.21	0.41	0.28	0.56	0.69
$max(\Delta z_1, \ldots, \Delta z_n)$	0.42	0.37	0.24	0.26	0.38

Detailed Investigation of Measuring Data

Figure 4 shows a plot of the measurement on P1 with the drawing of RP_l as circles. The plot of the z-plane (right) shows an irregular distribution of the reached positions (diagonal distribution). It seems like a drift in x- and y-direction. This drift may fit to a warm-up of the robot, but with a run-in of 20 min, this seems to be unlikely. Another cause of this drift may be insufficient preloads on the precision axis of the robots, which leads to bearing clearance. This was not investigated any further, because the effort to fix the drift behavior would be unreasonably high and the reached repeatability fulfills the requirements of the SAT.

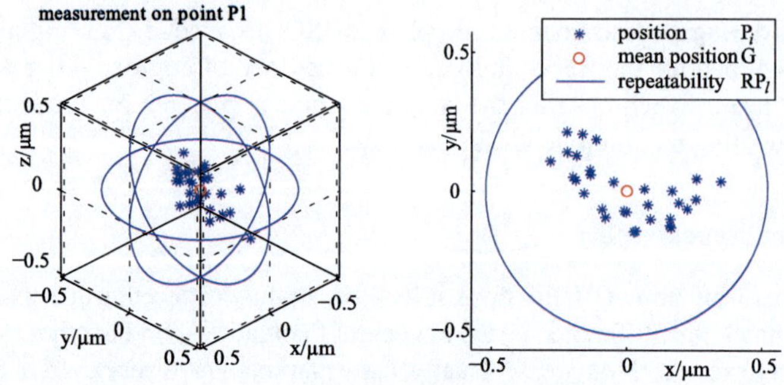

Fig. 4. Plot of reached positions and repeatability at P1

3.3 Position Stabilization Time and Overshoot

The measurement signal was continuously measured at each point. The data are very similar, so we only present P1 with the y-axis result. Figure 5 shows the plot of our signal. Additionally, the dwell time (6 s), RP_l and the specification limit are on display.

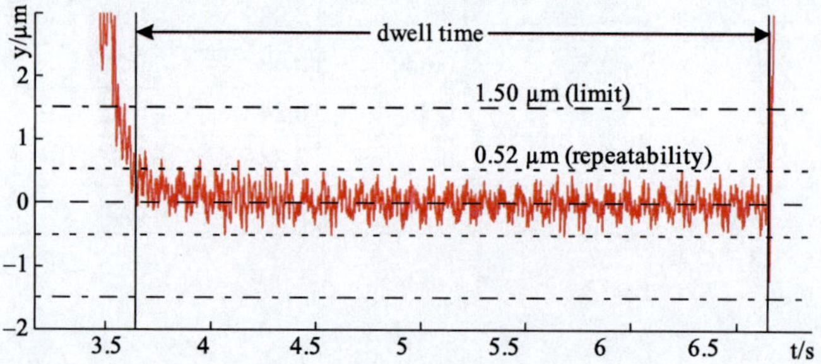

Fig. 5. Plot sensor data for position stabilization time (P1, y-axis)

Visible is the 55 Hz oscillation of the signal within an amplitude of approx. 0.28 µm. This is caused by the frequency of the mains voltage. The system behavior of the robot is overdamped. After entering the repeatability area, the system moves steadily toward the reached position and no (robot) oscillation is visible. As a result, there is no overshoot, which is favorable for a precision robot.

4 Discussion of the Performance Criteria According to ISO

ISO 9283 is very precise and the performance criteria are advantageous for the user. Key features are:

- Measurements take place at multiple poses in the workspace.
- Test load and test speed are at 100% of the maximal specification.
- The presented value is the least favorable of all measurements.
- All measurements are calculated with $3S$ (instead of $1S$).

The measurement comes with a high effort on metrology (see Sect. 4.1), time and knowledge about the ISO standards. This leads to insufficient use and incomplete datasheets (see Sect. 4.2). Over all, even with a given repeatability of a robot, calculation process accuracy is difficult due to the definition of *RP* (see Sect. 4.3).

With the focus on the measurement of accuracy and repeatability, several inconsistencies in the ISO cannot be discussed in this paper.

4.1 Practical Issues

Measuring the 6 DoF pose is very challenging. Since many robots have an accuracy between 10 μm and 50 μm, the measuring system needs to be highly accurate. Usually, laser tracking systems perform this task with good accuracy ≈20 μm and a wide range that also makes measurements of path accuracy possible [2, 6]. This is suitable for most calibration processes of industrial robots [7]. Nevertheless, even these 6 DoF tracking systems have a dependency on range, temperature and positioning, which reduced the suitability for high precision measurements. One additional challenge is to align the coordinate systems of the robot with the measuring system, which may lead to further slight inaccuracies due to transformation [8].

For precision robots, an accuracy below one micrometer is necessary and laser tracking systems reach their limits. Most sensors with (sub) micrometer performance have a small measuring range and only cover one DoF. The small range in combination with the position overshoot leads to the risk of collusion from load and sensors. For measuring multiple points, the systems must be repositioned in the working space. The measurement of the position repeatability is possible with an acceptable amount of effort. Making 6-DoF measurements is much more difficult, and measuring the accuracy and path repeatability is difficult to achieve for precision systems over a wide range.

4.2 Insufficient Use in the Industry

Robot manufacturers are interested in presenting the best possible performance in their data sheets. This is in contradiction to the elaborate measurements and the "worst case" conditions by ISO 9283.

Research shows that no robot manufacturer offers data sheets that meet ISO 9946. Some datasheets refer only to the axis accuracy and axis repeatability instead of the robot specification. This value is much easier to measure, because of the single degree of freedom. When a manufacturer implements an axis of a supplier, they can simply reference the specification without any (own) effort on researching robot accuracy.

Nevertheless, more and more manufacturers label their performance with the note "according to ISO 9283" [1] [2]. Information about the pose or path accuracy is hard to find. The position repeatability is common for many robots, partly extended by the single-axis repeatability of the additional rotation axis. Even for the data according to ISO 9283, some information is missing for the correct classification. For example, the position and size of the measuring cube are relevant, because the manufacturer has some leeway here. The definition of the assumed working space allows the selection of a measuring cube that offers better performance in a much smaller space.

Overall, the reference to ISO 9283 is questionable, as the majority of data sheets declare *RP* with $\pm$ *value*[1]. Since *RP* refers to the length of the deviation vector, this value can only be positive.

4.3 Challenges Posed by the Definition of Repeatability

A fundamental question is whether it makes sense to determine the repeatability according to ISO 9283. 30 measurements per point are very few to provide statistical reliability. In contrast to the 1990s, measurements will be performed and calculated automatically with computers now. An increase of the repetitions represents little more work in relation to the total effort of the measurement. We present the benefit of statistical reliability in Sect. 5.

The calculation of *RP* (1) is based on the $3S$ dispersion of the normal distribution. Corresponding to the three-sigma standard, users may assume that 0.27% of the approached poses are outside of *RP* (PooRP = Points outside of *RP*). This may be approximately true for the one-dimensional case. In the multidimensional case, the calculation by (1) is misleading, because the actual error rate is higher than suggested (see Table 2). With a normal distribution on a single axis, in the xy-plane, l has a Rayleigh distribution and with 3 DoF a Maxwell-Boltzmann distribution. The calculation of *RP* based on the normal distribution therefore is insufficient and results in a much higher error.

If the axes have significantly different repeatability, the factor of the inaccurate axis is significant to *RP*. In this case, *RP* only reflects the actual properties of the robot in a limited way. Some manufacturers have already decided to specify the repeatability for the xy-plane separately from the z repeatability (for SCARA robots).

5 Possible Adjustments on Repeatability Criteria

It makes sense to use *RP* to develop processes or to forecast assembly accuracies. Positioning errors could simply be summed up to calculate a predictive quality of a specific movement or operation. However, since *RP* (1) neither represents a normal distribution, nor a Rayleigh distribution, the addition formula for standard deviations cannot be applied. At this point, it would be appropriate to introduce a calculation function that better reflects the probability distribution and allows the calculation of the potential

[1] Found in data sheets and catalouges of KUKA, Yaskawa, Fanuc, Stäubli and Epson of different robot models in May 2023 (online research).

errors of several sequential handling operations. One way is to use the maximum likelihood estimation (MLE) method to determine the standard deviation of the Rayleigh distribution. This is reasonably accurate and is calculated by (2) [9]. By substituting the standard deviation into the density function of the Rayleigh distribution function (3) and setting it equal to the 3*S* specification (0.27% PooRP), we receive a *RP* value that corresponds to a 3*S* normal distribution:

$$\sigma \approx \hat{\sigma} = \sqrt[2]{\frac{1}{2n} * \sum\nolimits_{i=1}^{n} l_i^2} \tag{2}$$

$$1 - e^{-\frac{RP^2}{2\sigma^2}} = 0.9973 \tag{3}$$

$$RP = \left| \pm\sigma\sqrt{-2\ \ln(0.0027)} \right| \tag{4}$$

To verify the assumptions on statistical reliability and percentage distribution of a big number of values to the *RP*, we did simulations of a xy-cartesian robot. The simulation yields 10 million points, assuming that both axes have a normal error distribution (σ = 0.2). We calculated *RP* with (1) (ISO-RP) and (4) (MLE-RP) of the Rayleigh distribution for a sample of random 30, 100, 1000 and all simulated values. We determined the number of simulated positions that are outside of the *RP* value to get the percentage of PooRP. Each set was simulated 10 times and Table 2 shows the results with the mean values of ISO-RP/MLE-RP and the standard deviation within the 10 repetitions.

Table 2. Simulations on ISO-RP and MLE-RP

Meas. Points		mean Std. Dev ($\overline{\sigma_l}$)	mean PooRP (in %)	Std.Dev. PooRP (in %)	Std.Dev. PooRP (rel.*)
30	ISO-RP	0,1335	1,1861	1,0675	90%
	MLE-RP	0,1970	0,7264	0,7136	98%
100	ISO-RP	0,1275	0,8788	0,7808	89%
	MLE-RP	0,2010	0,3331	0,2383	72%
1000	ISO-RP	0,1321	0,5458	0,0799	15%
	MLE-RP	0,2004	0,2721	0,0402	15%
$10 \cdot 10^6$	ISO-RP	0,0066	0,5628	0,0016	0%
	MLE-RP	0,0100	0,2703	0,0018	1%

* relative percentage of the standard deviation to the mean value of PooRP

The results show, that 30 repetitions are not enough for a reliable calculation and even 100 repetitions still show a standard deviation of over 89% of the ISO-RP. Therefore, 1000 repetitions would be suitable for measuring repeatability.

On calculation with 30 and 100 simulated points, the amount of PooRP for ISO-RP is >0.87%. Even with 1000 or 10 million simulated points included in the ISO-RP calculation, the amount of PooRP only reaches 0.55%, which is significantly higher than the expected 0.27% (= $3S$). Comparing the MLE-RP calculation, the amount of PooRP is much lower and approaches 0.27% with 1000 simulated points. This proves that the *RP* calculation with the MLE provides a more reliable value for the three-sigma standard than ISO.

6 Conclusion and Outlook

ISO provides a series of standards for the performance criteria of robots. For many industrial robots, metrology exists to measure the performance criteria with sufficient accuracy. For precision robots, the measurement is more complex and some values are almost impossible to acquire.

For the manufacturers, there seems to be no need for adjustment of the ISO 9283, as many of them do not yet comprehensively follow the guidelines anyway. Only those values are presented that seem reasonable for the specific application and can be determined with reasonable effort. The current composition of data sheets reveals that many criteria are not currently represented. However, there are also good additions to the ISO standard, such as the separate repeatability in the xy-plane and z-axis.

In the calculation of repeatability, inconsistencies exist regarding the calculation. *RP* calculation is based on a normal distribution, although the reference quantity l corresponds to a Rayleigh or Maxwell-Boltzmann distribution. A calculation of the repeatability using the maximum likelihood estimate approach therefore seems more appropriate. A further scientific discussion and elaboration of possible calculation methods is necessary to establish more expressive performance criteria.

Acknowledgements. The research in this paper is funded by the Deutsche Forschungsgemeinschaft (DFG, German Research Foundation) under Germany's Excellence Strategy within the Cluster of Excellence PhoenixD (EXC 2122, Project ID 390833453). The authors thank the German Research Foundation (DFG) and Mr. Christoph Schumann, who made the measurements during his student thesis.

References

1. International Organization for Standardization: Manipulating industrial robots - Performance criteria and related test methods 25.040.30 (ISO 9283:1998-04)
2. Slamani, M., Nubiola, A., Bonev, I.: Assessment of the positioning performance of an industrial robot. Ind. Robot Int. J. (2012). https://doi.org/10.1108/01439911211192501
3. International Organization for Standardization: Robotics - Vocabulary 01.040.25, 25.040.30 (ISO 8373:2021-11)
4. International Organization for Standardization: Robots and robotic devices - Coordinate systems and motion nomenclatures 25.040.30 (ISO 9787:2013-05)
5. International Organization for Standardization: Manipulating industrial robots - Presentation of characteristics(ISO 9946:1999-04)
6. Jiang, Y., Yu, L., Jia, H., Zhao, H., Xia, H.: Absolute positioning accuracy improvement in an industrial robot. Sensors (Basel, Switzerland) (2020). https://doi.org/10.3390/s20164354

7. Young, K., Pickin, C.G.: Accuracy assessment of the modern industrial robot. Ind. Robot Int. J. (2000). https://doi.org/10.1108/01439910010378851
8. He, S., Ma, L., Yan, C., Lee, C.-H., Hu, P.: Multiple location constraints based industrial robot kinematic parameter calibration and accuracy assessment. Int. J. Adv. Manuf. Technol. (2019). https://doi.org/10.1007/s00170-018-2948-z
9. Walz, G.: Lexikon der Mathematik: Band 3. Springer, Berlin, Heidelberg (2017). https://doi.org/10.1007/978-3-662-53500-4

Framework for Industrial Cyber-Physical System of Systems and Application Towards an Effective Treatment of End-of-Life Products

Sandro Süß[1,2](✉) and Mark Mennenga[1,2]

[1] Institute of Machine Tools and Production Technology, Technische Universität Braunschweig, Langer Kamp 19b, 38106 Braunschweig, Germany
s.suess@tu-braunschweig.de

[2] Battery LabFactory Braunschweig, Langer Kamp 8, 38106 Braunschweig, Germany

Abstract. The amount of end-of-life products is increasing globally. Especially Waste of Electrical and Electronic Equipment (WEEE) is growing with more than 50 million metric tonnes annually. Effective treatment of end-of-life products aims to prevent landfilling and reduce the need for primary resources to manufacture new products. To contribute to sustainability targets, recycling must be energy- and resource-efficient. Industrial Cyber-Physical Systems are mechatronic systems that intend to improve efficiency within industrial infrastructure through near real-time digital modeling and simulation. In addition, coupling multiple Industrial Cyber-Physical Systems offers further potential to increase the overall system effectiveness by connecting and aligning digital models or simulations from different disciplines toward a common goal. The paper describes the concept of Industrial Cyber-Physical System of Systems and derives prospects and future research directions for its application within battery value networks.

Keywords: End-of-Life · Cyber-Physical System · System of Systems

1 Introduction

The average Electrical and Electronic Equipment consumption increases globally by 2.5 million metric tons annually. In 2019, this resulted in 53.6 Mt of globally generated Waste Electrical and Electronic Equipment discarded by their owners without the intent of reuse [1]. These investigations did not yet include any batteries, accumulators, electrical components of vehicles, or photovoltaic panels. On the contrary, climate mitigation drives the increasing demand for these technologies as, for example, solar energy is the mitigation option offering the highest potential contribution to net emission reduction by 2030 [2]. The development of automotive lithium-ion batteries can grasp the unprecedented ramp-up of these technologies, whose demand increased from about 330 GWh in 2021 to 550 GWh in 2022 [3]. With an expected lifetime of 10 to 15 years [4], solutions for End-of-Life (EoL) management of batteries will then be required at a similar scale as the current production capacities.

T. Bauernhansl et al. (Eds.): WGP 2023, LNPE, pp. 238–247, 2024.
https://doi.org/10.1007/978-3-031-47394-4_24

EoL management coordinates collection and decisions on the most suitable EoL scenario. The end-of-life treatment scenarios are maintenance, repair, product reuse, upgrading, downgrading, remanufacturing, component reuse, material recycling, incineration, and landfill. The choice for the most suitable scenario is based on the remaining useful life of products and components in relation to their physical and technological life [5]. Furthermore, strategies that induce higher environmental impacts than prevented are undesirable [6]. Although scenarios that ensure products remain at their highest possible value are preferable for higher sustainability, material recycling is necessary if no physical and technological life is left to reuse a component [7].

Material recycling treats waste materials, returning them to the supply chain. Its efficiency is measured by the output of the recycling process divided by its input such that material losses within the recycling process cause inefficiencies [8]. Combining different products in the same recycling stream increases the complexity of liberation within recycling. An approach to recycling that includes the whole cycle from the EoL phase to the production of renewed raw material can achieve improved recycling rates by modeling interactions within complex waste streams through technology and systems development [9].

In the case of batteries, production and recycling are both complex processes that pose multi-criteria optimization problems with a large set of adjustable process parameters toward various key performance indicators. In contrast to production, recycling induces dissociative material flows where one input is transformed into several outputs created along the process chain. Furthermore, basic procedures like mechanical treatment, thermal treatment, and hydrometallurgy can be combined in several different process routes [10]. Therefore, optimizing a recycling process focuses on a large set of quality parameters for different output fractions.

Cyber-Physical Systems (CPS) are used for such optimization problems. They are systems "based on a mechatronic core that is characterized by its networking with the Internet of Things and Services" [11]. This core is formed by one or more mechatronic system(s) such that the system can communicate and interact beyond its system boundaries. CPSs conduct processes in the physical and cyber worlds to observe and influence the physical environment [12]. Industrial Cyber-Physical Systems (ICPS) implement CPS in real-world industrial infrastructure, creating a corresponding representation in the cyber world to supervise, monitor, control, and manage the networked physical infrastructures [13].

Waste management systems for batteries that establish simple, mainly unmixed material streams suitable for final processing must be developed to achieve better liberation of compounds and, thus, higher recycling efficiencies.

Current research proposes different approaches to bringing forward and reverse supply chain actors together. Stock and Seliger assess the macro and micro perspective of Industry 4.0 for sustainable manufacturing. From the macro perspective, they argue for a strongly product-life-cycle-related point of view through value creation networks that link multiple phases of product life cycles [14].

A coordinated planning approach for closed-loop supply chains is proposed by Scheller et al., which identifies digitalization as additional support for cooperation and coordination in such systems [15].

The concept of Recycling 4.0 aims at the digitalization of processes at the EoL stage. Blömeke et al. propose a Recycling 4.0 information marketplace architecture to bridge the information gap between OEMs and recyclers [16].

A similar architecture was developed by Gu et al. in the context of extended producer responsibility, which is a widely implemented policy that gives producers significant obligations for the treatment or disposal of post-consumer products. The framework integrates information systems and product lifecycle management models focusing on information sharing [17].

Current literature proposes several frameworks to optimize closed-loop supply chains with a top-down approach on a macro level. As a remaining challenge they identify information gaps along the product life cycle, and digital technologies such as CPS are seen as a chance to overcome them. The proposed architectures are centralized structures focusing on supply chain management.

This work proposes a framework that supports the optimization of closed-loop supply chains in decentralized structures by coupling multiple Cyber-Physical Systems through interfaces. It focuses on Industrial Cyber-Physical Systems that enable the supervision, monitoring, control, and management of waste material streams within battery value networks. The paper is structured into five sections. Section 2 gives the theoretical background on value networks in the battery ecosystem and Industrial Cyber-Physical Systems. Section 3 introduces the general framework for Industrial Cyber-Physical System of Systems, whose implementation is discussed in Sect. 4. The final Sect. 5 concludes the work and discusses the potentials and challenges of the Industrial Cyber-Physical System of Systems.

2 Theoretical Background and State of Research

2.1 Value Networks in the Battery Ecosystem

In the battery ecosystem, value networks have become increasingly important. A Value Network is an alliance of several companies associated with risky up-front expenditures for the joint development of market potentials that could otherwise not be developed [18]. Within a Circular Economy, such alliances have the potential to build strong cooperation towards better environmental performance of waste management systems and value retention chains. This performance could be monitored by all involved companies with indicators such as Recyclability Benefit Rate, Retained Environmental Value, or Environmental Waste Utilization [19].

As new regulations in the European Union and other regions promote extended responsibilities of producers and distributing companies [20], these companies can no longer rely on their core business to implement all given requirements. Therefore, OEMs and battery producers tend to cooperate with organizations that can deliver end-of-life treatment for their batteries. Especially in the automotive industry, this is observable in joint ventures, technology partnerships, or similar cooperation forms between OEMs and recyclers like Volkswagen AG and Umicore AG & Co. KG [21], Mercedes-Benz Group AG and Primobius GmbH [22], or Stellantis N.V. and Galloo [23], to name a few. Apart from fulfilling regulations for end-of-life treatment, these partnerships often promote circular material flows. Assuming that one recycler exclusively recycles batteries from one producer and delivers raw materials back to this producer, a closed-loop supply chain is established. On the one hand, material composition, battery design, and manufacturing processes of the battery influence the options to process batteries at their end-of-life. On the other hand, the recyclate product quality influences the reusability of secondary raw materials in production. Therefore, interrelations in material cycles between production and recycling are present. These interrelations need to be understood for efficient recycling.

Value networks can be related to a scientific research direction that addresses the analysis, development, and improvement of so-called system of systems. A System of Systems (SoS) consists of multiple operationally and/or managerially independent systems whose aggregation produces results that are not achievable by the systems alone. Therefore, the behavior of the SoS depends on the behavior of the constituent systems and sub-systems [24, 25]. An SoS can be established within battery value networks by coupling systems from battery production and battery recycling.

2.2 Industrial Cyber-Physical Systems

In a production environment, Cyber-Physical Production Systems (CPPS) are CPS with a physical component, such as the machinery or machine tools, and a virtual component, like simulation or data-based models. The two worlds are connected via data acquisition from the physical world and some feedback or control of the physical world [26]. This approach is applied similarly to Cyber-Physical Recycling Systems (CPRS) to enable a model-based optimization of Recycling Systems [27]. In general, an Industrial Cyber-Physical System consists of four elements. The framework describes a cycle starting with the business understanding step in the physical world, similar to the one performed within the Cross-Industry Standard Process for Data Mining (CRISP-DM) cycle [28]. The Physical World (PW) represents the industrial infrastructure, such as production or recycling machines. Sensors generate data within and along the process chain and ingest that data via industrial protocols like OPC UA or Modbus into storage. Within storage, data engineering tasks like transformation and preparation steps are performed in the Transition Level Input (TLI) [29]. The data sets are accessed from the Cyber World (CW), where data analytics, machine learning, or simulation techniques are used to develop and evaluate models of the Physical World. Via the Transition Level Output (TLO), these models are deployed to adjust the Physical World or offer process decision support directly. Figure 1 shows the general framework of an ICPS with the corresponding information flows between its five parts Physical World, Transition Level Input, Cyber World, Transition Level Output, and the Human in the loop. The framework integrates steps from CRISP-DM, Data Engineering [29], and Cyber-Physical Production Systems [26, 30] to offer a general application framework for any industrial infrastructure.

The objective is to enable the assessment and control of recycling systems within value networks such that the overall performance of the waste management system is maximized. Cyber-Physical System of Systems enables stakeholders within value creation networks to couple Cyber-Physical Systems that optimize process efficiency within their facility with Cyber-Physical Systems in other facilities to improve effectiveness towards a common goal.

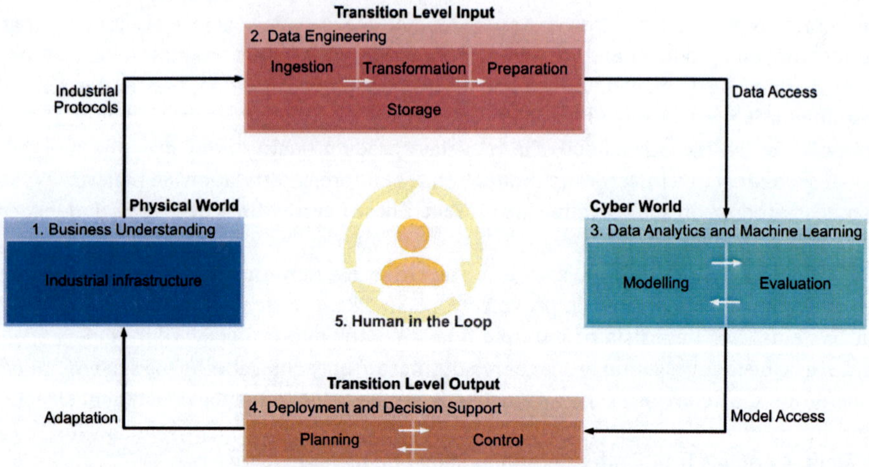

Fig. 1. General framework of an ICPS with its five elements

3 Industrial Cyber-Physical System of Systems

Building on the theoretical foundations, this section derives a general framework for an Industrial Cyber-Physical System of Systems.

A Cyber-Physical System of Systems (CPSoS) is a large, complex, and often spatially distributed SoS of at least two CPSs [25, 31]. Therefore, Industrial Cyber-Physical Systems of Systems (ICPSoS) are created by coupling at least two ICPS in an SoS. This means at least two ICPS interact, which are managerially and/or operationally independent. 4^n mathematical combinations of n ICPSs are possible if all four elements (PW, TLI, CW, TLO) connect between n ICPSs. As these connections are interpretable as input-output information flows, every combination must be considered separately.

Figure 2 illustrates all technically possible combinations within an ICPSoS with two constituting agents.

Data interfaces represent information flows from Transition Level Input to other TLIs or the Cyber World. In practice, the interface between two or more Transition Level Inputs results in some form of data exchange within data architectures, for example, exchanging acquired data with an external data storage system. Similarly, the interface from one Transition Level Input to an outer Cyber World practically represents models in the Cyber World accessing external data from another system.

Furthermore, information flows from one Cyber World to another can be developed via model interfaces. Thus, models can be coupled such that models of one system have direct access to the current model output of another system.

Decision support interfaces represent information flows to a Transition Layer Output. Therefore, a model can also support an external Transition Level Output by providing its model results to another system. An interface between Transition Level Outputs makes decision support transparent within connected systems.

All other mathematically possible combinations do not occur within the technical representation of the ICPSoS as they violate certain preconditions. One main precondition is that information flows are only possible successively from the Physical World to the Transition Level Input to the Cyber World to the Transition Level Output back to the Physical World. This originates in the general structure of an ICPS. An exception to this rule is information flows between elements of the same type such that information flows between two or more TLIs, CWs, or TLOs are possible.

An information flow between two Physical Worlds is excluded as this violates the SoS condition of managerially and/or operationally independent systems. By coupling two physical systems, operational independency no longer holds as physical parts or Programmable Logic Controllers of the machines are integrated. Furthermore, the SoS condition for managerially and/or operationally independent systems prevents information flows from a PW to an external TLI and a TLO to an external PW, as both would result in an operational dependency.

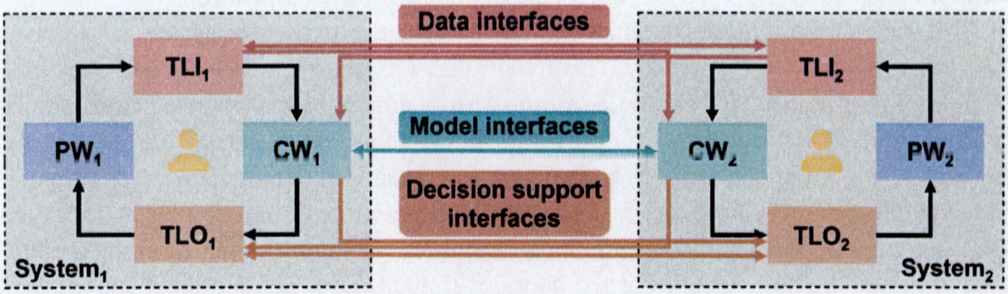

Fig. 2. ICPSoS with two constituting agents and their information flows; PW: Physical World, TLI: Transition Level Input, CW: Cyber World, TLO: Transition Level Output

4 Industrial Cyber-Physical System of Systems in Battery Value Networks

The application of the ICPSoS framework from Sect. 3 on battery recycling has the potential to digitally connect several actors within value networks enabling a connected data-based modeling approach that aims towards higher effectiveness of the overall system towards a circular battery economy. In its simplest form, such an ICPSoS in a battery value network consists of one Cyber-Physical Battery Production System (CPBPS) and one Cyber-Physical Battery Recycling System (CPBRS).

In the Physical World, raw and secondary material from the recycler is fed into the battery production process to manufacture the battery. This battery is then implemented within a product that is used to fulfill a particular task for a user. As the product reaches its EoL, in an optimal setting, it is collected and disassembled such that the battery reaches the recycler as post-consumer battery material (PCBM). The recycler then transforms the PCBM into secondary raw material that can be reused in production again.

Digital model representations for production and recycling are developed within the respective ICPSs in the Cyber World. Multiple interfaces between the CPBPS and

CPBRS are created to enhance these isolated models that can optimize the efficiency within their system boundaries.

Data interfaces enable access to or exchange of specific relevant datasets between partners. For example, the current recycling volumes could be shared within the SoS to empower the producer to react to changes in the availability of secondary materials. This could lead to preparatory steps for necessary adaptations in the mixing ratio of primary and secondary materials and, thus, the process parameters for the expected input material mixture. Similar use cases are possible where data exchange within such battery value networks poses significant advantages.

Model interfaces directly couple the two isolated models to each other. This usually is only possible if the partners work together to extend the current models such that the output of an external model improves the internal model. For example, a battery process model of the producer could be shared on demand by the recycler as soon as this exact battery reaches the recycling facility. The recycler can then use this model as an input for his recycling process model and therefore prepare the optimal process layout and setup for this battery type based on its design and material composition.

Decision support interfaces finally enable the optimization of the system's effectiveness through system transparency. This means that a recycler knows about the decisions made by the producer, for example, a long-term decision towards a particular battery cell layout or battery material composition. Furthermore, predictions from models are shared so the partner can adapt to specific expected changes within the partner's model. For example, a battery producer has a model of the lifetime of a particular battery type and uses this together with the number of batteries put onto the market at a specific time to model the number of returned batteries in the future. The producer can quickly assemble this data, which is highly valuable for the recycler. Vice versa, the recycler could provide models on the recyclability of certain battery types to aid the producer with his battery design for recycling.

All these interfaces generally support a collaborative and trustful exchange within a circular battery value network. Many new digital optimization potentials arise by combining two isolated ICPS for production and recycling. In this section, only a few examples were discussed. The following section discusses the prospects and challenges for ICPSoSs.

5 Conclusion: Prospects and Challenges for ICPSoS

In this work, a framework for the design of ICPSoS was developed. This framework was applied to a setting within a battery value network consisting of one recycling and one production system.

Industrial Cyber-Physical Systems of Systems offer several potential advantages within value networks, especially for recycling operations where cooperation is essential for a successful business model because simpler and less mixed material streams achieve higher recycling efficiencies. ICPSoS offer manageability of material input streams for recycling as they improve trustworthy information transparency of the products produced and used. This could be further enhanced by integrated Cyber-Physical Product Systems, which provide information from the use phase, and Cyber-Physical Logistic Systems, which support collection at the EoL. In this context, ICPSoS can also be a technological enabler for distributed recycling facilities and circular production. When distributed recycling facilities communicate with each other and a logistic system, they might have advantages by performing certain operations closer to the waste origin. The same holds for circular production, where production and recycling are integrated within one facility, and thus, especially the reverse material flow to the recycling needs to be well managed.

Furthermore, scenario analyses can be initiated within ICPSoSs which include more than one point of view on a systemic problem case. As stakeholders work together within ICPSoSs, they provide their points of view represented through their models and specialized datasets. In a shared scenario analysis, models could be coupled such that, for example, a specific production scenario triggers a corresponding recycling scenario, which discloses the production scenario's recyclability. With these insights, further steps like specialized technology development could be initiated.

ICPSoS can foster Industrial Symbiosis through open interface design where not only value networks and cooperation agreements enable access to data, models, and decision-making. By participating in a value ecosystem, these open interfaces can be utilized to create new business models and provide growth opportunities for all businesses. Furthermore, this could facilitate life cycle engineering as data on system interactions and behavior is provided.

One of the disadvantages of ICPSoS is that, on the one hand, the concept supports an effective and efficient treatment of EoL products. Still, on the other hand, implementation of such SoSs will further increase the amount of WEEE. Therefore, model-based systems engineering and a practical tool for designing minimal ICPSoS are crucial to prevent problem shifting and manage complexity. As each added system increases complexity, a robust system design is essential to operate ICPSoS successfully. Especially interfaces between ICPSs need to be designed for low maintenance and high cyber security while ensuring data integrity. Furthermore, as long as humans control the individual ICPSs, the concept must be framed as more than a technical system. Some recommendations from the decision support might not be trusted directly because the overall system effectiveness or temporal gap is not visible to an actor at an ICPS.

Lastly, the general application of the concept on complex mixed waste streams such as household solid waste is in doubt.

Therefore, the first step in further research is applying this approach to a small demonstrator SoS. In the next step, the application to value networks of products containing highly limited or valuable materials can be performed, starting with a proof of concept for a specific premium solution and later trying to propagate toward increasingly complex waste streams. Furthermore, robust system and interface design for ICPSoS must be conducted to address challenges such as cyber security and communication between ICPSs.

References

1. Forti, V., Balde, C.P., Kuehr, R., Bel, G.: The global e-waste monitor 2020: quantities, flows and the circular economy potential: united nations university/united nations institute for training and research. In: International Telecommunication Union, and International Solid Waste Association (2020)
2. Shukla, P.R., Skea, J., Reisinger, A. (eds.): Climate Change 2022: Mitigation of Climate Change. IPCC, Geneva (2022)
3. IEA, Ed., "Global EV Outlook 2023," Paris, 2023
4. Abdelbaky, M., Peeters, J.R., Dewulf, W.: On the influence of second use, future battery technologies, and battery lifetime on the maximum recycled content of future electric vehicle batteries in Europe. Waste Management (New York, N.Y.) **125**, 1–9 (2021). https://doi.org/10.1016/j.wasman.2021.02.032
5. Hauschild, M.Z., Kara, S., Røpke, I.: Absolute sustainability: challenges to life cycle engineering. CIRP Ann. **69**(2), 533–553 (2020). https://doi.org/10.1016/j.cirp.2020.05.004
6. Kara, S., Hauschild, M., Sutherland, J., McAloone, T.: Closed-loop systems to circular economy: a pathway to environmental sustainability? CIRP Ann. (2022). https://doi.org/10.1016/j.cirp.2022.05.008
7. Schmidt, S., Laner, D.: Material flow analysis of recycling systems. In: Brandão, M., Lazarevic, D., Finnveden, G. (eds.) Handbook of the Circular Economy. Edward Elgar Publishing (2020)
8. Worrell, E., Reuter, M.A.: Definitions and terminology. In: Worrell, E., Reuter, M.A. (eds.) Handbook of Recycling, pp. 9–16. Elsevier (2014)
9. Heiskanen, K.: Theory and tools of physical separation/recycling. In: Worrell, E., Reuter, M.A. (eds.) Handbook of Recycling, pp. 39–61. Elsevier (2014)
10. Doose, S., Mayer, J.K., Michalowski, P., Kwade, A.: Challenges in ecofriendly battery recycling and closed material cycles: a perspective on future lithium battery generations. Metals **11**(2), 291 (2021). https://doi.org/10.3390/met11020291
11. VDI/VDE 2206: Entwicklung mechatronischer und cyber-physischer Systeme, VDI/VDE-Gesellschaft Mess- und Automatisierungstechnik, 2021
12. Kopetz, H., Bondavalli, A., Brancati, F., Frömel, B., Höftberger, O., Iacob, S.: Emergence in cyber-physical systems-of-systems (CPSoSs). In: Bondavalli, A., Bouchenak, S., Kopetz, H. (eds.) Cyber-Physical Systems of Systems. LNCS, vol. 10099, pp. 73–96. Springer, Cham (2016). https://doi.org/10.1007/978-3-319-47590-5_3
13. Colombo, A.W., Karnouskos, S., Kaynak, O., Shi, Y., Yin, S.: Industrial cyberphysical systems: a backbone of the fourth industrial revolution. EEE Ind. Electron. Mag. **11**(1), 6–16 (2017). https://doi.org/10.1109/MIE.2017.2648857
14. Stock, T., Seliger, G.: Opportunities of sustainable manufacturing in industry 4.0. Procedia CIRP **40**, 536–541 (2016). https://doi.org/10.1016/j.procir.2016.01.129
15. Scheller, C., et al.: Coordinated planning in closed-loop supply chains and its implications on the production and recycling of lithium-ion batteries. Procedia CIRP **98**, 464–469 (2021). https://doi.org/10.1016/j.procir.2021.01.135
16. Blömeke, S., et al.: Recycling 4.0: an integrated approach towards an advanced circular economy. In: Proceedings of the 7th International Conference on ICT for Sustainability, Bristol United Kingdom, 2020, pp. 66–76 (2020)
17. Gu, F., Guo, J., Hall, P., Gu, X.: An integrated architecture for implementing extended producer responsibility in the context of Industry 4.0. Int. J. Prod. Res. **57**(5), 1458–1477 (2019). https://doi.org/10.1080/00207543.2018.1489161
18. Sommerville, R., Zhu, P., Rajaeifar, M.A., Heidrich, O., Goodship, V., Kendrick, E.: A qualitative assessment of lithium ion battery recycling processes. Resour. Conserv. Recycl. **165** (2021). https://doi.org/10.1016/j.resconrec.2020.105219

19. Schmidt, S., Laner, D.: Environmental Waste Utilization score to monitor the performance of waste management systems: a novel indicator applied to case studies in Germany. Resour. Conserv. Recycl. Adv. **18**, 200160 (2023). https://doi.org/10.1016/j.rcradv.2023.200160
20. Proposal for a regulation of the European parliament and of the council concerning batteries and waste batteries, repealing Directive 2006/66/EC and amending Regulation (EU) No 2019/1020, 2020
21. Umicore AG & Co. KG, Umicore und PowerCo gründen Joint Venture für die Produktion von Batteriematerialien in Europa. https://www.umicore.de/de/presse/news/umicore-und-powerco-grunden-joint-venture-fur-die-produktion-von-batteriematerialien-in-europa/
22. Mercedes Benz Group AG, Groundbreaking for battery recycling factory.: Mercedes-Benz plant Kuppenheim, Germany. https://group.mercedes-benz.com/company/news/recycling-factory-kuppenheim.html
23. Stellantis, N.V.: Stellantis und Galloo gründen Joint Venture für Recycling von Altfahrzeugen (2023). https://www.media.stellantis.com/de-de/corporate-communications/press/stellantis-und-galloo-gruenden-joint-venture-fuer-recycling-von-altfahrzeugen. Accessed 15 June 2023
24. Maier, M.W.: Architecting principles for systems-of-systems. Syst. Engin. **1**(4), 267–284 (1998). https://doi.org/10.1002/(SICI)1520-6858(1998)1:4%3C267::AID-SYS3%3E3.0.CO;2-D
25. Weilkiens, T., Lamm, J.G., Roth, S., Walker, M.: Model-Based System Architecture. Wiley, Hoboken, NJ (2022)
26. Thiede, S., Juraschek, M., Herrmann, C.: Implementing cyber-physical production systems in learning factories. Procedia CIRP **54**, 7–12 (2016). https://doi.org/10.1016/j.procir.2016.04.098
27. Blömeke, S., Rickert, J., Mennenga, M., Thiede, S., Spengler, T.S., Herrmann, C.: Recycling 4.0 – mapping smart manufacturing solutions to remanufacturing and recycling operations. Procedia CIRP **90**, 600–605 (2020). https://doi.org/10.1016/j.procir.2020.02.045
28. Chapman, P., et al.: CRISP-DM 1.0: step-by-step data mining guide (2000)
29. Reis, J., Housley, M.: Fundamentals of Data Engineering: Plan and Build Robust Data Systems. O'Reilly, Beijing, Boston, Farnham, Sebastopol, Tokyo (2022)
30. Turetskyy, A.: Data analytics in battery production systems, Dissertation, Technische Universität Braunschweig, Vulkan-Verlag GmbH
31. Engell, S., Paulen, R., Reniers, M.A., Sonntag, C., Thompson, H.: Core research and innovation areas in cyber-physical systems of systems. In: Mousavi, M., Berger, C. (eds.) Cyber Physical Systems. Design, Modeling, and Evaluation. CyPhy 2015. LNCS, vol. 9361, pp. 40–55. Springer, Cham (2015). https://doi.org/10.1007/978-3-319-25141-7_4

Model-Based Spindle Bearing Monitoring Using Vibration Sensors and Artificial Neural Networks

Magnus von Elling[1(✉)], Markus Weber[2], Viktor Berchtenbreiter[1], and Matthias Weigold[1]

[1] Department of Mechanical Engineering, Institute of Production Management, Technology and Machine Tools (PTW), Technical University of Darmstadt, Otto-Berndt-Str. 2, 64283 Darmstadt, Germany
m.vonelling@ptw.tu-darmstadt.de

[2] GMN Paul Müller Industrie GmbH & Co. KG, Äußere Bayreuther Str. 230, 90411 Nürnberg, Germany

Abstract. To ensure the longevity of the bearings of a motor spindle, it is advantageous to know the precise loads on the bearings during operation. Since sensor-based monitoring involves a great deal of effort due to the limited space available, and simulating the bearing load is not real-time capable, we investigated how the bearing loads can be estimated using machine learning methods. To estimate the bearing load, a co-simulation was first set up that generates large amounts of training data based on measured cutting forces and spindle vibration velocities. Measured and simulated quantities are then used to train artificial neural networks. The best-performing neural networks can estimate the surface pressure between rolling elements and bearing rings with an error of less than 2%. This deviation refers to the contact stress that was calculated for comparison with the simulation results.

Keywords: Condition Monitoring · machine learning · motor spindle

1 Introduction

The efficient and reliable operation of motor spindles is of great importance in various industrial applications, especially in machine tools. Motor spindle bearings, being critical components within these systems, are subject to significant mechanical stresses and vibrations during operation. Damaged bearings are one of the main reasons for spindle failure [1]. Monitoring bearing loads is essential to prevent premature failures, minimize downtime, and optimize maintenance strategies [2].

Wu et al. [3] presented a review of different sensors being used to monitor rolling bearings showing that mostly internal sensors with one-dimensional signal detection are used. However, they give limited insight into the observed structure and are mostly suitable for fault detection when used on their own and with analytical evaluation approaches. For load monitoring of rolling bearings more sophisticated sensory systems are used.

T. Bauernhansl et al. (Eds.): WGP 2023, LNPE, pp. 248–256, 2024.
https://doi.org/10.1007/978-3-031-47394-4_25

Conventional methods for monitoring spindle and bearing loads often rely on displacement sensors directly mounted on the bearings or surrounding structures [4]. While effective, these approaches can be complex, costly, and require intrusive modifications to the spindle assembly.

In recent years, advances in machine learning and computational modeling have opened new avenues for condition monitoring [5] and especially bearing vibration monitoring [6]. AI-based cutting force models have also been investigated [7]. However, with none of these methods can the load variables of the spindle's bearings be determined accurately, using only sensors that are already installed in many spindles, like vibration sensors.

In this paper, we present a novel approach that combines a multibody simulation model of the spindle with a neural network-based framework to estimate the bearing loads using vibration data acquired from built-in one-dimensional acceleration sensors. The proposed methodology uses the relationship between the mechanical loads experienced by the spindle bearings and the resulting vibrations captured by the acceleration sensors. By employing a model-based approach, we exploit the accurate representation of the spindle's structural dynamics, geometric parameters, and contact stresses in the bearings to establish a reliable mapping between the measured vibrations and the corresponding bearing loads.

Central to our approach is the use of a neural network architecture capable of learning the complex and nonlinear relationship between vibration data and bearing loads. Through training on a comprehensive dataset generated from the multibody simulation, the neural network can learn the patterns in the input-output domain, enabling accurate load estimation.

The primary objective of this study is to investigate the effectiveness and practical applicability of the proposed model-based monitoring system for motor spindle bearings. By comparing the estimated maximum bearing loads obtained from the neural network with the true loads derived from the multibody simulation, we aim to assess the accuracy, robustness, and real-time capabilities of the proposed approach. The simulation has been validated by a modal analysis of the real spindle and comparison with the virtual spindle.

Through the integration of advanced computational modeling, machine learning techniques, and sensor data, this research strives to enhance the understanding and monitoring of motor spindle bearing behavior, paving the way for improved reliability, reduced downtime, and optimized maintenance strategies in industrial applications.

2 Methodology

2.1 Multibody Simulation

The base for the described approach is a sophisticated multibody simulation of a motor spindle in a Hermle C32U dynamic machine tool. The simulation model was built in the Hexagon simulation software Adams. The full spindle has been rebuilt inside the simulation environment as described in [8]. The bearings themselves were modelled using so-called GFORCE elements which represent 3 translational and 3 rotational springs each and can therefore mimic the behavior of any real bearing. The stiffness of the spring elements depends on the rotational speed of the spindle since the hybrid angular and cylindrical ball bearings of the spindle behave different at different speeds. Increasing rotational speeds lead to a decrease in bearing stiffness due to elastic spring preload. Additionally, centrifugal forces cause the cylindrical roller bearing to become stiffer as the speed increases. Damping coefficients have been found that ensure a realistic bearing motion and stable simulation of the whole spindle-bearing assembly.

To accurately capture the dynamic behavior of the motor spindle system both rigid and finite element method (FEM) components were used. The rigid model of the spindle provided a simplified representation of the system, neglecting the flexibility of the tool, tool holder, and shaft. While this approach offered some insight into the overall behavior of the system, it did not fully capture the intricate spindle bearing load and deformations present in real-world conditions.

To overcome this limitation, a flexible model was developed using FEM techniques. The tool, tool holder, and shaft were modeled as flexible bodies in the software ANSYS, allowing for an accurate representation of their deformations and eigenfrequencies. The FEM model was integrated with the rigid components in ADAMS using .mnf-files, enabling a comprehensive simulation of the entire motor spindle system.

To validate the accuracy of the simulation results, a modal and order analysis was conducted comparing the simulated frequencies and mode shapes with those obtained from experimental measurements of the real spindle. The modal and order analysis of the rotating spindle served as a critical benchmark for evaluating the model and its ability to capture the dynamic characteristics of the system. Additionally typical frequencies like rotational and tool passing frequency were also clearly visible in the frequency spectrum of the flexible model and aligned closely with the measured data from the real spindle. These Frequencies cannot be seen with the rigid simulation (see Fig. 1). This observation confirmed that the flexible model represented the realistic dynamic behavior of the motor spindle system better.

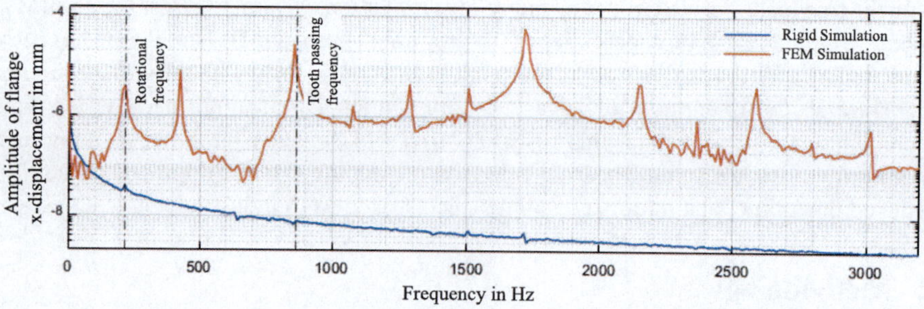

Fig. 1. Frequency analysis of the rigid and FEM motor spindle model. Rotational and tool passing frequencies are clearly visible in the FEM simulation.

The successful validation of the flexible model through modal and order analysis highlighted the importance of considering the flexibility of the tool, tool holder, and shaft in motor spindle simulations. Therefore, other simulation results, like the displacement and tilt of the bearings, can also be considered to be close to the real values.

2.2 Cutting Experiments

In order to investigate the cutting behavior and obtain a comprehensive dataset for simulation and subsequent training, aluminum cutting experiments were conducted using a design-of-experiment-approach (DOE). This methodology allowed for the exploration of a wide range of cutting conditions on the machine tool, providing valuable insights into the effects of various parameters on cutting forces and spindle vibrations. These especially included the angle at which the tool meets the workpiece as well as cutting speed and depths of cut.

To measure the cutting forces, a KISTLER 9257B dynamometer was employed. Simultaneously, the spindle's vibration velocity was recorded using the two built-in AI 100.012. Vibration sensors and SiViB 1001 preprocessing units of the company IBIS. The sensors are located at the lower end of the spindle close to the pre-loaded bearings. They are fixed at a distance of 90° around the circumference in the housing and point radially in the direction of the spindle center axis. See Fig. 2 for the experimental setup.

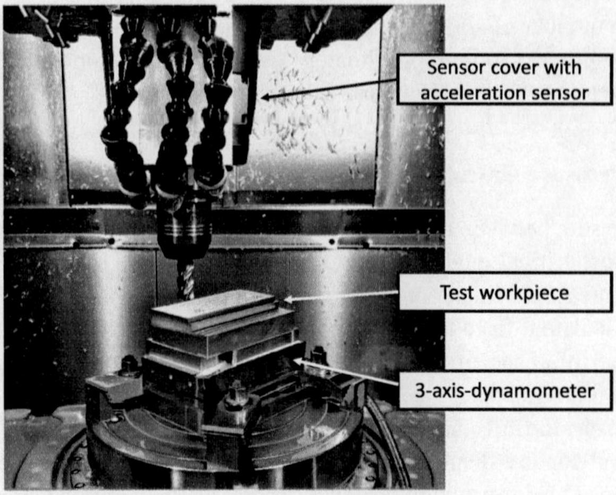

Fig. 2. Experimental setup with the dynamometer in the machine tool

To ensure synchronization and consistent time stamping of the data from multiple sensors, a National Instruments analog-digital converter was utilized to transform the sensor data to digital time series. The recorded signals were then processed using MATLAB code, enabling further analysis and data manipulation like coordinate transformation so the measured force data represents the force acting on the tool center point.

The 243 conducted aluminum cutting experiments, each representing a specific combination of cutting conditions, resulted in a total of 1215 time series. These time series, consisting of the 3 transformed cutting forces and the 2 corresponding spindle vibration velocities, constitute a rich and diverse dataset suitable for simulation and subsequent training.

2.3 Co-Simulation

To obtain a deep inside into the motor spindle system, the measured force data obtained from the aluminum cutting experiments was utilized to feed a co-simulation of the spindle in MATLAB. This co-simulation was achieved by exporting the validated spindle model from ADAMS to MATLAB Simulink. The exported model was then integrated into a MATLAB Simulink block diagram, allowing for the incorporation of experimental data and the automation of simulations through MATLAB code. The main motivation for employing this co-simulation approach was to facilitate the execution of a large number of simulations with minimal manual effort. This automated workflow provided insights into the spindle bearing's response under various cutting conditions.

The transformed cutting forces and the corresponding cutting parameters like rotational spindle speed, were input into the MATLAB Simulink block diagram. The co-simulation then ran automatically, with MATLAB code orchestrating the simulation process. The outputs of the simulation included forces, displacements, and tilts of all three bearings of the spindle. During the simulation process, the data was collected and saved automatically for subsequent analysis and investigation. It is worth noting that due to the large number of simulations performed and the complexity of the simulation itself, the entire process took approximately two weeks to complete. This multi-body simulation is therefore by no means real-time capable.

2.4 Surface Pressure Calculation

The surface pressure between the rolling elements and the rings of each spindle bearing is one of the most critical quantities to evaluate the bearing's load [9]. However, since they cannot be measured in an operating spindle assembly, an analytical approach was employed to calculate it from the simulation results. Notably, the analytical model took into account the influence of centrifugal forces, which play a significant role in the behavior of rotating systems.

By considering the tilt and displacements of the spindle bearings, the analytical model accounted for the displacement and deformations of the bearing components during operation. This information provided crucial insights into the contact conditions between the rolling elements and the rings, allowing for an estimation of the hertzian surface pressure distribution as described in [10]. The calculated surface pressure values are crucial for identifying potential areas of high stress or wear that can lead to premature failure of the bearing. The surface pressure is calculated using MATLAB code and can therefore seamlessly be connected to the co-simulation.

2.5 Training Neural Networks

Since the multibody simulation is time consuming efforts were made to enable real-time prediction of the simulation results, e.g. bearing tilt and force, during operation. Additionally, there was a need to replace the expensive and cumbersome dynamometer with a more practical solution. Machine learning methods were identified as suitable for mapping readily available vibration data to the previously simulated bearing variables.

Neural networks, trained using the Python library Keras, were employed to establish the relationship between the vibration data and the simulation data. In addition to vibration data, cutting parameters such as depth of cut, cutting speed, and rotational speed of the spindle, were included as additional input parameters. A variety of multi-layer perceptron neural networks were trained, and their hyperparameters optimized by comparing their mean absolute error (MAE) and coefficient of determination R^2, ensuring the best possible predictive performance. The schematic training procedure is depicted in Fig. 3.

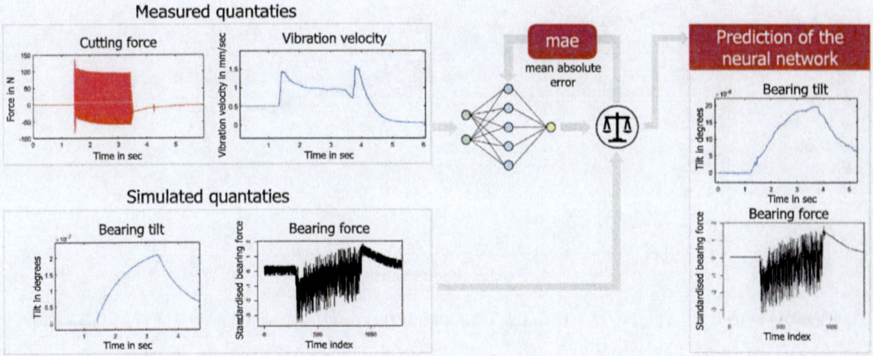

Fig. 3. Schematic representation of the training process how the vibration velocities are being mapped to the surface pressure.

Alternative approaches were also explored by training models using different input parameters, including measured forces and sheer velocities. However, the performance of these models was found to be inferior to the model trained with spindle vibration velocity and cutting parameters. Specifically, the model trained with forces as an input exhibited the poorest performance as can be seen in Table 1. It shows the maximum and mean deviation between the surface pressure calculated by simulation and estimated using the three different ML-models described above. The values were obtained by evaluating all 243 cutting experiments using both simulation and machine learning prediction.

Table 1. Maximum and mean deviation between the surface pressure calculated by simulation and estimated by means of three different ML models.

Type of input	Maximum deviation	Mean deviation
Cutting forces only	16.3%	3.3%
Vibration velocity only	9.8%	1.8%
Vibration velocity and cutting parameters	1.8%	0.6%

A key finding in the research was the benefit of using a rolling window time lag feature, which enhanced the predictive capability of the trained neural networks. By considering the temporal aspects of the vibration data, the models were able to capture the dynamic behavior and fluctuations in the bearing variables during operation better especially at the beginning and end of cut. Figure 4 shows the mean absolute error (MAE) and coefficient of determination R^2 depending on the size of the time lag window. It can be seen that the model with the largest R^2 and with the smallest error has the largest possible time lag. Small time lags tend to lead to a poorer prediction.

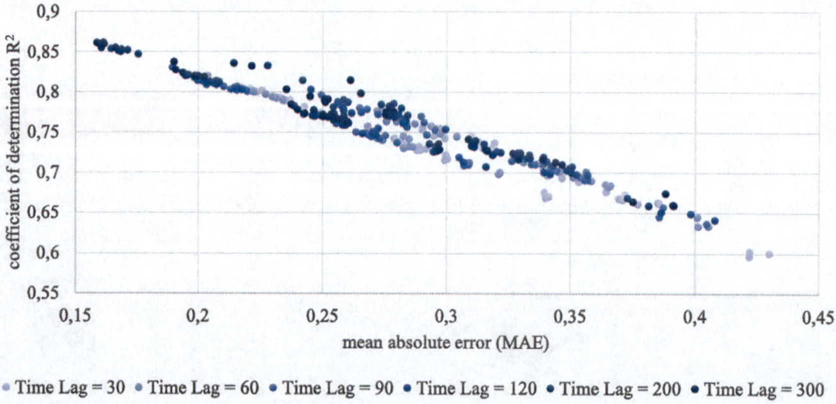

Fig. 4. Coefficient of determination and mean absolute error of trained neural networks depending on the size of the rolling window time lag feature.

In further investigations it was discovered that the trained neural networks could be executed efficiently in TwinCAT 3 on an Beckhoff industrial PC, serving as a soft PLC. By integrating the trained neural networks into the industrial PC and retrieving the sensor data using an analog input terminal, real-time predictions of the simulation results became possible.

3 Results

The comparison between the maximum surface pressure calculated using the simulation results and the maximum surface pressure estimated by the neural network revealed close agreement. The maximum deviation between the two was found to be smaller than 2%.

The execution time of the neural networks on the Beckhoff industrial PC was less than 70 μs, enabling real-time predictions within the machine's interpolation cycle of 1 ms. This capability allows for immediate monitoring and assessment of the surface pressures during operation, enabling timely adjustments and interventions to prevent excessive stresses on the spindle bearings. However, the prediction quality is not satisfactory since the signal processing using the applied Beckhoff EL3632 terminal is different to the IBIS SiViB 1001. The devices are using a digital and analog preprocessing respectively. Additionally, a significant time has passed between the data acquisition, training and deployment of the neural network on the Beckhoff PC. Wear in the machine and deterioration of sensor signals caused by aging and unwanted moisture are other reasons why the network is not working as well as during testing with the original data set.

4 Conclusion

Overall, the development of machine learning-based methods and the use of neural networks proved to be a practical and efficient approach for predicting surface pressures in real-time. By leveraging vibration data, cutting parameters and the utilization of a rolling window time lag feature it was possible to achieve accurate and timely predictions of maximum surface pressure. The integration of these models into a PLC shows the applicability for industrial application. These trained neural networks have the potential to enhance the monitoring and maintenance strategies of motor spindle systems by providing valuable insights into the surface pressure distribution inside the spindle's bearings.

To keep the prediction capabilities on a high level over a long period of time when deployed on an actual control unit further research is needed. The limitations in prediction accuracy can be addressed by adopting a continual learning approach. Continual learning allows the neural networks to adapt and improve over time by incorporating new data and experiences, thereby enhancing their prediction capability and robustness.

Currently, we are in the process of establishing and testing an edge-to-cloud environment that enables continuous transfer of data from the Beckhoff PLC to a virtual machine for simulation and training purposes. This environment also includes a mechanism to automatically deploy newly retrained neural networks when the prediction accuracy falls below a predefined threshold.

Given the time-consuming nature of the procedures described in this paper, we are also working on a solution that utilizes the currents of the feed axis to calculate cutting forces. These forces will then serve as inputs for a spindle simulation using the more widely used software MESYS. This approach allows to calculate bearing quantities related to the bearing load, using the currents the control unit can provide in real time.

To also ensure real-time capability of the MESYS simulation, neural networks will be trained on the simulation results, following the procedure outlined in Sect. 2.5 of this paper. This integration of neural networks with the MESYS simulation further enhances the efficiency and effectiveness of the overall monitoring and prediction system for motor spindle bearings.

Recently eddy-current sensors have been installed in the motor spindle to measure the displacement of the spindle shaft in all the directions. They will serve to further validate the simulation.

References

1. Neugebauer, R., Fischer, J., Praedicow, M.: Condition-based preventive maintenance of main spindles. Prod. Eng. Res. Dev. (2011). https://doi.org/10.1007/s11740-010-0272-z
2. Schuh, G., Kampker, A., Franzkoch, B., Wemhöner, N.: Intelligent Maintenance. Potenziale zustandsorientierter Instandhaltung (2010)

3. Wu, G., Yan, T., Yang, G., Chai, H., Cao, C.: A review on rolling bearing fault signal detection methods based on different sensors. Sensors (Basel, Switzerland) (2022). https://doi.org/10.3390/s22218330
4. Sarhan, A.A.D., Matsubara, A., Sugihara, M., Saraie, H., Ibaraki, S., Kakino, Y.: Monitoring method of cutting force by using additional spindle sensors. JSME Int. J. Ser. C (2006). https://doi.org/10.1299/jsmec.49.307
5. Surucu, O., Gadsden, S.A., Yawney, J.: Condition monitoring using machine learning: a review of theory, applications, and recent advances. Expert Syst. Appl. (2023). https://doi.org/10.1016/j.eswa.2023.119738
6. Sheng, H., Chen, Z., Xia, Y., He, J.: Review of artificial intelligence-based bearing vibration monitoring. In: 2020 11th International Conference on Prognostics and System Health Management (PHM-2020 Jinan). 2020 11th International Conference on Prognostics and System Health Management (PHM-2020 Jinan), Jinan, China, 23 October 2020–25 October 2020, pp. 58–67. IEEE (2020 - 2020) (2020). https://doi.org/10.1109/PHM-Jinan48558.2020.00018
7. Vaishnav, S., Agarwal, A., Desai, K.A.: Machine learning-based instantaneous cutting force model for end milling operation. J Intell Manuf (2020). https://doi.org/10.1007/s10845-019-01514-8
8. Abele, E., Weber, M., Daume, C.: Spindelwälzlagerbelastung beim trochoiden Fräsen*/Spindle bearing load during trochoidal milling. wt Werkstatttechnik online (2018). https://doi.org/10.37544/1436-4980-2018-10-94
9. Schall, D.: Spindelschwingungen beim Hochgeschwindigkeitsfräsen. Unter besonderer Berücksichtigung der Lagerbeanspruchung. Darmstädter Forschungsberichte für Konstruktion und Fertigung. Shaker, Aachen (2006)
10. Brändlein, J., Eschmann, P. (eds.): Die Wälzlagerpraxis. Handbuch für die Berechnung und Gestaltung von Lagerungen, 3rd edn. Vereinigte Fachverl., Mainz (1998)

Peak Shaving in a Common DC-Grid by Synchronization of High-Load Machine Operations

Andreas Wächter(✉), Magnus von Elling, Daniel Fuhrländer-Völker, Stefan Schmitt, and Matthias Weigold

Department of Mechanical Engineering, Institute for Production Management, Technology and Machine Tools (PTW), Technical University of Darmstadt, Otto-Berndt-Str. 2, 64283 Darmstadt, Germany

a.waechter@ptw.tu-darmstadt.de

Abstract. In the past decades, machine tools have been optimized for productivity, which has led to high feed rates and accelerations of drives. These high accelerations cause peak loads in the power grid. The electrical infrastructure for the machine supply must be designed to handle these peak loads even if peak loads occur simultaneously on several machines working in parallel. By reducing the resulting overall peak loads in the grid, components can be designed for lower currents and thus more resource-efficiently.

In this study we investigated how total peak loads in a shared DC-grid can be lowered by synchronizing high-load-operations. It is possible to slightly delay individual load-intensive NC steps via OPC UA so peak loads of the individual machines no longer occur simultaneously. The approach is implemented on two machine tools. In a field test we achieve a peak load reduction of up to 8.3%. In a simulation, we show that a reduction of almost 20% can be achieved by extending the approach to a network of six machine tools.

Keywords: DC-Network · OPC UA · Peak Load Shifting

1 Introduction

The realization of climate protection measures in line with the 2015 Paris Agreement to limit global warming to 1.5 °C is considered one of the key challenges of the coming years [1]. To pursue this goal, the German Climate Protection Plan formulates the intention to reduce net greenhouse gas emissions to zero by 2045 [2]. Energy supply and consequently industry as a major energy consumer deserve special attention in this context. The reference scenario assumes that the total energy consumption of German industry can be reduced from 735 TWh in 2008 to 498 TWh in 2050 [3]. Energy savings are expected in the provision of process heat and space heating [3]. Furthermore, the reference scenario assumes possible savings of 7.3% for mechanical energy in industry [3].

In machine tools, the required mechanical energy is usually provided by electric drives [4]. An approach to achieve savings in the context of such electrical drives are smart

T. Bauernhansl et al. (Eds.): WGP 2023, LNPE, pp. 257–267, 2024.
https://doi.org/10.1007/978-3-031-47394-4_26

direct current (DC) micro grids in industrial environments [5]. DC energy supply grids promise potentials to reduce the material effort – especially copper – in terms of wiring and device dimensioning and support the use of energy storage and photovoltaic systems in the production environment [6]. In addition, conversion losses between alternating current (AC) and DC grids can be reduced if consumers are supplied by DC [6]. Another central aspect of industrial DC supply is the optimized use of recuperated energy, which is fed into the machine's DC-link during deceleration activities [7]. Existing model plants suggest that savings of 20% are realistic through the reduction of conversion losses and recuperation of mechanical to electrical energy [8].

Electrical infrastructure is usually dimensioned to be able to supply short-term peak loads of connected systems [9]. However, load peaks only occur for a short period of time, so that partial load operation is predominant. The smoothing of the common load curve of the connected systems allows components such as inverters to be operated closer to the nominal point and thus more efficiently [10].

This paper presents an approach to operate two machine tools in a common DC grid in such a way that the overall load curve is smoothed, and maximum peak loads are reduced. For this purpose, NC (numerical control) steps and consequently accelerations of the drives are synchronized in such a way that simultaneous load peaks of the machines occur shifted in time. A smoothing of the overall load curve is enabled by this. The approach is implemented and evaluated on a test stand with two machine tools that are addressed via OPC UA (Open Platform Communication Unified Architecture). In addition, the operation of a network of six machine tools is simulated and the system's ability to feed recuperated energy back into the DC-link and the AC grid is investigated.

2 Fundamentals

In current machine tools and in industrial drive technology in general, both AC and DC conversion stages can be found. Their fundamentals in the context of the energy supply of machine tools are discussed below.

2.1 DC-Networks in Industry

AC network topologies have been the standard of our electricity system for many decades [10]. However, in recent years the focus on DC applications as an alternative has been increasing in various research projects [10]. One reason is that an increasing share of generators of electrical energy produce DC voltage - photovoltaics, for example [10]. The generated electrical energy is fed into the AC grid by means of rectifiers [10]. In addition, more and more devices are being used that require DC to operate anyway – for example computers, LED lighting and charging stations for electric cars [11].

In a modern production environment, many electric drives are supplied via frequency converters, which in turn are fed from a DC circuit [7]. Losses occur during the necessary conversion between AC and DC as well as when converting in the opposite direction [7]. At the same time, modern machine tools and industrial robots have high dynamics and consequently high proportions of acceleration and braking energy [10]. By using a common DC-bus at factory level, the recuperated energy could be used by other devices without additional conversion stages [10]. Furthermore, a balancing of the grid could take place due to the large number of frequency-controlled electric drives, so that energy losses in brake resistors can be reduced [10].

2.2 Peak Loads of Machine Tools

Cutting machine tools are mechanized and partly automated systems that produce a predefined shape on the workpiece by removing chips [4, 12]. In addition to control, safety and positioning functions, Mori et al. mention an energy supply as a characteristic feature of machine tools [12]. Typically, electric energy is used to operate spindles and feed-drives but also auxiliary units such as cooling systems or pumps for hydraulics and lubricant [4]. Variable speed drives such as spindles and feed drives are supplied via a DC link [13]. The frequency and voltage for the drive are modulated from this circuit by means of an inverter so that the required speed is achieved [13]. This machine internal DC-link is typically supplied by the AC mains connection via a rectifier [13].

During operations, short duration peaks of high amplitudes of power can be detected at the machine tools mains connection [14]. These peaks result from rapid positioning of the axes and acceleration of the spindle [14]. When decelerating axis or spindles, mechanical energy is converted back into electrical energy by the drives and fed into the DC link [14]. To maintain the level of voltage in the DC link, the energy is converted into heat by braking choppers or partially fed back to the AC grid via an inverter [7, 14].

2.3 NC-Control of Machine Tool Operations

The sequence of positioning, cutting and feed movements of the axis and spindles are controlled by the NC unit [4]. The basic commands, the syntax and the program structure for NC programs are standardized by DIN 66025 [15]. Geometric path conditions, such as the positioning of the axes, are described with G commands – for example rapid traverse operations are described by G00 [15].

2.4 Machine Communication via OPC UA

In the field of industrial communication, one standard has stood out in recent years. This communication standard is OPC UA, which is a platform-independent system interface that integrates all the functions of its predecessor OPC Classic and offers enhanced modelling capabilities [16]. Its functionality is completely independent of the underlying operating system and the hardware used [16]. This enables cross-manufacturer communication independent of the underlying hardware and software [17]. In addition to platform independence, further features have been included in the standard for communication between different systems [17]. These include scalability, robustness, fault tolerance, redundancy, integrated security mechanisms, interoperability, web compatibility and the possibility of sophisticated data modelling [17].

By using OPC UA, devices from different manufacturers can provide their data and services for interoperability [18]. A uniform interface such as OPC UA offers the advantage that different communication solutions can be united on one transmission mechanism, thus aiming at higher security in data transmission [18]. OPC UA provides two methods for data exchange, one of which uses the client-server model, which ensures secure information exchange based on the TCP (Transmission control protocol) and HTTPS (Hypertext transfer protocol secure) protocols [18]. However, the number of connections for this model is limited [18]. An unlimited number of devices can receive data from the OPC UA server when a publisher subscriber (pub sub) approach is used. In this case the server would function as a publisher and the clients are subscribers [19]. However, this standard is not readily available on most devices and has therefore no great relevance yet [19]. Communication in the client-server model makes use of standardized services, where the OPC UA Client as application requests the data provided by the OPC UA Server [17, 19]. A variety of devices can function as OPC UA servers, including field devices, controllers, desktop computers, and IT (Information Technology) servers [20]. In contrast to OPC Classic, applications in an OPC UA network can function as both server and client, whereby on the one hand the server can interact with several clients simultaneously and on the other hand each client can connect to several servers [20]. This way it is possible, for example, to write values on a PLC (Programmable Logic Controller) of a machine by using the OPC UA server on the controller and thus to change the behavior of the machine.

3 Method

To smooth the overall load curve of a machine tool network, an approach was pursued to coordinate and, if necessary, shift load-intensive operations. The procedure used is described in the following.

3.1 Identification of Load-Intensive NC Operations

The experimental setup of this study focuses on quantifying the peak load and identifying associated NC steps for subsequent peak shaving. For the peak shaving algorithm to operate effectively, we must categorize the peaks identified through experimental power measurements as relevant or prominent. For this purpose, we assigned all peaks to the corresponding NC steps of the NC-program used. We defined relevant peaks to encompass all load peaks with power consumptions exceeding 15% of the largest individual load peak of the example program executed on one of the two machines. On the other hand, we defined prominent peaks as those surpassing 40% of the peak load observed in the mentioned NC step. These values were determined empirically and proofed to give the best results when used with the algorithm. To facilitate analysis, these peaks can be presented in a matrix format with the second machine's peaks, allowing for an assessment of the combined total consumption when they occur simultaneously (see Fig. 1). This classification framework assists identify peaks where postponement strategies as part of the peak shaving algorithm explained in Sect. 3.2 are promising.

During the experimental analysis, it was observed that the occurrence of particularly high peaks was predominantly associated with two specific scenarios. Firstly, when the spindle of the machine started, significant power spikes were detected. Secondly, when the vertical axis of the machine rapidly traversed in an upward direction, we observed notable peaks in power consumption which can be attributed to the increased energy requirements for lifting the machine's spindle against gravity. Understanding the peaks that occur during these operations provides valuable insights for implementing targeted peak reduction strategies.

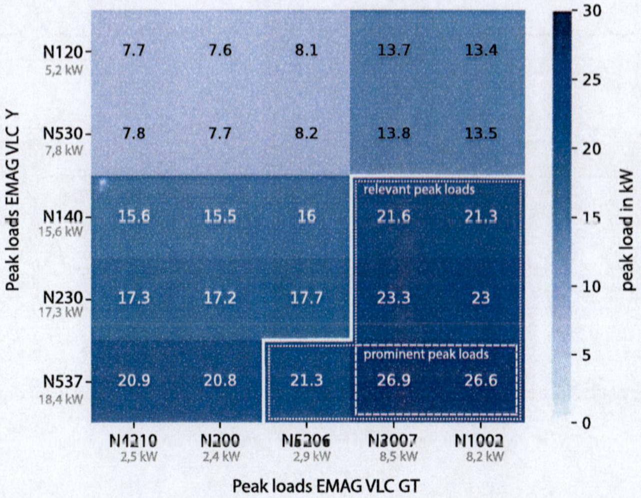

Fig. 1. Peak load matrix of two machines, showing relevant and prominent peaks and NC-steps that are targeted by the peak shaving algorithm.

3.2 Algorithm for Sequential Shifting

The peak load optimized operation of the machines in the connected grid is to be carried out via the sequential delay of individual NC steps. Menz describes sequential shifting as a possibility for peak shaving on machine tools [9]. In this process, individual sub-steps within an NC program are shifted in time to achieve an optimized overall load profile (see Fig. 2). Compared to a one-time shift, the sequential shift offers the advantage that it can also react to unscheduled events in the production process. To implement the measure, the NC code of the tool machine is read in line by line and interpreted according to the DIN 66025 standard. With the aid of a Python script, hold commands are inserted before the NC steps previously identified as load-intensive and the modified NC code is loaded into the machine tool. In the program sequence of the turning center, the hold variables are only set to active when there are simultaneous load peaks on the machine tools. If one of the two machines is in a load-intensive machining step, the holding variable is activated for the other machine tool. If a previously identified load-intensive machining step also occurs at other machining center, machining is extended by the time *t_wait*. As soon as the respective NC step is completed, the hold variable is deactivated again, and machining of the semi-finished products is continued. The maximum load peak can thus be reduced by ΔP, while at the same time the machining time is extended by the waiting times. The measure can be understood as a process interruption measure according to VDI guideline 5207 [21].

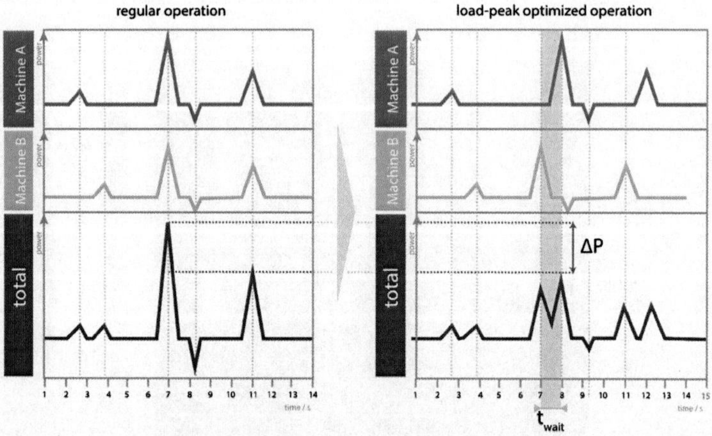

Fig. 2. Schematic figure of the desired effect on the overall load profile of a machine network

4 Implementation and Evaluation

The described method for reducing maximum peak loads in a network of machine tools was implemented and tested on a test stand with two real machine tools.

4.1 Test Stand with Two Machine Tools

The main components of the setup evaluated are two machine tools, one of which is an EMAG VLC 100Y lathing machining center and the second an EMAG VLC 100GT grinding machining center. In the setup, outlined in Fig. 3, both machines are connected to a common DC-link, which replaces the individual machine DC-links. The auxiliary units of the machines, for example hydraulic or cooling lubricant pumps, are connected to the AC-grid (not shown in Fig. 3) and are not supplied by the DC-link. Consequently, the DC side of the setup represents a local DC network from which the machine drives are fed. To quantify the power supplied to the individual machines, measuring points with QI-POWER-485 sensors are installed for each machine. At the AC mains connection, a Janitza UMG604 power analyzer is implemented to measure power fed from or to the surrounding AC grid.

4.2 Implemented Machine Communication Using OPC UA

The grinding machining center is the bottleneck of the process chain under consideration, so the grinding process is not interrupted, and the sequential shift is carried out on the turning machining center.

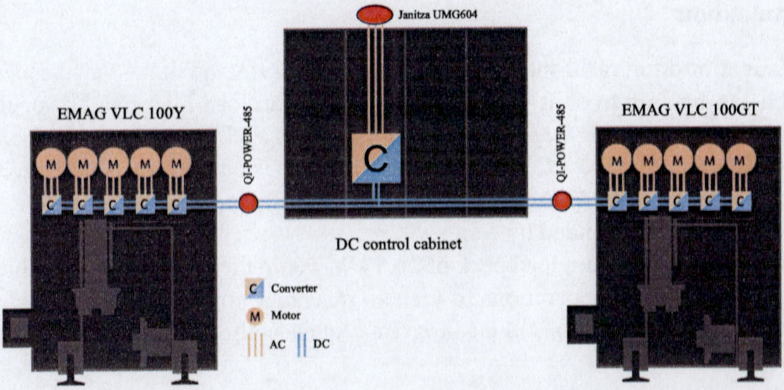

Fig. 3. Setup of two machine tools with a common DC-link and power measuring points.

To implement the peak shifting measures, an edge device is integrated in the network, in this case a desktop PC, and the connection to the OPC servers of the machine tools is established. The information relevant for machining, such as the current NC step, is made available on the machine tool controllers and read out by the edge device. To influence the braking and acceleration processes, the original NC code is extended by the corresponding CPL (customer programming language) commands. Wait commands are used to reduce peak loads. Here, the edge device monitors the respective valid NC steps of the machine tools and sets the hold commands as soon as a load-intensive step occurs at one of the machine tools. The edge device resets the waiting variable as soon as the load-intensive machining step is completed. A schematic depiction of how the edge device is communicating with the control units to activate the waiting variables is shown in Fig. 4.

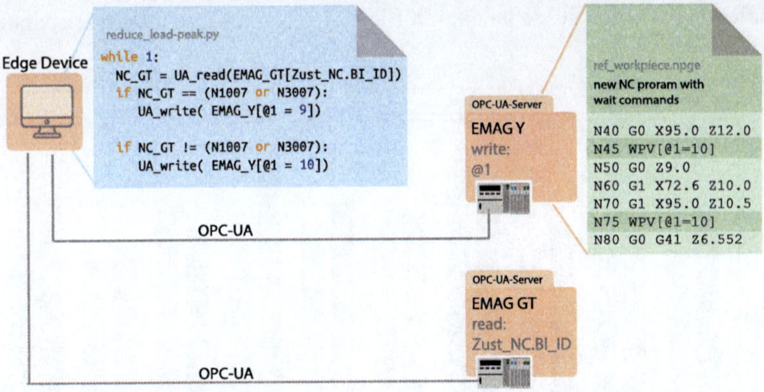

Fig. 4. Schematic representation of the peak load reduction process using an edge device and OPC UA communication.

4.3 Evaluation

An investigation during a 30-min production run on the test stand shows that the presented approach can be used to shift peak loads resulting from load-intensive NC operations using hold commands. For the test stand, a reduction of 1.85 kW or 8.5% in the maximum power consumption at the AC mains connection point is determined. Over the 30-min period, the program run of the turning center was postponed by a total of 1 s, so the processing time was extended by 1 s.

In addition, a maximum load peak of 24.5 kW would theoretically be possible if the two most load-intensive operations of the two machines coincided during production. Through targeted interruptions in the program sequence, this could be reduced by up to 29% to 17.4 kW.

4.4 Simulation of Larger Machine Networks

The extension of the described sequential load shifting in a DC grid to a larger machine group is being investigated by simulation. In a scenario, a network of six machine tools is considered over a period of eight hours. The three grinding and three lathing centers included are equivalent in their characteristics to the machines and machining programs described above. Equal to the implementation described above, only the NC-operations of the turning centers are shifted. For the load profile of the individual machines during processing, an average load profile is assumed based on several measurements on the real machines. The start times for the machining cycles on the individual machines are selected randomly.

Figure 5 shows the results of ten simulation runs, each corresponding to a production period of 8 h. On average, the maximum peak load is reduced by 4.9%. The largest reduction in the maximum peak load is observed in simulation run 7 (by 16.2%). The machining cycles are totally shifted by ΔT_total_average = 42.6 s (average of 10 simulations) during 8 h of production. For two simulation runs (4 & 8), the maximum peak load increases by 3.6% with sequential shifting. This is since load intensive operations of

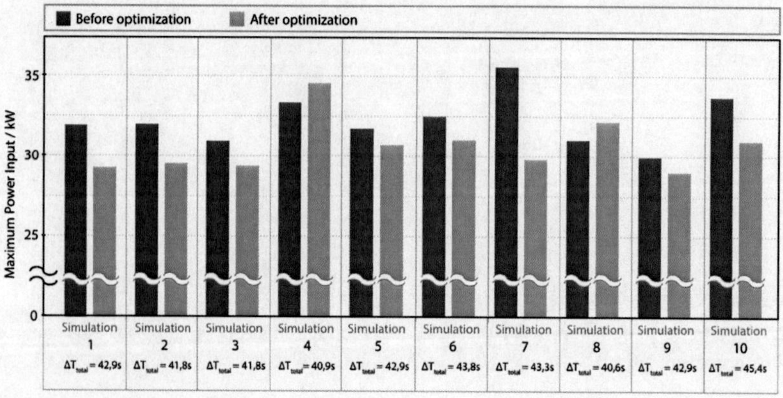

Fig. 5. Simulated maximum peak loads for ten 8 h intervals.

two machines, that were not enabled for sequential shifting, occur almost simultaneously during processing.

4.5 Recuperation of Braking Energy

In a further evaluation, the level of recuperation of braking energy between the individual conversion stages was considered. As a criterion, the amount of energy fed back was compared in relation to the amount of energy initially supplied at the respective measuring point (see measuring points in Fig. 3). For example, the amount of energy that was fed into the setup and recuperated for the operation of the grinding machine at the AC connection point, as well as the amount of energy at the machine-specific DC connection point, was used. For the second machine, the procedure was equal. We used an observation period of 30 min, during which the machines were in continuous operation.

The example of the lathing machining center (EMAG VLC 100Y) shows a share of 6.17% of recuperated energy into the DC-link (see Table 1). With an additional conversion stage, the share of energy recuperated into the AC grid is 1.40% which is 77% less than the recuperation into the DC-link. For the grinding machining center (EMAG VLC 100GT), the total share of recuperated energy is lower. However, the share of losses during recuperation into the AC grid shows a similar tendency and is even higher.

One possible cause for the large proportion of conversion losses is the dimensioning of the components in the function of a test stand. In the selected scenario, when operating only one machine tool at a time, the maximum load of the inverter is 10.8%. Efficiency losses are assumed due to partial load operation. The total time share of peak loads in the total operation is 0.9% (EMAG VLC 100Y) and 0.2% (EMAG VLC 100GT). Through better adapted dimensioning and smoother load profile, higher full-load shares and a better recuperation efficiency could be achieved.

Table 1. Comparison of recuperated energy between DC- and AC-Grid

	Share of recuperated energy (%)	
System Boundary	EMAG VLC 100Y	EMAG VLC 100GT
Drives to DC-link	6.17	0.90
Drives to AC-Grid	1.40	0.04

5 Discussion and Conclusion

The presented approach shows that machine-specific load peaks can be shifted by means of hold commands and thus a reduction of the maximum peak load in a machine network is possible. For implementation it is possible to use common machine communication like OPC UA. The overall delay in the production process remains low. In the presented case, this is 1 s over a period of 30 min.

Furthermore, the simulation of a larger machine group of six machines shows that a reduction of the maximum peak load can also be achieved. However, the reduction is lower on average in the simulation of the machine network. This is partly because only the turning centers in the simulation could shift their peak loads and, in some cases, peak loads of several grinding centers still occurred simultaneously. From this it is deduced that, to optimize the peak load reduction in the machine network through sequential shifts, all machines in the network should be enabled by hold commands. Overall, a smoothing of the load curve of the machine network can be achieved both on the test stand and in the simulation. In addition, the investigation of the regenerated energy on the test stand shows that a common DC-link between machine tools offers advantages in terms of energy recuperation.

Continued work aims to measure the machine's current at a higher frequency and transmit the data via OPC UA pub sub. This is to achieve a better resolution so no peak loads are missed in the measurement. Using the pub sub approach allows a high frequency data transfer in a standardized way.

Acknowledgements. The authors are grateful to the Federal Ministry for Economic Affairs and Climate Action (BMWK) for funding the presented research in the projects KI4ETA (grant number 03EN2053A) and ETA im Bestand (grant number 03EN2048A-I) and to Project Management Jülich for coordinating the projects.

References

1. Luderer, G., Kost, C., Sörgel, D. (ed.): Ariadne-Report: Deutschland auf dem Weg zur Klimaneutralität 2045. Potsdam Institute for Climate Impact Research (2021)
2. Deutscher Bundestag: Bundes-Klimaschutzgesetz (KSG) (2019)
3. Klaus, T., Vollmer, C., Werner, K., Lehmann, H., Müschen, K.: Energieziel 2050. 100% Strom aus erneuerbaren Quellen. https://www.umweltbundesamt.de/sites/default/files/medien/378/publikationen/energieziel_2050.pdf. Accessed 11 June 2023
4. Hirsch, A., Regel, J.: Werkzeugmaschinen und Vorrichtungen. Anforderungen, Auslegung, Ausführungsbeispiele, 4th edn. Springer, Wiesbaden (2022). https://doi.org/10.1007/978-3-658-37658-1
5. Weckmann, S., Sauer, A.: DC micro grid for energy efficient and flexible production. Procedia Manuf. **39**, 655–664 (2019)
6. Schaab, D.-A., Weckmann, S., Kuhlmann, T., Sauer, A.: Simulative analysis of a flexible, robust and sustainable energy supply through industrial smart-DC-grid with distributed grid management. Procedia CIRP **69**, 366–370 (2018)
7. Brosch, P.-F.: Moderne Stromrichterantriebe. Antriebssystem, Leistungselektronik, Maschinen, Mechatronik und Motion Control, Arbeitsweise drehzahlveränderbarer Antriebe mit Stromrichtern und Antriebsvernetzung, 5th edn. Vogel, Würzburg (2008)
8. BMW Group: Nachhaltige Fahrzeugproduktion mit smarter Gleichstromtechnik. Beteiligung des BMW Group Werks Dingolfing am Forschungsprojekt DC-INDUSTRIE 2. https://www.bmwgroup-werke.com/content/dam/grpw/websites/bmwgroup-werke_com/dingolfing/aktuelles/200610-dc-industrie2/PM_BMW_DGF_DC2.pdf. Accessed 09 June 2023
9. Menz, B.: Methode zur digitalen Vernetzung von Produktionsmaschinen zur Lastganggglättung im Betrieb. Dissertation. Shaker Verlag, Aachen (2017)

10. Sauer, A. (ed.): Die Gleichstromfabrik. Energieeffizient. Robust. Zukunftsweisend. Hanser, München (2020)
11. Juschkat, K.: AC vs. DC. Warum Gleichstrom eine echte Alternative ist. https://www.elektrotechnik.vogel.de/warum-gleichstrom-eine-echte-alternative-ist-a-625484/. Accessed 15 June 2023
12. Mori, M., Hansel, A., Fujishima, M.: Machine tool. In: Chatti, S., Laperrière, L., Reinhart, G., Tolio, T. (eds.) CIRP Encyclopedia of Production Engineering, 2nd edn., pp. 1093–1103. Springer, Berlin (2019). https://doi.org/10.1007/978-3-662-53120-4_6533
13. Hagl, R.: Elektrische Antriebstechnik, 3rd edn. Hanser, München (2021)
14. Dietmair, A., Verl, A.: A generic energy consumption model for decision making and energy efficiency optimisation in manufacturing. Int. J. Sustain. Eng. **2**(2), 123–133 (2009)
15. DIN German Institute for Standardization: DIN 66025-1: Programmaufbau für numerisch gesteuerte Arbeitsmaschinen; Allgemeines. Beuth, Berlin (1983)
16. Hoffmann, M.: Smart Agents for the Industry 4.0. Enabling Machine Learning in Industrial Production. Springer, Wiesbaden (2019). https://doi.org/10.1007/978-3-658-27742-0
17. Mahnke, W., Leitner, S.H., Damm, M.: OPC Unified Architecture. Springer, Heidelberg (2009). https://doi.org/10.1007/978-3-540-68899-0
18. Schleipen, M. (ed.): Praxishandbuch OPC UA. Grundlagen - Implementierung - Nachrüstung - Praxisbeispiele, 2nd edn. Vogel Communications Group, Würzburg (2020)
19. Langmann, R.: Vernetzte Systeme für die Automatisierung 4.0. Hanser, München (2021)
20. Pethig, F., et al.: Industrie 4.0 Communication Guideline Based on OPC UA. VDMA Guideline. Verband Deutscher Maschinen- und Anlagenbau e.V., Frankfurt (2017)
21. VDI Verein Deutscher Ingenieure: VDI-Standards 5207-2: energy-flexible factory - identification and technical evaluation. Beuth, Düsseldorf (2021)

Towards a Domain Specific Language for the Development of Distributed Real-Time Systems

Moritz Walker(✉), Marc Fischer, Michael Neubauer, Armin Lechler, and Alexander Verl

Institute for Control Engineering of Machine Tools and Manufacturing Units (ISW), University of Stuttgart, Stuttgart, Germany
moritz.walker@isw.uni-stuttgart.de

Abstract. Static configurations and slow adaption to changing requirements characterize today's production systems. In order to cope with variable external influences, these systems need to become more flexible. One basic building block to achieve more flexible automation systems from the software perspective are modular real-time architectures which are realized as distributed real-time systems. The improved modularity and flexibility come at the price of greater complexity. In order to enable developers to design robust systems regarding real-time properties, a domain specific language is proposed, which allows reasoning and validation of modules and composed distributed real-time systems.

Keywords: Real-Time · Distributed Systems · Domain Specific Language

1 Introduction

The volatility, uncertainty, complexity and ambiguity of the so called VUCA world has strong impact on the economy and thus on producing companies. These influences require flexible production systems to respond to changing demands. However, production systems are often characterized by static configurations and slow adaption to changing requirements, which make it difficult for such systems to compete in a VUCA world [1]. In order to react to the various external influences, production systems must therefore become more flexible. One way to improve the flexibility of production systems is by applying the concepts of Software-defined manufacturing (SDM) [1].

SDM, like other software-defined approaches, requires consistent abstraction and modularization. The main requirements for a software architecture enabling SDM revolve around flexibility and service orientation, which can be realized as Micro Service Architectures (MSAs). Virtualization, e.g., software containers and virtual machines, enables hardware-independent deployment. Instead of predetermined hardware, a software module can be placed on any suitable computing node. The flexibility of manufacturing systems can be increased through reconfiguration of the automation application

T. Bauernhansl et al. (Eds.): WGP 2023, LNPE, pp. 268–279, 2024.
https://doi.org/10.1007/978-3-031-47394-4_27

software. Depending on the requirements of the product to be manufactured, the corresponding software can be configured at runtime by selecting and composing the appropriate services. In SDM, a product and job description and knowledge of the underlaying manufacturing system are used to make these reconfiguration decisions.

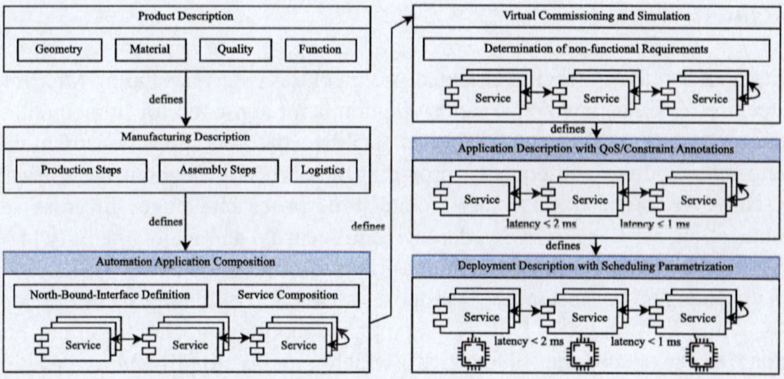

Fig. 1. Conceptual framework of SDM [1] extended with the deployment description. The areas of application of the proposed language are marked blue.

In [2] these decisions are made by the Software-defined. Part of this decision-making process is the specification of a suitable automation application composition, which enables the control of the underlaying manufacturing resources. As some of the involved services have real-time requirements, the assembled automation applications result in a distributed real-time system (DRTS) based on the MSA paradigm. These service compositions are given the process-related constraints such as cycle times and maximum permissible delays. Based on the requirements, an orchestrator decides on which compute node to deploy the services so that all real-time requirements are met (see Fig. 1).

In order to enable developers, IT-administrators and higher level SDM systems to design robust systems regarding real-time properties and to facilitate the integration of the various software systems involved; this work proposes the use of domain specific languages (DSLs). DSLs are one element of Model Driven Engineering (MDE) to abstract complexity to enable easier and faster handling of complex systems.

The following research question is therefore answered in this paper: "What must a DSL look like, to map the life cycle of a DRTS consisting of services?" The life cycle covers the description of real-time requirements, the service composition, the usage of different scheduling models, the reasoning and validation of modules and composed distributed real-time systems and the deployment. By answering this question, we present the Distributed Real-Time System Language (DRTSL) as the contribution of this paper. It allows reasoning and validation of services and composed distributed real-time systems without assuming a fixed scheduling model (see Fig. 1), enabling the user to select the most suitable scheduling model for the corresponding use case. Furthermore, it enables the description of constraints and QoS annotations, and the generation of deployment description files including the scheduling parametrization of individual services.

The remainder of the paper is structured as follows: Sect. 2 covers related work. Based on a requirements analysis in Sect. 3, DRTSL is proposed in Sect. 4 and qualitatively validated based on a use-case in Sect. 5. A conclusion, limitations and directions for future work are presented in Sect. 6.

2 Related Work

In contemporary times, Service-Oriented Architectures (SOA) including Microservices Architectures (MSAs) have emerged as the dominant approach for implementing distributed systems [3]. These architectures facilitate the decomposition of monolithic software into modular components, providing flexibility throughout the software life cycle. However, orchestrating such architectures poses challenges in areas such as communication, service discovery, deployment, security, and maintainability [4–6].

In the context of SOAs, several studies have examined the use of DSLs to address specific challenges. For instance, [7] proposes a DSL for validating security features in a MSA, while [8] presents a DSL for compliance monitoring. Similar to our research question, [6] introduces the Silvera DSL, which aims to simplify MSA development by generating boilerplate code and providing a unified description of service properties and interfaces. Similarly, the jHipster framework [9] facilitates code generation based on Java language descriptions. However, none of these works consider the life cycle of real-time systems within MSAs.

Although the modeling of DRTS is a well-explored topic in literature, our research aims to combine the modeling of DRTS with the life cycle considerations of MSAs. By doing so, we address a significant research gap and provide a comprehensive approach for incorporating real-time requirements into the life cycle management of MSAs.

3 Requirement's Analysis

To identify the requirements for the DRTSL, we conduct a detailed analysis of the life cycle of a DRTS built upon the SOA pattern. Figure 2 provides an overview of the crucial information that the involved software components require. The life cycle begins with an examination of the scheduling model employed and the composition of the Distributed Real-Time Application (DRTApp). Subsequently, a scheduler, which can either be integrated into a real-time orchestrator [10] or an external tool, assigns scheduling parameters to deployment units such as containers or virtual machines that encapsulate individual services. To draw conclusions about the real-time behavior of the DRTApp to be deployed, specific validators associated with the scheduling model and user-defined validators are utilized. Finally, service configuration files and deployment descriptions are generated for the runtime environment of a particular deployment unit, such as Kubernetes or Docker for containers.

The DRTApp description must be expressive enough to describe services and interconnections over a common communication system. In addition, the operating system level tasks of a service must be described to enable scheduling analysis, as this information cannot be abstracted away. Furthermore, constraints, QoS requirements, and simple resource requirements of services and their counterpart resource supplies of compute nodes, such as static memory allocations, must be described.

The scheduler analyzes the DRTApp description and determines a specific deployment strategy. Certain schedulers designed for specific applications can optimize the fulfillment of real-time constraints for the DRTApp. Note, that this high-level scheduler is not the same as the schedulers of the operating systems of compute nodes, which schedule real-time threads at runtime. However, some methods still require the validation of real-time properties even after scheduling parameters have been assigned. This is particularly necessary when application-specific requirements must be met, which are not satisfied by the default algorithm. For example, end-to-end (e2e) delays of service chains need to be considered, as described in more detail below. If a node assignment algorithm, such as [11], is employed, it may overlook communication dependencies between services and focus solely on the schedulability of services' real-time threads.

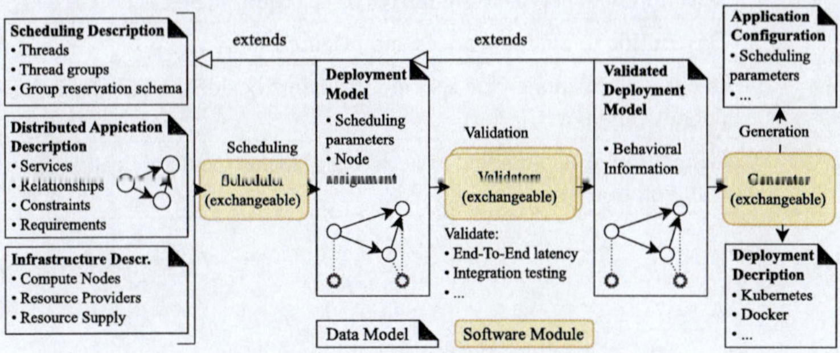

Fig. 2. Life cycle of a DRTapp's model.

Detailed conclusions about DRTApps require application models that depict the structures and communication patterns among services, as well as the scheduling of services at both compute node and application levels. A crucial metric for evaluating the composed application is the e2e latency, which indicates the time that elapses between the response of a service and its triggering. Various approaches have been developed for scheduling DRTApps with the objective of meeting end-to-end latencies. These methods encompass compositional performance analysis (CPA) [12], worst-case response time analysis [13], and decomposition-based algorithms [14]. CPA and worst-case response time analysis are analytical techniques used to calculate end-to-end latencies, while decomposition-based algorithms break down the distributed real-time application into parametrized independently scheduled tasks, thereby leveraging traditional schedulability tests for independent task sets [15].

However, each scheduling method has its own advantages and limitations, which makes it impossible to select a universally applicable method. Some methods are known to be pessimistic [16], i.e., overestimating the end-to-end latency or falsely claiming a DRTApp deployment to be not schedulable. Conversely, other methods are more accurate but have higher runtimes [15] or require node-level real-time operating system schedulers, which might not be available on the target operating system. Furthermore, as stated above, schedulers do not consider all relevant real-time properties [11]. Consequently, a validation step (see Fig. 2) becomes necessary, incorporating additional user-definable validation steps. These include simple plausibility tests, such as making sure, that a service is only assigned to exactly one compute node, as well as more complex tests, such as end-to-end-latency tests [17] for specific parts of the distributed real-time application. Table 1 summarizes the requirements.

Table 1. Requirements of the modelling and description language.

Requirement	Description
RQ1	Description of automation application compositions through communication links
RQ2	Description of service internal operating system level tasks
RQ3	Description of services' simple resource requirements
RQ4	Description of compute nodes and resource supplies
RQ5	Integration of application specific validators beside the scheduling models default validators
RQ6	Generation of deployment system (e.g., Docker) and application specific configuration files

4 Domain-Specific Language Design

This section presents a detailed description of the DRTSL based on the aforementioned requirements. First, the requirements are model and formalized in Sect. 4.1, after which the syntax of the DRTSL is presented in Sect. 4.2.

4.1 Internal Representation of Required Model Elements

Figure 3 presents a formalization of the requirements outlined in Table 1, providing a structured insight into the essential elements required by the software systems involved. For example, a real-time scheduler integrated into an orchestrator requires information about available nodes and services' tasks. In addition, an end-to-end delay analysis tool requires details of message transmission times and communication links between individual services.

A *DRTSystem* consists of multiple *DRTApps*, each composed of multiple *Services*, *Communication* links, and *Flows* (see below). *Services* and *Communication* links form graphs that naturally represent distributed applications. While many scheduling methods assume graph-based application models, other methods such as independent task model schedulers are not limited by this model, as they can be depicted as graphs with only one node. *Services* are composed of *ResourceRequirements*, corresponding deployment units (e.g., a container image), and multiple *TaskGroups*. *ResourceRequirements* at the *Service*

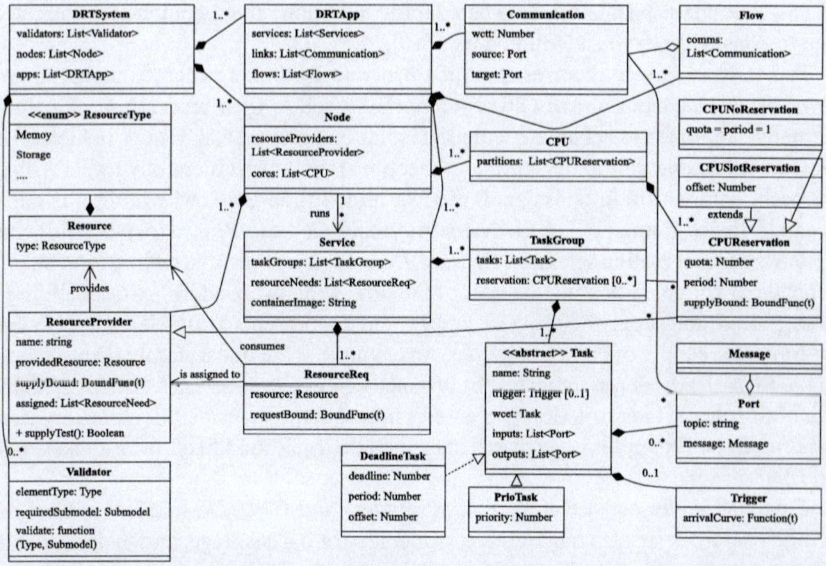

Fig. 3. Internal model representation necessary for various use cases regarding the modelling, validation and deployment of distributed real-time applications.

level align with resource management functions found in various operating systems, such as Linux control groups that regulate resource access at the services' levels. The resource model for *CPUs* is described in greater detail due to the focus on computing time in this work.

Multiple *TaskGroups* enable the definition of task groups with common scheduling properties, facilitating the modeling of hierarchical scheduling methods, such as Hierarchical Container-based Scheduling (HCBS) [18] or user-level hierarchical scheduling [16]. *Tasks* can be generic, of fixed priority, or of dynamic priority based on algorithms like earliest deadline first (EDF). *Tasks* communicate via *Ports*, which consist of a topic and a *Message* type. The publish-subscribe communication pattern is used as the basic pattern, capable of emulating other messaging patterns such as request-reply. *Message* descriptions rely on existing solutions like Protocol Buffers[1] for message description and serialization mechanisms. *Tasks* are typically triggered by messages from connected tasks. Some tasks, known as source tasks, are triggered by external event *Triggers* modeled as arrival curves [19], which describe the number of events that may occur within a specific time window as a function of time.

Certain scheduling methods offer temporal isolation for groups of tasks through scheduling servers such as the Constant Bandwidth Server (CBS) or the time-table driven hierarchical scheduling scheme [20]. Scheduling servers divide a processor (core) into multiple isolated partitions, each receiving a guaranteed quota of computing time cyclically for a fixed period. Hierarchical, server-based scheduling is advantageous for

[1] https://protobuf.dev/.

DRTSs as it enables isolation between deployment units (e.g., containers) without the need to consider every task within a unit [10].

Typically, a service receives multiple reservations that are shared among its tasks. However, tasks cannot utilize CPU time not assigned to their reservation. A common alternative approach to achieving temporal isolation between task groups in hierarchical scheduling is using a time-table driven schedule at the lowest hierarchy level. With this approach, each partition is assigned predetermined time windows within a cyclically repeating major frame [20]. *CPUReservation* and *CPUSlotReservation* are utilized to support these reservation schemes. *CPUSlotReservation* allows the definition of an offset in the major frame, indicating the start of the time window, in addition to specifying the quota, period, and supply bound. The supply bound represents a time-dependent function that limits the supply of computing time. An example is the linear supply bound function [11]. *CPUNoReservation* indicates the absence of a reservation. TaskGroups scheduled under this scheme lack isolation guarantees, necessitating consideration of every task in real-time tests. Examples of such schemes include traditional fixed-priority scheduling and EDF.

Compute Nodes consist of *ResourceProviders* and *CPUs*. *ResourceProviders* offer specific resources such as random-access memory or hard storage, and their availability is represented by a supply bound function. For simple static allocations, a single absolute number instead of supply bounds can be used to indicate a quantity of available resources. The adequacy of resources on a node can be validated by considering the *ResourceRequirements* of assigned *Services* to the node and using the provided *supplyTest()* method. E.g., for static allocations, the sum of request must be lower than the supply. A *CPU* represents a single compute core that can be divided into multiple partitions. In theory, a partition can be assigned to multiple *CPUs*.

Tasks are connected through *Communication* links form the graphs described at the beginning of this section. To allow users to describe requirements for end-to-end latencies in message propagation across multiple services, the concept of *Flows* is utilized [17]. A *Flow* represents a set of connections that form a coherent subgraph. Different response-time analysis methods can validate this metric. For hierarchically scheduled services, Compositional Performance Analysis (CPA) can be used, while for non-hierarchically scheduled service tasks, response time analysis techniques such as [13, 21] can be employed depending on whether fixed-priority scheduling or EDF is used. These analysis tools can be implemented as *Validators*, which are functions that operate on a subset of the model (*Submodel*) and validate a specific element type of the model (*elementType*).

4.2 DRTSL Syntax Design

Due to space limitations, this section provides a simplified description of the language for some of the necessary parts of DRTSL, to enable domain experts to quickly analyze and describe DRTApps. Code generation approaches are used to simplify the integration of software components, such as real-time orchestrators, while maintaining a human-readable representation of the model at all levels of extension described in Sect. 2. For the sake of clarity, the DSL is described using examples.

Figure 4 illustrates the description of a single service in DRTSL notation. It consists of two communicating tasks. The HCBS scheduler, with its two-tier model, determines the scheduling parameters. Thus, a server-based CPU reservation guarantees computing time for fixed-priority task groups. Both threads belong to group "Group1". The DRTApp description with the scheduling model extension specifies automatic or manual scheduling parameters, including quota, period and thread priority.

The service MyService1 described in Fig. 4 forms a cohesive DRTApp with MyServcie2 (not shown) through communication links. The textual representation in DRTSL is shown in Fig. 5. The worst-case transmission time (WCTT) is a constraint that must be provided by the underlying communication system. For that, network modelling approaches such as [22] could be used and integrated with DRTSL.

In addition to the communication links, a requirement for a flow is defined. Note, that there are two event chains because MyService1's Thread2 triggers MyService2's Thread1 and Thread2, but in the given example the user is only interested in the end-to-end delay of a particular flow, so only the e2e target for MyFlow1 is defined.

```
service MyService1 {
  thread MyThread1 {
      wcet=100
      port a input { topic="topic1" msg=Mt1 }
      port b output { topic="topic1" msg=Mt1 }
  }
  thread MyThread2 {
      wcet=200
      port a input { topic="topic1" msg=Mt1 }
      port b output { topic="topic1" msg=Mt1 }
  }

  conn conn1 MyThread1.b->MyThread2.a { wctt=10 }

  group Group1 as CPUReservation [ MyThread1 MyThread2 ]
  MyThread1.sched=PRIO
  MyThread2.sched=PRIO

  MyThread1.prio=89
  MyThread2.prio=70
  Group1.quota=10
  Group1.period=100
}
```

Distributed Application Description

Scheduling Model Extension

Deployment Model Extension

Fig. 4. A service model in DRTSL notation.

To validate the properties of the description and the determined scheduling parameters and node-assignments, validators as in Fig. 6 are used. The functions for validation are written in Python. In this example the validator is used to check the end-to-end delays of flows. For this purpose, the user-defined validation function uses the CPA method for HCBS, as described in [17]. The textual description for compute nodes, node assignment, partition description and other components follow a similar syntax.

```
comm MyService1.MyThread2->MyService2.MyThread1 { wctt=100 }
comm MyService1.MyThread2->MyService2.MyThread2 { wctt=100 }
                                                          Description
flow MyFlow1 MyService1.MyThread1->MyService1
             .MyThread2->MyService2.MyThread2 { e2eTarget=600 }
```

Fig. 5. A distributed real-time application model in DRTSL notation.

```
validator MyValidator2 validates flow {
  submodel=DRTApp
  function={
    def validate(flow: Flow, app: DRTApp):
      flow_e2es = runCPA(app)                          Description
      if flow_e2es[flow['id']] > flow['e2eTarget']:
        exit_w_error("Target e2e-response time exceeded!")
  }
}
```

Fig. 6. Examples of validators in DRTSL notation.

5 Preliminary Validation: Towards a DRTSL-Based Engineering Framework

In this section, the prototypical implementation of the DRTS language is described. The validation is done qualitatively through the use case of an engineering framework that supports two different methods of scheduling DRTApps.

The grammar is written in the meta language textX[2]. In addition to the textual language, a graphical representation has been designed. Communication links and flows are modelled as edges, while compute nodes, CPUs, tasks and services are modelled as graphical nodes. The graphical modelling language is based on the Graphical Language Server Platform (GLSP)[3], an open-source framework for building diagram editors. The user-defined validators have been integrated into the language server protocol for diagrams, which provides direct feedback to the user. We choose a Visual Studio Code extension as the platform for the graphical editor. This allows the user to switch seamlessly between the textual and graphical representations of the model. Figure 7 shows the graphical representation with error messages from the validators highlighted. The priority of MyThread2's exceeds the maximum possible priority, while the target end-to-end delay of MyFlow1 cannot be met, based on the result of the CPA.

In addition to HCBS-based DRTApps as shown in Fig. 7, DRTApps without scheduling hierarchy can also be modelled using DRTSL. For example, the DRTApp in Fig. 8, which is used to control an automation system consisting of several stations connected by conveyors and pushers. The latter move workpieces between stations and conveyor belts. The validators are the response time analysis of Pathan [13] and Melani [21]. Melani's test is applicable, if only the final end-to-end latency is of interest. The CPUNoReservation was used as the only partition of cores and every thread was given its own thread

[2] http://textx.github.io/textX/3.1/.

[3] https://www.eclipse.org/glsp/.

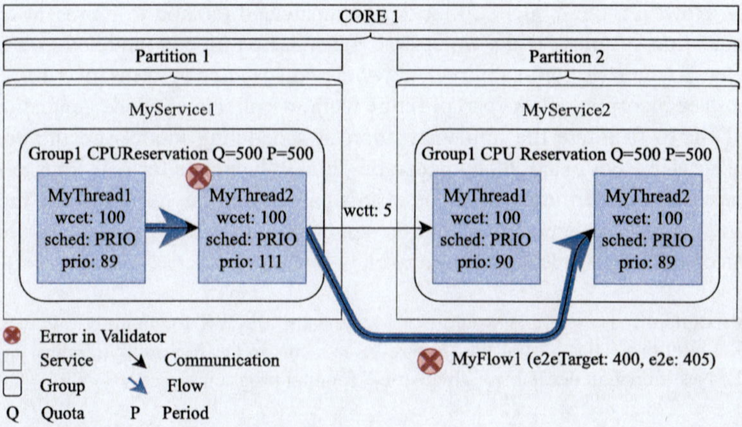

Fig. 7. Graphical representation of a distributed real-time app with responses from validators. The scheduling model of choice is HCBS.

group. All thread groups assigned to the same core, share the same CPUNoReservation partition. Figure 8 (right) shows the computed response times and the distribution of measured response times of the single-threaded services on the left. The response times were computed based on the worst-case execution times of the threads and their precedence constraints. Evaluation was done by running the DRTApp for six hours. The computing platform used was Linux 4.19 with the PREEMPT RT-PATCH.

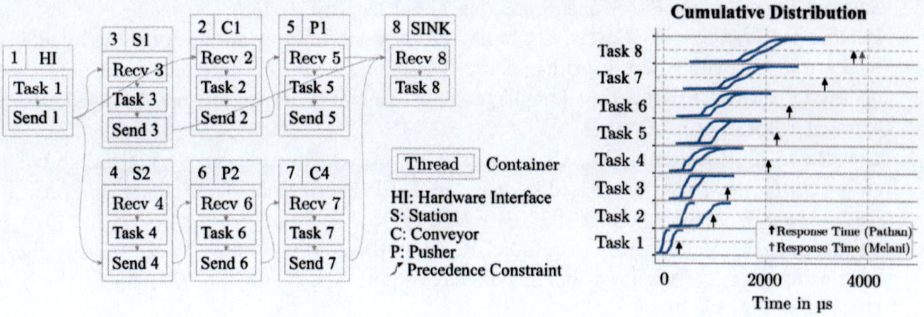

Fig. 8. Fixed priority scheduling-based distributed real-time application, which can be implemented using the described framework, with integrated response time tests [23].

6 Conclusion, Limitations and Future Work

The DRTSL was used to model two different execution models, which shows the suitability of the approach to provide domain experts with necessary abstractions, to implement heterogeneous real-time scheduling models. i.e., HCBS and non-hierarchical fixed-priority graph scheduling and the corresponding validators, i.e., CPA and response time analysis. However, the model still lacks a sophisticated method to represent real-time communication systems, as the worst-case transmission time is modeled as a constant parameter. To improve applicability in networked distributed systems, network calculus [22] could be incorporated. As part of future work, we plan to use code generation based on the DSL to facilitate the implementation of scheduling model-specific orchestrators and services. For example, by generating boilerplate code for real-time schedulers of Kubernetes. Furthermore, the focus of this paper is on the design of the modelling language, the full implementation and the application to use cases spanning the entire lifecycle of DRTApps is left for future work.

Acknowledgment. The authors would like to thank the BMWK for funding this work in the SDM4FZI joint project as part of the "Future Investments in the Automotive Industry (german: Zukunftsinvestitionen in der Fahrzeugindustrie)" funding program.

References

1. Neubauer, M., Reiff, C., Walker, M., et al.: Cloud-based evaluation platform for software-defined manufacturing. at-Automatisierungstechnik **71**, 351–363 (2023). https://doi.org/10.1515/auto-2022-0137
2. Potok, M., Chen, C.-Y., Mitra, S., et al.: SDCWorks: a formal framework for software defined control of smart manufacturing systems. In: 9th ACM/IEEE International Conference on Cyber-Physical Systems: ICCPS 2018, Porto, Portugal, 11–13 April 2018, Proceedings, pp. 88–97. IEEE, Piscataway (2018)
3. Salah, T., Zemerly, M.J., Yeun, C.Y., et al.: The evolution of distributed systems towards microservices architecture. In: 11th International Conference for Internet Technology and Secured Transactions, Barcelona, Spain, pp. 318–325. IEEE (2016)
4. Belafia, R., Jeanjean, P., Barais, O., et al.: From monolithic to microservice architecture: the case of extensible and domain-specific IDEs. In: ACM/IEEE International Conference on Model Driven Engineering Languages and Systems Companion, pp. 454–463. IEEE Computer Society (2021)
5. Granchelli, G., Cardarelli, M., Di Francesco, P., et al.: Towards recovering the software architecture of microservice-based systems. In: IEEE International Conference on Software Architecture Workshops, pp. 46–53. IEEE (2017)
6. Suljkanović, A., Milosavljević, B., Inđić, V., et al.: Developing microservice-based applications using the silvera domain-specific language. Appl. Sci. **12**, 6679 (2022). https://doi.org/10.3390/app12136679
7. Kosar, T., Gaberc, S., Carver, J.C., et al.: Program comprehension of domain-specific and general-purpose languages: replication of a family of experiments using integrated development environments. Empir. Softw. Eng. **23**, 2734–2763 (2018). https://doi.org/10.1007/s10664-017-9593-2
8. Mulo, E., Zdun, U., Dustdar, S.: Domain-specific language for event-based compliance monitoring in process-driven SOAs. SOCA **7**, 59–73 (2013). https://doi.org/10.1007/s11761-012-0121-3
9. Team J: JHipster - Full Stack Platform for the Modern Developer! (2023). https://www.jhipster.tech/. Accessed 12 Jun 2023
10. Struhar, V., Craciunas, S.S., Ashjaei, M., et al.: REACT: enabling real-time container orchestration. In: Proceedings 2021 26th IEEE International Conference on Emerging Technologies and Factory Automation (ETFA), Västerås, Sweden, pp. 1–8 (2021)

11. Struhár, V., Craciunas, S.S., Ashjaei, M., et al.: Hierarchical resource orchestration framework for real-time containers. ACM Trans. Embed. Comput. Syst. (2023). https://doi.org/10.1145/3592856
12. Henia, R., Hamann, A., Jersak, M., et al.: System level performance analysis-the SymTA/S approach. IEEE Proc.-Comput. Digit. Tech. **152**, 148–166 (2005)
13. Pathan, R., Voudouris, P., Stenstrom, P.: Scheduling parallel real-time recurrent tasks on multicore platforms. IEEE Trans. Parallel Distrib. Syst. **29**, 915–928 (2018)
14. Saifullah, A., Ferry, D., Li, J., et al.: Parallel real-time scheduling of DAGs. IEEE Trans. Parallel Distrib. Syst. **25**, 3242–3252 (2014). https://doi.org/10.1109/TPDS.2013.2297919
15. Baruah, S., Baker, T.: Schedulability analysis of global EDF. Real-Time Syst. **38**, 223–235 (2008). https://doi.org/10.1007/s11241-007-9047-9
16. Tang, Y., Feng, Z., Guan, N., et al.: Response time analysis and priority assignment of processing chains on ROS2 executors. In: Nelissen, G. (ed.) 2020 IEEE 41st Real-Time Systems Symposium, 1–4 December 2020, Proceedings, pp. 231–243. IEEE, Piscataway (2020)
17. Dasari, D., Becker, M., Casini, D., et al.: End-to-end analysis of event chains under the QNX adaptive partitioning scheduler. In: Nelissen, G. (ed.) 28th IEEE Real-Time and Embedded Technology and Applications Symposium: RTAS 2022: Proceedings, Milan, Italy, 4–6 May 2022, pp. 214–227. IEEE, Piscataway (2022)
18. Abeni, L., Balsini, A., Cucinotta, T.: Container-based real-time scheduling in the Linux kernel. SIGBED Rev. **16**, 33–38 (2019). https://doi.org/10.1145/3373400.3373405
19. Moy, M., Altisen, K.: Arrival curves for real-time calculus: the causality problem and its solutions. In: Esparza, J., Majumdar, R. (eds.) Tools and Algorithms for the Construction and Analysis of Systems: 16th International Conference, TACAS 2010, Held as Part of the Joint European Conferences on Theory and Practice of Software, ETAPS 2010, Paphos, Cyprus, March 20-28, 2010. Proceedings, pp. 358–372. Springer, Heidelberg (2010). https://doi.org/10.1007/978-3-642-12002-2_31
20. Amurrio, A., Azketa, E., Gutierrez, J.J., et al.: Response-time analysis of multipath flows in hierarchically-scheduled time-partitioned distributed real-time systems. IEEE Access **8**, 196700–196711 (2020). https://doi.org/10.1109/ACCESS.2020.3033461
21. Melani, A., Bertogna, M., Bonifaci, V., et al.: Response-time analysis of conditional DAG tasks in multiprocessor systems. In: 2015 27th Euromicro Conference on Real-Time Systems: ECRTS 2015: Proceedings, pp. 211–221. IEEE, Piscataway (2015)
22. Zhang, J., Chen, L., Wang, T., et al.: Analysis of TSN for industrial automation based on network calculus. In: 2019 24th IEEE International Conference on Emerging Technologies and Factory Automation (ETFA): Paraninfo Building, University of Zaragoza, Zaragoza, Spain, 10–13 September, Proceedings, pp. 240–247. IEEE, Piscataway (2019)
23. Walker, M., Tasci, T., Lechler, A., Verl, A.: Analysis of real-time execution models for container-based control applications. In: Kiefl, N., Wulle, F., Ackermann, C., Holder, D. (eds.) Advances in Automotive Production Technology – Towards Software-Defined Manufacturing and Resilient Supply Chains: Stuttgart Conference on Automotive Production (SCAP2022), pp. 14–24. Springer International Publishing, Cham (2023). https://doi.org/10.1007/978-3-031-27933-1_2

The Role of Value Co-creation and Top Management Involvement in the Development of Data-Based Services

L. Schild(✉), S. Sandikçi, and A. Stumpp

Technical University of Munich, Leopoldstraße 145, 80804 Munich, Germany
lukas.schild@tum.de

Abstract. Data-based services play a major role in reaching sustainability targets as they contribute to optimizing the use of production resources and enable sustainability improvements in production processes. The development of data-based services poses a challenge for companies of the manufacturing industry due to lack of knowledge regarding value creation and top management involvement. This paper provides a view on related work on development processes for data-based services and sheds light on proposed development methods. Interviews are conducted with large sized companies of the manufacturing and logistics industries to qualitatively investigate practices and lessons learned regarding the development of data-based services. Applying Gioia methodology, the paper derives recommendations for data-based service development and proposes implications for further research on the involvement of top management and the use of value co-creation.

Keywords: data-based service · value co-creation · top management involvement

1 Introduction

Manufacturing companies have been increasingly shifting their offering from physical products to services in the past years. This phenomenon, often labelled as "servitization", is a result of increased global competition and shorter product life cycles. The increased need for sustainable solutions to meet challenges due to global warming and resource scarcity as well as the emergence of new technologies such as the Internet of Things (IoT) and artificial intelligence further reinforce this development [1]. In this context, many companies change their business models and seek to gain an advantage by offering data-based services [2]. This opportunity arises from the fact that most products are increasingly becoming complex systems consisting of different elements such as hardware, sensors, microprocessors, data storage, and software [3]. While data-based services enable the optimization of products and the creation of new business models, their development is challenging due to the limited applicability of the traditional product development process [4]. Rather, a specific approach to developing data-driven services

T. Bauernhansl et al. (Eds.): WGP 2023, LNPE, pp. 280–289, 2024.
https://doi.org/10.1007/978-3-031-47394-4_28

is needed. Our findings show that the involvement of top management and value co-creation play a role in this context. Against this background, this paper aims to provide recommendations for the development of data-based services. This paper is structured into six sections. Following the introduction, the relevant terminology is discussed in the second section. To build on existing findings in research, the third section deals with related work in the context of the development of data-based services. Representing the empirical basis of this paper, the fourth section focuses on the research design and setting, data collection and data analysis. Our findings from the qualitative interviews are outlined in the fifth section. Finally, the conclusion and implications for further research are given in the last section.

2 Terminology

To provide a common understanding about the terminology used in this paper, the main term "data-based service" in the context of development processes is explained in this section. First, it is necessary to distinguish between the term "product" and the term "service". Products are generally associated with tangible items [5]. GOEDKOOP et al. define a product as "a tangible commodity [...] capable of falling onto your toes and of fulfilling a user's need" [6]. SHOSTACK uses a similar definition, emphasizing that products are tangible objects that exist in both time and space and are characterized by the fact that they should create added value to customers [7]. In contrast, services tend to be heterogeneous and perishable, and while the service provider produces the service the customer simultaneously consumes it [8, 9]. Therefore, services cannot be stored but only experienced or participated in [10]. According to these definitions, a car is a product, and a car insurance is a service. The connection between services and products is complicated. In reality, the supply of products depends on the completion of an extensive chain of services, whereas the provision of services encompasses both tangible and intangible components. Therefore, there is often a close and symbiotic relationship between products and services [11]. As the term suggests, "data-based service" falls under the category of the term "service", with "data" being in its core. The opportunities to collect substantial amounts of data and use them to develop data-based services arise from recent technological advances. The data can be generated by the customers, service providers or partner organizations [12]. Product-service systems (PSS) are market offerings that pair products and services together and provide solutions for the customer. The service in a PSS increases the consumer's value in consumption as opposed to conventional services linked to items such as a warranty. For instance, a customer may be able to get a service warranty while purchasing a washing machine. The engagement or experiences the consumer has with the washing machine, however, are not impacted by this warranty. An example of a PSS is the purchase of a bicycle that includes a two-year maintenance contract. This allows the customer to bring in the bicycle at any time during the two years to have a tire repaired free of charge, for instance [6]. Industrial PSS take the industrial aspect of product and service bundles into account [13]. Smart PSS, as opposed to smart products, combine service with the product to jointly answer the demands of consumers. Most of these services are electronic services that provide users with value in an electronic format. With regards to the example of the washing machine, the machine itself with its sensors represents the smart product, whereas remotely booking the washing machine with an app comprises the smart service [14]. ZHENG et al. define smart PSS as an "[information & communication technology] based dynamic ecological smart PSS network, which integrates customers, smart product service systems, smart service platform and product service suppliers for value co-creation and customer experience improvement, by means of smart interaction, mutual cooperation, resource sharing and optimal configuration" [3]. In this paper, data-based services refer to an integrated combination of data combined with a service and a product to enhance the customer's benefit.

3 Related Work

Within the literature, there are different approaches with recommendations for value co-creation and top management involvement in development processes of data-based services. Value co-creation plays an important role within service literature. It enables manufacturing companies to generate new value propositions [15]. The general idea is that a company collaborates with its customers, suppliers and other business partners to develop new data-based services [16]. This is in contrast to the traditional business where customers are passively involved in the value creation process, rather being value consumers [17]. Value co-creation especially calls for two things. Firstly, it requires the provision of appropriate information or education and establish adequate operational links with customers. Similarly, when engaging in co-creation activities with suppliers, the company needs to understand their goals, competencies and directions for growth [16]. One way to engage in value co-creation is to collaborate with start-ups. In these settings, established firms often provide required resources such as reference products for experimentation, while start-ups bring in a base technology for validation on a specific application [18]. Within the viewpoint of service-dominant logic, AKAKA & VARGO point out the relevance of value co-creation in the context of service. They deduct, that a consideration of skills and competencies of customers and other actors has to be taken during the development of service offerings [19]. Based on literature research, OERTZEN et al. explore an investigative framework for the co-creation of services. The authors identify involvement, engagement, and participation as prerequisites of co-creation processes and distinguish regenerative and operative co-creation phases in the service process [20]. In the context of data-based services, ASHRAFI et al. and ERIKSSON & HEIKKILÄ emphasize the importance of the involvement of customers in the innovation process [21, 22]. HEINZ et al. include the degree of value co-creation into their maturity model for smart PSS as a deciding component [23]. EIGNER & STARY view the role of value co-creation especially crucial in the development of additional services in the field of smart PSS [24]. LIU et al. develop a value co-creation framework for smart PSS with four stages: co-exist stakeholders and their requirements, co-design service functions, co-implement service systems and co-evaluate the service systems [17]. LENKA et al. focus on the necessary digitalization capabilities and mechanisms to facilitate value co-creation within the servitization of manufacturing companies. The authors identify perceptive and responsive mechanisms for the integration of customers into the service development process and link them to required digitalization capabilities [25].

From the presented approaches, the relevance of value co-creation during the development processes for data-based services becomes apparent. While the approaches address frameworks for its application, clear instructions for the operative layout of the co-creation process misses in the existing literature. With regards to this research gap, the first research question is derived as follows:

(1) How is value co-creation conducted in the development process for data-driven product and service solutions?

Apart from value co-creation, top management decisions play an essential role in the development process of data-based services. SCHNEUING & JOHNSON stress its importance especially in the authorization step of a development process, as the commitment of corporate resources is necessary for the further development of a service idea [26]. FIRK et al. address the role of top management teams fostering digital innovation. The authors assess the necessary characteristics for the CEO, the importance of a CDO role and link flat hierarchical structures to the success of digital innovation [27]. KIMITA et al. propose a maturity model based on 46 distinctive capabilities that are required for successful servitization. They state, that the commitment of the top management as well as the organization as a whole play an important role in the design of services [16]. To further investigate the role of top management during the development of data-based services, the following research question is formulated:

(2) How should the top management be involved in the development process for data-driven services?

4 Methodology

The following section explores the empirical research approach that is taken in this paper. Sect. 4.1 addresses the research design and setting. The data collection in qualitative expert interviews is elaborated in Sect. 4.2. Finally, Sect. 4.3 sheds light on the analysis of the data from the expert interviews.

4.1 Research Design and Setting

In this paper, grounded theory building through expert interviews is applied. This research strategy is taken due to a lack of existing literature regarding practical guidelines for value co-creation and top management involvement in the development of data-based services. The empirical study is set in the manufacturing and logistics industry. The shift towards offering data-based services is a major trend in the manufacturing and logistics sector, due to the relative ease of data-generation in running equipment and the thereby possible generation of use cases for data-based services on the basis of the collected data. This paper draws on multiple expert interviews to allow a cross-interview analysis. The study design comprises four expert interviews with managers from three different medium and large-scale companies. The selection of interviewees was based on theoretical sampling [28]. One criterion for the selection of the interviewees was their expertise in the field of data-driven solutions or the topic of IoT. Moreover, each interviewee needed to have experience in one of the following categories: product development for data-driven solutions, service development for data-driven solutions, or software development for data-driven solutions. The interviewees are key managers in the field of data-based services and hold senior positions such as head of development or group lead. The first interview was conducted with a medium scale subsidiary company in the machine manufacturing sector, which this paper refers to as SubCo. In its role as a subsidiary company, it provides software solutions for internal and external customers. The service offering includes data-driven analyses, simple, automated documentation, and decision-making tools. The company is located in Germany and comprises more than 100 employees. For the second interview, a large-scale logistics company was selected, which will be referred to as ParcelCo. The company offers a wide range of services including international express shipping, freight transport, supply chain management, and e-commerce solutions. Through the application of local sensors, the company uses real time data to provide services like parcel localization as well as monitoring, e.g. temperature, humidity, shock, light, and air pressure. ParcelCo operates globally with over 500.000 employees worldwide. Finally, the last two interviews were conducted with a large-size manufacturing company, referred to as DrillCo. The globally operating company provides products, system solutions, software and services to its customers and has a relatively long-standing history in the provision of data-based services. For chartered machinery, the company offers data-based services such as machinery localization, fleet cockpits and predictive maintenance. DrillCo operates in more than 120 countries with more than 30.000 employees worldwide.

4.2 Data Collection

The study uses expert interviews with company leaders and open access company data as its data sources. The primary source are semi-structured interviews that were conducted online via Zoom and MS Teams with a duration ranging from 30 to 60 min. The interviews were recorded and transcribed. Prior to the interview, the topic and purpose of the study was described to each informant. Furthermore, the company as well as their offered product and service portfolios were analysed in advance as preparation for the interview. Newly generated insights from previous interviews were included in the interview process. For example, identified methods for user involvement were raised as a topic in further interviews to capture further data on their application. However, information from prior interviews was not shared with interviewees to protect anonymity and encourage openness. The interview guideline for the semi-structured interviews was organized in three sections. The first section addresses background information regarding the company and the informant. This section includes questions regarding the expertise and role of the informant as well as the role of data-based services in the company's current business as well as strategy. The second section focuses on the development of data-based services. The interviewee was asked questions regarding the development process and its stages including relevant tasks during development. Lastly, the third section is built around value co-creation and top management involvement during development.

4.3 Data Analysis

In line with inductive research, the expert interviews were firstly analysed separately, before a cross-interview analysis was conducted [29]. The study applies the content analysis strategy from GIOIA et al. as method for the coding of interviews. This analysis is divided into first-order codes which are then processed into second-order codes. Lastly, those second-order codes are accumulated into higher-order dimensions [30]. Firstly, the audio recordings of the interviews were listened to repeatedly and thoughts on possible codes were made. Subsequently, existing literature was scanned and, based on existing concepts in the literature, the first-order codes were combined and grouped into themes. Some first order codes were merged, deleted, or renamed. These themes represent a more abstract level, which constitutes the second-order codes. [31] Lastly, the developed themes were again grouped and abstracted to another dimension. Again, with the help of existing concepts in the literature, "aggregate dimensions" were formed. These dimensions serve as building blocks for answering and describing the phenomena investigated in this paper.

5 Findings

The interviews yield practical guidelines for the application of value co-creation and top management involvement in the development of data-based services – two aspects that are rather underrepresented in the extant literature. This chapter sheds light on the findings of the expert interviews. While Sect. 5.1 addresses value co-creation methods and guidelines, Sect. 5.2 focuses on top management involvement in the development process.

5.1 Value Co-creation

The informants proposed different methods for the interaction with different types of customers in dependence with the development stage of data-based services. Between the informants there is the agreement, that customer co-creation is valuable during the whole development process. This implication aligns with the framework by LENKA et al. Who propose a chain of co-creation steps along the whole development process [25]. The informant from SubCo states regarding the timing and frequency of customer involvement:

> *… sometimes not at step zero, but maybe at step one. […]. So, we are actually interviewing and testing customers in every phase.*

The informants from SubCo and DrillCo stress the limitations of value co-creation due to the high degree of required time and money. Value co-creation is facilitated through lead users. However, the lead user approach also poses challenges. The informant from ParcelCo shared, that the selection of the right users is an important criterion for the successful development of the solution. He proposes to focus on lead users that have already formed an emotional connection to the offering of the company:

… it's very difficult because selecting lead users out of 80 million people within Germany whom all use [our offering]. […] we notice that the emotional connection to our services consistently has a high value with many people and we hope for [their] honest feedback.

The informants proposed various methods of interaction within the value co-creation process. Several methods are based on feedback and testing approaches, and some enable the customers to co-develop the solution with the company. Between the informants, user research, customer surveys, live demos, A/B testing, as well as analytics and observation are proposed. The informant from DrillCo addresses the observation of customers as powerful method for the testing of service developments:

We tend to observe customer behavior and how it changes when we change something about [our offering]. We do this, however, so that we don't ignore the customer, but rather focus on exactly what the customer needs and don't do everything that is technically possible, but rather what ultimately represents the core and the value for the customer.

5.2 Top Management Involvement

Beside value co-creation, the analysis of the interviews also results in guidelines regarding the occurrence and responsibilities of top management during the development process. Both informants from DrillCo describe the occurrence of top management involvement largely during the gates for the development process. The informant from SubCo emphasizes that top management is only involved subsequently to a larger product release:

…we would already do a launch party and there also the […] CEOs will […] give their feedback, but they are not an integral part of the development process. You don't really need senior management for that, unless it has to be scaled or something […].

The informant from ParcelCo adds to this, stressing that top management is only ever involved during the development of entirely new products:

… when it comes to the introduction of entire products, at the end of which there are also service obligations, for example, service level agreements, then the corporate board of the Group, also decides.

The responsibilities of top management vary from coordination and consultation activities for the development team to approving budgets, ideas, and concepts. An informant from DrillCo describes the responsibilities to give the final go for launching a new offering and to release the budget:

[Top management is presented] with a business opportunity description and […] an idea for a new project. It depends on the budget and the major milestones. […] these main milestones, these gates, they are approved by the [top] management.

Top management acts thereby as the gatekeeper of the budget and builds steering committees to take the decision regarding project continuation at each of the development gates. The second informant from DrillCo adds to this describing the strategy definition process:

There are also times when we do strategy, right? [We go to the] executive board basically to say these are the things that we want to do and this is why. [We do it] at a very higher level [not] down [at] the project level. [We call our] executive management team [...] to get that approval and do check-ins, that kind of thing.

A similar description of the responsibilities of top management is laid out by the informant from ParcelCo. From these descriptions, the involvement of top management during the development of data-based services seems not to play a major role. While the occurrence of interaction seems to be low in all portrayed companies, top management acts mostly as the decision-making authority at clearly defined phases during the development. This result may is consistent with the findings of HARMANCIOGLU et al., who identify a negative relation of the level of top management involvement and innovation productivity during development processes of new products and therefore propose less involvement [32]. Against this background, this study proposes a careful evaluation of the level of top management involvement during the development process of data-based services.

6 Conclusion

With the emergence of new technologies, data-based services enable growth opportunities for companies. Drawing on qualitative interviews with experts from manufacturing and logistics industries, this paper provides practical implications concerning the development of data-based services. By focusing on the aspects of value co-creation and top management involvement, it contributes to the growing literature on data-based services. From the interviews it became clear, that the involvement of the customer via value co-creation such as an application of lead users matters above all during the development process of data-based services. It is hereby important to address the different stages of the development process with fitting co-creation methods. The process of co-creation though increases the required time and monetary resources of the development. A careful consideration of the right level of co-creation and the selection of fitting customers to involve is therefore critical for success. In the case of lead users, for example, their emotional connection to the brand seems to be especially relevant. The involvement of top management within the development process was considered to be necessary by the interviewees, foremost in the role as a gatekeeper of budgets and in a deciding role for crossing gates in the process. In the companies of the included informants, top management mostly assumed the role of a steering committee. It further became clear in the interviews, that the involvement of top management is required to kick-off new service development projects and during strategy definition. The interviews yielded the assessment that top management is mostly involved at distinct phases of the development process. An extended involvement in the development process was not proposed by the informants and should therefore be carefully considered. Despite these findings, our investigation is subject to limitations. For instance, the scope of the interviews conducted is limited. Conducting more interviews with industry experts or even case-study analyses can help bring out details that we merely touch on. Moreover, the cross-sectional character of our study only provides a snapshot. A longitudinal study can help identify valuable insights for both research and practice. Given the increasing relevance of data-based services in the global economy, future research is encouraged to extend our investigation.

References

1. Lerman, L.-V., Benitez, G.-B., Müller, J.-M., de Sousa, P.-R., Frank, A.-G.: Smart green supply chain management: a configurational approach to enhance green performance through digital transformation. SCM **27**(7), 147–176 (2022)
2. Momeni, K., Vaittinen, E., Jähi, M., Martinsuo, M.: Introducing smart services: requirements and interconnections in multi-actor cooperation. JBIM **38**(13), 105–121 (2023)
3. Zheng, M., Ming, X., Wang, L., Yin, D., Zhang, X.: Status review and future perspectives on the framework of smart product service ecosystem. Procedia CIRP **64**, 181–186 (2017)
4. Schubert, V., Kuehner, S., Krauss, T., Trat, M., Bender, J.: Towards a B2B integration framework for smart services in Industry 4.0. Procedia Comput. Sci. **217**, 1649–1659 (2023)
5. Scholz, U., Pastoors, S., Becker, J.-H., Hofmann D., van Dun, R.: Praxishandbuch nachhaltige Produktentwicklung: Ein Leitfaden mit Tipps zur Entwicklung und Vermarktung nachhaltiger Produkte, Springer, Heidelberg (2018). https://doi.org/10.1007/978-3-662-57320-4
6. Goedkoop, M.-J., van Halen, C.-J., te Tiele, H.-R., Rommens, P.-J.: Product service systems. Ecological and Economic Basics (1999)
7. Shostack, L.-G.: How to design a service. EJM **16**(1), 49–63 (1982)
8. Levitt, T.: Marketing intangible products and product intangibles. Cornell Hotel Restaurant Adm. Q. **22**(2), 37–44 (1981)
9. de Brentani, U.: Success factors in developing new business services. EJM **25**(2), 33–59 (1991)
10. Cao, L., Mohan, K., Xu, P., Ramesh, B.: A framework for adapting agile development methodologies. Eur. J. Inf. Syst. **18**(4), 332–343 (2009)
11. Cooper, T., Evans, S.: Products to services: a report for friends of the earth. Sheffield Hallam University (2000)
12. Lim, C., Kim, M.-J., Kim, K.-H., Kim, K.-J., Maglio, P.-P.: Using data to advance service: managerial issues and theoretical implications from action research. JSTP **28**(1), 99–128 (2018)
13. Cedeno, J.-M.-V., Papinniemi, J., Hannola, L., Donoghue, I.: Developing smart services by internet of things in manufacturing business. Logforum **14**(1), 59–71 (2018)
14. Bulut, S., Anderl, R.: Towards ecosystems with smart product-service systems. Procedia CIRP **109**, 221–226 (2022)
15. Reinhold, J., Koldewey, C., Dumitrescu, R.: Value creation framework and roles for smart services. Procedia CIRP **109**, 413–418 (2022)
16. Kimita, K., McAloone, T.-C., Ogata, K., Pigosso, D.-C.: Servitization maturity model: developing distinctive capabilities for successful servitization in manufacturing companies. JMTM **33**(9), 61–87 (2022)
17. Liu, Z., Ming, X., Song, W., Qiu, S., Qu, Y.: A perspective on value co-creation-oriented framework for smart product-service system. Procedia CIRP **73**, 155–160 (2018)
18. Sandikçi, S.: Organisatorische Gestaltung von Kooperationen etablierter Unternehmen mit Start-ups, 1st edn. TCW Transfer-Centrum, München (2022)

19. Akaka, M.-A., Vargo, S.-L.: Extending the context of service: from encounters to ecosystems. J. Serv. Mark. **296**(7), 453–462 (2015)
20. Oertzen, A.-S., Odekerken-Schröder, G., Brax, S.-A., Mager, B.: Co-creating services—conceptual clarification, forms and outcomes. JOSM **29**(4), 641–679 (2018)
21. Ashrafi, A., Zare Ravasan, A., Trkman, P., Afshari, S.: The role of business analytics capabilities in bolstering firms' agility and performance. Int. J. Inf. Manage. **47**, 1–15 (2019)
22. Eriksson, T., Heikkilä, M.: Capabilities for data-driven innovation in B2B industrial companies. Ind. Mark. Manage. **111**, 158–172 (2023)
23. Heinz, D., Benz, C., Silbernagel, R., Molins, B., Satzger, G., Lanza, G.: A maturity model for smart product-service systems. Procedia CIRP **107**, 113–118 (2022)
24. Eigner, A., Stary, C.: The role of internet-of-things for service transformation. SAGE Open **13**(1), 215824402311592 (2023)
25. Lenka, S., Parida, V., Wincent, J.: Digitalization capabilities as enablers of value co-creation in servitizing firms. Psychol. Mark. **34**(1), 92–100 (2017)
26. Scheuing, E.-E., Johnson, E.-M.: A proposed model for new service development. J. Serv. Mark. **3**(2), 25–34 (1989)
27. Firk, S., Gehrke, Y., Hanelt, A., Wolff, M.: Top management team characteristics and digital innovation: Exploring digital knowledge and TMT interfaces. Long Range Plan. **55**(3), 102166 (2022)
28. Eisenhardt, K.-M., Graebner, M.-E.: Theory building from cases: opportunities and challenges. Acad. Manag. J. **50**(1), 25–32 (2007)
29. Eisenhardt, K.-M.: Building theories from case study research. Acad. Manag. Rev. **14**(4), 532–550 (1989)
30. Gioia, D.-A., Corley, K.-G., Hamilton, A.-L.: Seeking qualitative rigor in inductive research. Organ. Res. Methods **16**(1), 15–31 (2013)
31. Corbin, J.-M., Strauss, A.-L.: Basics of Qualitative Research: Techniques and Procedures for Developing Grounded Theory. 4th edn. SAGE, Los Angeles (2013)
32. Harmancioglu, N., McNally, R.-C., Calantone, R.-J., Durmusoglu, S.-S.: Your new product development (NPD) is only as good as your process: an exploratory analysis of new NPD process design and implementation. R&D Manage. **37**(5), 399–424 (2007)

Production Technologies for a Circular Economy

Ecology Analysis of Machines and Plants During Product Development

C. Ruschitzka[1(✉)], M. Kuhn[1], M. Riesener[1], and G. Schuh[1,2]

[1] Laboratory for Machine Tools and Production Engineering WZL, RWTH Aachen University, Campus Boulevard 30, 52074 Aachen, Germany
c.ruschitzka@wzl.rwth-aachen.de

[2] Fraunhofer Institute for Production Technology IPT, Steinbachstraße 17, 52074 Aachen, Germany

Abstract. The linear economic model of the manufacturing industry requires a fundamental transformation towards a circular economy. To support this transformation in machinery and plant engineering, transparency about the environmental impacts along the entire product life cycle is necessary. Since more than 80% of the environmental impact is determined during product development, it is important to create a basis for decision-making that makes the consequences of different circular economy strategies visible and clarifies the sphere of influence of design decision. In this paper, an ecology analysis for machines and plants is introduced, taking into account different circular economy strategies, in order to enable a holistic comparison of product and process variants. For this purpose, the relationships between ecology, product life cycle and product architecture are investigated. A mathematical calculation procedure of the interrelationships is presented to determine the main levers for improving the overall sustainability of machines or plants.

Keywords: Environmental Assessment · Circular Economy · Product Architecture

1 Introduction

The development of sustainable products and solutions is more important than ever [1]. Manufacturing companies have the responsibility to reduce their high greenhouse gas emissions and consumption of raw materials [2]. The intensifying signs of climate change, the growing competition for dwindling natural resources and an environmentally conscious society are increasing the pressure on companies to develop new products according to sustainability aspects. [3, 4] Following, the dominating linear "take-make-dispose" patterns of production and consumption cannot continue indefinitely [5]. Circular Economy is a promising economic system to decouple economic growth from the extraction of new raw materials. As an alternative to the current linear system, it aims to return resources to the value creation system instead of disposing them. [6, 7] This transformation is especially important for manufacturing companies in machinery and plant engineering, due to the high complexity of machines and plants, resource-intensive processes in the usage phase, and long product life cycles [8, 9].

T. Bauernhansl et al. (Eds.): WGP 2023, LNPE, pp. 293–303, 2024.
https://doi.org/10.1007/978-3-031-47394-4_29

Product development takes on an important role, as it defines up to 80% of a product's environmental impact [10]. In this phase, decisions regarding environmental and circularity aspects are made [5]. Among other things, the basic components of the product are designed, the materials and resources are selected, and the required manufacturing processes are predefined [11, 12]. In order to develop sustainable products or solutions, transparency about the environmental impact and the consequences of design decision are necessary. There are numerous methods and approaches, such as the Life Cycle Assessment (LCA), that focus on evaluating the company and its products in terms of sustainability [13]. However, there is a lack of more in-depth analysis considering the actual functional executions of a machine during the entire product lifetime, the replacement needs of individual product scopes with corresponding resource consumption, and recovery opportunities following different ReX-strategies. [14].

The methodology presented within this paper builds on the authors' previous publication [14]. The overall objective is the development of an ecology analysis to guide product development in machinery and plant engineering towards more sustainable and circular solutions. To this end, it is important to increase the usability of LCA for product development through functional orientation and detailed analysis of the usage and recovery phase. After the introduction in this section, section two gives an overview of the fundamentals for this paper. The third section describes existing approaches and identifies the research gap. Section four presents the developed methodology, whereas the last section gives a conclusion and provides the need of further research.

2 Fundamentals

Life Cycle Assessment. LCA according to DIN EN ISO 14040 [15] and DIN EN ISO 14044 [16] is the most widely used method for analyzing and evaluating environmental sustainability [17] and is structured in a four-step procedure: goal and scope definition, life cycle inventory analysis, life cycle impact assessment and interpretation of the results [15]. As the life cycle inventory requires many data, computational structures are introduced. A process matrix P lists the collected inputs and outputs of processes, giving each process unit its own column, shown in Fig. 1. Negative values indicate inputs, while positive values indicate outputs. The process matrix can be distinguished in two parts: one representing the flows within the economic system (technology matrix A), and one representing the flows from and into the environment (intervention matrix B). Besides the documentation of the processes, it is necessary to specify the required performance of the system, documented in the final demand vector f. The inventory vector g consists all environmental flows associated with the reference flow under consideration. In order to calculate those, it is necessary to include a scaling as the values of the final demand matrix and the output of the processes do not always add up. Given that the technology matrix A and the final demand vector f are known, the balance equation can be solved to yield the scaling vector s. As the scaling of a process affects both the economic flows and the environmental flows, the inventory matrix B also needs to be multiplied by the scaling vector s resulting in the inventory vector g. [18].

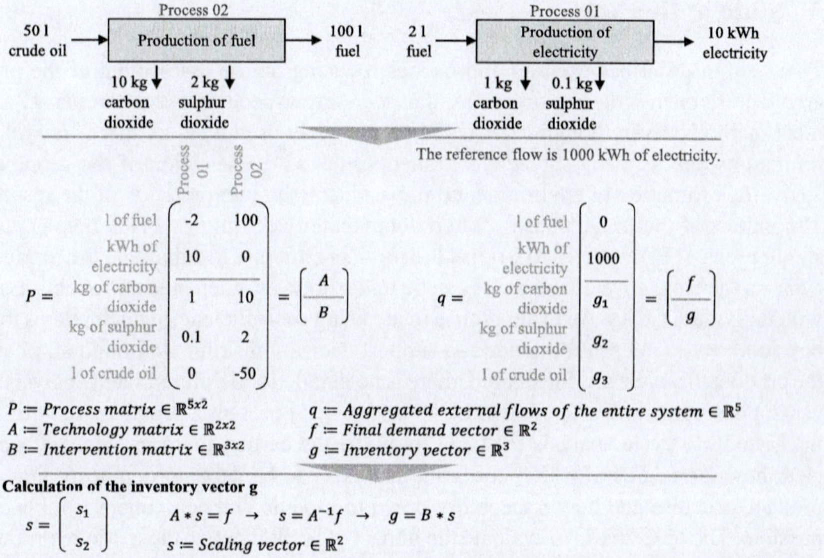

Fig. 1. Exemplary calculation of the inventory vector g according to HEIJUNGS AND SUH [18]

Product Architecture. Complex products are characterized by a large number of interrelated functions and components [11]. Transparency can be created by a product architecture consisting of the functional structure, product structure and connection of both by transformation relations. Product functions derive from customer requirements and can be decomposed into sub functions. The result is a hierarchical representation, the functional structure. The product structure describes the physical composition of the product consisting of assemblies and components. Within a structural bill of materials (BOM) each product structure element has an unique product identification number and the quantities of elements included in an assembly are listed. The linkage between both structures answers the question of how and with which modules or components individual sub functions are realized. [19].

Product Life Cycle. In literature, several definitions of the product life cycle exist. Most approaches distinguish between beginning of life (BoL), middle of life (MoL), and end of life (EoL) [1]. In the context of circular economy, the redistribution of product scopes, the reprocessing and return to production as well as recycling for the restoration of raw materials are to be pursued [20]. Circularity strategies within the production chain are also discussed under the term ReX-strategies. Often named strategies are reduce, reuse, refurbish, remanufacture and recycle. [21] Within this paper the four phases development, production, usage and recovery are defined as a product life cycle, with the last phase comprising different ReX-strategies according to POTTING ET AL. [21]. This is important, because it is only through reassembly in an industrialized process that a value-added, life-extending circular economy can be realized [22, 23].

3 State of the Art

This section describes existing approaches focusing on the assessment of the product architecture or the integration of circular economy aspects into environmental assessment methods. Even though LCA is the most widely used method, there are still some limitations due to many subjective elements such as the selection of the scope of the study, the estimation of environmental impacts, and the interpretation of the results. [3] The impact of the usage phase is often documented in form of a black box as just one single process. LOW ET AL. [24] use the product structure to subdivide the product life cycle so the respective components can be understood as independent products assessed with individual LCAs. An evaluation is made using eco-efficiency, which relates the carbon footprint to net profit intended to support decision making. Although all phases of the product life cycle are considered, there is no detailed and differentiated analysis of the usage phase. In another approach, LOW ET AL. [25] present a product structure-based integrated life cycle analysis (PSILA) to model the carbon footprint. Here it becomes clear how the results of a CO_2 footprint analysis can be decomposed according to the product structure and how traceability down to module and component level becomes possible. The objective is to evaluate the impact of implementing closed-loop production systems. However, there is still no further analysis regarding the usage phase or which product scopes have the greatest potential for improvement.

DEVANATHAN ET AL. [26] use a so-called Function-Impact-Marix (FIM) to integrate a sustainability assessment into the early phase of product development. A LCA of the product structure elements of an existing product is performed. The transformation relationships between product and functional structure are then used to assign the environmental impact to each function within the framework of the FIM. This enables statements regarding the improvement potential of new design developments. Although this enables a function-oriented assessment, the usage phase is only considered to a limited extent. BEGER [27] has developed a methodology for assessing the environmental impact of existing products along their life cycle to develop optimized sustainable solutions. The procedure is divided into the following steps: recording the functional and product structure; analyzing the environmental impact according to LCA; identifying weak points and potentials; automated recombination of the most sustainable components into a new variant and optimization of weak points. The procedure shows the potential for reducing the environmental impact based on existing possibilities and provide ideas for new designs. BEGER places the product architecture at the center of the methodology and describes it as the link between product design and LCA. However, an analysis of function execution in the usage phase does not take place. At the same time, only recycling and disposal of product components are considered.

KAMP ALBAEK ET AL. [5] analyze the literature regarding tools for product circularity evaluation and develop their own concept circularity evaluation tool (CCET) to be applied during product development. The goal is to overcome the obstacle that circularity aspects are yet seldom considered in decision-making procedures of product development. Even though the concept covers relevant aspects for circular product design and displays a circularity potential score, the evaluation is still based on a qualitative questionnaire with subjective elements. Moreover, individual elements of the functional or product structure are not investigated.

The existing literature underlines the usefulness to analyze individual functional and product structure elements in order to compare different variants and ReX-strategies. However, function- and component-based assessment are scarcely present in eco-design so far [27]. The approaches presented focus either on the assessment of design alternatives based on the product structure or on the assessment of circular economy perspectives. Furthermore, the usage phase, which represents the longest life cycle phase in machinery and plant engineering, is often not considered in detail.

4 Methodology for the Ecology Analysis

The approaches described in section three emphasize the need for an ecology assessment of machines and plants to support design decision in product development, especially taking into account function executions in usage phase as well as different recovery processes. Therefore, the main research question within this paper is formulated as follows: *How can the ecology of the functional and the product structure in the product life cycle be analyzed for product development?* The presented methodology to address the main research question is structured in five steps, which are described in the following. In doing so, this builds on the existing LCA framework and further develops the Life Cycle Inventory.

Step 1: Structuring of Machines or Plants. At the beginning, it is necessary to structure the product using the product architecture. To describe the functional and product structure mathematically, the direct demand matrix can be used. This matrix shows which sub function contribute to the main functions (D_{FSE}) or which and how many components an assembly consists of (D_{PSE}) [28, 29]. Also the transformational relationship between functional and product structure can be displayed in the transformation matrix (T). In order to be able to allocate the environmental impact measured for product structure elements (PSE) to the corresponding functional structure elements (FSE) and vice versa, a percentage is introduced in what way, e.g., a component contributes to the fulfillment of a specific sub function.

Step 2: Analyzing the Production Phase. Within the production phase (PP), each PSE is produced from raw materials, further processed, and assembled. The occurring processes can therefore be assigned to the PSE that is manufactured within this process, described as process owner. In order to be able to clearly assign and analyze the environmental impacts caused by these processes, certain data must be recorded for each process step. This includes, among other things, the assignment of a unique process number, the naming of the specific process step, the specification of process owners, and the naming of the number of processed parts *n*. In addition, information on various environmental impacts is provided by naming the respective ecology indicator, the quantity and the unit based on existing procedures like LCA. Before the collected environmental impacts can be assigned to the process owners, an allocation and normalization of the determined ecological values must be carried out. If a process contains several process owners, it is necessary to carry out an allocation analogous to the LCA (AF_{PP}). For example, an injection mold can include the molds for two different components. To enable the environmental impact to be reported separately, the material share could serve as an allocation factor. In addition, a normalization is necessary (N_{PP}). This means dividing the environmental impact of each process step by the number of processed parts n. If e.g. ten rails were punched at the same time in a punching process, but only one was installed in the product, the environmental impact must be distributed among the ten rails. With the collected input, a modified technology and intervention matrix can be derived, as displayed in Fig. 2. The modification allows the later calculation of individual environmental impacts for each PSE.

Structuring ecology indicators (OEK)

No.	Ecology category	Ecology indicator	Unit	⋮
OE01	Auxiliary materials	Cooling lubricant	Liter	⋮

Structuring production processes (PPZ)

No.	Process step	Time	Process owner			Processed parts			Ecology value			Allocation factor		
			P01	P02	⋮	n01	n02	⋮	OE01	OE02	⋮	AF01	AF02	⋮
PP01	Milling	0,75 h	x			10			1,2	0		1		
PP02	Drilling	0,25 h	x			1			0	3		1		
			P_{PP}			N_{PP}			OEV_{PPZ}			AF_{PP}		

$P_{PP} := PPZ\ of\ PSE \in \mathbb{R}^{pxs}$ $\quad OEV_{PP} := Ecology\ values\ in\ PPZ \in \mathbb{R}^{pxo}$

$N_{PP} := Number\ processed\ PSE\ in\ PPZ \in \mathbb{R}^{pxs}$ $\quad AF_{PP} := Allocation\ factors\ for\ PSE\ in\ PPZ \in \mathbb{R}^{pxs}$

s: number of PSE; p: number of PPZ; o: number of ecology indicators

Calculation of technology and intervention matrices

$$A_{PP} = \begin{array}{c} \\ PSE \\ PPZ \end{array} \left(\begin{array}{c|c} PSE & PPZ \\ D_{PSE} & \\ \hline D_{PSE-PPZ} & I_{PPZ} \end{array} \right) \qquad B_{PP} = OEK \left(\begin{array}{c|c} PSE & PPZ \\ U_{PSE} & OEV_{PP} \end{array} \right)$$

$A_{PP} := Technology\ matrix\ of\ PP \in \mathbb{R}^{(s+p)x(s+p)}$ $\quad B_{PP} := Intervention\ matrix\ of\ PP \in \mathbb{R}^{ox(s+p)}$

$D_{PSE-PPZ} := Number\ of\ PPZ\ required\ for\ one\ PSE \in \mathbb{R}^{pxs}$

$$(D_{PSE-PPZ})_{ij} := d_{PSE-PPZ,ij} = \frac{af_{PP,ij} * n_{PP,ij}}{\sum_{i=1} af_{PP,ij} * n_{PP,ij}} \ ;\ i = 1 \ldots np;\ i = 1 \ldots t$$

$I_{PPZ} := Unit\ matrix\ of\ PPZ \in \mathbb{R}^{pxp}$

$U_{PSE} := Environmental\ impact\ of\ PSE\ in\ PP \in \mathbb{R}^{oxs}$

Fig. 2. Derivation of modified technology and intervention matrices

Step 3: Analyzing the Usage Phase. For analyzing the environmental impact within the usage phase (UP), a function-oriented view is necessary. Functions are executed by a machine or plant, so the processes within the UP can be assigned to FSE, instead of PSE before. The environmental impact is measured and documented similar to the PP, with small changes described in the following. Instead of the number of processed parts, the number of function executions within each process is listed. Since the intensity of the execution of individual processes and functions can vary, it is necessary to work out different usage profiles for the product. For this purpose, it is useful to divide the entire life span of a machine according to VDI 3423 [30] into time of use, organizational downtime, technical downtime and maintenance time. Within the time of use, the focus lies on dynamic functional elements that cause changing loads on product elements during function execution. The processes in the UP can then be documented according to the defined usage profile. An additional column h_{UP} represents the number of the same process executions within the specific usage profile.

Step 4: Analyzing the Recovery Phase. The usage profile of a machine also allows conclusions to be drawn with regard to the different lifetimes of individual PSE. For example, the nozzles for supplying the cooling lubricant are replaced every four years, while the machine itself is in use for more than twenty years. In general, the lifetimes of components can be calculated in several ways. Using Mean Time Between Failure (MBTF) [31] as an example, the replacement probabilities for one usage profile can be calculated as shown in Fig. 3. MBTF refers to the operating time between two failures of a repairable component [32]. By working out the replacement probabilities (rp_{PSE}) of the PSE individual lifetimes can be taken into account as part of the calculation of the overall environmental impact.

Calculation of the loading duration for each PSE

$(\vec{f}_{hd})_i := f_{hd,i} = \sum_{i=1} t_{UP,i} * h_{UP,i} \;\; ; i = 1 \ldots t;$

$\overrightarrow{f_{hd}}$:= *Function execution time for one usage profile* $\in \mathbb{R}^t$

$\overrightarrow{t_{UP}}$:= *Time of UPZ* $\in \mathbb{R}^{up}$ *up: number of usage processes*

$\overrightarrow{h_{UP}}$:= *number of usage process executions within the usage profile* $\in \mathbb{R}^{up}$

$\vec{b} = T^T * \overrightarrow{f_{hd}}$ $\vec{b}$:= *Loading duration for each PSE* $\in \mathbb{R}^s$

MTBF of different PSE

No.	PSE	MTBF [h]	⋮
P01	Drive System	15000	⋮
P02	Tool drive	9000	
...	...	...	

Calculation of the replacement probability for each PSE within one usage profile

$(\overrightarrow{rp_{PSE}})_i := rh_{PSE,i} = \frac{b_i}{MTBF_i}$

$\overrightarrow{rp_{PSE}}$:= *Replacement probability for each PSE* $\in \mathbb{R}^s$

Fig. 3. Calculation of the replacement probability for each PSE within one usage profile

In the recovery phase (RP) the processes with their environmental impacts can be assigned to the PSE analogous to the PP. However, different recovery processes can occur depending on the condition of the PSE at the end of its specific usage phase and the company's strategy. Therefore, it is possible to define several process chains following different ReX-strategies for each PSE. In order to allocate the environmental impacts in the right proportion, a recovery key is introduced. This provides information on the probabilities that a PSE is redistributed, repaired, refurbished, remanufactured, recycled or disposed at the end of its useful life. The ratios can be defined based on the company's experience, e.g. by analyzing service reports.

Step 5: Allocation and Determination of Fields of Action. Within the last step of the methodology, the objective lies on the derivation of the interdependencies within the product life cycle. In doing so, the processes with their recorded environmental impacts must be calculated according to the developed usage profile, lifetimes with replacement probability, and recovery key. To avoid miscalculation for components that need to be exchanged in the usage phase, the PP counts as initial production whereas following reproduction processes fall into the RP counting as recurring life cycles. The environmental impact in the PP can be calculated as shown in Fig. 4. In the UP, the technology matrix A_{UP} needs to be multiplied according to the necessary numbers of usage profiles within the entire lifetime of e.g. 20 years. For the RP, the replacement probability needs to be multiplied by the corresponding recovery keys, so the environmental impact for the different recovery processes can be calculated according to the replacement frequency.

The definition of the final demand matrix F consisting of the unit matrix allows the calculation of the environmental impact for each PSE individually. Following, a comparison of different PSE or solutions with one another becomes possible. To this point, the environmental impact of the PSE in production and recovery phase as well as FSE in usage phase are calculated separately. The next step is the consolidation of those impacts to take on a holistic view. The transformational relationships between FSE and PSE can be used to calculate all environmental impacts for the other consideration side as well. With all environmental impacts calculated for FSE as well as PSE in all life cycle phases, the total environmental impact can be formed as the sum of the environmental impacts over the individual phases. The developed database thus enables a holistic analysis in the functional context. This provides an effective way for developers to interpret results and generate new ideas [27]. For example, it becomes possible to compare different product scopes that fulfill the same function and make decisions based on their impact.

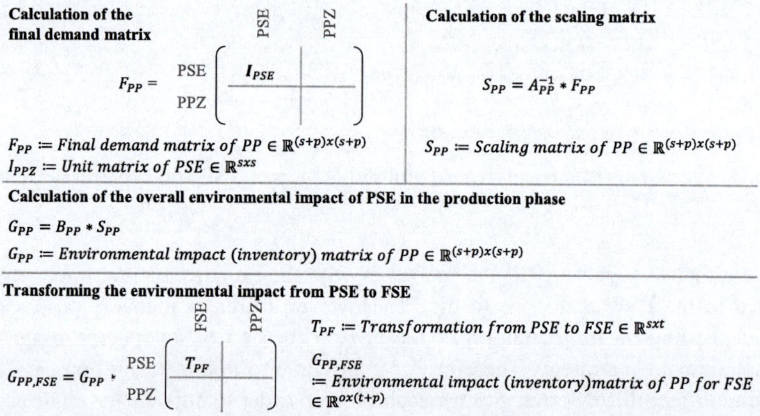

Fig. 4. Calculation of the environmental impact in PP for PSE and FSE

5 Conclusion and Further Work

This paper introduces a methodology for the ecology analysis in machinery and plant engineering to support product development in creating more sustainable and circular products. For this purpose, the environmental impact along the entire product life cycle is calculated and allocated to the functional and product structure. By this, different lifetimes of all PSE as well as recovery processes are taken into account, which creates a holistic view over the entire environmental impact of a machine or plant.

Right now, the authors are working on the realization of the mathematical calculation in a software tool using python as programming language. The methodology itself was already discussed with sustainability experts and product development from manufacturing companies. They are currently collecting data to validate the methodology using the new software tool. In addition, different calculation and allocation possibilities exist on a detailed level within the product architecture and elaboration of usage profiles, so the conduction of a sensitivity analysis is planned to avoid possible losses of accuracy. Moreover, it is planned to discuss different preparation forms of the ecology analysis in the form of dashboards with a group of product developers. In conclusion, it can be stated that the developed methodology enables a more in depth analysis of different product scopes and functions and supports essential decisions within product development.

Funded by the Deutsche Forschungsgemeinschaft (DFG, German Research Foundation) under Germany´s Excellence Strategy – EXC-2023 Internet of Production – 390621612.

References

1. Riesener, M., Kuhn, M., Ruschitzka, C., Becker, D., Schuh, G.: Identification of fields of action for the development of ecologically sustainable products. In: 2022 Portland International Conference on Management of Engineering and Technology (PICMET), Portland, OR, USA, pp. 1–10 (2022)
2. Hodgson, D., Vass, T., Levi, P., Hugues, P.: Industry: sectoral overview. https://www.iea.org/reports/industry. Accessed 20 Jan 2023
3. U. Sailer, Nachhaltigkeitscontrolling: Was Controller und Manager über die Steuerung der Nachhaltigkeit wissen sollten. 3rd ed. München: UVK Verlagsgesellschaft mbH; UVK/Lucius (2020)
4. Schuh, G., Dölle, C.: Sustainable Innovation: Nachhaltig Werte Schaffen. 2nd edn. Springer, Heidelberg (2021). https://doi.org/10.1007/978-3-662-61910-0
5. Kamp Albæk, J., Shahbazi, S., McAloone, T.C., Pigosso, D.C.: Circularity evaluation of alternative concepts during early product design and development. Sustainability **12**(22), 9353 (2020). https://doi.org/10.3390/su12229353
6. Europäische Kommission. Closing the loop - an EU action plan for the circular economy. In: Communication from the Commission to the European Parliament, the Council, the European Economic and Social Committee and the Committee of Regions COM (2015) 614/2 (2015). http://eur-lex.europa.eu/legal-content/EN/TXT/?uri=CELEX:52015DC0614. Accessed 27 May 27 2023
7. Niero, M., Rivera, X.C.S.: The role of life cycle sustainability assessment in the implementation of circular economy principles in organizations. Procedia CIRP **69**, 793–798 (2018). https://doi.org/10.1016/j.procir.2017.11.022
8. Ponn, J., Lindemann, U.: Konzeptentwicklung und Gestaltung Technischer Produkte. Springer, Heidelberg (2011). https://doi.org/10.1007/978-3-540-68563-0
9. Herterich, M.M., Uebernickel, F., Brenner, W.: Industrielle Dienstleistungen 4.0. Springer, Wiesbaden (2016). https://doi.org/10.1007/978-3-658-13911-7
10. Scholz, U.: Chancen der nachhaltigen Produktentwicklung. In: Praxishandbuch Nachhaltige Produktentwicklung, pp. 257–264. Springer, Heidelberg (2018). https://doi.org/10.1007/978-3-662-57320-4_21
11. Watz, M., Hallstedt, S.I.: Integrating sustainability in product requirements. In: Proceedings of the DESIGN 2018 15th International Design Conference, pp. 1405–1416 (2018)

12. Koen, P., et al.: Providing clarity and a common language to the "fuzzy front end." Res. Technol. Manage. **44**(2), 46–55 (2001)
13. Sheldrick, L., Rahimifard, S.: Evolution in ecodesign and sustainable design methodologies. In: Nee, A., Song, B., Ong, SK. (eds.) Re-engineering Manufacturing for Sustainability, pp. 35–40. Springer, Singapore (2013). https://doi.org/10.1007/978-981-4451-48-2_6
14. Riesener, M., Kuhn, M., Ruschitzka, C., Schuh, G.: Concept for a function-oriented ecology analysis in machinery and plant engineering. Procedia CIRP **116**, 53–58 (2023). https://doi.org/10.1016/j.procir.2023.02.010
15. DIN EN ISO 14040:2021-02, Umweltmanagement- Ökobilanz- Grundsätze und Rahmenbedingungen: (ISO_14040:2006_+ Amd_1:2020); Deutsche Fassung EN_ISO_14040:2006_+ A1:2020, 14040, Deutsches Institut für Normung e. V., Berlin (2021)
16. DIN EN ISO 14044:2021-02, Umweltmanagement- Ökobilanz- Anforderungen und Anleitungen: (ISO_14044:2006_+ Amd_1:2017_+ Amd_2:2020); Deutsche Fassung EN_ISO_14044:2006_+ A1:2018_+ A2:2020, 14044, Deutsches Institut für Normung e. V., Berlin (2021)
17. Muralikrishna, I.V., Manickam, V.: Life cycle assessment. In: Muralikrishna, I.V., Manickam, V. (eds.) Environmental Management, pp. 57–75. Elsevier (2017)
18. Heijungs, R., Suh, S.: The Computational Structure of Life Cycle Assessment. Springer, Dordrecht (2002). https://doi.org/10.1007/978-94-015-9900-9
19. Schuh, G., Riesener, M.: Produktkomplexität Managen: Strategien - Methoden - Tools, 3rd edn. Hanser, München (2018)
20. Europäisches Parlament: Kreislaufwirtschaft: Definition und Vorteile (2021). https://www.europarl.europa.eu/news/de/headlines/economy/20151201STO05603/kreislaufwirtschaft-definition-und-vorteile. Accessed 7 Mar (2022)
21. Potting, J., Hekkert, M. P., Worrell, E., Hanemaaijer, A.: Circular Economy: Measuring Innovation in the Product Chain. The Hague 2544 (2017)
22. Schuh, G., et al.: Green re-assembly upgrade factory (2023)
23. Schuh, G., Schmitz, S., Lukas, G., Niwar, L., Welsing, M., Calchera, R.: Framework for circular production economy (2023)
24. Low, J.S.C., Lu, W.F., Song, B.: Methodology for an integrated life cycle approach to design for environment. KEM **572**, 20–23 (2013). https://doi.org/10.4028/www.scientific.net/KEM.572.20
25. Low, J.S.C., Tjandra, T.B., Lu, W.F., Lee, H.M.: Adaptation of the product structure-based integrated life cycle analysis (PSILA) technique for carbon footprint modelling and analysis of closed-loop production systems. J. Clean. Prod. **120**, 105–123 (2016). https://doi.org/10.1016/j.jclepro.2015.09.095
26. Devanathan, S., Ramanujan, D., Bernstein, W.Z., Zhao, F., Ramani, K; Integration of sustainability into early design through the function impact matrix. J. Mech. Des. **132**(8) 2010. https://doi.org/10.1115/1.4001890
27. Beger, A.-L.: Methodik für eine nachhaltigkeitsoptimierte Produktentwicklung. Dissertation, Rheinisch-Westfälische Technische Hochschule Aachen, Aachen (2020)
28. Möller, H.P., Zimmermann, J., Hüfner, B.: Erlös- und Kostenrechnung. Pearson Studium, München (2007)
29. Menges, A.: Bewertung von Produktvarianten mittels Process Data Mining. Dissertation, Apprimus Verlag (2022)
30. VDI 3423 - Verfügbarkeit von Maschinen und Anlagen: Begriffe, Zeiterfassung und Berechnung, VDI 3423, Verein Deutscher Ingenieure e.V. (2021)

31. Bertsche, B., Lechner, G.: Zuverlässigkeit im Fahrzeug- und Maschinenbau: Ermittlung von Bauteil- und System-Zuverlässigkeiten. 3rd edn. Springer, Heidelberg (2004). https://doi.org/10.1007/3-540-34996-0_10. http://digitale-objekte.hbz-nrw.de/webclient/DeliveryManager?pid=1495106&custom_att_2=simple_viewer
32. DIN 40041:1990-12: Zuverlässigkeit; Begriffe, 40041: 1990-12, Deutsches Institut für Normung e. V., Berlin

Towards a Sustainable Industrial Society – Critical Capabilities for the Transformation to a Circular Economy in Manufacturing Companies

P. Burggräf, F. Steinberg, A. Becher, C. R. Sauer, and M. Wigger(✉)

International Production Engineering and Management, University of Siegen, Siegener Straße 152, 57223 Kreuztal, Germany
marius.wigger@uni-siegen.de

Abstract. The global earth overshoot day occurs earlier each year, highlighting the fact that industry and society consume more resources than the Earth can regenerate within a year. Nevertheless, the current focus on combating climate change is primarily on renewable energies and energy efficiency measures. To address resource scarcity as well, a shift from a linear economy to a circular economy is necessary. The adoption of circular strategies, which aim to minimize waste and maximize resource utilization, provides not only ecological benefits but also economic advantages. However, only a small number of manufacturing companies have fully integrated the circular economy into their business units. To identify the associated barriers and challenges, an exploratory literature review was conducted. Based on this, the critical capabilities of circular business modelling, circular design, transition management, circular networking as well as digital transformation were developed, to pave companies the way towards a circular economy.

Keywords: Sustainable Production · Resource efficiency · Closed-loop systems · Circular business strategies · Literature review

1 Introduction

In recent years, the global Earth Overshoot Day, which marks the point in the year when humanity has consumed more natural resources than the Earth can regenerate in one year, has been occurring earlier and earlier, highlighting the pressing need for a shift towards a more sustainable economy [1, 2]. The current linear economic model with its wasteful "take, make, dispose" mindset, is leading to the degradation of ecosystems and the depletion of natural resources, which are unable to provide essential services to a growing population and jeopardising profitability [3]. In 2021, approximately 37.9 billion tonnes of CO2 were emitted worldwide, with 2.8 billion tonnes in the European Union [4]. The supply of energy and its consumption (including energy for building, energy for transportation and energy systems) is responsible for 55% of global greenhouse gas

T. Bauernhansl et al. (Eds.): WGP 2023, LNPE, pp. 304–315, 2024.
https://doi.org/10.1007/978-3-031-47394-4_30

emissions (in CO_2 eq.), while the remaining 45% are directly linked to the production of goods (including agriculture, forestry, and other land use as well as material production) [5]. These staggering emission levels highlight the need to explore additional strategies to reduce carbon emissions and transition to a more sustainable economy. Faced with this challenge, climate action has focused primarily on the role of renewable energy and energy-efficiency measures [5], but addressing the emissions associated with production is critical to meet the European Green Deal targets of a 55% reduction in greenhouse gas emissions by 2030 compared to 1990 as well as climate neutrality by 2050 [6]. Society and industry should therefore focus on a circular economy, which aims to retain the value of products, materials, and resources for as long as possible by returning them to the product cycle at the end of their use while minimizing the generation of waste [7]. Circular economy is a comprehensive approach that addresses several global challenges, including climate change, biodiversity loss, waste, and pollution, by decoupling economic activity from the consumption of finite resources [8, 9]. However, only a few manufacturing companies have fully integrated circularity into their business operations [10] and those companies that incorporate circular principles struggle with its implementation [11].

Therefore, based on the challenges and barriers faced by manufacturing companies in their transformation to a circular economy, we define critical capabilities required for successful implementation. By understanding and developing these capabilities, companies can overcome barriers and challenges and effectively transform to a circular economy. The findings of this study will provide valuable insights and guidance for manufacturing companies seeking to embrace circular economy principles and achieve sustainable and resource-efficient practices. This paper is structured as follows: Sect. 2 provides an overview of the theoretical foundations of the circular economy, followed by the barriers and challenges. Building up on this, Sect. 4 introduces the critical capabilities. Conclusions and the outlook are presented in Sect. 5.

2 Fundamentals of a Circular Economy

The concept of circular economy has gained significant attention and recognition in science and industry as a viable solution to the pressing environmental and economic challenges of our time [12, 13]. As an industrial system that is restorative or regenerative by intention and design, it replaces the traditional linear economic model in which resources are extracted, processed, and disposed of at the end of their life cycle [14]. The circular economy is based on three main principles: eliminating waste and pollution, circulating products and materials at their highest value, and regenerating natural systems [8, 13]. The principles and components of circular economy can be depicted within the butterfly diagram (Fig. 1).

The biological cycle represents the flow of materials that can biodegrade and safely return to the earth and focuses primarily on products that are consumed, such as food and other organic matter [15]. Biological materials are considered renewable and operate in an open-loop system of resources that cascade through the subsequent steps of extraction, production of bio-based materials, energy recovery and return of nutrients to the biosphere to feed the next cycle of primary products [16]. The biological cycle includes the concepts of restoration, which is the process of revitalising degraded ecosystems, intending to enhance biodiversity and ecosystem health, as well as the concept of regeneration, which focuses on maintaining and improving existing conditions, thereby continuously strengthening the natural capital of ecosystems [17]. In the biological cycle, resources are extracted from ecosystems and cascaded through several material applications. At the end of their life, these resources decompose and return to the biosphere, helping to restore natural capital [18]. In addition to the biological cycle, circular economy also incorporates the concept of the technical cycle, which is designed to maximise product life, keep materials in use, and minimise waste. It consists of several stages that allow continuous use and re-use of materials, promoting resource efficiency, and reducing environmental impact [19]. The technical cycle emphasises strategies such as recycling, remanufacturing, reuse, and maintenance to extend the life and value of products, thereby reducing the demand for finite materials and minimising environmental impact [14]. The concept of the inner loops surrounded by larger outer loops in the butterfly emphasises the importance of maintaining the integrity of products and materials to maximise their value and minimise waste and capturing the most value, as they preserve the product in its entirety and thus retain more of its inherent value [19].

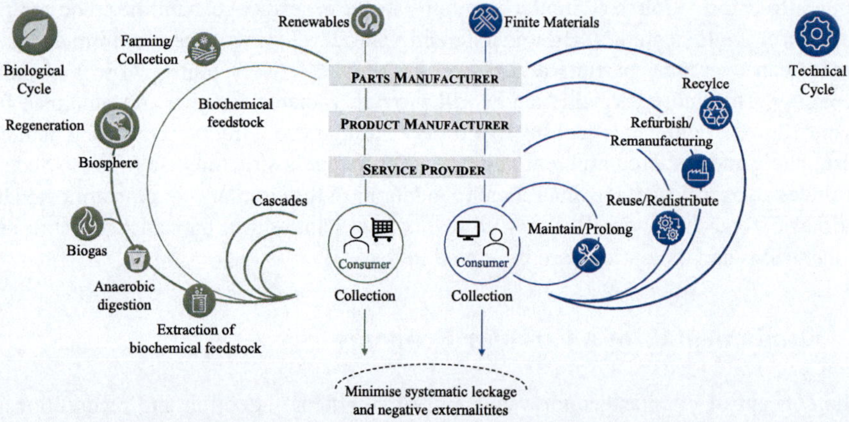

Fig. 1. Circular Economy Butterfly Diagram according to [20]

The butterfly model serves as a valuable framework for companies, providing a holistic understanding of material stocks and flows in the circular economy. It highlights the importance of understanding the linkages between different elements, embracing circular economy principles and implementing strategies such as recycling, remanufacturing and reuse to optimise value, minimise waste, improve resource efficiency and reduce environmental impact.

3 Methodological Approach

To initially identify the challenges and barriers associated with the transformation to a circular economy, an exploratory literature review was conducted, based on the approach of Arksey and O'Malley [21], which comprises five distinct stages: (1) identify the research question, (2) identify relevant studies, (3) study selection, (4) charting the data as well as (5) collating, summarising, and reporting the results.

The exploratory literature review aimed to synthesise and identify existing literature concerning the research question: "*What are the key challenges and barriers hindering the transformation to a circular economy in manufacturing companies, and which critical capabilities are needed to effectively address these barriers*?". We deliberately focused on individual relevant empirical studies by utilizing academic databases such as IEEE Xplore, EBSCO, ScienceDirect, and SpringerLink with a combination of keywords such as circular economy, manufacturing companies, challenges, barriers, transition, and transformation. The initial focus was on the barriers and challenges, leading us to conduct targeted searches for studies, expert interviews, and surveys that specifically addressed these aspects. Rather than considering challenges and barriers in circular economy in general, we carefully targeted and selected literature that specifically addressed the challenges unique to manufacturing companies during their transformation to a circular economy. Afterwards, we synthesized and interpreted the key information obtained (charting the data) into different categories. Based on this, we defined the critical capabilities required in the context of the research question.

Subsequently, we conducted further literature research to explore and gather more insights related to these critical capabilities. Therefore, the above-mentioned databases were used again with a particular focus on keywords that are directly related to the defined critical capabilities identified in the research question.

4 From Linear to Circular: Barriers and Challenges in the Transformation to a Circular Economy

The transformation from linear to circular economy presents many challenges that need to be addressed to successfully implement circular strategies. This section, therefore, provides an overview of the main challenges and barriers associated with adopting a circular economy. Based on a large N-study on challenges in the circular economy in the European Union with 208 survey respondents and 47 expert interviews, these are grouped into four categories: cultural, market, regulatory and technological factors [22].

4.1 Cultural Challenges

In contrast to earlier literature, which focused primarily on technical barriers as the main obstacles to the adoption of a circular economy, companies and policymakers have recognised that resistance to the adoption of a circular economy is often rooted in cultural barriers within companies [22, 23]. These barriers include a lack of consumer interest and awareness, a hesitant company culture, operating within a linear business model and a limited willingness to collaborate across the value chain [22]. These cultural challenges also contribute to structural barriers, such as a lack of leadership commitment and the lack of a clear governance framework [24] as well as a lack of information sharing and unclear accountability [25]. Circular initiatives are often implemented in isolation from core business operations, resulting in limited integration and effectiveness. There is also a general lack of senior management commitment and prioritisation of circular economy strategies [23]. Cultural challenges highlight the need for a shift in mindset and practices to promote a circular economy. Companies should prioritise circular economy strategies, foster leadership commitment and establish clear governance frameworks to integrate circular initiatives into core business operations.

4.2 Market Challenges

Market barriers pose significant challenges to the adoption of a circular economy, resulting from non-existent or ill-functioning of markets and economic barriers as financial arguments [23]. These barriers mainly revolve around difficulties in securing financing for circular business models and the high up-front investment costs, which can indicate a hesitant business culture [22, 23]. In addition, the pricing of raw materials, especially the low price of virgin materials, affects the transition to a circular economy [22]. Sustainable products are often priced at or above the upper limit of consumer affordability [26], e.g., “fossil-fuel based plastics are much cheaper than our bio-based plastics” which would undermine the affordability of the respondent’s recycled products [22]. Uncertainty due to price volatility can reduce the willingness to invest in recycled material markets, while the uptake of recycled materials remains limited [23]. Market challenges highlight the need to overcome financial and economic barriers and therefore require the exploration of innovative circular business models and collaborative partnerships to create demand and stable markets for recycled materials.

4.3 Regulatory Challenges

Both, market and societal challenges are often linked to regulatory barriers. These barriers arise from the lack of synergistic government policies and regulations as well as supportive incentives needed to accelerate the transition to a circular economy, thus hampering the development of circular markets and the adoption of circular business models [22]. Institutional and regulatory barriers to the adoption of a circular economy include inconsistent policy messages and inadequate institutional infrastructure. These barriers are particularly important for large companies operating in globally integrated supply chains, as policy consistency between countries is crucial for the effective adoption of a circular economy [23]. Addressing regulatory challenges is critical, especially for large companies operating in globally integrated supply chains, and requires synergistic government policies, and consistent regulations across countries.

4.4 Technological Challenges

Technical barriers include mainly the circular design as well as the ability to supply high-quality remanufactured products. Lack of data also plays a role, e.g., due to missing knowledge about impacts or the product life cycle [22]. It influences factors such as the product's durability, reparability, recyclability, the proportion of recycled and renewable materials used, and its suitability for refurbishment or remanufacturing. By incorporating circular design principles, such as designing for durability, ease of repair, efficient disassembly and the use of sustainable materials, products can be optimised for circularity [27]. However, the knowledge and understanding of practices, methods and tools focused on circular product design is currently limited. Despite the importance of product design in driving the transition to a circular economy, there is a lack of widespread development in this specific area [28]. Technical barriers require advances in circular design practices and the availability of high-quality remanufactured products. Companies should invest in circular design research and development, and integrate circularity considerations into product development processes while encouraging cross-sector collaboration and government support.

5 Critical Capabilities for the Transformation Towards a Circular Economy

To successfully implement circular strategies and to overcome the challenges, manufacturing companies need to develop critical capabilities, which are defined as "those dynamic capabilities which enable a company to shape, adapt and renew business models to create value in a sustainable way" [29]. Based on a literature review and assessment conducted by [30], a total of 21 critical capabilities for competitive manufacturing have been identified and categorized into six dimensions. One of these dimensions is the sustainability dimension, encompassing the capabilities of "product sustainability" and "process sustainability". The former refers to "the ability to provide sustainable products", while the latter pertains to "the ability to manufacture products in a sustainable manner" [30]. However, in addition to process and product-related capabilities, companies need a systematic integration across their value stream including the implementation of business models and the involvement of stakeholders and policymakers [31]. Therefore, the adoption and implementation of critical capabilities become imperative to effectively address the demands and challenges of a changing economic landscape. Companies need to broaden their perspectives and operational practices to achieve full integration across their entire value network. In light of this, and to address the challenges outlined above, we propose the critical capabilities of *Circular Business Modelling, Circular Design, Transition Management, Circular Networking, and Digital Transformation* (Fig. 2). These are not just related to one area of the above challenges but are intended to serve as a holistic approach to enable manufacturing companies their transformation to a circular economy.

Circular Business Modelling involves rethinking and redesigning business models to incorporate circular principles such as product-as-a-service, closed-loop systems, and circular supply chains [32]. One of the aspects of developing circular business models is the implementation of product service systems (PSS) that enable customer-centric value creation [33, 34]. PSS go beyond the traditional model of selling products and instead focus on providing a bundle of products and services to meet customer needs [35]. By offering circular services, companies provide customers with additional options beyond

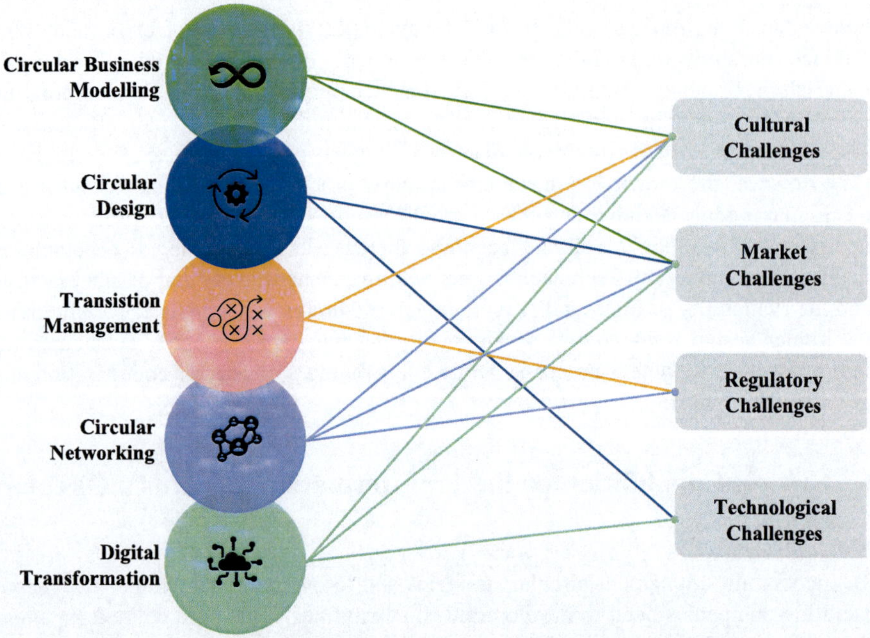

Fig. 2. Proposed Critical Capabilities to Overcome Barriers and Challenges

traditional product ownership, creating value for both the customer and the company. These services typically involve upgrading, refurbishing, or repairing products and are often combined with leasing or take-back schemes to ensure that products are returned to the manufacturer at the end of their useful life [36]. To implement circular business models, companies need to adopt tailored management practices that correspond to the three dimensions of the business model: value creation, value transfer and value capture. By doing so, companies can reduce costs, generate new revenue streams, and improve their environmental and social performance [37].

Circular Design focuses on designing products, systems, and materials to maximise their useful life, enable reuse and minimise waste [38]. This capability requires companies to consider the entire product lifecycle, from material selection to end-of-life disposal including the economic, environmental, and social impacts [39]. The development of a circular design includes strategies to close loops such as design for a technological cycle, design for a biological cycle as well as design for dis- and reassembly [40]. In addition, circular materials should be designed to be fully recyclable, ensuring that the materials can be regenerated without toxic precursors or by-products. These need to be processed locally to meet local needs and encourage a decentralised approach [38]. Designing for circularity also means considering waste as a valuable resource in the design process. Instead of treating waste as a burden, it can be seen as a potential raw material for new products or as an input for other industrial processes [41]. This concept, often referred to as "upcycling" or "cradle-to-cradle" design, promotes the idea that resources should be transformed in such a way that they maintain their value and functionality over time [42]. By designing products with waste materials in mind, manufacturers can reduce their dependence on virgin resources, minimise waste generation and contribute to the circular economy. These strategies can bring significant environmental and economic benefits, including stimulating innovation in the development of greener product designs [27].

Transition Management is a governance approach that addresses long-term, complex societal challenges, particularly those related to sustainable development. It involves specific networking and decision-making processes and focuses on the implementation of strategies and tools to facilitate societal transitions [43]. The transition to a circular economy requires effective communication of environmental benefits to both internal stakeholders, such as employees, and external stakeholders, such as customers, as well as improved stakeholder relations extensive stakeholder engagement and a strong awareness of issues such as ecosystem protection, health protection and resource conservation [44]. Transition Management also includes the implementation and development of regulatory requirements, e.g., by proposing policy recommendations to enable a circular economy for plastics including the role of plastic-related chemicals [45]. Governments and regulators play a crucial role in shaping the business environment and promoting circular practices [46]. Policymakers can do this by further developing regulations that facilitate the recyclability of components and by introducing new regulations on waste management, recycling, product standards and environmental protection [47]. These regulations impose obligations on companies' businesses and encourage them to align their operations with circular economy principles.

Circular Networking focus on creating networks and partnerships to facilitate the flow of resources, materials, and information [48], and to integrate data throughout the product lifecycle [49]. This involves partnerships with suppliers, customers, and other stakeholders and the sharing of resources such as raw materials, components, and equipment to improve the flow of materials, products, and services. In addition, the sharing of data and information, as well as expertise and knowledge, helps to ensure transparency and traceability [48].

Digital Transformation has the potential to significantly enhance and accelerate the transformation towards a more sustainable circular economy [49]. This involves initially the integration of data and technologies such as Big Data and the Internet of Things, which plays a crucial role in enabling companies to track and monitor the entire lifecycle of products, materials, and resources [49]. These technologies also facilitate the identification of design improvements, product, and process optimisation, technical support, and customer targeting [50], while providing accurate data on materials and products to support monitoring, remote maintenance and opportunities for reuse, refurbishment, and repair [51]. The integration of digital technologies, in particular Artificial Intelligence, supports eco-design by reducing bias in testing and prototyping, and facilitating real-time data transformation and analysis for massive data analytics, leading to reduced energy consumption and waste in the design process [51]. Data integration also facilitates transparency and traceability, giving customers and stakeholders insight into the circularity and sustainability of products and services [52].

By implementing the proposed critical capabilities, companies are enabled to shape, adapt, and renew their business models, incorporate circular principles, and foster collaboration to create sustainable value, minimize waste, and achieve resource efficiency.

6 Conclusion and Outlook

The transformation to a circular economy allows businesses to reduce their dependence on finite resources, resulting in cost savings and the creation of new revenue streams through material recovery and reuse [53]. Despite the growing interest in a circular economy and the obvious benefits, there has been limited progress in its implementation due to several challenges and barriers faced by manufacturing companies [22]. Therefore, the transformation to a circular economy supports innovation and a rethinking within companies to overcome barriers and challenges. With the proposed critical capabilities, we provide a holistic approach to help manufacturing companies on their way to a circular economy.

The further development of these critical capabilities, as well as the development and integration of potential additional capabilities, is important to meet future challenges. As more and more companies master the transition to a circular economy, this can lead to a domino effect, encouraging more companies to adopt a circular supply chain and circular business models. The development of strategies and capabilities will thus accelerate the transformation to a circular economy.

References

1. World Economic Forum. Action needed as Earth Overshoot Day gets earlier each year (2022). https://www.weforum.org/agenda/2022/08/earth-overshoot-day-climate-crisis/. Accessed 09 May 2023
2. Earth Overshoot Day. Past Earth Overshoot Days - #MoveTheDate of Earth Overshoot Day. https://www.overshootday.org/newsroom/past-earth-overshoot-days/. Accessed 09 May 2023
3. Sariatli, F.: Linear economy versus circular economy: a comparative and analyzer study for optimization of economy for sustainability. Visegrad J. Bioeconomy Sustain. Dev. **6**(1), 31–34 (2017). https://doi.org/10.1515/vjbsd-2017-0005
4. EDGAR - Emissions Database for Global Atmospheric Research. CO2 emissions of all world countries (2022). https://edgar.jrc.ec.europa.eu/report_2022. Accessed 12 May 2023
5. Ellen Mac Arthur Foundation. Completing the picture: How the circular economy tackles climate change (2021). https://ellenmacarthurfoundation.org/completing-the-picture. Accessed 11 May 2023
6. European Commission. A European Green Deal (2019). https://commission.europa.eu/strategy-and-policy/priorities-2019-2024/european-green-deal_en. Accessed 11 May 2023
7. Eurostat. Overview - Circular economy. https://ec.europa.eu/eurostat/web/circular-economy. Accessed 09 May 2023
8. Ellen MacArthur Foundation. What is a circular economy?" https://ellenmacarthurfoundation.org/topics/circular-economy-introduction/overview. Accessed 11 May 2023
9. Scheel, C., Aguiñaga, E., Bello, B.: Decoupling economic development from the consumption of finite resources using circular economy. A model for developing countries. Sustainability **12**(4), 1291 (2020). https://doi.org/10.3390/su12041291
10. World Business Council for Sustainable Development. DNV Viewpoint survey 'Circular Economy. How are companies transitioning? (2021)

11. Takacs, F., Brunner, D., Frankenberger, K.: Barriers to a circular economy in small- and medium-sized enterprises and their integration in a sustainable strategic management framework. J. Clean Prod. **362**, 132227 (2022). https://doi.org/10.1016/j.jclepro.2022.132227
12. Kirchherr, J., Yang, N.H.N., Schulze-Spüntrup, F., Heerink, M.J., Hartley, K.: Conceptualizing the circular economy (revisited): an analysis of 221 definitions. Resour. Conserv. Recycl. **194**, 107001 (2023). https://doi.org/10.1016/J.RESCONREC.2023.107001
13. Velenturf, A.P.M., Purnell, P.: Principles for a sustainable circular economy. Sustain. Prod. Consum. **27**, 1437–1457 (2021). https://doi.org/10.1016/J.SPC.2021.02.018
14. Ellen MacArthur Foundation. Towards the circular economy - economic and business rationale for an accelerated transition (2013)
15. Ellen MacArthur Foundation. The biological cycle of the butterfly diagram. https://ellenmacarthurfoundation.org/articles/the-biological-cycle-of-the-butterfly-diagram. Accessed 15 May 2023
16. Velenturf, A.P.M., Archer, S.A., Gomes, H.I., Christgen, B., Lag-Brotons, A.J., Purnell, P.: Circular economy and the matter of integrated resources. Sci. Total Environ. **689**, 963–969 (2019). https://doi.org/10.1016/J.SCITOTENV.2019.06.449
17. Morseletto, P.: Restorative and regenerative: exploring the concepts in the circular economy. J. Ind. Ecol. **24**(4), 763–773 (2020). https://doi.org/10.1111/JIEC.12987
18. Navare, K., Muys, B., Vrancken, K.C., Van Acker, K.: Circular economy monitoring – how to make it apt for biological cycles? Resour. Conserv. Recycl. **170**, 105563 (2021). https://doi.org/10.1016/J.RESCONREC.2021.105563
19. Ellen MacArthur Foundation. The technical cycle of the butterfly diagram. https://ellenmacarthurfoundation.org/articles/the-technical-cycle-of-the-butterfly-diagram. Accessed 15 May 2023
20. Ellen MacArthur Foundation. The Butterfly Diagram: Visualising the Circular Economy. https://ellenmacarthurfoundation.org/circular-economy-diagram. Accessed 15 May 2023
21. Arksey, H., O'Malley, L.: Scoping studies: Towards a methodological framework. Int. J. Soc. Res. Meth. Theory Pract. **8**(1), 19–32 (2005). https://doi.org/10.1080/1364557032000119616
22. Kirchherr, J., et al.: Barriers to the circular economy: evidence from the European Union (EU). Ecol. Econ. **150**, 264–272 (2018). https://doi.org/10.1016/j.ecolecon.2018.04.028
23. Grafström, J., Aasma, S.: Breaking circular economy barriers. J. Clean. Prod. **292**, 126002 (2021). https://doi.org/10.1016/j.jclepro.2021.126002
24. Droege, H., Raggi, A., Ramos, T.B.: Overcoming current challenges for circular economy assessment implementation in public sector organisations. Sustainability (Switzerland) **13**(3), 1–22 (2021). https://doi.org/10.3390/su13031182
25. Ritzén, S., Sandström, G.Ö.: Barriers to the circular economy - integration of perspectives and domains. Procedia CIRP **64**, 7–12 (2017). https://doi.org/10.1016/J.PROCIR.2017.03.005
26. Tan, J., Cha, V.: Innovation for circular economy. In: Liu, L., Ramakrishna, S. (eds.) An Introduction to Circular Economy, pp. 369–395. Springer, Singapore, (2021). https://doi.org/10.1007/978-981-15-8510-4_19
27. European Environmental Agency. Circular by design — products in the circular economy (2017)
28. Aguiar, M.F., Mesa, J.A., Jugend, D., Pinheiro, M.A.P., Fiorini, P.D.C.: Circular product design: strategies, challenges and relationships with new product development. Manage. Environ. Qual. Int. J. **33**(2), 300–329 (2022). https://doi.org/10.1108/MEQ-06-2021-0125
29. Achtenhagen, L., Melin, L., Naldi, L.: Dynamics of business models - strategizing, critical capabilities and activities for sustained value creation. Long Range Plann. **46**(6), 427–442 (2013). https://doi.org/10.1016/J.LRP.2013.04.002
30. Sansone, C., Hilletofth, P., Eriksson, D.: Evaluation of critical operations capabilities for competitive manufacturing in a high-cost environment. J. Glob. Oper. Strate. Sourcing **13**(3), 229–250 (2020). https://doi.org/10.1108/JGOSS-10-2019-0055

31. European Environment Agency. A framework for enabling circular business models in Europe (2021)
32. Lüdeke-Freund, F., Gold, S., Bocken, N.M.: A review and typology of circular economy business model patterns. J. Ind. Ecol. **23**(1), 36–61 (2019). https://doi.org/10.1111/jiec.12763
33. Burggräf, P., Wagner, J., Heinbach, B., Wigger, M.: Design of a methodological framework for adaptive remanufacturing-based business models. Procedia CIRP **98**, 547–552 (2021). https://doi.org/10.1016/J.PROCIR.2021.01.149
34. Kjaer, L.L., Pigosso, D.C.A., Niero, M., Bech, N.M., McAloone, T.C.: Product/service-systems for a circular economy: the route to decoupling economic growth from resource consumption? J. Ind. Ecol. **23**(1), 22–35 (2019). https://doi.org/10.1111/JIEC.12747
35. Barquet, A.P.B., de Oliveira, M.G., Amigo, C.R., Cunha, V.P., Rozenfeld, H.: Employing the business model concept to support the adoption of product–service systems (PSS). Ind. Mark. Manage. **42**(5), 693–704 (2013). https://doi.org/10.1016/J.INDMARMAN.2013.05.003
36. Guldmann, E., Huulgaard, R.D.: Barriers to circular business model innovation: a multiple-case study. J. Clean Prod. **243**, 118160 (2020). https://doi.org/10.1016/j.jclepro.2019.118160
37. Centobelli, P., Cerchione, R., Chiaroni, D., Del Vecchio, P., Urbinati, A.: Designing business models in circular economy: a systematic literature review and research agenda. Bus. Strategy Environ. **29**(4), 1734–1749 (2020). https://doi.org/10.1002/BSE.2466
38. Dumée, L.F.: Circular materials and circular design—review on challenges towards sustainable manufacturing and recycling. Circ. Econ. Sustain. **2**(1), 9–23 (2021). https://doi.org/10.1007/s43615-021-00085-2
39. Hapuwatte, B.M., Jawahir, I.S.: Closed-loop sustainable product design for circular economy. J. Ind. Ecol. **25**(6), 1430–1446 (2021). https://doi.org/10.1111/JIEC.13154
40. Bocken, N.M.P., de Pauw, I., Bakker, C., van der Grinten, B.: Product design and business model strategies for a circular economy. J. Ind. Prod. Eng. **33**(5), 308–320 (2016). https://doi.org/10.1080/21681015.2016.1172124
41. Perey, R., Benn, S., Agarwal, R., Edwards, M.: The place of waste: changing business value for the circular economy. Bus. Strategy Environ. **27**(5), 631–642 (2018). https://doi.org/10.1002/BSE.2068
42. Braungart, M., McDonough, W., Bollinger, A.: Cradle-to-cradle design: creating healthy emissions – a strategy for eco-effective product and system design. J. Clean. Prod. **15**(13–14), 1337–1348 (2007). https://doi.org/10.1016/j.jclepro.2006.08.003
43. Loorbach, D.: Transition management for sustainable development: a prescriptive, complexity-based governance framework. Governance **23**(1), 161–183 (2010). https://doi.org/10.1111/j.1468-0491.2009.01471.x
44. Salvioni, D.M., Almici, A.: Transitioning toward a circular economy: the impact of stakeholder engagement on sustainability culture. Sustainability (Switzerland) **12**(20), 1–30 (2020). https://doi.org/10.3390/su12208641
45. Aurisano, N., Weber, R., Fantke, P.: Enabling a circular economy for chemicals in plastics (2021). https://doi.org/10.1016/j.cogsc.2021.100513
46. Towards a more resource-efficient and circular economy. The role of the G20 (2021)
47. European Commission Directorate-General for Communicatio. Circular economy action plan - for a cleaner and more competitive Europe (2020). https://ec.europa.eu/environment/pdf/circular-economy/new_circular_economy_action_plan.pdf. Accessed 11 May 2023
48. Nogueira, A., Ashton, W., Teixeira, C., Lyon, E., Pereira, J.: Infrastructuring the circular economy. Energies (Basel) **13**(7), 1805 (2020). https://doi.org/10.3390/en13071805
49. Antikainen, M., Uusitalo, T., Kivikytö-Reponen, P.: Digitalisation as an enabler of circular economy. In: Procedia CIRP, pp. 45–49. Elsevier B.V. (2018). https://doi.org/10.1016/j.procir.2018.04.027

50. Bressanelli, G., Adrodegari, F., Perona, M., Saccani, N.: Exploring how usage-focused business models enable circular economy through digital technologies. Sustainability **10**(3), 639 (2018). https://doi.org/10.3390/su10030639
51. Ghoreishi, M., Happonen, A.: New promises AI brings into circular economy accelerated product design: a review on supporting literature. In: E3S Web of Conferences, EDP Sciences (2020). https://doi.org/10.1051/e3sconf/202015806002
52. Davari, S., Jaberi, M., Yousfi, A., Poirier, E.: A traceability framework to enable circularity in the built environment. Sustainability **15**(10), 8278 (2023). https://doi.org/10.3390/su15108278
53. Patwa, N., Sivarajah, U., Seetharaman, A., Sarkar, S., Maiti, K., Hingorani, K.: Towards a circular economy: an emerging economies context. J. Bus. Res. **122**, 725–735 (2021). https://doi.org/10.1016/j.jbusres.2020.05.015

Structuring Ecological Indicators in Machinery and Plant Engineering for Product Development

C. Ruschitzka[1](✉), S. Reimann[1], M. Kuhn[1], M. Riesener[1], S. C.-Y. Lu[2], and G. Schuh[1,3]

[1] Laboratory for Machine Tools and Production Engineering WZL, RWTH Aachen University, Campus Boulevard 30, 52074 Aachen, Germany
c.ruschitzka@wzl.rwth-aachen.de
[2] Viterbi School of Engineering, University of Southern California, 3650 McClintock Ave., Los Angeles, CA 90089, USA
[3] Fraunhofer Institute for Production Technology IPT, Steinbachstraße 17, 52074 Aachen, Germany

Abstract. A necessary precondition for a sustainability transformation is creating transparency about the environmental impact of products along their entire life cycle. However, possibilities for influencing sustainability are not directly apparent for product development using existing ecology metrics or methodologies such as Life Cycle Assessment, especially when dealing with complex machines and plants. This is a major challenge because up to 80% of a product's environmental impact is defined during development. Therefore, this paper presents a methodology for structuring and selecting ecological indicators for the sustainability assessment of products within product development. Thus, a systematic literature review was conducted to identify relevant ecological indicators and elaborate a generic ecology structure. Simultaneously, the life cycle of machines and plants was analyzed and a generic process structure was developed to describe it from product development's perspective. By identifying the relationships between processes and ecological indicators, a product-specific selection of indicators is made possible.

Keywords: Ecological Indicator · Life Cycle Assessment · Product Development

1 Introduction

Over time, the topic of sustainability not only has become increasingly important but one of the most relevant challenges of our time. Due to high energy and resource consumption, environmentally conscious actions have become increasingly important for manufacturing companies [1, 2]. By designing more sustainable products, there is a high potential for the manufacturing industry to influence the environmental impact. Besides the long life cycle of machines and plants, the environmental impact of the components produced by these products must be considered, too [3]. Over 80% of the environmental

T. Bauernhansl et al. (Eds.): WGP 2023, LNPE, pp. 316–325, 2024.
https://doi.org/10.1007/978-3-031-47394-4_31

impact is determined in product development, which gives this phase a high significance [4]. Among other things, functionalities of the machines and plants are elaborated, materials of the product's components are selected, and required manufacturing processes are predefined [5, 6]. To develop sustainable products, companies first need to be aware of their environmental impact and additionally need to know how they can influence or improve it [7]. There are already methodologies measuring the environmental impact of a product, such as Life Cycle Assessment. However, most are very subjective, costly, and time consuming [8]. This includes especially the elaboration or selection of relevant ecological indicators [9]. Moreover, it is not clear, which indicators can be influenced by product development [10]. Since machines and plants consist of numerous components and functions of high mechatronic complexity, the result is an unfortunate ratio of benefit and effort of using a methodology like LCA [11, 12]. To increase the ratio and to help the machinery and plant engineering industry improve the environmental impact, a suitable solution is needed. Therefore, this paper presents a structuring and selection of ecological indicators for the sustainability assessment of products from a product development perspective. Thus, a systematic literature review is conducted to identify relevant ecological indicators and elaborate a generic ecology structure. Simultaneously, the product life cycle is analyzed, and a generic process structure is developed that can be used within product development. By identifying the relationships between processes and ecological indicators, a product-specific selection of relevant indicators is possible.

After the introduction in Sect. 1, Sect. 2 gives an overview over the relevant terminology and existing approaches. Section 3 represents the methodology for structuring and selecting ecological indicators within product development. The last section states the conclusion and the need for further research.

2 Related Research

Product Life Cycle. A product life cycle describes the phases or stages through which an object passes during its lifetime. A rough division of the phases into the beginning of life (BOL), middle of life (MOL) and end of life (EOL) is common [13]. The specific subdivision into individual phases differs depending on the author. Examples in the literature can be found in VDI 2221 [14], SCHUH [15], DIN EN 60300-3-3 [16] and GROTE ET AL. [17]. JAWAHIR ET AL. [18] take a division into the four stages pre-production, production, use and post-use. In the context of the circular economy, the reduction of waste quantities and, in additional to the disposal of resources, different ReX-strategies to close the material flow are aimed at [19, 20] POTTING ET AL. [21] emphasize nine ReX-strategies, which are recover, recycle, repurpose, remanufacture, refurbish, repair, reuse, reduce, rethink and refuse. In this paper, the phases development, production, usage and recovery are used, with different ReX-strategies included in the last phase.

Life Cycle Assessment. This paper focuses on ecological sustainability, which means the focus on nature or the environment in general with the goal of preserving the natural foundations of life [22]. Because of the focus, sustainability and ecological sustainability are used synonymously within this paper. The most common methodology to measure sustainability is the Life Cycle Assessment (LCA) according to DIN EN ISO 14040

[23] and DIN EN ISO 14044 [24]. It provides a systematic procedure for measuring environmental impact throughout the entire product life cycle [25] and uses a cradle-to-grave-approach [26]. The methodology is divided into four steps. In the first step, the goal and the scope of the study must be defined and aligned with the intended application. In this step, it is important to define which inputs, like mass contribution, energy, and environmental relevance, are included in the LCA [24]. In the second step, a life cycle inventory is prepared iteratively. Life cycle inventories comprise data collection and calculation procedures with the aim of quantifying relevant input and output flows of a product system. The next step is to use the results of the life cycle inventory to assess the significance of the potential environmental impacts. This is done by linking the data from the life cycle inventory to impact categories and impact indicators [23]. At the end, a detailed assessment is carried out, taking into account the objectives set at the beginning [24].

Ecological Indicators. To measure sustainability, ecological indicators can be used. An indicator is defined as a qualitative or quantitative measurement or fact that can be used to display the state or condition of the criterion as a single parameter [27, 28]. In literature, indicators are categorized in several ways. The European Union developed an Eco-Management and Audit Scheme (EMAS) which divides the environmental performance indicators in the six key areas of energy efficiency, material efficiency, water, waste, biodiversity, and emissions [29]. Other divisions that go more into detail can be found in ISSA ET AL. [30] who use similar categories but further subcategories like recyclable, recycled, renewable or hazardous materials. Also LANZ ET AL. [31] elaborated subcategories like the CO_2 footprint or the recycling input material. JOUNG ET AL. [9] specify eleven indicator sets and give examples of sustainability indicators. DIN EN ISO 14031 [32] gives a large number of operation performance indicators. FRISCHKNECHT [33] describes widely used environmental impact indicators, such as enhancement of greenhouse effect, photochemical ozone creation, eutrophication, particle formation, and so on. SARKAR ET AL. [28] present a sustainability indicator repository, which is an integration and extension of thirteen already developed and established sustainability indicator sets. This is intended to address the challenge that there are already many indicator sets to analyze and to assess sustainable manufacturing.

To pursue an ecology assessment, the selection of suitable indicators for the company's products is necessary. AHMAD ET AL. [11] give a good insight into various assessment approaches in different industries and countries and make clear that indicators are often used very specifically, which makes uniformity and the associated comparability difficult. However, they do not elaborate in detail how indicators can be selected. ISSA ET AL. [30] propose a five-step approach to assist in the selection of environmental performance indicators (EPIs) for measuring the environmental performance of a product during its life cycle. The whole product life cycle with the stages of pre-manufacturing, manufacturing and design, distribution and packing, use and maintenance, end of life, and general activities is examined. Moreover, they address general environmental aspects as energy, solid waste, wastewater, gaseous emissions, and energy losses. Nevertheless, an approach to structure ecological indicators for a specific target group such as product development is not explained in detail.

A connection between processes and ecological indicators is made by KIM ET AL. [34]. They examine existing research and development efforts related to machining processes in discrete parts manufacturing to measure the sustainability of these processes. The focus is on machining processes, so neither the whole product life cycle, nor the view or influence of the product development is considered. It turns out that there are already first approaches in the literature that encompass some aspects that allow a specific selection of indicators within product development. However, there is not yet a methodology that covers all described aspects. So far, too little attention has been paid to the interrelationships between product, process and ecology.

3 Methodology

The approaches described above underline the need for the development of a methodology for a product-specific selection of ecological indicators within product development. To do so, the dependencies of ecological indicators, processes along the product life cycle and the sphere of influence of product development need to be investigated in more detail. Therefore, the main research question within this paper is formulated as follows: *How can ecological indicators that can be influenced by product development be derived based on dependencies between ecology and processes of the entire product life cycle?* To answer this question, the methodology is divided into four main steps, described in the following.

Step 1: Structuring and Description of Ecological Indicators
The goal of the first step is to elaborate a generic structure that classifies ecological indicators into different categories. Primarily, the scope of consideration is defined. KIM AND LEONG [34] differentiate between the hierarchy level from product to process, company, sector, and country level up to the world. Within this paper, the focus lies on the level of product and process, since they can be influenced by product design decisions directly e.g. by choosing another material or interface between components to be connected. Moreover, the quantifiability of indicators for an easy measurement is pursued. Within this scope, indicator sets are collected based on the work of JOUNG ET AL. [9] and expanded with other sets from the literature. To categorize the indicators collected from these sets, a systematical clustering procedure is pursued following the steps of consolidation, harmonization, and evaluation by reviewing the indicators with different criteria. The European Commission recommends that indicators should fulfill the RACER (Relevant, Accepted, Credible, Easy and Robust) criteria [29]. In addition, existing literature on structuring ecological indicators is considered, such as KRAJNC ET AL. [35] or SARKAR ET AL. [36]. For this purpose, different papers and proposals for clustering the indicators into different main categories are among other things compared by their frequency of usage. Important and frequently cited proposals can be seen in Fig. 1.

Taking everything into account, the final main categories elaborated are material, energy, and water as process inputs and air emission, solid waste, water effluent and others as process outputs. The category others includes further indicators that do not really fit into the other ones, such as energy loss in form of heat. Within each main

Main Category	AHMAD ET AL. [11]	EMAS [29]	ISSA ET AL. [30]	KIM ET AL. [34]	KRAJNC ET AL. [36]	LANZ ET AL. [31]	SARKAR ET AL. [28]	SINGH ET AL. [37]	Sum	Selection
Biodiversity						x	x	x	3	No
Environmental atmosphere							x	x	2	No
Environmental oceanic seas								x	1	No
Material		x	x	x	x	x	x	x	7	Yes
Water	x	x	x		x	x	x	x	7	Yes
Energy		x	x		x	x		x	5	Yes
Emissions in the air	x		x		x		x	x	5	Yes
Energy losses			x						1	No
Effluents and waist		x	x			x	x	x	5	Yes

Fig. 1. Exemplary indicator classification from different authors

category, the described procedure is repeated, dividing the ecology indicators further into subcategories based on the frequency of the categorization used in literature and the content-related closeness of the indicators among each other. All categories with their subcategories form the generic ecology structure and can be graphically presented in tree diagrams, shown in Fig. 2. In doing so, the indicators of the longlist are structured into the newly defined subcategories. The scientific approach allows further indicators to be added to the categories at any time.

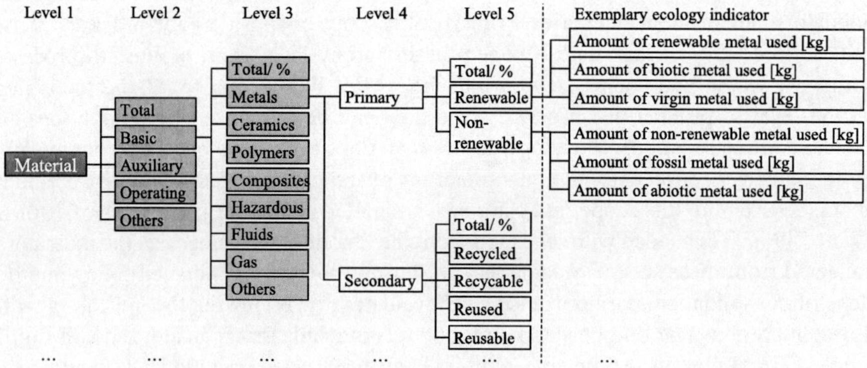

Fig. 2. Ecology structure and exemplary indicators for the category Material

Step 2: Classification and Description of Processes in the Product Life Cycle

The goal of the second step is to generate a generic structure of processes in the product life cycle and its phases. First, different approaches for structuring the product life cycle from different authors are compared [11, 28–31, 35–37]. The focus is on operational processes, management and auxiliary processes are not considered. By analyzing the ability of the product development to influence those phases and processes, the product life cycle is subdivided. Those processes, which can be influenced directly or indirectly by product development, are considered in the following. If design decisions have active influence on a life cycle phase e.g. choosing different materials allowing additive manufacturing instead of potting, a direct influence possibility is present. An indirect or partial possibility of influence is present, if the activities of the product development have influence on the life cycle phases, but the product development is not the sole decision maker and/or still other goals must be considered. For example, if the shape of the product is changed so that it no longer fits on an Euro pallet, new transport methods or packaging materials will be required, which may not have been defined by the product designer. Based on the described procedure, the main phases production, usage and recovery are divided into several subcategories, described in the following. The *production phase* is subdivided into procurement and manufacturing. Following the UMWELTBUNDESAMT [38] the second level of subcategories for procurement are mining, conveying, conditioning and refinement. For manufacturing there are different possibilities of classification, which bring different advantages. The categorization within this paper follows DIN 8580 [39], which classifies this phase according to forming, transforming and material property. As a result, the category for forming is original forming, the ones for transforming are transforming, separating, joining, coating and finishing and the one for material property is change of material properties. In the *usage phase*, the technical product takes on the active role of an operator, thus fulfilling its intended purpose by performing the product functions. The clustering of this phase of the product life cycle is based on MATTMANN'S [40] structure, which results in a subdivision into setup, actual usage, and maintenance. The structure of the *recovery phase* considers different ReX-strategies, as they can also be found in POTTING ET AL. [21] and VDI 2343 [41]. As there are various overlaps between those ReX-strategies, each process group is preceded by the process groups re-assembly and cleaning. The result is a generic structure of operating processes in the product life cycle.

Step 3: Determination of the Dependencies Between Processes and Indicators
In this step, the results of the two previous steps are combined. The assumption is made that the relationship between product and environment can be displayed transparent via processes. Processes along the entire product life cycle occur to, e.g., manufacture a product component or execute a function in the usage phase. During those processes, ecological indicators can be measured. Therefore, this step determines which ecology subcategory is relevant for which process subcategory. Dependencies between them are systematically determined using process definitions from literature and expert knowledge. For this purpose, a matrix is created in which the ecology subcategories are contrasted with the process subcategories. A distinction is made between a strong (value 2), a general (value 1) and no dependence (value 0). Criteria, like the change of aggregate state (IV), are established that can be used to assess the strength of the dependency. For example, a drilling machine uses energy to work. The same applies to an injection molding machine. However, the energy input is higher as the material is melted to take the new desired shape. Since the phases of the product life cycle differ greatly from one another, the criteria are set up anew for each life cycle phase or main process category. The criteria used for the different phases can be found in Fig. 3.

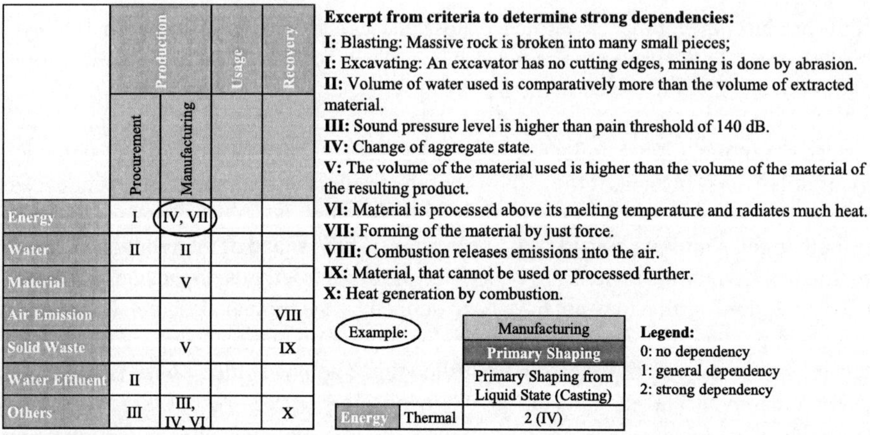

Fig. 3. Overview of criteria to determine strong dependencies

Step 4: Selection of Product-Specific Ecological Indicators
The last step focuses on the selection of product-specific ecological indicators. Since a product life cycle can vary greatly from product to product, product development can describe the individual product life cycle using processes from the predefined generic process groups. An automated calculation counts the number of generic process groups consisting in the described product life cycle. As the dependencies between generic process group and generic ecology subcategories consisting of several ecological indicators are determined, the overall sum for each ecology subcategory or process category can be calculated. The normalized sum then helps product development to select those indicator groups with the highest relevance for their product depending on the specific life cycle processes. In order to be able to additionally take into account individual goals or targets of the companies or product developers, a weighting option is introduced for the processes and indicator groups. A determined longlist of specific example indicators further supports the detailing of the selected ecology subcategories. Thus, the number of indicators with which a LCA can subsequently be performed can been significantly reduced and it becomes easier to conduct a LCA. The Life Cycle Initiative, which is part of the UN environment program proposes a number of six to ten indicators [42]. The result is a decision guiding matrix that helps the product development selecting ecological indicators for LCA.

4 Conclusion and Further Research

Sustainability has become increasingly important, especially for manufacturing companies. Methods such as LCA, which uses ecological indicators to measure environmental impact, are used to analyze sustainability. One challenge consists in the high number of indicators available and the lack of methodologies supporting a specific selection of those. Therefore, this paper presents a methodology for structuring ecological indicators in machinery and plant engineering within product development. In particular, the setup of a generic structure for the categorization of ecological indicators and processes over the entire product life cycle are described. Based on this, the interdependency between these two fields are investigated. This provides a solid basis for the specific selection of ecological indicators by product development. For more usability of the methodology, a future implementation in the form of a software application, should be pursued. Against the background of the high complexity of machinery and plant engineering and the fact that sustainability must always be considered as a property of the overall system, the methodology still has much potential for further development. To reach the goal of reliable software tool, broader research should be done on the sphere of influence of product development. The tool could also include the linking of the indicators and processes to the product architecture to facilitate the product development while still considering sustainability aspects as described in further publications of the authors [43, 44]. Finally, the focus on environmental sustainability suggests that the two other dimensions of social and economic sustainability should also be included in the scope of consideration. In conclusion, the methodology developed in this paper provides a solid basis for the specific selection of ecological indicators by product development.

References

1. Schuh, G., Dölle, C.: Sustainable Innovation. Springer, Heidelberg (2021). https://doi.org/10.1007/978-3-031-18751-3
2. Bundesregierung, Ziele der nachhaltigen Entwicklung: Nachhaltig produzieren und konsumieren. https://www.bundesregierung.de/breg-de/themen/nachhaltigkeitspolitik/produzieren-konsumieren-181666. Accessed: 9 Mar 2023
3. Heinzle, E., Hungerbuhler, K.: Integrated process development: the key to future production of chemicals (1997)
4. Scholz, U., Pastoors, S.: Nachhaltige produktentwicklung. In: Praxishandbuch Nachhaltige Produktentwicklung, pp. 63–77. Springer, Heidelberg (2018). https://doi.org/10.1007/978-3-662-57320-4_7
5. Koen, P., et al.: Providing clarity and a common language to the "fuzzy front end." Res. Technol. Manag. **44**(2), 46–55 (2001). https://doi.org/10.1080/08956308.2001.11671418
6. Watz, M., Hallstedt, S.I.: Integrating sustainability in product requirements. In: Marjanovic, D., Storga, M., Skec, S., Bojcetic, N., Pavkovic, N. (eds.) DESIGN, Proceedings of the DESIGN 2018: 15th International Design Conference, pp. 1405–1416 (2018)
7. Schäfer, P., Apostolov, H.: Nachhaltige Produktentwicklung. In: Eigner, M., Roubanov, D., Zafirov, R. (ed.) Modellbasierte virtuelle Produktentwicklung, pp. 369–392. Springer, Heidelberg (2014)
8. Sheldrick, L., Rahimifard, S.: Evolution in ecodesign and sustainable design methodologies. In: Nee, A., Song, B., Ong, S.K. (eds.) Re-engineering Manufacturing for Sustainability, pp. 35–40. Springer, Singapore (2013). https://doi.org/10.1007/978-981-4451-48-2_6
9. Joung, C.B., Carrell, J., Sarkar, P., Feng, S.C.: Categorization of indicators for sustainable manufacturing. Ecol. Ind. **24**, 148–157 (2012). https://doi.org/10.1016/j.ecolind.2012.05.030
10. Clarke-Sather, A.R., Hutchins, M.J., Zhang, Q., Gershenson, J.K., Sutherland, J.W.: Development of social, environmental, and economic indicators for a small/medium enterprise. Int. J. Account. Inf. Manag. **19**(3), 247–266 (2011). https://doi.org/10.1108/18347641111169250
11. Ahmad, S., Wong, K.Y., Rajoo, S.: Sustainability indicators for manufacturing sectors: a literature survey and maturity analysis from the triple-bottom line perspective. JMTM **30**(2), 312–334 (2018). https://doi.org/10.1108/JMTM-03-2018-0091

12. Röser, M.: Produktbasierte Bewertung von Kundenbeziehungen im Maschinen- und Anlagenbau. Wiesbaden. Springer, Wiesbaden (2021). https://doi.org/10.1007/978-3-658-35952-2
13. Fend, L., Hofmann, J.: Digitalisierung in Industrie-, Handels- und Dienstleistungsunternehmen. Springer, Wiesbaden (2018). https://doi.org/10.1007/978-3-658-35950-8
14. Verein Deutscher Ingenieure e.V., VDI 2221 - Blatt 1: Entwicklung technischer Produkte und Systeme Modell der Produktentwicklung. Berlin: Beuth Verlag GmbH
15. Schuh, G.: PLM (Product Lifecycle Management). https://wi-lex.de/index.php/lexikon/inner-und-ueberbetriebliche-informationssysteme/sektorspezifische-anwendungssysteme/plm-product-lifecycle-management/. Accessed 29 June 2022
16. Deutsches Institut für Normung e.V., DIN EN 60300-3-3: Zuverlässigkeitsmanagement
17. Grote, K.-H., Engelmann, F., Herbst, S., Beitz, W.: Grundlagen der Produktentwicklung. In: Hennecke, M., Skrotzki, B. (eds.) HÜTTE - Das Ingenieurwissen. 35th edn., pp. 1–16. Springer, Berlin (2020)
18. Jawahir, I.S., Rouch, K.E., Joshi, K.J., Kardekar, A.: Total life-cycle considerations in product design for sustainability: a framework for comprehensive evaluation (2006)
19. Stark, R., Seliger, G., Bonvoisin, J.: Sustainable Manufacturing. Springer, Cham (2017). https://doi.org/10.1007/978-3-319-48514-0
20. Europäisches Parlament, Kreislaufwirtschaft: Definition und Vorteile. https://www.europarl.europa.eu/news/de/headlines/economy/20151201STO05603/kreislaufwirtschaft-definition-und-vorteile. Accessed 7 Mar 2022
21. Potting, J., Hekkert, M., Worrell, E., Hanemaaijer, A.: Circular economy: measuring innovation in the production chain. In: PBL Netherlands Environmental Assessment Agency (2017)
22. Stoll, M.: 3-Säulen-Modell der Nachhaltigkeit. https://www.michael-stoll.info/glossar-3-saeulen-modell/
23. Technisches Komitee ISO/TC 207. DIN EN ISO 14040: Umweltmanagement - Ökobilanz-Grundsätze und Rahmenbedingungen (2021)
24. Technisches Komitee ISO/TC 207. DIN EN ISO 14044: Umweltmanagement - Ökobilanz - Anforderungen und Anleitungen (2021)
25. Sartal, A., Bellas, R., Mejías, A.M., García-Collado, A.: The sustainable manufacturing concept, evolution and opportunities within Industry 4.0: a literature review. Adv. Mech. Eng. **12**(5), 168781402092523 (2020). https://doi.org/10.1177/1687814020925232
26. Hermann, B.G., Kroeze, C., Jawjit, W.: Assessing environmental performance by combining life cycle assessment, multi-criteria analysis and environmental performance indicators. J. Clean. Prod. **15**(18), 1787–1796 (2007). https://doi.org/10.1016/j.jclepro.2006.04.004
27. Hallstedt, S.I.: Sustainability criteria and sustainability compliance index for decision support in product development. J. Clean. Prod. **140**, 251–266 (2017). https://doi.org/10.1016/j.jclepro.2015.06.068
28. Sarkar, P., Joung, C.B., Carrell, J., Feng, S.C.: Sustainable manufacturing indicator repository. In: Proceedings of the ASME 2011 International Design Engineering Technical Conferences and Computers and Information in Engineering Conference. August 28–31, 2011, Washington, DC, USA, vol. 2, 31st Computers and Information in Engineering Conference, Washington, DC, USA, pp. 943–950 (2011)
29. Weiß, D., Müller, R., Lössl, S.: Umweltkennzahlen in der Praxis: Ein Leitfaden zur Anwendung von Umweltkennzahlen in Umweltmanagementsystemen mit dem Schwerpunkt auf EMAS (2013)
30. Issa, I.: Product-related environmental performance indicators: a guide to support the selection of product-related environmental performance indicators (2013)

31. Lanz, M., Järvenpää, E., Nylund, H., Tuokko, R., Torvinen, S., Georgoulias, K.: Sustainability and performance indicators landscape. In: Proceedings of the 24th International Conference on Flexible Automation & Intelligent Manufacturing, pp. 283–290 (2014)
32. Frischknecht, R.: Lehrbuch der Ökobilanzierung. Springer, Heidelberg (2020). https://doi.org/10.1007/978-3-662-54763-2
33. Deutsches Institut für Normung e.V. DIN EN ISO 14031: Environmental performance evaluation – Guidelines. Deutsches Institut für Normung e.V. (2013)
34. Kim, D.B., Leong, S.: An overview of sustainability indicators and metrics for discrete part manufacturing. In: 32nd Computers and Information in Engineering Conference, Parts A and B, Chicago, Illinois, USA, vol. 2, pp. 1173–1181 (2012)
35. Krajnc, D., Glavic, P.: Indicators of sustainable production. Clean Technol. Environ. Policy **5**(3–4), 279–288 (2003). https://doi.org/10.1007/s10098-003-0221-z
36. Singh, R.K., Murty, H.R., Gupta, S.K., Dikshit, A.K.: An overview of sustainability assessment methodologies. Ecol. Ind. **15**(1), 281–299 (2012). https://doi.org/10.1016/j.ecolind.2011.01.007
37. Kim, D.B., Leong, S., Chen, C.-S.: An overview of sustainability indicators and metrics for discrete part manufacturing, vol. 2012, pp. 1173–1181 (2012). https://doi.org/10.1115/DETC2012-71020
38. Umweltbundesamt, Abiotische Rohstoffe schonend gewinnen. https://www.umweltbundesamt.de/themen/abfall-ressourcen/ressourcenschonung-in-produktion-konsum/abiotische-rohstoffe-schonend-gewinnen. Accessed 6 Mar 2023
39. Deutsches Institut für Normung e.V., DIN 8580. Fertigungsverfahren
40. Mattmann, I.: Modellintegrierte Produkt- und Prozessentwicklung. Springer, Wiesbaden (2017). https://doi.org/10.1007/978-3-658-19409-3
41. VDI 2343 - Blatt 1: Recycling elektrischer und elektronischer Geräte, Grundladen und Begriffe, 2343, Verein Deutscher Ingenieure e.V. (2001)
42. Life Cycle Initiative by the UN environment program. Environmental life cycle impact assessment indicators. https://www.lifecycleinitiative.org/activities/phase-iii/global-guidance-on-environmental-life-cycle-impact-assessment-indicators/. Accessed 13 Mar 2023
43. Riesener, M., Kuhn, M., Ruschitzka, C., Becker, D., Schuh, G.: Identification of fields of action for the development of ecologically sustainable products. In: 2022 Portland International Conference on Management of Engineering and Technology (PICMET), Portland, OR, USA, pp. 1–10 (2022)
44. Riesener, M., Kuhn, M., Ruschitzka, C., Schuh, G.: Concept for a function-oriented ecology analysis in machinery and plant engineering. Procedia CIRP **116**, 53–58 (2023). https://doi.org/10.1016/j.procir.2023.02.010

Study on Scrap Reduction in Cold Forging During Ramp-Up Phases Through Actuator Control

J. Böhm(✉), M. Liewald, and P. Clauß

Institute for Metal Forming Technology, University of Stuttgart, Holzgartenstraße 17, 70174 Stuttgart, Germany
jonathan.boehm@ifu.uni-stuttgart.de

Abstract. Ramp-up phases right at the beginning of batch processing in cold-forging of high volumes are accountable for a significant amount of scrap due to transient process conditions. To understand the emerging impacts of transient effects during ramp-up phases on the final dimensional product properties more deeply, process conditions like forming load, part temperature and part geometry were monitored exemplarily in this study.

Using measured data will enable the application of a sensor-based actuator control system to adjust some process parameters in cold forging.

A newly developed adjustable wedge device is introduced allowing for continuous adjustment of the punch in-between subsequent strokes and a precise positioning of the punch in the working direction. Therefore, a reliable way to control transient process conditions is proposed, aiming to reduce the amount of scrap production and shorten the ramp-up phase timewise.

Keywords: reduced scrap production · transient effects · adjustable wedge device · product quality

1 Introduction

The avoidance of scrap in cold forging processes is one of their generally accepted features resulting in economic and ecological advantages compared to chipping technologies, such as turning or milling. This feature, until now, is limited by transient process conditions and their undesirable influences on component properties and tolerances, especially during the process ramp-up phase after any change of tools. A stable production status is characterized by disclosing stationary process conditions after each ramp-up phase. During this phase of production, in practice, a considerable number of workpieces are produced that do not fulfil desired quality requirements. The economic efficiency of cold forming could be significantly improved if producers could compensate for unstable process conditions during the ramp-up phase. Though the common compensation strategy by experienced, manual work conducted by a machine operator lacks the fast, precise and repeated rate of necessary adjustment. On top of that adjustment is only possible during production stops. In contrast, automated mechanical devices

T. Bauernhansl et al. (Eds.): WGP 2023, LNPE, pp. 326–334, 2024.
https://doi.org/10.1007/978-3-031-47394-4_32

enable precise forming parameter compensation if deep knowledge about the technological correlations between properties of produced components is available. At the same time, the mechanical and temperature behaviour of the press machine and tools is essential for realising automated solutions. To decrease scrap during the ramp-up phase, this study investigates the mechanical and temperature behaviour of the press machine and tools during the ramp-up phase by conducting experiments and simulations.

2 State of the Art

Investigating the topic of ramp-up phases, the entire metal forming system consisting of press drive, gearboxes, ram lever designs, ram structure and clamped forming tool etc. are important. The dimensional and shape accuracy of cold forged components is not only depicted by the mechanical cavity of both parts of the forming tool. The main factors influencing the behaviour of the overall system include fluctuations in the mechanical properties of raw parts, the stiffness of the clamped press/forming tool system and thermal and tribological process conditions. In the case of fluctuating mechanical properties of the billet, evidently, the level of flow stress in particular, reveals a major influence on the forging process and on the dimensional and shape accuracy of the component. Fluctuating process forces in fact do lead to varying heat dissipation in the forming tool [1, 2]. Variations in the raw part material's microstructural properties also lead to a change in thermal conductivity [3]. An assessment of the forming system's sensitivity to temperature changes shows the effect of the elastic compression of the punch during forming. Punch compression is reduced by the thermal expansion of the punch that results from friction and energy used for plastic deformation. These effects lead to changes not only of the tribological conditions of the forming system but also do influence the part's dimensional accuracy, risking exceeding tolerance limits. During the ramp-up phase of the cold forging process, a transient thermal state is established within the die, in which cyclic temperature changes occur as depicted on the left side of the diagram in Fig. 1. After numerous forging cycles, a saturation of temperature and thus a stationary thermal state is reached, as shown on the right side of the diagram in Fig. 2 [3–5].

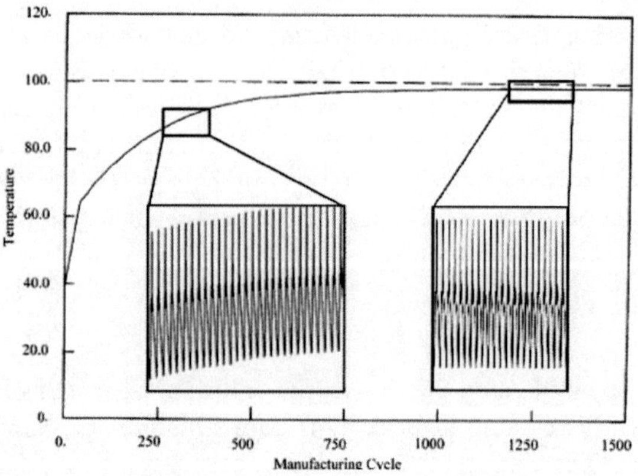

Fig. 1. Temperature cycles measured close to the inner die cavity due to thermomechanical punch length extension [4]

3 Investigation of a Ramp-Up Phase of an Industrially Established Process

In the framework of this paper to further investigate different ramp-up processes in cold forming, a comparison between a cold forging process performed in a laboratory and an industrially established process with a high production rate was made. Arnold Umformtechnik GmbH & Co KG, provided the opportunity to monitor a six-stage serial process consisting of extrusion and several heading operations for approximately 100 min during fasteners production. Forming forces as well as part and tool temperatures were measured during this time frame. To ensure that the start of the experiment represents the beginning of the ramp-up phase, the press and tool configurations were set up and kept unchanged since the last batch production one day before. Press and forging tool sets cooled down overnight. The manufacturing frequency was given to 110 parts per minute and the billet material was 38B2 (1.5515) steel. Combining an extrusion process with several upsetting stages, the fastener production was chosen to exemplarily show the effect of ramp-up phases especially on length dimension accuracy. Consequently, the first parts were not found according to quality requirements due to unsuitable tool settings during the start phase of batch production.

In contrast to the well-known zic-zac temperature course according to Fig. 2 in cold forging, the measured data in temperature emerging in the production line did not show the same characteristics due to the high production frequency. Instead, the part temperature at the fastener's head, captured by a thermal camera, correlates to the well-known temperature growth. Further short interruptions are indicated as blue marks (I and II) in Fig. 2, caused by the operator, were observed due to fluctuating dimensions of the pressed parts. A rapid drop of part temperature by 25 °C after two concurrent wedge adjustments in different stages after 4400 forged parts lead to a production stop by 5 min, which consequently lowers the part temperature down to 132 °C. Contrary to this observation, temperatures measured at the tool block in 250 mm distance behind the part ejector did not show any decrease in temperature but continued to increase as shown in

Fig. 2 (c). Here it can be seen that during the production stop the heat is distributed to adjacent tool parts and thus the measured temperature near the forming zone decreases.

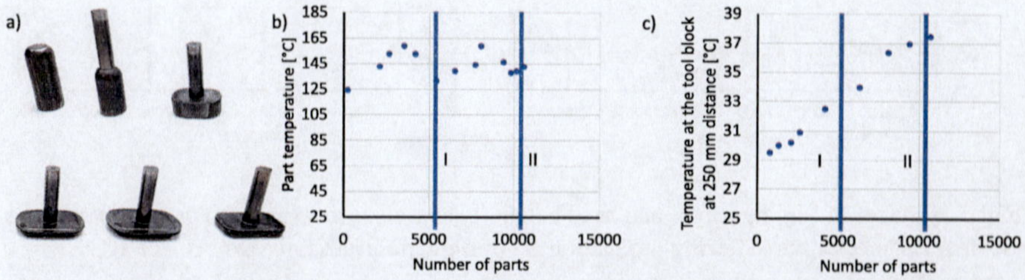

Fig. 2. (a) Different forming stages of investigated fastener component, (b) part temperature course of the finished product and (c) tool block temperature course measured at a distance of 250 mm away from the forming zone behind the part ejector. Production stops are shown with vertical blue lines.

Correlating the measured temperature courses with corresponding press force courses during the ramp-up phase, an increase in maximum force was observed during the rapid initial increase of temperature in all forming stages. This trend was interrupted by changes in punch position in the extrusion forming stage, referred to as stage two, and the first one of three upsetting stages referred to as stage four in the sequence of forming stages. Initiated by the machine operator to maintain product quality, the distance of the punch to the die in both stages was increased by 0.2 mm and 0.15 mm respectively. The resulting increase of peak force can be explained by the superimposition of thermal tool length expansion, changes in flow stress and influencing factors caused by previous forming stages. Thus, during the ramp-up phase in this process, the thermal expansion has a dominant influence on the captured force courses in all forming stages.

The changes in temperature and peak force do result in a fluctuating product geometry. The length of the parts depends on the acting forces and temperatures due to considerable changes in punch length. This leads to manually adjusting the settings by the machine operator (I and II). In this study, the height of the fastener's head was investigated and it clearly shows a linear decrease during the first phase (up to stop I) of the ramp-up process resulting in a reduced height of more than 3% after 4400 strokes compared to its initial height. Although the tool adjustment resulted in a closer distance of 0.15 mm between the punch and die, so, the consequential increase in the product's head height ranges around 0.1 mm. Head Height is measured by calculating the distance between a parallel flat surface and a mount while pushing the part into a mount so the part head protrudes. In the following production phase (after stop I) after the wedge adjustment, a further reduction of the head height was noted. In contrast to the initial reduction before changes in the tool settings, the gradient of head height changes was less and leads to a decrease of 1% over the next 3300 forged parts.

The time span of scrap production was measured to be approximately five minutes, corresponding to 550 parts. However, defining the end of the ramp-up phase at this stage is misleading because part temperature and temperatures at the tool block are still increasing significantly. For this specific product quality, the tolerances are acceptable

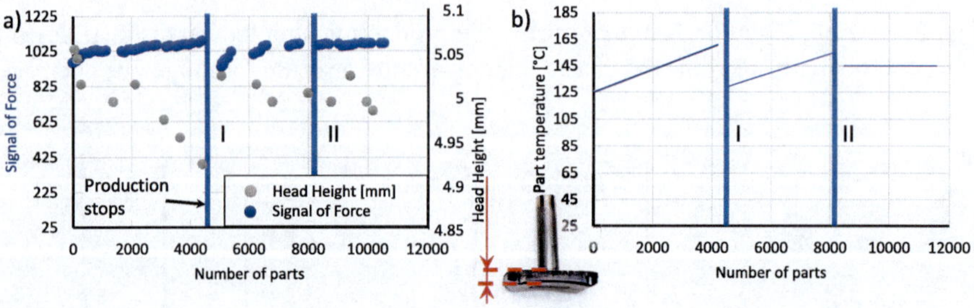

Fig. 3. Courses of forging forces and head height (a) as well as measured (d) part temperature of the final fastener captured during production at Arnold Umformtechnik GmbH CO. KG

at this stage of the production process though. Due to the machine operator's adjustment of the tool settings, a specific time during production to specify the end of the ramp-up phase is difficult to determine but expected at around 4500 parts when the force and temperature signals converge to a maximum value.

4 Investigation of a Cold Forging Ramp-Up Phase at the Institute

At the Institute for Metal Forming Technology in Stuttgart the continuous, but slower production, by forming backward cup extrusion specimens, was established and monitored during the ramp-up phase. Compared to the high production efficiency of the above-mentioned forming process, this process ramp-up phase allowed for a detailed investigation of changes in tool temperature and forming forces affecting the process. Single-stage forming of C15 cups at the production speed of nine parts per minute allows for individual monitoring of varying die temperatures and press forces simultaneously. Starting with an initial temperature level of 23 °C in press, tools and billets, a temperature sensor, embedded 2 mm beneath the inner tool cavity surface into the die and the die shrink ring, enables the inline capturing of a temperature gradient field in radial die direction close to the forming zone. The sensor within the die clearly shows the characteristic zic-zac temperature profile of each stroke as well as the superimposed rise in temperature up to 73 °C during the progress of the ramp-up phase as shown in Fig. 4. Due to the delay between strokes, also a considerable drop in temperature was measured at the inner sensor leading to a slow overall increase of temperature levels. Product geometry changes in correlation to the increase in temperature resulting in a decrease in cup bottom thickness of 1,5%, as shown in Fig. 4. Connecting both the temperature, force and the cup bottom thickness figure, the end of the ramp-up phase can be determined at approximately 250 parts.

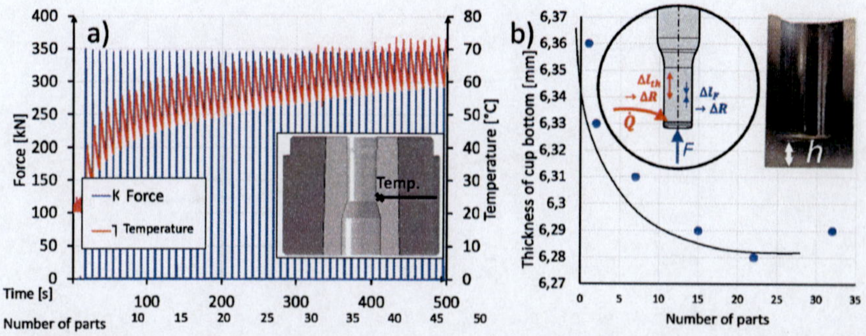

Fig. 4. a) Forging forces and temperature over time during backwards cup extrusion, measured in the die (temperature measurement points marked) and die shrink ring and b) thickness of the cup bottom

5 Strategies to Stabilize Ramp-Up Processes

In principle the authors recognize two different strategies to shorten the time of any ramp-up phases: 1. Quickest process stabilization or 2. Continuous adjustment of the process conditions. Both strategies can also be used simultaneously. The quickest stabilization of ramp-up phases strategy aims to avoid transient process conditions during the ramp-up phase by any means. To achieve this, there are several ways, for example, pre-heating of tools, pre-heating of billets or the smart combination of tool materials which may reduce any thermal expansion during production. Tight control of all process conditions with minimal tolerances and a small margin for errors due to statistical effects as well as minimised human impact can lead to a short ramp-up phase.

The strategy of continuous adjustment approaches the challenge of transient conditions by recognizing the effects of the ramp-up phase as well as statistical effects in the process and material as well as human influences and tries to adjust to any changes affecting product quality. Mechanically or electrically driven actuators and intelligent data processing can lead to short ramp-up phases and stable processes ensuring shortened ramp-up phases. Different kinds of actuator systems that can compensate for thermo-mechanical tool expansion are known, for example, actuators, that induce stresses with the help of elastomer inlays in tools or by hydraulic pressure. Tool parts based on the effects of shape memory alloys or dies with active thermal management are ways to continuously affect process conditions as well [6–9].

With the goal to get deeper knowledge into the ramp-up phase and comparing both strategies, at the Institute of Metal Forming Technology in Stuttgart, experiments with different lubricants and different pre-heated billets were conducted similar to the experiments shown in Fig. 4. Pre-heated parts at different final temperatures show a clear tendency towards smaller cup bottom thickness as shown in Fig. 5.

Several different effects are superimposing each other during the ramp-up phase: A slight decrease in material flow stresses and less viscose lubricants due to higher temperatures lead to improved material flow. Additionally, a higher difference between tool and billet temperature leads to faster initial heat transfer between billet and tools. Adding the constant forming energy that transforms mostly in heat, too, the maximum

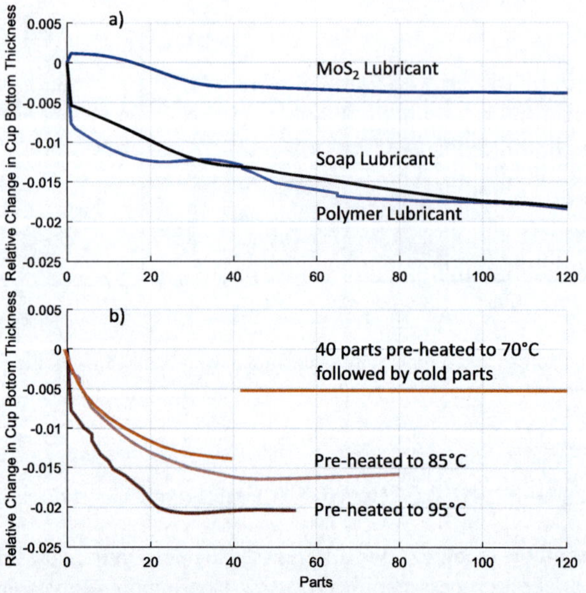

Fig. 5. a) Changes in cup bottom thickness during batch production with different lubricants and b) pre-heated billets at different temperatures

part temperature increases, overall leading to reduced cup bottom thickness in the pressed part. No clear result in ramp-up phase time reduction was recognized, although ramp-up phases with higher pre-heated part temperatures seem to show a faster decline in cup bottom thickness and seemingly faster horizontal progression in the graph.

Using only a limited number of pre-heated parts followed by cold parts, the experiments show an immediate increase in cup bottom thickness after changing to cold billets again. The heat seems to dissipate before leading to a more permanent increase in tool core temperature. These experiments show a remarkably constant median trend after the initial increase in cup bottom thickness offering possibilities to adjust the pre-heating temperature according to the final product part tolerances. On top of that, a precise number of parts are scrap, allowing machine operators to easily separate scrap from good parts. The useful application depends mainly on the required tolerance limits.

To investigate the possible impacts of continuous adjustment strategies actuators are important to influence the process. At the Institute of Metal Forming Technology, Stuttgart a wedge adjustment device based on a self-locking system of different wedges as shown in Fig. 6 was developed by the authors to achieve fast adjustment of the punch position. The wedge device will be installed between the press ram and the die punch with little space requirements and changes to the tool specifications. The wedge adjustment is controlled in a closed-loop system to precisely position the punch under load. With the idea in mind to combine numerical data gained in advance and human expert knowledge of the ramp-up phase of the process as well as continuous product geometry data, the wedge adjustment device is designed to control the product quality during changing process parameters. The adjustment is developed to be fast and in-between subsequent strokes or with short production stops enabled by the automatic self-locking system. As shown in Fig. 3, the tool wedge adjustment could intervene at several precise timings to shorten ramp-up phases and reduce scrap during production. Powered by a stepper motor that powers a spindle and pre-sets the final punch position, several wedges (I -III in Fig. 6) are in contact with each other. The connection of these wedges allows precise positioning of the punch with positioning tolerances of less than 1 μm. Based on the position of the central wedge (I) that is connected to the spindle and powered by the motor, the second wedge (II) is moved perpendicular to the movement of the wedge I on the same horizontal plane. Connected to the third wedge (III) a vertical movement, again perpendicular to the second wedge (II), now in the third dimension, is realized. With the fixed mechanical connection with wedge III the system is now able to position the punch on the vertical press with the stepper motor. The maximum change of the punch position in the vertical direction is 0.5 mm. Usual changes necessary in production are assumed to be less than 0.1 mm, allowing fast changes in the punch position according to the speed of the motor. The advantages of the system, as shown in Fig. 6, are the compact design with a self-locking mechanism.

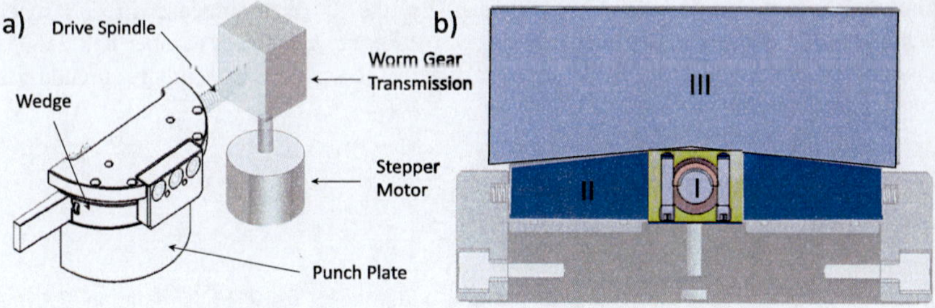

Fig. 6. a) Wedge adjustment system consisting of several wedges connecting a stepper motor via a spindle with the punch, b) detailed assembly of the wedge adjustment device

6 Summary

During ramp-up phases in cold forging batch production, transient thermomechanical process conditions lead to fluctuating part geometry. Resulting scrap parts currently are attributed to slow adaptations of forming parameter changes as well as a general lack of knowledge of the process effects. Aiming for a scrap-free production, two general strategies were proposed. The first is the quickest stabilization of ramp-up phases to optimize all forming parameters as well as press, tool and adjacent device settings. The second strategy is to continuously adjust the forming parameters. In pursuing this strategy, the final part geometry parameters are monitored during the ramp-up phase and adjustments are initiated with the help of an actuator device. To effectively use the actuator and quickly achieve stable process conditions, general knowledge about the ramp-up phase serves as a foundation for the control.

In this paper, the overall increase in part and tool temperature was measured in different forming stages in an industrially established process to investigate the impact of high production rate forging on the ramp-up phase. The influence of the tool adjustments by the machine operator and the impact of production stops were shown. To understand the zic-zac-shaped increase and decrease of the tool temperature captured close to the forming zone, a backwards cup extrusion process at the Institute for Metal Forming Technology Stuttgart was investigated. Expanding this basic process, the effect of pre-heated parts on the ramp-up was shown. Therefore, in the future additional simulation data can be used to predict the time and values of necessary tool adjustments, implemented by the machine operator. On top of that, they can lead to specific control knowledge to implement the actuator. Challenges like the numerical calculations of the heat flow in the tool parts need further investigation to correlate the effects of thermal extension and mechanical compression, especially to predict part geometry precisely. To shorten ramp-up phases and to continuously adjust process conditions, a closed-loop controlled wedge device was introduced in this paper. The goal of this device is the reduction of ramp-up phase process time by continuous adjustments of the punch position during the beginning of each serial process to reduce scrap.

Acknowledgement. This research is founded by the "Arbeitsgemeinschaft industrieller Forschungsvereinigungen" (AiF, Industrial Collective Research), project number IGF 21635 N. The authors would like to thank the Industrial Collective Research Foundation for founding and supporting this research project.

References

1. Long, H.: Dimensional errors of cold formed components using FE simulation. J. Mater. Process. Technol. **151**(1–3), 355–366 (2004)
2. Kuzman, K.: Problems of accuracy control in cold forming. J. Mater. Process. Technol. **113**(1–3), 10–15 (2001)
3. Rosochowska, M., Balendra, R., Chodnikiewicz, K.: Measurements of thermal contact conductance. J. Mater. Process. Technol. **135**(2–3), 204–210 (2003)
4. Qin, Y., Balendra, R., Chodnikiewicz, K.: Method for the simulation of temperature stabilisation in the tools during multi-cycle cold-forging operations. J. Mater. Process. Technol. **107**(1–3), 252–259 (2000)
5. Volke, P., Groche, P.: Interfacial heat transfer coefficients in cold forging of stainless steel. Int.J. Mater. Form. **15**, 32 (2022)
6. Qin, Y.: Forming-tool design innovation and intelligent tool-structure/system concepts. Int. J. Mach. Tools Manuf **46**(11), 1253–1260 (2006)
7. Koch, J., Kroiß, D.T.: Abschlussbericht der Studie: Aktive Matrize, Erlangen (2010)
8. Baumgarten, J.: Aktiver Dehnungsausgleich. Mit neuartigem Werkzeugkonzept die elastische Matrizenaufweitung bei der Kaltmassivumformung kompensieren (2002)
9. Veldhuis, M., Heingärtner, J., Krairi, A., et al.: An industrial-scale cold forming process highly sensitive to temperature induced frictional start-up effects to validate a physical based friction model. Procedia Manuf. **47**, 578–585 (2020)

Investigation of the Influence of Static Recrystallization on Void Evolution After Hot Compression Forming

A. Heiser(✉), J. Gerlach, D. Bailly, and G. Hirt

Institute of Metal Forming (IBF), RWTH Aachen University, Intzestr. 10, 52072 Aachen, Germany

annemarie.heiser@ibf.rwth-aachen.de

Abstract. Damage that exists in the form of microvoids in a material can be influenced by various process parameters during forming. Depending on these parameters, the static recrystallization (SRX) occurring during the holding time between two hot forming steps can also lead to a reduction in damage. Therefore, this work aims to investigate the influence of SRX on ductile damage after hot compression forming in the case hardening steel 16MnCrS5. First, stress relaxation tests are conducted to characterize the kinetics of static recrystallization in the material. Based on that, hot compression tests are performed at different temperatures and holding times. The void area fraction (VAF) after the hot compression and consecutive static recrystallization is determined based on combined image processing of scanning-electron microscopy (SEM) and energy-dispersive X-ray spectroscopy (EDX). The results show a decrease in VAF with an increasing recrystallized volume fraction, which indicates a positive effect of static recrystallization on void healing.

Keywords: Hot Forming · Damage · Static Recrystallization

1 Introduction

Hot working processes aim to form material into desired geometries while reaching desired mechanical properties through influencing the microstructure via recrystallization and grain refinement. In addition to these main objectives, hot forming processes can also be designed to reduce damage in terms of microvoids present in the semi-finished product during deformation, thereby improving its properties. Considering and controlling damage offers a means to modify the process design, enhancing the product's performance and ultimately utilizing its lightweight potential. While damage-reducing compressive stresses dominate in typical hot forming processes such as forging, damage can still occur in tension-loaded regions of the components [1]. The damage occurs through decohesion of non-metallic inclusions and second-phase particles from the matrix [2]. The recovery and recrystallization processes during hot forming provide means to delay the development of damage by reducing internal stresses [2] or, if static recrystallization (SRX) occurs after deformation, to heal voids [3].

T. Bauernhansl et al. (Eds.): WGP 2023, LNPE, pp. 335–344, 2024.
https://doi.org/10.1007/978-3-031-47394-4_33

Ganguly et al. [4] simulated the reverse rolling of steel using a finite element (FE) model. The influence of void size, void location in the workpiece and temperature on void healing was investigated. Additionally, the effect of softening due to recrystallization during and between individual rolling passes was considered. It was demonstrated that the voids in the center of the plate during the rolling process could be completely closed, unlike the voids at the edge of the sheet. Increased temperature resulted in faster void closure. The addition of recrystallization in the material model also had a positive effect on the simulated void closure. Feng et al. [5] demonstrated that interfaces can be completely dissolved through recrystallization after the welding of a void. In their study, two blocks of a high Cr ferritic steel were axially compressed at temperatures ranging from 900 °C to 1200 °C, followed by isothermal holding for varying time periods. They showed that higher temperatures lead to an enhanced surface bonding behavior. Similar findings to Feng et al. were obtained by Gao et al. [6] during the rolling of a carbon steel with artificially introduced interfaces acting as microcracks. Recrystallization can occur at the free surfaces of the crack, and the newly formed grains can gradually close the crack and ultimately heal it completely.

However, none of the mentioned studies establishes a direct correlation between the damage evolution and SRX following a hot forming process. Therefore, this study aims to investigate the influence of SRX on the void area fraction (VAF) after hot compression tests.

2 Material and Methods

In this study, the case hardening steel 16MnCrS5 is investigated. The chemical composition of the steel was determined using an optical emission spectrometer and is given in Table 1. The material contains various forms of non-metallic inclusions including MnS, Al_2O3, and SiO_2. All tests are performed on cylindrical compression samples with a height of 10 mm and a diameter of 5 mm.

Table 1. Chemical composition of 16MnCrS5 steel.

Element	C	Si	Mn	S	Cr	Ni	Cu
Wt. %	0.200	0.173	1.22	0.032	1.05	0.135	0.162

2.1 Experimental Setup

Stress relaxation tests are performed to model the recrystallization kinetics of the investigated steel. Afterwards holding times corresponding to desired amounts of recrystallization within the material are determined based on these kinetics. In the hot compression tests, the samples are quenched after those defined holding times to prevent the microstructure from changing further and thus to investigate the influence of SRX on damage evolution. SEM and EDX measurements are then used to quantify the void area fractions of the samples.

The stress relaxation and hot compression tests are carried out on a quenching and deformation dilatometer "DIL805A/D/T" from TA Instruments. For all tests, the samples are heated to the desired temperatures, ranging between 900 to 1200 °C, at a heating rate of 10 K/s, followed by a homogenization period of 60 s. Following, the samples are compressed in both test methods. In the initial stress relaxation tests for determining the SRX kinetics, the subsequent isothermal hold time is designed long enough to allow for complete recrystallization. In contrast, in the hot compression tests the samples are quenched after defined holding times to obtain microstructures with no recrystallization, partial recrystallization (50%) or complete recrystallization (95%).

Stress Relaxation Tests

The stress relaxation tests are performed with two different compressive true strains of 0.1 and 0.2 and a strain rate of 0.01 s^{-1} at temperatures of 900, 1000, 1100 and 1200 °C. Subsequently, an isothermal holding period of 30 min at the same temperature follows. The position of the tooling stamp, and thus the attained deformation, is maintained constant during this time, while the force response of the specimen is measured over time.

The derivation of the SRX kinetics from the force-time curves is performed using the method proposed by Karjalainen [7], which assumes that the material consists of both a hardened and a softened fraction at any given time. This allows the utilization of a mixing rule to determine the recrystallized fraction X of the microstructure at each time point t. The recrystallized fraction X is then fitted to the Johnson-Mehl-Avrami-Kolgomorov (JMAK) equation, where n stands for the Avrami exponent [8, 9]:

$$X(\mathrm{t}) = 1 - \exp\left[-0,693\left(\frac{t}{t_{50}}\right)^{n}\right] \tag{1}$$

The time t_{50} represents the time at which 50% of the microstructure is recrystallized. It can be determined using the following equation

$$t_{50}=A \cdot \varphi^{q} \cdot d_0^{r} \cdot \dot{\varphi}^{s} \cdot \exp\left(\frac{Q}{RT}\right) \tag{2}$$

with the dimensionless material-dependent parameters A, q, r, and s, the true strain φ, the strain rate $\dot{\varphi}$ and the initial grain size d_0. Q represents the activation energy for recrystallization, R the universal gas constant and T the temperature in Kelvin [10].

To avoid dynamic recrystallization during deformation, a critical strain ε_c below the peak strain ε_p must not be exceeded. For C-Mn steels, a value of $0.8 \cdot \varepsilon_p$ is often assumed for the critical strain [11]. If the critical strain is not exceeded, only static recrystallization occurs. Since t_{50} is assumed independent of the strain rate if the conditions for static recrystallization are met and the initial grain size is the same for all samples, the equation is simplified to:

$$t_{50}=A \cdot \varphi^{q} \cdot \exp\left(\frac{Q}{RT}\right) \tag{3}$$

Hot Compression Tests

The hot compression tests follow the heating and deformation scheme shown in Fig. 1.

The samples are heated to a temperature of 900 or 1100 °C at a heating rate of 10 K/s, followed by a homogenization period of 60 s. The two temperatures were selected to examine the temperature effect on damage evolution while simultaneously reducing the experimental effort required for damage characterization. By conducting the experiments at 900 °C and 1100 °C, two objectives are achieved: firstly, a substantial temperature difference is established; secondly, recrystallization occurs at a slower rate at lower temperatures, resulting in a larger time interval between 50% and 95% recrystallization. This ensures two significantly distinct recrystallization fractions even in the event of recrystallization persisting briefly during quenching following the holding period. The subsequent compression is performed with a true strain of 0.1 and a strain rate of 1 s^{-1}in all experiments. Flow curves published in [12] show that the chosen combination of true strain and strain rate ensures that the condition $\varepsilon < 0.8 \cdot \varepsilon_p$ is met at all temperatures, allowing only static recrystallization to occur. This is important in order to exclusively investigate the influence of static recrystallization, excluding the effects of dynamic recrystallization. After the compression, the samples are held isothermally for various holding times corresponding to a recrystallized fraction of 0%, 50%, and 95%. At the end of the holding time, the samples are quenched by flooding the dilatometer with protective gas to prevent further recrystallization.

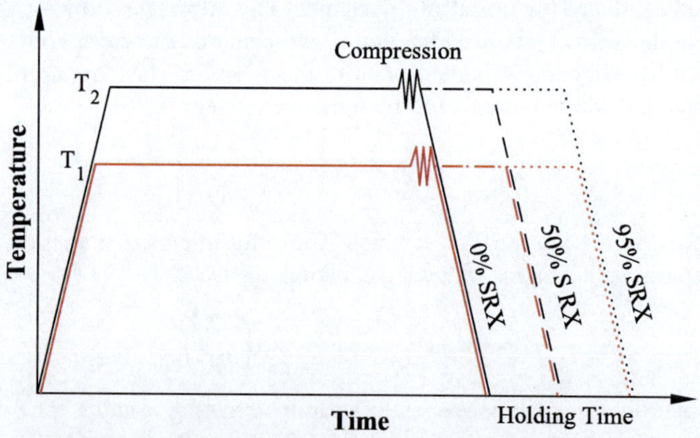

Fig. 1. Schematic sequence of hot compression tests.

2.2 Damage Characterization

In this paper, ductile damage is understood as the nucleation, growth and coalescence of microvoids. Therefore, a field emission gun scanning electron microscope "JEOL JSM 7000F" from Schottky is used to quantify the number and size of microvoids within the material. An automated energy-dispersive X-ray spectroscopy (EDX) quantification method with the EDAX Genesis (Version 6.53) software allows for the distinction between matrix material, inclusions and voids. Figure 2b shows the backscattered electron (BSE) image of a void and a manganese sulfide inclusion in the sample compressed at 900 °C. The varying grayscales allow differentiation between the matrix and the inclusion. Since voids and preparation residue or contamination exhibit similar grayscale values, energy-dispersive X-ray spectroscopy (EDX) is employed for their identification. The used method is presented in detail in [13].

As shown in Fig. 2a, the void area fraction (VAF) and void size distribution is investigated at half the height of the samples and approximately 1 mm below the surface. There are two adjacent measurement fields with a width of 2 mm and a height of 0.75 mm resulting in a total analyzed area of 3 mm^2. In the bulging zone of the compressed sample the highest level of damage is expected due to present circumferential tensile stresses.

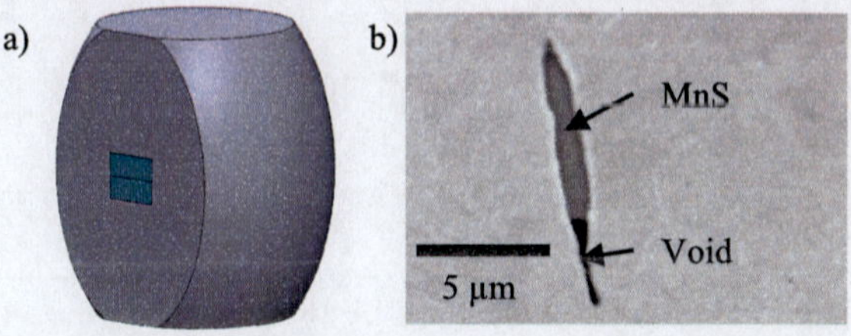

Fig. 2. a) Schematic view of compression sample with measurement fields b) BSE image of void at a MnS inclusion.

3 Results

3.1 Modelling of SRX Kinetics

The experimental and fitted data of the stress relaxation tests are presented in Fig. 3. When analyzing the recrystallization kinetics, it is important to note that the stress relaxation tests were conducted at a strain rate of 0.01 s^{-1} since lower strain rates benefit the procedure. Flow curves of the studied material published in [12] indicate that for this strain rate at a true strain of 0.1, only static recrystallization is expected to occur. However, this is not the case for the higher true strain of 0.2 at such a low strain rate. Therefore, only the results of the stress relaxation tests with the lower true strain of 0.1 are used to determine the Avrami exponent n and activation energy Q for modelling the static recrystallization kinetics.

The derived material parameters for the recrystallization kinetics of the investigated 16MnCrS5 are shown in Table 2. They were used in the simplified Eq. 3 to calculate the holding times necessary for 50% and 95% static recrystallization for the hot compression tests.

Table 2. Material parameters for recrystallization kinetics.

Avrami exponent n	Activation energy Q [kJ/mol]	A	q
1.04	102.462	0.001318	0.4698

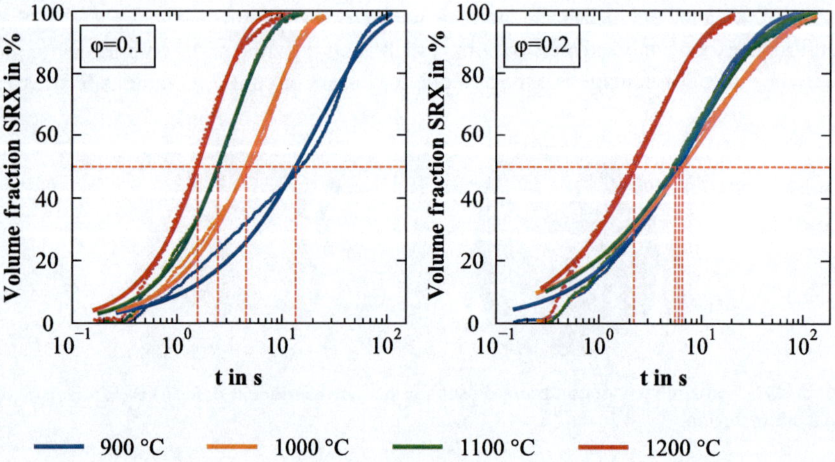

Fig. 3. Kinetics of static recrystallization determined via stress relaxation tests at $\varphi = 0.1$ and $\varphi = 0.2$.

3.2 Influence of SRX on Damage Evolution

As displayed in Fig. 4, the VAF decreases with an increasing recrystallized volume fraction. Based on the available data points, a linear relationship between the VAF and the recrystallized volume fraction can be inferred. This assumption should suffice for a general assessment of the impact of static recrystallization on damage. To verify this assumption, further data points need to be collected. The slope of the line is dependent of the temperature, at which the recrystallization occurs. The initial VAF of the samples at room temperature is measured to be 0.00499%. When directly quenching the samples after deformation, the VAF decreases to a lower value at 1100 °C (0.00224%) compared to 900 °C (0.00304%). It is indicated, that the phase transformation of the ferritic-perlitic microstructure to austenite during the heating and homogenization process prior to deformation contributes to reducing damage. Additionally, the results demonstrate an effect of different temperatures during homogenization and subsequent deformation on the reduction in damage. In addition to enhanced grain growth at higher annealing temperatures resulting in an already lower VAF before compression, the temperature of 1100 °C also leads to a lower flow stress compared to 900 °C. A lower flow stress of the matrix material is beneficial for the void closure during the compression. These results are consistent with findings of Imran et al. [2], who investigated the influence of dynamic recrystallization on damage evolution.

The varying magnitudes of standard deviations can be explained by the void distributions in Fig. 5 and Fig. 6. At 900 °C, a few, albeit larger, voids are present. With a low number of voids, there is a higher likelihood of their uneven distribution across the two measurement areas, resulting in an increase in the standard deviation.

At 50% of recrystallization the sample at a temperature of 900 °C still shows a higher level of damage (0.00250% VAF) in comparison to the sample at 1100 °C (0.00215% VAF). However, after full recrystallization, the samples show nearly the same amount of damage at both temperatures (900 °C: 0.00163% VAF, 1100 °C: 0.00170% VAF). The higher impact of the SRX at 900 °C, resulting in a nearly 50% decrease of VAF compared to a 25% decrease at 1100 °C, might be attributed to the initial levels of damage at 0% recrystallized volume fraction. At a temperature of 1100 °C, the sample already contains less damage, and the lower flow stress leads to a better void closure during deformation. The total number of voids for the non-recrystallized samples is 499 for 900 °C and 407 for 1100 °C. Consequently, the amount of voids that can potentially heal during subsequent static recrystallization is lower compared to the sample annealed at 900 °C.

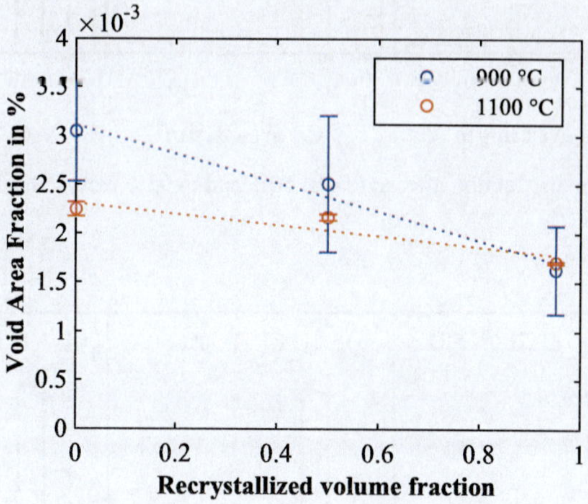

Fig. 4. Void area fraction (VAF) as a function of the recrystallized volume fraction.

The analysis of the void size distribution in Fig. 5 and Fig. 6 provide a more comprehensive understanding of the damage evolution during static recrystallization. At a temperature of 900 °C, the VAF consistently decreases from the non-recrystallized to the fully recrystallized state. The total number of voids remains nearly constant between 50% and 95% recrystallized volume fractions. The distribution of void sizes shifts towards fractions of voids with smaller size, indicating that existing voids decrease in size and can be healed through static recrystallization.

At 1100 °C, both the VAF and the total number of voids consistently decrease most probably due to the closure and healing of voids. The shift towards fractions of voids with smaller size is less significant. However, the relative proportion of small voids, up to a void area of 0.132 mm2, to the total count of voids increases with the progressing recrystallization from 60% to 74%.

Both the decrease in VAF over the progressing recrystallized volume fraction as well as the shift in the void size distributions to fractions containing smaller voids demonstrate that SRX has the capability to close and heal voids. Consequently, materials with a recrystallized microstructure exhibit lower levels of damage compared to materials that undergo direct quenching following hot working processes.

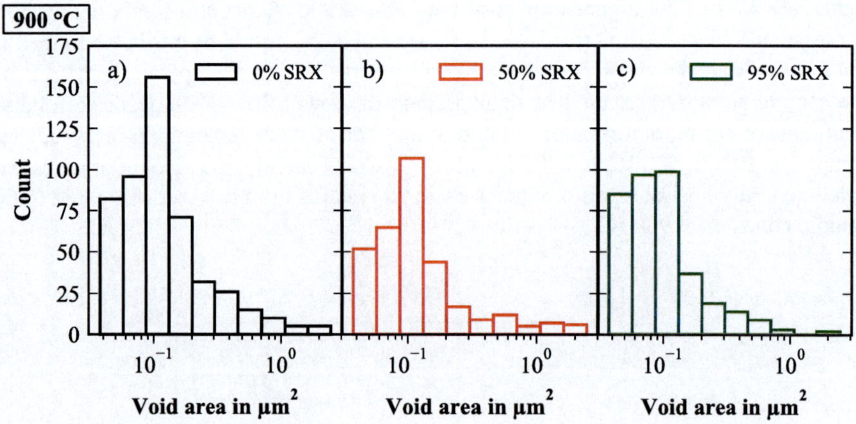

Fig. 5. Void size distribution after a) 0% b) 50% and c) 95% recrystallization at 900 °C.

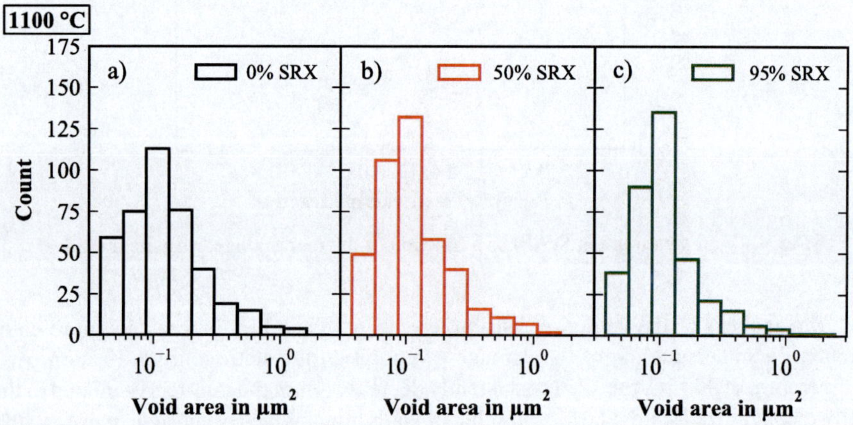

Fig. 6. Void size distribution after a) 0% b) 50% and c) 95% recrystallization at 1100 °C.

4 Summary and Outlook

Using stress relaxation tests, the recrystallization kinetics for 16MnCrS5 were determined. The results were utilized to select holding times for a partial and a full static recrystallization in hot compression tests. In an SEM and EDX based damage analyses following the hot compression tests, it was demonstrated that the void area fraction decreases with an increasing recrystallized volume fraction. The influence of SRX on damage was more pronounced in the samples processed at 900 °C due to a higher number of voids at the beginning of SRX, resulting in an equal VAF after complete recrystallization. It can be presumed that static recrystallization positively influences damage by reducing it and contributing to the healing of voids.

To further confirm the accuracy of the recrystallization kinetics model and to gain a more in-depth understanding of how the microstructure, particularly grain growth and grain size distribution during SRX, influence the damage evolution a microstructure analysis is planned. Furthermore, the model for recrystallization kinetics is then to be extended with terms considering the grain growth following the SRX and its corresponding influence on damage.

Acknowledgement. This research was funded by the Deutsche Forschungsgemeinschaft (DFG, German Research Foundation) - Project number 278868966 - TRR 188, Project A07.

The support of this work by the Gemeinschaftslabor für Elektronenmikroskopie (GFE) of RWTH Aachen University and Anthony Dunlap is gratefully acknowledged, who provided the SEM measurements.

References

1. Duan, X.W., Liu, J.S.: Research on damage evolution and damage model of 316LN steel during forging. Mater. Sci. Eng. A **588**, 265–271 (2013)
2. Imran, M., Szyndler, J., Afzal, M.J., Bambach, M.: Dynamic recrystallization-dependent damage modeling during hot forming. Int. J. Damage Mech **29**(2), 335–363 (2020)
3. Li, X., Liebsch, C., Hirt, G., et al.: Modelling of void healing in hot rolling due to recrystallization. Prod. Eng. Res. Dev. **14**(1), 43–52 (2020)
4. Ganguly, S., Wang, X., Chandrashekhara, K., et al.: Modeling and simulation of void closure during steckel mill rolling for steel plate. Steel Res. Int. **92**(2), 1611–3683 (2021)
5. Feng, C., Cui, Z., Liu, M., et al.: Investigation on the void closure efficiency in cogging processes of the large ingot by using a 3-D void evolution model. J. Mater. Process. Technol. **237**(1), 371–385 (2016)
6. Gao, H., Ai, Z., Yu, H., et al.: Analysis of internal crack healing mechanism under rolling deformation. PLoS ONE **9**(7), 1–6 (2014)
7. Karjalainen, L.P.: Stress relaxation method for investigation of softening kinetics in hot deformed steels. Mater. Sci. Technol. **11**(6), 557–565 (1995)
8. Johnson, W.A., Mehl, R.F.: Reaction kinetics in processes of nucleation and growth. Trans. Am. Inst. Min. Metall. Eng. **135**(8), 416–442 (1939)
9. Avrami, M.: Kinetics of phase change. I general theory. J. Chem. Phys. **7**(12), 1103–1112 (1939)
10. Sellars, C.M., Whiteman, J.A.: Recrystallization and grain growth in hot rolling. Metal Sci. **13**(3–4), 187–194 (1979)

11. Karjalainen, L.P., Maccangno, T.M., Jonas, J.J.: Softening and flow stress behaviour of Nb microalloyed steels during hot rolling simulation. ISIJ Int. **35**(12), 1523–1531 (1995)
12. Wang, S., Dunlap, A., Möhring, K., et al.: Torsion plastometer trials to investigate the effect of non-proportional loading paths in caliber rolling on damage and performance of metal parts. Prod. Eng. Res. Dev. **14**(1), 17–32 (2020)
13. Hering, O., Dunlap, A., Tekkaya, A.E., et al.: Characterization of damage in forward rod extruded parts. Int. J. Mater. Form. **13**(6), 1003–1014 (2020)

Analysis of Semi-Hot Forming Behaviour of AA7075 Sheet Metal

D. Vasquez Ramirez(✉), H. Wester, D. Rosenbusch, and B.-A. Behrens

Institute of Forming Technology and Machines, Leibniz University Hannover, An der Universität 2, 30823 Garbsen, Germany
vasquezramirez@ifum.uni-hannover.de

Abstract. 7xxx series aluminium alloys have a high potential for lightweight constructions in automotive applications due to lower weight, similar strength and higher recyclability compared to steel. However, formability of such alloys at room temperature is relatively low. Elevated temperatures are one way of improving formability, e.g. by semi-hot forming. This work is therefore concerned with the analysis of strain rate and temperature dependent behaviour of AA7075 in T6 state. For this, experimental characterisation has been performed using uniaxial and shear tension, layer compression, plane strain compression and Nakajima tests over temperatures of 20 to 275 °C and strain rates of 0.01 to 1 s^{-1}. The experimental results are used to calibrate and build a temperature and strain rate dependent plasticity model using Hockett-Sherby hardening law. For failure modelling, a temperature dependent forming limit curve was calculated. Additionally, various yield criteria were applied to predict a temperature dependent yield locus for a constant strain rate.

Keywords: sheet forming · aluminium alloy · semi-hot forming · strain rate · temperature dependency

1 Introduction

Lightweight automotive and aerospace components are becoming increasingly important to improve fuel efficiency and reduce pollution, in particularly in the context of e-mobility and its range extension. For example, 7xxx-series aluminium alloys have a greater strength to weight ratio than some high strength steels [1]. Moreover, recyclability and corrosion resistance of aluminium is much higher than of steel. However, a drawback of such alloy is the limited room temperature formability, which is why semi-hot forming, also known as warm forming, offers a promising approach [2].

In contrast to hot forming, semi-hot forming of aluminium alloys up to 300 °C offers improved formability by reducing energy as well as process duration. However, suitable process design requires an appropriate material behaviour analysis to meet the requirements in terms of temperatures, strain rates and stresses. Few studies have considered semi-hot forming of pre-aged 7xxx-series aluminium alloys. Lin et al. [1] investigated AA7075 alloy in T6 state at temperatures between 25–230 °C using uniaxial

T. Bauernhansl et al. (Eds.): WGP 2023, LNPE, pp. 345–355, 2024.
https://doi.org/10.1007/978-3-031-47394-4_34

and biaxial tensile tests. However, neither strain rate dependence nor shear stresses were considered. Österreicher et al. [3] analysed AA7075 T6 sheet behaviour during warm forming, however, only for tensile stresses. Rader et al. investigated semi-hot forming of AA7075 sheet in T6 state with the aim of retrogression forming, again using tensile tests only [4]. Based on this literature review, few works were found that analyse AA7075 sheet in T6 state regarding various stress states, temperatures and strain rates. For a realistic numerical representation of semi-hot forming, a detailed experimental material characterisation was therefore carried out on a 2 mm thick AA7075 T6 sheet by means of temperature and strain rate dependent miniaturised uniaxial and shear tensile tests, temperature dependent layer and plane strain compression tests and temperature dependent Nakajima tests. Each test was repeated five times for statistical certainty.

2 Experimental Setup

2.1 Tensile Tests

Miniaturised tensile tests on uniaxial and shear specimens were performed using a quenching and forming dilatometer DIL 805A/D + T. The dilatometer as well as the specimen geometries are shown in Fig. 1 (a) and (b), respectively, as presented in [5].

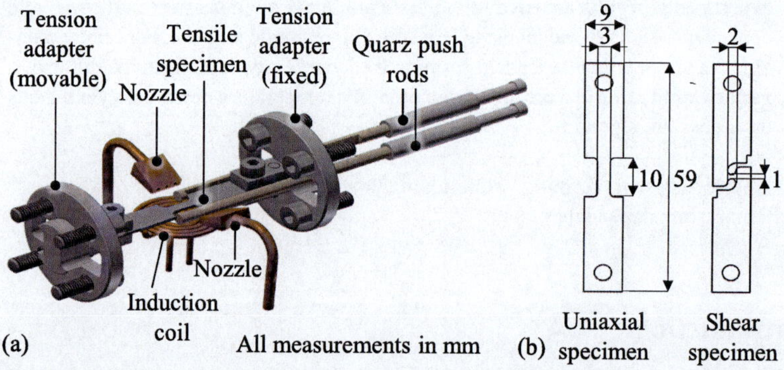

Fig. 1. Dilatometer test setup for miniaturised tensile tests (a), miniaturised specimens (b)

Uniaxial tensile tests were conducted on specimens in 0°, 45° and 90° to sheet's rolling direction (RD), while shear tensile tests were performed on specimens in 0° and 90° to RD. Testing temperatures applied via inductive heating with 15 K/s heating rate for both specimen geometries were 20, 125, 175, 225 and 275 °C at 0.01 s^{-1}. Rates of 0.1 and 1 s^{-1} were considered solely on uniaxial tensile specimens in 0° to RD to determine temperature and strain rate dependent flow curves. The experimental flow curves were then extrapolated by means of Hockett-Sherby hardening model

$$\sigma_{tensile} = \sigma_s - \left(\sigma_s - \sigma_y\right) \cdot exp(-N\varepsilon)^P \tag{1}$$

with σ_s as steady-state flow stress in MPa, σ_y as yield stress in MPa, ε as plastic strain and N and P as material constants using the least square method [6]. Experimental plane strain compression tests were used to extend the flow curves from miniaturised tensile tests, as presented in 2.3, and subsequently to select the extrapolation method with the best fit by means of least square method, resulting in the selection of Hockett-Sherby. The uniaxial tensile tests in all three angles to RD were also applied to model a temperature dependent yield locus at 0.01 s^{-1} by using the flow stress of each experiment for the principal stress calculation according to [7], while the shear tensile tests in 0° and 90° to RD were used as yield locus validation.

2.2 Layer Compression Tests

Layer compression tests were used to determine the biaxial point in the yield locus curve. For this, a servo-hydraulic forming simulator DYNSJ5590 was used, in which the layer compression test setup was mounted, shown in Fig. 2 (a). The test setup consisted of two compression punches and an insulated container. The tests were conducted at temperatures 20, 125, 175, 225 and 275 °C with a heating rate of 5 K/s at 0.01 s^{-1}. As solution heat treatment of AA7075 takes place at approximately 480 °C and thus no significant change in grain structure is observed below 300 °C, the difference in heating rate between layer compression and tensile tests was neglected [3]. A furnace was used to heat up the test setup. The temperature was measured using thermocouples. After the test setup reached the set temperature, it was removed from the furnace, the specimens were promptly placed on the lower punch and the entire set up was once again placed and held in the furnace until the set temperature was reached.

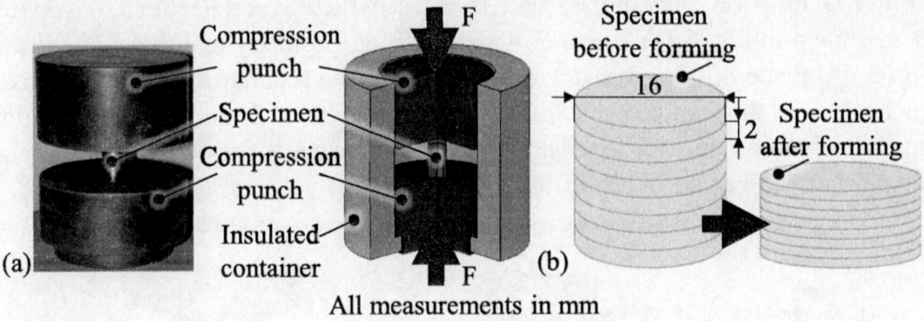

Fig. 2. Test setup of layer compression test (a), layered specimens before and after forming (b)

An assembly of eight specimen layers was built resulting in an approximate ratio between the specimen's diameter and the total height of 1:1, as shown in Fig. 2 (b). The relative position to the rolling direction of the layered specimens was held constant. A graphite lubricant was applied between the contact surface of the compression punch and the specimen as well as between the individual layers. The punch stroke of the forming simulator was set to 8 mm, which compressed the layered specimen by half of its initial height, as shown in Fig. 2 (b). The test evaluation was conducted under isotropy assumption. Resulting force-stroke curve was used to calculate initiation of biaxial flow stress for each considered temperature.

2.3 Plane Strain Compression Tests

The plane strain compression test (PSCT) was used for two purposes in this research. Firstly, the resulting flow curves served to extend the data base of uniaxial tensile flow curves to higher degrees of deformation and thus to achieve the best fit of the flow curve extrapolation method. For this purpose, the determined biaxial flow stress had to be converted into an equivalent von Mises stress with the following equation assuming no friction due to applied graphite lubrication was present

$$\sigma_{uni,psct} = \sigma_{bia,psct} \cdot \frac{\sqrt{3}}{2} = \frac{F}{a \cdot b} \cdot \frac{\sqrt{3}}{2} \quad (2)$$

with F as measured Force in kN, a as contact width of 2 mm and b as contact height of 20 mm [8]. Subsequently, using the plastic deformation work equivalence condition between biaxial W_{psct} and uniaxial $W_{tensile}$ loading

$$W_{psct} = W_{tensile} = \int \sigma_{uni,psct} \cdot d\varepsilon_{uni,psct} = \int \sigma_{tensile} \cdot d\varepsilon_{tensile} \quad (3)$$

with an integral calculation of the respective flow stresses $\sigma_{uni,psct}$ and $\sigma_{tensile}$ over each equivalent plastic strain $\varepsilon_{uni,psct}$ and $\varepsilon_{tensile}$, the uniaxial flow curves were extended and extrapolated using the least square method, as presented in 4.1 [9]. Secondly, the PSCT was used to validate yield locus criteria. For this reason, the temperatures and strain rates considered in the PSCT at 0° to RD direction were 20, 125, 175, 225 and 275 °C with a heating rate of 15 K/s at 0.01, 0.1 and 1 s^{-1}. Identical temperatures were applied for tests at 90° to RD, however, only at 0.01 s^{-1}. To perform the PSCT, Gleeble 3800-GTC was used, as shown in Fig. 3 (a), together with specimen geometry and the applied stamps in Fig. 3 (b), made of a hardened 1.2367 material. The total stroke of the stamps was set to 1 mm. For the stamp, radius of 0.5 mm and width of 2 mm was selected due to comparatively low values of friction and thrust components for the considered plastic strain and sheet thickness of 2 mm, as described in [8].

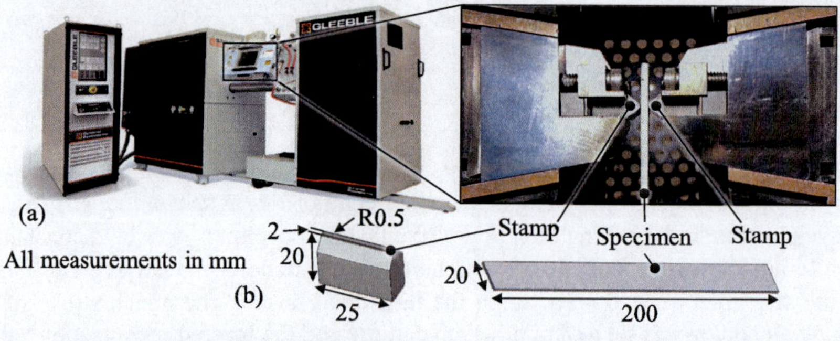

Fig. 3. PSCT setup (a), applied testing stamp with specimen geometry (b)

2.4 Nakajima Tests

The goal of Nakajima tests was to build a temperature dependent flow limit curve (FLC). The FLC test setup in Fig. 4 (a) is applicable for temperatures up to 1100 °C and is shown exemplarily at 700 °C. The test setup was made of 1.2367 hardened material and mounted into a servo-hydraulic forming simulator DYNSJ5590, equipped with a punch drive, a pneumatic plate and a digital image correlation (DIC) device Aramis. The punch drive served to move the Nakajima punch of ∅ 100 mm inside the FLC setup and thus to form the sheet, which was placed between the drawing die and the blank holder. The pneumatic plate built a blankholder force of 60 kN on the sheet. Thermocouples were used to measure the temperature during the heating process in a furnace and during the testing. The scaled Hasek specimens, designed analogous to the ISO 12004 norm regarding their bridge width-diameter ratio so the strain ratios of the respective specimen geometry remained identical to the ISO standard, are shown in Fig. 4 (b) and vary between uniaxial, plane strain and biaxial stress states [10].

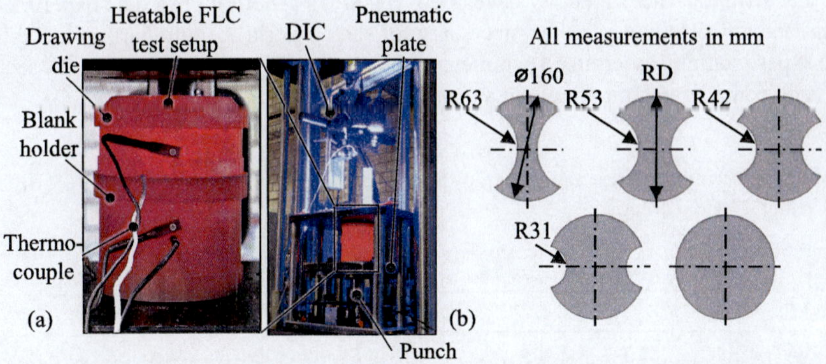

Fig. 4. Mounted heatable FLC test setup (a) scaled Hasek FLC specimens (b)

Prior to testing, each specimen was coated with white varnish and a stochastic pattern of black grid to prevent reflection and allow an online DIC measurement. Following the ISO 12004 standard, the tests were performed on AA7075 sheet material in 0° to RD. Prior testing, both the test setup and the specimen were heated in a furnace separately until the considered temperature was reached. After reaching the temperature, the specimen was placed on the test setup between drawing die and blank holder and transferred to the forming simulator, where it was mounted on the pneumatic plate. The temperatures tested were 125, 175, 225 and 275 °C with a heating rate of 5 K/s, in analogy to layer compression tests, at a punch speed of 1 mms^{-1}. As a lubrication, Raziol NTO 300G was used together with a thin graphite foil to reduce friction between the Nakajima punch and the specimen. The DIC was used to determine the major and minor strains. The DIN EN ISO 12004 line intersection method was applied to evaluate the results and determine major and minor strains for each temperature and specimen geometry to build a temperature dependent FLC.

3 Yield Locus Determination

The yield locus were determined using experimental data from miniaturised tensile, layer compression and FLC tests. By means of uniaxial miniaturised tensile tests in 0°, 45° and 90° to RD, yield stresses $\sigma_{uni,0}$, $\sigma_{uni,45}$, $\sigma_{uni,90}$, anisotropic r-values $r_{uni,0}$, $r_{uni,45}$ and $r_{uni,90}$ and uniform elongation A_g were ascertained. Layer compression tests were used for the determination of biaxial yield stress σ_{bia}. The biaxial r-value

$$r_{bia} = \frac{\varepsilon_y^{pl}}{\varepsilon_x^{pl}} = \frac{e^{\varepsilon_y} - 1}{e^{\varepsilon_x} - 1} \tag{4}$$

was calculated using FLC results on a biaxial specimen by means of major ε_x and minor strain ε_y application [1]. Shear tensile and plane strain compression tests were used for validation by determining the deviation between experimental principal stresses and each yield locus. Various yield criteria were used to build and compare temperature dependent locus for the considered AA7075 sheet material, such as the common Mises, Yld'03, BBC'05 and a recently developed Vegter'17 criterion [11]. The experimental stress and anisotropy parameters necessary for each criteria are summarised in Table 1. The experimentally ascertained parameters of each model are shown in the results section for each considered temperature at a quasi-static strain rate.

Table 1. Experimental stress and anisotropy parameters necessary for von Mises, Yld'03, BBC'05 and Vegter'17 yield criteria

	$\sigma_{uni,0}$	$\sigma_{uni,45}$	$\sigma_{uni,90}$	$r_{uni,0}$	$r_{uni,45}$	$r_{uni,90}$	σ_{bia}	r_{bia}	A_g
Von Mises	x								
Yld'03	x	x	x	x	x	x	x	x	
BBC'05	x	x	x	x	x	x	x	x	
Vegter'17	x			x	x	x			x

4 Results

4.1 Temperature and Strain Rate Dependent Plasticity

The extrapolated temperature and strain rate dependent flow curves, determined by means of dilatometer tensile tests on uniaxial miniaturised specimens and extended using plane strain compression tests, are presented in Fig. 5 (a).

The flow curves show a comparatively high strain rate dependence at elevated temperatures, whereas at temperatures of 20 and 125 °C the strain rate influence on the flow curves is noticeable starting at a forming degree of 0.1. However, the AA7075 flow curves present a clear dependence on all considered temperatures by a decreasing flow stress at increasing temperatures. Figure 5 (b) shows an additional detail view of the

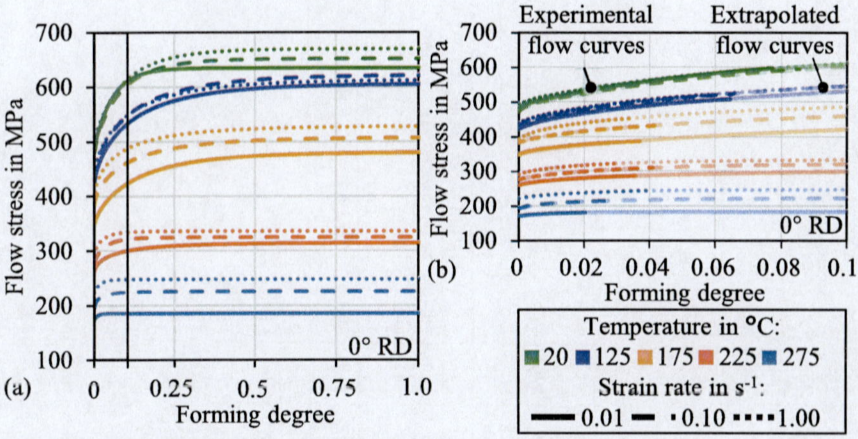

Fig. 5. Extrapolated temperature and strain rate dependent flow curves at 0° to RD (a), detail view of the transition between the experimental and extrapolated flow curves (b)

transition between the extended experimental and extrapolated flow curves. It should be noticed, that the fracture strain reached by the experimental flow curves decreases with increasing temperature. This is due to material's hardening, which has been shown to be strongly temperature dependent [2].

4.2 Temperature Dependent Forming Limit Curve

A temperature dependent FLC was built by above presented Nakajima tests, shown in Fig. 6 with the corresponding values in Table 2. Both minor and major strains of each FLC geometry show a significant dependence on temperature by an increasing major strain with increasing temperature and by a predominantly increasing minor strain with increasing temperature. Compared to all temperatures considered, the FLC at 175 and 225 °C shows the least response to temperature change, similarly to the research in [12].

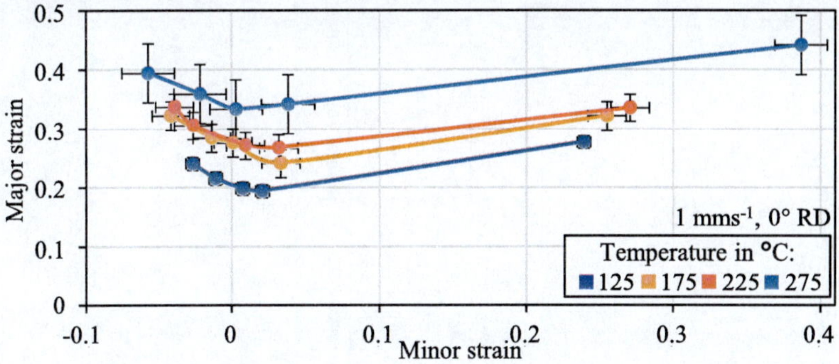

Fig. 6. Temperature dependent FLC curve at 0.01 s^{-1} and 0° to RD

Table 2. Temperature dependent experimental FLC values of minor and major strains

T (°C)	Minor strain					Major strain				
	R63	R53	R42	R31	R0	R63	R53	R42	R31	R0
125	−0.03 ± 0.004	−0.01 ± 0.006	0.007 ± 0.005	0.02 ± 0.002	0.24 ± 0.006	0.24 ± 0.01	0.22 ± 0.01	0.19 ± 0.01	0.19 ± 0.008	0.28 ± 0.005
175	−0.04 ± 0.01	−0.01 ± 0.015	0.001 ± 0.002	0.03 ± 0.003	0.26 ± 0.04	0.32 ± 0.03	0.28 ± 0.04	0.27 ± 0.004	0.24 ± 0.01	0.32 ± 0.04
225	−0.04 ± 0.01	−0.03 ± 0.01	0.01 ± 0.01	0.03 ± 0.01	0.27 ± 0.03	0.34 ± 0.02	0.3 ± 0.03	0.27 ± 0.05	0.27 ± 0.01	0.34 ± 0.02
275	−0.06 ± 0.02	−0.02 ± 0.02	0.003 ± 0.014	0.04 ± 0.013	0.4 ± 0.03	0.4 ± 0.02	0.36 ± 0.1	0.33 ± 0.06	0.34 ± 0.05	0.44 ± 0.03

4.3 Temperature Dependent Yield Locus

The aforementioned experiments were used to determine temperature dependent stress and anisotropy parameters and to build the considered yield locus. The experimentally determined parameters are shown in Table 3, followed by Table 4, which presents stress values of shear σ_{sh} and plane strain compression σ_{psct} tests in 0° and 90° to RD at 0.01 s^{-1}. The latter stress values were used to validate the considered yield criteria. First and second principal stresses σ_I and σ_{II} of each experimental test were determined by correlations in [7] using characteristic stresses in Table 3 and Table 4. With this approach, a temperature dependent yield locus was successfully built by means of von Mises, Yld'03, BBC'05 and Vegter'17 criteria in Fig. 7. Each experiment used for either the parametrization or validation of the individual yield locus is indicated in the figure. It can be seen, that principal stresses are strongly influenced by temperature. Von Mises criterion shows the largest deviation from the principal stresses of the shear tensile and PSCT validation experiments at all temperatures. However, for the parametrisation of this criterion only one characteristic stress value of $\sigma_{uni,0}$ was needed.

Table 3. Temperature dependent experimental stress and anisotropy values necessary for parametrization of von Mises, Yld'03, BBC'05 and Vegter'17 yield criteria

T (°C)	$\sigma_{uni,0}$ (MPa)	$\sigma_{uni,45}$ (MPa)	$\sigma_{uni,90}$ (MPa)	$r_{uni,0}$	$r_{uni,45}$	$r_{uni,90}$	σ_{bia} (MPa)	r_{bia}	A_g
20	483.1 ± 0.3	467.2 ± 7.2	474.6 ± 2.8	0.61 ± 0	0.95 ± 0.07	0.82 ± 0	475.5 ± 9.5	0.97 ± 0.01	0.09 ± 0.01
125	421.6 ± 12.4	423.9 ± 1.8	423.6 ± 0.7	0.57 ± 0.05	0.98 ± 0.05	0.83 ± 0.02	415.8 ± 0.2	0.85 ± 0	0.07 ± 0
175	346.3 ± 12.9	336.3 ± 6.4	345.7 ± 14.2	0.69 ± 0	0.92 ± 0.04	0.71 ± 0.01	338.3 ± 2.7	0.77 ± 0.03	0.04 ± 0.02
225	260.5 ± 0.5	255.1 ± 7.4	244.9 ± 7.4	0.77 ± 0.07	0.91 ± 0.02	0.71 ± 0	258.9 ± 6.4	0.78 ± 0.06	0.03 ± 0.01
275	172.5 ± 6.5	173.8 ± 3.8	179.1 ± 1.2	0.74 ± .0.3	0.76 ± 0.01	0.67 ± 0	172.5 ± 3.5	0.85 ± 0.05	0.01 ± 0

Table 4. Temperature dependent experimental stress values used to validate yield criteria

T (°C)	$\sigma_{sh,0}$ (MPa)	$\sigma_{sh,90}$ (MPa)	$\sigma_{psct,0}$ (MPa)	$\sigma_{psct,90}$ (MPa)
20	278.1 ± 1.2	264.1 ± 6.1	535.8 ± 22.1	256.7 ± 9.3
125	231.9 ± 7.6	228.2 ± 15.4	469.5 ± 29.3	230.8 ± 14.3
175	198.3 ± 6.7	187.5 ± 3.1	377.1 ± 18.6	183.9 ± 5.9
225	152.2 ± 12.5	133.8 ± 0.9	298.9 ± 1.3	133.4 ± 14.7
275	106.6 ± 7.6	87.3 ± 2.6	182.2 ± 2.2	102.5 ± 3.3

Yld'03 and BBC'05 yield criteria show nearly identical courses and a very good agreement with the experimental principal stresses of the validation tests at all considered temperatures. This is due to the fact that the parameterisation of both criteria requires the largest number of characteristic stress and anisotropy values, and thus the most extensive experimental analysis, compared to von Mises or Vegter'17 criteria. Vegter'17 yield criterion also shows an excellent agreement to experimental shear and PSCT principal stresses with a simultaneously reduced number of experimental tests necessary for its parametrisation compared to Yld'03 and BBC'05 criteria.

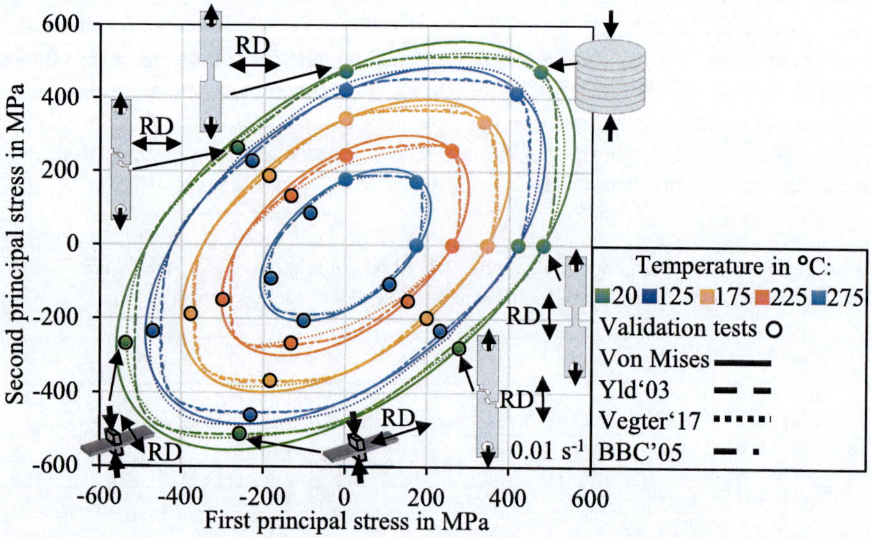

Fig. 7. Temperature dependent von Mises, Yld'03, Vegter'17 and BBC'05 yield criteria

5 Conclusion and Outlook

7xxx series aluminium has become popular due to its high potential for lightweight constructions. However, since its formability at room temperature is comparably low, it is important to investigate materials behaviour under semi-hot forming conditions. The presented research encompasses an experimental AA7075 sheet material analysis by varying strain rate, temperature and stress state. Following conclusions are drawn:

- Uniaxial flow behaviour is strongly temperature dependent, however, strain rate dependency is only noticeable starting at 175 °C.
- Due to temperature dependent material's hardening, the maximum fracture strain prior to necking in uniaxial tension decreases with increasing temperature.
- Minor and major strains of FLC show a strong temperature dependence by increasing major strain with increasing temperature and predominantly increasing minor strain with increasing temperature.

- The considered yield criteria were successfully validated by shear tensile and plane strain compression tests. Von Mises criterion shows lowest agreement with experimental principal stresses. Yld'03 and BBC'05 yield criteria show nearly identical courses and best agreement with experimental principal stresses. Vegter'17 criterion shows an excellent agreement to experimental principal stresses with reduced number of tests necessary for its parametrisation compared to Yld'03 and BBC'05.

Further research will include implementation of material data in an FE simulation, which will be verified by post-simulations of experimental tests. Subsequently, a semi-hot forming process will be implemented in an FE simulation based on material data and validated by experimental tests on deep drawing of an automotive component.

Acknowledgements. The authors thank the funders, Federal Ministry for Economic Affairs and Energy. The funding was organised by German Federation of Industrial Research Associations (AiF) part of a program for Industrial Collective Research (IGF) under grant number 21645N.

References

1. Lin, J., Bao, X., Hou, Y., Min, J., Qu, X., Tao, Z.: Investigation on yield behavior of 7075–T6 aluminum alloy at elevated temperatures. CJME **33**, 76 (2020)
2. Pandya, K.S, Roth, C.C., Mohr, D.: Strain rate and temperature dependent plastic response of AA7075 during hot forming. Mater. Sci. Eng., 12100 (2019)
3. Österreicher, J.A., et al.: Warm-forming of pre-aged Al-Zn-Mg-Cu alloy sheet. Mater. Des. **193**, 108837 (2020)
4. Rader, K.E., Carter, J.T., Hector, L.G.: Plastic deformation and ductility of AA7075 and AA6013 at warm temperatures suitable to retrogression forming. Metall. Mater. Trans. A **52**, 4003–4017 (2021)
5. Behrens, B.-A., Brunotte, K., Wester, H., Kock, C., Kildonaviciute, D.: Determination oftemperature dependence in modified-Mohr-Coulomb failure model for process simulation of shearcutting, In: IOP Conference Series: Materials Science and Engineering, vol. 12028 (2022)
6. Rahmaan, T., Butcher, C., Worswick, M.J.: Constitutive response of AA7075-T6 aluminum alloy sheet in tensile and shear loading. J. Exp. Appl. Mech., 115–122 (2017)
7. Isik, K., Soyarslan, C., Richter, H., Tekkaya, A.: Analysis of formability of advanced high strength steel sheets with phenomenologically based failure criteria with separate treatment of instability, shear and normal fracture. Mater. Sci. (2011)
8. Behrens, B.-A., Hübner, S., Bouguecha, A., Grbic, N., Vucetic, M.: Comparison of determination of biaxial true stress – true strain curves by the use of plane strain compression test and bulge test. In: AIP Conference Proceedings, pp. 650–653 (2013)
9. Chermette, C., Unruh, K., Peshekhodov, I., Chottin, J., Balan, T.: A new analytical method for determination of the flow curve for high-strength sheet steels using the plane strain compression test. Int. J. Mater. Form., 269–292 (2020)
10. Behrens, B.-A., Rosenbusch, D., Wester, H., Stockburger, E.: Material characterization and modeling for finite element simulation of press hardening with AISI 420C. JMEP 825–832 (2022)
11. Abspoel, M., Scholting, M.E, Lansbergen, M.: An, A., Vegter, H.: A new method for predicting advanced yield criteria input parameters from mechanical properties. J. Mater. Process. Technol., 161–177 (2017)
12. Reuther, F., Lieber, T., Heidrich, J., Kräusel, V.: Numerical investigations on thermal forming limit testing with local inductive heating for hot forming of AA7075. J. Mater., 1882 (2021)

Characterization of the Surface Error During Peripheral Milling of Thin-Walled, Near-Net-Shaped Structures

Lasse Evers(✉), Melchior Blühm, Sebastian Junghans, Carsten Möller, and Jan Hendrick Dege

Institute of Production Management and Technology (IPMT), Hamburg University of Technology (TUHH), Denickestraße 17, 21073 Hamburg, Germany
lasse.evers@tuhh.de

Abstract. Lightweight aerospace structural components are usually milled out of plate-material, converting up to 95% of the material into difficult to recycle chips. Therefore, additive-manufacturing holds great resource saving potential through the near-net-shape production of thin-walled parts. However, these parts pose new challenges. For example, previous methods such as the "waterline"-path approach, which uses the residual stiffness of the solid-block to reduce the deformation of the thin-walls due to process forces, can only be applied to a very limited extent. This results in a greater deflection of the near-net-shape, additively-manufactured structure. In this paper the surface error is investigated using milling experiments. An analytical approach to qualitatively predict the surface characteristics is provided. The results get used to discuss whether an adjustment to the tool path is suitable to compensate for the dimensional error. Such a compensation method has the potential to efficiently machine near-net-shape semi-finished parts within the required tolerances.

Keywords: milling · thin-walled structures · surface error · additive manufacturing

1 Introduction

Due to the social and political movement towards sustainability, additive manufacturing of lightweight components is steadily moving closer to the industry's focus. Currently, lightweight components, e.g. for aviation, are milled out of plate-material, converting up to 95% of the material into chips [1]. These are contaminated with cooling lubricants and show surface oxide layers, which makes recycling difficult and thus preventing a circular economy. Additive manufacturing offers the potential to reduce this waste by directly producing a near-net-shape semi-finished product. Only small allowances have to be removed from the component by machining operations.

However, the final machining of the near-net-shape semi-finished products poses new challenges for the manufacturing technology. The slender, thin-walled structures are prone to elastic deformation during machining, which remains as a dimensional

T. Bauernhansl et al. (Eds.): WGP 2023, LNPE, pp. 356–366, 2024.
https://doi.org/10.1007/978-3-031-47394-4_35

surface error on the component. For manufacturing out of plate-material, methods, like the waterline-path approach have been developed, that utilize the residual stiffness of the solid-block to reinforce the surfaces to be machined. The surface error therefor results exclusively from the tool bending and has been extensively studied [10, 11]. For the finishing of additive, slender parts new solutions must be found. Various approaches have been tested in the past:

One way to reduce the elastic deformation and thus the surface error is to use less tool engagement [2], but since this limits the productivity, alternative approaches are preferable. Another approach is to reinforce the thin structures with fixtures [3]. However, these have the disadvantage of requiring assembly time, induce additional risk of collision and add extra cost. A different way to compensate the surface error is to adjust the tool path, e.g. by increasing the width of cut [4], or adjust the tool angle [5] to compensate for the bending line of the slender part. However, it has been noted [5, 6], that not every surface error can be compensated by adjusting the tool path. The characteristic of the surface error depends on the engagement conditions and the tool geometry. In addition, for additive parts, the slenderness of the component is expected to further influence the shape of the surface error. In order to find the ideal compensation strategy, an estimation of the resulting surface error is necessary so that a choice can be made between the different strategies. Typical approaches to predict the surface error use a combination of analytical force and numerical deformation calculation [6, 7, 9]. This method requires a time-consuming set up of the simulation, unknown cutting coefficients and are computation heavy, but offer the most accurate results. However, if the exact quantitative surface error is not needed, [5] provides an analytical prediction, that accelerates the calculation process significantly and needs less input parameters. The result can be used to pre-select tools and process parameters that lead to surface errors that are compensable by tool path adjustment. The quantitative amount of needed adjustment has to be gained another way, e.g. by online monitoring [8].

The aim of this paper is to provide a similar analytical qualitative surface error prediction that needs a minimal number of input parameters and provides fast calculation. The current model gets expanded to calculate not only the maximal, but the full shape of the qualitative surface error. The field of application gets enlarged for thin walls without a fixed clamping at the depth of cut and can thus be used for machining of near-net-shaped parts in multiple axial feeds, as required for additive parts. The results can get used to decide, whether an angular adjustment or increase in width of cut is suitable for surface error compensation.

2 Analytical Characterization of the Surface Error

With the assistance of an analytical model, it is possible to estimate the shape of the surface error without having to perform FE-simulations or experiments. With the help of a few parameters known to the NC-programmer, it is thus possible to predict the surface error and select a suitable tool, process parameters and compensation strategy in a computationally efficient manner. This makes the model attractive for industry.

2.1 Calculation Approach

Due to the cutting forces at the contact lines between tooth and workpiece, forces generated during climb milling push workpiece and tool away from each other. Figure 1 (a) shows a flank milling process of a slender fillet that has a wall height h_w below the depth of cut, which is typical for additive manufactured components. A typical surface error due to the displacement is shown in Fig. 1 (b). The error can be characterized via the following points:

- Point A – Dimensional error at the highest depth of cut
- Point B – Dimensional error at the lowest depth of cut
- Point C – Maximum dimensional error
- Point D – Minimum dimensional error

[4] introduces the Error-Linearity-Factor ζ to describe the linearity of the dimensional error:

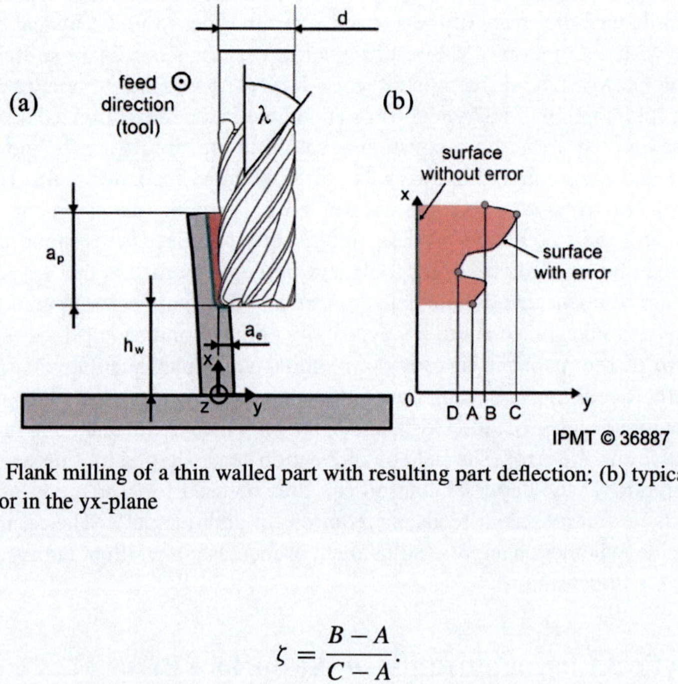

Fig. 1. (a) Flank milling of a thin walled part with resulting part deflection; (b) typical resulting surface error in the yx-plane

$$\zeta = \frac{B - A}{C - A}. \tag{1}$$

A higher Linearity-Factor indicates a linear surface error, which can be compensated to a large extent by angular adjusting the toolpath. Manufacturing of slender parts results in surface errors, that can be compensated to an extend by increasing the width of cut. To describe, which fraction of the surface error can be compensated by increasing the width of cut, a new factor, the Error-Compensation-Factor ξ, is introduced:

$$\xi = \frac{D}{C}. \tag{2}$$

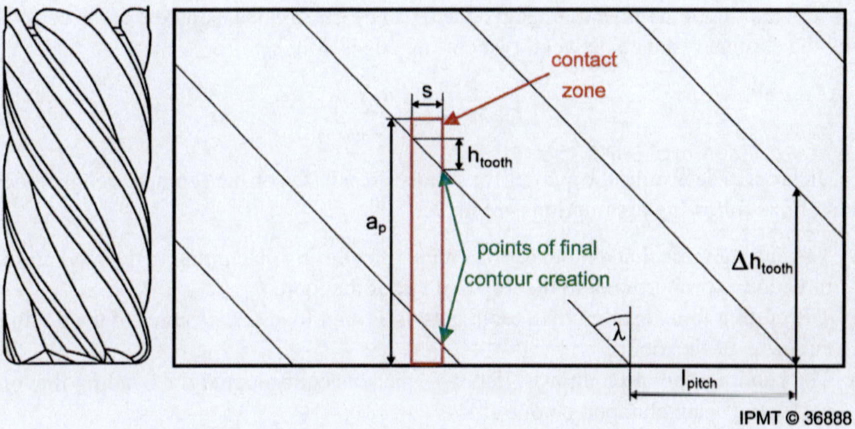

Fig. 2. Uncoiled tool teeth and contact zone with the workpiece.

To predict the qualitative surface error, derive the points A-D and Error-Factors an analytical approach is provided:

If the area of the tool that is in engagement is projected onto the uncoiled cutting profile, the cutting edge areas that are in engagement at a given time can be identified (see Fig. 2). Based on the geometric correlations, the required parameters for the analytical model can be calculated from known engagement and tool parameters:

2.2 Geometric Engagement Conditions

The contact angle φ_i can be derived from the width of cut a_e and tool diameter d:

$$\varphi_i = \cos^{-1}\left(1 - \frac{2a_e}{d}\right). \tag{3}$$

By using the contact angel φ_i the immersion arc s can be calculated:

$$s = \frac{\varphi_i}{360} \cdot \pi \cdot d. \tag{4}$$

The arc length between two teeth l_{pitch} can be obtained with the tool diameter d and the number of teeth z:

$$l_{pitch} = \frac{d \cdot \pi}{z}. \tag{5}$$

The cutting depth of a tooth, that's in full engagement h_{tooth} can be calculated using the immersion arc s and tool helix angle λ:

$$h_{tooth} = \frac{s}{\tan(\lambda)}. \tag{6}$$

The vertical distance between two cutting edges Δh_{tooth} is equally calculated from the arc length between two cutting edges l_{pitch} and the tool helix angle λ:

$$\Delta h_{tooth} = \frac{l_{pitch}}{\tan(\lambda)}. \tag{7}$$

The maximum number of engaged teeth can be determined using the depth of cut a_p and the vertical distance between two cutting edges Δh_{tooth}:

$$n_{tooth} = \lceil \frac{a_p}{\Delta h_{tooth}} \rceil. \tag{8}$$

In order to determine the qualitative surface error based on the geometric engagement conditions following assumptions are made:

- The cutting force that a tooth exerts on the fillet during machining in the feed normal direction is proportional to the depth of cut of the tooth h_{tooth}.
- The cutting force exerted by a tooth acts as a point load at the center of the depth of cut h_{tooth} of the tooth.
- The bending line of the thin-walled workpiece is equivalent to the bending line of a cantilever beam clamped on one side.
- The tool compliance is negligible compared to the workpiece compliance.

During climb milling, the outline of the part and thus the dimensional error is created during the disengagement of the cutting edge (see Fig. 2). The surface error at a specific height position can therefore be predicted by looking at the moment at which a cutting edge exits at this position. In the following, therefore, this point in time and the forces and positions of the forces acting on the wall during this moment are determined for each height point of the wall. Based on this information, the resulting surface error can be determined with the use of the bending line.

2.3 Surface Error Over the Wall Height

The surface error $w_{ges}(x)$ at every height position x is calculated individually. The error results from the displacement of the structure by several teeth in contact. For the model, the resulting surface error $w_i(x)$ for each tooth i, is calculated separately. Subsequently, the individual errors are summed up for the total error according to the superposition principle:

$$w_{ges}(x) = \sum_{i=1}^{2 \cdot n_{tooth}} w_i(x). \tag{9}$$

The maximum number of teeth in contact n_{tooth} is already determined. Therefore, only n_{tooth} cutting edges above and n_{tooth} cutting edges below the searched height point x are considered for the calculation, ensuring that all possible cutting edges in contact are taken into account:

$1 \leq i \leq n_{tooth}$ Teeth below the searched height point.

$(n_{tooth} + 1) \leq i \leq (2 \cdot n_{tooth})$ Teeth above the searched height point.

The resulting surface error $w_i(x)$ at a height point x, which is caused by the tooth i and the force F_i acting at the position a_i, is known for a cantilever beam of length l:

$$w_i(x) = \begin{cases} \frac{F_i}{6EI}(3a_i x^2 - x^3), 0 \leq x < a_i \\ \frac{F_i}{6EI}(3a_i{}^2 x - a_i{}^3), a_i \leq x \leq l \end{cases}. \tag{10}$$

The area moment of inertia I and elastic modulus E are material constants in the equation that influence the quantitative result. However, the qualitative outcome is independent of these constants, thus they are neglected for the model:

$$EI = 1. \tag{11}$$

According to the assumptions the at a tooth i acting force F_i can be substituted by the tooth cutting depth $h_{tooth,i}$:

$$F_i \triangleq h_{tooth,i}. \tag{12}$$

The cutting depth of a tooth that is completely in contact has already been determined by h_{tooth}. However, if the tooth is currently in the entry, exit, or outside the cutting area, the cutting depth must be recalculated. For the cutting edges below the searched height point $1 \leq i \leq n_{tooth}$ the following applies:

$$h_{tooth,i} = \begin{cases} 0, & x - i \cdot \Delta h_{tooth} + h_{tooth} < 0 \\ h_{tooth}, & x - i \cdot \Delta h_{tooth} + h_{tooth} \geq h_{tooth} \\ x - i \cdot \Delta h_{tooth} + h_{tooth}, & elsewhere \end{cases}. \tag{13}$$

For the cutting edges below the searched height point $(n_{tooth} + 1) \leq i \leq (2 \cdot n_{tooth})$ the following applies:

$$h_{tooth,i} = \begin{cases} 0, & x + (i - n_{tooth} - 1) \cdot \Delta h_{tooth} + h_{tooth} \geq a_p + h_{tooth} \\ h_{tooth}, & x + (i - n_{tooth} - 1) \cdot \Delta h_{tooth} + h_{tooth} < 0 \\ a_p - (x + (i - n_{tooth} - 1) \cdot \Delta h_{tooth}, & elsewhere \end{cases}. \tag{14}$$

According to the assumptions, the position a_i, at which the force of the tooth is applied, occurs at the middle of the engagement height $h_{tooth,i}$ of the tooth and can therefore be calculated as follows:

$$a_i = h_w + \begin{cases} x - i \cdot \Delta h_{tooth} + \Delta h_{tooth} - \frac{h_{tooth,i}}{2}, & 1 \leq i \leq n_{tooth} \\ x + (i - n_{tooth} - 1) \cdot \Delta h_{tooth} + \frac{h_{tooth,i}}{2}, & (n_{tooth} + 1) \leq i \leq (2 \cdot n_{tooth}) \end{cases}. \tag{15}$$

2.4 Deformation Damping Due to Tooth Engagement

The preceding calculation describes the expected surface error assuming that the structure reacts freely and immediately to the changing cutting forces. However, because the tool is in contact with the structure during cutting, the contact dampens the movement of the structure. During climb milling, a tooth initially enters the workpiece at a high depth of cut. Due to the helix, the contact area of the tooth moves toward a lower depth of cut during tool rotation. As a time sequence in which the surface error is generated, it can be deduced that first the surface error is generated at a high depth of cut, and with the rotation of the tool the surface error follows at a lower depth of cut.

Due to the damping effect of the tool contact, the generated surface error at a height position is thus dependent on the surface error below this position, since the workpiece can only change its deflection to a limited extent. A downward moving average filter is therefore applied to the calculated surface error to represent this dependence. An interval $size = 2\text{mm}$ has yielded results close to reality for the investigated parameters:

$$w_{ges,damped}(x) = \left[w_{ges}(x) + w_{ges}\left(x - \frac{size}{2} \right) + w_{ges}(x - size) \right] / 3 \quad (16)$$

3 Experimental Setup

Milling tests were carried out on a FOOKE ENDURA 711 LINEAR to validate the model. Sheets made of Ti-6Al-4V Grade 5 with 2.5 mm wall thickness, 150 mm length and variable heights were clamped and flank milled. In addition to the wall height, the depth of cut, width of cut and tool were varied. An overview of all test parameters of the full-factorial test plan and cemented carbide tools can be found in Table 1. The profile after machining was recorded with a Mahr GD 120 surface profiler at 75 mm length.

Table 1. Cutting parameters and end mills used in full factorial experiments.

Process Parameter						
Cutting Speed v_c	Feed per Tooth f_z	Depth of Cut a_p	Width of Cut a_e	Start Wall Thickness b_w	Wall Height h_w	End Mills
70 $\frac{m}{min}$	0.07 mm	10 mm	0.2 mm	2.5 mm	0 mm	1–3
		20 mm	0.4 mm		20 mm	
					40 mm	
End Mills						
Tool Number	Diameter d	Teeth z	Helix Angle λ		Corner radius r_β	
1	16 mm	5	40°		2.5 mm	
2	16 mm	6	55°		3.5 mm	
3	16 mm	8	45°		3.5 mm	

4 Results and Discussion

4.1 Qualitative Surface-Error-Prediction

Figure 3 shows the measured workpieces after machining and the qualitatively predicted surface error according to Eq. (1) to (16). Figure 3 (a) displays the influence of varying the width of cut. It can be seen that the qualitative, S-shaped progression of the measured dimensional error is well reproduced by the prediction. Figure 3 (b) shows that at different cutting speeds, the resulting surface error is constant. This confirms the prediction by not including the cutting speed in the model. Figure 3 (c) displays the surface error at different tooth feed rates. The feed per tooth also does not appear in the model. However, a higher feed leads to higher cutting forces and thus to a quantitatively larger surface error, as can be seen in the measurement. The qualitative progression remains constant and is well represented by the prediction over the different tooth feeds rates. Figure 3 (d) shows the surface error at different wall heights. It must be noted that the qualitative prediction agrees best with reality at medium wall heights. A deviation at low wall heights can be explained mainly by the fact that the stiffer workpiece increases the influence of tool displacement on the surface error, which is not included in the model. The deviation at greater heights can be explained by the fact that the actual width of cut decreases due to the deflection and thus the effective cutting force and intensity of the S-Curve decreases. This dependency of the effective cutting force to the deflection is not included in the model.

Figure 3 (e) displays the dimensional error when using the same tool as in (d), but with a smaller depth of cut. It can be noted, that the S-Shaped surface error is shortened with less turning points and consequently less linear. This can be traced back to less tooths in engagement at the same time. The surface error is therefore strongly dependent on the depth of cut and this dependency is well represented by the model. Figure 3 (f) shows the surface error when using tool 2, which uses more tooth and a higher helix angle compared to (d) that results in more teeth in contact with the workpiece. Same can be found with tool 3. They produces therefore an S-curve with more turning points and less pronounced minima and maxima that leads to a more linear surface error. The model again shows good agreement with the measurement.

4.2 Error-Linearity-Factor Und Error-Compensation-Factor

To validate the model, the characteristic points A, B, C and D of the machining tests were measured and the resulting Error-Linearity-Factor ζ and Error-Compensation-Factor ξ were calculated. The plot of the results together with the predicted values according to Eq. (1) to (16) are shown in Fig. 4.

Figure 4. (a) plots the data points for the Error-Linearity-Factor ζ. The linear regression shows a coefficient of determination of $R^2 = 0.86$ and thus a good prediction of the surface error linearity by the provided model. It is therefore suitable for estimating whether the surface error can be compensated by angular adjustment to the tool.

However only a weak correlation between model and reality for the Error-Compensation-Factor ξ, which describes the ratio of minimum to maximum dimensional error, can be seen from Fig. 4. (b). This is confirmed by a low coefficient of determination $R^2 = 0.38$ obtained from a linear regression. A possible reason for this is the neglect of tool deflection in the model. As a result, the model tends to underestimate the minimum dimensional error for stiffer components ($h_w = 0$ mm), since this is largely caused by tool deflection (compare Fig. 3. (f)). On the other hand, large deformations, that leads to a significant reduction in width of cut get overestimated, because the dependency of deformation and less cutting forces, due to the reduced width of cut is not included in the model either. Therefore, the model is only conditionally able to estimate which part

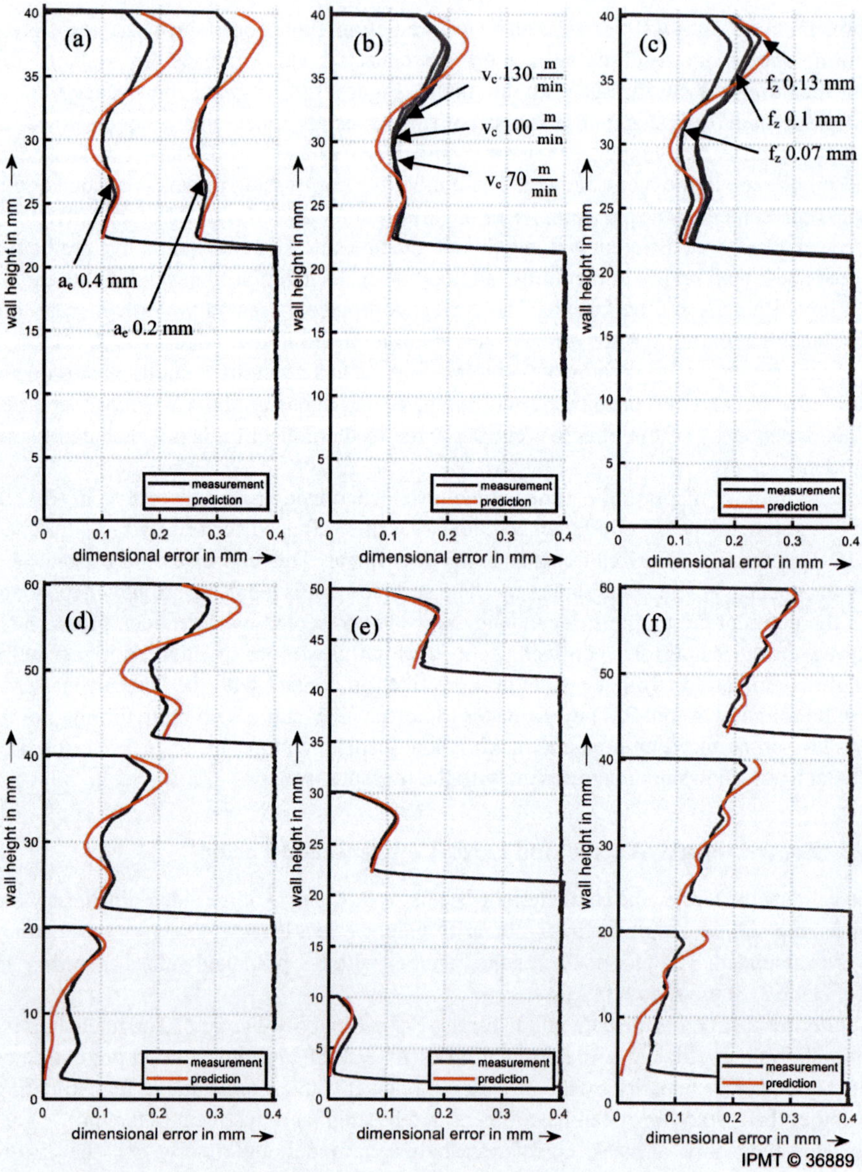

Fig. 3. Measured and predicted surface error with different cutting parameters:
(a) a_p 20 mm; a_e variable; h_w 20 mm; f_z 0.07 mm; v_c 70 m/min; End Mill 1
(b) a_p 20 mm; a_e 0.4 mm; h_w 20 mm; f_z 0.07 mm; v_c variable; End Mill 1
(c) a_p 20 mm; a_e 0.4 mm; h_w 20 mm; f_z variable; v_c 70 m/min; End Mill 1
(d) a_p 20 mm; a_e 0.4 mm; h_w variable; f_z 0.07 mm; v_c 70 m/min; End Mill 1
(e) a_p 10 mm; a_e 0,4 mm; h_w variable; f_z 0.07 mm; v_c 70 m/min; End Mill 1
(f) a_p 20 mm; a_e 0,4 mm; h_w variable; f_z 0.07 mm; v_c 70 m/min; End Mill 2.
Other Parameters according to Table 1 and Chapter 3.

of the dimensional error can be compensated by an increased width of cut and should be optimized in further research.

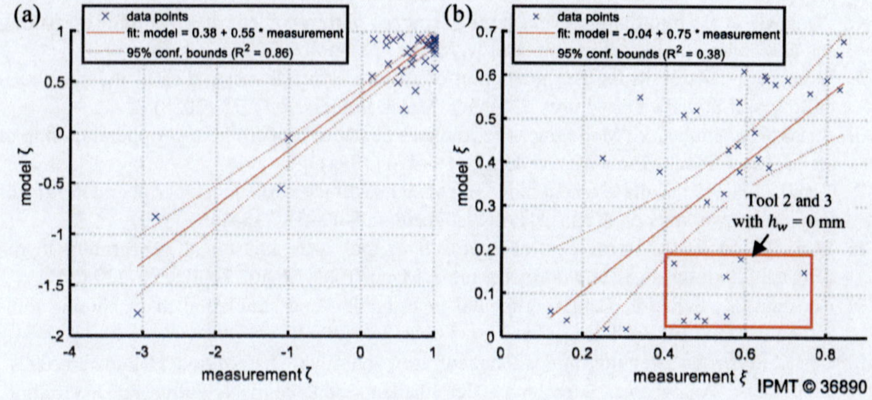

Fig. 4. Experimental validation of the Prediction Model with ζ and ξ

5 Summary and Outlook

An analytical model was provided, that can be used to qualitatively predict the surface error after machining of slender, additively manufactured parts based on engagement and tool data without the need of FE-simulations or experiments. The model limits are:

- too stiff workpieces, where the surface error is mostly a result of the tool bending.
- too compliant workpieces, where the width of cut gets significantly changed due to the workpiece deformation.

Within its limits the model is able to determine, whether more radial engagement or angular adjustment to the tool is a suitable compensation strategy to reduce the resulting surface error after flank milling. To further open the field of use for the model, a quantitative calculation is needed. Whether such an enlargement in the needed parameters and computing power is in proportion to the gained accuracy is dependent on the application. Further optimization of the model without a quantitative approach could also include different deformation damping intervals based on tooth clearance angle, that is assumed to influence the surface error.

Acknowledgments. The authors gratefully acknowledge the financial support of the Federal Ministry for Economic Affairs and Climate Action, Ministerium für Wirtschaft und Klimaschutz (BMWK) within the research project AMAvia, project number 20W1902F.

References

1. Herranz, S., et al.: The milling of airframe components with low rigidity: a general approach to avoid static and dynamic problems. Proc. Inst. Mech. Eng. Part B: J. Eng. Manuf. **219**(11), 789–801 (2005)

2. Budak, E.: Analytical models for high performance milling. Part I: cutting forces, structural deformations and tolerance integrity. Int. J. Mach. Tools Manuf. **46**(12–13), 1478–1488 (2006)
3. Fei, J., et al.: Investigation of moving fixture on deformation suppression during milling process of thin-walled structures. J. Manuf. Process. **32**, 403–411 (2018)
4. Wang, G., et al.: Improving the machining accuracy of thin-walled parts by online measuring and allowance compensation. Int. J. Adv. Manuf. Technol. **92**, 2755–2763 (2017)
5. Wimmer, S., Zaeh, M.F.: The prediction of surface error characteristics in the peripheral milling of thin-walled structures. J. Manuf. Mater. Process. **2**(1), 13 (2018)
6. Budak, E., Altintas, Y.: Modeling and avoidance of static form errors in peripheral milling of plates. Int. J. Mach. Tools Manuf **35**(3), 459–476 (1995)
7. Tsai, J., Liao, C.: Finite-element modeling of static surface errors in the peripheral milling of thin-walled workpieces. J. Mater. Process. Technol. **94**(2–3), 235–246 (1999)
8. Wan, X., et al.: An error control approach to tool path adjustment conforming to the deformation of thin-walled workpiece. Int. J. Mach. Tools Manuf **51**(3), 221–229 (2011)
9. Denkena, B., Schmidt, C.: Experimental investigation and simulation of machining thin-walled workpieces. Prod. Eng. Res. Dev. **1**, 343–350 (2007)
10. Gey, C.: Prozessauslegung für das Flankenfräsen von Titan. Dissertation, Hannover (2002)
11. Klobasa, I.: Analytische Berechnung der Flankengestalt beim Nutenfräsen. Dissertation, Hannover (2007)

Characterisation and Investigation of the Mechanical Material Properties of Recycled Glass-Fibre Reinforced Polymers for Sustainable Reusability

Jonas Ludwig[1](✉), Toni Hocke[1], Mitchel Polte[1,2], and Eckart Uhlmann[1,2]

[1] Institute for Machine Tools and Factory Management IWF, Technical University of Berlin, Berlin, Germany
jonas.ludwig@iwf.tu-berlin.de

[2] Fraunhofer Institute for Production Systems and Design Technology IPK, Berlin, Germany

Abstract. Waste products from glass-fibre reinforced polymers (GFRP), which are used for wind turbine rotor blades, represent a substantial environmental and disposal challenge. According to the state of the art, thermal or chemical recycling methods require high energies and resource consumptions. Based on a new recycling approach, this study analyses the potential of mechanical recycling as a sustainable alternative for the management of GFRP waste. This paper presents a novel solution for mechanical recycling of GFRP waste and analyses the mechanical properties of recycled GFRP materials manufactured with mechanical recycled and processed GFRP waste. The study investigates the impact of particle size classifications on the mechanical properties of recycled GFRP materials, providing insights for optimising the recycling process. Tensile strength tests were carried out to measure the Young's modulus E and ultimate tensile strength Rm to compare the properties of recycled GFRP specimens with newly manufactured GFRP materials and thus identify the suitability of these materials for specific applications. Furthermore, the GFRP material samples were analysed using computer tomography (CT) to determine the fibre lengths l_f in order to analyse their influence on the mechanical material properties. First results showed a correlation between the particle length l_p and the ultimate tensile strength Rm of the investigated GFRP material samples. The mechanical recycling approach enables more sustainable GFRP materials compared to the current state of the art, which significantly reduces the environmental impact of GFRP waste disposal.

Keywords: glass-fibre reinforced polymers · GFRP · mechanical recycling · tensile strength test · material characterisation

1 Introduction

GFRP, composed of glass-fibres embedded in a polymer matrix, represent a significant factor in industries such as wind energy, aviation and automotive [1]. The processing and recycling of GFRP waste, including end-of-life components and production-related

T. Bauernhansl et al. (Eds.): WGP 2023, LNPE, pp. 367–375, 2024.
https://doi.org/10.1007/978-3-031-47394-4_36

residues, are subject to challenges in terms of economy and efficiency [2]. Significant amounts of recyclable waste are expected to be generated during wind turbine decommissioning and maintenance. In this context, wind turbine rotor waste is expected to reach a number of 2 million tons by 2050 [3].

To manage these waste volumes V_w, new recycling strategies and reuse options are in high demand. One potential approach involves incorporating GFRP waste into new GFRP materials as recycled content. Reusing GFRP waste as secondary raw materials for developing new products is one potential application.

In order to establish a market for recycled fibre-reinforced materials, their mechanical properties need to be ensured and cost-effective compared to newly manufactured fibres. However, the fragmentation and processing of GFRP waste for recycling purposes is characterised by cost-intensive processes, which represent a challenge for small and medium-sized enterprises with existing industrial facilities. Furthermore, the individual components of GFRP, especially glass-fibres and the resin matrix, lead to major challenges for an effective recycling. The standard manufacturing process of GFRP materials includes an irreversible hardening process in thermosetting resins, which completely prevents the resin from melting back into its original shape [4, 5, 6]. An improved recycling process would involve a complete separation of fibres from the resin without compromising their mechanical properties, which is challenging according to the current state-of-the-art due to the thermoset properties of the resin. Current solutions for these challenges are various recycling methods, including mechanical, thermal, chemical and electrical processes. Thermal recycling requires high energy inputs and may compromise the mechanical properties of glass-fibres compared to newly manufactured GFRP materials. Chemical recycling is limited as certain chemical reagents are not recyclable or reusable, leading to waste disposal issues. However, mechanical recycling of residual waste produces particles that can be used as filler additives in various products and is considered as the most energy-efficient approach for GFRP waste [7, 8].

This paper presents a novel solution for mechanical recycling of GFRP waste and analyses the mechanical properties of recycled GFRP materials manufactured with mechanical recycled and processed GFRP waste. The study investigates the impact of particle size classification on the mechanical properties of recycled GFRP materials, providing insights for optimising the recycling process. The findings contribute to advancing sustainable practices in the industry and addressing the technological limitations of GFRP recycling.

2 Materials and Methods

2.1 Manufacturing of GFRP Test Specimens

For the following tensile strength investigations of recycled GFRP material, GFRP test specimens were manufactured with a composition of 70% recycled material embedded in 30% UP resin. The processing of the recycled material is shown in Fig. 1. For the 70% recycled material, GFRP waste material from KRUK FASERVERBUND GMBH, Lichtenau, Germany, as shown in Fig. 1a is used. The base material was manufactured in a handcraft lamination process with four internal tissue layers enclosed in spray fibre and gelcoat layers. This waste material was chopped in a specifically developed mechanical shredding system, depicted in Fig. 1b. This system consists of a rotating breaking tool arrangement with two rotating axes, which allows the GFRP waste material to be fed and shredded. The rotating axes are equipped with several blunt tools, which interlock in an offset manner and thus induce mechanical shearing forces F_s in the inserted GFRP panels, causing them to be broken up and shredded. This is shown schematically in Fig. 1b. Figure 1c shows the mechanical chopped particles. Based on this, the shredded particles were dry classified in a sieving system to extract specific particle lengths $l_p =$ 7 mm, $l_p = 20$ mm and $l_p = 80$ mm, see Fig. 1d. The recycled particles are embedded in a 30% UP resin matrix with a Young's modulus E = 3.7 GPa. The classification process is necessary to reach specified particle distributions in order to investigate the influence of the particle lengths l_p on the Young's modulus E as well as the ultimate tensile strength Rm. The particle lengths l_p were applied into the UP resin in a mixture process to achieve a stochastically distributed particle orientation. This allows non-directional mechanical properties of the GFRP material. To mix the shredded fibers for embedding in the UP resin matrix, 70% shredded fiber flakes of recycled GFRP waste material from KRUK FASERVERBUND GMBH, Lichtenau, Germany, were filled into the disperser Kniele Konus Mischer KKM 500/750, KNIELE GMBH, Bad Buchau, Germany, and 30% UP resin DISTITRON 429 ESX20Q, POLYNT SPA, Italy, was added, which was then homogeneously mixed. To reduce the appearance of air pores and voids the chamber was put under vacuum. Including the three particle lengths l_p, three different types of recycled GFRP test specimen were manufactured. For each type five test specimens were analysed in tensile strength tests to identify statistical differences.

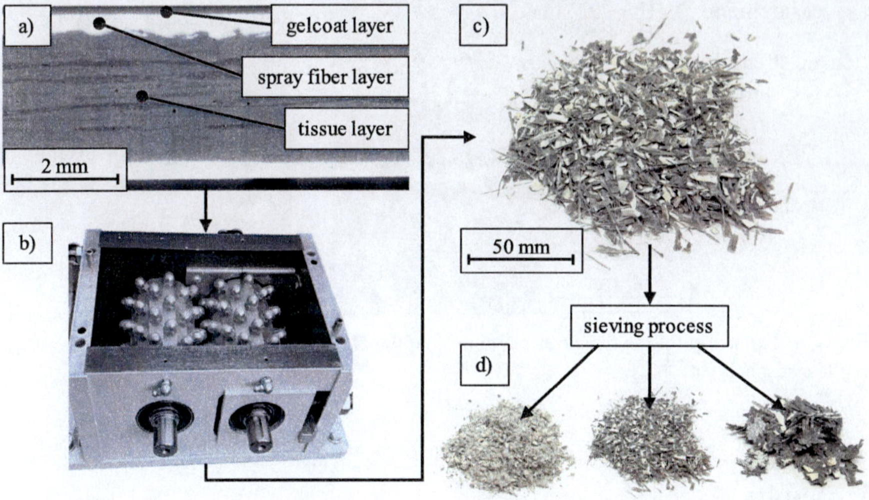

Fig. 1. Processing of the recycled material, a) GFRP waste as base material, b) mechanical chopping, c) chopped particles, d) dry classified particle distributions after the sieving process.

2.2 Characterisation of GFRP Test Specimens

The fibre lengths l_f, fibre orientation o_f and fibre density ρ_f were analysed using the computer tomograph CT Metrotom 800, CARL ZEISS GMBH, Oberkochen, Germany. The analysed test specimen with dimensions of 30 mm × 25 mm × 14 mm were exposed to the X-Rays in the CT using an analysis time duration of $t_{at} = 30$ min.

2.3 Tensile Strength Tests

The three different GFRP types with each five test specimens were analysed for their mechanical properties in a tensile strength analysis. The tensile strength tests were carried out using the material testing machine of the type T1-FR150SN.A4K from ZWICK-ROELL GMBH & CO. KG, Ulm, Germany. Within the investigations, the Young's modulus E and an ultimate tensile strength Rm for each test specimen were analysed. The test setup is shown in Fig. 2a and a cross-section of the three GFRP types are visualised in Fig. 2b.

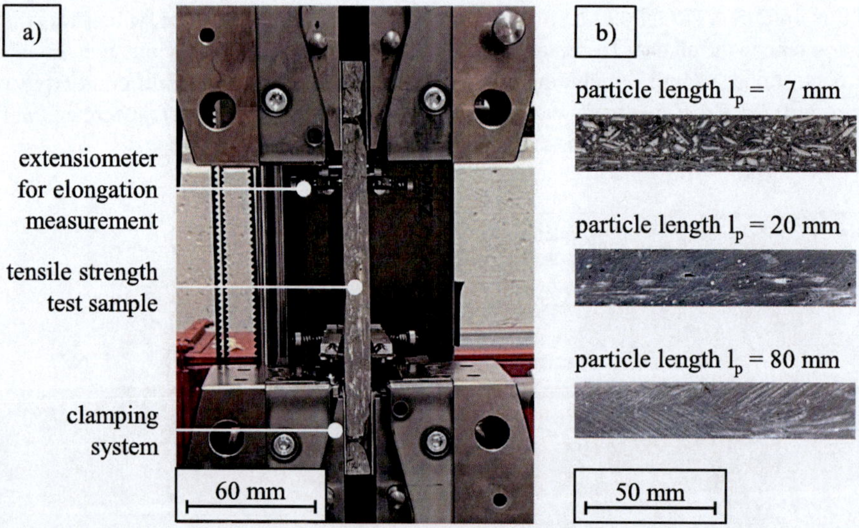

Fig. 2. a) Test setup for the tensile strength tests of the GFRP specimens, b) Types of the GFRP tensile test specimen.

3 Results

3.1 Mechanical Material Properties of the GFRP Test Specimens

Within the investigations, the mechanical properties of the test specimens concerning the Young's modulus E and the ultimate tensile strength Rm were evaluated. The results of the tensile strength analysis are shown in Table 1 and Fig. 3. The Young's modulus E reaches its maximum with $E = 5.05$ GPa for a particle length $l_p = 20$ mm. The occurrence of this maximum is due to smaller air cavities and pores in the material with the particle length $l_p = 20$ mm, see Fig. 4b, as air inclusions have a diminishing effect on the Young's modulus E. In addition, the results show a relatively high standard deviation of 21.4%, which can be explained to deviations in sample preparation or manufacturing process. Compared to the pure UP resin used for the tensile test specimens, the added recycled particles show an improvement of $i_{min} = 13.6\%$ and $i_{max} = 36.4\%$ to the mentioned Young's modulus E. The ultimate tensile strength Rm increases with longer recycled particles with its maximum of $Rm = 11.48$ MPa for particle lengths $l_p = 80$ mm. This shows that increasing particle lengths l_p correlate with growing ultimate tensile strengths Rm. This can be explained by a higher adhesion area A_a between fibres and the UP resin. The standard deviation of the Young's modulus E for the particle lengths $l_p = 20$ mm is significantly lower, which could be due to variations in the manufacturing process of the test samples or due to the sample homogeneity. For shorter particle lengths $l_p = 7$ mm, a more uneven orientation of the fibers in the sample are visible, see Fig. 4a. This could lead to local differences in material behavior, which in turn leads to a higher standard deviation. With particle lengths $l_p = 80$ mm on the other hand, mechanical interactions between the fibers could occur and in addition larger air cavities, see Fig. 4c, lead to higher standard deviations.

Table 1. Results of the tensile strength tests.

Particle length l_p in mm	Young's modulus E in GPa		Ultimate tensile strength Rm in MPa	
	average	standard deviation	average	standard deviation
7	4.3210	0.8537	6.5800	0.8070
20	5.0455	0.3384	9.9732	2.9528
80	4.2014	0.9002	11.4808	4.5676

Figure 3 shows exemplary stress-strain curves for each investigated GFRP material type. The depicted stress-strain curves show some nonlinearity. For the derivation of Young's modulus E, gradients were formed between the zero points of the measurement curves and the maximum values before the specimens failed. The Young's modulus E was calculated from these gradients. The curve for the particle length $l_p = 7$ mm shows a nonlinearity with a degressive course, which could be attributed to a less effective adhesion between the fibres and the resin matrix leading to an increased deformation. Also, shorter fibers have more different orientations, see Fig. 4a, which leads to anisotropic behavior and nonlinear responses. The similar gradients of all curves in Fig. 3 indicate similar Young's moduli E of the test specimens, whereby higher gradients of the curves indicate larger Young's moduli E. The specimen with a particle length $l_p = 80$ mm show interruptions in the stress-strain curve, which result from the failure of individual glass-fibres or the beginning of brittle fractures in the UP resin.

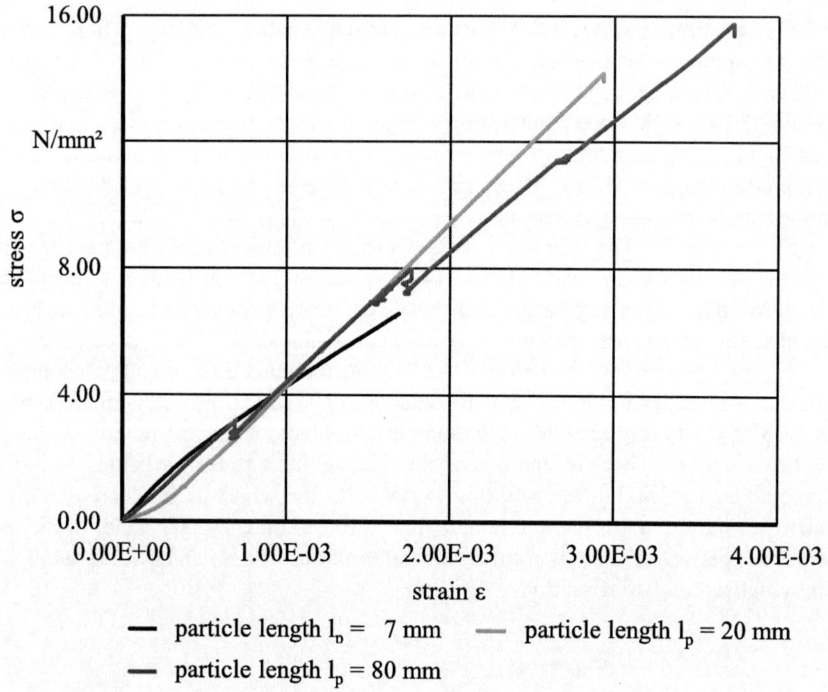

Fig. 3. Exemplary stress-strain curves for tensile test specimens with particle lengths of l_p = 7 mm, l_p = 20 mm and l_p = 80 mm.

However, the differences of the Young's moduli E between the samples are relatively low considering the standard deviations. This leads to the assumption that the mechanical strength is mainly provided by the resin and the fibers with their different particle length l_p have a minor influence. Compared to the Young's modulus of the resin with E = 3.7 GPa, the Young's modulus of the samples is on average 22% higher. The ultimate tensile strength Rm is measured at the maximum values of the curves before the mechanical failure of the test specimens. As depicted in Fig. 3, the minimum ultimate tensile strength $Rm_{,min}$ was found at particle length l_p = 7 mm and the maximum ultimate tensile strength $Rm_{,max}$ could be identified at particle length l_p = 80 mm respectively, which shows a significant influence of the particle lengths l_p on the ultimate tensile strength Rm.

3.2 CT Analysis

The CT analyses performed on the GFRP specimens provided valuable insights concerning the distribution and characteristics of fibre lengths l_f, visualised in Fig. 4. The fibre lengths l_f were identified in the CT analysis software myVGL, VOLUME GRAPHICS GMBH, Heidelberg, Germany, by measuring the maximum length of the individual fibers whereas the scaling is carried out automatically by the analysis software.

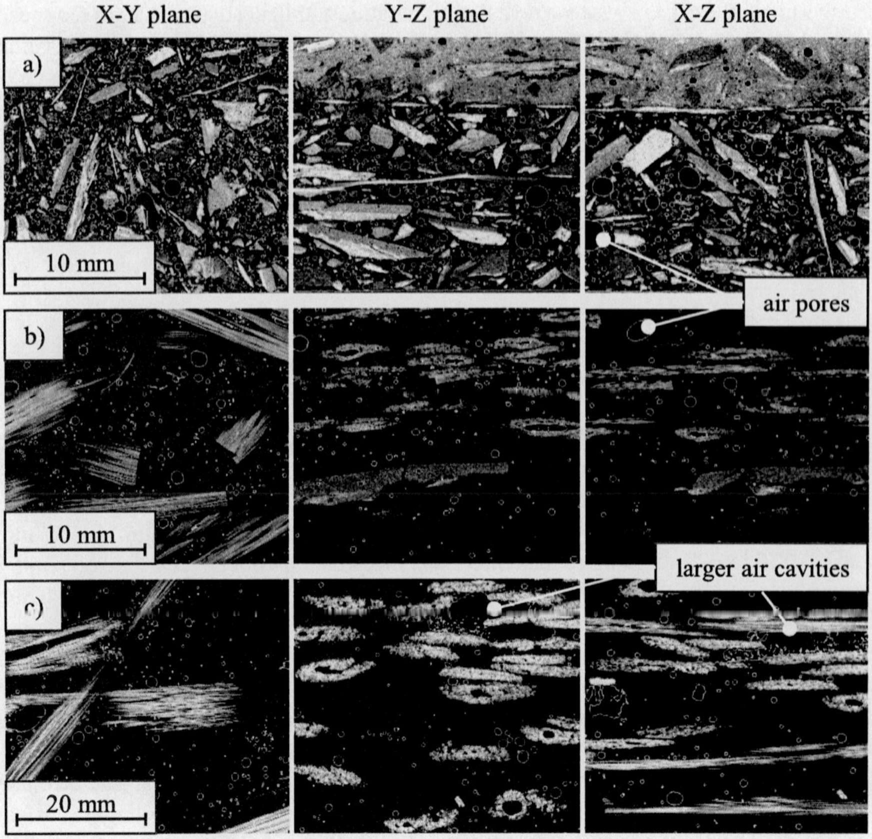

Fig. 4. CT cross-sections of the recycled GFRP material types, a) $l_p = 7$ mm, b) $l_p = 20$ mm, c) $l_p = 80$ mm.

For the sieved particle lengths $l_p = 7$ mm, the measured fibre lengths could be identified for $l_f = 4.77 \pm 1.54$ mm. With increasing sieved particle lengths $l_p = 20$ mm and $l_p = 80$ mm, the fibre lengths also increased, measuring $l_f = 10.12 \pm 5.37$ mm and $l_f = 19.83 \pm 4.53$ mm, respectively. A comparison between the sieved particle lengths l_p and the corresponding fibre lengths l_f showed significant variations. These variations can be explained by multiple shortened fibres that overlap within the sieved particles, a result of the mechanical fragmentation process during recycling.

A notable finding from the analysis was the predominant alignment of longer fibres towards the X-Y plane within the material matrix, see Fig. 4. This alignment pattern suggests a reinforcement effect primarily along the longitudinal direction of the tensile specimens. The pulling force F_p of the tensile strength tests was carried out in the X-direction. In the CT images of specimens with particle lengths $l_p = 20$ mm and $l_p = 80$ mm, areas with higher fibre densities ρ_f within the material matrix were observed. These regions indicate enhanced structural integrity and potential reinforcement effects.

Furthermore, the CT images also revealed air cavities within the material matrix, appearing as black areas in Fig. 4. These cavities comprised both small pores and larger voids, with the voids predominantly located at the interfaces between fibres. The presence of these air cavities and voids raises concerns about potential structural weaknesses and their influence on the overall material performance.

The results highlight the complex relationship between fibre lengths l_f, particle lengths l_p and their influence on the structural characteristics of composite materials. The findings emphasize the importance of considering fibre alignment and the occurrence of air cavities, when assessing the mechanical properties and performance of such materials.

4 Conclusions and Outlook

In this study, the mechanical behaviour of recycled GFRP materials was investigated through tensile strength tests. The GFRP materials consisted of an UP matrix and mechanically fragmented dry-classified recycled particles. Three different GFRP types were analysed and are characterised by recycled particle lengths $l_p = 7$ mm, $l_p = 20$ mm and $l_p = 80$ mm. According to the results, the investigated GFRP types showed similar Young's moduli E, measuring E = 4.32 GPa for $l_p = 7$ mm, E = 5.05 GPa for $l_p = 20$ mm and E = 4.20 GPa for $l_p = 80$ mm. Compared to new GFRP materials with a Young's modulus E = 11.21 GPa the recycled materials have a 55% reduced Young's modulus E. The new GFRP material without any recycling content consists of 4 layers of fiberglass fabric surrounded by a spray fiber layer and an external gelcoat layer which was manufactured in the hand lamination process. The fabric layers consist of glass fibers woven at a 90°-angle. The ultimate tensile strength of this material is Rm = 53.24 MPa. The ultimate tensile strength Rm of the analysed recycled materials increased with larger particle lengths l_p. The maximum ultimate tensile strength Rm with an average value Rm = 11.48 MPa was achieved for the particle length $l_p = 80$ mm which is 78% lower than the new GFRP material. This shows the demand for the development of new manufacturing and recycling processes which will be analysed within this project. Additionally, the GFRP materials were subjected to CT analysis. The CT images showed significantly differences between the recycled particle lengths l_p and the corresponding fibre lengths l_f. The investigated fibre lengths l_f of the used particle length l_p are shorter compared to the belonging particle lengths l_p. The results show that the fibre lengths l_f of $l_p = 7$ mm is shorter up to 4.6%. The fibre lengths l_f of $l_p = 20$ mm up to 49.4% and the fibre lengths l_f of $l_p = 80$ mm up to 75.2%. Furthermore, the presence of air pockets based on pores and voids in the material matrix was detected, which reduce the mechanical properties of the test specimens due to decreased material densities ρ in the material volume v_m. This results in a small amount of material to absorb the mechanical forces F_m. In conclusion, this study provides valuable insights into the mechanical behaviour of recycled GFRP materials. By understanding the relationship between particle lengths l_p, fibre lengths l_f and mechanical properties, future research focus on optimising the manufacturing process to enhance the performance of these materials. This leads to a more flexible use of sustainable composites as well as enable the reduction of energy E and resources R in various industries. Future research work addresses the reduction of these air inclusions during the manufacturing process to improve the mechanical properties of the GFRP materials. This includes further investigations into suitable particle lengths l_p and fibre lengths l_f to achieve enhanced mechanical performance. Based on this, the developed GFRP materials can be used in applications that require increased mechanical properties, resulting in energy E and resource savings.

Acknowledgements. The authors would like to thank Robert Kruk from KRUK FASERVERBUND GMBH, Lichtenau, Germany, for the manufacturing of the tensile test specimens and Bernd Weißenborn from WEIßENBORN MASCHINENBAU GMBH, Einbeck, Germany, for his support concerning the development of the mechanical shredding system.

References

1. Woidasky, J.: Weiterentwicklung des recyclings von faserverstärkten verbunden. In: Recycling und Rohstoffe, vol. 6, pp. 241–259. TK Verlag Karl Thomé-Kozmiensky, Neuruppin (2013)
2. Seiler, E., Teipel, U.: Recycling von polymeren verbundstrukturen aus rotorblättern. In: Recycling und Rohstoffe, vol. 11, pp. 395–414. TK Verlag Karl Thomé-Kozmiensky, Neuruppin (2018)
3. Liu, P., Barlow, C.Y.: Wind turbine blade waste in 2050. Waste Manage. **62**, 229–240 (2017)
4. Ludwig, C.: Glasfaserverstärkte kunststoffe unter hoher thermischer und mechanischer Belastung. In: Forschungsberichte aus dem Institut für Tragkonstruktionen und konstruktives Entwerfen der Universität Stuttgart. ITKE, Dissertation, Stuttgart (2009)
5. Witten, E., Avk, I.V.K.: Handbuch Faserverbundkunststoffe : Grundlagen Verarbeitung Anwendungen. Vieweg + Teubner, Wiesbaden (2010)
6. Shuaib, N.A., Mativenga, P.T.: Effect of process parameters on mechanical recycling of glass fibre thermoset composites. Procedia CIRP **48**, 134–139 (2016)
7. Khalid, M.Y., Arif, Z.U., Ahmed, W., Arshad, H.: Recent trends in recycling and reusing techniques of different plastic polymers and their composite materials. Sustain. Mater. Technol. **31**, 382 (2022)
8. Mazzoli, A., Moriconi, G.: Particle size, size distribution and morphological evaluation of glass fibre reinforced plastic (GFRP) industrial by-product. Micron **67**, 169–178 (2014)

Development of a Damped Toolholder for Vibration-Assisted High-Precision Milling Using High Damping Metals

Julius Tschöpel[1(✉)], Toni Hocke[1], Mitchel Polte[1,2], Bernfried Fleiner[3], Clemens Dicke[4], and Eckart Uhlmann[1,2]

[1] Institute for Machine Tools and Factory Management IWF, Technical University of Berlin, Berlin, Germany
julius.tschoepel@iwf.tu-berlin.de

[2] Fraunhofer Institute for Production Systems and Design Technology IPK, Berlin, Germany

[3] Exeron GmbH, Oberndorf, Germany

[4] Son-X GmbH, Aachen, Germany

Abstract. Ultrasonic vibration-assisted machining (UVAM) is one of the most promising innovations in high precision machining of advanced materials such as titanium alloys for aerospace applications or silicon carbide materials for the semiconductor industry. UVAM improves the process efficiency by increasing tool life through lower cutting forces and temperatures. However, in vibration-assisted milling applications, it is necessary to isolate the spindle system from the vibrations induced by the oscillation actor in order to maintain the long-term accuracy of the machine. This paper presents an innovative approach based on the integration of high damping materials into the toolholder to improve its damping capacity while maintaining the necessary stiffness. Therefore, the tool holder was developed with cylindrical damping inserts in a radial pattern with shrink fit connections. According to the results it was shown that the damping capability of a conventional toolholder can be improved without unacceptably reducing its radial stiffness compared to a standard toolholder geometry.

Keywords: Ultrasonic vibration-assisted machining (UVAM) · ultra-precision machining · material damping

1 Introduction

Ultrasonic-assisted manufacturing is one of the most important innovations in modern high- and ultra-precision machining in the past decades. The use of vibration assistance offers improved process conditions to machine surfaces with highest optical requirements such as mould inserts for injection moulding processes as well as mirrors and lenses in aerospace applications [1]. Vibration assistance in milling represents a particularly innovative field of technology. In this application case, the rotation of the cutting tool is superimposed with an axial oscillating motion, induced by an actuator system. This reduces the tool cutting time t_c, which decreases the resulting temperatures ϑ and

T. Bauernhansl et al. (Eds.): WGP 2023, LNPE, pp. 376–386, 2024.
https://doi.org/10.1007/978-3-031-47394-4_37

process forces f_{pr} and leads to an increased tool life and improves process results such as surface quality [2]. Furthermore, the effect of the interruption of contact between the tool and the workpiece increases the critical chip thickness d_c. This allows ductile machining of hard-brittle materials [2]. However, these systems place high demands on the machining setup - particularly the spindle bearings - to resist the vibrations induced by the ultrasonic actuator. This is critical because these components are largely responsible for the precision of a machine tool [3]. The spindle bearings and the linear guides are stressed by these vibrations [4]. This is due to the mechanism of material fatigue, where periodic loading favours the propagation of microcracks in the inner and outer bearing rings, which in extreme cases can lead to chipping in the bearing raceway [5]. Due to the natural vibration behaviour and the possible overrunning of resonance frequencies f_n of the bearing balls, even a periodic force f_{vib} with a small amount can cause contact deformation of the bearing raceway. To exceed the long-term accuracy of the spindle a_{sp} and the axis system, these parts should be isolated from high-frequency vibrations. Due to its proximity to the ultrasonic actuator, the toolholder is the most reasonable part to be used for damping. Various damping approaches were investigated by comprehensive research works. MÖHRING ET AL. [6] gave a comprehensive review of materials for the manufacture of machine tool frame structures and their damping properties. JEHRING [7] developed and experimentally validated damping elements consisting of a hollow sphere partially filled with particles, which could be integrated into vibrating machine parts. The resulting substantial vibration suppression was based on the internal friction of the particles set in motion by the oscillating movement of the hollow sphere. Due to the centrifugal forces F_{rot} occurring in the fast-rotating milling toolholder, the particles could not be set in motion by the vibration. VOGEL ET AL. [8] extended this approach based on the development of a turning holder with integrated powder-filled cavities. Therefore, a toolholder with an ellipsoid hollow element was additively manufactured and filled with metallic particles from various materials. Accordingly, particle-based approaches can only be used for applications without superimposed rotary motions. TEIGE [9] investigated another friction-based damping strategy, which is only applicable to fast rotating machine parts. Boring bars are used as the test object, since they are vulnerable to torsional and bending vibrations. The primary reason is based on their large diameter to length ratio L_s/d_s, compared to conventional toolholders. Therefore, boring bars are equipped with concentric tubes that are compressed to convert the vibrational energy E_{vib} to heat through frictional loss. Due to the constructive implementation of the friction damping-effect, the toolholder should provide a greater length l than a conventional used toolholder for precision milling applications. Furthermore, active damping systems were investigated to extinct vibrations using phase-shifted oscillations. BAUER [10] added a damping system to the machine spindle, consisting of an electromagnetic actuator and an acceleration sensor, which was acting as a vibration absorber. The experimental results showed a decrease in the maximum amplitude a_{max} at low frequencies f. To eliminate machine vibrations in ultrasonic frequency ranges, a damping system with a high energy input would be necessary. In addition, the vibrations still reach the spindle bearings even when the damping system is attached to the spindle housing. The implementation of magnetorheological fluids inside machine parts represents a further approach. WERKLE ET AL. [11] presented specially prepared bearing seats that can contain magnetorheological fluids. When these fluids are exposed to an electric field, a change in their viscosity as well as damping properties occurs. The main disadvantage of this solution approach is related to its complex structure. An integrated electromagnet is needed, which requires an inductive or direct electric power supply at the spindle nose. The increased energy E consumed by the system and the more complex construction shows a negative effect on the overall efficiency of the machine tool η. The innovative approach presented in this paper addresses the need for a defined damping behaviour of a toolholder used for vibration-assisted high-precision machining. When using a conventional toolholder, high-frequency vibrations can be transmitted to the machine system, resulting in reduced long-term accuracy a_{sp} and machine service life l_m. The presented damping method is based on the combination of different metals with different stiffness coefficients S_b and damping ratios ζ. Conventional toolholders are made from tempered steels (e.g. 31CrMoV9). These materials show suitable mechanical characteristics in terms of stiffness and elasticity, but the damping coefficients are reduced. Therefore, materials with improved damping properties are selected and inserted into the toolholder. The connection between the damping inserts and the toolholder is made securely with a shrink fit connection. In order to set the desired damping properties, simulations were performed and a modal analysis of the manufactured specimens were carried out. Due to the way the piezo element is integrated into the oscillation system, a negative influence of the damping element on the resulting amplitude a_{max} is not to be expected. The structure of the ultrasonic system consists of the front mass, which is represented by the sonotrode, and the back mass, which is realised by a screw connection that braces the piezo element against the top of the sonotrode.

2 Methods and Materials

2.1 Sample Design

The design of the samples used in this paper is based on the geometry of a standard HSK-40 toolholder. This type of toolholder is one of the most commonly used approaches to connect the tool to the machine spindle in high- and ultra-precision machining setups. In order to maintain a high level of integrability of the presented damping solution with the toolholder, a cylindrical geometry with an outer diameter $D_{out} = 40$ mm and a length $l = 80$ mm was chosen. The considered damping elements show a diameter $D_i = 6$ mm, $D_i = 8$ mm and $D_i = 10$ mm and a length $l_i = 5$ mm. The inserted elements are arranged in a radial pattern with six inserts per rotation and different number of elements. The structure of the specimens and an exemplary implementation of the damping elements are shown in Fig. 1.

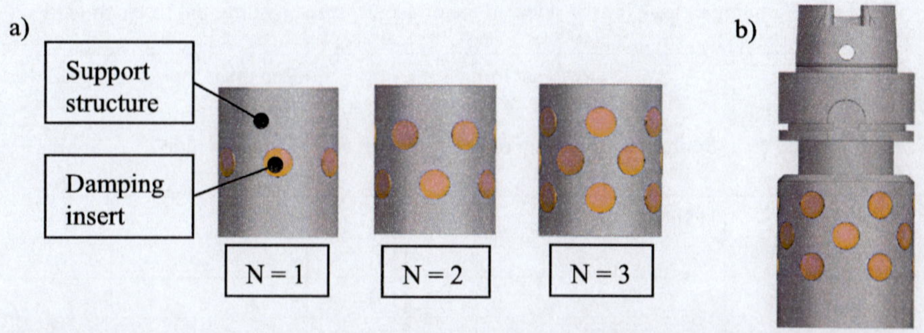

Fig. 1. Implementation of the damping elements with a) specimen geometries with different number of elements and b) exemplary implementation into HSK-40 toolholder.

The main criteria for the selection of suitable materials for the damping inserts is the damping capability. Since mechanical stiffness properties, e.g. bending stiffness S_b, were also important for the application shown, only metallic materials were considered. The effect leading to energy dissipation and consequently material damping is based on atomic dislocations in the crystal structure and a consequently thermal energy conversion [12]. Temperature leading processes in the metallic phase structure between regions of different strain are also cited as a key mechanism leading to inner damping effects [13]. Resulting from these dependencies, small changes in grain structure can result in large variances in damping properties. These modifications in the microscopic structure can be induced by fatigue processes due to periodic loading of the specimen [13]. In addition, the determination of the damping ratio ζ can be associated by significant uncertainty, depending on the employed estimation methodology and the specifics of the experimental configuration. This is due to the impossibility to distinguish between the damping effects induced by the specimen and the effects caused by the experimental setup [13].

The used materials are 99,9%-copper type Cu-EPT, grey cast iron with lamellar graphite type EN-GJL-250 and red brass type GC-Rg7. Their damping properties, determined by different research efforts, are shown in Table 1. As a result of the reasons already mentioned, considerable deviations in the damping ratio ζ can be observed. The cylindrical support structure, with radial holes in the pattern described, is made of structural steel type S235JR. A shrink-fit connection was used to ensure a reliable connection of the damping inserts to the support structure that could resist the high centrifugal forces f_{rot} caused by the high rotational speeds V_{rot}.

Table 1. Damping ratios ζ for the selected materials measured by different research works

Copper Cu-EPT		Grey cast iron EN-GJL-250		Red brass GC-Rg7	
Damping ratio ζ ($*10^{-3}$)	Source	Damping ratio ζ ($*10^{-3}$)	Source	Damping ratio ζ ($*10^{-3}$)	Source
1	[13]	1.5–15	[14]	1.5–3	[13]
-	-	6.5–15	[13]	-	-
-	-	20	[15]	-	-
-	-	50	[16]	-	-

2.2 Simulative Results

Simulations, based on the finite element method, were performed to determine the sample geometry that provides the highest damping ratio ζ without significantly reducing the radial stiffness S_b. The simulations were carried out using ANSYS® Mechanical 2022 R2, ANSYS INC., Canonsburg, USA. To allow an objective evaluation of the obtained simulation results, the performance value α was introduced. Its calculation is shown in Eq. 1.

$$\alpha = \frac{1}{S_b} * \zeta \tag{1}$$

To estimate the damping ratio ζ of the samples, harmonic analysis was executed based on the results of modal analysis. In order to determine the influence of the insert geometry independently of the damping material used, the same damping ratios ζ were used for each simulation. Therefore, the material for the damping inserts was defined with a specific damping ratio $\zeta_d = 0.02$, which represents a high-damping metal. The damping ratio of the support structure was defined as $\zeta_S = 0{,}01$. A periodic force with a maximum value of $f_{ex} = 1$ kN acting on the specimen's surface in the axial direction was used as the excitation. The resulting amplitudes a_{max} at the first natural frequencies f_n show a decrease in the maximum amplitudes a_{max} with the increase of the percentage of damping material (Fig. 3). The damping ratio ζ was estimated using the "Half-Power Bandwidth Method" from the frequency response data. This calculation approach uses the gradients at the rising and falling edges of the frequency response curve to derive the damping ratio ζ. The procedure is presented in Eq. 2–4. The variables $f_{1,2}$ representing the two corresponding frequency values, which are at half of the maximum amplitude a_{max}. Q is defined as the amplification factor. Equation 4 establishes the mathematical relation between the amplification factor Q and the resulting damping ratio ζ.

$$\Delta f = f_2 - f_1 \tag{2}$$

$$Q = \frac{f_n}{\Delta f} \tag{3}$$

$$\zeta = \frac{1}{2 * Q} \tag{4}$$

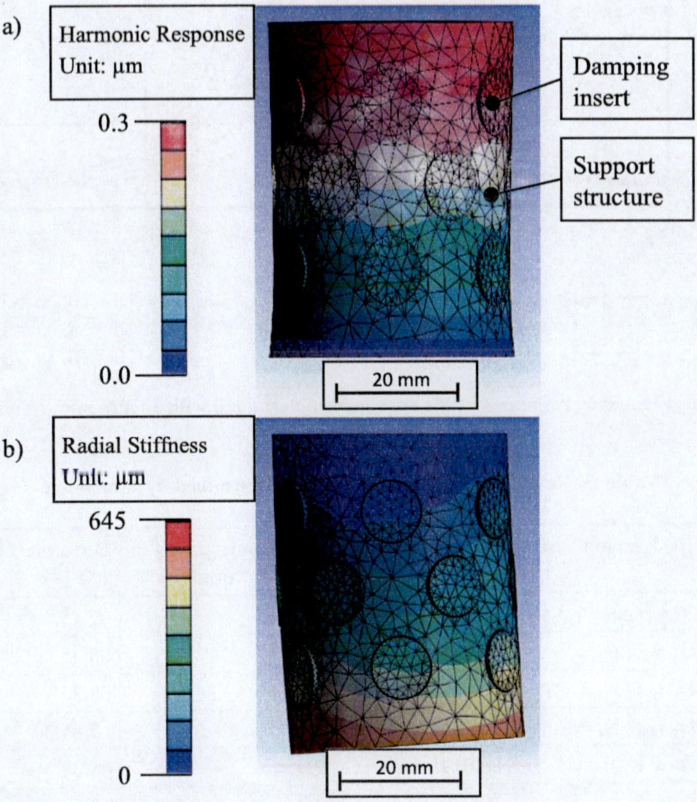

Fig. 2. Simulative determination of the stiffness S_b and the damping ratio ζ. a) resulting eigenmode from the harmonic response (total deformation displayed) and b) radial stiffness S_b (total deformation displayed).

The radial stiffness S_b was simulated by defining a force $F_{rad} = 10$ kN acting on the cylinder top surface in radial direction and measuring the resulting deflection at the end of the specimen (Fig. 2 b)). In order to choose the part with the best application-specific damping ratio ζ and lowest reduced radial stiffness S_b, the performance value α was calculated. The results are shown in Table 2.

The simulation results prove that the test assembly can be equipped with damping inserts without significantly reducing its overall radial stiffness S_b. When comparing the reference part (sample No. 1) without any inserts with sample No. 10, it was observed that the target value α was increased by a factor of 3.49, while the radial stiffness was reduced by $S_b = 27\%$. Based on the results presented, the geometry related to sample 10 was used for the experimental verification of the simulation.

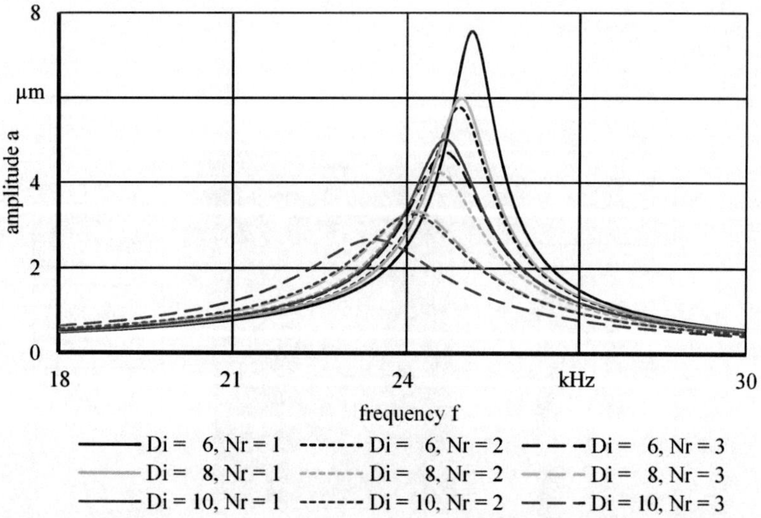

Fig. 3. Frequency responses of the specimens determined with a harmonic simulation.

Table 2. Relevant mechanical properties determined by simulation.

No Sample	Designation Sample	Damping ratio ζ	Radial stiffness S_b in $1*10^{-3}$ * mm/kN	Performance value α 1*10–3
1	Reference	0.011	2.02	05.45
2	$D_i = 6, N_r = 1$	0.015	2.32	06.45
3	$D_i = 6, N_r = 2$	0.020	2.39	08.36
4	$D_i = 6, N_r = 3$	0.026	2.42	10.85
5	$D_i = 8, N_r = 1$	0.019	2.35	08.09
6	$D_i = 8, N_r = 2$	0.028	2.42	11.59
7	$D_i = 8, N_r = 3$	0.038	2.48	15.29
8	$D_i = 10, N_r = 1$	0.024	2.38	10.06
9	$D_i = 10, N_r = 2$	0.039	2.48	15.70
10	$D_i = 10, N_r = 3$	0.049	2.57	19.07

2.3 Experimental Results

Due to the high uncertainty regarding the damping ratio ζ of the materials used, the simulations were conducted with idealised mechanical properties. To transfer the simulated damping characteristics to the real application, an experimental verification is required. Therefore, three samples with the damping materials copper Cu-EPT, grey cast iron EN-GJL-250 and red brass GC-Rg 7 were manufactured based on the described simulation results. The manufactured reference specimen was equivalent to the sample No. 1 in

Table 2. To measure their damping characteristics, an experimental modal analysis was performed using an impulse hammer excitation. The resulting oscillations were measured with a triaxial ICP® acceleration sensor, Model HT356A44, PCB PIEZOTRONICS, USA, and recorded with a measuring frontend VibPlot 8-channel, M + P INTERNATIONAL, Germany. By hitting the specimen's top surface with the tip of the impulse hammer, the first natural frequency f_n was excited. With the experimental setup shown in Fig. 4, the time-resolved vibration signal was recorded. After performing a Fast Fourier transform (FFT), the damping ratios ζ were calculated from the frequency-resolved vibration signal using the Half-Power Bandwidth Method. The obtained results are shown in Table 3, as well as the measured frequency responses in Fig. 5. The specimen made of the material type EN-GJL-250 shows the highest damping ratio ζ. The determined damping ratios ζ of the other metals investigated vary by only 18% (red brass GC-Rg 7) and 37% (copper Cu-EPT) related to the specimen made of grey cast iron. It can be also seen that the recorded measurement results deviate the simulated damping ratio by $\zeta = 34\%$ for EN-GJL-250, $\zeta = 30\%$ for GC-Rg 7 and $\zeta = 25\%$ for Cu-EPT.

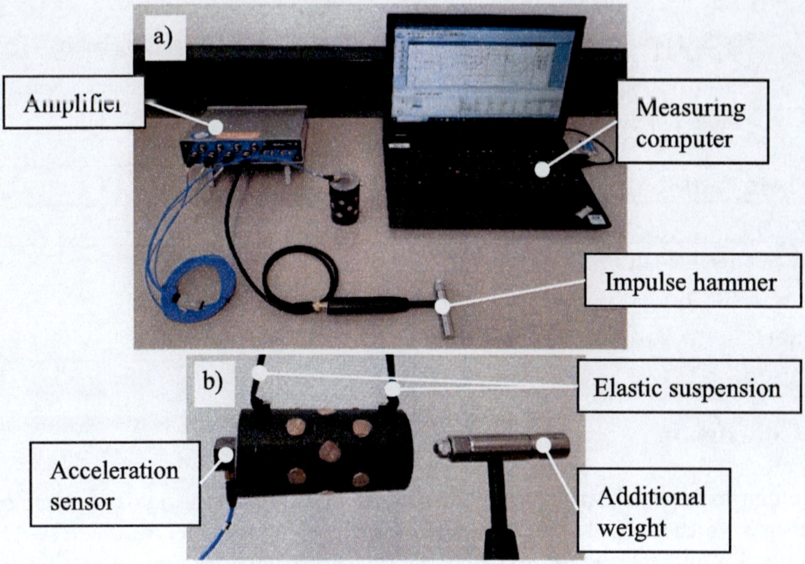

Fig. 4. Estimation of the samples damping ratio ζ with a) experimental setup and b) conduction of the modal analysis.

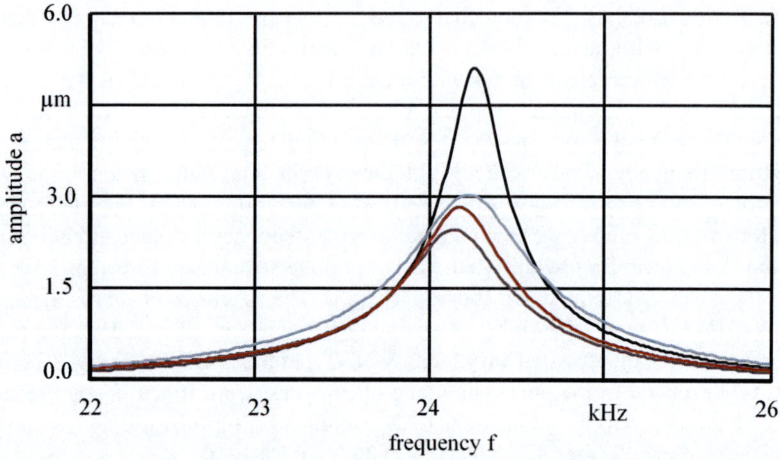

Fig. 5. Frequency responses of the specimens equipped with damping inserts.

Table 3. Resulting damping ratios ζ from the experimental modal analysis.

Damping material	Damping ratio ζ ($*10^{-3}$)
Reference	11
Grey cast iron EN-GJL-250	16.6
Red brass GC-Rg7	14.6
Copper Cu-EPT	12.5

3 Conclusion

According to the results, it could be demonstrated that the damping ratio ζ of a developed geometry, which is based on the structure of an HSK-40 toolholder, was increased by a factor of 1.5 by implementing damping inserts made from grey cast iron EN-GJL-250 into its structure. Based on the damping ratios ζ described in Table 1, a relatively small deviation of the samples could be observed. It could be demonstrated that there are other factors besides material damping that influence the decay behaviour. This is particularly evident in the large deviation of the simulation results from the actual measured damping ratios ζ. This discrepancy is largely due to the limitations of damping modelling in the contact area of traditional FE approaches. For example, the surface quality, wetting condition and surface pressure in the contact area would have to be taken into account to realistically represent the damping properties. These relevant parameters need to be further investigated. Subsequent research works will address the effect of the interface damping on the frequency response [18, 19]. This could be done by manipulating the geometry of the inserted elements, including their corresponding interfaces, which opens up the possibility of distinguishing the material damping from the interface damping. In addition, different offsets at the shrink-fit connection will be investigated to estimate the impact of contact pressure on the damping characteristics. The presented solution offers several notable advantages, including its seamless integration into standard toolholders, its passive approach, without the requirement for an external energy supply as well as its ability to operate at high rotational speeds V_{rot}. As a result, the developed toolholder with integrated damping inserts is nearly as mechanically rigid as a conventionally used toolholder.

Acknowledgements. This work is supported by the funding program *Zentrales Innovationsprogramm Mittelstand* (ZIM) by the Federal Ministry for Economic Affairs and Climate Action (BMWK), Berlin, Germany.

References

1. Zhang, C., et al.: An investigation of the high-frequency ultrasonic vibration-assisted cutting of steel optical moulds. MDPI Micromachines **12**, 460 (2021)
2. Bulla, B.: Ultrapräzisionszerspanung von Nanokorn-Hartmetall mit Monokristallinen Diamantwerkzeugen. Apprimus Verlag, Aachen (2013)
3. Abele, E., Altintas, Y., Brecher, C.: Machine tool spindle units. CIRP Ann. Manuf. Technol. **59**, 781–802 (2010)
4. Homepage SMB Bearings. https://www.smbbearings.com/technical/bearing-noise-vibration.html. Accessed 21 Aug 2023
5. Lee, H.W., et al.: Modelling ultrasonic vibration fatigue with unified mechanics theory. Int. J. Solids Struct. 236–237 (2022)
6. Möhring, H.-C., et al.: Materials in machine tool structures. CIRP Ann. Manuf. Technol. **64**, 725–748 (2015)
7. Jehring, U.: Schwingungsdämpfung mit partikelgefüllten Hohlkugeln. Technische Universität Dresden, Dresden (2019)
8. Vogel, F.A.M., et al.: Vibration supressing in turning TiAl6V4 using additively manufactured tool holders with special structured, particle filled hollow elements. In: 19th Machining Innovations Conference for Aerospace Industry, pp. 32–37. Elsevier, Amsterdam (2019)
9. Teige, C.: Entwicklung gedämpfter Werkzeugaufnahmen für lang auskragende rotierende Werkzeuge. In: Institut für Fertigungstechnik und Werkzeugmaschinen Leibnitz Universität Hannover, Hannover (2018)
10. Baur, M.: Aktives Dämpfungssystem zur Ratterunterdrückung an Spanenden Werkzeugmaschinen. Technische Universität München, München (2014)
11. Werkle, K.T., et al.: Additively manufactured, particle-filled damping structures with magnetorheological fluids. In: 54th CIPP Conference on Manufacturing Systems, pp. 1418–1423. Elsevier, Amsterdam (2021)
12. Kollmann, F.G.: Praktische Maschinenakustik. Springer, Heidelberg (2005)
13. Cremer, L., Heckl, M.: Körperschall, 3rd edn. Springer, Heidelberg (2010)
14. Beards, C.F.: Structural Vibration: Analysis and Damping. Butterworth-Heinemann, Oxford (1996)
15. Grothe, K.-H., Feldhusen, J.: Dubbel-Taschenbuch für den Maschinenbau, 22nd edn. Springer, Heidelberg (2007)

16. Lazan, B.J.: Damping of Materials and Members in Structural Mechanics. Pergamon Press, Oxford (1968)
17. Ilschner, B., Singer, R.F.: Werkstoffwissenschaften und Fertigungstechnik, 4th edn. Springer, Heidelberg (2005)
18. Hildebrand, M., Adams, R.D.: Vibration damping. In: Adhesive Bonding, 2nd edn. Woodhead Publishing, Sawston (2021)
19. Niehus, K.K.: Identifikation Linearer Dämpfungsmodelle für Werkzeugmaschinenstrukturen. Technische Universität München, München (2015)

Hybrid Models for Waste Reduction and Sustainable Manufacturing: A Comparative Study of Stoneware Tile Polishing

Marcel Wagner[1](✉), Fábio J. P. Sousa[2], Max Werrel[1], Moritz Glatt[1], and Jan C. Aurich[1]

[1] Institute for Manufacturing Technology and Production Systems, RPTU Kaiserslautern, 67663 Kaiserslautern, Germany
marcel.wagner@rptu.de

[2] School of Sciences and Technology, Laboratory of Instrumentation and Robotics, Federal University of Rio Grande do Norte, Natal, Brazil

Abstract. Hybrid models which combine data-driven and physics-based models have emerged as a trend in manufacturing for improving process efficiency, reducing costs, and enhancing product quality. Hybrid models can deliver greater performance benefits than either model by itself. As sustainability becomes an increasingly important issue for the manufacturing industry, hybrid models show potential to assist the transformation of manufacturing processes towards increased sustainability. Hence this paper investigates the potential for reducing resource and energy consumption through the application of hybrid models, based on a use case of polishing stoneware tiles. A comparative analysis between a physics-based and a hybrid model-based approach is presented. For both approaches, the ratio of defective parts is calculated. Furthermore, the environmental impacts are estimated with a life cycle assessment. The findings suggest that the hybrid model-based approach enables reductions in energy and resource consumption, thereby improving the sustainability of the manufacturing process.

Keywords: Hybrid model · Sustainable manufacturing · Industrial polishing

1 Introduction

Digital technologies have emerged as the primary enabler in manufacturing for improving process efficiency, reducing costs, or enhancing product quality. Hybrid models, which combine physics-based and data-driven models, are among the technologies that show promise for further improvement. Previous research has presented various improvement possibilities for different systems using hybrid models. For example, Langlotz et al. reported improvements in energy management in a factory using a hybrid model [1]. Sousa et al. pointed out the potential for improving the prediction of gloss patterns of stoneware tiles produced in an industrial polishing process [2]. Li et al. showed that their hybrid model for continuous damping control has higher accuracy and lower computational cost compared to existing physics-based models [3].

T. Bauernhansl et al. (Eds.): WGP 2023, LNPE, pp. 387–396, 2024.
https://doi.org/10.1007/978-3-031-47394-4_38

However, in addition to traditional optimization goals such as efficiency, cost, and quality improvement, companies increasingly focus on improving the environmental sustainability of manufacturing processes and the resulting products. Therefore, this paper investigates the potential of hybrid models for the reduction of environmental impacts when the models are applied to manufacturing processes.

In order to do so, we will present the state of the art in research and technology and present how hybrid models are able to reduce the environmental impact of manufacturing processes. The case study presents how a hybrid model can reduce the number of defective parts and the environmental impact in the stoneware tile polishing process.

2 State of Research and Technology

2.1 Digital Models in Manufacturing

Digital models in manufacturing can be broadly categorized into two types: physics-based models and data-driven models.

Physics-based models, also known as white box models, aim to simulate and represent specific physical phenomena of real manufacturing systems using established knowledge. These models use functional relationships to accurately capture the behavior of a system. For example, finite element methods are widely used to simulate complex structures and physical interactions within manufacturing processes. A specific example is the prediction of forces in drilling [4]. Another application of physics-based models is the multi-body simulation, which allows the analysis of dynamics and kinematics of mechanical systems, for example, to simulate machine tool motion [5]. The advantage of physics-based models is that they are based on existing and validated knowledge, providing a clear understanding of the underlying principles. However, to develop an accurate model, this deep understanding of the real system is required. Furthermore, the computational requirements can be significantly high and modeling accuracy can be compromised if real-world complexities are not adequately addressed, which leads to uncertainties in modeling results. This is due to the fact that stochastic influences are often not taken into account in physics-based models, and scattering is difficult to represent [6].

Data-driven models, also known as black box models, use computational algorithms to process historical and real-time operational data. These models use techniques such as statistics, machine learning, pattern recognition, and data mining to extract insights from real-time and historical data [7]. Examples of applications include failure prediction [8] or fault diagnosis [9], where patterns are identified to detect and diagnose system failures. Data-driven models offer several benefits. They provide general insights into trends, patterns, and sources of inefficiency without requiring extensive expert knowledge of the physical processes. These models are flexible, generic, and adaptable to different manufacturing systems. In addition, with the advent of Industry 4.0, an increasing amount of data is becoming available, which can be used by companies to improve manufacturing processes. However, data-driven models have disadvantages as well. They require large amounts of training data to build accurate models. Furthermore, the interpretability and transparency of data-driven models can be challenging, as they do not provide explanations of the relationships between variables [6].

In summary, digital models in manufacturing encompass both physics-based and data-driven approaches. Each approach has its own set of advantages and disadvantages, and the choice between them depends on the specific requirements of the application. It is also possible to combine both physics-based and data-driven approaches to take advantage of the strengths of both types of models.

2.2 Hybrid Models

Hybrid models, also known as gray box models, are a combination of physics-based and data-driven models [10]. Using this combination allows manufacturers to benefit from the advantages of both approaches while reducing their respective disadvantages mutually [6]. This has been successfully tested in several applications, such as predictive maintenance for CNC machine tools [11], simulation of bond strength of overmolded thermoplastic composites [12], or prediction of fuel consumption [13]. The current impacts of hybrid models in manufacturing are mostly related to goals such as cost reduction, product quality improvement, or failure prevention. So far, little insight has been gained on the potential of hybrid models in terms of reducing the environmental impacts of manufacturing processes.

3 Impact of Hybrid Models on Environmental Sustainability

As shown above, hybrid models offer many ways to improve processes by simulating, predicting or analyzing them. Accordingly, there are also different ways in which hybrid models can be used for the improvement of the sustainability of manufacturing processes. For example, hybrid models can be part of digital twins that are utilized to optimize energy consumption by controlling manufacturing parameters of machines. In this case the physics-based model can be used to describe the machining process, while the data-driven model uses the parameters of the physics-based model to predict energy consumption. Another option is to use hybrid models for predictive maintenance, e.g. to use tools for as long as possible without compromising the quality of manufactured products. In this case the physics-based model can be used to estimate the remaining useful life using tool wear rates, while the data-driven part of the hybrid model can detect patterns in vibrations, temperature or other data to extend the physics-based approach. A last example is the usage in quality management to increase the number of good parts and reduce carbon emissions by reducing the number of defective parts.

This last case is the subject of this paper. To elaborate on the research objective of this study, which is to investigate whether hybrid models have the potential to reduce waste and the environmental impact of the manufacturing process, this paper builds on the results of a previous work by Sousa et al. [2]. In this work, the polishing process of an industrial polishing train was studied using a hybrid model that combines kinematic equations (physics-based model) with an artificial neural network (ANN) (data-driven model) to predict gloss values in sections of stoneware tiles. The hybrid model is used in this application because the most used physics-based model for predicting gloss values by Hutchings et al. [14] (referred to as classical model) does not consider the highly stochastic nature of the polishing process. The application of this hybrid model resulted in a higher prediction power of gloss values compared to the classical model.

This work is the foundation for further investigations on how the prediction can be used to achieve more accurate simulations and thus be able to select a better set of operating parameters for the given polishing train. As a result, a more uniformly distributed glossiness can be achieved. Such uniformity has a positive impact on the number of polished tiles expected to be within the specified tolerance range. The higher the predictive power of the simulation model, the lower the level of losses that can be expected to occur due to poorly polished tiles. This in turn results in savings of energy and resources during the manufacturing process.

Hybrid models therefore seem to be a promising tool to simulate the gloss values, adjust the operating parameters, improve the product quality, and potentially reduce the number of defective parts. Potential improvements in environmental impacts can be seen, for example, in the reduction of required energy, lubricants or polishing tools per unit of produced floor tiles. By studying the environmental impact of minimizing the number of defective parts, valuable insight can be gained into the potential of hybrid models to reduce the environmental impact of manufacturing processes.

4 Case Study

The hybrid model developed by Sousa et al. is used for predicting gloss values in specific sections of stoneware tiles [2]. The investigation focuses on an industrial polishing train, which consists of numerous tangential polishing heads arranged in a row. These polishing heads are equipped with abrasive blocks of decreasing grit size. The purpose of the polishing process is to achieve a uniform distribution of gloss levels across the tiles and to achieve a target mean gloss level within a specified tolerance range.

The hybrid model combines a physics-based component that models the kinematics of the abrasive contacts with a data-driven model that uses an ANN to predict gloss values at specific sections of the tile. Firstly, the physics-based model uses established equations to simulate the kinematics of the abrasive contacts between the polishing heads and the stoneware tiles. It describes the number of abrasive contacts (N), the mean scratch speed (μ), and the variability of the scratch angle (σ) at each location on the tile. Secondly, the values from the physics-based model are used in the data-driven component of the hybrid model that predicts gloss values at different pixels on the tile surface. The input parameters for the ANN are $\{N, \mu, \sigma\}$. To obtain the inputs for the ANN, a 17x17 matrix is used to divide two adjacent tiles into 289 pixels each. This results in a total of 578 pixels for which gloss values are determined. The hybrid model improves the prediction accuracy, increasing it from a determination coefficient of $\rho = 0.08$ (achieved by the classical model) to $\rho = 0.45$. The details of the development and application of the hybrid model can be found in Sousa et al. [2].

4.1 Application of a Hybrid Model for Waste Reduction in the Polishing Process

The evolution of gloss experienced by a ceramic tile during the polishing process can be predicted and visualized with simulation tools. Such tools can be based on the classical model of gloss-gaining curve available in literature [14] and more recently, on a hybrid model [2]. The improvement offered by a simulation tool is twofold. Firstly, such a tool can recommend the set of operational parameters that optimizes the polishing kinematics, so that the action of the myriad of abrasive particles becomes more uniformly distributed over the tile surface. Secondly, with a defined quality control target, a simulation tool can be used for purposes of quality control, detecting significant variations of gloss during the polishing process, and accordingly performing corrections, such as recommending a new set of operational parameters [15].

Regarding the spatial uniformity of the distribution of abrasive contacts to be promoted by polishing process, in a long run, both classical and hybrid model are expected to deliver similar optimized sets of operational parameters to be implemented in a polishing train. On the other hand, the capability of in-line detecting significant variations of gloss during polishing depends directly on the prediction power of each simulation tool. The present study performs a quantitative analysis of the performance expected from both the classical and the hybrid model in terms of production control. This situation is sketched in Fig. 1, which indicates that the use of the hybrid model is expected to result in a lower rate of defective tiles due to the greater prediction power and the according adjustment of the set of operational parameters.

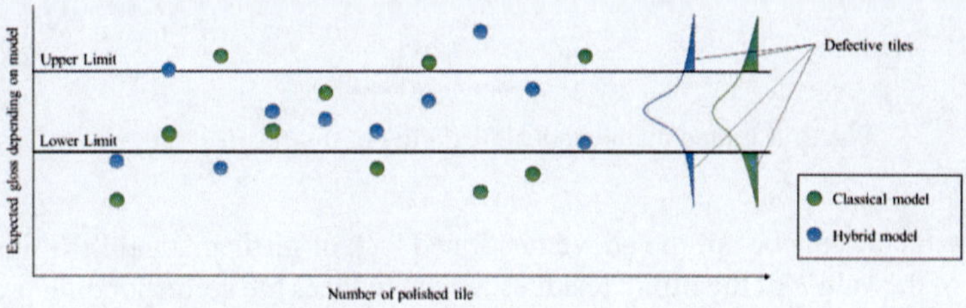

Fig. 1. Simulation tools used for quality control purposes.

As previously mentioned, the classical model to predict the evolution of gloss at a specific region of ceramic tile during the polishing time was conceived by Hutchings et al. [14]. The model assumes that the instantaneous rate of gloss gaining of a finite region remains inversely proportional to its current level of gloss, which can be described mathematically as:

$$G_t = G_\infty - (G_\infty - G_0) \cdot e^{\frac{-t}{k}} \tag{1}$$

G_t is the value of gloss at a given time t, G_∞ is the maximum possible level of gloss, which depends exclusively on microstructure and optical properties of the material, G_0 is the level of gloss before the polishing process, and k is known as characteristic time, which is a constant that has to be empirically calibrated. Mathematically, k represents the value of polishing duration leading to approximately 63% of the total enhancement.

In a more physical sense, k encompasses the phenomenological aspects of the polishing process altogether. Therefore, despite the reasonable assumption and the pioneer approach, a certain lack of mathematical flexibility of the Eq. (1) becomes evident. As a result, several nuances of the polishing phenomenon cannot be accurately represented. Figure 2 exemplifies this impossibility.

The correct adjustment of the classical model based on the measured values is presented in the graph by the gray line, considering $G_\infty = 67.5\%$, $G_0 = 2.5\%$, and $k = 237$ s. The residues of such adjustment for the entire dataset (578 points) sum up to nearly zero, indicating the desired absence of bias. Alternatively, different approaches could also be tried, and some of them may even yield better mathematical fittings, such as those two cases also presented in Fig. 2, focusing on the higher and on the lower values of gloss. Although other fitting approaches could also be mathematically possible. However, in view of the constraints imposed by the physical meaning of the parameters G_∞, G_0, and k, such mathematical flexibility vanishes, and the data adjustment becomes limited to that presented by the classical model (gray curve).

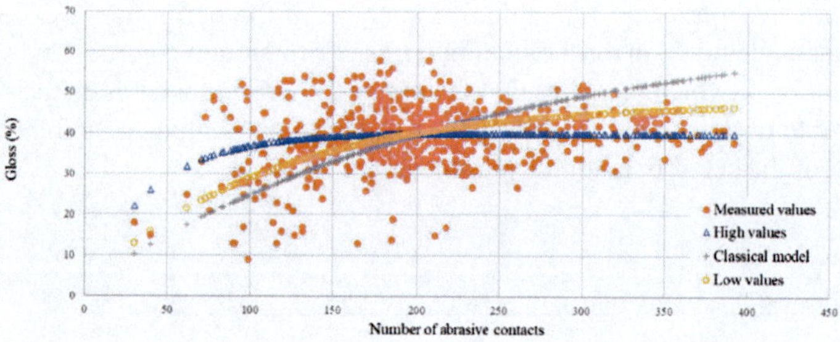

Fig. 2. Limited mathematical flexibility of the classical model.

A direct consequence of the abovementioned lack of mathematical flexibility is the increase of the values of the fitting residues, which reduces the corresponding prediction power. Based on the sum of the square deviations of the residues, the resulting standard deviation of the classical model can be estimated as $\sigma_{CM} = 9.3\%$ (in gloss units).

In contrast, the much larger mathematical flexibility of the ANN can be seen in Fig. 3. As a result, much smaller values of residues are expected for each simulation. The resulting standard deviation of the hybrid model, also based on the sum of the square deviations of the residues, is calculated as $\sigma_{HM} = 4.52\%$ (in gloss units).

Some typical parameters of quality control in industry are the acceptable range of gloss from 65% to 70%, centered in 67.5%, and the usual checking process that consists of taking measurements of gloss on 12 different regions equally spaced over the polished surface [16]. Furthermore, as the residues of deviations from both models were analyzed and found to be normally distributed, the capability of detecting variations of gloss within the tolerance range can be quantified by the proportion of tiles correctly evaluated. This in turn can be estimated by Eqs. (2), and (3), which for both models simply represent the proportion of tailings outside of the tolerance range, as previously indicated in Fig. 1.

$$T_{CM} = \frac{2 \cdot \sigma_{CM}}{\sqrt{12}} = 0.35 \tag{2}$$

$$T_{HM} = \frac{2 \cdot \sigma_{HM}}{\sqrt{12}} = 0.06 \tag{3}$$

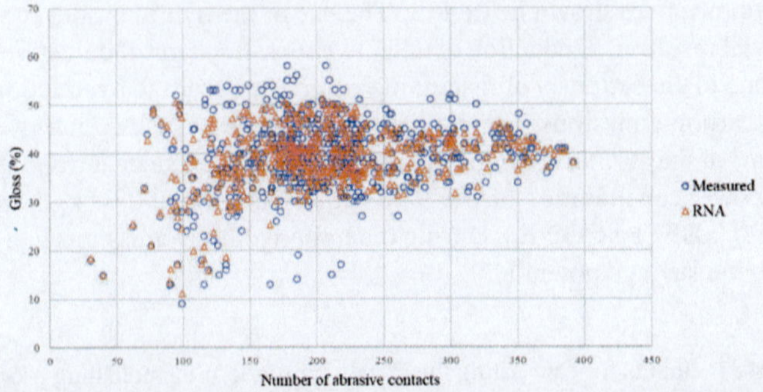

Fig. 3. Mathematical flexibility of the hybrid model in view of the ANN (adapted from [2]).

The proportion of losses due to defects in the polishing process reported by literature is of about 5% [17]. For the current level of tile production worldwide, this represents a carbon footprint of approximately 11 million tons CO2-Eq. Resulting from the production of defective tiles [18, 19]. No reports on the use of simulation tools for reducing this level were found in literature so far, so that the absolute value of carbon emission, that can be correspondingly saved cannot be properly estimated. In relative terms, however, the use of hybrid models has shown a substantial advantage when compared to the classical physics-based model. Admitting an industrial polishing train operating under typical conditions of quality control, the proportion of tiles to be rejected by a simulated tool equipped with a hybrid model is expected to be nearly six times smaller than that based on the classical model of gloss gaining curve.

4.2 Assessment of the Environmental Impacts of Tile Polishing

To quantify the possible savings of environmental impacts, a life cycle assessment of the polishing process was performed, using the LCA software Umberto LCA + and the database ecoinvent (version 3.8). The functional unit of the assessment is 1 m^2 of stoneware tiles. The polishing process, as shown in Fig. 4., uses 1 m^2 of blank tiles [18], 1.47 kWh electricity [20], 30 l water [21] and 0.55 kg of polishing tool silicon carbide [16] as inputs, while one m^2 of polished tiles represents the output. Silicon carbide is the most common abrasive used in polishing processes [16].

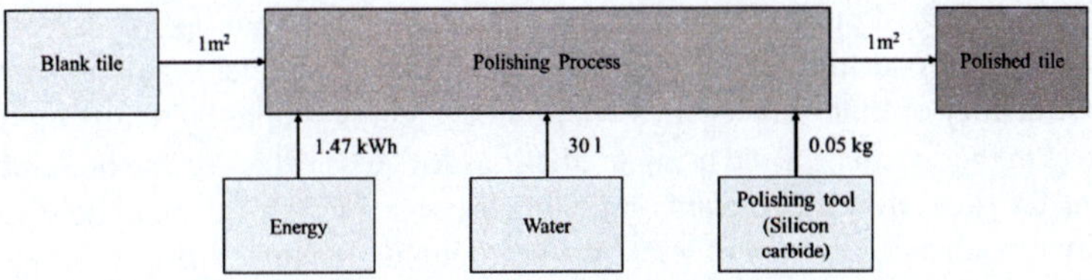

Fig. 4. Illustration of the polishing process.

The overall carbon footprint to produce 1 m^2 of polished tiles is 20.31 kg CO2-Eq. Using the impact assessment method IPCC 2013 GWP 100a. Details used for calculating the carbon footprint are shown in Table 1. The use of the hybrid model compared to the classical model results in a reduction of defective tiles. Each m^2 of defective tiles that can be reduced due to the better set of machining parameters leads to a reduction of 20.31 kg CO2-Eq. in carbon emissions. For the sake of completeness, the energy consumption of the training of the ANN should be considered. However, since a training duration of 347 s and a power consumption of 149.9 W with a CO2-Eq. of 0.69 kg/kWh only cause emissions of 0.00997 kg CO2-Eq. During the training, these emissions can be neglected in relation to the savings potential.

Table 1. Inputs for calculating the carbon footprint of the polishing process.

Input	Carbon Footprint (in kg CO_2-Eq.)
Blank Tile	15.02
Silicon Carbide	4.98
Energy (Electricity)	0.30
Water	$7.33 * 10^{-3}$

5 Summary and Outlook

The findings of this study highlight the potential of utilizing a hybrid model to reduce the carbon footprint in tile manufacturing. By comparing the hybrid model simulation with the classical model of gloss gaining curve, it was observed that the proportion of tailings falling out of the tolerance range is nearly six times lower for the hybrid model compared to the classical model. This reduction in defective tiles offers a promising avenue for minimizing carbon emissions associated with the manufacturing process.

Considering that approximately 13.5 billion m^2 of tiles were manufactured in the year 2017 [19] and with a carbon footprint of 20.31 kg CO2-Eq./m^2 of polished tiles, the impact of reducing the number of defective tiles by accurate gloss predictions and subsequent adjustments of operational parameters can be substantial. By employing the hybrid model, manufacturers can enhance their ability to perform in-line predictions and improvements directly in the process, thereby reducing the overall carbon emissions.

Future research should not only focus on reducing the number of defective tiles but also explore opportunities to optimize operational parameters, aiming to produce the same quantity of high-quality tiles with reduced energy input by adjusting machine parameters in the simulation and using predictions for gloss values on the one and carbon emissions for this set of parameters on the other hand side. This avenue holds promise for achieving enhanced energy efficiency within the tile manufacturing.

Furthermore, this study serves as a compelling example of how hybrid models can contribute to enhancing the environmental sustainability of manufacturing processes. It is recommended to extend the application of hybrid models to manufacturing processes within different industries. By conducting further investigations, it would be possible to assess the broader applicability of hybrid models in improving the efficiency and environmental performance of manufacturing processes.

Overall, the findings presented in this research paper emphasize the potential of hybrid models to reduce carbon emissions in the tile manufacturing industry. By implementing accurate gloss prediction and adjusting operational parameters accordingly, manufacturers can reduce the number of defective tiles, leading to environmental benefits. The implications extend beyond the specific case studied here, as hybrid models have the capacity to support sustainability efforts across other manufacturing sectors as well. The integration of such models in industrial practices therefore seems to represent one possible step towards achieving a more sustainable future.

Acknowledgements. The German Research Foundation (DFG) and the Brazilian Coordination for the Improvement of High Education Personnel (CAPES) partly fund this work within the program "Investigating the applicability of hybrid digital models for manufacturing systems under varying model input conditions", in the scope of the Collaborative Research Initiative – PIPC 8881.473092/2019–1 (DFG funding number AU 185/72–2).

References

1. Langlotz, P., Klar, M., Yi, L., Hussong, M., Sousa, F.J., Aurich, J.C.: Concept of hybrid modeled digital twins and its application for an energy management of manufacturing systems. Procedia CIRP **112**, 549–554 (2022)
2. Sousa, F.J.P., Halla, R., Souza, A., Langlotz, P., Glatt, M., Aurich, J.C.: Fusion of physical principles and data-driven based models: an industry 4.0 perspective for improving the polishing process of stoneware tiles. Prod. Eng. **14**(5–6), 639–654 (2020)
3. Li, T., Liao, Z., Cai, Y.: Physical modeling for digital twin of Continuous Damping Control damper. J. Manuf. Process. **99**, 96–104 (2023)
4. Strenkowski, J., Hsieh, C., Shih, A.: An analytical finite element technique for predicting thrust force and torque in drilling. Int. J. Mach. Tools Manuf. **44**(12–13), 1413–1421 (2004)
5. Zaeh, M., Siedl, D.: A new method for simulation of machining performance by integrating finite element and multi-body simulation for machine tools. CIRP Ann. **56**(1), 383–386 (2007)
6. Langlotz, P., Glatt, M., Sousa, F.J.P., Aurich, J.C.: Modelle als grundlage für den digitalen zwilling. Zeitschrift für wirtschaftlichen Fabrikbetrieb **115**(5), 340–343 (2020)
7. Chen, C.L.P., Zhang, C.-Y.: Data-intensive applications, challenges, techniques and technologies: a survey on Big Data. Inf. Sci. **275**, 314–347 (2014)
8. Leukel, J., González, J., Riekert, M.: Adoption of machine learning technology for failure prediction in industrial maintenance: a systematic review. J. Manuf. Syst. **61**, 87–96 (2021)
9. Dogan, A., Birant, D.: Machine learning and data mining in manufacturing. Expert Syst. Appl. **166**, 114060 (2021)
10. Oussar, Y., Dreyfus, G.: How to be a gray box: dynamic semi-physical modeling. Neural Netw. **14**(9), 1161–1172 (2001)
11. Luo, W., Hu, T., Ye, Y., Zhang, C., Wei, Y.: A hybrid predictive maintenance approach for CNC machine tool driven by digital twin. Robot. Comput. Integr. Manuf. **65**, 101974 (2020)
12. Hürkamp, A., et al.: Combining simulation and machine learning as digital twin for the manufacturing of overmolded thermoplastic composites. J. Manuf. Mater. Process. **4**(3), 92 (2020)

13. Ma, Y., Zhao, Y., Yu, J., Zhou, J., Kuang, H.: An interpretable gray box model for ship fuel consumption prediction based on the SHAP framework. J. Marine Sci. Eng. **11**(5), 1059 (2023)
14. Hutchings, I.M., Xu, Y., Sánchez, E., Ibáñez, M.J., Quereda, M.F.: Development of surface finish during the polishing of porcelain ceramic tiles. J. Mater. Sci. **40**(1), 37–42 (2005)
15. Sousa, F.J.P., Aurich, J.C., Weingaertner, W.L., Alarcon, O.E.: Analytical determination of the distribution of polishing time over the surface of polished tiles. J. Am. Ceram. Soc. **90**(11), 3468–3477 (2007)
16. Hutchings, I.M., Adachi, K., Xu, Y., Sánchez, E., Ibáñez, M.J., Quereda, M.F.: Analysis and laboratory simulation of an industrial polishing process for porcelain ceramic tiles. J. Eur. Ceram. Soc. **25**(13), 3151–3156 (2005)
17. Vázquez, M.F.Q.: Porcelain tile polishing. A study of process variables and materials characteristics. Universitat Jaume I de Castelló, Castellon (2008)
18. Peng, J., Zhao, Y., Jiao, L., Zheng, W., Zeng, L.: CO2 Emission calculation and reduction options in ceramic tile manufacture-the Foshan case. Energy Procedia **16**, 467–476 (2012)
19. Ferrer, S., Mezquita, A., Aguilella, V.M., Monfort, E.: Beyond the energy balance: exergy analysis of an industrial roller kiln firing porcelain tiles. Appl. Therm. Eng. **150**, 1002–1015 (2019)
20. Wang, Y., et al.: Comparative life cycle assessment of different fuel scenarios and milling technologies for ceramic tile production: a case study in China. J. Clean. Prod. **273**, 122846 (2020)
21. Hutchings, I.M., Adachi, K., Xu, Y., Sánchez, E., Ibáñez, M.J.: Laboratory simulation of the industrial ceramic tile polishing process. In: Proceedings of Qualicer 2004 in Castellon, Spain, pp. 19–30 (2004)

Artefacts for the Implementation of Continuous Innovation for Technical Products

S. Perau[1(✉)], P. Rixen[1], M. Riesener[1], M. Kuhn[1], and G. Schuh[1,2]

[1] Laboratory for Machine Tools and Production Engineering (WZL), RWTH Aachen University, Aachen, Germany
s.perau@wzl.rwth-aachen.de

[2] Fraunhofer Institute for Production Technology IPT, Aachen, Germany

Abstract. The increasing importance of sustainability leads to changing conditions and new requirements for the manufacturing industry. Thereby, the concept of the circular economy plays a decisive role in meeting these challenges. Continuous Innovation (CI) focuses on the ability of a company to continuously improve or renew products or features during the product usage phase. Thereby, it addresses the growing demand for a circular economy through the upgradability of the products, while at the same time increasing customer satisfaction. Although manufacturing companies are aware of the opportunities, the implementation is currently not successful. This paper attempts to close this gap by introducing artefacts for the implementation of CI for technical products. Thereby, the systematic identification of the artefacts focus on the dimensions product, process, business model and organization. Thus, the identified implementation artefacts support manufacturing companies in the implementation and operative realization of continuous innovation.

Keywords: Innovation Management · Continuous Innovation · Implementation of Innovation · Upgradeability · Circular Economy

1 Motivation

Climate change creates major challenges for society. It affects the ecological balance of the earth and endangers the basis of life [1]. The manufacturing industry has a central role within managing climate change, due its high impacts regarding resource consumption, waste and pollution as well as greenhouse gas emissions [2]. The concept of circular economy provides a promising solution to address the sustainability challenges by using resources efficiently and for as long as possible [3, 4]. According to the Ellen MacArthur Foundation a Circular Economy is a restorative or regenerative industrial system by intent and design [5, 6]. Despite of these challenge, the increasing competition in the international market and ever more specific customer requirements must also be faced with innovative products to maintain customer satisfaction and market share [7]. Thereby, the concept of continuous innovation of technical products in the usage phase presents a way to successfully support a value-enhancing circular economy by implementing new

T. Bauernhansl et al. (Eds.): WGP 2023, LNPE, pp. 397–406, 2024.
https://doi.org/10.1007/978-3-031-47394-4_39

features and components (upgrades) into existing products [8, 9]. It is an interdisciplinary approach considering different dimensions such as product, process, business model and organization [10]. The concept of continuous innovation is already well known within the software industry. Due to the multidimensionality however, there is no guideline to implement the concept of continuous innovation for technical products [11, 12]. Figure 1 illustrates the possible value enhancement through continuous innovation for technical products.

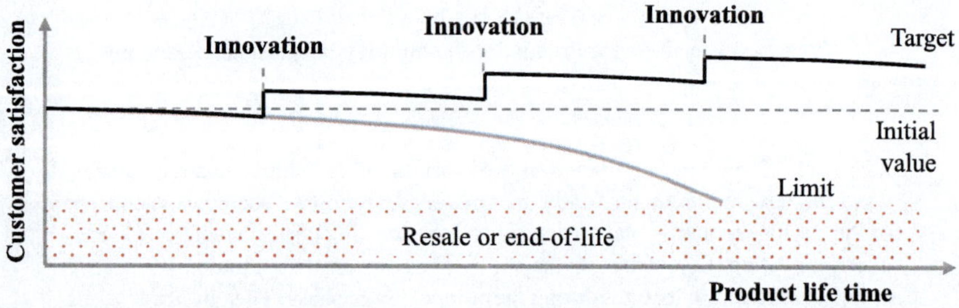

Fig. 1. Schematic concept for a possible value enhancement of customer satisfaction over time through continuous innovations

This paper aims to provide a base for the implementation for manufacturing companies by identifying relevant artefacts as a partial solution for the implementation of continuous innovations.

2 Fundamentals

In this chapter, relevant fundamentals in the context of this paper are described. The fundamentals mainly focus on continuous innovation and circular economy.

2.1 Continuous Innovation

First, it is necessary to define the term innovation and distinguishes it from the term idea. Based on Thushman an innovation is defined as a product that fulfills market need. The term innovation can be distinguished from the term idea according to Schumpeter regarding its market maturity [13, 14]. Thereby, an idea can only be described as an innovation when it has successfully been launched on the market [15]. In the context of continuous innovation, there is currently no widely accepted definition in scientific literature. Lianto et al. have categorized the existing definitions of continuous innovation, defining it as an ongoing process, that is carried out continuously, repeatedly and regularly over a longer period of time [11]. This stimulates a learning culture, which aims to continually develop and improve the company to adapt to the changing needs of present and future customers [11]. Based on Lianto et al., Riesener et al. provide a more technical definition, describing continuous innovation as a holistic approach for continuously identifying customer needs during the product usage phase and integrating and delivering new functions, components, technologies and services [10].

2.2 Circular Economy

Motivated by publications like "The limits to Growth" by the Club of Rome in 1972 several debates about the compatibility of economic growth and preservation of worldwide resources began [16]. Out of this, the concept of a circular economy evolved and presents a promising approach to face the challenges of the 21st century [16]. In comparison to a linear production economy, the circular economy aims to keep the products and its material in the economic system for as long as possible [17, 18]. The Ellen McArthur Foundation describes the circular economy as a system that is regenerative by design, aiming to ensure that products, components and materials retain their highest utility and value at all times, distinguishing between technical and biological cycles [5]. Circular Economy is usually presented as a seamless interaction of loops of recovery, named the R-strategies. The amount of the "R's" in literature might differ between 3R-10R. However, cycles such as repair, reuse, refurbish, remanufacture including upgradeability or recycle are common categories [5, 17, 19]. A value-enhancing circular economy in which regular and life-extending upgrades (continuous innovations) are integrated into products provides the concept for an industrialized circular economy for the manufacturing industry [2, 6].

3 Related Research

In this section, related scientific research regarding the implementation of continuous innovation is examined. There are many approaches that partly focus on the implementation of continuous innovation or capabilities for the realization of continuous innovation.

Inspired by the DevOps approach in the software industry Schuh et al. develop an approach for linking the use and development phases of cyber-physical systems. Derived from the software approach mentioned above, fields of action are presented [20]. Lianto et al. generate a comprehensive definition of continuous innovation, emphasize the need for continuous innovation and describe essential elements of continuous innovation capabilities. Moreover, six strategies for the development of continuous innovation are presented [11]. Furthermore, Lianto et al. present a model to measure continuous innovation capabilities in the manufacturing industry. Despite the focus of the author on continuous innovation no specific implementation artefacts are described [21]. Schuh et al. focus on the interdependencies between customer-centric continuous innovation in the subscription business and underline the importance of continuous innovation presented by Lianto et al. In the paper a management model is presented in which continuous innovation functions as the interface between a subscription business model and an organization [22]. Boer et al. as well as Xu focus on knowledge management in the context of continuous innovation [23, 24]. Boer et al. developed a model to analyze how companies can achieve a significant competitive edge by broadening their innovation activities past the product development phase as well as facilitating knowledge transfer and learning not only within the company but also with other partner organizations [23]. In addition, Xu et al. examine the mechanisms of knowledge management for innovation and present a method for companies on how to use knowledge management in the context of continuous innovation [24]. The two approaches presented above consider only parts of the scope necessary for implementing continuous innovation and do not take into account the aspects of the product and the business model. Steiber presents management principles and skills critical for the implementation of continuous innovation and discusses the details of implementation in the case of Google. The corporate culture, structure and management of Google are described in more detail. Furthermore, five steps to a management concept of continuous innovation are presented [12]. Nevertheless, the author does not display a holistic framework for the operative support of the implementation process in manufacturing companies. Riesener et al. address challenges for the realization of continuous innovation in the context of manufacturing companies [10]. These challenges can be categorized into the domains of product, process, culture and business model. Moreover, fields of action are described. Nevertheless, no concrete implementation artefacts are presented. With the Continuous Innovation Framework (COIN), van Spijker focusses on the implementation of continuous innovations in a more operative way [25, 26]. Thereby, innovations are described on three levels: Innovation level, portfolio level and company level. Although the model is specifically designed for the implementation of continuous innovations in companies, it does not consider the specific requirements of technical products and relevant implementation artefacts regarding product, organization, business model and organization.

In summary, none of the described approaches provides artefacts that can be used as a base for the implementation of technical products in the usage phase.

4 Methodology and Research Results

The approaches and research gap described above as well as the mentioned motivation address the need for a systematic identification of relevant artefact to support the implementation of continuous innovation (CI) for technical products in the usage phase. Therefore, the research question is "How can relevant artifacts for the implementation of continuous innovation in the context of value-enhancing circular economy be identified?" Three steps are performed in order to answer the question:

- Step 1: Identification of areas of interest for a literature research
- Step 2: Identification of relevant literature
- Step 3: Identification and description of artefacts

4.1 Identification of Areas of Interest for a Literature Research

The aim of the first step is the identification of areas of interest for the systematic literature research following Noblit and Hare [27]. For the holistic identification of the areas of interest, two different methods are used. First, the management model of Schuh et al. is used, as it is a suitable framework for the overarching interrelationships in the context of CI [22]. Within the model, the interrelationships between business model, product, organization, and the process of CI are displayed. These interrelationships result in requirements for CI and its implementation in the domains product, process, organization and business model. For example, the process has to come up with single embeddable innovations, while the product must be able to integrate those innovations.

The analysis of the management model results in 18 different areas of interest. Additionally, the method of networked thinking is being used for exploration of additional areas of interest from new perspectives as well as for the verification of already identified fields [28]. Finally, a list of 19 areas of interest for the systematic literature review is generated. Figure 2 shows the result in the form of the identified fields.

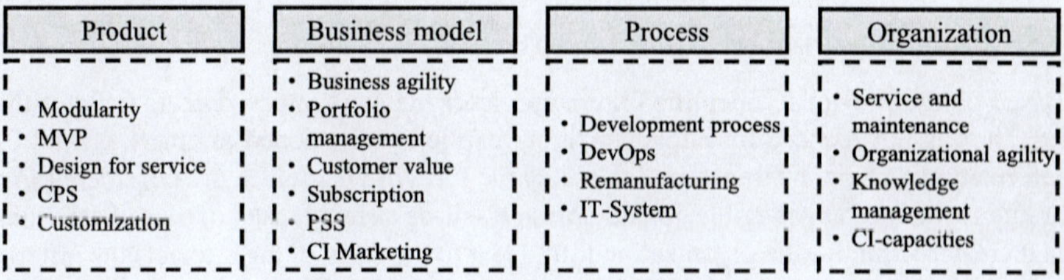

Fig. 2. Overview of identified fields of interests

4.2 Identification of Relevant Literature

The aim of the second step is the systematic literature review based on the previously identified areas of interest. The procedure follows the process suggested by Xiao and Watson [29]. The search string for the literature review includes the area of interest as well as the terms "continuous innovation", for specifying CI, and "Implementation", "Realization" and "Realisation" for specifying the implementation. The terms are connected by Boolean operators and searched for in the literature databases Web of Science, Google Scholar, IEEE Xplore, Science Direct and EconBiz. For every database and area of interest the most relevant titles are reviewed and, depending on the total number of results, the search string is detailed or extended. Figure 3 summarizes the quantitative result of the individual steps of the literature search.

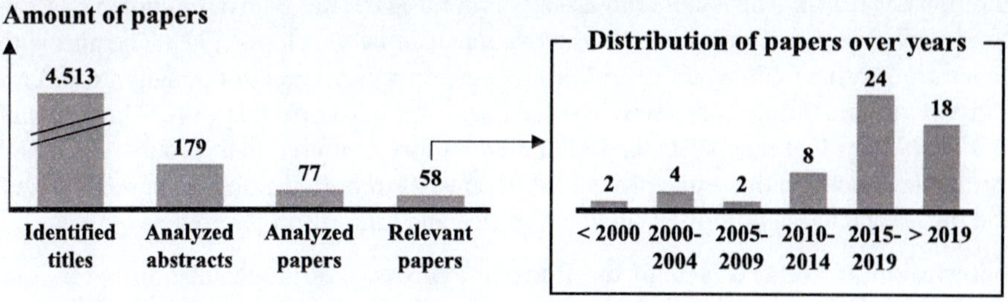

Fig. 3. Overview of analyzed literature

By this a total of 4513 titles are identified. The identified titles are reviewed by title and, in case of a positive fit, the abstracts and correspondingly following the contents are analyzed. The review is analyzed regarding the expected usability of the literature for the identification of artefacts for the implementation of CI. Thereby, German and English literature are considered. Thus, all areas of interest are analyzed and the literature should remain manageable, the goal is a small amount of literature with the highest possible usability for the synthesis of artefacts. After analyzing 179 abstracts and 77 contents, the result of the second step is a list of 58 relevant papers that is used for the identification of artefacts.

4.3 Identification and Description of Artefacts

Based on the previously identified literature, artefacts are synthesized. As most of the literature is not focused on implementation, explicitly mentioned artefacts as well as information that can be transferred to match the form of an artefact, are considered. An artefact is defined as a tangible, identifiable and self-describing result of a set of activities that creates value for the organization [30]. Next to the form of the artefacts the criteria correctness of content, freedom from contradictions, functionality, manageability and simple formulation are used to identify and evaluate artefacts from the literature [31]. Through eliminating duplicates and evaluation according to the criteria, a list of 114 sub-artefacts is identified. The list can be clustered into 21 artefacts for better operational manageability and usability. Here, the proximity of the artefacts to each other in terms of content was compared and similar artefacts were summarized. The artefacts for each dimension are described in more detail below.

Description of the Artefacts of the Domain Product. In the domain product, the following five artefacts should be implemented for the successful implementation of CI: *modularity, service-integrated design, cyber-physical features and minimal marketable product*. The *modularity* of the initial product is necessary in order to integrate new components into an existing product [32]. The modules themselves as well as the associated interfaces and architectures must be considered with regard to upgradeability and technology development [33]. The *service-integrated design* enables and simplifies the replacement of modules, as well as the realization of additional services in product-service systems, by integrating the corresponding requirements into the design of the product [32]. The *cyber-physical features* enable the product to gather usage data and transfer it to the development. Based on the usage data, innovation potentials can be identified and customer-centered innovations can be developed [32]. The aim is to guarantee physical connectivity and product cyber security and, optionally, to offer a digital twin and further connectivity-based services based on this [34]. The *minimal marketable product* represents the initial product that is offered in the course of CI and further developed in the usage phase [35]. It consists of both the physical product itself and the associated definition of product, service and innovation offerings.

Description of the Artefacts of the Domain Process. For the realization of CI, four overlapping and iterative process phases, *idea generation, idea validation, experimentation process and scaling process*, as well as a *process for portfolio management* should be implemented. The ideas for new innovations can be generated through a general creative idea generation process or through analysis of the products in the usage phase [25]. Idea validation ensures the efficient, continuous and transparent processing of ideas and is necessary in order to ensure the increase of the customer benefit by the new idea [25]. The development of modules requires experiments that explore technical possibilities and thus generate new knowledge about the unknown field of the innovation option, as well as subsequent functional tests that validate the functionality and safety of the innovations [35]. This process step requires various supporting tools such as project plan, prototyping infrastructure and modularisation process [32, 36]. After the successful development, the innovation is integrated into existing products in the process of scaling. Therefore, the necessary coordination of updates and upgrades requires not only a dedicated product update management, but also release, delivery and integration processes, ensuring innovation performance, a problem-solving concept, usage validation and a concept for the end-of-life of the replaced modules [37]. *Portfolio management* receives innovation options in the form of ideas and prototypes and decides, in alignment with the business model, how the options should be prioritized and, if necessary, adapts the business model [38]. This requires, among others, a detailed and transparency roadmap of future innovations and technologies [25].

Description of the Artefacts of the Domain Business Model. The successful realization of CI requires a *new business model, an adapted finance and accounting system, business agility, success controls*, as well as a *communication strategy*. The basis for CI is a *business model* adapted to the new value creation, such as a subscription model, whose economic functionality must be ensured by sufficient feasibility analyzes [20]. In this course *finance and accounting* requires adaptation, since the sale of individual innovations, especially in the context of the subscription model, is likely to involve smaller but more frequent transactions compared to the sale of an entire product [39]. Due to the continuous flow of innovations, the incremental changes to the product add up and demand active management of the initial business model and the adaption of success control to optimally align it in the long term [12, 25]. Finally, the *communication of changes* resulting from CI and information on CI itself, like information for use, functionality of the product and options for upgrades, is essential in order to minimize the resistance of all stakeholders, such as customers, suppliers and employees [10].

Description of the Artefacts of the Domain Organization. In the domain of organization, seven artefacts are required: *IT system, flexible production capacities, organizational structure for innovation, process speed management, knowledge management, service and logistics infrastructure*, and an *implementation organization*. The *IT system* accelerates and simplifies the CI process [36]. It includes the IT system for digital knowledge management, communication technology, customer and service structure and cyber security, among others [35]. Since CI requires simultaneous production of existing and innovative modules, *flexible and modern production technologies* are necessary [36]. Furthermore, for the subsequent regular and reliable delivery, a defined service and logistics infrastructure is required [40]. The *organizational structure* decisively determines how innovative employees are and how quickly innovations are implemented [36]. To achieve an organizational structure for innovation, it is necessary to implement, among others, an innovation culture, organizational agility and interdisciplinary innovation teams dealing with both development and operations of their team-specific innovation [36]. In addition, in order to specifically improve the innovation frequency of CI, an additional *process speed management* is needed that systematically removes hurdles in the innovation process and improves the process of realising CI itself [41].

Moreover, *knowledge management* is a particular challenge in CI, as knowledge from new innovations, existing products and its customer usage and the corporate environment must be coordinated at the same time, in order to achieve maximum customer benefit [20, 35]. Here, gathering and analyzing usage data of the existing product is essential. Finally, the implementation of CI itself requires a *implementation organizational*, consisting of an implementation team and plan, which carry out and coordinate the implementation operationally [42].

5 Conclusion and Further Research

Within the paper, artefacts for implementing continuous innovation for technical products in the usage phase were identified. The identified artefacts serve as a base to support manufacturing companies regarding the dimensions product, process, business model and organization. In further research activities, these artefacts needs to be validated with companies in the manufacturing industry. Furthermore, the artefacts must be prioritized in terms of their implementation relevance and rearranged into a sequential order. In addition, it is necessary to assign relevant roles and responsibilities for the realization of the artefacts. Furthermore, it must be taken into account that companies face different implementation challenges and that the necessity of individual artefacts must be analyzed with regard to the corporate context and the corporate target system.

Acknowledgement. Funded by the Deutsche Forschungsgemeinschaft (DFG, German Research Foundation) under Germany's Excellence Strategy – EXC-2023 Internet of Production – 390621612.

References

1. The Intergovernmental Panel on Climate Change: Global Warming of 1.5 °C: Special Report. https://www.ipcc.ch/sr15/. Accessed 16 June 2023
2. Schuh, G., et al.: Green Re-Assembly Upgrade Factory (2023)
3. Baumgartner, R.J.: Nachhaltiges Produktmanagement durch die Kombination physischer und digitaler Produktlebenszyklen. In: Redlich, T., Moritz, M., Wulfsberg, J.P. (eds.) Interdisziplinäre Perspektiven zur Zukunft der Wertschöpfung, pp. 347–360. Springer Fachmedien Wiesbaden, Wiesbaden (2018)
4. den Hollander, M.C., Bakker, C.A., Hultink, E.J.: Product design in a circular economy. J. Ind. Ecol. **21**(3), 517–525 (2017)
5. Ellen MacArthur Foundation: Towards a circular economy: Business rationale for an accelerated transition (2015)
6. Schuh, G., et al.: New Modularity and Technology Roadmapping (2023)
7. Dereli, D.D.: Innovation management in global competition and competitive advantage. Procedia. Soc. Behav. Sci. **195**, 1365–1370 (2015)
8. Chen, S.-H.: The influencing factors of enterprise sustainable innovation: an empirical study. Sustainability **8**(5), 425 (2016)
9. Aarikka-Stenroos, L., Welathanthri, M.D., Ranta, V.: What is the customer value of the circular economy? Cross-industry exploration of diverse values perceived by consumers and business customers. Sustainability **13**(24), 13764 (2021)

10. Riesener, M., Kuhn, M., Perau, S., Mertens, M., Schuh, G.: Fields of action for the realization of continuous innovation. In: 10th International Conference on Mass Customization and Personalization - Community of Europe, pp. 138–146 (2022)
11. Lianto, B., Dachyar, M., Soemardi, T.P.: Continuous innovation: a literature review and future perspective. Int. J. Adv. Sci. Eng. Inf. Technol. **8**(3), 771–779 (2018)
12. Steiber, A.: The Google Model. Springer, Heidelberg, New York, Dordrecht, London (2014)
13. Tushman, M., Nadler, D.: Organizing for innovation. Calif. Manage. Rev. **28**(3), 74–92 (1986)
14. Schumpeter, J.A.: Business Cycles: A Theoretical, Historical and Statistical Analysis of the Capitalist Process. McGraw-Hill Companies Inc., New York, Toronto, London (1939)
15. Schuh, G. (ed.) Innovationsmanagement. Springer Vieweg, Berlin, Heidelberg (2012)
16. Acatech: Deutschland auf dem Weg zur Circular Economy (2019)
17. Ekins, P., Domenech Aparisi, T., Drummond, P., Bleischwitz, R., Hughes, N., Lotti, L.: The Circular Economy: What, Why, How and Where (2020)
18. Triguero, Á., Moreno-Mondéjar, L., Sáez-Martínez, F.J.: Circular economy and firm performance: the influence of product life cycle analysis, upcycling, and redesign. Sustain. Dev. (2023)
19. Potting, J., Hekkert, M., Worrell, E., Hanemaaijer, A.: Circular Economy: Measuring Innovation in the Product Chain (2017)
20. Schuh, G., et al.: Always Beta: DevOps für cyber-physische Produkte. Internet Prod. Turn. Data Into Value, 114–137 (2020)
21. Lianto, B., Dachyar, M., Soemardi, T.P.: A holistic model for measuring continuous innovation capability of manufacturing industry: a case study. IJPPM **71**(6), 2061–2086 (2022)
22. Schuh, G., Gützlaff, A., Cremer, S., Lammersmann, J., Liu, Y.: Business model and organization: interdependencies for customer-centric continuous innovation in subscription business. In: 2020 IEEE International Conference on Industrial Engineering and Engineering Management (IEEM), pp. 260–264, Singapore, Singapore (2020)
23. Boer, H., et al.: Knowledge and continuous Innovation: the CIMA methodology. Int. J. Oper. Prod. Manag. **21**(4), 490–503 (2001)
24. Xu, J., Houssin, R., Caillaud, E., Gardoni, M.: Fostering continuous innovation in design with an integrated knowledge management approach. Comput. Ind. **62**(4), 423–436 (2011)
25. van 't Spijker, A.: Continuous innovation: how successful organizations continuously develop, scale, and embed innovations to lead tomorrow's markets. Technics Publications, Basking Ridge, New Jersey, USA (2019)
26. van 't Spijker, A.: COIN - Continuous Innovation Framework. https://continuousinnovation.net/. Accessed 17 Aug 2022
27. Beck, C.T.: Mothering multiples: a meta-synthesis of qualitative research. Am. J. Child Maternal/Child Nurs. **27**(4), 214–221 (2002)
28. Probst, G.J., Gomez, P.: Vernetztes Denken: Unternehmen ganzheitlich führen. Gabler, Wiesbaden (1989)
29. Xiao, Y., Watson, M.: Guidance on conducting a systematic literature review. J. Plan. Educ. Res. **39**(1), 93–112 (2019)
30. Liu, R., Bhattacharya, K., Wu, F.Y.: Modeling business contexture and behavior using business artifacts. In: King, R. (ed.) Notes on Numerical Fluid Mechanics and Multidisciplinary Design Ser, vol 141: Active Flow and Combustion Control 2018, pp. 324–339. Springer International Publishing AG, Cham (2019)
31. Patzak, G.: Systemtechnik: Planung Komplexer Innovativer Systeme. Grundlagen, Methoden, Techniken. Springer, Berlin, Heidelberg (1982)
32. Khan, M.A., Wuest, T.: Towards a framework to design upgradable product service systems. Procedia CIRP **78**, 400–405 (2018)
33. Cebon, P., Hauptman, O., Shekhar, C.: Product modularity and the product life cycle: new dynamics in the interactions of product and process technologies. IJTM **42**(4), 365–386 (2008)

34. Alabadi, M., Habbal, A., Wei, X.: Industrial internet of things: requirements, architecture, challenges, and future research directions. IEEE Access **10**, 66374–66400 (2022)
35. Bartezzaghi, E., Corso, M., Verganti, R.: Managing knowledge in continuous product innovation. In: 5th International Product Development Management Conference, pp. 1–15 (1998)
36. Lianto, B., Dachyar, M., Soemardi, T.P.: Modelling the continuous innovation capability enablers in Indonesia's manufacturing industry. JM2, **17**(1), 66–99 (2022)
37. Chierici, E., Copani, G.: Remanufacturing with upgrade PSS for new sustainable business models. Procedia CIRP **47**, 531–536 (2016)
38. van 't Spijker, A.: Portfolio Management: COIN - Continuous Innovation Framework
39. Ebi, M.: Implementierung von Subskriptionsmodellen im Maschinen- und Anlagenbau, 1st edn. Apprimus Verlag, Aachen (2021)
40. Raddats, C., Burton, J., Ashman, R.: Resource configurations for services success. J. Serv. Manag. **26**(1), 97–116 (2015)
41. Denning, S.: Reinventing management: the practices that enable continuous innovation. Strategy Leadersh. **39**(3), 16–24 (2011)
42. Meyers, D.C., Durlak, J.A., Wandersman, A.: The quality implementation framework: a synthesis of critical steps in the implementation process. Am. J. Commun. Psychol. **50**(3–4), 462–480 (2012)

Planning of Hybrid Portfolios for Industrial Solution Providers in Machinery Engineering

C. Boßmann[1(✉)], M. Kuhn[1], M. Riesener[1], and G. Schuh[1,2]

[1] Laboratory for Machine Tools and Production Engineering (WZL) of RWTH Aachen University, Campus-Boulevard 20, 52074 Aachen, Germany
c.bossmann@wzl.rwth-aachen.de

[2] Fraunhofer Institute for Production Technology IPT, Steinbachstraße 17, 52074 Aachen, Germany

Abstract. Increasingly difficult technical differentiation from the competition and growing demand for integrated customized solutions require companies to make strategic adjustments to their portfolios and business models. Due to the increasing possibilities of service and software applications in the provision of solutions, a holistic view of the portfolio is required, in contrast to the widespread focus on physical products. The transformation into a solution provider can only succeed through systematic combinations of product elements with value-added services to create customized solutions. These contribute to longer, more frequent and more intensive use of the physical product, as this is tailored specifically to customer requirements, and thus also to better resource utilization through a longer life cycle. Therefore, this paper presents a methodology that enables the planning of hybrid portfolios consisting of products and value-added services in machinery engineering. Furthermore, a procedure is described how customized solutions can be created under complexity-reducing aspects.

Keywords: Portfolio Management · Solution Provider · Servitization

1 Introduction

In the current economic environment, companies in the machinery engineering sector are mastering numerous challenges in order to remain successful in the long term [1]. In addition to increasing regulatory requirements in terms of collaboration in the digital era as well as doing business in general and their uncertain developments, it is also important not to lose the competitive advantage in the value creation system [2]. However, it is becoming increasingly difficult to differentiate oneself from competing companies in this respect [3]. Among other things, this is due to better technology availability, general technological progress in the global market and the resulting increase in the number of competitors, some of whom are able to realize significantly lower prices due to different cost structures [4]. As a consequence of this development, products are emerging that are becoming increasingly similar in terms of technology and quality. As a result of the larger offer of similar products, a so-called buyer's market arises, in which prices are determined more strongly by the customer and the importance of the price as a purchase-deciding criterion increases [5].

T. Bauernhansl et al. (Eds.): WGP 2023, LNPE, pp. 407–416, 2024.
https://doi.org/10.1007/978-3-031-47394-4_40

Nevertheless, the buyer's market also offers many opportunities for future competitiveness, since customers are increasingly demanding individual solutions to their individual problems [6]. The solution provider must act as a single point of contact for the customer. Close customer contact enables the compilation of the necessary components of the solutions in line with the customer's requirements [7]. It is irrelevant whether the components can be provided entirely by the company or whether they are provided by partners in a value network [8]. However, the increased use of services and software applications, which can be characterized as value-added services, remains a necessary prerequisite [9]. Only when these are combined with the physical products is it possible to create the customized solutions that are ultimately responsible for customer value. Furthermore, the business model will be changed from a transactional to a service-based model, such as the subscription model or the pay-per-use model, in order to embed the given value proposition of a solution provider for both the customer and the provider [10]. The transformation to a solution provider thus encompasses all process-related changes in a manufacturing company that must inevitably be addressed along the value creation process. In particular, this also includes adapting the portfolio of all products and value-added services [11]. The increasing importance of value-added services for the provision of solutions will inevitably lead to a further increase in their number (see Fig. 1). In order to prevent an excessive increase in complexity in the portfolio, a systematic approach is required at this point, which derives technical functions for the respective customer requirements and decides how these can be implemented. This offers the potential to reduce product complexity in the portfolio [11, 12].

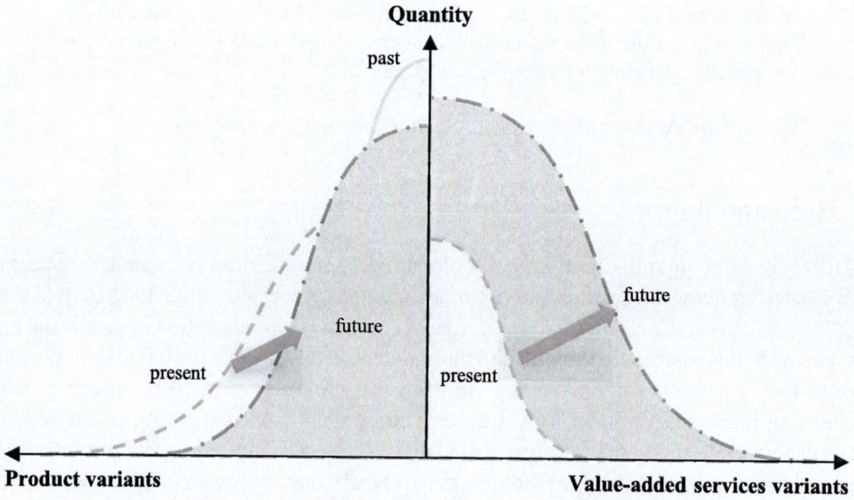

Fig. 1. Changes in variance of products and value-added services

However, the transformation to a solution provider not only contributes to increasing the differentiation potential of companies, intensifying customer relationships or opening up new sources of revenue, but it also supports compliance with sustainability targets and the enabling of a circular economy in the manufacturing industry [13]. By tailoring solutions to the individual customer, which can be adapted to changing needs in the course of use, an intensified user experience is generated with an associated increase in value. The physical product can be used longer and more frequently by the customer, since the value-added services and the responsibility for productivity creation transferred to the manufacturer guarantee constant availability. In addition, the product can be used more intensively, since it has been selected precisely for the customer's needs [14]. Thus, the product life cycle is extended, e.g. by upgrading some physical components, repairs and downtimes are prevented by the intensive monitoring and service, and additional revenue potentials are generated in the context of the circular economy [15].

The aim of this paper is to present a methodology for planning the necessary hybrid portfolios consisting of products and value-added services in order to enable solution providers in the machinery engineering sector to create a competitive portfolio. For this purpose, after the introduction, necessary basics are explained as well as existing approaches from the scientific literature are presented, evaluated and the literature deficit is derived. This is followed by a detailed description of the methodology and an exemplary application based on a case study. Finally, there is a summary and an outlook on further research activities in this field. The paper thus follows the research process with theoretical and practical relevance according to [16]. The process is characterized in particular by the identification of problems relevant in practice, the investigation of existing scientific approaches, the derivation of rules for design and the return of the results to practice.

2 Related Work

In this section, related scientific work is examined. Selected exemplary approaches are presented briefly and evaluated according to their suitability to contribute substantially to the portfolio planning of industrial solution providers.

[17] present a methodology for the implementation of subscription models in machinery and plant engineering. In addition to a description of the characteristics of a subscription model in machinery and plant engineering, activities for implementation are identified. Furthermore, a process for implementing subscription models in machinery and plant engineering is defined. This encompasses an initial assessment of possible risks and benefits of an implementation of subscription business models. In the approach of [18], a concept for the application of complexity management mechanisms in the process of servitization is presented. A theory adaptation approach is used for this purpose in order to identify synergies between complexity reduction and absorption. In addition to the identification and classification of complexity management mechanisms, dedicated design recommendations are derived at a management level in order to improve the servitization process from the perspective of complexity management. [19] present an approach which derives different trajectories for companies that want to start the process of servitization under the prerequisite of offering product, service and software elements. The subdivision of the solution portfolio into product, service, and software systems allows profound classification of the portfolio in an underlying business strategy. Regarding the interdependencies of the elements among themselves, eight norm strategies are identified, which were substantiated by case studies. This approach

enables companies to select the most suitable path for themselves. The methodology of [20] presents an approach for the digitalization of products in which a special focus is placed on the digital elements to enable servitization. The aim is to develop a digitalization roadmap for the further development of the portfolio on the basis of the three design dimensions of product, service and business model using generic innovation principles. Thus, the methodology takes into account the current digitization of the products, possible future scenarios and corresponding digitization strategies. [21] present an approach to updateable and reconfigurable product-service systems in order to enable customization and individualization. To this end, the requirements for the product structure regarding the modular design as well as for the business model regarding the use instead of the possession of the product are first established and operationalized. After articulating the basic requirements, the structure of the product is described and the interdependencies between the components are evaluated. A functional model of the product is then created. In this model, those functions are identified, which are to be realized by updateable product elements. Finally, a simulation is carried out to verify the established system.

[22] present a methodology how global companies can optimize their product portfolio management in the context of converting very early product ideas and heterogeneous customer requirements into marketable products. Firstly, a framework is presented how this process can be designed over the life cycle of the product. Central in this step is the development of a structure in which service and software components can be integrated in addition to physical product components. In the second step of the methodology, the goals of product portfolio management are monitored with KPIs. In contrast, the approach of [23] presents activities for the portfolio management of solutions. The fields of action of solution portfolio management are divided into alignment with corporate strategy, portfolio balancing and value maximization. The focus is on a strategic, dynamic portfolio development, which strives for a balance between customer interests and internal efficiency. For this purpose, the objectives of portfolio management are first operationalized and decomposed into sub-objectives. Subsequently, the relevance of these goals is examined on the basis of exemplary case studies before individual activities are determined for each objective. In the article of [24], a methodology is presented for merging the portfolios of different companies into one portfolio. This is particularly relevant in the context of the transformation into an industrial solution provider, as it enables different companies to contribute their competencies as well as their services and thus increase the overall customer value of an integrated solution. To this end, the authors follow a five-step procedure from a uniform description and structuring, enabling commonalities, identifying system solutions and composing a portfolio from the required components based on the customer benefits and the system solutions. [25] address portfolio management in the context of New Service Development. In the course of this, they conduct a practical study to investigate the handling of New Service Portfolios in industry. Findings in the study are classified into the dimensions of portfolio management goals, processes, and tools. While the portfolio management goals differ only slightly between product and service development, the decision process regarding service development is much more complex. This distinction can also be applied to the use of tools in service development.

In conclusion, there are many approaches in the areas of the transformation into industrial solution providers and portfolio management for companies in the sector of machinery engineering, some of which have been presented as examples. It is evident from the presented approaches that the scientific literature acknowledged the shift to selling individual solutions instead of products, the rising importance of value added services for realizing individual solutions and the advantages of service-based business models in machinery engineering. However, there are also certain deficits in the existing scientific work. Many approaches for portfolio management do not focus on solutions consisting of products and value-added services. Furthermore, they do not consider the entire portfolio including every element a company offers to their customers. In addition, the existing approaches do not take the specific challenges of companies in the sector of machinery engineering into account. In the applied procedures of the approaches it should be stated that a evaluation of the existing portfolio is usually neglected. Furthermore, a systematic planning of the necessary variance in the portfolio is not conducted. This likewise applies to the strategic linkage of products and value-added services to create integrated solutions. Finally, it should be noted that existing approaches to portfolio planning neglect the premise of the portfolio's conformity to company targets. A few approaches can be identified, which cover individual aspects, such as an examination of portfolio and corporate goals, show. However, the majority of approaches have a singular focus on individual sub aspects, such as solutions or portfolio management. However, no approach focuses on an integrated addressing of the deficits outlined above. As a consequence, the approaches presented here do not adequately address central aspects of the research project. The identified deficits in the related work constitute the research demand addressed by the presented research project.

3 Methodology

The methodology for planning portfolios of solution providers in machinery engineering consists of four sub-steps. First, a procedure is described for determining the company-specific targets of a solution provider. Then, relevant characteristics for structuring the portfolio are identified. In the third step, solution packages consisting of products and value-added services are to be derived by means of a solution staggering, which are then to be anchored in the portfolio in the fourth step.

3.1 Identification of Company-Specific Targets for the Portfolio of Solution Providers

The planning of the portfolio of a solution provider should be in line with the respective company-specific targets in order to enable long-term success in the transformation to a solution provider. Since the targets of the companies cannot be generalized, a company-specific selection of possible targets must be viable. In order to aggregate a selection of possible targets for a solution provider's portfolio, corresponding target dimensions were first determined. In the context of portfolio management, the target dimensions according to [26] are still relevant today: value of the portfolio, balance within the portfolio, strategic fit of the portfolio, and allocation of portfolio-relevant resources. Nevertheless, the changed economic environment and the expected changes in the next years have to be taken into account, especially by integrating the target dimension sustainability fit of the portfolio. Through an analytical-deductive analysis of relevant scientific literature and an empirical-inductive analysis of existing case studies, corresponding targets were extracted and aggregated. An excerpt of this is shown in Fig. 2.

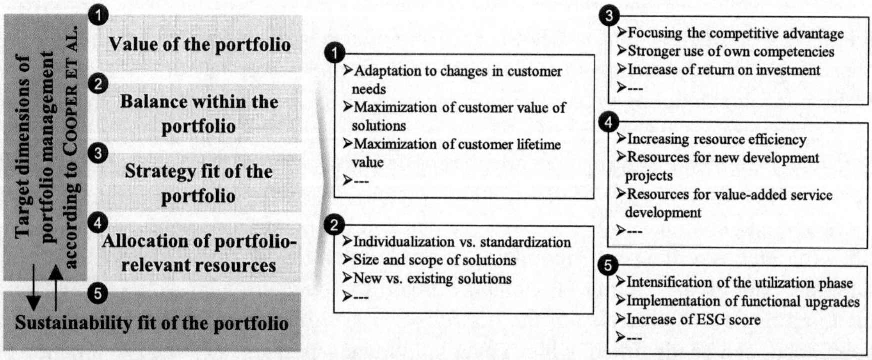

Fig. 2. Target dimensions and targets for solution providers in machinery engineering

Since not all of the identified targets are relevant to every company to the exact same degree, the next step is to create a company-specific target system. To this end, the targets are first weighted and then assigned to the internal or external perspective of the company. In step four, this allows a selection of the solutions that are valuable from both the company's and the customer's perspective. This division and the weightings result in the company-specific target system.

3.2 Identification of Relevant Characteristics for Structuring the Portfolio

For the planning of the portfolio and in particular the determination of the necessary portfolio elements for the realization of the intended value propositions, relevant characteristics for the description and structuring of the portfolio must first be identified. Since a solution provider must aim for a high level of customer focus to solve their problems [11], the determination of the features must also be conducted with a sufficiently high level of customer participation. This ultimately ensures that the customer problems can be abstracted and incorporated into the portfolio planning. To determine the characteristics for structuring the portfolio, the Value Proposition Canvas according to [27] can be used. With the participation of lead users and target customers, the gains and pains in the use of the company's solutions are jointly recorded. A clustering procedure is then used for systematization, so that the recorded pains and gains are aggregated into features. Since the features were only recorded from a purely external customer perspective, a relevance evaluation follows in order to ensure the practicability of these features. For this purpose, the frequency of use of the features in the individual customer groups is determined, the contribution to the productivity of the solutions in terms of performance, availability and quality is determined and the strategic relevance for the company is defined. Using a three-level scale (1, 3, 9) and multiplying the values for each characteristic, an exponentially descending sum is derived for all values in the relevance evaluation. By defining a termination criterion, only a sub-quantity of the identified features is used to structure the portfolio. In addition, it is ensured that the features are meaningful from both the external customer perspective and the internal company perspective. Subsequently, the necessary variance in the portfolio can be determined by defining feature characteristics and the combination of these characteristics.

3.3 Derivation of Solution Packages in Hybrid Portfolios

Following the determination of the features for structuring the portfolio and the necessary relevance for realizing the value propositions, the portfolio elements required must then be determined for portfolio planning. To do this, a generic architecture of the solutions in the portfolio is first created. Based on a product architecture, the functions of the solutions are defined and systematically documented in a functional structure [28]. The existing product structure is then extended to include the elements of the value-added services and linked to the corresponding functions. This architecture then indicates which elements of the solutions are needed to realize the functions.

However, portfolio planning requires a subsequent step, because the generic solution architecture does not provide an overview of the existing variance in the products or value-added services, nor of the variance at the level of the individual solution elements. To plan the portfolio at the product and value-added service level, the relationships between the solution elements and the features of the portfolio must first be determined. Since the relationships between functions and features can be easily determined [8], the corresponding solution elements for each feature can also be identified via the generic solution architecture that has been created. Then, these insights form the basis for the following solution staggering. For this purpose, all technically possible combinations of product and value-added service elements are determined for each combination of feature value and function. In particular, a focus has to be placed on the potential of value-added services to realize the functions, as this helps to achieve the necessary complexity reduction in the portfolio mentioned above [8]. This investigation systematically identifies potentials for complexity reduction, which have a positive effect on the company's monetary goals, since the range of solutions is not limited. The next step is to aggregate this investigation at the level of products and value-added services for staggering offers. Taking into account the restrictions of the required variance and the already existing portfolio, corresponding packages of product and value-added services can be derived for realization for each required solution variant. Furthermore, relevant key variants of the products must be determined at this point, and attention must be paid to reasonable differentiation between the individual solutions.

3.4 Consolidation of the Results in the Portfolio Planning

Following the solution staggering and the associated evaluation of the technical realization, the final step is to assess the target conformity of the specific packages. As only packages with a high level of target conformity should be included in the portfolio, this contributes to the company achieving its overall targets more effectively. First, the internal and external target conformity is evaluated for each package by determining a corresponding value for each target from the target system established in the first step. By multiplying these values with the corresponding weightings, one value each for internal and external target conformity can then be indicated. By visualizing these values in a two-axis portfolio representation, it becomes clear which package has the better values for target conformity for each solution to be offered and is therefore to be preferred. The elements of the other packages are no longer required for the solution provision. This results in a streamlining of the portfolio and a reduction in complexity, but without the loss of target-oriented variance for the customer. Besides the differentiation in the individual packages, however, the overall classification should also be checked, since packages or solutions that have both a low internal and a low external target conformity should not be included in the portfolio. In the case of solutions for which only one target conformity is to be rated as high, this decision must be made selectively. The findings from this evaluation must finally be anchored in the portfolio, so that the portfolio is restructured by means of the identified features and is composed of the necessary solution components. Furthermore, the described procedure enables companies to use as many components of the existing portfolio as possible also as a solution provider, thus preventing a complete redevelopment of the portfolio in the best possible way.

4 Conclusion

Companies in the machinery engineering sector increasingly have to deal with the transformation to a solution provider to ensure the long-term success of the company. This is becoming necessary in particular due to the increasingly difficult differentiation by means of purely technical products and the demand for customer-specific individualized solutions. The potential of this transformation and its implications have already been proven in numerous instances. Nevertheless, scientific literature is missing an approach that enables companies planning of the future portfolio. The objective should not be to develop all solutions from new, but to use existing solution components if possible. To this end, this paper presents a methodology for portfolio planning that takes into account company-specific goals, a reduction in complexity, and consideration from both an external and internal perspective.

The further detailing of the methodology and in particular the interrelationships within the solution staggering are the subject of the authors' current research. Limitations of this paper are set in the investigation of the influence of partners from the value network on the provision of solutions. Furthermore, especially in the case of a large portfolio and its numerous possible combinations, the application of an automated solution staggering has to be examined.

Acknowledgment. Funded by the Deutsche Forschungsgemeinschaft (DFG, German Research Foundation) – 513121577.

References

1. Arlinghaus J., Antons, O.: Management for digitalization and industry 4.0. In: Frenz, W. (ed.) Handbook Industry 4.0: Law, Technology, Society, pp. 927–948. Springer, Berlin (2022). https://doi.org/10.1007/978-3-662-64448-5_49
2. Frenz, W.: Industry 4.0 and Competition Law. In: Frenz, W. (ed.) Handbook Industry 4.0: Law, Technology, Society, pp. 275–288. Springer, Berlin (2022). https://doi.org/10.1007/978-3-662-64448-5_14
3. Gassmann, O., Ferrandina, F.: Die win-win-win-formel. Harvard Bus. Manager **6**, 1–11 (2021)
4. Schuh, G., Dölle, C.: Sustainable Innovation. Springer, Berlin (2021)
5. Schuh, G., et al.: Geschäftsmodell-Innovation. In: Reinhart, G. (ed.) Handbuch Industrie 4.0, pp. 3–30. Hanser, München (2017)
6. Wilken, R., Jacob, F.: Vom Produkt- zum Lösungsanbieter. In: Backhaus, K., Voeth, M. (eds.) Handbuch Business-to-Business-Marketing, pp. 147–164. Springer, Wiesbaden (2015)
7. Huikkola, T., Kohtamäki, M., Ylimäki, J.: Becoming a smart solution provider: reconfiguring a product manufacturer's strategic capabilities and processes to facilitate business model innovation. Technovation **118**, 1–13 (2022)
8. Kreß, J.M.: Erbringung hybrider Leistungsangebote in Wertschöpfungsnetzwerken. Dissertation (2022)
9. Thomassen, M.K., Alfnes, E.: Mass customization challenges of engineer-to-order manufacturing. In: Bellemare, J., Carrier, S., Nielsen, K., Piller, F.T. (eds.) Managing Complexity, pp. 27–40. Springer, Cham (2017)
10. Tukker, A.: Eight types of product–service system: eight ways to sustainability? Experiences from SusProNet. Bus. Strat. Env. **13**(4), 246–260 (2004)
11. Forschungsbeirat der Plattform Industrie 4.0/acatech - Deutsche Akademie der Technikwissenschaften, Ed., "Themenfelder Industrie 4.0 - Forschungs- und Entwicklungsbedarfe für die erfolgreiche Umsetzung von Industrie 4.0 (2022)
12. Riesener, M., Kuhn, M., Boßmann, C., Schuh, G.: Concept for the portfolio management of industrial solution providers in machinery and plant engineering. Procedia CIRP **33**, 1–6 (2023)
13. Schmitt, R., Bergs, T., Brecher, C., Schuh, G. (eds.) Empower Green Production (2023)
14. Schuh, G., Schmitz, S., Lukas, G., Niwar, L., Welsing, M., Calchera, R.: Framework for circular production economy. In: Schmitt, R., Bergs, T., Brecher, C., Schuh, G. (eds.) Empower Green Production, Aachen, pp. 286–313 (2023)
15. Riesener, M., Kuhn, M., Perau, S., Mertens, M., Schuh, G.: Fields of action for the realization of continuous innovation. In: 10th International Conference on Mass Customization and Personalization (MCP - CE 2022), pp. 138–146 (2022)
16. Ulrich, P., Hill, W.: Wissenschaftstheoretische Grundfragen der Wirtschaftswissenschaften. Vahlen, München (1979)
17. Riesener, M., Dölle, C., Ebi, M., Perau, S.: Methodology for the implementation of subscription models in machinery and plant engineering. Procedia CIRP **90**, 730–735 (2020)
18. Eloranta, V., Ardolino, M., Saccani, N.: A complexity management approach to servitization: the role of digital platforms. Int. J. Oper. Prod. Manag. **41**(5), 622–644 (2021)
19. Hsuan, J., Jovanovic, M., Clemente, D.H.: Exploring digital servitization trajectories within product–service–software space. Int. J. Oper. Prod. Manag. **41**(5), 598–621 (2021)
20. Echterfeld, J., Gausemeier, J.: Digitising product portfolios. Int. J. Innov. Manag. **22**(5), 1–27 (2018)
21. Marques, M., Agostinho, C., Zacharewicz, G., Jardim-Goncalves, R.: Reconfigurable and updatable product-service systems. In: Mod4Sim 2017: Proceedings of the Symposium on Model-driven Approaches for Simulation Engineering, pp. 1–12 (2017)

22. Lahtinen, N., Mustonen, E., Harkonen, J.: Commercial and technical productization for fact-based product portfolio management over lifecycle. IEEE Trans. Eng. Manage. **68**(6), 1826–1838 (2021)
23. Hakanen, T., Jähi, M.: Central activities of solution portfolio management. Int. J. Serv. Technol. Manage. **27**(1/2), 104–128 (2021)
24. Mustonen, E., Tolonen, A., Härkönen, J., Haapaslo, H.: Constructing a joint product portfolio to support commercial cooperation. Int. J. Manag. Knowl. Learn. **8**(1), 3–23 (2019)
25. Aas, T.H., Breunig, K.J., Hydle, K.M.: Exploring new service portfolio management. Int. J. Innov. Mgt., **21**(6) (2017)
26. Cooper, R.G., Edgett, S.J., Kleinschmidt, E.J.: Portfolio Management for New Products. Perseus, Cambridge (2001)
27. Osterwalder, A., Pigneur, Y., Bernarda, G., Smith, A.: Value Proposition Design. Wiley, Hoboken (2014)
28. Riesener, M., Dölle, C., Kreß, J., Boßmann, C., Schuh, G.: Methodology for structuring and bundling product-service systems. In: DS 112: Proceedings of the 23rd International DSM Conference (DSM 2021), pp. 1–10 (2021)

Evaluation of the Cold Atmospheric Plasma Metallization of Bare Dies with Copper Through Life Cycle Assessment Through Life Cycle Assessment

Manuela Ockel(✉), Felix Funk, Lucas Janisch, and Jörg Franke

Friedrich-Alexander University Erlangen-Nuremberg, Erlangen, Germany
manuela.ockel@faps.fau.de

Abstract. The cold atmospheric plasma metallization (CAPM) with copper is an alternative coating process to electroplating for applying a thermo-mechanical buffer onto bare dies to enable more powerful top-side interconnection technologies for power electronics. A complete life cycle assessment (LCA) is to be performed in order to evaluate improvements to increase sustainability of the copper metallization. Since life cycle assessment results strongly depends on the set system boundaries and availability of primary data, the scope of this study is determined by the availability of the latter. Since the coating properties are dependent on the plasma spray parameters, a variation of set parameters is assessed, which are currently used for the coating of the bare dies. The evaluation showed that electricity, plasma gas and powder production have major impacts on the sustainability. The necessary heat treatment is responsible for on third of the emissions if bare die production and plasma coating are included into the assessment. Potential improvements are the reduction of total conveyed powder and carrier gas, as well as increasing the number of bare dies treated in one annealing process.

Keywords: Cold Atmospheric Plasma Metallization · Life Cycle Assessment · Power Electronics · Copper

1 Introduction

In order to keep up with the trend towards electromobility the development of power modules must be intensified. New packaging and interconnection technologies are necessary to fulfill the increased requirements regarding efficiency, reliability and performance. Additionally, with the introduction of wide band gap semiconductors, the surrounding connection technologies should not be the limiting factor for their application in power modules. This translates in the substitution of materials like aluminum (Al) through copper (Cu), in order to ensure the functionality at higher temperatures and harsh environments. By substituting Al with Cu several challenges occur. In order to apply top-side-interconnections like heavy Cu wire bonds through ultrasonic wire bonding or leadframes via laser bonding, the top-side of the bare die, which is typically provided

T. Bauernhansl et al. (Eds.): WGP 2023, LNPE, pp. 417–427, 2024.
https://doi.org/10.1007/978-3-031-47394-4_41

with an Al metallization, must be coated with Cu. The Cu layers serve as a buffer for thermal or mechanical forces applied during the interconnection process and prevent the destruction of the semiconductor. They also enhance lifetime by evenly distributing heat during operation across the bare die surface. The Cu metallization can be applied by several technologies (e.g. electroplating, foil sintering) which result in different, partially complex, materials setups and process chains [1].

The plasma spray coating represents an alternative metallization technology (see Fig. 1), provided the coating properties are of suitable quality, to the already industrialized Cu coatings. The main advantage is the short process chain and currently simple material setup.

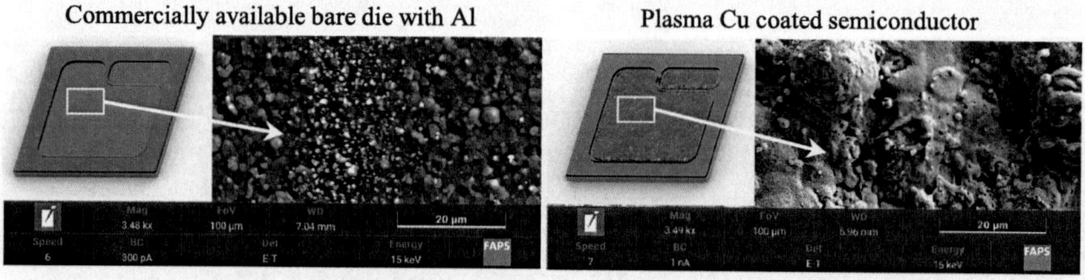

Fig. 1. Surface analysis of Al metallization and plasma Cu metallization

[2] indicates, that thermal spray processes may not be as sustainable as galvanic electroplating, even when no wet chemical etching procedures are used. In order to assess the cold atmospheric plasma metallization on commercially available bare dies, life cycle assessment (LCA) is used as a methodology in this paper. It serves as a bases for the comparison of cold atmospheric plasma spray processes with other plasma spray processes. LCA is an acknowledged methodology to quantify environmental impacts (e.g. regarding climate change, human health and ecotoxicity) of products along their life cycle and thus enables the informed development of more sustainable production practices [3]. Nevertheless, the results of LCAs heavily depend on the chosen system boundaries, which makes it difficult to ensure comparability between different production processes. To create a foundation for further comparisons and analysis a first life cycle assessment of cold atmospheric plasma spray process is performed and discussed in this paper.

2 Basics of Plasma Sprays

Thermal spray processes are characterized by high energy consumption and harmful gas and particle emissions directly into the atmosphere, which is challenging for improving the eco-friendliness of the process. Additionally, the gases cause harm to the environment during their production, powder production is energy intensive and powder remainders are difficult to recover, besides the changed particle properties after they have been injected into a heat source [4]. Since the coating geometry (thickness and covered area) and material properties directly correlate with energy consumption and deposition time the required electricity is strongly dependent on the chosen parameters, which makes optimizations by reducing energy and gas consumption difficult. Accordingly, an improved deposition efficiency shows the most potential for reducing the environmental impact [5]. The main influencing factors on deposition efficiency are depicted in the following chapter.

2.1 Cold Atmospheric Plasma Spray Metallization

Cold atmospheric plasma sprays are thermal spray processes where a source of heat, e.g. an arc produced by a direct current between cathode and anode, ionizes a gas which melts the material in-flight an propels the particles to the workpiece. The plasma is an ionized gas with the same number of ions and electrons (electrically neutral) and its properties are defined by the used plasma forming gases. Typical primary gases are heavy gases like Argon (Ar) and Nitrogen (N_2) for flow and particle entrainment and secondary gases like Hydrogen (H_2) and Helium (He) for heat transfer. Plasma based on noble gases are classified as cold and fast (direct ionization $Ar \rightarrow Ar^+$), whereas plasma gases based on molecules are slow and hot (two stages: first dissociation $N_2 \rightarrow 2N$, later ionization $2N \rightarrow 2N^+$) [6]. Typical plasma spray parameters are significantly higher than the parameters used in the low voltage plasma spray technology which is evaluated in this paper (see Table 1) [7 9].

Table 1. Parameter comparison of typical plasma sprays and the adapted plasma spray process

	Flow rate [g/min]	Powder flow rate [kg/h]	Particle velocity [mm/s]	Power [kW]
Typical plasma spray	0.8–2	3–6	300–1500	20–80
Adapted plasma spray	0.5	0.4	150–200	8

The different technologies, dependent on the used plasma gas, are combinations of Ar or N_2 as a primary heavy gas with at least 80% and He, H_2 or a combination of He and H_2 [9]. One of the main advantages of cold spray is usually the limited oxidation of the melted particles during flight (<100 ppm for Cu) [7]. To enhance adhesion on the bare dies, the process has been modified which leads to further oxidation of the Cu particles and reduced electrical conductivity [10].

2.2 Adaption for Copper Coating of Bare Dies

The cold atmospheric plasma metallization (CAPM) technology has been adapted for the application of Cu coatings in previous works. Firstly, the applied voltage is limited to 24 V in order to keep the electrical properties of the chip. The coating parameters have been adjusted for low particle velocities in order to prevent abrasive effects on the surface while still applying enough energy to melt the Cu particles with an average grain size of 12 μm. The substrate is preheated in order to minimize the temperature difference between the surface and the arriving particles. The spherical Cu particles solidify as a splat. The mechanical stress within the substrate is reduced by increasing the substrate temperature and thereby prolonging the time until solidification. Hotter coating parameters, which result in higher gas flow rates and energy consumption, may induce higher thermo-mechanical stresses, while improving the adhesion of the particles significantly. The exact interaction between the coating parameters and coated bare dies is still under research. Additionally, the preheating of the substrates leads to a strong oxidation of the Cu coating. In order to achieve electrical conductivities comparable to pure Cu, an annealing procedure under a CuO reducing N_2H_2-atmosphere is performed, which also alters the coating properties by reducing the Cu hardness, increasing its density and promoting a diffusion layer between Cu and Al, if a semiconductor with an aluminum surface finish has been used. Since CAPM is used for covering large areas, the desired coating geometry is depicted by a laser cut steal mask [1, 8, 10] Klicken oder tippen Sie hier, um Text einzugeben. This enables the coating of individual chips as well as a possible coating of complete wafers.

3 Life Cycle Assessment

Life cycle assessment provides a framework for identifying, comparing and improving environmental impacts associated with the product and the production processes and is defined by the ISO-standard 14040-44 [11]. The goal of this study is the compilation of relevant energy and material input flows as well as environmental flows (life cycle inventory) and the calculation of several impact factors of the cold atmospheric plasma spraying used for the Cu coating of bare dies as a baseline to identify optimization potentials regarding environmental friendliness.

3.1 Goal and Scope Definition

The experiment aims to determine the main environmental impacts. Therefore, three of the most important and variable steps are assessed: resource consumption during plasma coating (variation of standard parameters and gases), post-treatment (soft annealing) and the efficiency dependent on lot size. According to DIN EN ISO 14040 the goal of the study has to be defined first, system boundaries are determined after that. An overview of the process including the chosen boundaries is depicted in see Fig. 1. It includes the consumables that are needed for the plasma spray process. Periphery, wear materials, transport and storage are not considered, since the evaluation is focusing on the variable parameters. The flows are separated into input, output and production sequences or materials. The outputs are classified as emission to air, product for waste treatment and products for further use. According to the set system boundaries the assessment is classified as cradle-to-gate. The functional unit is a Cu coating on a 100 cm^2 surface with 60 μm coating thickness which results in a volume of 600 mm^3. This represents the applicated metallization after repeating the coating sequence three times.

3.2 Data Collection of Inputs and Outputs

The experiment was conducted in a direct current plasma coating unit “PlasmaCoat PCU3D” with Ar as a primary, ArH_2 as a secondary plasma gas and an ignition voltage of 50 V. The 60 μm thick coating is based on Cu powder and is applied onto preheated commercially available silicon bare dies with a 34 mm^2 closing Al metallization. For heat treatment and CuO-reduction the tube furnace ©Gero Carbolite GHA 12/300 with a N_2H_2 atmosphere was used. The average powder consumption is determined offline via precision scale and is set at 3 g/min for one vibration conveyor.

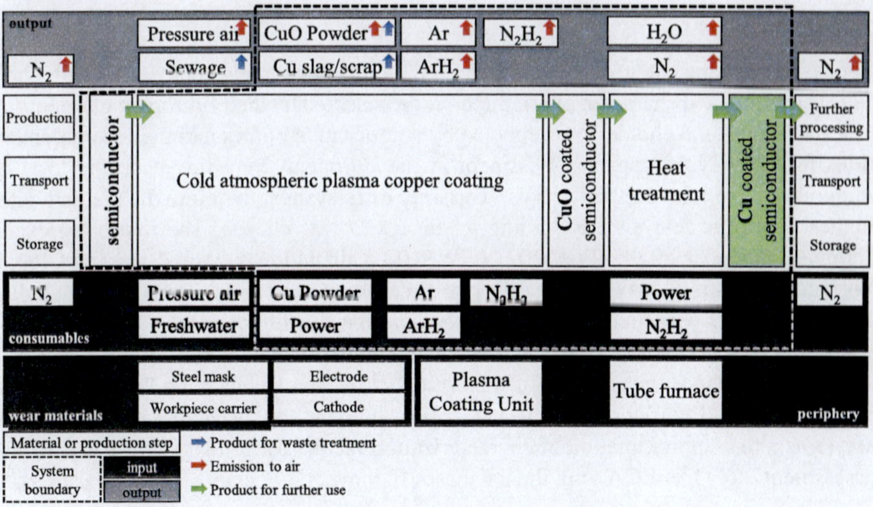

Fig. 2. Overview of production steps and materials with chosen system boundaries

The system boundaries of Fig. 2 serve as a reference for the data collection of input and output products and are chosen based on the large availability of primary data. The ingoing power, gas and powder consumption are measured. The effective power P_{eff} [W] of the plasma coating unit is measured via the energy measuring kit ©Sentron Pac 4200, which is positioned between the main power supply lane and the power grid. The power consumption of the plasma burner unit is monitored by the integrated true voltage sensor system and current logging. The total power consumption of the tube furnace is roughly calculated based on the annealing parameters and is estimated around 16.7 MJ. The power consumption during one coating and annealing procedure is depicted in Fig. 2 and shows the varying power consumption dependent on the production step.

The gas consumption is measured via integrated flow measurements (integrated thermal mass flow meters) for the plasma gases (Ar and ArH_2), carrier gas of conveyor unit one and two (Ar) and shroud gas (N_2H_2), resulting in Q_{gas} [l/min]. A typical ratio and graph progression of the gases, dependent on the production step, is depicted in Fig. 3. The gas consumption of the furnace oven is computed from the gas cylinder’s pressure before and after the soft annealing process and average flow rate during recorded during annealing, which results in a total gas consumption of 2.57 kg N_2H_2. For the life cycle inventory the logged power and gas consumption are integrated over the production time (see 1 and 2), resulting in the total energy consumption E_t [MJ] with effective Power P_{eff} and total gas consumption $m_{gas\,t}$ [g] with the logged gas consumption Q_{gas} and gas dependent density ρ_{gas}.

$$P_t = \int_{t_{start}}^{t_{end}} P_{eff} \tag{1}$$

$$m_{gas\,t} = \left(\int_{t_{start}}^{t_{end}} Q_{gas} \right) \cdot \rho_{gas} \cdot 10^3 \tag{2}$$

3.3 Results

For assessment the categories climate change, human health and ecotoxicity were chosen. The associated characterization methods are selected by their fulfilment of scientific criteria, overall stakeholders' acceptance by authoritative bodies and the recommendations for Life Cycle Impact Assessment in the European context. According to [12] midpoint or endpoint methods vary in certainty or relevance, therefore the method with highest scientific relevance according to the ILCD was chosen. The first category is "climate change" with the IPCC 2021 (GWP100) calculation method. Based on Global Warming Potentials (GWP) with different timeframes this midpoint method includes 20-, 100- or 500-year timeframes. Since the Kyoto protocol applies the 100-year timeframe, the IPCC 2021 is classified as a level I method. The method ReCiPe was selected, since it uses the most recent state-of-the-art model for human health on European scale. The method UseTox is applied in the categories "human toxicity" and "ecotoxicity" and serves as a first approximation of characterization factors for primary PM10 [3]. As an assessment tool OpenLCA with the databases Ecoinvent 391 cutoff upr N2 was used.

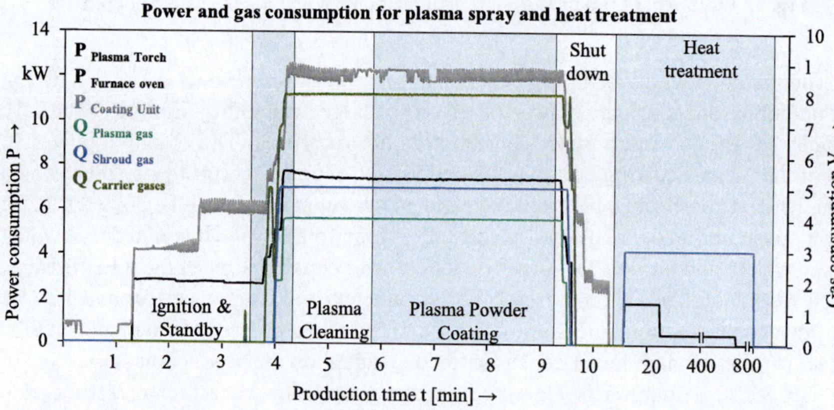

Fig. 3. Power and gas consumption of one sequence of plasma spray and heat treatment

As an overview the GWP impact contributions [kg CO_2eq] of each production step are compared dependent on the manufactured lot size. The main processes are the manufacturing of the bare die (production of surface-mounted diodes), the plasma coating procedure and the controlled atmosphere heat treatment. The coated lot size during one plasma spraying sequence is determined by the maximum area which can be covered without changing the production parameters and results in the simultaneous coating of 196 bare dies. The maximum capacity of the furnace oven is estimated around 400 plasma coated bare dies.

For the production of one single bare die the annealing procedure makes up for almost 100% of the total impact (see Fig. 4). With increase of the number of bare dies annealed in one sequence the ratio significantly improves but is still responsible for most of the emissions. Electricity makes up for at least 40% (heat treatment), whereas N_2 usage is responsible for almost 0% to 19% of the impact per semiconductor, if the lot size is decreased. The detailed impact factors of the plasma coating procedure are shown in Fig. 5. Since electricity has already been identified as the main source of CO_2 emission, a comparison between a standard energy mix (total of 1.27 kg CO_2eq) and photovoltaic energy (total of kg CO_2eq) is depicted as well.

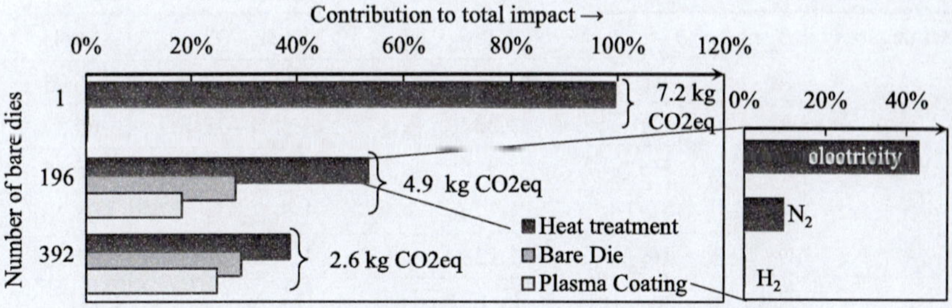

Fig. 4. Percentages of total impact dependent on manufactured lot size

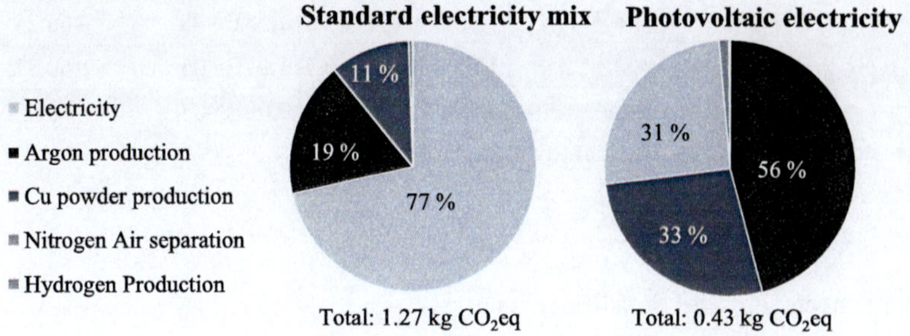

Fig. 5. GWP impact factors [kg CO_2eq] compared between standard and "green" energy

Besides electricity the Ar and Cu powder production are main impact drivers. Ar additionally accounts for the highest proportion of gas consumption by weight during the plasma spray procedure. If green energy is used Ar production is the main driver of the CO_2-emissions and the total foot print is reduced significantly.

The main influencing factors of the plasma spraying sequence, which also alter the coating properties, are assessed in detail by varying the kind of primary plasma gas and the combination of flow rates which results in different particle temperatures and coating properties and are used for different applications (see Table 2). The variation of different kinds of secondary plasma gases hardly changes the impact on climate change (ca. 1.35 kg CO_2eq for Ar as primary plasma gas). Switching to N_2 as a primary plasma gas would reduce the climate change impact by 16.4% percent. With increasing total gas consumption m_{tot} the impact corelates directly. The same pattern is recognizable in the category human health. The only significant improvement would be switch to a N_2-based plasma. The lowest impact on human health has experiment 2. The calculations for ecotoxicity show an improvement of the impact on ecotoxicity with all plasma gases and parameter combinations.

Table 2. Results of impact analysis dependent on the plasma gas constitution

Variation	Category (method)		Climate change (IPCC 2021)	Human Health (ReCiPe)	Ecotoxcitiy (UseTox)
	Primary	Secondary	kg CO_2eq	DALYs[a]	CTU[b]
Plasma gases	Ar	–	1.385	1.61E−05	−66.05
	Ar	ArH_2	1.358	1.58E−05	−66.25
	Ar	H_2	1.363	1.58E−05	−66.23
	Ar	He	1.355	1.57E−05	−66.35
	Ar	He H_2	1.370	1.58E−05	−66.30
	N_2	H_2	1.128	1.31E−05	−67.78
	Ratio	m_{tot} [g]	kg CO_2eq	DALYs[a]	CTU[b]
Flow rate combination	3:1	124.5	1.00	1.51E−05	−63.50
	3:1	201.7	1.137	1.32E−05	−63.40
	1:1	238.7	1.358	1.58E−05	−66.25
	1:1	342.1	1.694	1.97E−05	−60.57
	5:3	395.4	1.798	2.09E−05	−60.49

[a]disease-adjusted life years [b]comparative toxic unit

3.4 Interpretation and Sensitivity Analysis

There are several limitations and results which need to be discussed and assessed in detail. The sensitivity of the study was assessed by the parameter variation in Table 2. The chosen system boundaries exclude any wear materials and technical equipment, as well as transportation and N_2 usage during storage of the powder and bare dies, which is necessary to prevent oxidation of the material. These factors have to be considered in future analysis. The impact of the bare dies used in this experiment was substituted with the process "diode for surface mounting" which is an already existing process in the Ecoinvent database. The calculated impact may be different dependent on the production process and used metals for potential diffusion barriers (e.g. Ti) and finishing metallization (e.g. Ag, Au, Pd, Al). The impact of the heat treatment is based on an estimated calculation, detailed measurements in the future should provide more reliable data. The different quantities of coated bare dies are extrapolated. The experiments were conducted with four bare dies per coating procedure. The annealing and reduction process of several hundred bare dies may require higher N_2H_2 flow rates in order to ensure proper reduction of the CuO formed during plasma spraying, which would increase the impact due to higher resource consumption. The available data for the Cu powder is limited as well, the provided data is limited to the exhausted CO_2 permission per kg powder (2.7 kg). Since powder production is not yet included into the database the results may deviate strongly. Additionally, the powder properties, which are adjusted by their way of production, could influence the impact as well. The spray process also emits fine Cu particles into the atmosphere, which may strongly influence the category human health. Neither the emission of the process gases nor the emission of fine Cu particles is monitored at the moment. Besides, the database only provides data for N_2 and Ar in a liquidous state – the gases used in production are in a gaseous state. The results of ecotoxicity are to be used with caution. Since the harmful emissions of the process are not monitored, the results of this category have to be verified in further experiments, since they show an overall reduction of the ecotoxicity. This is based on the output processes in the analysis tool, where remaining oxides Cu and Cu scraps are reintroduced into the material cycle.

4 Identification of Optimization Potentials

The focus of this evaluation was to identify potentials of the technology to reduce the environmental impact. Besides using a green energy mix, there are other optimization potentials. Firstly, Ar production is responsible for a large portion of the emissions. A reduction of the overall Ar consumption would reduce the impact. The process curves show that the Ar carrier gas is active during the plasma cleaning procedure, although no powder is conveyed. A removal of the additional Ar would alter the gas constitution in the plasma flame and could change the cleaning effect, but is one option to be evaluated. If stronger plasma spray parameters are required to achieve specific coating properties through higher temperatures – specifically better adhesion – one could differentiate between an initial bonding layer with a higher temperature and application of additional layers with lower temperatures. Since the later also results in less gas consumption the overall use of Ar could be reduced. Secondly, the powder consumption must be optimized. Only one quarter of the total conveyed powder is used to form the Cu coating. The first step is to reduce the powder feeding time by stopping the powder injection between the different layers. Other ways to improve the deposition rate need to be evaluated, since the recovery of remaining Cu particles in the experimental setup is difficult. Powder which is not used to form the coating is collected by a wet scrubber and forms a Cu slag which is currently treated as waste. Additionally, the properties of the particles are altered during the spraying process through oxidation and agglomeration. Figure 5 shows the particles before and after the spraying processes. The particles were collected from different parts of the chamber and show high differences in grain size distribution and surface roughness. Even if unused particles could be collected the powder has to be processed again. Regarding resource depletion this is an option which has to be further evaluated. In consideration of human health and ecotoxicity the emission of process gases and powder particles have to be monitored inside the laboratory and at the outlet, to determine the correct impact.

The heat treatment should be optimized as well. It is mainly used to reduce thermo-mechanical stresses at the interface, alter the material hardness and reduce the CuO to improve electrical conductivity. A minimization of oxidation during the plasma spray coating could reduce the necessary time and temperature for a complete reduction. The oxidation largely takes place during the slower solidification of the particles due to the preheating of the substrate. Lower preheating temperatures should lead to less reduction. If the resulting material properties are suitable for further applications has to be evaluated.

5 Conclusion

The evaluation of the plasma spray process and adjacent heat treatment allowed for the identification of several optimization potentials to reduce the total life cycle impact. The impact reduction through parameter variation is limited. But, gaps in the process monitoring were identified, especially regarding the monitoring of process emissions. The assessment showed, that dependent on the available data and set boundaries, the impact calculation differs strongly. In order to allow a comparison between completely different metallization processes to identify the buffer technology with the lowest environmental impact these boundaries have to be considered carefully. Future analysis should also be lined with as much primary data as possible.

Acknowledgements. The work was funded by the German Federal Ministry for Economic Affairs and Energy (BMWi) as part of "FlaMe", project FKZ:03EN4008A-G. The scanning electron microscope is funded by the Deutsche Forschungsgemeinschaft (DFG, German Research Foundation) – 442921285.

References

1. Ockel, M., Hensel, A., Stegmeier, S., Franke, J.: Plasma Powder Copper Coating on silicon substrates for copper wire bonds in comparison to state-of-the-art top-side interconnection technologies. In: Pan Pacific Microelectronics Symposium (2023)
2. Moign, A., Vardelle, A., Legoux, J.G., Themelis, N.J.: LCA comparison of electroplating and other thermal spray processes. In: International Thermal Spray Conference (2009)
3. Sala, S.: International reference life cycle data system (ILCD) handbook: general guide for life cycle assessment: provisions and action steps (EUR, Scientific and technical research series 24571). Publications Office, Luxembourg (2011)
4. Liu, J., et al.: Evaluation and improvement of the greenness of plasma spraying through life cycle assessment and grey relational analysis. Int. J. Life Cycle Assess. **26**(8), 1586–1606 (2021). https://doi.org/10.1007/s11367-021-01910-4
5. Moign, A., Vardelle, A., Themelis, N.J., Legoux, J.G.: Life cycle assessment of using powder and liquid precursors in plasma spraying: the case of yttria-stabilized zirconia. Surf. Coat. Technol. **205**(2), 668–673 (2010). https://doi.org/10.1016/j.surfcoat.2010.07.015
6. Bach, F.-W., Möhwald, K., Laarmann, A.: Moderne Beschichtungsverfahren, 2. Auflage. Wiley (2004)

7. Boulos, M.I., Fauchais, P.L., Pfender, E.: Handbook of Thermal Plasmas. Springer, Cham (2023)
8. Hensel, A., Mueller, M., Franke, J., von Platen, K.K.: Additive copper metallization of semiconductors for enabling a copper wire bonding process. In: 40th International Spring Seminar, pp. 1–6 (2017)
9. Fauchais, P.L., Heberlein, J.V., Boulos, M.I.: Thermal Spray Fundamentals. Springer, Boston (2014)
10. Hensel, A., Stoll, T., Stegmeier, S., Franke, J.: Additive manufacturing for power electronics substrates: comparison of plasma- and laser-based metallization technologies. In: Pan Pacific Microelectronics Symposium (2022)
11. DIN, EN ISO 14040 Umweltmanagement - Ökobilanz - Grundsätze und Rahmenbedingungen
12. Bare, J.C., Hofstetter, P.: Life cycle impact assessment workshop summary midpoints versus endpoints: the sacrifices and benefits (2000)

Energy Reduction in Powder Atomization by Novel Production Process for Recycled Titanium Electrodes

Berend Denkena[1], Benjamin Bergmann[1], Florian Pyzcak[2], Marcus Willi Rackel[2], and Jonas Matthies[1](✉)

[1] Institute of Production Engineering and Machine Tools (IFW), Leibniz University Hannover, An der Universität 2, 30823 Garbsen, Germany
matthies@ifw.uni-hannover.de
[2] Helmholtz-Zentrum Hereon, Max-Planck-Straße 1, 21502 Geesthacht, Germany

Abstract. The established process chain for fabrication of titanium powder for additive manufacturing (AM) consists of the production of titanium sponge, the forging of electrodes and the atomization. A high energy demand of about 85% of the overall process energy consumption is taken by the production of the electrodes, resulting in a high carbon dioxide footprint and costs. This paper introduces a novel process chain to recycle titanium chips from milling and powder residues directly for the use in fabrication of electrodes. A methodology to increase the recyclability of the chips by adapting the process parameters of the milling process is shown. Cleaning and compaction to electrodes by utilizing hot isostatic pressing enables the possibility to use titanium chips and powder residues as base material for atomization. By using this novel recycling process, the energy demand for the production of titanium powder can be significantly reduced by approximately 72% in comparison to standard powder production.

Keywords: resource efficiency · sustainable manufacturing · recycling · titanium

1 Introduction

Titanium is used in the aerospace industry because of its excellent properties in terms of weight, strength and load capacity. The manufacturing process of structural components made of titanium for the aircraft industry is characterized by high cutting rates due to its complex shape. With a length of up to 4 m and a thickness of about 2 mm, it is necessary to manufacture the components from solid hammer-forged products. Typically, material removal rates from 90% occur in the cutting process. This results in up to 400 t of titanium chips for the production of an aircraft [1]. Usually, the chips are used for substandard products in the steel industry. A reuse of titanium chips for the production of high-grade titanium alloys is rather unusual and cost intensive. This is manly related to process-induced contaminations. Chips are contaminated with coolant, which are used for cutting processes. These fluids consist of hydrocarbon, alcohol and water and lead to elevated carbon and oxygen contents within the chips. The second source of contaminations

T. Bauernhansl et al. (Eds.): WGP 2023, LNPE, pp. 428–437, 2024.
https://doi.org/10.1007/978-3-031-47394-4_42

is the temperature generated in the machining process, which results in an increased reactivity of titanium with the surrounding media (air and coolant) leading to elevated oxygen and nitrogen contents again. Finally, adhesion and diffusion between workpiece and cutting tool could result in left over tool particles in the chips [2]. According to standards for titanium alloy grade 5 ASTM B 265a Ti-6Al-4V, a maximum oxygen content of 2,000 ppm, a maximum nitrogen content of 500 ppm and a maximum carbon content of 800 ppm are allowed for high-class applications [3]. However, even within these ranges, the impurities directly influence the mechanical properties of the alloy preventing recycling [4]. Thus, the high-grade titanium alloys are mainly produced from primary material (titanium sponge). The production process of titanium sponge includes generally the Kroll process, which is a very energy-intensive process. Approximately 80% of the overall energy consumption for titanium production is required for this [5–7].

For this purpose, methods aiming to realize a recycling of titanium chips under economic and environmental aspects have been developed and evaluated in various research projects. Ahn et al. presented a recycling process for the production of titanium hydride, which is used as a foaming agent [8]. Levinskii et al. and Mtsariashvili et al. described the production of titanium carbide and silicide from titanium scrap [9, 10]. More recent studies are concerned with an optimized melting process in order to reduce contaminations [11, 12]. In contrast to these studies one holistic approach from Denkena et al. aims to reduce impurities beforehand by optimizing the cutting process [13, 14]. It is shown that with adapted process parameter and dry machining recycling rates of 85% are possible for titanium chips by utilizing a melting with admixture of raw titanium. Applying these recycling rates for the production of structural airplane parts reduce CO_2 emission by 75% [15, 16].

In addition to traditional machining, innovative additive manufacturing processes are increasingly used in the production sector for high cost titanium parts. One of these manufacturing processes is Laser Powder Bed Fusion (LPBF), which is used in industry because it can enable new possibilities for functional integration, lightweight construction and resource conservation. However, the powder required for LPBF must first be produced from titanium electrodes from titanium rods by atomization in an energy-intensive process. In addition, the atomization process produces up to 70% scrap due to the high quality requirements and as it is not possible to recycle powder residues back into the process [17]. From an ecological point of view, a recycling strategy of titanium chips and scrap powder would significantly improve energy efficiency for additive manufactured parts. Therefore, in this paper a new developed process chain from titanium scrap to finished additive manufactured part is presented.

A methodology to increase the recycling rate of chips by adapting the influencing cutting and process parameters on chip quality under cooling conditions is given. Additionally, an approach to directly produce titanium powder without any cost and energy intensive remelting by utilizing Hot Isostatic Pressing (HIP) is pointed out. Furthermore, a comparison of the LPBF process and the machinability of additively manufactured components made from standard powder and recycled powder is compared.

2 Methodology

The new developed approach and the different manufacturing steps for the recycling of chips to powder are depict in Fig. 1 The conventional approach for the production of additive manufactured parts from powder consists of the initial fabrication of electrode from raw titanium followed by the atomization to powder followed up by the LPBF process and a finalization of the part by machining the functional surfaces with finishing process, following the top row of images in Fig. 1. The new developed process chain starts with a substitution of the initial electrodes and uses the residues from conventional production as well as the unusable powder residues from the atomization (Bottom line of images in Fig. 1).

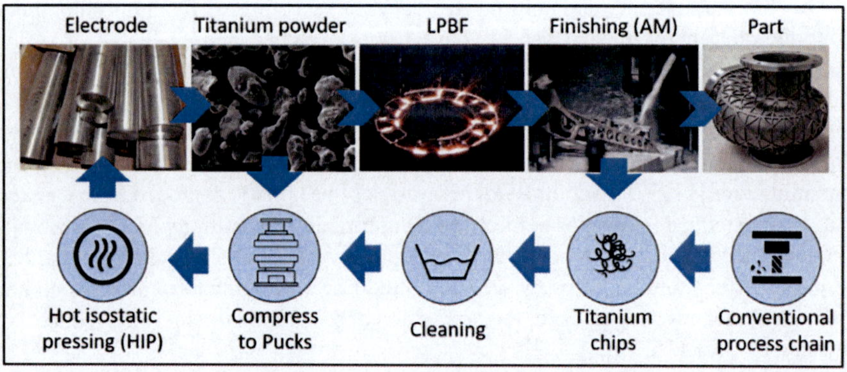

Fig. 1. Overview of the titanium chip and titanium powder scarp recycling approach

The individual process steps start with subtractive manufacturing and the resulting chips as the starting material for the new electrodes. In the next step, these are cleaned of adhering cooling lubricant residues and compacted. The subsequent hot isostatic pressing with residues from the powder production leads to a new electrode made of recycled titanium, which in turn is the beginning of the additive production of complex titanium components. The difficulties and challenges that arose in the process chain discussed above are described in the next sections.

2.1 Initial Chip Contamination

The initial product for the new developed recycling process are chips from conventional manufacturing. As shown before [13, 14], the chemical properties of produced titanium chips are dependent on process parameters of the cutting process. As a result, the machining parameters directly influence the recycling rate. The chemical analysis of the chips shows that the measures described to minimize the energy consumption (short process time through high feed and cutting speed as well as minimal use of coolant [18]) of the cutting process are partly contrary to maximizing the recyclability. With regard to chemical chip quality machining at elevated cutting speeds leads to extended wear progression and elevated chip oxidation. Additionally high cutting speeds lead to large chipping of the tools and therefore contamination with tool fragments. Furthermore, the relative ratio of chip area A_{ch} to chip volume V_{ch} defines a surface for natural oxidation of chips leading to a high percentage of natural uptake of oxygen of the titanium for chips with low undeformed chip thickness h. Due to the natural oxidation finishing chips exceeded the permitted oxygen contents. Additionally, carbon uptake is also highly dependent on relative chip volume V_{ch}. Therefore, concerning chip quality, machining at moderate cutting speed v_c with high feed per tooth f_z and high width of cut a_p is most beneficial for chip recycling rate limiting oxidation and maximizing und the chip volume to chip area ratio.

2.2 Cleaning, Compaction and Solidifying of Chips and Powder Scarps

In order to be able to use the chips, they must first be cleaned of deposits that lead to unwanted impurities. A washing process wars implemented on a Mafac Java industrial cleaning machine with a capacity 105 l. In laboratory tests of three industrial cleaning media the SurTec 001 102 SFT with a cleaner concentration of 2.5% showed the best results and wars further used in the cleaning of the chips. A comparison of the washing results of the different roughing and finishing chips showed a significantly higher Carbon content in chips from the finishing process compared to chips from the roughing process, due to the higher ratio of surface area to volume as well as a higher density of cracks in the chips. The characteristic chip segmentation of titanium shows that with increasing cutting speed and with lower undeformed chip thickness, the chip segment size increases. Fractures are formed at the roots of the segments act as traps for adhering coolant. This coolant is difficult to remove in downstream cleaning processes and contaminates new material during melting. Therefore, it has been found that a cleaning process directly after the machining process is necessary, because the coolant quickly dries inside these cracks. In addition, a denser packing of the finishing chips increases the likelihood of dead zones within the wash baskets, due to the more compact shape.

Due to the poorer cleanability of the finishing chips, the permitted carbon content mostly is exceeded. The oxygen content was not significantly changed by the washing process. The mean carbon and oxygen content of the chips after cleaning is illustrated in Fig. 3. Based on the findings, the washing process for the chips was adapted and a washing procedure was determined for the starting material of the electrodes. This is carried out with a washing medium temperature of 70 °C and on batches of 45 kg of chips. The soaking time is 300 s. During this process, the chips are repeatedly sprayed with the washing medium and mixed. This process is followed by a 60 s dry blowing with compressed air to remove the impurities dissolved in the cleaning medium from the chips. After that chips are rinsed clear with water for 180 s. This is followed by another dry blowing process of 1,080 s. Chips from rough machining with cutting speed $v_c = 30$ m/min and high feed $f_z = 0.2$ mm have been further used for the production of abomizable electrodes because of most desirable chemical conditions. The measurements of these chips depict in Fig. 3 reveal a relative carbon increase of 227% (274 ppm) and a relative oxygen increase of 6% (115 ppm) in comparison to the initial raw material.

After cleaning process, chips are compacted to pucks, illustrated in Fig. 2, in a hydraulic press. In this process the density of chip material increases from about

0.17 g/cm^3 for lose chips to pucks of a density of 2.98 g/cm^3. This reaches 67% of the density of classic raw Ti6Al4V material. Therefore, the lower limit of the required properties for hot isostatic pressing (60% of final density) is reached. In preparation for the HIP process serval pucks of diameter 55 mm and approximately 50 mm height are encapsulated in a metal tube with 60 mm inner diameter, fitted with a suction valve and hermetically sealed. Different powder scarp contents are mixt in between the two pucks and the space in between the inner wall and the puck. The remaining atmosphere inside the capsule is then evacuated. The titanium chip pucks and powder samples are than treated with a HIP process at a temperature of 850 °C and argon pressure inside the camber of 150 MPa. A sample after this process is shown in Fig. 2. The texture of the resulting sample from HIP process is globules as expected for a heat treatment of around 850 °C, Fig. 2. The density of the sample is 4.35 g/cm^3 and therefore reaches nearly 97% of Ti6Al4V. The sections of former transition from puck to powder and to the surrounding metal tube have negligible influences on the mechanical stability of the finished electrode. Nevertheless, the straightness of the finished HIP sample is not within the necessary requirements for the atomization process. A sufficient straightness of the electrode must be achieved, in order to produce a uniform melt jet during melting in the induction coil. Therefore, a rotary finishing process is applied to that extent that the electrodes have a symmetrical rotation. The chemical measurements of the electrode shown in Fig. 3. Reveals a relative carbon increase of 211% (274 ppm) and a relative oxygen increase of 5% (95 ppm) in comparison to the initial raw material. Lower values compared to the pucks are explained by the admixture of more pure powder from primer production process. It is therefore demonstrated that chips can be used for the production of electrodes, which fulfil the mechanical integrity and chemical requirements of the atomization processes.

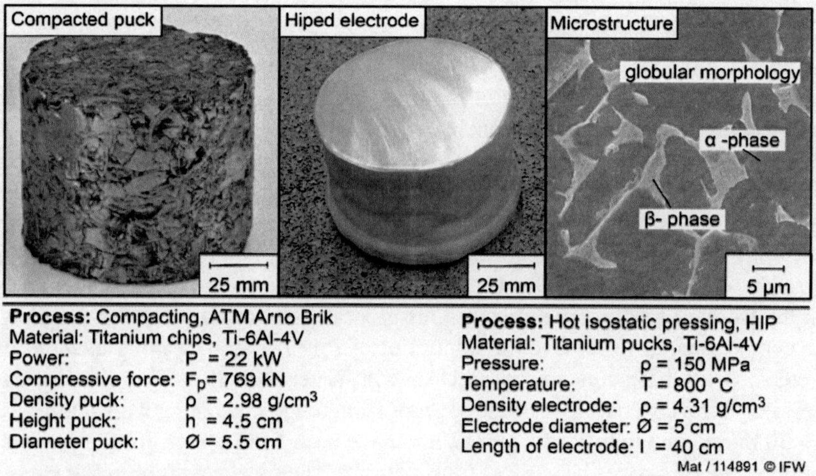

Fig. 2. Titanium-chip-puck, hot isostatic pressing sample of the electrode and microstructure of the finished electrode.

2.3 Powder Atomization

The atomization with the Induction Melting Inert Gas Atomization (EIGA) process was successfully carried out at the Helmholtz Zentrum Hereon at a research atomization plant. The electrodes have been shaped in a turning process to straight specimens which consistently show no cracks and are mechanically stable. Nevertheless, during melting it is noticeable that there is a difference in the longitudinal structure of the electrode. During the atomization process the transition areas of the individual pucks to areas that have been filled with powder melt at different rates. This requires a permanent adjustment of the atomization parameters, especially the high-frequency power. In some cases a melt sac formation occurs. This melt sac formation always occurs when areas of different packing density in the electrode melt at the same time. This can lead to increased molten material falling into the nozzle, resulting in an unfavourable material to atomization gas ratio. In this cases, the melt sac formation has a negative effect on the powder yield in the desired powder particle range between 25–63 μm. This behavior can also be observed in the standardized production of powders and is therefore constantly monitored and adjusted in terms of regulations. Following atomization under argon gas, the atomized powder was separated into the respective powder fractions using ultrasonic sieves. A comparison of recycled and conventional powder fraction is shown in Fig. 3. Figure 3 shows, that minor differences regarding the shape and size occur. A chemical analysis of the produced powder show an increase of 20 ppm with regard to the process-related increase in oxygen.

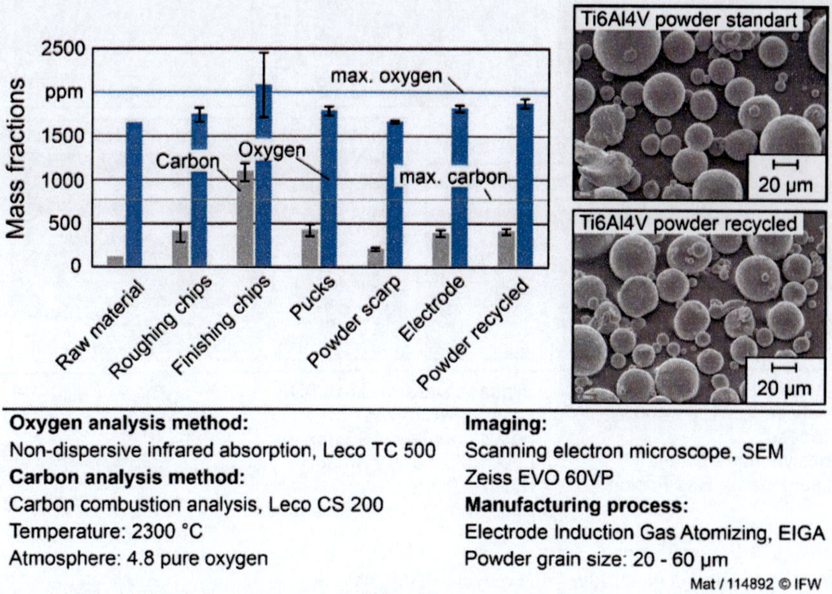

Fig. 3. Mass fractions of oxygen and carbon in the course of the production chain and raster electron images of conventional and recycled Ti6Al4V powder

2.4 Laser Powder Bed Fusion and Finishing of Recycled Parts

In order to evaluate the influence of the laser parameters and the powder quality on the component properties, tensile specimens and Small cubes are additively manufactured on a Realizer lasertec SLM125. The cubes are produced under variation of laser power and hatch distance to vary the energy input into the powder via the surface fluencies (cf. Fig. 4). The sample cubes were analysed with regard to density and microstructure. Subsequently, the 144 W laser power with a 90 μm hatch and a layer thickness of 30 μm showed the best results for recycled and conventional powder in the grain fraction of 25–60 μm. Nevertheless, a high density of at least 99.23% was achieved for all powders used and selected laser powers, which is in the industrially usable range. The tensile strength R_m of additively manufactured Ti6Al4V from recycled and conventional powder is almost constant with $R_{m,conventional} = 1{,}304.5$ MPa and $R_{m,recycled} = 1{,}308.5$ MPa. Therefore, no quantifiable difference in powder quality can be observed. With the production of the dental bridge (cf. Fig. 4), a component was produced for the first time by using the new developed process chain. The chemical values show that there is an increase of approx. 250 ppm oxygen and 370 ppm carbon over the entire process chain fulfilling the chemical requirements for titanium of the grade 5.

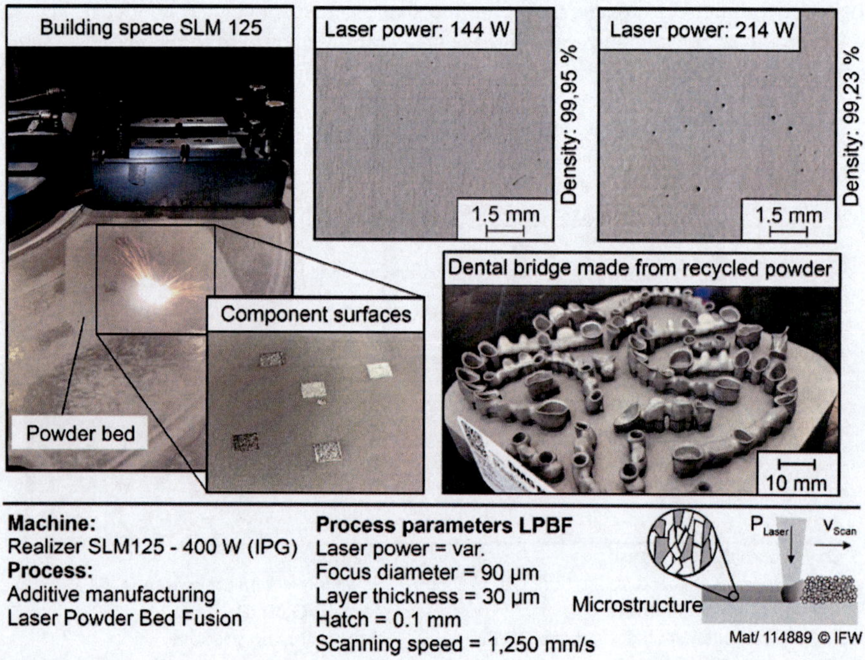

Fig. 4. LPBF manufacturing process, raster electron images from laser process parameter determination and dental bridge from recycled Ti6Al4V powder.

In order to investigate machinability of the recycled additivity manufactured parts, milling experiments were carried out on a Heller H5000 machine tool with parameters depict in Fig. 5. Blocks of the dimensions 50 × 30 × 70 mm have been manufactured with a laser power of 144 W with the same parameters as the dental bridge in Fig. 4. For the machining investigation the cutting speed v_c has been varied. An internal coolant supply with 11% water based emulsion (Blaser Vasco 6000) was used. As indicators for the workpiece quality the surface roughness was measured by means of a tactile measurements on a Mahrsurf LD130. As depict in Fig. 5 no significant difference in tool wear to conventional parts could be observed during machining of the workpieces made by recycled Ti6Al4V powder. With respect to the workpiece quality, a slight reduction of the mean roughness from $Ra = 0.24 \pm 0.13$ µm on conventional to $Ra = 0.21 \pm 0.1$ µm on recycled parts could be observed thought the mean values of all cutting speeds.

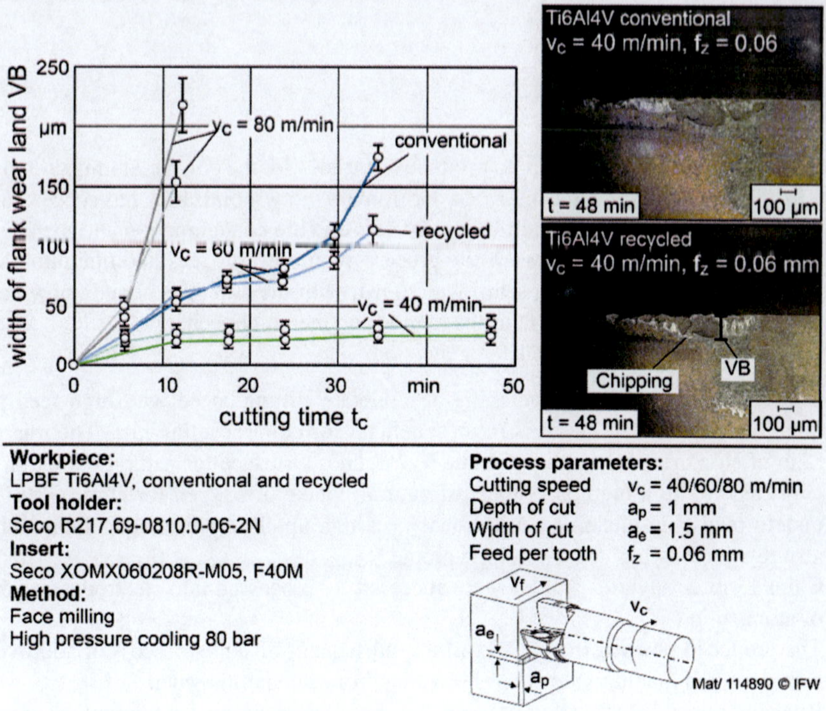

Fig. 5. Tool live time and ware mechanism when machining additively manufactured parts form conventional and recycled Ti6Al4V powders.

3 Energy Savings

Within the framework of the process chain presented, energy savings by maximizing the recyclability of produced titanium chips are possible. The new process chain for producing titanium powder involves the consumption of various resources: Electrical energy is consumed during the pressing of the chips into pucks, during the HIP process and during the subsequent EIGA process. A significant energy consumption also occurs during the production of primary powder (by the use of primary material for the production of powder by the EIGA process). The use of primary material in the recycling process is caused by the admixture of powder scarps and by the use of a titanium sleeve in the HIP process. The energy consumption wars calculated on the basis of data and experience of the research partners for the energy required. A distinction is made for production of electrodes from raw material and from chips and powder via HIP process. A distinction is made between production from primary material and production from recycled material. Overall, the production of recycled titanium powder is more favorable in ecological terms than production from primary material out of the Kroll process and can save up to 72% of overall energy consumption. This can be explained due to the high-energy input required for the production of primary titanium whose share is significantly reduced in the recycling process.

4 Conclusions

In this paper, a new process chain to produce electrodes for the powder atomization process from titanium chips and excess powder from previous atomisation has successfully been exhibit. The influence of cutting parameters on chip contamination and the development and optimization process of the process chain towards recycled titanium parts from powder out of production scarp was shown with the aim to generate knowledge about the acting mechanisms and influences during the process chain.

The following findings have been generated.

- Concerning chip quality, machining at moderate cutting speed with high feed per tooth and high width of cut is most beneficial for chip recycling rate. The relative ratio of chip area A_{ch} to chip volume V_{ch} defines a surface for natural oxidation of chips leading to a high percentage of natural uptake of oxygen for chips with low undeformed chip thickness. Additionally, carbon uptake from cooling lubricant is also favoured by high relative chip volume V_{ch}.
- Chips from a roughing process were successfully processed into electrodes capable of atomization.
- The produced powder, the LPBF and the subsequent finishing process of additively manufactured titanium parts does not varies from standard powder.
- By using this novel recycling process, the energy demand for the production of titanium powder can be significantly reduced by approximately 72% in comparison to standard powder production.

Acknowledgements. The authors thank the German Federal Ministry for Economic Affairs and Energy (BMWi) for its financial and organizational support of the project Return II (03EN2032 A-E)). The authors are grateful for the support of their partner within this project. Special thanks are due to the Helmholtz-Zentrum Hereon, Eckard TLS and CRONIMET for atomization and the chemical analyses of the material and DMG Mori for the Production of the additively manufactured parts.

References

1. Younossi, O., Michael, K., Graser, C.: Military Airframe Costs: The Effects of Advanced Materials and Manufacturing Processes. RAND Corporation, Santa Monica. MR-1370-AF (2001). As of April 2017
2. HOESCH Metallurgie GmbH: Schließung des Stoffkreislaufes von Titan – Entwicklung eines Recyclingverfahrens für Schrotte der Gießereien und mechanischen Bearbeitungszentren. Abschlussbericht 2008, BMBF 01 RW 0413
3. ASTM B265-13ae1. Standard Specification for Titanium and Titanium Alloy Strip, Sheet, and Plate, ASTM International, West Conshohocken, PA (2013). www.astm.org
4. Donachie, M.: Titanium – A Technical Guide. ASM International, 2nd edn. (2000). ISBN 0871706865
5. Titanium 2014: Aerospace Supply Chain and Raw Material Outlook, ICF International, USA (2014)
6. Goonan, T.G.: Titanium Recycling in the United States in 2004; Flow Studies for Recycling Metal Commodities in the United States, U.S. Department on the Interior, U.S. Geological Survey (2004)
7. Boyer, R., Welsch, G., Collings, E.W.: Materials Properties Handbook: Titanium Alloys, 4th edn. ASM International (2007)
8. Ahn, D.K., Kim, S.Y., Park, S.H., Bur, B.Y.: A study on titanium hydride formation of used titanium scrap for metal foaming agents. In: Proceedings of the 65th World Foundry Congress, Gyeongju, Korea, pp. 579–586 (2002)
9. Levinskii, Y.V., Petrov, A.P., Zolotova, O.S.: Preparation of Titanium Silicide Powders from Titanium Chips. Inorganic Mater. **34**(5), 452–455 (1998)
10. Mtsariashvili, L.V., Dzhalobadze, N.V.: Sintered alloy obtained from titanium waste. Poroshkovaya Metallurgiya **2**, 97–101 (1988)
11. Friedrich, B., Morscheiser, J., Reitz, J., Lochbichler, C.: Recycling of titanium-aluminide scrap. Intermetallics **6**, 762–768 (2011)
12. Rotmann, B., Lochbichler, C., Friedrich B.: Challenges in titanium recycling - do we need a new specification for secondary alloys? In: Proceedings of EMC 2011, pp. 1–15 (2011)
13. Denkena, B.: RETURN - Prozesskette Recycling von Titanspänen. Abschlussbericht, BMBF 03 ET 1174 (2016)
14. Denkena, B., Krödel, A., Pyzcak, F., Rackel, M.W., Kettelmann, S., Matthies, J.: Recycled titanium chips as initial product for the atomisation process of powders for additive manufacturing to increase resource efficiency.In: Proceedings of the Machining Innovations Conference for Aerospace Industry (2021)
15. Denkena, B., Jacob, S.: Approach for increasing the resource efficiency for the production process of titanium structural components. In: Machining Innovations Conference (2015)
16. Denkena, B., Dittrich, M.A., Jacob, S.: Energy efficiency in machining of aircraft components. Procedia CIRP **48**, 479–482 (2016)
17. Capus, J.: Titanium powder developements for AM – a round up. Met. Powder Rep. **72**(12), 384–388 (2017)
18. Denkena, B., Helmecke, P., Hülsemeyer, L.: Energy efficient machining of Ti–6Al–4V. CIRP Ann. **64**(1), 61–64 (2015)

Software-Defined Control Cabinet - Reprogramming Instead of Replacing

Florian Frick, Manuel Weiss(✉), and Armin Lechler

Institute for Control Engineering of Machine Tools and Manufacturing Units, University of Stuttgart, Seidenstr. 36, 70174 Stuttgart, Germany
manuel.weiss@isw.uni-stuttgart.de

Abstract. Flexible and reconfigurable production systems are composed of modules which fulfil a defined functionality and can operate within predetermined use cases. Some of these modules are fully integrated cyber-physical systems, others require electrical components located in control cabinets. Exchanging or adapting their functionality is limited, complex and typically requires manual effort due to hard-wired hardware and application-specific software. A new concept for control cabinets, designed as cyber-physical systems and based on reconfigurable hardware (FPGA) with flexible interfaces, is presented in this paper. The new approach overcomes existing limitations, offers more adaptability, reusability and extends the lifetime of the modules. Furthermore, the concept allows a seamless integration in a software-defined manufacturing environment. First results and experiences are based on a servo-controller which is extended with new features for an adaptive PWM by following the new design approach.

Keywords: Software-defined Manufacturing · Field Programmable Gate Array · Control Cabinet · Modular Production

1 Introduction

Constantly changing market demands and ever shorter product life cycles require increasingly flexible and adaptable production systems. Flexible production systems are designed for a certain range of variants or products and enable a fast and often software-defined adjustment of the production. The downside of these systems are strict limitations defined during the system design combined with high complexity, high cost and typically a low utilization due to a systematic overprovisioning. Adaptable production systems are designed to be reconfigured, typically using a modular approach. These modules are increasingly being designed as cyber-physical systems (CPS), combining mechanics, electronics, software and intelligent algorithms in a fully integrated system. They can be combined to form cyber-physical production systems (CPPS) [1].

These systems can easily be used with highly digitalized manufacturing approaches like software-defined manufacturing (SDM) [2], adjusting and changing the intelligent algorithms over the life time of the CPS. However, when CPS reach their physical limits, mechanical and electrical adjustments are necessary [3].

T. Bauernhansl et al. (Eds.): WGP 2023, LNPE, pp. 438–447, 2024.
https://doi.org/10.1007/978-3-031-47394-4_43

Furthermore, the decentralized and highly integrated approach of CPS is also a significant cost, size and energy factor. This further limits their usability when there is a lack of space, energy or specific environmental conditions like extreme temperatures or mechanical vibrations.

For all these reasons, traditional control-cabinet-based systems are still state of the art and required for the foreseeable future. Unfortunately, producing and adjusting control cabinets require significant effort, leading to economical issues in high-wage countries and must be carried out by skilled workers, which are also lacking [4]. Furthermore, an integration in software-defined manufacturing is complex.

In order to combine the advantages of control cabinets with the benefits of CPS, a novel concept aiming to reduce the effort required to build and adapt control cabinets and their functions is presented in this paper, eventually enabling a seamless integration with SDM.

2 State of the Art

In addition to the components for the power supply, a control cabinet essentially contains components for signal processing and logic (Fig. 1). Terminal blocks are used to connect external wires of the inputs and outputs with the internal ones. Control PCs are connected to the bus terminals via bus systems and bus couplers. The bus terminals are modules with individual functions for input and output signals, which have their own logic units. Dedicated modules like servo amplifiers with their own controller structures are used to drive servo motors. Most components are mechanically fixed in the control cabinet via rails.

Nowadays, the planning and engineering is done using electronic computer-aided design (ECAD), including 3D models and increasingly digital twins that describe the structure and behavior of the control cabinet and its components [5].

The digital twin and, in particular, the 3D design allows for a partially automatic placement of the components on the rails and derivation of the cabling. The cabling itself can also be automated to some extent by robots [6]. At present, however, there are still significant technological limitations. In particular, there is no automated solution available to modify existing control cabinets.

The effort for the construction, assembly and wiring of control cabinets is still very expensive, as shown in [7]. The duration for the construction of a control cabinet is on the average twenty hours. That amount of time is not acceptable for the reconfiguration of production systems within an operating factory.

Control cabinets that do not require cabling, such as systems with a backplane [8] are limited in their flexibility, since only components compatible with the backplane can be used. Likewise, the wiring to the modules themselves is not eliminated in such systems.

The state of the art shows that the cost of building control cabinets is still very high and represents a serious limitation in the efficient reconfiguration of production systems.

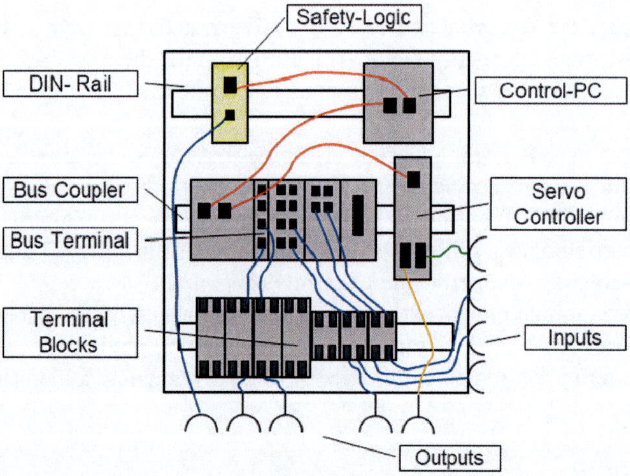

Fig. 1. Typical components of a control cabinet. Components for power supply, relays and fuses are not shown.

3 Software-Defined Control Cabinet

To address the present limitations of control cabinets, a novel architecture is required. For its design, multiple guiding principles are used: (i) The control cabinet should be an integrated system, based on modules. (ii) It should be a digital and intelligent system itself. (iii) As much of the logic as possible should be software-definable. (iv) Physical components should be flexible and provided in sufficient quantities. (v) Interfaces should be flexible and standardized.

In consequence, the dedicated logic components of a control cabinet, as described in the state of the art, must be replaced by software-definable solutions, including in particular bus terminals and drive amplifiers. As a result, a multi-layer virtualization platform for control cabinets is defined.

3.1 Modular Architecture

The architecture (Fig. 2) gives a software-defined cabinet a new structure which is mapped on three layers: physical hardware, reconfigurable hardware (FPGA) and software. All existing functions according to the state of the art as well as novel functions can be mapped to the new structure.

The details regarding the architecture are presented in the next sections.

3.2 Components of a Software-Defined Control Cabinet

According to the CPS-approach, components are not necessarily purely physical or digital but are defined on a more abstract level. They can be implemented on any or multiple of the available layers.

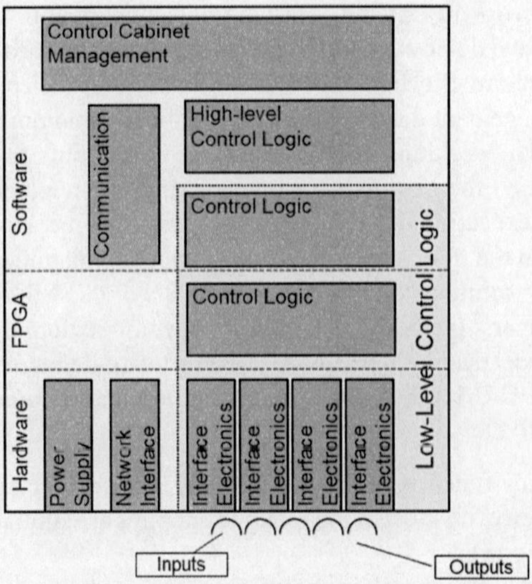

Fig. 2. Software-Defined Control Cabinet architecture.

Low-Level Control Logic. The low-level control logic includes all components interfacing external devices like digital and analog inputs and outputs, protocol-based interfaces (sensor- and field-buses), hard wired logic (relays, etc.), user interfaces but also specialized devices like drive control systems. A detailed description is provided in Sect. 3.3.

High-Level Control Logic. High-level control logic relates to controllers like programmable logic controllers (PLC) or computerized numerical controls (CNC). According to the state of the art, these controllers were either implemented on a dedicated device with a field bus interface or located on an IPC platform as a software solution. For the software-defined control cabinet, the logic is separated from the hardware-dependent low-level logic and is exclusively represented in software. Depending on the actual available compute resources, high-level control logic can potentially be relocated to external compute infrastructure.

Control Cabinet Management. State-of-the-art control cabinets are mainly seen as an electrical system and its mechanic housing. The cabinet itself has no digital part except potential artifacts from the design and planning stage. The "intelligence" of a control cabinet is the sum of the individual devices located in the cabinet.

In the context of the software-defined cabinet approach, the control cabinets become entities in the real and digital world. They must be self-aware, must be included in higher-layer processes, e.g. in the context of SDM, and able to configure their internal function according to the requested behavior. These tasks are handled by an internal management unit, deploying software, setting parameters and configuring the FPGA.

Communication Across all Layers. The control cabinet requires a broad connectivity (IT-based services, field devices, other cabinets etc.) and therefore a corresponding communication infrastructure. Since various types of traffic must be considered – field data, controller data, management data – converged real-time communication is required.

Time-Sensitive Networking (TSN), the emerging real-time extensions to standard Ethernet, is considered the best fit. It allows a seamless integration with IT-systems, enables the parallel execution of multiple distributed real-time application and furthermore direct access to the field level when using it in combination with OPC UA FX or "fieldbus-over-TSN" solutions like Profinet-over-TSN or CC-Link IE TSN.

On the higher layers, including data models, a combination of protocols and technologies is used to meet the requirements. For the software-defined control cabinet itself, a combination of OPC UA and Asset Administration Shells is used on top of standard protocols like TCP/IP etc.

Power Supply. A key function of control cabinets is the power supply of internal and external modules. However, most of the required supplies are similar across applications and only vary in the power.

Therefore, a modular approach is chosen for software-defined cabinets with various compatible options, each providing different power levels.

3.3 Low-Level Control Logic Based on the Open Automation Platform

A common challenge within the research of industrial control systems are platforms to implement and evaluate innovative solutions. As an enabler, the Open Automation Platform (OAP) was developed and utilized across numerous projects, focusing on drive control, novel sensors, complex algorithms or industrial communication [9]. The key idea of the OAP is openness on three layers: a modular hardware layer, a reconfigurable FPGA layer and software layer. The system design is conducted using a component-based approach, with each component potentially being partially implemented on any layer. Each layer offers dedicated advantages and is connected through standardized interfaces with the adjacent layers. The overall structure is depicted in Fig. 3. The three open layers are described in detail in the following sections.

Hardware Layer. The lowest layer is nonprogrammable hardware, consisting of a base and extension modules. The base module is designed around a commercially available FPGA, which is extended with standard components like memory, programming interfaces etc. The General-Purpose Input/Output (GPIO) pins of the FPGA are mapped to standardized connectors on a backplane.

A multitude of modular extension boards can be connected to the backplane. These boards provide functions that cannot be configured within the FPGA including drivers, ADC, power electronics, network interfaces or sensor interfaces for motor encoders. Memory modules on the extension boards allow an automatic detection of the modules.

FPGA Layer. Reconfigurable hardware combines the performance and determinism of hardware with the flexibility of software. Therefore, the FPGA layer is used for time critical functions, such as control loops or signal processing. The desired functions

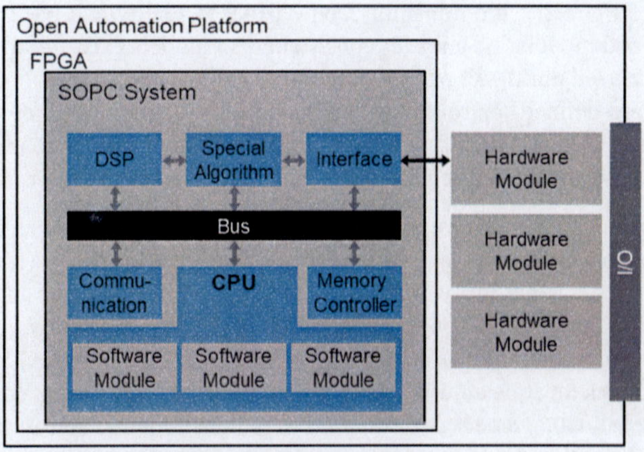

Fig. 3. Architecture of the OAP

are implemented in VHDL and wrapped in a component, allowing to use System-on-a-Programmable-Chip (SOPC) design approaches. The self-developed SOPC-components can be supplemented by IP-Cores of other sources, e.g. specialized field bus or encoder protocols.

The SOPC components have three types of interfaces: First, I/Os are mapped to the GPIO pins and consequently to the physical hardware. Second, they are connected to other modules on the FPGA-layer. Third, all SOPC components are connected to the software layers using a memory mapping approach, thereby exposing registers.

Software Layer. The third layer is the Software layer located on a CPU. The CPU can either be a Soft CPU generated as a SOPC component or a Hard Processing System (HPS), which is integrated in the same silicon as the FPGA. In both cases, the CPU is connected to the SOPC through a bus. The bus enables changing parameters, as well as tapping data of SOPC components.

The software consists of a generic base design and modules that can be integrated. These modules are written in high-level languages and fulfil functions like parametrizations, complex but not-time-critical algorithms, stacks or user interaction.

3.4 System Design in the Context of Software-Defined Manufacturing

Defined processes for system design, configuration, updates and adaption are an integral part of the software-defined control cabinet concept. The high level of digitalization requires an integration of the concept in an appropriate environment like SDM in order to exploit the benefits and unlock the full potential. Nevertheless, dedicated tools can also be used to have a direct human involvement in the processes.

Software-Defined Manufacturing (SDM). SDM is a new paradigm attempting to enable the dynamics and innovativeness known from the IT in manufacturing. The approach combines a layered resource-based model for the required technologies with a new understanding of processes. A key aspect is a continuous view which is applied to the processes in general, the transition from virtual to reality and back as well as for the used data and models. As a result, consistent data models exist for all entities in the system and are used during all phases of the life-cycle.

The software-defined control cabinet approach allows a seamless integration in SDM. Control cabinets can be handled as resources in the system, can be managed and their exact function and the related requirements can be derived from the overall model.

System Design. The system design of a software-defined control cabinet is based on the assumption that a well-defined specification of the required functions is (digitally) available. In a first step, the hardware modules must either be composed or an existing setup fulfilling all requirements must be selected. Based on the physical hardware and specified functions, the reconfigurable hardware and software can be composed and generated. External interfaces must be adjusted as well as the external converged communication. Also, the internal management is configured as well as required self-descriptions. The control cabinet can then be integrated with the rest of the system.

4 Proof of Concept: Adaptive PWM

In the context of various research activities in the field of drive control systems, a servo controller based on the OAP was developed [9]. All three previously presented layers of the OAP are used: on the hardware layer, extension boards are used for power electronics, encoder and network interfaces. On the FPGA layer, a cascaded control loop and the signal generation for the power electronics as well as the sensor and encoder logic are implemented, providing high-performance and determinism. The software layer is used for the initialization of the system and the configuration of parameters on the FPGA layer, monitoring of drive data and for the communication. The OAP is used in combination with a PC-based numerical control (NC), which is connected using OPC UA Pub/Sub over TSN. It is operating an axis with a servo drive for evaluation of controller concepts, as shown in Fig. 4.

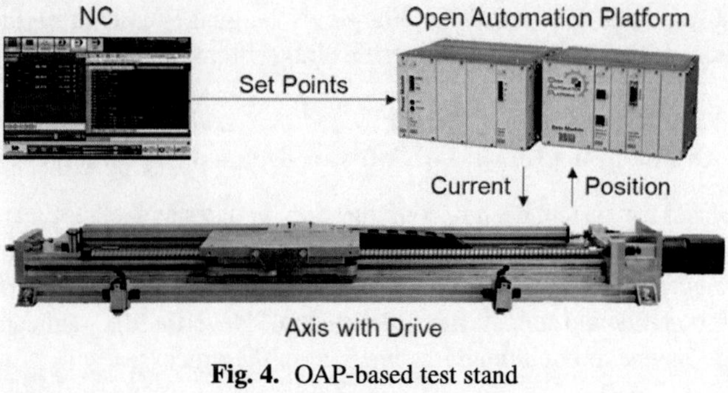

Fig. 4. OAP-based test stand

With the modular approach of the OAP, a flexible adaption of the servo controller is possible. For example, the extension board for the power electronics can be changed to other power levels. Similarly, motor specific hardware could be added, for example an extension board for interfacing the motor temperature or for switching the motor break. The FPGA and software layer can also be adapted with new components, for example to monitor the power consumption or DC voltage, complemented by safety functions.

System Adaption. In contrast to state-of-the-art drive systems, the modular approach allows for an easy integration of new functions to extend the conventional servo control loop. This is utilized in an ongoing research project investigating an adaptive pulse-width modulation (PWM) to increase the energy efficiency of servo drives in tooling machines.

The approach is based on the fact that the PWM frequency of a drive is selected based on the highest requirements of the process. To achieve the required dynamic behavior of feed drives, a high PWM frequency is typically set for the complete manufacturing process, since high switching frequencies of the PWM reduce the dead-time of the system and increases the control bandwidth [12]. However, high PWM frequencies result in significant switching losses in the power electronics, which increase the overall energy consumption of the tooling machine. Furthermore, the lifetime of the semiconductors is reduced [13].

To overcome these issues, the PWM frequency could be reduced in manufacturing steps in which a high accuracy is not required. Typical scenarios are for example traverses to the tool changer, standstill of the axis, as well as processes which don't manufacture the final finish of the product. A reduced PWM frequency results in a reduced switching frequency of the power electronics and therefore in lower switching losses and an overall better energy efficiency of the tooling machine.

The functionality of the adaptive PWM is mainly implemented on the FPGA layer of the OAP. For controlling the PWM frequency, the control loop is extended by a switching frequency controller. The switching frequency controller implements different algorithms to determine the actual PWM frequency. Furthermore, the switching frequency controller adapts the control parameter of the control loop to the currently active PWM frequency [12]. Studies show that the adaptive PWM has the potential to reduce the energy consumption of feed drives [13]. For the adaptive PWM, the existing PWM module is extended to support different PWM frequencies and a dynamic switching between them.

The presented adaptive PWM features can extend a conventional servo controller via reconfiguration of the FPGA. Therefore, the conventional PWM module can then be exchanged with the new adaptive PWM module and the switching frequency controller can be added as a SOPC component to the design, as depicted in Fig. 5.

Using a reconfiguration of the FPGA, no mechanical or electrical modifications on the hardware platform are necessary. The presented OAP enables the flexible implementation and evaluation of new control structures.

The approach can be used in a similar way for the entire control logic within a control cabinet, thereby enabling a software-based approach.

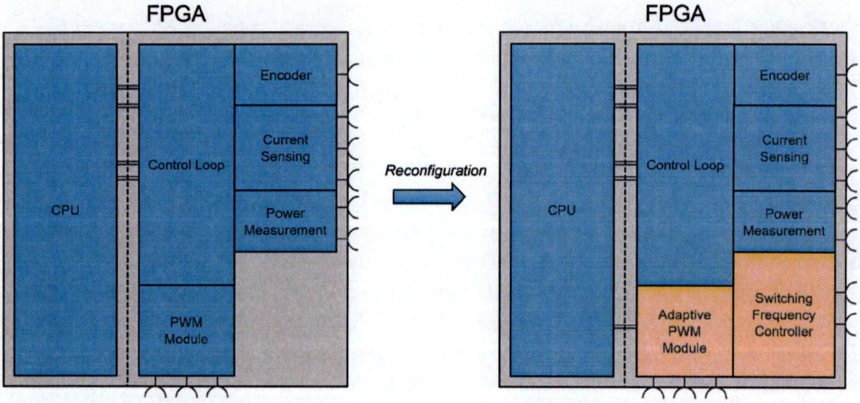

Fig. 5. Reconfiguration of the FPGA to integrate new functions.

5 Summary and Outlook

Control cabinets are essential for traditional and CPS-based production systems. However, their design, adaption and operation have a noticeable lower level of digitalization and automation than the components they are integrating and the systems they are used in. In consequence, control cabinets are a significant bottle neck when it comes to fully digitalized approaches like software-defined manufacturing.

The presented concept maintains the idea of control cabinets but changes their architecture to a more modular and standardized approach which allows a software-defined configuration of its function. In summary, the concept defines a multi-layer virtualization platform for control cabinets – from low-level hardware to sophisticated control algorithms.

Based on the proposed architecture, further research focusses on optimal design of the low-level hardware, more standardized interfaces, an efficient reconfigurable hardware layer and the seamless integration with software-defined manufacturing – including the required models and communication.

References

1. Monostori, L., et al.: Cyber-physical systems in manufacturing. CIRP Ann. Manuf. Technol. **65**(2), 621–641 (2016)
2. Lechler, A., Verl, A.: Software defined manufacturing extends cloud-based control. In: 12th International Manufacturing Science and Engineering Conference, vol. 3. American Society of Mechanical Engineers (2017)
3. Koren, Y., Gu, X., Guo, W.: Reconfigurable manufacturing systems: principles, design, and future trends. Front. Mech. Eng. **13**, 121–136 (2018)
4. Wiechers, R., Scholl, F.: Mechanical engineering – figures and charts 2022. VDMA Economics and Statistics, Frankfurt (2022)
5. Lukac, D., Montiel, S., Kilburn, J., Mulherrin, S., Telles, T., Omari, A.: Digital twins in control cabinet construction. In: 2022 Intermountain Engineering, Technology and Computing (IETC), pp. 1–5 (2022)

6. Spies, S., Bartelt, M., Kuhlenkötter, B.: Wiring of control cabinets using a distributed control within a robot-based production cell. In: 19th International Conference on Advanced Robotics (ICAR), pp. 332–337. IEEE (2019)
7. Tempel, P., Eger, F., Lechler, A., Verl, A.: Control Cabinet Manufacturing 4.0 Study. ISW University of Stuttgart (2017)
8. Beckhoff MX-System. https://www.beckhoff.com/en-gb/products/mx-system/. Accessed 16 May 2023
9. Frick, F., Zahn, P., Lechler, A., Verl, A.: Modular design approach for model-based drive control systems on reconfigurable logic. Appl. Mech. Mater. **704**, 380–384 (2015)
10. Böcker, J.: On the control bandwidth of servo drives. In: 13th European Conference on Power Electronics and Applications. IEEE, Piscataway (2009)
11. Wei, L., McGuire, J., Lukaszewski, R.A.: Analysis of PWM frequency control to improve the lifetime of PWM inverter. IEEE Trans. Ind. Appl. **47**(2), 922–929 (2011)
12. Braband, M., Frick, F., Lechler, A., Verl, A.: Process requirements based adaptive PWM for improved efficiency of machine tool feed drives. PCIM Europe, Nuremberg (2016)
13. Henke, T., Frick, F., Lechler, A., Verl, A.: Adaptive PWM modulation for an increased energy efficiency of industrial drives. In: Symposium on Power Electronics, Electrical Drives, Automation and Motion (25th SPEEDAM) (2020)

Disassembly Graph Generation and Sequence Planning Based on 3D Models for the Disassembly of Electric Motors

Malte Hansjosten(✉) and Jürgen Fleischer

wbk Institute of Production Science, Kaiserstrasse 12, 76131 Karlsruhe, Germany
Malte.Hansjosten@kit.edu

Abstract. Disassembly is an important first step for new and innovative end-of-life solutions for critical components and materials on the way to a circular economy. Generating possible disassembly sequences and finding optimal sequences for a given disassembly target is a major challenge in disassembly. Using electric traction motors as an example, this paper presents an approach to automatically generate possible disassembly sequences and calculate the optimal path to a given disassembly state. This is based on geometric data in the form of polygon meshes enriched with additional information about non-geometric joints (e.g. adhesive joints). Geometric joints are automatically detected by analysing the contacts between parts. All parts, joints and resulting constraints are modelled in an undirected graph. By analysing the joints and movement constraints, a directed graph is generated. This graph contains all possible disassembly sequences, which can then be searched for an optimal sequence.

Keywords: Disassembly · Graph generation · Sequence planning

1 Introduction

In response to a growing demand for resources, volatile prices as well as the challenges posed by climate change, waste and pollution, the concept of a circular economy is raising in popularity [1]. This concept focuses on moving from a linear economy, where new resources are used to manufacture products which are ultimately discarded as waste, to a closed-loop system that relies heavily on concepts like recycling, remanufacturing or reuse to limit resource consumption and waste generation [2]. The necessity for this transition has also been recognized by legislative bodies, such as the European Union, who actively promote this development [3]. To accomplish this, it is therefore necessary to develop innovative end-of-life solutions for critical parts and materials.

A major overlapping trend is the electrification of personal transport, leading to an increase in the production and variety of electric motors, primarily permanent magnet synchronous motors (PSM). More than 70% of the value of these motors lies in the materials used, such as copper and rare earth metals, which are considered to have a critical supply chain [4, 5]. PSMs are therefore an ideal case for the development of innovative end-of-life (EOL) solutions.

T. Bauernhansl et al. (Eds.): WGP 2023, LNPE, pp. 448–457, 2024.
https://doi.org/10.1007/978-3-031-47394-4_44

PSMs typically consist of a housing, a stator with a laminated core and copper windings, a rotor with a shaft, permanent magnets and additional elements such as bearings and sensors [6]. In the context of a circular economy, different EOL solutions can be considered for these components. Rotors could be directly reused due to the minimal wear, or stators could be remanufactured by replacing the copper windings [7, 8]. In addition, sorting materials after disassembly can improve recycling efforts by increasing the purity and therefore the value of the recycled material.

Regardless of the cycling method, the motor must first be disassembled into its main components. However, cost effective disassembly, particularly in high wage areas, requires automated solutions.

In [9] we present a concept for an autonomous adaptive disassembly system for PSM based on industrial robots with process specific end effectors. To achieve the necessary level of autonomy the system must not only be able to automatically carry out disassembly actions but also to plan possible disassembly sequences. A suitable product representation/model are in turn essential for successful disassembly sequence planning (DSP) as is the subsequent planning of the individual disassembly actions [10].

Since manual creation of such models and the subsequent generation of possible disassembly sequences is associated with considerable effort, automated solutions are necessary. One common approach is the usage of CAD-Data as a basis for automated DSP. In this paper we present a solution specifically for the automated DSP of electric traction motors based on 3D-geometry models. The software solution is implemented as a standalone python application and uses simple mesh models (e.g. .stl or .obj files) It is therefore independent from proprietary CAD-software or file formats. In addition, we expand on the purely geometric DSP by also considering the different types of joints present in common electric motors. This is based on findings from the manual disassembly of real, state-of-the-art motors and analysis of the present joints.

In the following paragraphs we present an overview of the state of the art for CAD-based DSP, before going into detail on the implemented motor model, its automatic generation and the generation of possible disassembly sequences.

2 State of the Art

DSP has been an active research area for many years. Early approaches were considered in the context of assembly, rather than disassembly planning (assembly by disassembly), with examples for automatic, computer-aided approaches reaching back as far as the 1980s. Friedmann [11] for example presented an approach to automatically generate precedence graphs and exploded drawings based on an analysis of 3D-Models and additional, formalized knowledge of assembly processes. Similar approaches, also explicitly for DSP, can still be found in current research.

Münker and Schmitt [12] propose a system to automatically generate AND/OR-Graphs for (Dis-)assembly planning. They use neutral STEP-Files as input for a CATIA interface to extract both connectivity between individual parts and collision-free movement directions for part-pairs. However, they only consider movements in the three main coordinate axis and remove all connector parts (e.g. screws, nuts, etc.) from the assembly.

Prioli et al. [13] present a similar implementation for a disassembly sequence planner based on collision analysis of STEP-Files. They also only consider the three main coordinate axis and did not include additional information about fasteners or joints.

Neb and Göke [14] combine collision and gravity analysis with feature-based information, both generated automatically from CAD-models using a SolidWorks Macro. Using a genetic algorithm, they were able to generate feasible assembly sequences.

Another example that expands the geometric analysis of 3D CAD-models with the integration of technological information can be found in [15]. The authors propose a system that also uses a SolidWorks API to extract geometric constraints from CAD-models. They also include additional information about components, like elasticity, to generate more realistic disassembly sequences, without however including information about joints which might necessitate specialized disassembly processes (e.g. glue joints) and also don't consider destructive disassembly.

While these approaches show the feasibility of 3D-geometry analysis as well as the benefits of supplementing these analyses with selected additional technological information, for the specific use case of the disassembly of electric motors an equally specific motor model is necessary. Additionally, while some current approaches already utilize neutral data formats, like STEP-files, they still build on commercial CAD systems, which further sets the approach described in this publication apart.

3 Motor Model

The basis for the automatic generation of possible disassembly sequences is a suitable digital model of the object to be disassembled, in this case the motor. The approach chosen for this work was the modelling of an *Assembly* (the motor) as a collection of individual *Parts* and the existing *Connections* between those parts, each with an individual set of properties. The following subsections first go into detail on the different types of *Connections* found in representative electric traction motors, before describing the developed motor model and the mostly automatic approach for generating such models based on simple 3D models.

3.1 Motor Analysis

Before a suitable digital model for the disassembly planning of electric traction motors could be build it was necessary to analyse such motors and determine how the individual parts are joint together and how they can be disassembled. Therefore, two used motors procured on the open market were disassembled manually (see Fig. 1) and occurring joint types according to DIN 8593 [16] were identified and classified by their ease of disassembly.

Although the examined motors were constructed differently the used joining processes were generally similar and mainly fall in the three *groups putting together, press-on/press-in* and *bonding.*

This classification served as the basis for further subdivision of the relevant joints in the following.

Putting Together

This group includes joints in which "[...] the cohesion of the parts to be joined is brought

Fig. 1. Manually disassembled motors. Left: VW ID.3. Right: e.go

about by gravity (friction), by form fit, by spring force or by a combination of these" [17]. All identified joints in this group are non-destructively detachable.

Relevant sub-groups here are first of all the laying-on and pushing-into-one-another, which mainly includes housing components as well as some bearings. When not regarding other joints, they can be dismantled by simply grasping them. Depending on accessibility, the difficulty of disassembly is therefore rated as *easy to medium.*

Another relevant subgroup of assembly is spring-loaded expansion. This includes the retaining rings that were used in all analysed motors to secure the bearing of the rotor shaft axially. Since loosening the retaining rings proved to be a kinematically complex process with often very limited accessibility, especially in the manual disassembly tests, the difficulty in loosening these joints is rated as *medium to difficult.*

Press-on/Press-in

This group includes joints in which "[...] the parts to be joined and any auxiliary parts are essentially only elastically deformed and unintentional loosening is prevented by frictional locking" [16].

The first relevant subgroup here is bolting. This concerns a large part of the housing components, as well as some additional components, such as built-in sensors. Although screw joints can generally be loosened non-destructively, this may no longer be the case for used products, due to corrosion or wear. In this case, destructive methods must be used. The difficulty of loosening these joints is therefore rated as *easy to medium*, depending on the condition of the components.

The second relevant subgroup is press joints. In the context of the analysed motors, this concerns bearings with interference fits. These can be loosened non-destructively with a suitable pulling tool, so the difficulty for loosening is rated as *easy*.

Bonding

This group includes "joining using an adhesive (see DIN 16920), i.e. a non-metallic material that can join parts by surface adhesion and internal strength (adhesion and cohesion)" [18].

During the manual disassembly tests, it was found that adhesive bonding could occur between the parts due to the sealant applied to various housing parts. While bonding is not the primary function of the sealant, it still prevents easy removal of the components and classifies this as a bonded joint in the context of disassembly. Although the gasket or sealant must be destroyed when loosening these joints, the remaining components can be separated non-destructively. Therefore, the loosening of the joint is classified as basically non-destructive. Due to the very poor accessibility of the actual adhesive, the difficulty of loosening is additionally rated as *medium to difficult*.

This analysis was the basis for building a suitable digital motor model, which will be described in the following subsection.

3.2 Model Architecture

As stated above, the motor was modelled as an *Assembly* which in turn contains all *Parts* of the motor and the *Connections* between those parts. Using python, this was implemented as a corresponding class structure.

Each *Part* object contains a unique part name, the geometry of the part as a 3D-triangle mesh and, derived from the mesh, the center-of-mass of the part. A basic *Connection* object contains the unique name of the two *Parts* it connects, the connection type and the movement constraints the connection imposes on the connected parts. To model the unique properties of the different joints identified in the motor analysis (see Sect. 3.1) four different sub-classes of the *Connection* class were implemented:

- Geometric Connections: These *Connections* represent all instances, where direct contact between two components imposes movement constraints on these components. These *Connections* solely result from the geometry of the individual parts and can be determined automatically by analysing a suitable 3D model (see Sect. 3.3).
- Threaded Connections: These *Connections* represent all instances, where the movement of a part with an external thread (i.e. a screw) is constrained in all directions by a part with an internal thread. In addition to the movement constraints this type of *Connection* object also contains information on the type of screw-head.
- Glue Connections: These *Connections* represent all instances, where the movement of two *Parts* is constrained by an adhesive.
- Retaining-Ring Connection: These *Connections* are used to specify a connection of a retaining ring with the part containing the slot, in which the retaining ring sits. This connection doesn't impose additional constraints on the involved components, it is however necessary to retain the information which part can be deformed relative to which other part during a disassembly process.

The term *Connection* rather than joint was chosen to avoid confusion, since a joint as described above can consist of multiple connections. A screw-joint consisting of three parts held together by two screws would for example be represented by multiple *Geometric Connections* and two *Screw Connections.*

By combining individual *Parts* and *Connections* in an *Assembly* object a complete model containing all necessary information for the planning of possible disassembly sequences is generated. An *Assembly* can be represented as a unidirectional graph, where

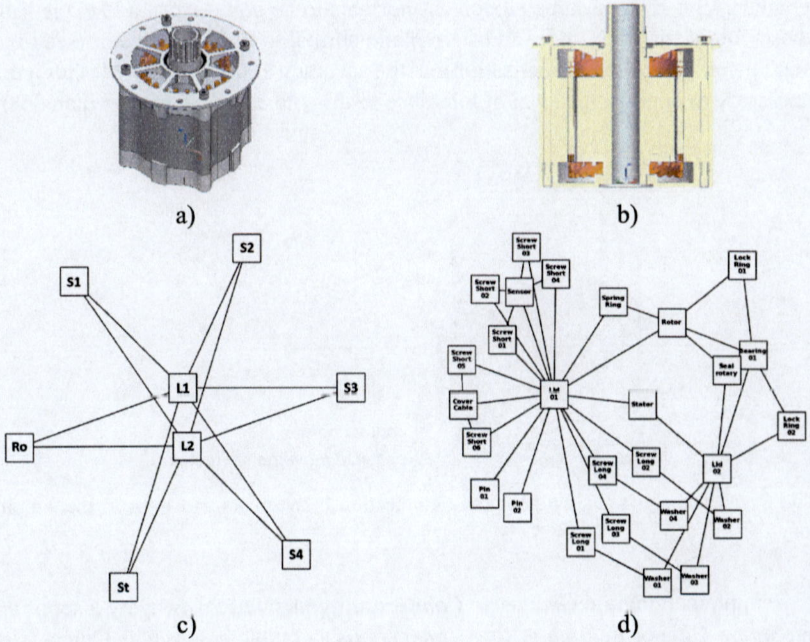

Fig. 2. Motor model examples. a) Simplified model consisting of four screws, two lids, rotor, stator. b) Cross section of simplified motor. c) graph for simplified motor. d) graph for real motor consisting of 25 *Parts*

each node is a *Part* and each edge represents a *Connection* between those parts. This is shown for two examples in Fig. 2.

3.3 Model Generation

To minimize the manual modelling effort to generate a specific *Assembly* Object an automated solution was implemented based on a geometric model of the motor and a list of non-geometric *Connections*. To keep the approach as widely applicable as possible simple 3D-Meshes (.stl or .obj files) of the individual motor parts were chosen as input format.

To generate *Geometric Connection* Objects, it is necessary to determine the movement constraints (or possible movement directions (PMD) as their inverse) imposed by individual parts on each other. To allow for efficient calculations the problem has to be discretised. This can be achieved by first defining all possible PMD-unit-vectors as equidistant points on a unitary sphere. For each possible pair of *Parts* contact points and contact normals (also as unit vectors) are then calculated. A possible PMD is considered valid, if the angle between the PMD-vector and each contact normal equal or greater than 90°, meaning that no contact point exists that would block movement in the direction of the possible PMD. All invalid PMDs are then considered movement-constraints.

It is also possible to directly check, if the dot-product of PMD-vector and contact-normal is equal or lesser than 0. A more illustrative way to describe this is to look at it geometrically: If we assume all contact-normals to be points, then a PMD is valid if all these points are on or under an origin plane normal to the PMD-vector (see Fig. 3). Depending on the chosen discretisation and the accuracy of the used 3D-Meshes it might be necessary to apply some level of tolerance to this (i.e. an angle greater than 85°).

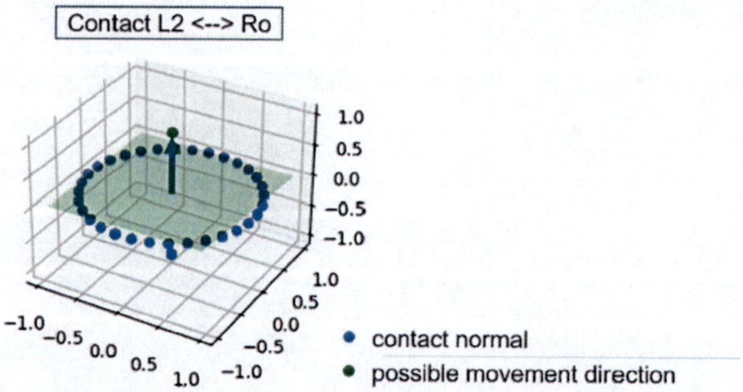

Fig. 3. Contact analysis for the geometric connection between lid and rotor in the simplified motor

By supplementing the *Geometric Connections* generated this way by a set of additional *Screw, Glue* or *Retaining-Ring Connections* a complete *Assembly* Object is generated. Since these *Connections* can be described as a list of part-pairs, connection types and little additional information like the type of screw-head the necessary user input is minimal.

4 Disassembly Sequence Generation

Using the model described in Sect. 3 it is now possible to generate possible disassembly sequences. These possible sequences can then be searched for an optimal sequence. The following subsections describe the current state of the implemented approaches, as well as planned expansions to include alternative processes, especially destructive processes.

4.1 Sequence Graph Generation

For this a recursive algorithm was implemented that generates a directed graph, containing the individual disassembly states, represented by individual *Assembly* objects, as nodes and the performed disassembly processes as edges.

To reduce the complexity of the resulting graph as well as the necessary computational effort the algorithm runs under the assumption, that it is beneficial to remove easy to remove parts (especially loose parts) first so that harder to remove parts are better accessible. Based on this assumption and the evaluation performed in the motor analysis (see Sect. 3.1) the algorithm performs the following steps, starting from the fully assembled motor:

1. Determine the discrete PMD for each *Part* as intersection of the PMD of all *Connections* the part is involved in.

2. If one or more *Parts* with PMD exist, remove the part from a copy of the *Assembly* currently considered, add a corresponding edge to the graph and if the resulting *Assembly* is not already part of the graph also a new node.
3. If no *Part* with PMD exists "dismantle" *Screw Connections* and add corresponding edges and nodes (if necessary)
4. If no *Screw Connection* is removable "dismantle" analogues first *Glue*, then *Retaining-Ring Connections.*
5. If no disassembly action is possible, terminate.

Every time a new node is added to the graph the algorithm is called again for the corresponding *Assembly* object. The result for the two example motors mentioned in Sect. 3.2 is depicted in Fig. 4.

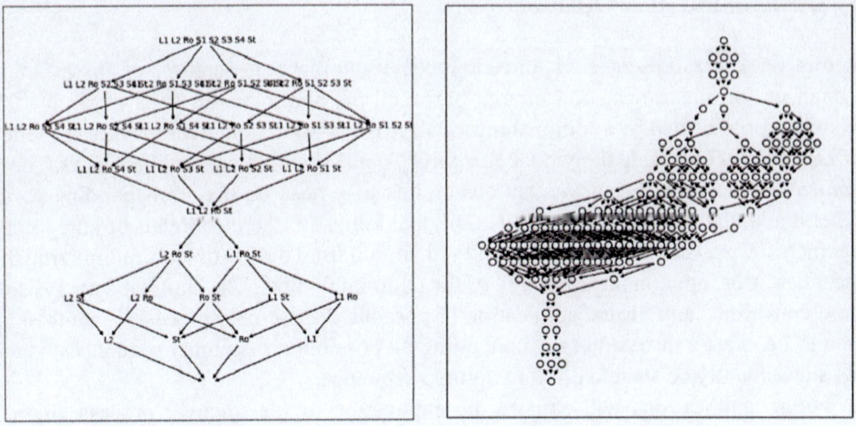

a) b)

Fig. 4. Disassembly sequence graph. a) simplified motor. b) real motor.

While not yet implemented, this approach can easily be expanded to include alternative (destructive) processes, by adding additional edges and nodes to the graph where necessary. Especially when considering destructive process alternatives these alternatives might lead to different disassembly states, even if applied to loosen the same joint. If we consider for example the loosening of a screw joint, this might be accomplished by unscrewing the screw, but also by milling of the screw head. Since part of the screw remains in the *Assembly* this is a different disassembly state and must therefore be added as an individual node to the graph. Since a disassembly state represented by an individual *Assembly* object and thereby in turn by its included *Parts* and *Connections* it is distinctly described by the resulting movement restrictions posed by the connections.

4.2 Sequence Planning

The generated disassembly sequence graphs can subsequently be searched for an optimal disassembly sequence, for a given disassembly target state. This can be done by applying common graph search algorithms, for example Dijkstra's algorithm [19]. A target disassembly state does not necessarily have to be the complete motor disassembly, but could also be the disassembly of a single part or a subset of parts. In such cases the algorithm terminates as soon as it reaches a state in which all target parts are removed.

In the current implementation of the algorithm the cost for each disassembly process in the sequence graph is calculated during the graph-search as the distance from the center-of-mass of the part to be removed to the previously removed part. For the simplified motor model mentioned in Sect. 3.2 this results in the following disassembly sequence: Removal of the screws, followed by the upper lid, the rotor, the stator and the lower lid. Other cost metrices may include the necessity for tool changes, process time or the process destructiveness (i.e. measured by the destroyed part volume). While these are not currently implemented they will be part of future works.

5 Conclusion and Outlook

In this work we present an approach for the automatic generation of disassembly sequences for electric traction motors based on the automatic analysis of simple 3D models supplemented by additional information on existing joints. Motors are modelled as *Assemblies* containing individual *Parts* and *Connections* between pairs of parts. These *Connections* contain the movement constraints they pose on the corresponding parts, either due to the part geometry or by additional joints, like screw threads or glue joints. Geometric *Connections* are automatically identified from the 3D models minimizing the necessary user input on the definition of the additional joints. The automatic analysis of these constraints and virtual application of possible disassembly processes can then be used to generate a directed graph containing all possible disassembly sequences which can subsequently be searched for an optimal sequence.

Future publications will explore the application of (destructive) process alternatives in more detail as well as the implementation of additional cost metrices for the sequence planning. In a greater context, this work is part of an effort to develop an adaptive disassembly system for electric motors using industrial robots and process specific endeffectors as described in [9]. The inclusion of the automatic planning module with the physical disassembly hardware currently under development, as well as an overall system validation will also be addressed in future publications.

References

1. Calisto Friant, M., Vermeulen, W.J., Salomone, R.: A typology of circular economy discourses: navigating the diverse visions of a contested paradigm. Resour. Conserv. Recycl. **161**, 104917 (2020). https://doi.org/10.1016/j.resconrec.2020.104917
2. Geissdoerfer, M., Pieroni, M.P., Pigosso, D.C., Soufani, K.: Circular business models: a review. J. Clean. Prod. **277**, 123741 (2020). https://doi.org/10.1016/j.jclepro.2020.123741
3. European Commission. Closing the loop: Commission adopts ambitious new Circular Economy Package to boost competitiveness create jobs and generate sustainable growth. Brussels (2015)
4. Stanek, R., et al.: Wertschöpfungspotenziale von E-Motoren für den Automobilbereich in Baden-Württemberg (2021)

5. Reuter, B., Hendrich, A., Hengstler, J., Kupferschmid, S., Schwenk, M.: Rohstoffe für innovative Fahrzeugtechnologien: Herausforderungen und Lösungsansätze (2019)
6. Nordelöf, A., Grunditz, E., Lundmark, S., Tillman, A.-M., Alatalo, M., Thiringer, T.: Life cycle assessment of permanent magnet electric traction motors. Transp. Res. Part D: Transp. Environ. **67**, 263–274 (2019). https://doi.org/10.1016/j.trd.2018.11.004
7. Tiwari, D., Miscandlon, J., Tiwari, A., Jewell, G.W.: A review of circular economy research for electric motors and the role of industry 4.0 technologies. Sustainability **13**(17), 9668 (2021). https://doi.org/10.3390/su13179668
8. Tang, Q., Shu, X., Zhu, G., Wang, J., Yang, H.: Reliability study of BEV powertrain system and its components—a case study. Processes **9**(5), 762 (2021). https://doi.org/10.3390/pr9050762
9. Hansjosten, M., Fleischer, J.: Towards autonomous adaptive disassembly of permanent-magnet synchronous motors with industrial robots. Manuf. Lett. **35**(37), 1336–1346 (2023). https://doi.org/10.1016/j.mfglet.2023.08.084
10. Guo, X., Zhou, M., Abusorrah, A., Alsokhiry, F., Sedraoui, K.: Disassembly sequence planning: a survey. IEEE/CAA J. Autom. Sinica **8**(7), 1308–1324 (2021). https://doi.org/10.1109/JAS.2020.1003515
11. Friedmann T.: Integration von Produktentwicklung und Montageplanung durch neue, rechnergestützte Verfahren [Dissertation]. Karlsruhe: Universität Karlsruhe (1989)
12. Münker, S., Schmitt, R.H.: CAD-based AND/OR graph generation algorithms in (dis)assembly sequence planning of complex products. Procedia CIRP **106**, 144–149 (2022). https://doi.org/10.1016/j.procir.2022.02.169
13. Prioli, J.P.J., Alrufaifi, H.M., Rickli, J.L.: Disassembly assessment from CAD-based collision evaluation for sequence planning. Robot. Comput.-Integrated Manuf. **78**, 102416 (2022). https://doi.org/10.1016/j.rcim.2022.102416
14. Neb, A., Göke, J.: Generation of assembly restrictions and evaluation of assembly criteria from 3D assembly models by collision analysis. Procedia CIRP **97**, 33–38 (2021). https://doi.org/10.1016/j.procir.2020.05.201
15. Issaoui, L., Aifaoui, N., Benamara, A.: Modelling and implementation of geometric and technological information for disassembly simulation in CAD environment. Int. J. Adv. Manuf. Technol. **89**(5–8), 1731–1741 (2017). https://doi.org/10.1007/s00170-016-9128-9
16. DIN 8593-0:2003-09, Fertigungsverfahren Fügen_- Teil_0: Allgemeines; Einordnung, Unterteilung, Begriffe. Beuth Verlag GmbH, Berlin. https://doi.org/10.31030/9500684
17. DIN 8593-1:2003-09, Fertigungsverfahren Fügen_- Teil_1: Zusammensetzen; Einordnung, Unterteilung, Begriffe. Beuth Verlag GmbH, Berlin. https://doi.org/10.31030/9500685
18. DIN 8593-8:2003-09, Fertigungsverfahren Fügen_- Teil_8: Kleben; Einordnung, Unterteilung, Begriffe. Beuth Verlag GmbH, Berlin. https://doi.org/10.31030/9500692
19. Dijkstra, E.W.: A note on two problems in connexion with graphs. Numerische Mathematik 269–271 (1959). https://doi.org/10.1145/3544585.3544600

Approach for Advanced Mechanical Recycling Strategies of Rare Earth Magnets Applied in Traction Drives

Thorsten Ihne(✉), Roman Hahn, Nico Wieprecht, Jörg Franke, and Alexander Kühl

Friedrich-Alexander University Erlangen-Nuremberg, Institute FAPS, 90429 Nuremberg, Germany

thorsten.ihne@faps.fau.de

Abstract. Rare Earth (RE) Magnets are crucial for the green transition of the European Union (EU). This applies especially for e-mobility where RE magnets are used for highly efficient traction drives. Permanent magnetic motors offer the best efficiency, but the use of Rare Earth Elements (REEs) is related to several drawbacks. The magnets and the required raw materials are almost exclusively imported. This causes problems for European industry due to supply chain risks and continuously increasing prices. In addition, primary production is associated with a poor ecological footprint. Therefore, the EU has declared REs as critical raw materials with highest supply chain risk and plans a compulsory recycling share of 15% until 2030. Currently, no industrially applicable processes exist for systematic recovery of RE magnets. In this context the following paper should present approaches for an automated process chain starting with an end-of-life traction drive until the extracted and demagnetized magnets.

Keywords: Rare earth magnets · Recycling · Traction drives · Critical raw materials

1 Introduction

The consequences of climate change are confronting global politics, industry and society with major challenges. In this context the European Union (EU) has set the target of climate neutrality until 2050 by the "European Green Deal" [1]. To achieve the related emission targets, security of supply for relevant raw materials is essential. A key role is played by rare-earth elements (REE) used for high-performance magnets such as Neodymium (Nd), Samarium (Sm), Praseodymium (Pr), Terbium (Tb) and Dysprosium (Dy). Rare earth (RE) magnets are crucial for many green technologies, with neodymium–iron–boron (NdFeB) magnets dominating (market share > 95%) [2–6]. Relevant applications include for example e-mobility, renewable energy generation, domestic infrastructure and applications in electronics, automatization and aerospace. As a result, high growth in demand is expected, which could exceed the production despite productivity increases [4, 7–10]. Forecasts suggest that demand for relevant RE materials may reach 83 kt in 2030 while estimated production in the same year is only 50 kt (2020: 24 kt) [4]. REE

T. Bauernhansl et al. (Eds.): WGP 2023, LNPE, pp. 458–468, 2024.
https://doi.org/10.1007/978-3-031-47394-4_45

free alternatives such as hard ferrites and Aluminum-Nickel-Cobalt (AlNiCo) can hardly substitute RE magnets because they provide much lower magnetic properties [6, 8, 11].

The European demand for REE magnets and the required raw materials is almost exclusively based on imports. Although rare earth material reserves occur globally, mining is profitable only in a few locations causing poor market diversification. Even more critical is the concentration of the further processing. Currently the PR China is dominating the global market for REE magnets with a market share of 94% due to a leading position in mining and refining of REE [10–13].

This situation is causing several drawbacks for the European industry. The market of relevant REE oxides and REE magnets is very dynamic with high price volatility driven by speculation and continuously increasing prices due to expanding demand. Furthermore, there is an additional supply chain risk resulting from political influence, as most suppliers are controlled by the Chinese state and the PR China uses its market power as leverage in political and economic negotiations. An example of this strategy is the REE crisis in 2010 and 2011 caused by a Chinese embargo against Japan [3, 10, 14, 15]. Finally, the primary production in China is related to a poor ecological footprint including the dispersion of radioactive elements into the environment [3–5, 15, 16].

Due to the combination of strategic relevance and supply risk the EU is declaring REE as critical raw materials (CRM) of the highest category since 2011 as shown in Fig. 1 [17]. Several measures are planned by the proposed European Critical Raw Materials Act including the establishment of recycling capacity for 15% of the annual consumption of the EU [2].

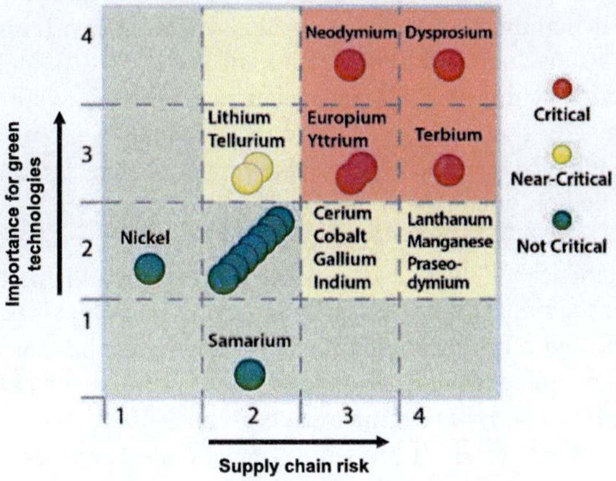

Fig. 1. Categorization of critical raw materials [6]

End-of-life (EOL) products represent a major reserve of RE materials, as the amount of Nd and Dy in within in-use devices exceeds the primary production [18]. However, no industrially applicable processes currently exist that allow systematic recovery of rare earth magnets from e-waste and EOL products [19, 20]. Factors affecting the complexity of the recycling process include the type of EOL product, the magnet composition and type (sintered or resin bonded) and the applied protective coating [12]. Therefore less than 1% of rare earth elements are recycled in Europe at the moment [4, 9, 10, 21]. In addition, a small share of magnetic scrap in Europe is collected and exported to Asia for metallurgical processing [22]. Accordingly, it is essential to develop sufficient technologies enabling establishment of a European closed raw material cycle for REE. Moreover an effective EOL product management is required [23].

2 Usage of REE Magnets for Automotive Traction Drives

2.1 Relevance of Permanent Magnets for Electric Mobility

A central part of the mobility agenda of the EU is the expansion of electric mobility to achieve decarbonization. Currently the European electric vehicle market is almost growing exponentially [10, 24]. By 2030 at least 30 million zero-emission cars should be operated in the EU which will mostly be driven by electric traction drives [25].

The automotive industry is already today a major application field for rare earth magnets where they are used for various devices such as electric motors, actuators, sensors, pumps and speakers [14, 26, 27]. By the expansion of e-mobility this share will significantly grow mainly based on permanent magnetic traction drives. In 2030 a share of approx. 25% of the NdFeB consumption is expected to be used for electric vehicles (EV) with more than 90% utilized for traction drives [10, 14]. This corresponds to a growth in demand for rare earth permanent magnets from between 6,000 to 9,000 tons in 2020 to between 40,000 and 70,000 tons in 2030 [10].

Automotive traction motors are built in various configurations based on the specific strategy of the OEM and the installation position. Current commercial use focuses on three motor types including the permanent excited synchronous machine, the externally excited synchronous machine and the induction machine [28–30]. Only the permanent excited synchronous machine contains permanent magnets but is by far the most relevant motor type with market shares reaching up to more than 90% [7, 10, 12, 31, 32]. The reason for this dominance is based on favorable properties which are mostly related to the absence of excitation windings in the rotor and the associated losses. As result the motor combines very high efficiency, gravimetric torque density, reliability and positive packaging characteristics [33–37]. Due to their high efficiency, the permanent magnetic drives contribute to maximized vehicle ranges and reduced battery sizes. Hybrid vehicles are usually equipped with PM machines due to packaging restrictions [20, 36]. For all-wheel drive vehicles, permanent excited synchronous machines and induction machines are often combined as both types complement each other [28].

Despite the excellent technical properties of permanent magnet motors, there are various activities by automotive OEMs and suppliers to substitute this type of motors due to economic and strategic reasons related to the usage of REE. In addition to supply chain risks, the main reason are the high magnet costs, which can account for around 50% of total motor costs [14].

Currently for series use available alternatives generally offer lower performance than permanent magnet drives. In addition, new alternative motor types are under development that have no or reduced magnet content [38]. Axial flux motors have the potential of reduced magnet content. Current developments without permanent magnets include switched reluctance machines and optimized externally excited synchronous machines with inductive transmitters. However, the development maturity of these motors is still not sufficient for full series applicability. The same applies for potential magnet materials with no or reduced RE content. In this context, the establishment of a circular economy for RE permanent magnets offers an attractive alternative to achieve cost stabilization and reliable supply chains of permanent excited synchronous machines.

2.2 Potentials of Magnet Extraction from Automotive Traction Drives

The recovery of RE magnets from traction drives has not been attractive in the past due to lack of economic competitiveness against primary production in China, high organizational and technological effort and non-sufficient volumes of EOL vehicles [21, 23]. As result many research activities have been stopped after normalization of markets following the REE crisis in 2010 and 2011. Meanwhile this situation has significantly changed by commercial reasons and political regulations.

Assuming an average lifetime of an electric vehicle (EV) of 150,000 km or 8 to 12 years significant amounts of EOL EVs can be expected in the near future, as many relevant high-volume models have been introduced in Europe since around 2013. For example, the number of registrations of vehicles with PM drive in 2015 was more than 155,000 which are now gradually reaching the recycling process [39]. In the period between 2015 and 2021, registration numbers of EVs (battery electric and plug-in hybrids) in the EU-27 increased by a factor of more than 13 [24]. This growth must also be reflected in a rapid expansion of recycling capacities and scalability of recycling technologies.

Traction drives are a promising source for the recovery of REE due to the absolute amount of magnets and the applied quality grade [40]. In general hybrid and full electric vehicles contain between 1.0 and 5.0 kg of sintered RE magnets depending on the specific drive architecture with typical values between 2.5 and 3.0 kg [12, 14, 20, 41]. The applied NdFeB alloy usually consists of iron (approx. 60–70 w%), boron (approx. 1 w%) and 28–35 wt% REE, with varying composition and Nd as dominating element [6, 23, 39, 42]. The addition of heavy rare earths (HREE) like Dy and Tb to the alloy optimizes the properties of the material but also significantly affects the material cost [14, 41–44]. The most relevant HREE are Tb improving the coercivity and Dy affecting the temperature resistance. The oxides of both elements exceed the price of neodymium by a factor of approx. three for Dy and by a factor of twelve for Tb [14]. In contrast, the less expensive praseodymium can be used to reduce the neodymium content without affecting the magnetic properties too much [6].

Due to the high requirements for traction drives regarding the magnetic properties and the prevailing operating temperatures, they have a comparatively high content of HREE, which increases their attractiveness for recycling [12, 27]. As result of these characteristics traction drives are worthwhile for the implementation of a specific disassembly process without shredding, which is standard for most EOL products including electric motors.

3 Approach for an Extraction Strategy of Magnets from Traction Drives and Related Challenges

Political regulations within the EU define mandatory recycling shares for EOL vehicles, so that they are already being recycled relatively efficiently today (target recycling rate of 95 w%) [45, 46]. In general, the recycling process of end-of-life vehicles (ELV) can be divided into the steps depollution, manual dismantling, shredding and post-shredding sorting [16, 45].

Already in today's standard recycling process of ELVs, it is usual to extract specific assemblies for separate recycling due to environmental and safety requirements or commercial reasons. This usually applies also for the powertrain of conventional vehicles as well as for hybrids and battery electric vehicles (BEV). Electric traction motors are often shredded for material extraction. The resulting material mixture is usually sorted into the main fractions steel, aluminum and copper, which is the main value carrier [12, 17, 47]. As the implemented sorting processes do not differentiate between soft and hard magnetic materials the magnet fragments are collected and processed together with the steel scrap. As result, the REE materials cannot be recycled because economic recovery from the slags resulting from steel recycling is currently not economically feasible. The separation of soft and hard magnetic components of a granulate in industrial scale is only possible with very high effort, if at all, so that a potential process for the REE recovery should avoid shredding magnets [9, 42]. As result a sequential process chain is required for dedicated magnet recycling starting with the already existing removal of the drive as shown in Fig. 2. This process is complex and faces various challenges [40]. Important requirements are high flexibility and scalability.

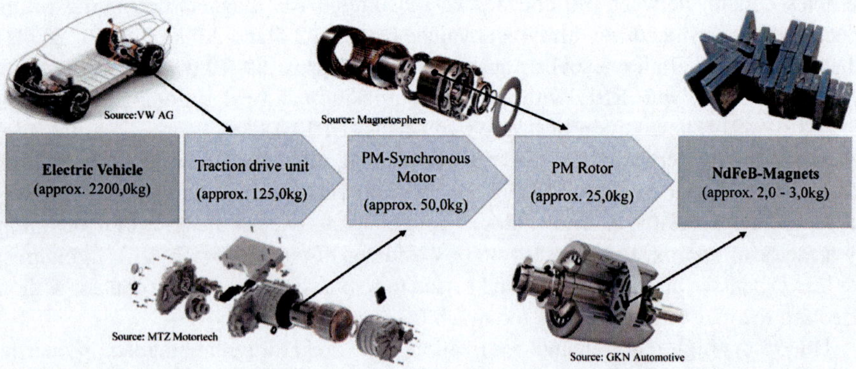

Fig. 2. Overall process chain for the extraction of NdFeB magnets from traction drives

Recycling of EOL vehicles usually takes place across manufacturers at specialized facilities [46]. Therefore, a lack of information regarding the individual EOL product and the related disassembly process can be expected. As not all traction drives are containing magnets an identification of the motor type must take place. The simplest way is to match the vehicle type with a database even before the drive train is removed.

While drives without magnets can be processed by shredding PM drives have to be feed into the dismantling process. Afterwards the electric motor has to be isolated before further dismantling. Traction drives of BEVs are usually a functional unit consisting of the motor, a gearbox and the power electronics. In case of a hybrid vehicle the electric drive, the ICE and the gearbox have to be separated. In both cases it can be assumed that the disassembly will be predominantly manual because of the variance of the drives and the complexity of the disassembly task. Afterward the motor housing has to be opened and the rotor and the stator have to be separated due to the installation position of the magnets. According to existing processes this operation can be carried out semi-automatically using hydraulic tools. To handle the expected amount of EOL drives an increased degree of automatization is required, which can be realized by the combination of handling robots and vision systems [48].

The isolated rotors have to be transferred to a disassembly station for further processing. In this context, the particular rotor design represents an important boundary condition for the extraction of the magnets. A corresponding dismantling process is currently being developed at the institue FAPS, whereby a focus is set on adaptability to the rotor topology. The consideration concentrates on interior permanent magnets (IPM), which are dominating for traction drives [49]. The magnets of IPM rotors are located in cavities in the laminated core and fixed by adhesive bonds, plastic overmoldings or press fits, which must be destroyed accordingly. A further challenge results from the common design of PM traction drives with twisted rotor stacks reducing torque ripple. As result the single magnets often cannot be directly removed from the cavities.

In addition, the magnetic reaction forces between the individual magnets and between the magnets and the laminated core have to be considered [47, 50].

As result the integration of a demagnetization operation into the overall process is mandatory because the interfering reaction forces are impeding the magnet extraction. The simplest method is heating the magnets to temperatures above the specific Curie temperature. In general, two potential process chains can be differentiated. On the one hand post-demagnetization allows energy efficient demagnetization with a wide range of methods but the acting reaction forces complicate the previous handling and magnet extraction. Furthermore, the magnets have to be handled separately to prevent collisions. This can be realized by spatial or temporal separation [50]. On the other hand, in-situ demagnetization can be established offering simplified handling. The main disadvantage of this procedure is that the complete rotor with a high thermal capacity has to be heated up.

Preliminary tests at insitute FAPS based on a specific EOL traction drive have shown that inductive in-situ heating of complete rotors in several steps is a promising option. During the experiments, single rotor disks were heated with an inductor. The target value of a surface temperature of 400° at the magnet surface was achieved within a few seconds. In a later process, the induction coil and rotor can be moved relatively to each other for this purpose. As side effect decomposition of the epoxy resin often used for magnetic fixation occurs at the applied temperatures [51].

Afterwards the rotor shaft is pressed out to alllow separation of the lamination stacks. Trials have shown that both steps can be done in parallel by using a profiled pressing tool in combination with a hydraulic press causing also deformation of the iron core.

As alternative the single stacks can be sheared-off by transverse forces. After separating the stacks, the magnets can be directly forced out of the cavities. This process step is supported by the enbrittlement of the adhesive bonds. As shown in Fig. 3, in-situ heating ensures a significant reduction in the required press-out forces, so that several magnets could be pressed out simultaneously.

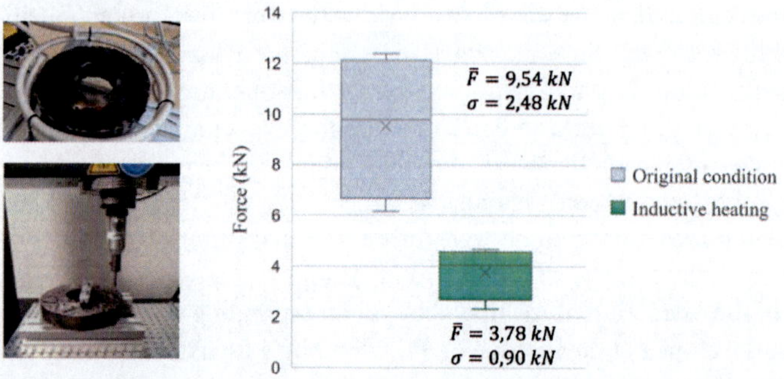

Fig. 3. Press-out forces without and with inducitve heating for single magnets of a rotor stack

Increased flexibility can be realized by a vision system for identification of the magnet positions in combination with an adaptive pressing tool. Based on the results of the investigations at the institute FAPS, the process chain visualized in Fig. 4 is proposed for the dismantling of IPM rotors.

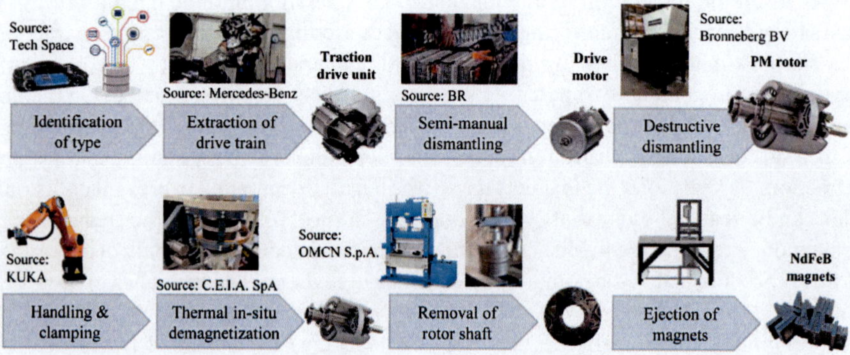

Fig. 4. Potential overall process chain for the extraction of magnet from IPM rotors

The collected magnets can be transferred to further processing. Direct reuse could be an option but is not preferred based on quality issues like mechanical defects and organizational difficulties like compatibility issues [39, 50]. As result reuse is only an option if the OEMs are processing their own production scrap or take back their EOL products. Instead, hydro- and pyrometallurgical processing is preferred allowing the recovery of the used REE and the production of new magnets. Hydrometallurgical processes allow REE to be recovered by chemical dissolution of the magnets. As result the further processing is very flexible, but the environmental footprint is high. Pyrometallurgical methods are based on remelting and allow the subsequent extraction of pure REE or RE-containing alloys [32, 39, 42]. Both process chains are currently available and are continuously optimized.

4 Conclusion and Outlook

In the present paper the potentials and the challenges of REE recovery from automotive traction drives are discussed. With the strong growth of e-mobility, electric traction motors are a major driver of demand for REEs such as Nd, Dy, and Tb, as well as a promising source for the recovery of these critical raw materials. Currently, no systematic recovery of REE is existing in the automotive industry, although economic and political constraints are increasing the pressure to act. Accordingly, there is high need for flexible and scalable processes that allow the establishment of a closed material cycle. However, there are various technological and organizational obstacles that need to be overcome.

In this context, general approaches for the mechanical extraction of RE magnets from traction drives are presented. The required preprocessing includes the extraction of the drive train from the EOL vehicle, the isolation of the electric motor and the extraction of the PM rotor. Afterwards the magnets are thermally demagnetized, the rotor shaft is removed and the rotor stacks are separated. Finally, the magnets are forced out and collected for further hydro or pyrometallurgical processing.

As part of the EU-funded REEPRODUCE project, the complete process chain from extraction of the motor to production of new magnets is currently being examined and experimentally validated. As a proof of function, the construction of test machinery is currently ongoing which should extract significant quantities of magnets from EOL drives.

Acknowledgements. The presented results have been obtained in context of research activities at the Institute for Factory Automation and Production Systems (FAPS) of the Friedrich-Alexander University Erlangen-Nuremberg (FAU). This research has been funded by the European Union Grant Agreement N° 101057733.

References

1. European Commission. The European Green Deal, Brussels (2019)
2. European Commission. Regulation of the European Parliament and of the Council: establishing a framework for ensuring a secure and sustainable supply of critical raw materials and amending Regulations (EU) 168/2013, (EU) 2018/858, 2018/1724 and (EU) 2019/1020, Brussels (2023)
3. Barakas, G., Mischo, H., Gutzmer, J.: An outlook on the rare earth elements mining industry (2016)
4. Golroudbary, S.R., Makarava, I., Kraslawski, A., Repo, E.: Global environmental cost of using rare earth elements in green energy technologies. Sci. Total. Environ. **832**, 155022 (2022)
5. Patil, A.B., Struis, R.P.W.J., Ludwig, C.: Opportunities in critical rare earth metal recycling value chains for economic growth with sustainable technological innovations (2022)

6. Binnemans, K., Jones, P.T., Müller, T., Yurramendi, L.: Rare earths and the balance problem: how to deal with changing markets? **4**, 126 (2018)
7. Bobba, S., Carrara, S., Huisman, J., Mathieux, F., Pavel, C.: Critical raw materials for strategic technologies and sectors in the EU: a foresight study. Publications Office of the European Union, Luxemburg (2020)
8. Glöser-Chahoud, S., Kühn, A., Tercero Espinoza, L.: Globale Verwendungsstrukturen der Magnetwerkstoffe Neodym und Dysprosium: Eine szenariobasierte Analyse der Auswirkung der Diffusion der Elektromobilität auf den Bedarf an Seltenen Erden, Karlsruhe (2016)
9. Binnemans, K., Jones, P.T., Blanpain, B., van Gerven, T., et al.: Recycling of rare earths: a critical review. **51**, 1 (2013)
10. Gauß, R., et al.: Rare Earth Magnets and Motors: A European Call for Action: A report by the Rare Earth Magnets and Motors Cluster of the European Raw Materials Alliance, Berlin (2021)
11. Binnemans, K., McGuiness, P., Jones, P.T.: Rare-earth recycling needs market intervention. **6**, 459 (2021)
12. U.S. Department of Energy. Rare Earth Permanent Magnets: Supply Chain Deep Dive Assessment (2022)
13. München, D.D., Stein, R.T., Veit, H.M.: Rare earth elements recycling potential estimate based on end-of-life NdFeB permanent magnets from mobile phones and hard disk drives in Brazil. **11**, 1190 (2021)
14. Gielen, D., Lyons, M.: Critical materials for the energy transition: rare earth elements (2022)
15. Sprecher, B., Xiao, Y., Walton, A., Speight, J., et al.: Life cycle inventory of the production of rare earths and the subsequent production of NdFeB rare earth permanent magnets. Environ. Sci. Technol. **48**, 3951 (2014)
16. Jin, H., Afiuny, P., McIntyre, T., Yih, Y., et al.: Comparative life cycle assessment of NdFeB magnets: virgin production versus magnet-to-magnet recycling. **48**, 45 (2016)
17. Blengini, G.A., et al.: Study on the EU's list of critical raw materials (2020): Final report. Publications Office of the European Union, Luxembourg (2020)
18. Du, X., Graedel, T.E.: Global rare earth in-use stocks in NdFeB permanent magnets. **15**, 836 (2011)
19. Rademaker, J.H., Kleijn, R., Yang, Y.: Recycling as a strategy against rare earth element criticality: a systemic evaluation of the potential yield of NdFeB magnet recycling. Environ. Sci. Technol. **47**, 10129 (2013)
20. Reimer, M., Schenk-Mathes, H., Hoffmann, M., Elwert, T.: Recycling decisions in 2020, 2030, and 2040—when can substantial NdFeB extraction be expected in the EU? **8**, 867 (2018)
21. Ciacci, L., Vassura, I., Cao, Z., Liu, G., et al.: Recovering the "new twin": analysis of secondary neodymium sources and recycling potentials in Europe. **142**, 143 (2019)
22. Elwert, T., Schwarz, S., Bergamos, M., Kammer, U.: Entwicklung einer industriell umsetzbaren Recycling-Technologiekette für NdFeB-Magnete – SEMAREC. In: Recycling und Rohstoffe, Thomé-Kozmiensky Verlag GmbH, Neuruppin, p. 253 (2018)
23. Maani, T., Mathur, N., Singh, S., Rong, C., et al.: Potential for Nd and Dy recovery from end-of-life products to meet future electric vehicle demand in the U.S. **98**, 109 (2021)
24. European Environment Agency. New registrations of electric cars, EU-27. https://www.eea.europa.eu/data-and-maps/daviz/new-electric-vehicles-in-eu-2#tab-chart_3
25. European Commission. Sustainable and Smart Mobility Strategy: Putting European transport on track for the future, Brussels (2020)
26. Raghuraman, B., Nategh, S., Sidiropoulos, N., Petersson, L., et al.: Sustainability aspects of electrical machines for E-mobility applications part I: a design with reduced rare-earth elements. In: 47th Annual Conference of the IEEE Industrial Electronics Society, IECON 2021, p. 1. IEEE (2021)

27. Goodenough, K.M., Wall, F., Merriman, D.: The rare earth elements: demand, global resources, and challenges for resourcing future. Generations **27**, 201 (2018)
28. Flemming, J., et al.: Wertschöpfungspotenziale von E-Motoren für den Automobilbereich in Baden-Württemberg: Themenpapier Cluster Elektromobilität Süd-West
29. Kampker, A., Heimes, H., Dorn, B.: Elektromotoren-Produktion: Die Wertschöpfungskette im Spannungsfeld von "Market Pull" und "Technology Push", Aachen (2022)
30. Tschöke, H., Gutzmer, P., Pfund, T.: Elektrifizierung des Antriebsstrangs. Springer, Heidelberg (2019)
31. Alves Dias, P., Bobba, S., Carrara, S., Plazzotta, B.: The role of rare earth elements in wind energy and electric mobility: an analysis of future supply/demand balances. Publications Office of the European Union, Luxembourg (2020)
32. Rizos, V., Righetti, E., Amin, K.: Developing a supply chain for recycled rare earth permanent magnets in the EU: CEPS In-Depth Analysis, Brussels (2022)
33. Hwang, M.-H., Han, J.-H., Kim, D.-H., Cha, H.-R.: Design and analysis of rotor shapes for IPM motors in EV power traction platforms. **11**, 2601 (2018)
34. Al-Qarni, A., EL-Refaie, A.: On eliminating heavy rare-earth PM elements for high power density traction application motors. In: 2021 IEEE International Electric Machines & Drives Conference (IEMDC), p. 1. IEEE (2021)
35. Du, Z.S., Lipo, T.A.: High torque density and low torque ripple shaped-magnet machines using sinusoidal plus third harmonic shaped magnets. **55**, 2601 (2019)
36. Sato, E.: Permanent magnet synchronous motor drives for hybrid electric vehicles. **2**, 162 (2007)
37. Henneberger, S., Van Haute, S., Hameyer, K., Belmans, R.: Design of a 45 kW Permanent Magnet Synchronous Motor for a Hybrid Electric Vehicle, Leuven (1997)
38. Smith, B.J., Eggert, R.G.: Multifaceted material substitution: the case of NdFeB magnets, 2010–2015. **68**, 1964 (2016)
39. Elwert, T., Goldmann, D., Roemer, F., Schwarz, S.: Recycling of NdFeB magnets from electric drive motors of (hybrid) electric vehicles. **3**, 108 (2017)
40. Jönsson, C., Awais, M., Pickering, L., Degri, M., et al.: The extraction of NdFeB magnets from automotive scrap rotors using hydrogen. **277**, 124058 (2020)
41. Gielen, D.: Critical minerals for the energy transition (2021)
42. Yang, Y., Walton, A., Sheridan, R., Güth, K., et al.: REE recovery from end-of-life NdFeB permanent magnet scrap: a critical review. **3**, 122 (2017)
43. Glöser-Chahoud, S., Pfaff, M., Tercero Espinoza, L., Faulstich, M.: Dynamische Materialfluss-Analyse der Magnetwerkstoffe Neodym und Dysprosium, Tutzing (2016)
44. Leader, A., Gaustad, G.: Critical material applications and intensities in clean energy technologies. **1**, 164 (2019)
45. Sakai, S., Yoshida, H., Hiratsuka, J., Vandecasteele, C., et al.: An international comparative study of end-of-life vehicle (ELV) recycling systems. **16**, 1 (2014)
46. Martens, H., Goldmann, D.: Recyclingtechnik. Springer Fachmedien Wiesbaden, Wiesbaden (2016)
47. Klier, T., Risch, F., Franke, J.: Disassembly, recycling, and reuse of magnet material of electric drives. In: 2013 IEEE International Symposium on Assembly and Manufacturing (ISAM), p. 88. IEEE (2013)
48. Fleischer, J., Gerlitz, E., Rieβ, S., Coutandin, S., et al.: Concepts and requirements for flexible disassembly systems for drive train components of electric vehicles. **98**, 577 (2021)
49. Van Haute, S., Henneberger, S., Hameyer, K., Belmans, R., De Temmerman, J., De Clercq, J.: Design and control of a Permanent Magnet Synchronous Motor Drive for a Hybrid Electric Vehicle, Leuven (1996)

50. Bast, U., et al.: Recycling von Komponenten und strategischen Metallen aus elektrischen Fahrantrieben: Abschlussbericht zum Verbundvorhaben (2014)
51. Buch, X., Shanahan, M.E.R.: Influence of the gaseous environment on the thermal degradation of a structural epoxy adhesive. **76**, 987 (2000)

Sustainable Computing Through Open Standard ISAs: Leveraging Tailor-Fit Hardware Designs for Circular Economies

J. C. Mariscal-Melgar(✉), Manuel Moritz, Tobias Redlich, and Jens P. Wulfsberg

Helmut-Schmidt University/University of the Federal Armed Forces, Hamburg, Germany
jc.mariscal@hsu-hh.de

Abstract. The increase in demand for computational power along with society's reliance on proprietary processor architectures, raise concerns about processor sovereignty and prompts the exploration of open standard instruction set architectures (ISAs) such as RISC-V. Concurrently, the impact of e-waste in the environment and society's resource consumption highlights the need for circular economy principles and sustainable production methods within the electronic industry. In this study, computational power, open standard ISAs are explored to survey recent hardware implementations, particularly of RISC-V processors with a focus on power efficiency gains. We argue that integrating open standard ISAs, circular economy strategies, and sustainable production practices may contribute to resilient and responsible products. This approach addresses some of the questions in resource sustainability and invites further exploration of the potential of open-source processor implementations to improve the way that electronic devices are produced and consumed.

Keywords: RISC-V · Circular Economy · Open-Source Hardware · Open Standard ISA

1 Introduction

Current geo-political events are changing the landscape of production and consumption in times of climate-neutral agendas. Global manufacturing susceptibility to supply shocks, exacerbated by commodity shocks [1], pandemics [2], financial crises [3], environmental disasters [4], and geopolitical tensions [5] substantiate the importance of regional and local economic self-sufficiency strategies [6]. The need of global manufacturers to reduce the environmental damage caused by processes within product life cycles [7] is an urgent matter. Circular economies, where closed loops of resources flow in production and consumption activities with minimal waste [8] is a major strategic objective in EU countries [9], and potentially a future topic of discussion at a global policy scale.

Within the reality of today's manufacturing environment, semiconductors are a hot topic of discussion. National security concerns have led to an industry with geographical and geopolitical challenges where export controls dictate supply-chain deals [10],

T. Bauernhansl et al. (Eds.): WGP 2023, LNPE, pp. 469–480, 2024.
https://doi.org/10.1007/978-3-031-47394-4_46

and where local semiconductor manufacturing is now a critical strategy [11]. Concerns about security have let to cancellation of technological infrastructure deals, such as with Huawei's 5G [12]. In this context, US and the EU policy makers, have incentivized the set-up of future semiconductor fabs in major cities, to provide an alternative to production networks in Asia for processor sovereignty [13].

As the supply chain challenge is alleviated with more production sites, the need for computational power continues to increase. This shift in policy, and production networks, may offer the opportunity to tackle new challenges, where we go from chip-consumers to tailor-fit hardware designs.

This paper surveys latest innovations in open standard instruction sets (ISAs) processor designs and relates them to circular economic frameworks for sustainable computing for the future.

2 Methodology

The methodology begins with an analysis of trends in computational power, utilizing data from the Stanford VLSI Dataset [14] and micro-controller trend dataset [15] as illustration of current trends, to explain growth patterns and determine factors influencing the form in which computing capabilities are advancing. Subsequently, Open-Source Instruction Set Architectures (ISAs) are explored, to gain further insight into the current state of open-standard ISA processor designs.

Following the analysis and discussion of open-standard ISAs, a survey of tailor-made processor designs is done, with a focus on those papers demonstrating significant improvements in power or performance. Finally, the potential of custom-made processor designs is discussed, in the context of circular economy.

3 Demand of Computational Power

Data generation and processing requirements across domains in the industry, governments and research institutions is growing. Cloud computing, AI, Internet of Things (IoT) require increasing amount of computational power to power search engines, social networks, servicing video or providing API to large language models (LLM).

Computing has been able to keep up with faster, cheaper, and lower power solutions that society is accustomed. Technology in cars, smartphones and smart embedded devices contain processing power that is closely following the down-scaling consistency of Moore's law exponential growth.

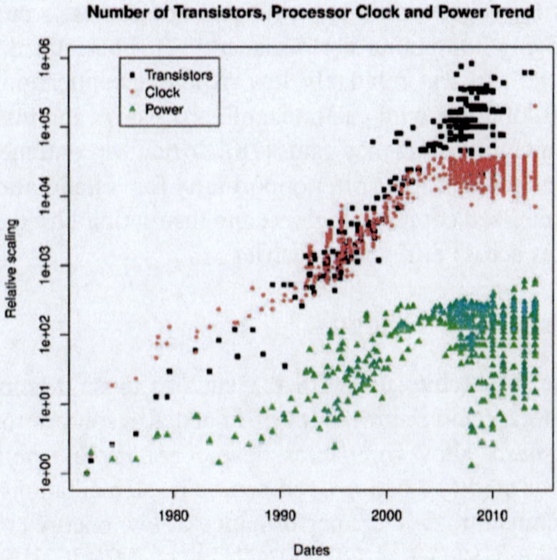

Fig. 1. Processor Scaling trends in clock and power (Data obtained from CPU-DB Dataset [14])

However, the power density of processors has not scaled in the same way as the quest to scaling clock frequencies, despite Dennard's suggestion of scaling power supply voltages at a constant field rate [16]. The result is an increase of power consumption of processors and challenges in maintaining electric field constant. Figure 1 illustrates the trend in scaling clock and power as transistor numbers increase. This deviation has prompted an alternative strategy since uniprocessors got into this power struggle, by including more processors on each die with multi-core architectures as shown in Fig. 2.

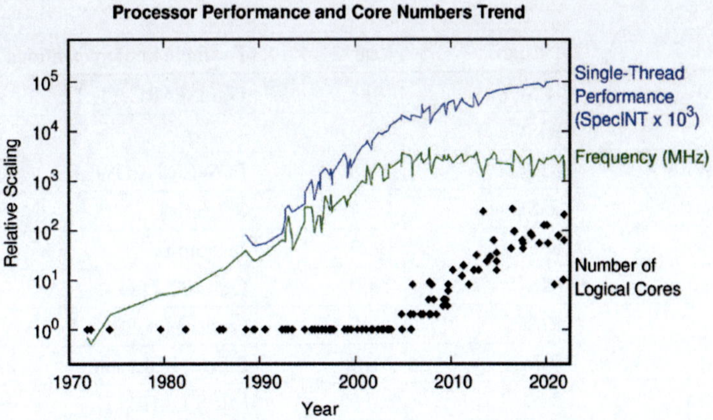

Fig. 2. Shift to multi-core architectures (Data obtained from '50 years of microprocessor data' Dataset [15])

Hence, there is the suggestion that power and not transistor numbers, is the limiting factor in improving computing performance [17]. Specialized processors may be optimized for special tasks at relatively low expense, application of FPGAs enable application-specific solutions with customizable processors and may introduce significant power and computation efficiency gains [18]. So that we continue to scale computing performance - Open Standard ISAs offer opportunity for collaboration in exploring new design process, specialized compute engines, and instruction-based extensions to create new innovative chips across different industries.

4 Tailor-Fit Processor Designs

Modern Instruction Set Architectures (ISAs), such as those developed by IBM, Intel or ARM, are proprietary and require a lot of financial resources to license and adapt. ISAs that are open source allow companies, developers, and independent researchers to implement designs or modify existing open-source implementations.

RISC-based architecture delivers performance at low energy consumption comparable to CISC designs [19]. ARM [20], SPARC [21], MIP [22], POWER [23] introduced processors as an alternative to CISC-based x86 processors. ARM has been able to dominate the smart-phone market and is gaining traction in data-centers usage.

RISC based open-standard ISAs architectures were introduced in the 80's and are still an ongoing development as illustrated in Table 1. RISC-V, introduced by a team at UC Berkeley [24], is allowing designers to integrate custom instruction sets to specific application domains in different fields. The RISC-V community has been working on the standardization of a wide range of instruction set extensions, including cryptography, bit manipulation and vector extensions, to enable hardware designers to build specialized processor cores or SoCs.

Table 1. Open-Source ISAs

Architecture	Design	Year	Example implementations
OpenSPARC	RISC	1985	UltraSPARC T1, T2
DLX	RISC	1990	–
OpenPOWER	RISC	1990	POWER8, POWER9
SuperH	RISC	1994	SH-series
MMIX	RISC	1999	fpgammix
OpenRISC	RISC	2000	OR1200, OR1K
MICO32	RISC	2006	LatticeMico32
RISC-V	RISC	2010	SiFive, PULPino
MIPS Open	RISC	2018	MIPS I, MIPS II

RISC or CISC chips with comparable performance levels have trivial differences in energy characteristics [19]. However, it is at low power/performance applications, where RISC architecture may provide advantages.

Table 2 provides a summary of latest implementations of RISC-V processors for different applications. Various aspects of RISC-V processors and their applications in the context of ultra-low power, high-performance, and energy efficient computing in various domains, including IoT, deep learning and HPC. Many of the projects showcase the versatility of RISC-V designs as tailor-made designs deliver specific solutions.

Table 2. Tailor-fit hardware implementations

Name/Desc	Summary
PsPIN	RISC-V in-network accelerator for high-performance low-power packet processing [25]
Low-power multimedia RISC	Battery powered multimedia systems with low-power chip and special instruction set [25]
Wallace Tree Multiplier	Low power and high-speed Wallace tree multiplier for RISC processor [26]
Zero-riscy	Ultra-low-power RISC-V cores for IoT applications with mixed arithmetic/control tasks [27]
PULP-NN	Parallel ultra-low-power RISC-V processors for quantized neural network inference [28]
Arnold	eFPGA-augmented RISC-V SoC for flexible and low-power IoT end nodes [29]
RISC-V ISA Extension	Ultra-low power IoT wireless signal processing with custom instructions [30]
Day-Night Architecture	Ultra-low power processor for wearable devices with anomaly detection [31]
HULK-V	Heterogeneous Linux-capable RISC-V SoC for complex DSP and ML tasks [32]
ColibriES	Milliwatt RISC-V embedded system with neuromorphic and neural network hardware accelerators [33]
Cheshire	Lightweight Linux-capable RISC-V host platform for domain-specific accelerator plug-in [34]
Vector operation unit	RISC-V-based vector operation unit for embedded processor [35]
AllWinner D1	Benchmarking the AllWinner D1 with vectorisation speed-ups [36]
Dustin	16-cores parallel ultra-low-power cluster with flexible bit-precision and vector lockstep execution mode [37]
BARVINN	Arbitrary precision DNN accelerator controlled by a RISC-V CPU [38]

(*continued*)

Table 2. (*continued*)

Name/Desc	Summary
Vitruvius+	Area-efficient RISC-V decoupled vector coprocessor for HPC applications [39]
Quark	Integer RISC-V vector processor for sub-byte quantized DNN inference [40]
HALO	Hardware-software co-designed processor for brain-computer interfaces [41]
FAUST	Pipelined RISC-V vector floating-point unit for vector processing [42]

5 Power Savings Through Tailor-Fit Designs

Among the projects listed, many share focus on RISC-V processor designs with emphasis on ultra-low power consumption without performance debt. Table 3 shows the various power related improvements.

The applicability of RISC-V powered devices may not be limited to embedded devices or ultra-low-power applications. RISC-V processors are suitable for HPC applications as vector computing advantages are used to run supercomputers such as Fugaku, a TOP500 leader [43].

The versatility and adaptability of custom processor designs for different applications in different domains may be combined with libre licensing and open co-development arrangements to support circularity and empower users and engineers.

6 Discussion: RISC-V Designs, Open-Source and Circular Economies

Open sourcing hardware, as exemplified by the numerous certified projects in the Open-Source Hardware Association (OSHWA) [44], might support the harmonization of project designs with circular economy policies. Digital fabrication machines, such as the case of 3D printers, have innovated the tinkering space of product customization and inspired new ideas of mass customization markets. Product customization may not be limited to mechanical customization, PCB re-layout or firmware tweaking. In the future, semiconductor chips may be part of the trend of miniaturization of product customization.

Advances in processor designs that use Open ISAs such as RISC-V improve resource efficiency, as shown in Table 3. This flexibility could translate to new design paradigms that include circularity as a requirement in product ideas in the industry, where issues such as repairability, re-usability, recyclability, modularity and openness in the design align with government policies.

Table 3. Power saving considerations

Name/Desc.	Category	Power savings
PsPIN	Network Accelerator	400 Gbit/s processing 26 ns latency for 64 B packets 18.5 mm^2 area [25]
Low-power multimedia RISC	Multimedia	Low-power chip with fast 32-bit parallel multiply-adder [25]
Wallace Tree Multiplier	Processor Implementation	44.4% faster 11% power reduction compared to conventional CMOS [26]
Zero-riscy	IoT	2.2× smaller 100 μW power envelope 1.4× more energy-efficient for control code [27]
PULP-NN	Neural Networks	14.1× higher efficiency than STM32L4 39.5× higher efficiency than STM32H7 [28]
Arnold	eFPGA-Augmented RISC-V SoC	3.4× better performance 2.9× better energy efficiency compared to other SoCs [29]
RISC-V ISA Extension	IoT Wireless Signal Processing	<1% power 28% area overhead 380 μW for Bluetooth LE 225 μW for LoRa [30]
Day-Night Architecture	Anomaly Detection	Day part off most of the time night part runs with low power [31]
HULK-V	IoT	250 mW power envelope 2× energy efficiency compared to LPDDR memories [32]
ColibriES	Low-Latency Closed-loop Control	7.7 mJ energy consumption 164.5 ms latency per inference [33]
Cheshire	Linux-capable RISC-V Host Platform	325 MHz frequency 300 mW total power on data-intensive workloads [34]
Vector operation unit	Embedded Processor	1.2× performance improvement compared to scalar processor [35]

(*continued*)

Table 3. (*continued*)

Name/Desc.	Category	Power savings
AllWinner D1	HPC	Reasonable vectorization speedup favorable performance compared to other processors [36]
Dustin	Integer RISC-V Vector Processor	58 GOPS peak performance 1.15 TOPS/W peak efficiency [37]
BARVINN	DNN Accelerator	8.2 TMACs computational power scalable throughput for quantized models [38]
Vitruvius+	HPC	1.4 GHz frequency 1.3 mm^2 area 920 mW maximum estimated power [39]
Quark	Sub-Byte Quantized DNN Inference	2× smaller 1.9× more power-efficient compared to Ara supports 1-bit and 2-bit quantized models [40]
HALO	Brain-Computer Interfaces	Low-power hardware Processing Elements [41]
FAUST	Pipelined RISC-V Vector FPU	Floating Point Units support high performance at low powers [42]

Figure 3 illustrates a diagram of production and consumption, based on the ideas of the Ellen MacArthur Foundation's butterfly diagram [45], to incorporate the role of the product designer and the openness of circular economic life cycle of designs that consider a component-wise breakdown of products, in modular systems.

In the industry, there have been attempts at creating products that fit the criteria of repairability, modularity, and in some degree, circularity of components such as the Project Ara, Phonebloks [46], Framework Laptop [47] or the MNT Reform [48]. Such projects have promising economic, ecological and societal benefits but their success is not guaranteed [46]. Nonetheless, the inclusion of tailor-fit open-source processors may prove useful in creating practical solutions that fit into component-level life cycles or even provide new ways for components re-use in electronic products.

The role of software longevity with the support of Free and Libre Open-Source Software (FLOSS) matters [49], as well as the co-design advantages of Open-Source Hardware (OSH) which allow for new ways of interacting with users. The role of open-standard ISAs comes into play with the future of product customizations. New projects will be able to include custom designed processors which the community and circular economic cycles support.

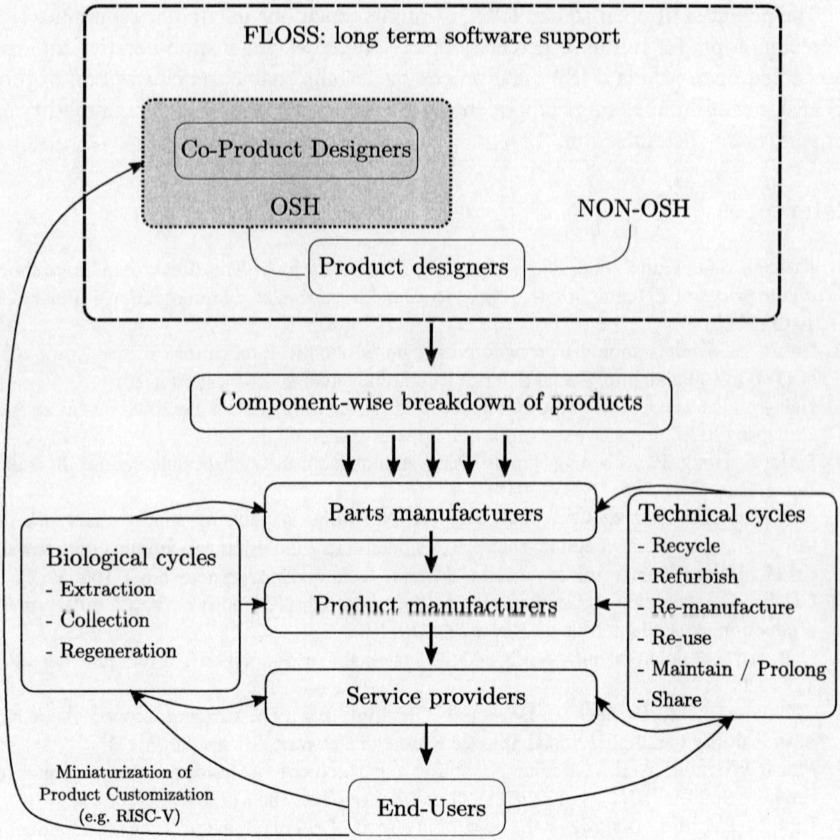

Fig. 3. Circular economic cycle (Based on [45])

7 Conclusion

The changing landscape of production and consumption, together with the urgency to reduce environmental impact and achieve circular economies, may need a shift towards local economic self-sufficiency. The semiconductor industry, at the forefront of geopolitical and national security concerns, presents an opportunity to explore ISAs and custom-made processor designs. This paper surveyed the latest innovations to gauge the potential of open-standard ISAs, such as RISC-V, in achieving performance and power efficiency gains for tailor-fit implementations. Particularly, RISC-V can leverage vector-operations, and flexible ISA extensions to improve processor designs for specific applications.

As global demand for computational power continues to grow, the development and adoption of custom-made processor designs based on open-standard ISAs, may play a role in addressing challenges of supply chain security, environmental impact, and technological sovereignty.

The potential of open-source RISC-V implementations in circular economies is an interesting topic for future research, as open-source designs inspire iterative improvements, and open-standard ISAs and processor designs enable miniaturization of product customization, fostering component-wise modularity, re-usability, repairability and perhaps, mass-customization.

References

1. Krokida, S.-I., Lambertides, N., Savva, C.S., Tsouknidis, D.A.: The effects of oil price shocks on the prices of EU emission trading system and European stock returns. Eur. J. Financ. **26**, 1–13 (2020)
2. Barua, S.: Understanding Coronanomics: The Economic Implications of the Coronavirus (COVID-19) Pandemic. Social Science Research Network, Rochester (2020)
3. Baldwin, R.: The Greater Trade Collapse of 2020: Learnings from the 2008–09 Great Trade Collapse (2020). https://voxeu.org/article/greater-trade-collapse-2020
4. Park, Y., Hong, P., Roh, J.J.: Supply chain lessons from the catastrophic natural disaster in Japan. Bus. Horiz. **56**, 75–85 (2013)
5. Karmaker, C.L., Aziz, R.A., Palit, T., Bari, A.B.M.M.: Analyzing supply chain risk factors in the small and medium enterprises under fuzzy environment: Implications towards sustainability for emerging economies. Sustain. Technol. Entrepreneurship **2**, 100032 (2023)
6. Cutcher-Gershenfeld, J., Gershenfeld, A., Gershenfeld, N.: The promise of self-sufficient production. MIT Sloan Manag. Rev. **62** (2020)
7. O'Brien, C.: Global manufacturing and the sustainable economy. Int. J. Prod. Res. **40**, 3867–3877 (2002)
8. Geissdoerfer, M., Savaget, P., Bocken, N., Hultink, E.J.: The Circular Economy - A New Sustainability Paradigm? Social Science Research Network, Rochester (2017)
9. Mazur-Wierzbicka, E.: Circular economy: advancement of European Union countries. Environ. Sci. Eur. **33**(1), 1–15 (2021). https://doi.org/10.1186/s12302-021-00549-0
10. Park, D.-J., Liu, S.: A study on the economic effects of US export controls on semiconductors to china. J. Int. Trade Commer. **19**, 129–142 (2023)
11. Varas, A., Varadarajan, R., Goodrich, J., Yinug, F.: Government incentives and US competitiveness in semiconductor manufacturing. Boston Consulting Group (2020)
12. Krolikowski, A., Hall, T.H.: Non-decision decisions in the Huawei 5G dilemma: policy in Japan, the UK, and Germany. Jpn. J. Polit. Sci. 1–19 (2023)
13. Dieter, H., Biedermann, J.: The new advocacy for autarky: self-sufficiency is now once again becoming popular for geopolitical reasons. Economists' Voice **19**, 125–145 (2023)
14. Stanford VLSI Group: CPU-DB: Looking at 40 Years of Processor Improvements (2023). http://cpudb.stanford.edu/
15. Rupp, K.: Microprocessor Trend Data (2023). https://github.com/karlrupp/microprocessor-trend-data
16. Dennard, R.H., Cai, J., Kumar, A.: A perspective on today's scaling challenges and possible future directions. Solid-State Electron. **51**, 518–525 (2007)
17. Horowitz, M.: 1.1 Computing's energy problem (and what we can do about it). In: 2014 IEEE International Solid-State Circuits Conference Digest of Technical Papers (ISSCC), San Francisco, CA, USA, pp. 10–14. IEEE (2014)
18. Kalapothas, S., Flamis, G., Kitsos, P.: Efficient edge-AI application deployment for FPGAs. Information **13**, 279 (2022)
19. Blem, E., Menon, J., Sankaralingam, K.: Power struggles: revisiting the RISC vs. CISC debate on contemporary ARM and x86 architectures. In: 2013 IEEE 19th International Symposium on High Performance Computer Architecture (HPCA), Shenzhen, pp. 1–12. IEEE (2013)

20. Patterson, D.: Past and future of hardware and architecture. In: SOSP History Day 2015, Monterey, California, pp. 1–63. ACM (2015)
21. Slater, M.: A Guide to RISC Microprocessors. Elsevier Science, Oxford (1992)
22. Kane, G., Heinrich, J.: MIPS RISC Architecture. Prentice Hall, Englewood Cliffs (1992)
23. Gara, A., et al.: IBM power architecture. In: Padua, D. (ed.) Encyclopedia of Parallel Computing, pp. 900–907. Springer, Boston (2011). https://doi.org/10.1007/978-0-387-09766-4_13
24. Chen, T., Patterson, D.A.: RISC-V geneology. EECS Department, University of California, Berkeley, Technical report. UCB/EECS-2016-6 (2016)
25. Nadehara, K., Kuroda, I., Daito, M., Nakayama, T.: Low-power multimedia RISC. IEEE Micro **15**, 20–29 (1995)
26. Vinoth, C., et al.: A novel low power and high speed Wallace tree multiplier for RISC processor. In: 2011 3rd International Conference on Electronics Computer Technology, Kanyakumari, India, pp. 330–334. IEEE (2011)
27. Schiavone, P., et al.: Slow and steady wins the race? A comparison of ultra-low-power RISC-V cores for Internet-of-Things applications. In: 2017 27th International Symposium on Power and Timing Modelling, Optimization and Simulation (PATMOS), Thessaloniki, pp. 1–8. IEEE (2017)
28. Garofalo, A., Rusci, M., Conti, F., Rossi, D., Benini, L.: PULP-NN: accelerating quantized neural networks on parallel ultra-low-power RISC-V processors. Phil. Trans. R. Soc. A **378**, 20190155 (2020)
29. Schiavone, P.D., et al.: Arnold: an eFPGA-augmented RISC-V SoC for flexible and low-power IoT end nodes. IEEE Trans. VLSI Syst. **29**, 677–690 (2021)
30. Amor, H.B., Bernier, C., Prikryl, Z.: A RISC-V ISA extension for ultra-low power IoT wireless signal processing. IEEE Trans. Comput. **71**, 766–778 (2022)
31. Park, J., Choi, E., Lee, K., Lee, J.-J., Han, K., Lee, W.: Developing an ultra-low power RISC-V processor for anomaly detection. In: 2023 Design, Automation & Test in Europe Conference & Exhibition (DATE), Antwerp, Belgium, pp. 1–2. IEEE (2023)
32. Valente, L., et al.: HULK-V: a heterogeneous ultra-low-power Linux capable RISC-V SoC. In: 2023 Design, Automation & Test in Europe Conference & Exhibition (DATE), Antwerp, Belgium, pp. 1–6. IEEE (2023)
33. Rutishauser, G., Hunziker, R., Di Mauro, A., Bian, S., Benini, L., Magno, M.: ColibriES: a milliwatts RISC-V based embedded system leveraging neuromorphic and neural networks hardware accelerators for low-latency closed-loop control applications (2023)
34. Ottaviano, A., Benz, T., Scheffler, P., Benini, L.: Cheshire: a lightweight, Linux-capable RISC-V host platform for domain-specific accelerator plug-in (2023)
35. You, Z., Zhang, X., Han, L., Zhao, R.: Design and implementation of RISC-V-based vector operation unit for embedded processor. In: Loskot, P., Yan, S. (eds.) Eighth International Symposium on Advances in Electrical, Electronics, and Computer Engineering (ISAEECE 2023), Hangzhou, China, p. 18. SPIE (2023)
36. Lee, J.K.L., Jamieson, M., Brown, N., Jesus, R.: Test-driving RISC-V Vector hardware for HPC (2023)
37. Ottavi, G., et al.: Dustin: a 16-cores parallel ultra-low-power cluster with 2b-to-32b fully flexible bit-precision and vector lockstep execution mode. IEEE Trans. Circuits Syst I **70**, 2450–2463 (2023)
38. Askarihemmat, M., Wagner, S., Bilaniuk, O., Hariri, Y., Savaria, Y., David, J.-P.: BARVINN: arbitrary precision DNN accelerator controlled by a RISC-V CPU. In: Proceedings of the 28th Asia and South Pacific Design Automation Conference, Tokyo Japan, pp. 483–489. ACM (2023)
39. Minervini, F., et al.: Vitruvius+: an area-efficient RISC-V decoupled vector coprocessor for high performance computing applications. ACM Trans. Archit. Code Optim. **20**, 1–25 (2023)

40. AskariHemmat, M., et al.: Quark: an integer RISC-V vector processor for sub-byte quantized DNN inference (2023)
41. Sriram, K., et al.: HALO: a hardware-software co-designed processor for brain-computer interfaces. IEEE Micro **43**, 64–72 (2023)
42. Kovač, M., et al.: FAUST: design and implementation of a pipelined RISC-V vector floating-point unit. Microprocess. Microsyst. **97**, 104762 (2023)
43. Perotti, M., Cavalcante, M., Wistoff, N., Andri, R., Cavigelli, L., Benini, L.: A "New Ara" for vector computing: an open source highly efficient RISC-V V 1.0 vector processor design. In: 2022 IEEE 33rd International Conference on Application-specific Systems, Architectures and Processors (ASAP), Gothenburg, Sweden, pp. 43–51. IEEE (2022)
44. OSHWA certified projects list (2023). https://certification.oshwa.org/list.html
45. Ellen MacArthur Foundation: The butterfly diagram: visualising the circular economy. https://ellenmacarthurfoundation.org/circular-economy-diagram. Accessed 13 June 2023
46. Hankammer, S., Jiang, R., Kleer, R., Schymanietz, M.: Are modular and customizable smartphones the future, or doomed to fail? A case study on the introduction of sustainable consumer electronics. CIRP J. Manuf. Sci. Technol. **23**, 146–155 (2018)
47. Framework Computer Inc.: Framework Laptop. https://web.archive.org/web/20230000000000*/. https://frame.work/de/en/about. Accessed 12 June 2023
48. MNT Research GmbH: MNT Reform: Open Hardware Laptop. https://web.archive.org/web/20230000000000*/. https://mntre.com/. Accessed 12 June 2023
49. European Commission. Directorate General for Communications Networks, Content and Technology: The impact of open source software and hardware on technological independence, competitiveness and innovation in the EU economy: final study report. Publications Office, LU (2021)

Concept for Process Parameter-Based Inline Quality Control as a Basis for Pairing in a Production Line

A. Geiser(✉), F. Stamer, and G. Lanza

Karlsruhe Institute of Technology (KIT), 76131 Karlsruhe, Germany
alexander.geiser@kit.edu

Abstract. This paper presents a general concept for an in-line quality control system and the basis for component pairing using the example of a pressure valve. The aim is to improve the functional quality of the product while increasing dimensional tolerances to reduce waste. The proposed system requires an end-of-line (EOL) functional test and collects pre-existing sensor data from the production line. This data is used to train machine learning models to identify correlations between measurements and EOL test results. The system uses this information to predict future EOL test results. Anomaly detection and root cause analysis is performed by comparing predicted results with actual measurements. To improve the data set, additional sensors are integrated into the identified production steps. Once parameters with a high influence on the product function have been identified, these should be used to find ideal pairs of components with favorable parameter combinations in order to improve functionality. The EOL test is then used for validation.

Keywords: Functional quality control · pairing · In-Process

1 Introduction and Motivation

To maintain competitiveness, ensuring product quality is crucial in modern manufacturing processes. [1] Often, tight tolerances are chosen to guarantee optimal product performance. [2] However, for manufacturers operating at the limits of their production capabilities, achieving and sustaining these narrow tolerances can be a daunting task. [3] Accepting a certain level of initial rejection or scrap at the beginning of the value chain becomes a necessary trade-off to prevent costly waste at the end of the production line, where significant resources such as time and materials have already been invested. To avoid waste, various quality control strategies have promising approaches. [4] By leveraging knowledge about the process or the components, often geometric features which have an impact on the product quality are identified. Those are measured in order to detect deviations. Then, a quality control loop can use the measurements and adjust machine parameters to stay inside the given tolerances. In order to do so efficiently, in-process or at least in-machine measurement technology is preferable. These are the final stages according to the degree of integration as defined by Lanza et al. 2019 [5] (see

T. Bauernhansl et al. (Eds.): WGP 2023, LNPE, pp. 481–487, 2024.
https://doi.org/10.1007/978-3-031-47394-4_47

Fig. 1). However, measuring dimensions with tight tolerances this way can be challenging. [6] In practice, samples are selected for offline or even off-shop floor inspections to circumvent this challenge. Individual components of the valve are randomly sampled after production to manually check characteristic dimensions.

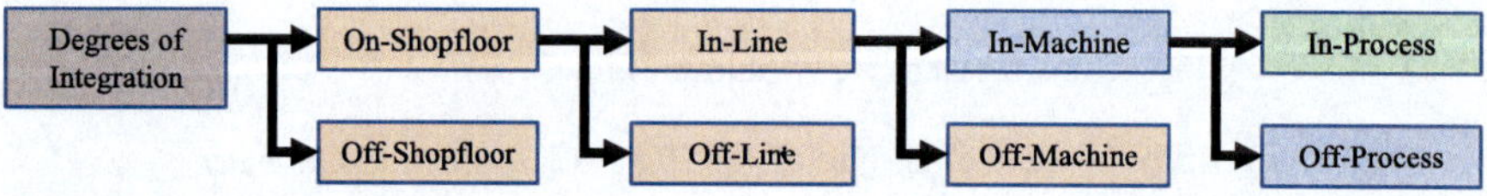

Fig. 1. Degrees of integration of production-integrated measurement technology (according to Lanza [5]) The aim is to use only In-Machine data such as machine parameters and In-Process measurements.

The results of the manual measurements are recorded by an employee in a handwritten test plan. This additional step in the production line introduces delays while parts between tests that do not meet the required tolerances may still pass through further production steps, leading to rejection or scrap only detected during the final functional testing stage. In a production process with long-standing experience, manufacturers rely on their accumulated knowledge and expertise. The manufacturing parameters, including machine settings and process specifications, are well-established based on years of experience. Consequently, the dimensional tolerances of the parts can be tightly controlled and product knowledge can be integrated into quality control strategies, ensuring the desired functionality. [7] However, the fuel cell valve under consideration is a new technology and production is accordingly in the ramp-up phase. During the new production ramp-up phase, where the manufacturer aims to increase production capacity or introduces modifications to the production line, maintaining the same level of dimensional tolerances becomes increasingly challenging. The existing knowledge and experience might not be directly transferable to the new production setup. This situation necessitates the need for an alternative approach that focuses on functional rather than on dimensional tolerances.

This paper presents a combination of methods to help ensure and improve the product functional quality without integrating time-consuming and costly measurement technology for a valve used in a fuel cell system. Instead, existing process data should be used to perform quality control, make functional predictions and enable individual assembly strategies.

2 Methods

2.1 Functional Orientation

While meeting dimensional tolerances often is a crucial aspect of production, the end customer's primary concern lies in the functional performance of the product rather than the adherence to specific dimensional specifications. For instance, at a valve used in an automotive engine. The primary function of this valve is to regulate flow. If the flow falls below the specified threshold, the engine's performance is compromised, whereas excessive flow leads to increased fuel consumption. For this purpose, for example, inline measuring devices that measure flow are preferred over devices that measure geometric features of a valve. [8] An end-of-line (EOL) function test bench is therefore a prerequisite. With this, quality-critical characteristics can be recorded in order to be able to build up a functional orientated quality control system. [7].

2.2 Data Acquisition

The relevant data includes machine parameters and in-process data for each production step. Each component goes through an individual number of production steps. (see Fig. 2) The individual components are then assembled step by step. The first step is to extract as much information as possible from the existing process data. This should save both time and money when integrating new sensor technology. Above all, however, the aim is to avoid having to use any additional measurement technology that is not in-process.

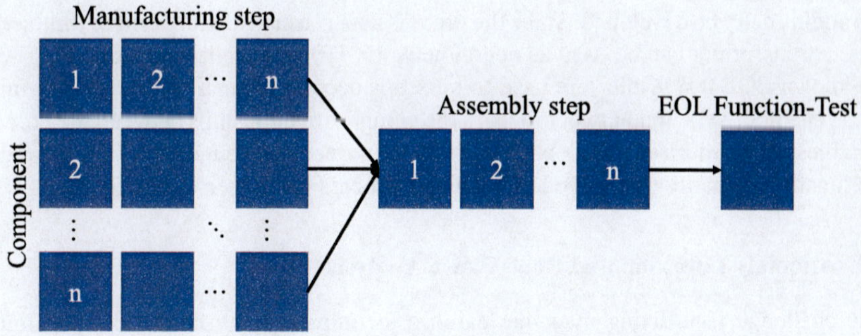

Fig. 2. Production process. The production of the final unit is divided into the step-by-step manufacturing of the individual parts and assembly before the final functional test

Each production step already contains a range of data. These can be parameters, referred to as P or measured values, referred to as M. (see Fig. 3). The manufacturing of the valve is based on CNC machines. In CNC machines, for example, the cutting speed can be used as a parameter and the machine current as a measured value. In an assembly process such as a pressing operation, the press-in depth can occur as a parameter and the pressure as a measured value.

2.3 Correlation Analysis

To utilize the existing data and unlock its potential for inline quality control, correlation analysis and machine learning techniques are employed. The objective is to establish correlations between the collected data and the functional quality measurements obtained from the EOL test. Initially, correlation analysis is performed to identify the process variables that exhibit strong relationships with the functional quality measurements. This analysis helps to uncover the critical parameters that directly influence the valves functionality. [9] By understanding these correlations, the manufacturer can focus their

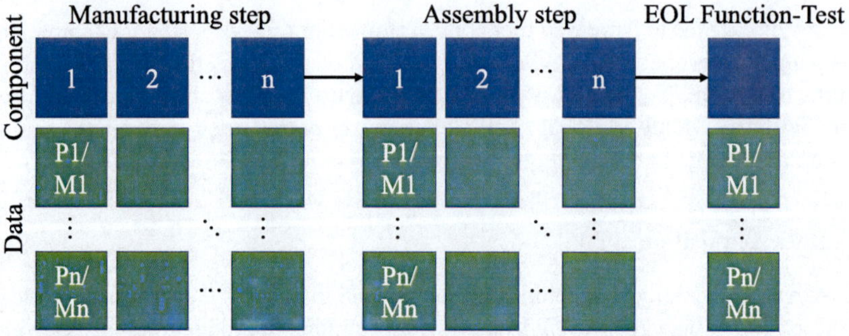

Fig. 3. Production data. The individual production steps from manufacturing and assembly to EOL testing provide data in the form of machine parameters and in-process measured values.

efforts on monitoring and controlling the key parameters during the production process. With the insights into the key parameters, a predictive model for the flow rate of the valve can additionally be developed. Since the process data is usually in the form of time series data, a transformer can be used as neural network. [10] The collected data, along with the known EOL test results, are used to train this deep-learning network. This training phase enables the neural network to learn the complex relationships between the process variables and the desired functional quality. Once trained, the neural network can predict the functional quality of a valve based on new process parameter values. [11].

2.4 Anomaly Detection and Root-Cause-Analysis

One challenge in utilizing machine learning for inline quality control is the limited availability of historical data, especially during the early stages of production ramp-up. Changes in the production process of the valve may occur, and these changes can significantly impact the correlation between process variables and functional quality. Without sufficient historical data capturing these changes, the predictive accuracy of the model may deteriorate. The prediction may continue to assume the production of units within the function tolerances even though the actual measured function of the units in the EOL test is no longer within the limits. (see Fig. 4).

To address this issue, an anomaly detection approach is employed in addition to the neural network. The predicted functional quality values from the trained neural network are compared with the actual measurements obtained from the EOL test. Deviations between the predicted and actual values indicate potential anomalies in the production process. Machine learning methods can also be used for this purpose. [12]. Furthermore, a root cause analysis is performed to understand the underlying factors contributing to the detected anomalies. This analysis helps identify the specific areas or process steps that require additional data collection or intervention. By targeting these areas, the manufacturer can focus their resources on gathering more data to enhance the accuracy and reliability of the predictive model.

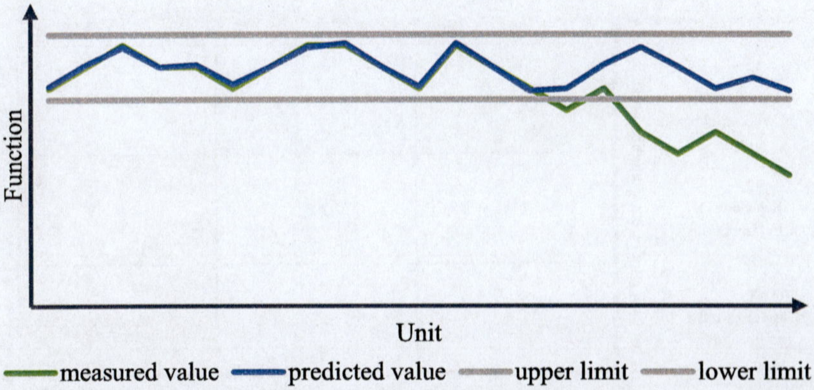

Fig. 4. Deviation between predicted and measured value. Predicted functional quality of the units still remains OK, although the measured units are NOK due to unknown changes in the production process.

2.5 Component Pairing

Once the predictive model demonstrates reasonable robustness, and the critical parameters that significantly influence the desired functionality are identified, these parameters can be used to improve production with classical quality control loops. In addition, with the gathered knowledge about the parameters, component pairing can be employed to further improve functionality. For example, components can be selectively or individually paired based on different features to achieve better results in terms of product functionality using the same manufacturing methods. [13] While the features are often a mixture of geometric measurements, in-line functional measurements and specific assembly process parameters, [9] the aim of this approach is to work with the identified parameters from the previous steps as much as possible to save intermediate steps such as geometric measurements. Therefore, the adherence to dimensional tolerances should no longer be in the foreground. It should now be possible to manufacture components to nominal dimensions and to pair and assemble them individually on the basis of the process data (3 A in Fig. 5). If pairings with good functional quality are not possible, the machine parameters can be adapted statistically. (3 B in Fig. 5) To achieve ideal component pairings, the manufacturer has to employ multidimensional optimization techniques. These optimization methods aim to identify the optimal combinations of parameters that maximize the desired functionality while adhering to any constraints or limitations. By formulating the component pairing problem as an optimization problem, the manufacturer can efficiently search the parameter space for the most favorable combinations. Therefore, the parameter space needs to be reduced to a minimum of the most influential parameters identified beforehand. Otherwise, the complexity of the optimization problem becomes too high to be solved in the cycle time of production. This would lead to undesirable delays.

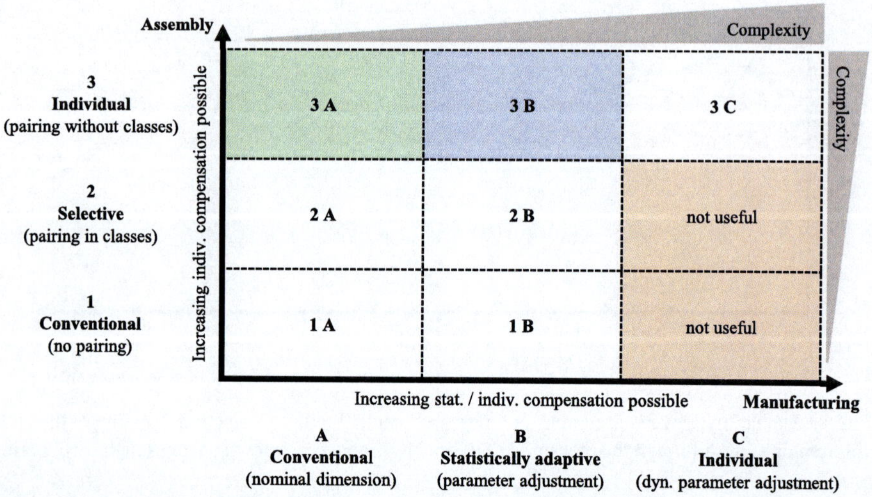

Fig. 5. Framework of possible strategies for quality control (according to Wagner [13] and Silbernagel [14]). The aim is to enable individual pairing of components based on parameters and in-process data without having to manufacture components individually.

3 Conclusion

Overall, this concept offers an approach for inline quality control and component pairing in a production line. By leveraging existing data, utilizing machine learning, and prioritizing functional quality, the manufacturer can enhance product performance, reduce waste, and improve overall customer satisfaction. Future research can focus on implementing and validating this concept in real-world manufacturing environments to assess its effectiveness and scalability. By successfully implementing the proposed concept for inline quality control and component pairing, it becomes possible to gradually widen dimensional tolerances while monitoring the system's response. This approach could lead to a future in which dimensional tolerance-free series production is possible. Tolerances can therefore relate more to function and, in the broadest sense, to parameters to ensure functional quality.

References

1. Friedli, T., Lanza, G., Remling, D.: Global Manufacturing Management. Springer, Cham (2021). https://doi.org/10.1007/978-3-030-72740-6
2. Mease, D., Nair, V.N., Sudjianto, A.: Selective assembly in manufacturing: statistical issues and optimal binning strategies. Technometrics **46**(2), 165–175 (2004). https://doi.org/10.1198/004017004000000185
3. Heling, B., Aschenbrenner, A., Walter, M., Wartzack, S.: On connected tolerances in statistical tolerance-cost-optimization of assemblies with interrelated dimension chains. In: Procedia CIRP, vol. 43, pp. 262–267 (2016). https://doi.org/10.1016/j.procir.2016.02.031. https://www.sciencedirect.com/science/article/pii/S2212827116003097

4. Lanza, G., Haefner, B., Kraemer, A.: Optimization of selective assembly and adaptive manufacturing by means of cyber-physical system based matching. CIRP Ann. **64**(1), 399–402 (2015). https://doi.org/10.1016/j.cirp.2015.04.123
5. Lanza, G., et al.: In-line measurement technology and quality control. In: Gao, W. (ed.) Metrology (Precision Manufacturing), pp. 1–35. Springer, Singapore (2019). https://doi.org/10.1007/978-981-10-4912-5_14-1
6. Schmitt, R., Damm, B.: WIE SIE MIT INLINE-MESSTECHNIK IHRE WERTSCHÖPFUNG MAXIMIEREN : Prüfen und Messen im Takt," 2008. https://www.semanticscholar.org/paper/WIE-SIE-MIT-INLINE-MESSTECHNIK-IHRE-WERTSCH%C3%96PFUNG-%3A-Schmitt-Damm/e3bbece3142f8e78c4ac4627d56f9fcdf1374490
7. Wagner, R., Haefner, B., Lanza, G.: Function-oriented quality control strategies for high precision products. Procedia CIRP **75**, 57–62 (2018). https://doi.org/10.1016/j.procir.2018.04.069
8. Weckenmann, A., Werner, T.: Messen und Prüfen. In: Pfeifer, T., Schmitt, R. (eds.) Masing Handbuch Qualitätsmanagement, pp. 677–724. Carl Hanser Verlag GmbH & Co. KG, München (2021). https://doi.org/10.3139/9783446466210.029
9. Wagner, R., Fischer, J., Gauder, D., Haefner, B., Lanza, G.: Virtual In-line inspection for function verification in serial production by means of artificial intelligence. Procedia CIRP **92**, 63–68 (2020). https://doi.org/10.1016/j.procir.2020.03.126
10. Liu, M., et al.: Gated Transformer Networks for Multivariate Time Series Classification (2021). https://arxiv.org/pdf/2103.14438
11. Haefner, B., Biehler, M., Wagner, R., Lanza, G.: Meta-model based on artificial neural networks for tooth root stress analysis of micro-gears. Procedia CIRP **75**, 155–160 (2018). https://doi.org/10.1016/j.procir.2018.04.031
12. "Sonderdruck_Ott_BI_02_2018". https://www.consist.de/export/sites/www.consist.de/.content/downloads/_downloads/Sonderdruck_Ott_BI_02_2018.pdf
13. Wagner, R.: Strategien zur funktionsorientierten Qualitätsregelung in der Serienproduktion. Dissertation. https://doi.org/10.5445/IR/1000124008
14. Silbernagel, R.M.: Funktionsorientierte Qualitätsregelung in Produktionsnetzwerken. Dissertation. https://doi.org/10.5445/IR/1000154569

Resilient Value Chain Systems

A Signal-Based Approach to Switching Between Micro and Macro Material Flow Models for Production Systems

A. Kienzlen(✉) and A. Verl

University of Stuttgart, Seidenstraße 36, 70174 Stuttgart, Germany
annika.kienzlen@isw.uni-stuttgart.de

Abstract. In order to enhance resource efficiency and minimize waste in production, effective planning and monitoring of material flow is crucial, which can be facilitated by material flow simulations. There are various material flow models on different scales with different advantages and disadvantages. With a significant quantity of piece goods, transitioning or switching between different model scales becomes essential for balancing accuracy and computational efficiency. However, determining the appropriate switching points in space or time can be challenging. This paper presents a signal-based methodology for identifying appropriate switching points between different models, which can be customized by the user. We concentrate on the switch from a physics-based model on a micro scale to a flow model on a macro scale. By defining switch conditions based on necessary information, signals can trigger switches between models. The proposed approach provides a flexible solution for simulating complex material flow systems with different scales.

Keywords: Material Flow · Virtual Commissioning · Multi-Scale

1 Introduction

1.1 Motivation

In the era of digital transformation, simulation and digital engineering have become integral tools in production, revolutionizing the design, optimization, and analysis of complex processes and systems for sustainable production. Virtual commissioning (VC), the simulation of a production process before the actual implementation in a physical production system [1], can reduce the debugging time by 73% in the development of control code for production [2]. It offers several significant advantages over traditional commissioning approaches: VC enables engineers to identify and fix potential issues and bottlenecks early in the design phase, reducing the need for expensive and time-consuming modifications in the physical production environment. The movement of piece goods

This work is funded by the Deutsche Forschungsgemeinschaft (DFG, German Research Foundation) – Project FlowVC, 508338261.

T. Bauernhansl et al. (Eds.): WGP 2023, LNPE, pp. 491–501, 2024.
https://doi.org/10.1007/978-3-031-47394-4_48

in the production (material flow) is crucial for the production. Therefore, the material flow should not be neglected in the VC. By simulating and analyzing material flow processes in a virtual environment, engineers can optimize operations to reduce rejects and improve the quality of the process. The optimization and fine-tuning of material flow layouts enhance the overall efficiency, saving time and energy and optimizing resource allocation. VC facilitates the exploration of different scenarios and analyses without any real prototypes which helps to create sustainable and cost-effective productions. It provides engineers with a powerful toolset for designing, optimizing, and validating production systems to make informed decisions, improve system performance, and ensure smooth and efficient material flow operations.

Several material flow models exist with different strengths and weaknesses. One single model is not enough to use in VC of production plants [3]. By utilizing two models on different scales, engineers can gain a comprehensive understanding of the material flow system. The ability to switch between models on two scales within a unified simulation framework would be an important progress [3]. This allows to optimize processes and increase the productivity and efficiency.

1.2 State of the Art

For material flow simulation of internal material flow systems in VC, two types of material flow models are most relevant: a physics-based model on micro scale and a flow model on macro scale [4]. Both use a continuous time representation unlike an event-discrete model. A physics-based model (cf. Fig. 1a) uses kinetics and describes individual piece goods as discrete entities based on their geometry and properties according to Newton-Euler [5]. A kinematic model would consider only velocities and no other properties. Depending on the solver, the real geometry is used in a physics-based model as a mesh or a simplified version such as a convex hull. The movements of the piece goods are calculated based on the forces, torques, etc. from the environment and their interaction with each other. These models can be very accurate and close to reality but the computational effort is proportional to the number of collisions and therefore can get large [4]. This level of detail enables a thorough understanding of the interactions and dynamics between individual items, facilitating precise control of the material flow.

The flow model (cf. Fig. 1b) describes a set of piece goods in terms of a density distribution based on a partial differential equation and does not consider individual piece goods. The environment is expressed as velocity fields that cause the density distribution to move. When the highest possible density is reached, which corresponds to a collision of the piece goods, the density distribution is distributed by a convolution. This leads to a lower - and more importantly constant - computational effort in case of many collisions [4]. However, it is not possible to consider individual piece goods in this model. A multi-scale material flow model including the flow model only gets relevant if the restrictions of the flow model are met. There has to be a large quantity of piece goods where it is not important to track specific piece goods. Up until now, the piece goods all have to be equal with a circular or quadratic base area otherwise the collisions would behave differently. Moreover, the flow model is only developed for two dimensions; the density distribution is two-dimensional and there are two velocity fields, in each direction one. The focus shifts towards analyzing the overall distribution and density of piece goods in the system which allows to gain insights when transporting large amounts of piece goods, to identify bottlenecks and to optimize the distribution. Since the flow model is differentiable, unlike the physics-based model, it is very useful for optimization.

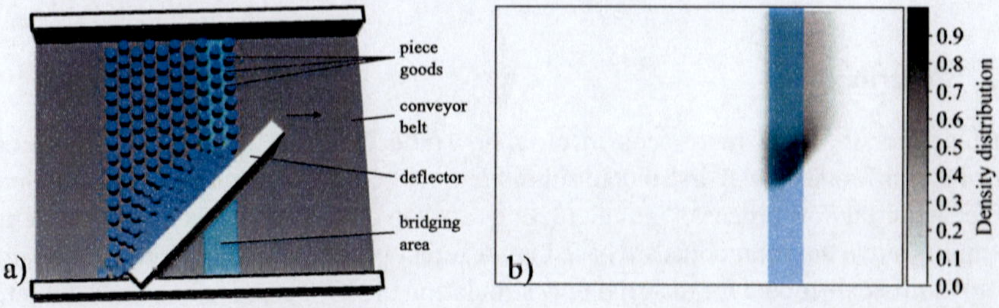

Fig. 1. Simulation of piece goods on a conveyor belt in a) a physics-based submodel and b) a flow submodel with switch from physics-based to flow in the bridging area (light blue).

To combine the advantages of both models, the concept of a multi-scale material flow model was presented [3]. In this idea, the physics-based model and the flow model are simulated simultaneously. To switch between the models, the description of the piece goods has to be transformed. Göttlich et al. [6] show that the switch from micro to macro is straightforward while the opposite direction requires optimization.

Models on multiple scales are used in several domains. They are quite common in biology [7, 8], materials [9, 10] and chemistry [11, 12]. Besides science, multi-scale modelling has also reached the engineering domain [10, 13]. In particular for cyber physical systems, combining multiple disciplines is of increasing importance [14–16]. However, the term is ambiguous. A multi-scale model can mean that the same things are modelled on different scales parallelly, e.g. [14, 15], or that different parts of the subject are modelled on different scales, e.g. [8, 17, 18]. Often, the distribution between the submodels remains the same. It is also possible to switch at specific times or locations [19, 20]. The scales can be either spatial or temporal [10]. Sometimes, even the coupling of different disciplines is called multi-scale modelling, e.g. [14]. Regarding the computations on different scales, they can either be solved separately exchanging the results (sequential or hierarchical coupling), or all computations can be integrated in one system and solved simultaneously (concurrent coupling or hybrid models) [12, 21, 22].

There are several methods for scale bridging in sequential coupling which transform one submodel into another. The use of these methods depends strongly on the domain and, more importantly, on the formulation of the coupled submodels. It is of utmost importance to understand the principles of the submodels and to identify their coupling aspects [9, 10]. For spatial scales, a handshake or bridging region is often defined [8, 22, 23], which represents the interface between the submodels. For example, in the bridging domain method, the different submodels overlap in the bridging region [24]. Furthermore, it is not enough to be able to convert the submodels, they can also have effects on each other. In dynamic simulations, the bridging region between the submodels on different scales can pose certain challenges. These challenges can include artificial wave reflections caused by differences in the properties of the models used on different scales (impedance mismatches) [24, 25]. Additionally, the coupling of local and non-local systems can introduce ghost forces. Ghost forces are undesired artefacts that arise from the integration of these different systems, potentially affecting the accuracy and reliability of the simulation results [11, 22, 26].

1.3 Contribution

In this paper, the term "multi-scale material flow model" is used for simulating one model in which different spatial distributed submodels are combined to simulate different parts of the material flow process. The calculations are solved separately equaling a sequential coupling since the submodels are based on different principles. To our knowledge, there is no multi-scale model for material flow simulation in VC. There is a coupled simulation which creates a new model out of a kinematic and a kinetic model [27] but these are neither different scales nor do both models persist at the same time. To realize a multi-scale model for internal material flow simulation which improves VC of production plants, different aspects have to be considered:

1. Scales respectively submodels to take into account
2. Time and place to change between the scales (dynamically or statically)
3. Boundary conditions
4. Method to change between the scales

For the first aspect, we decided on physics-based and flow submodels because of their high accuracy. For a multi-scale material flow model, a switching logic is needed to decide where or when to switch between the submodels on different scales. In addition to the switching logic, the change itself has to be realized. So far, research has focused on changing the whole model, not just parts of it, and not during simulation. Therefore, boundary conditions have not been considered. This paper focuses on the second aspect, with a conceptual approach to the third. A method for changing between these models (fourth aspect) has been theoretically designed [6].

2 Concept of Switching Logic for Material Flow Models

The switching logic of a multi-scale material flow model should be decided before the simulation model is designed but after the required signals are known.

2.1 Differentiation Between Signal Types

In VC, the goal of the simulation is to test and improve the production system. Concerning the domain of control engineering, the focus is on the behavior resulting from the control system. Signals in this context are all information that is needed by and available to the control system. Most of the signals regarding the material flow are based on sensors. The signals can be separated in four categories (cf. Table 1): A presence signal only detects whether there is any piece good, which is possible for example with a magnetic sensor.

It is also called a "flow-no flow" signal. An area signal can detect the amount of piece goods in a specific area e.g. with a scale or a vision system. It should be mentioned that all the required signals can be solved with a vision system if the surroundings are clear and light (in the table only for the most commonly used signal). However, the vision system is sometimes too much effort. A point signal detects where exactly a piece good is placed for example with a light curtain. In comparison to the specific object signal, it detects the position of any piece good not a specific one. In the pharmaceutical sector, for example, it is important to be able to track individual piece goods which requires a specific object signal like a RFID or code reader.

Table 1. Relevant signals in material flow simulation.

Preferred model	Signal	Sensors
Flow model	presence	Microwave, Capacitive, Magnetic, Inductive, Photoelectric (light)
	area	Scale, Vision systems
Physic-based	point	Laser profilers, Light curtain, Ultrasonic distance
	specific object	RFID, Code reader, Color

The physics-based model is very detailed depicting every single piece good with its individual properties. Therefore, it is possible to generate all the signals based on the physics-based model. However, some signals can be generated just as good by the flow model needing less effort. The generation of the signals from the flow model is described in [28]. The flow model is convenient for generating area signals. It is also able to generate presence signals as long as the area under consideration is big enough. However, the flow model is not able to show the exact position of piece goods and, since it uses a continuous density distribution, it is impossible to follow any specific object.

2.2 Signal-Based Switching Criteria

Since collisions increase the computational effort on the micro scale, it would be beneficial to use the flow model in case of many collisions. The submodel for the area or time of interest can be selected based on the required signals (Fig. 2, signals in grey).

If information about specific objects is required anywhere or anytime later, the physics-based model must be used the entire time (1). If a point signal is required, the physics-based model must be selected in this area or at this time (2). For an area signal, it depends on whether the flow model is currently being used. If so, it is advantageous to keep the flow model (3). If this is not the case, but a large amount of piece goods will also be considered afterwards, it is advantageous to switch to a flow model (4). Otherwise, the changeover effort may be too great. Finally, if a presence signal is needed in a small area, a physics-based model should be chosen (5). If a presence signal is needed in a large area (6), it again depends on whether a flow model is currently used (3) and eventually whether a large amount of piece goods will be needed later.

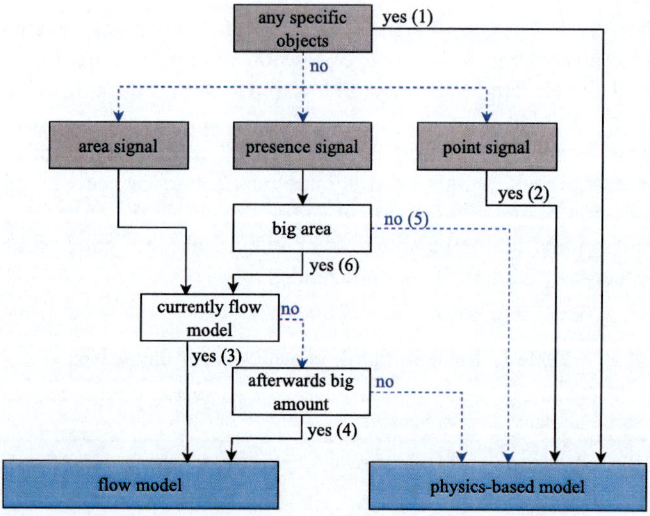

Fig. 2. Criteria to choose the appropriate submodel.

3 Concept to Switching Between Material Flow Models

Having decided where or when to switch between the submodels, the next point is how to change. The transformation between the physics-based model and the flow model has been described in [6]: To transform the geometric piece goods of a physics-based model into the density distribution of a flow model, each piece good is represented by a two-dimensional Gaussian bell curve with variance σ_0 using its position $(X_{o,j}, Y_{o,j})$ and radius r. All the separate bell curves are summed up to one density distribution ρ in space (x, y):

$$\rho(x, y) = \sum_{j=1}^{N} \left(\frac{\sqrt{3}r^2}{\pi\sigma_0^2} \exp\left(-\frac{1}{2\sigma_0^2} \left(\|x - X_{o,j}\|_2^2 + \|y - Y_{o,j}\|_2^2 \right) \right) \right) \tag{1}$$

To transform the density distribution of the flow model to the positions of the piece goods in a physics-based model, an optimization problem as nonlinear least squares problem has to be solved where the density distribution of the flow model ρ_{flow} should equal the retransformed density distribution ρ_{physic} of the positions p

$$\|\rho_{flow} - \rho_{physic}(p)\|_2^2 \tag{2}$$

with the constraints of the total number of piece goods, the available area and the restriction of overlapping.

A time-based switch for an entire submodel is straightforward to realize. To implement a multi-scale material flow model with a switch for a certain area, the requirements of the flow model have to be considered for this area as well: The amount of piece goods in this area has to be large. There can only be piece goods of one type on one level (third dimension irrelevant). If independent sections of the material flow are modeled in different submodels, they will not affect each other. If not the whole submodel or separate sections are transformed, but only a part, the submodels can influence each other at an interface. This happens either for a location-based switch or for a time-based switch within a certain area. The latter could be, for example, in front of a clocked machine. We will focus on these cases in the following subchapters. It is important to avoid artificial wave reflections and ghost forces at this interface.

3.1 Concept of Bridging Areas

A material flow usually has one direction, which simplifies the switch by requiring the transformation in only one direction. For a switch in a certain area, a bridging area can be defined where the description of the piece goods is already transformed to that of the second submodel, but is still considered in the first submodel. The required size of the bridging area depends on the size of the piece goods, the velocity of the material flow and the simulation step size. If the piece goods do not collide with each other in this area and there is no backlog of the following area, the switch is straightforward without the danger of ghost forces. In the following, only backlogs leading to a stop of piece goods and not a backwards velocity are considered.

Considering a location-based switch from the physics-based submodel to the flow submodel, the individual piece goods have to be transformed to a density distribution in the bridging area (exemplary in light blue in Fig. 1). The flow model has one density distribution for its entire area. When a piece good enters the bridging area, its density is calculated as a bell curve and added to the density distribution of the flow model. However, as long as the piece good is inside the bridging area, its individual representation in the physics-based model is maintained. On the one hand, this individual piece good is influenced by the other individual piece goods and the environment in the physics-based submodel. On the other hand, this individual piece good is influenced by the density distribution in the flow submodel accounting for backlogs. As long as the piece goods are not densely packed, the flow submodel has no effect. If the density distribution in front of the individual piece good is almost maximally densely packed (assuming one direction), it cannot move there. Consequently, there is a force resulting from the flow submodel that acts on the individual piece good in the physics-based submodel. The direction and size of the force result from the distribution in the density. The process is quite similar for a time-based switch in a certain area. Not individual piece goods are transformed but all new piece goods in the bridging area when the switch gets active. The next steps are the same as for the location-based switch.

Considering a location-based switch from the flow submodel to the physics-based submodel, the bridging process consists of several steps. In the first step, the individual piece goods of the physics-based submodel in the bridging area are transformed into a density distribution and added to the density distribution of the flow submodel. Then a simulation step is calculated in both submodels. Afterwards, the individual piece goods of the physics-based submodel in the bridging area are again transformed in a density distribution, but this time they are subtracted from the density distribution of the flow submodel. The remaining density distribution of the flow submodel in the bridging area is transformed into individual piece goods for the physics-based submodel. The transformation only occurs if a certain sum of density has entered the area. Only density that is equal to a complete piece good can be transformed. The consideration of the individual piece goods of the physics-based submodel in the flow submodel is the consideration of backlogs. This backlog density distribution is taken into account for reactions to a pile of maximal density. For a time-based switch in a certain area, it is not necessary to add and subtract the transformed piece goods each simulation time step. They are added once when the switch gets active and maintained until the next time.

3.2 Implementation of Bridging Areas

As an example, a conveyor belt with a deflector is simulated in a physics-based submodel with AGX Dynamics [29] with a switch to a flow submodel implemented in Python (Fig. 1). The piece goods are first processed individually and then arrive at the deflector in bulk after which an area signal measures the number of piece goods. Therefore, the individual piece goods of the physics-based submodel are transformed in the density of the flow submodel. There is a check for high density which leads to a force on the remaining piece good in the physics-based submodel.

To evaluate the effect of the transformation, the influence of various starting points for bridging areas from physics-based to flow model are tested with constant size of the area (Fig. 3b). The curve of number of piece goods over time should be similar regardless of the starting point. The baseline of the results, the physics-based and the flow submodel (solid lines), differ mainly due to the different representation of piece goods: In the physics-based submodel the position of the center of the piece goods accounts for the signal, in the flow submodel the density enters gradually. Consequently, the comparison seems promising with similar curves (Fig. 3a), but the starting point of and the forces in the bridging area need further evaluation.

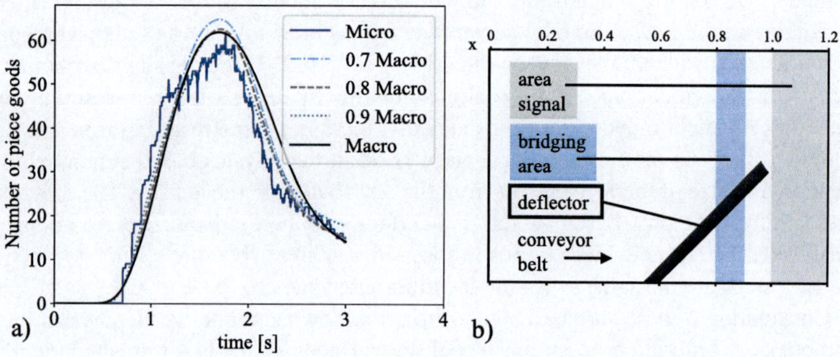

Fig. 3. Number of piece goods in area signal after deflector with various bridging areas(micro – only physics-based, x macro – switching at point x to flow model, macro – only flow model).

4 Conclusion and Outlook

A multi-scale model is required to enable a material flow model in VC of an entire production plant, which is capable of reducing waste and improving the quality and efficiency of the production. Models for VC in control engineering have to provide the signals used by the control system, making them a suitable indicator for selecting a submodel. While specific objects and point signals ask for a physics-based submodel on micro scale, presence and area signals can be depicted with a macro flow submodel depending on the area size and later required signals. We introduced bridging areas to avoid ghost forces at the interfaces between different scale submodels. In the bridging area, one submodel is transformed in another but it can also be affected by the other one in case of a backlog.

The concept of bridging areas needs more evaluation in the context of various scenes regarding the effect of the size of the piece goods and their environment on the required size and forces. A method for aligning the submodel types can be introduced as baseline. A material flow in multiple directions has to be regarded additionally. Moreover, the transformation from a flow submodel to a physics-based submodel is computation intensive and therefore needs further research. All of the concepts have to be included in one implementation to increase the productivity and efficiency with VC.

References

1. VDI Verein Deutscher Ingenieure. Virtuelle Inbetriebnahme - Modellarten und Glossar. Beuth, Berlin, Berlin 25.040.40, 35.240.50 (2016)
2. Schamp, M., et al.: Impact of a virtual twin on commissioning time and quality. IFAC-PapersOnLine **51**, 1047–1052 (2018). https://doi.org/10.1016/j.ifacol.2018.08.469
3. Scheifele, C., Lechler, A., Verl, A. Materialflussmodelle für die HiL-Simulation. wt Werkstattstechnik online, vol. 106, pp. 119–124 (2016)
4. Göttlich, S., et al.: Modeling, simulation and validation of material flow on conveyor belts. Appl. Math. Model. **38**, 3295–3313 (2014). https://doi.org/10.1016/j.apm.2013.11.039
5. Bender, J. Impulsbasierte Dynamiksimulation von Mehrkörpersystemen in der virtuellen Realität. Dissertation. Universität Karlsruhe (2007)
6. Göttlich, S., Pfirsching, M.: A micro-macro hybrid model with application for material and pedestrian flow. Cogent Mathem. Stat. **5**, 1–20 (2018). https://doi.org/10.1080/25742558.2018.1476049
7. Meier-Schellersheim, M., Fraser, I.D.C., Klauschen, F.: Multiscale modeling for biologists. Wiley Interdisc. Rev. Syst. Biol. Med. **1**, 4–14 (2009). https://doi.org/10.1002/wsbm.33
8. Quarteroni, A., Veneziani, A.: Analysis of a geometrical multiscale model based on the coupling of ODE and PDE for blood flow simulations. Multiscale Model. Simul. **1**, 173–195 (2003). https://doi.org/10.1137/S1540345902408482
9. Weinan, E.: Principles of multiscale modeling. Cambridge Univ. Press, Cambridge (2011)
10. Hoekstra, A., Chopard, B., Coveney, P.: Multiscale modelling and simulation: a position paper. Philos. Trans. R. Soc. Math. Phys. Eng. Sci. **372**(2021), 20130377 (2014). https://doi.org/10.1098/rsta.2013.0377

11. Chamoin, L., et al.: Ghost forces and spurious effects in atomic-to-continuum coupling methods by the Arlequin approach. Int. J. Numer. Meth. Engng **83**, 1081–1113 (2010)
12. Horstemeyer, M.F.: Multiscale Modeling: A Review. In: Leszczynski, J., Shukla, M. (eds) Practical Aspects of Computational Chemistry. Springer, Dordrecht (2009). https://doi.org/10.1007/978-90-481-2687-3_4
13. Fish, J.: Multiscale Methods: Bridging the Scales in Science and Engineering. OUP Oxford (2009)
14. Demkovich, N., Yablochnikov, E., Abaev, G. Multiscale modeling and simulation for industrial cyber-physical systems. In: Proceedings Industrial Cyber-Physical Systems (ICPS), pp. 291–296. IEEE, Piscataway, NJ (2018)
15. Neumann, M., Constantinescu, C., Westkämper, E.: Method for multi-scale modeling and simulation of assembly systems. Procedia CIRP **3**, 406–411 (2012). https://doi.org/10.1016/j.procir.2012.07.070
16. Schönemann, M., et al.: Multiscale simulation approach for production systems. Int. J. Adv. Manuf. Technol. **102**, 1373–1390 (2019). https://doi.org/10.1007/s00170-018-3054-y
17. Luan, B.Q., et al.: Multiscale modeling of two-dimensional contacts. Phys. Rev. E Stat. Nonlinear Soft Matter Phys. **74**, 046710 (2006). https://doi.org/10.1103/PhysRevE.74.046710
18. Talebi, H., Silani, M., Rabczuk, T.: Concurrent multiscale modeling of three dimensional crack and dislocation propagation. Adv. Eng. Softw. **80**, 82–92 (2015). https://doi.org/10.1016/j.advengsoft.2014.09.016
19. Tiwary, P., van de Walle, A.: A review of enhanced sampling approaches for accelerated molecular dynamics. In: Weinberger, C.R., Tucker, G.J. (eds.) Multiscale materials modeling for nanomechanics. SSMS, vol. 245, pp. 195–221. Springer, Cham (2016). https://doi.org/10.1007/978-3-319-33480-6_6
20. Ingólfsson, H.I., et al.: Machine learning-driven multiscale modeling: bridging the scales with a next-generation simulation infrastructure. J. Chem. Theory Comput. **19**, 2658–2675 (2023). https://doi.org/10.1021/acs.jctc.2c01018
21. Lee, J.D., Li, J., Zhang, Z., Wang, L.: Sequential and concurrent multiscale modeling of multiphysics: from atoms to continuum. In: Meguid, S.A., Weng, G.J. (eds.) Micromechanics and Nanomechanics of Composite Solids, pp. 1–38. Springer, Cham (2018). https://doi.org/10.1007/978-3-319-52794-9_1
22. Liu, X., Furrer, D., Kosters, J., Holmes, J.: Vision 2040: A Roadmap for Integrated, Multiscale Modeling and Simulation of Materials and Systems. Accessed Jun. 2023. https://ntrs.nasa.gov/citations/20180002010
23. Abraham, F.F., et al.: Spanning the continuum to quantum length scales in a dynamic simulation of brittle fracture. Europhys. Lett. **44**, 783 (1998). https://doi.org/10.1209/epl/i1998-00536-9
24. Xiao, S.P., Belytschko, T.: A bridging domain method for coupling continua with molecular dynamics. Comput. Methods Appl. Mech. Eng. **193**, 1645–1669 (2004). https://doi.org/10.1016/j.cma.2003.12.053
25. Lee, Y., Basaran, C.: A multiscale modeling technique for bridging molecular dynamics with finite element method. J. Comput. Phys. **253**, 64–85 (2013). https://doi.org/10.1016/j.jcp.2013.06.039
26. Li, X., Ming, P.: On the effect of ghost force in the quasicontinuum method: dynamic problems in one dimension. Commun. Comput. Phys. **15**, 647–676 (2014). https://doi.org/10.4208/cicp.250213.270813a
27. Hoher, S.: Ein gekoppeltes Materialflussmodell zur durchgängigen Entwicklungsunterstützung von Materialflusssteuerungen. Stuttgarter Beiträge zur Produktionsforschung. Dissertation. Universität Stuttgart, Stuttgart (2017)

28. Kienzlen, A., Verl, A.: Konzept zur Integration eines kontinuierlichen Materialflussmodells in die Virtuelle Inbetriebnahme durch Signalgenerierung, pp. 585–594 in Simulation in Produktion und Logistik 2021, 1st ed. J. Franke and P. Schuderer, eds. Cuvillier Verlag, Göttingen (2021)
29. Algoryx Simulation AB, Umea. AGX Dynamics v2.36.1.2 (2023). www.algoryx.se/products/agx-dynamics

Enhancing the Resilience of Global Production Networks Against Unpredictable Disruptions: A Systematic Literature Review

Benedict Janssen[1,3](✉), Martin Welsing[1,3], Seth Schmitz[1,3], and Günther Schuh[1,2,3]

[1] Laboratory for Machine Tools and Production Engineering (WZL), Aachen, Germany
b.janssen@wzl.rwth-aachen.de

[2] Fraunhofer Institute for Production Technology IPT, Aachen, Germany

[3] RWTH Aachen University, Campus-Boulevard 30, 52074 Aachen, Germany

Abstract. Global production networks have become a dominant feature of the global economy, with companies relying on dispersed and closely linked production activities in different regions. However, various disruptive events have highlighted the vulnerability of such networks. Companies that are robustly positioned against these events have a significant competitive advantage. To understand this strategic advantage better, a systematic literature review is necessary to compile elementary factors that describe the resilience of networks and identify further research gaps. The main objective is to identify (1) the central weaknesses in global networks, (2) how the vulnerability of the network can be determined and (3) which approaches make global production networks more resilient. The paper provides a comprehensive analysis of the current research and highlights the implications for global manufacturing companies.

Keywords: Global Production Networks · Resilience · Literature Review

1 Motivation

The beginning decade is marked by a series of disruptive crises that are having a global impact. According to the World Economic Forum's annual risk report, companies face an unforeseen combination of familiar risks, such as rising inflation and cost of living, and new risks, such as de-globalization and declining market growth. [1].

Manufacturing companies have increasingly internationalized their global production networks (GPN) throughout the past decades. The share of traded intermediate products has tripled worldwide since 2000. [2] This trend has also led to companies becoming increasingly exposed to the risks of global crises [3]. For example, supply chain disruptions of more than one month increased, so they now occur on average every 3.7 years [2]. This leads to substantial financial losses for manufacturing companies. Todays GPN are characterized by complexity due to a worldwide network of production sites, customers and suppliers and dynamics as the aforementioned disruptions [4]. As a result, future events are becoming more and more difficult to predict and

T. Bauernhansl et al. (Eds.): WGP 2023, LNPE, pp. 502–513, 2024.
https://doi.org/10.1007/978-3-031-47394-4_49

control [5]. Incomplete knowledge of these developments and the risk companies are facing can lead to deviations between target and actual values [6].

The diversification of value creation in a GPN can help companies react better to unforeseen events. Through many different locations and partners, know-how and capacity can be distributed if companies are consciously resilient to these events.[7]. This shows that companies can gain a strategic competitive advantage by ensuring the resilience of their GPN [4].

For this reason, this paper conducts a systematic literature review to capture elementary factors of resilient production network design. Understanding key weaknesses, assessing vulnerability, and identifying methods to increase robustness in production networks are crucial for companies to enhance their resilience and gain a strategic competitive advantage in an unpredictable and challenging global environment. This systematic overview aims to survey the accumulated knowledge and identify new issues in the field of resilience of GPN. With the help of the systematic literature research, the following three research questions are considered, evaluated and analyzed with the help of the literature.

1. How to identify key weaknesses in the GPN?
2. How to determine the vulnerability of the GPN?
3. Which approaches increase the resilience of GPN?

The remainder of this paper is structured as follows. First, an introduction to the current state of research on resilience in GPN is given. Based on this, the research structure of the systematic literature review is presented. Subsequently, the above three research questions are answered and summarized with the help of the evaluation of the identified literature. Finally, a summary and classification takes place.

2 Prior Research on the Topic of Resilience in Global Production Networks

In the following, a brief introduction to the basics of GPN and their resilience is given in order to provide the focus for the systematic literature review.

2.1 Comprehension of the Terminology Agility, Robustness and Resilience

The terms agility, robustness and resilience are often used almost synonymously [8]. For this reason, all three are included in the search string of the subsequent literature analysis. In detail, however, the three terms differ from each other. *Agility*, similar to flexibility, describes the ability of a system to recognize changes at an early stage and to adapt quickly to the new circumstances [9]. *Robustness* describes the ability of a system to continuously maintain functions and processes despite changes, hence its stability. In contrast to agility, robustness does not require any short-term adjustments [10, 11]. The concept of *resilience* unites these two concepts into the ability to react flexibly to changes in the system through preparation. Resilience can thus also be understood as robustness through agility or as sensitivity to changes [11, 12].

Since this understanding of resilience as the union of flexibility and agility is not shared by all publications, the analysis covers all three terms equally. In terms of the research questions raised, all three concepts are relevant in dealing with unforeseen events, as all offer leverage to provide the company with a competitive advantage.

2.2 Resilience in Global Production Networks

Due to the extensive structural changes brought about by globalization, companies have increasingly adopted a new approach to organizing their production. This involves the formation of GPN in recent decades. GPN are characterized by the presence of multiple production units located around the world, which are interconnected through material and information flows. These networks can also involve collaboration with suppliers and customers. The main responsibilities within the GPN can be categorized into three areas: establishing the production strategy, determining the network's configuration, and coordinating the production network itself. Alongside tasks such as defining business models and product portfolios, these responsibilities encompass activities such as allocating product mixes, developing capabilities, and overseeing supply and order management.[13, 14].

As described in the introduction the outset, a lack of resilience in GPN can lead to significant economic consequences for manufacturing companies. The resilience of GPN can be enhanced at all three levels: network strategy, network coordination and network configuration.[11].

However, since not all publications considered follow this distinction, not all can be clearly assigned. Therefore, all approaches that fundamentally deal with the issues of global production networks are considered equally.

3 Research Methodology

The structure of the systematic literature analysis follows the structure of the systematic review according to LIBERATI ET AL. They divide the structure of a systematic literature analysis into four phases. In the first phase of *identification*, all relevant contributions from the selected databases are collected. This is followed by a *screening* phase, where duplicates are removed. In the third phase the literature is filtered using defined *filter criteria.* In the fourth phase of the *suitability process*, the remaining articles are manually filtered for their suitability with regard to the research question. During the final stage of the inclusion process, all pertinent sources, texts, or articles are now documented within a designated solution space.[15].

The databases used for the present systematic literature search are Google Scholar and Springer Link for German search queries. For English-language searches, the databases used are Web of Science, Scopus and Google Scholar. The databases were selected because they are among the largest scientific databases for English or German language literature with several million articles [16]. Only literature in German or English language was considered.

The search string used to filter the literature from the databases mentioned is visualized in Fig. 1. The search criteria were formed along the topic of "Resilience in global production networks" and serve to identify the largest possible number of relevant literature.

Language	Component Term	Search String	Connecting Operator
English	Resilience	(Robustness OR Agility OR Resilience OR Flexibility OR Durability OR Strength)	AND
	Production network	("Production Networks" OR "Manufacturing Networks" OR "Global Production Networks" OR "Global Manufacturing Networks" OR "Global Production Systems")	AND
	Methods	(Methods OR Influences OR Drivers OR Factors OR Processes OR Approaches)	AND
German	Resilienz	(Robustheit OR Agilität OR Resilienz OR Flexibilität OR Strapazierfähigkeit OR Stärke)	AND
	Produktions-netzwerk	("Produktionsnetzwerke" OR "Fertigungsnetzwerke" OR "Globale Produktionsnetzwerke" OR "Globale Fertigungsnetzwerke" OR "Globale Produktionssysteme")	AND
	Methoden	(Methoden OR Einflüsse OR Triebkräfte OR Einflussfaktoren OR Prozesse OR Vorgehensweisen)	AND

Fig. 1. Search string for the systematic literature search

Figure 2 shows the results of the four-stage research process. From original 7.854 contributions that could be found with the help of the composite search string 34 relevant contributions to be analyzed are identified in the literature using the following method. In the present systematic literature search, non-German or English language articles were excluded and duplicates removed during the *screening* phase. In the *filtering* phase, filters and a blacklist were used to focus on the application area, so that only articles from the subject area of engineering and economics were considered. Afterwards the titles of the articles were filtered manually. Finally, the *suitability* phase identified a selection of relevant articles to answer the questions based on the abstracts. The number of articles during the identification phase at Springer Link is upwardly biased due to the search engine construction because the search string could not be entered as a whole. Thus, more literature was considered than would have been relevant according to the search string.

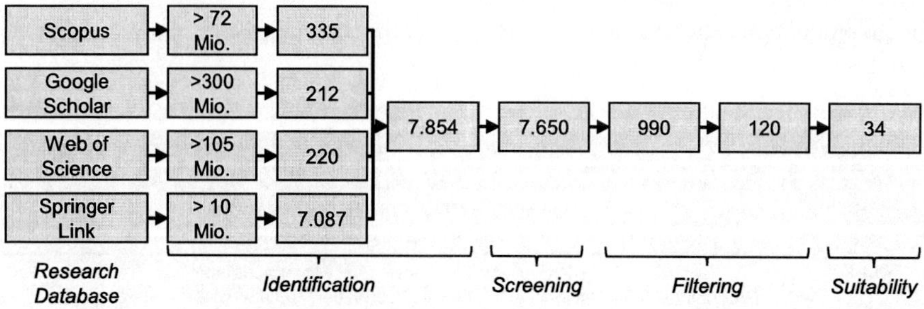

Fig. 2. Number of identified articles in the systematic literature research

4 Identified Research Focus Areas

To answer the three research questions addressed in chapter one, the 34 identified articles were examined and categorized in terms of the three research questions.

4.1 Central Weaknesses in Global Production Networks

The systematic literature review resulted in the identification of different approaches that elaborate weaknesses in global production networks.

The identified weaknesses were categorized by the majority of authors into external risks and internal weaknesses, as it is presented in Fig. 3. Internal weaknesses such as the loss of production resources at a site are those that can be influenced by the company, since they occur along the company's production process. External risks such as natural disasters or geopolitical conflicts have an impact on the company from a global macro perspective through the interaction of the company with the environment and can be hardly influenced. [17].

According to STICH ET AL. and AYS ET AL., internal weaknesses occur in particular at the interfaces of the company with suppliers on the procurement side and customers on the sales side. These are cause due to sourcing problems or demand fluctuations, for example. In addition, disruptions at individual production sites, such as production shutdowns, can lead to extended lead times and thus reduced profitability of the company. Thus, procurement-side, sales-side and internal site factors are the main causes of internal weaknesses in GPN.[4, 5] BUTOLLO ET AL. furthermore identify that management strategies such as single sourcing or just-in-time production in particular have increased internal weaknesses. Especially during disruptive crises, such as the corona pandemic, these weaknesses became apparent. [18]. The potential risks arising from these weaknesses are examined in detail by NIKNEJAD ET AL., SONI ET AL. and LUND ET AL. With the help of concrete probabilities of occurrence for defined scenarios, the effects of internal weaknesses are quantified in order to be able to derive measures for the most likely occurrence scenarios and to determine possible losses [2, 19–21].

These internal weaknesses are offset by global influences that lead to external risks that are difficult to control. AYS ET AL., BALDWIN ET AL., BUTTOLO ET AL. and SONI ET AL. list various macro-influencing factors and risks on the GPN. The mainly distinguished factors are political instabilities, economic crises, socio-cultural changes, legal barriers and environmental catastrophies. [4, 18, 20, 22] The Corona pandemic and the war between Russia and Ukraine are two examples that are discussed in this context by NIKNEJAD ET AL. as well as LUND ET AL. or BUTOLLO ET AL. [2, 17–19, 21].

These weaknesses, both internal and external, can occur continuously, such as through a decline in orders, or disruptively, such as through a collapsed supply chain. Disruptive events occur unpredictably and lead to a significant impairment of a company's ability to do business. In particular, SCHUH ET AL., AYS, ERLACH ET AL., MOURTZIS ET AL. and HU ET AL. discuss these disruptive events and derive adaptations for the GPN [4, 11, 23–25]. These consist of either long-term adjustments to the strategy and configuration of the network [e.g. NIKNEJAD ET AL., AYS, RODEMANN ET AL.] or short-term reactions by means of network coordination [PEUKERT] [4, 6, 19, 26].

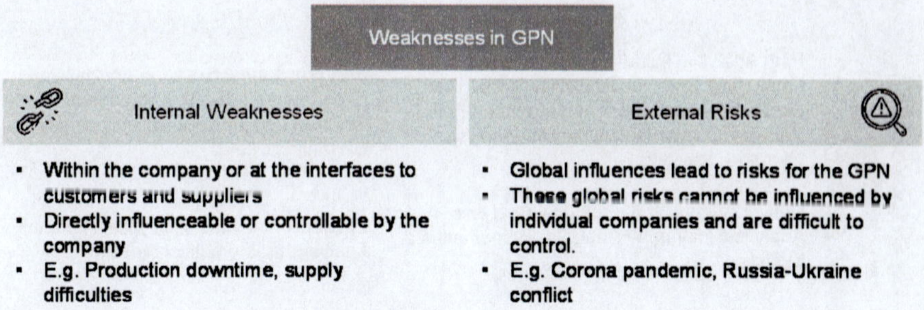

Fig. 3. Weaknesses in GPN

The analysis of both internal and external weaknesses serves companies in that it allows them to draw conclusions regarding the susceptibility of the GPN and establish an initial point which entities in the GPN need a more resilient network design. Based on the identified literature, various internal weaknesses and external risks of the network have been identified. These weaknesses occur either continuously or disruptively and can be addressed in the long term through GPN strategy and design or through short-term network management. However, approaches that have quantitatively assessed the probabilities of occurrence may differ from each other in individual application cases.

4.2 Determination of Network Vulnerability

Based on the general weaknesses in the network from chapter 4.1, the systematic literature review was furthermore analyzed on how to determine the vulnerability of the global production network.

In order to implement resilience-enhancing measures in the GPN that are commensurate with risks and efforts, a thorough identification of vulnerabilities in the network is necessary [21]. The categorization into internal and external weaknesses identified in chapter 4.1 was used to analyze which approaches can be used to identify either weaknesses. The identification of specific vulnerabilities in the GPN is conducted in the literature through either quantitative or qualitative means, as it is shown in Fig. 4 [4].

The literature presents a variety of qualitative approaches for identifying vulnerabilities in the network. Ays et al., Rodemann et al., and Stich et al. propose literature-based and expert knowledge-based methods for identifying fundamental vulnerabilities [4, 5, 26]. Wonsak initially defines the central core elements of the production site and derives scenario-based vulnerabilities through a stress test [27]. Baldwin et al. and Häntsch et al. follow risk management processes as proposed by the World Economic Forum to identify risks in the network [22, 28]. Notably, the level of abstraction differs in their respective approaches. While literature-based processes provide a broad overview, applied risk management processes can focus specifically on the considered company.

		Weaknesses	
		Internal	External
Determination Methods	Qualitative	Path Analysis, Redundancy Classification, Risk Monitoring, Scenario Analysis, Stress Test, Expert Interviews, Risk Management, FMEA, Pairwise Comparison, Criticality Analysis, Literature Reviews	Risk Monitoring, Scenario Analysis, Expert Interviews, Risk Management, Pairwise comparison, Criticality analysis, Literature reviews
	Quantitativ	Network Simulation, Empirical Analyses, Analytical Hierarchy Analysis, Mathematical Models	Empirical Analyses, Analytical Hierarchy Analysis, Mathematical Models

Fig. 4. Methods to Determinate the Vulnerability in GPN

In addition to qualitative methods for identifying vulnerabilities in the GPN, there are quantitative approaches that can provide further insights into the vulnerability analysis through simulation-based or metric-based techniques. Witthohn et al. present an approach to quantify the probability of an event or vulnerability occurrence, thus assigning a probability value to each identified weakness [29]. Similarly, Ays et al. determine the volatility of event occurrence using a Key Performance Indicator (KPI), based on the information content used to calculate the probability of occurrence [30]. Massari et al. and Putnik et al. introduce simulation-based approaches. Putnik et al. simulate the correlation between the robustness of the GPN and its size, identifying an optimal network size [31]. Massari et al. link the question of network resilience to the challenges of the circular economy and simulate the impact of circular economy practices on the resilience of GPN structures [32]. Schurig and Niknejad et al. also present data-driven approaches for identifying GPN vulnerabilities by analyzing the failure probabilities of GPN scenarios through simulation [19, 33].

Therefore, vulnerabilities in the GPN can be identified, evaluated, and categorized through qualitative and quantitative methods. It can be seen that there are currently more methods for qualitative analysis of vulnerabilities than quantitative. In particular, simulation methods for analysing external risks are currently hardly considered due to the complex and multi-layered influencing factors. This forms the challenge for decision-making and implementing appropriate resilience-enhancing methods in the GPN.

4.3 Resilience Improvement Methods in the Network

Building upon the general weaknesses of the GPN and the approaches to analyse the vulnerability of the GPN, the literature was examined for different approaches and measures to enhance the resilience of the network.

Figure 5 lists 22 articles from the systematic literature review that present a method or approach for increasing resilience in the GPN. The identified articles can be differentiated in whether they seek to whether they analyse risks, develop methods to strengthen the GPN or characterise the actual handling during an incident. The phases of the Fraunhofer EMI resilience cycle, as shown in Fig. 5, serve to classify the identified methods for increasing resilience. It is applied because it is already widely used as an established interdisciplinary framework and therefore forms a basis for categorisation.

The Fraunhofer EMI resilience cycle is used as a framework to divide the identified resilience approaches into five different phases: *Prepare, Prevent Protect, Respond* and *Recover*. In the Prepare phase, the focus is on assessing risks before they have an impact on the company. In the Prevent phase, these risks are identified at an early stage and countered with measures. The Protect phase serves to effectively protect the company in the event of a disruptive event. The Respond and Protect phases deal with the handling of disruptive events as soon as they occur. Respond deals with effective strategies for crisis management, while Recover deals with the return to the initial state after the occurrence of a disruptive event. [34] In the course of the literature research, the first three phases in particular were identified. These deal with measures that can be taken before a disruptive event has occurred.

SCHURIG, AYS, EPURENAU ET AL., ERLACH ET AL. and STICH ET AL. focus their methods on enhancing agility in the GPN. AYS emphasizes the design of economic measures to increase agility through a cost-benefit evaluation of the identified measures [30]. They therefore analyse the process from preparing for possible disruptive events to preventing and protecting the GPN from the identified risks and vulnerabilities. SCHURIG and STICH ET AL. concentrate on data-driven approaches to enhance agility in the production network [5, 33]. SCHURIG'S approach advocates for a stress test that identifies the most agile configuration through simulated GPN configuration scenarios [33]. They therefore also addresses how the GPN can be protected from disruptive events within the framework of the resilience cycle. EPURENAU ET AL. and ERLACH ET AL. also develop their methods with the aim of enhancing agility in the production network [23, 35]. In particular, EPURENAU ET AL. focus on the flexible design of critical production areas, motivated by the production of medical devices during the Corona pandemic. The focus of the method is primarily on preventing the effects of a disruptive event by strengthening the network through agility measures. It specifically aims at flexible configuration of production capacity within the network. In this regard, EPURENAU ET AL. and HÄNTSCH ET AL. agree, as HÄNTSCH ET AL. also pursue the goal of manufacturing flexibility. The focus here is particularly on achieving the ability to theoretically manufacture the entire product portfolio at all locations through technological flexibility. STICH ET AL. prioritize data integration for systematic network design, aiming to make decisions on a more agile and robust network configuration based on data [5]. MOURTZIS also pursues the idea of achieving transparency and resilience in the GPN through extensive digitization [24]. It can be seen that methods that focus on increasing agility measures in

GPN primarily address the "Prevent" point of the resilience cycle. This is because agility measures in GPN mitigate the effects of disruptive events in advance.

While the previously discussed approaches significantly pursue the goal of increasing agility in the GPN, BALDWIN ET AL. WINDT ET AL. and LUND ET AL. focus on impacts and strengthening the robustness of the network [2, 22, 36]. Various measures for increasing robustness are presented here. BALDWIN ET AL. and WINDT ET AL. focus on the possibility of making the network more robust by building redundancies [22, 36]. Redundancies on the supplier side are considered, for example through multi-sourcing, as well as redundancies within the GPN, through the allocation of products to several locations in order to distribute manufacturing competencies, similar to what HÄNTSCH ET AL. present. With the help of these redundancy measures, WINDT et al. focus on protecting the network through robustness measures. BALDWIN ET AL. also introduces the possibilities of increased inventories and standardized supplier parts to strengthen robustness against failures on the supplier side [22]. With her method, PEUKERT presents a simulation-based possibility to investigate the interdependencies of disturbances in the network and their causes in order to derive measures for the GPN configuration on this basis [14]. A similar idea was also pursued by HU ET AL., who formalized the GPN as a mathematical model in order to optimize the stock level under consideration of disruptions [25]. Using these simulation-driven approaches, both PEUKERT and HU ET AL. present methods that focus on both prevention and protection from disruptive events.

The methods and measures presented so far have defined themselves with increasing robustness and agility through network design. In contrast, CARAVEO ET AL. present a method that develops contingency plans to coordinate a flexible selection of suppliers in response to unforeseen events. Thus, the ability to react to possible events is increased without restructuring the network [37].

BUTOLLO ET AL. have examined how global value chain networks recovered from the impacts of the COVID-19 pandemic. Additionally, they methodically present measures that can be taken to mitigate the economic effects of such disruptions. In doing so, BUTOLLO ET AL. are already addressing the Recover phase of the previous disruption and deriving actions for an upcoming one.

In the context of examining approaches and methods for enhancing the resilience of GPNs, it was observed that the "Protect" phase is primarily addressed. Resilience measures do not prevent the impacts of disruptive events; however, they do strengthen the GPNs ability to withstand such events, for example, through repulsive actions like the offloading of processes as described by PEUKERT.

In summary, most of the methods focus on how risks can be assessed and, based on this, either prevented or the GPN can be designed in such a way that it is resilient to the occurrence of disruptive events. The identified literature thus addresses mainly the first three phases of the resilience cycle in particular.

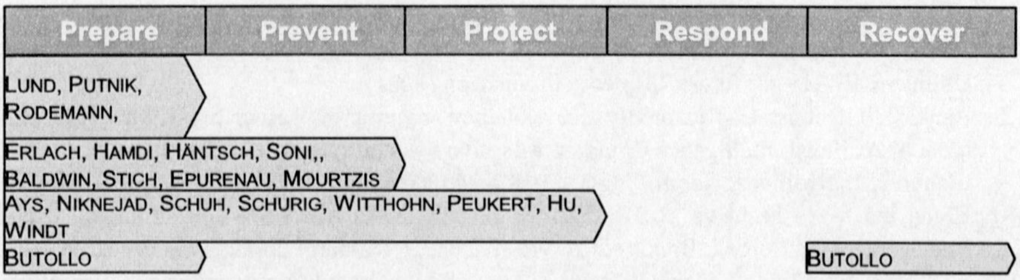

Fig. 5. Identified literature clustered along the resilience cycle

5 Conclusion and Further Research

This study highlights the importance of strengthening GPN to withstand global crises and uncertainties. Through a systematic literature review, it identifies key weaknesses, classifies network vulnerability, and proposes and classifies existing approaches to enhance resilience. The review emphasizes the need to address both internal vulnerabilities within the company's production process and external risks influenced by global factors. Various qualitative and quantitative approaches for vulnerability analysis are discussed, including expert knowledge-based methodologies and simulation-based techniques. Additionally, the study outlines measures to improve agility and robustness in production networks, such as adaptable design, data integration, and digitization. Using the resilience cycle, these methods were integrated into an existing resilience approach and categorized.

Building on the identified existing research, a method can be developed in the future to identify the individual weaknesses in the GPN of companies based on the configuration of their GPN and to be able to prepare recommendations for action. This method will enable companies to strengthen their network independently of disruptions that have already occurred. Furthermore, the approaches can be analysed and evaluated in the future according to their strategic, configurative or coordinative orientation. A distinction can make a contribution to the timeliness of resilience decisions.

Acknowledgement. Funded by the Deutsche Forschungsgemeinschaft (DFG, German Research Foundation) under Germany's Excellence Strategy – EXC-2023 Internet of Production – 390621612.

References

1. World Economic Forum (Hrsg.): The Global Risks Report 2023 (2023)
2. Lund, S., Manyika, J., Woetzel, J., et al.: Risk, resilience and rebalancing in global value chains. Executive summary (2020)
3. Maihold, G.: Die neue Geopolitik der Lieferketten. »Friend-shoring« als Zielvorgabe für den Umbau von Lieferketten (2022)
4. Ays, J.: Gestaltung agiler Produktionsnetzwerke. Design of agile production networks (2021)
5. Stich, V., Schröer, T., Linnartz, M., et al.: Wertschöpfungsnetzwerke in Zeiten von Infektionskrisen (2021)

6. Peukert, S.K.: Robustheitssteigerung in Produktionsnetzwerken mithilfe eines integrierten Störungsmanagements (2021)
7. Uhlmann, E. (Hrsg.): futur: Resiliente Produktion (2021)
8. Bankes, S.: Robustness, adaptivity and resiliency analysis. In: Association for the Advancement of Artificial Intelligence Complex Adaptive System Symposium (2010)
9. Eichholz, J.: Resilienz. Controlling 33(2), 83–84 (2021)
10. Kleemann, F.C., Frühbeis, R.: Resiliente Lieferketten in der VUCA-Welt. Supply Chain Management für Corona, Brexit & Co. Wiesbaden, Heidelberg: Springer Gabler 2021
11. Schuh, G., Friedli, T., Lanza, G., et al.: Zukunftsfähige Produktionsnetzwerke in disruptiven Zeiten. Zeitschrift für wirtschaftlichen Fabrikbetrieb **117**(12), 794–798 (2022)
12. Frank Ritz: Organisationale Resilienz - Paradigmenwechsel, Konzeptentwicklung und Anwendung (2015)
13. Friedli, T., Lanza, G., Remling, D.: Global Manufacturing Management. From Excellent Plants Toward Network Optimization. Springer, Cham (2021)
14. Peukert, S., Lohmann, J., Haefner, B., et al.: Towards increasing robustness in global production networks by means of an integrated disruption management. Procedia CIRP **93**, 706–711 (2020)
15. Moher, D., Liberati, A., Tetzlaff, J., et al.: Preferred reporting items for systematic reviews and meta-analyses: the PRISMA statement. BMJ (Clin. Res.) **339**, b2535 (2009)
16. Martín-Martín, A., Thelwall, M., Orduna-Malea, E., et al.: Google scholar, microsoft academic, scopus, dimensions, web of science, and opencitations' COCI: a multidisciplinary comparison of coverage via citations. Scientometrics **126**(1), 871–906 (2021)
17. Hamdi, F., Masmoudi, F., Dupont, L.: Supply Chain Risk Management. Conceptual Framework. . In: Haddar, M., Chaari, F., Benamara, A., Chouchane, M., Karra, C., Aifaoui, N. (eds.) Design and Modeling of Mechanical Systems—III. CMSM 2017. LNME (2017). https://doi.org/10.1007/978-3-319-66697-6_72
18. Butollo, F., Staritz, C.: Deglobalisierung, Rekonfiguration oder Business as Usual? COVID-19 und die Grenzen der Rückverlagerung globalisierter Produktion. Berl. J. Soziol. **32**(3), 393–425 (2022)
19. Niknejad, A., Petrovic, D., Popplewell, K., et al.: A New Framework for Strategic Risk Analysis in a Global Pump Manufacturing Network. In: Mertins, K., Jardim-Gonçalves, R., Popplewell, K., Mendonça, J. (eds) Enterprise Interoperability VII. Proceedings of the I-ESA Conferences, vol 8, pp. 13–23. Springer, Cham (2016). https://doi.org/10.1007/978-3-319-30957-6_2
20. Soni, U., Vaishnavi, A.: Analyzing Risk in Dynamic Supply Chains Using Hybrid Fuzzy AHP Model. In: Malik, H., Srivastava, S., Sood, Y., Ahmad, A. (eds) Applications of Artificial Intelligence Techniques in Engineering. Advances in Intelligent Systems and Computing, vol. 697, pp. 411–420. (2019). https://doi.org/10.1007/978-981-13-1822-1_38
21. Schuh, G., Potente, T., Varandani, R.M., et al.: Consideration of risk management in global production footprint design. Procedia CIRP **17**, 345–350 (2014)
22. Baldwin, R., Freeman, R.: Risks and global supply chains: what we know and what we need to know. Ann. Rev. Econ. **14**, 153–180 (2022)
23. Erlach, K., Berchtold, M.-A., Kaucher, C., et al.: Design of Resilient Production Networks with Agility Enablers Site Roles as a Solution in Factory Planning (2023)
24. Mourtzis, D., Panopoulos, N.: Digital Transformation Process Towards Resilient Production Systems and Networks. In: Dolgui, A., Ivanov, D., Sokolov, B. (eds.) Supply Network Dynamics and Control. Springer Series in Supply Chain Management, vol. 20, pp. 11–42 (2022). https://doi.org/10.1007/978-3-031-09179-7_2
25. Hu, Y., Li, J.S., Holloway, L., et al.: Achieving Resilience for a Class of Serial Production Networks.In: 2010 American Control Conference, pp. 5326–5331 (2010)

26. Rodemann, N., Ays, J., Gützlaff, A., et al.: Influencing factors for the design of agile global production networks: Einflussfaktoren für die Gestaltung agiler globaler Produktionsnetzwerke, pp. 563–571 (2019)
27. Wonsak, I.G.A.: Evaluation der Zukunftsrobustheit von Produktionssystemen durch Stresstests (2022)
28. Häntsch, M., Huchzermeier, A.: Identifying, analyzing, and assessing risk in the strategic planning of a production network: the practical view of a German car manufacturer. J. Manage. Control **24**, 125–158 (2013)
29. Witthohn, C.: Konfiguration globaler Produktionsnetzwerke im Spannungsfeld von Kosten- und Risikominimierung (2019)
30. Ays, J., Gützlaff, A., Thomas, K., et al.: Determination of a Dedicated. Cost-Effective Agility in Manufacturing Networks. In: Behrens, BA., Brosius, A., Hintze, W., Ihlenfeldt, S., Wulfsberg, J.P. (eds) Production at the leading edge of technology. WGP 2020, pp. 699–706 (2020). https://doi.org/10.1007/978-3-662-62138-7_70
31. Putnik, G.D., Škulj, G., Varela, L., Butala, P.: Simulation study of large production network robustness in uncertain environment. CIRP Ann. **64**(1), 439–442 (2015)
32. Massari, G.F., Giannoccaro, I.: Simulating the network structures in the circular economy and their impact on resilience. IFAC-PapersOnLine **55**(10), 2863–2868 (2022)
33. Schurig, M.: Methodology to evaluate the agility of a production network using a stress test approach (2016)
34. Hiermaier, S., Scharte, B.: Ausfallsichere systeme. In: Neugebauer, R. (ed.) Digitalisierung, pp. 295–310. Springer, Heidelberg (2018). https://doi.org/10.1007/978-3-662-55890-4_17
35. Epureanu, B.I., Li, X.Y., Nassehi, A., et al.: An agile production network enabled by reconfigurable manufacturing systems. CIRP Ann. Manuf. Technol. **70**(1), 403–406 (2021)
36. Windt, K., Hütt, M.-T., Meyer, M.: A Modeling Approach to Analyze Redundancy in Manufacturing Systems. In: ElMaraghy, H. (eds) Enabling Manufacturing Competitiveness and Economic Sustainability. Springer, Heidelberg (2011). https://doi.org/10.1007/978-3-642-23860-4_81
37. Caraveo, A.; Krishnamoorthy, S.; Lawas, J. M. et al.: Mutual Resiliency and Lean Analysis for Supply Chain Management in a COVID-19 Ventilator Production Case Study. In: Valle, M., et al. Advances in System-Integrated Intelligence. SYSINT 2022. LNNS, vol. 546, pp. 299–309 (2023).https://doi.org/10.1007/978-3-031-16281-7_29

Check for updates

Software-Defined Value Networks: Motivation, Approaches, and Research Activities

David Dietrich(✉), Manuel Zürn, Colin Reiff, Michael Neubauer, Armin Lechler, and Alexander Verl

Institute for Control Engineering of Machine Tools and Manufacturing Units, University of Stuttgart, Seidenstr. 36, 70174 Stuttgart, Germany
david.dietrich@isw.uni-stuttgart.de

Abstract. Globalization has led to cost efficient but highly fragile value chains in industrial manufacturing, as demonstrated by the corona pandemic. This paper proposes a rethinking of value chains towards software-defined value networks (SDVN) to achieve resilience advantages in a VUCA world. To unveil these advantages, a holistic view of the causes of global value chain disruptions was set up. In a novel approach, an initial set of causes was extracted from ChatGPT and combined, structured, and expanded with an automated analysis of news and research literature. On this basis, levels of SDVN adaptivity and their corresponding application-oriented requirements are deduced. From this, current barriers in digitalization at the value-added level are derived. To reflect the contribution made by current developments, these are assigned to the barriers. Finally, open research questions towards the transition to SDVN are identified. The result holistically shows how the flexibility, redundancy, scalability, and data efficiency of SDVNs lead to success.

Keywords: software-defined value network · production networks · value chain · VUCA world · resilience · data spaces

1 Introduction

The corona pandemic was not the only demonstration of the fragility of global value chains (GVCs) due to highly interconnected industrial markets. Wars, shortages, sanctions, or unstable shipping costs and energy supply make for a very volatile market [1, 2]. Looking at a single production system, the trend towards software-defined manufacturing reduces the impact of production disruptions by increasing the flexibility and adaptability in production systems, but has its system boundary at the factory gate [3].

Contrawise, current digitalization developments aim beyond the factory gate, such as the standardization of digital twins (DT) and data spaces. They tend to enable automated and standardized linkages at the value-added level. More fundamentally, however, the advantages of such linkages need to be discussed under current causes of GVC disruptions.

T. Bauernhansl et al. (Eds.): WGP 2023, LNPE, pp. 514–524, 2024.
https://doi.org/10.1007/978-3-031-47394-4_50

The aim of this paper is to clarify the holistic requirements for value chain linkages, identify current barriers, and relate recent developments to open research questions. To take these linkages into account and cope with current challenges of GVC, we propose a rethink of GVC towards software-defined value networks (SDVN). The scope of the SDVN is based on thinking about a broad network of suppliers, manufacturers, distributors or retailers, customers, service providers, R&D, and regulatory or standardization organizations that collectively create value, where software-defined approaches aim to create a dynamic, diversified, and adaptable ecosystem that can respond quickly to change. This transformation is schematically shown in Fig. 1.

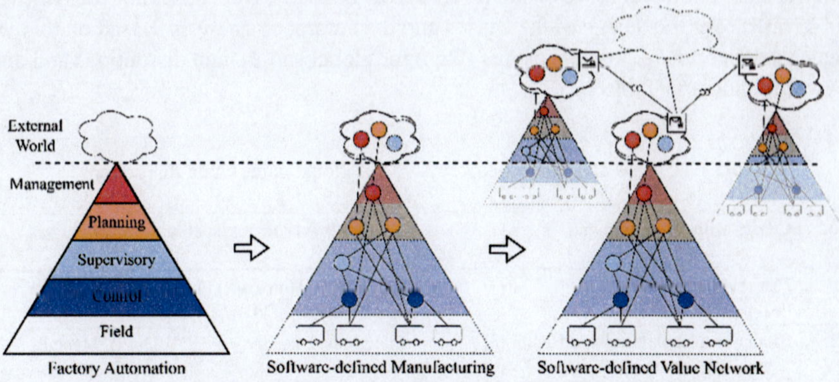

Fig. 1. Transformation of automation to software-defined value network.

This paper is structured according to Fig. 2 as follows: After the introduction in this section, a holistic overview of *causes* for value chain disruptions is determined in Sect. 2. Hierarchical *levels* are introduced in Sect. 3 to classify the range of impact of changes in value-added systems ranging from specific balancing of supply and value-adding services to product releases. They demonstrate the advantages of SDVN, and result in application-oriented *requirements*. From this, underlying *barriers* of digitalization at the value creation level are derived in Sect. 4. The developments of *enablers*, such as digital twins and data spaces are introduced and discussed, by which their contribution to a structural concept of SDVN and open *research questions* are determined. In the last section a conclusion and an outlook are given.

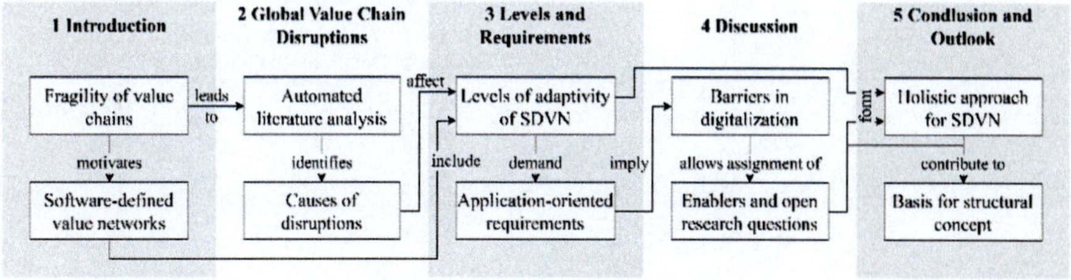

Fig. 2. Structure of the paper.

2 Automated Cause Extraction for Global Value Chain Disruptions

We asked ChatGPT (trained with data up to September 2021) [4] to identify the 20 most likely causes of supply and value chain disruptions. We then reduced the causes to achieve a consistent level of abstraction. For the resulting 15 main causes, we generated synonyms, associations, and sub-features as keywords. In the next step, we checked whether at least one of the keywords was present in the title or snippet (~200 chars) of international research papers (past 5 years) and news articles (past month) requested via Google Scholar API [5] and a news API [6] according to the query: "(value chain OR supply chain OR global market) AND (disruption OR challenge)". To ensure a comprehensive collection, we manually analyzed the articles for causes not yet identified and iterated over the design of the causes and the automatic analysis. Based on this, we created a collection of possible causes (C…) for global value chain disruptions and their keywords shown in Table 1.

Table 1. Causes and associated keywords of global value chain disruptions.

No.	Causes with top cited paper	Keywords (synonyms, spellings and plural excluded)
C1	Natural Disasters [7]	Earthquake, Hurricane, Flood, Tsunami, Wildfire, Serve Weather, Drought, Landslide
C2	Pandemics, Epidemics, and Health [8]	Infectious, Disease (e.g. COVID-19), Mental Health
C3	Political Factors [9]	Political Instability, Political Unrest, (Civil) Wars, Terror, Government Policy, Regulation, Law, Legislation, Trade Dispute, Trade Policy (e.g. Brexit), Trade Barrier, (Geo)Political Tension or Conflict, Legal Challenge (e.g. Data and Privacy Protection), Taxation, Sanction
C4	Economic Factors [10]	Recession, Inflation, Currency Fluctuation, Financial Crisis, Fluctuating Costs (e.g. Energy and Material), Bank Failure, Trade Imbalance
C5	Supplier and Vendor Issues [11]	Bankruptcy, Insolvency, Financial Instability, Quality Issue, Production Delay, Dependence, Labor Issue
C6	Transportation Disruptions [12]	Congestion, Port Closure, Strike, Fuel or Container Shortage, Accident, Lack of Capacity, Customs, Border Closure
C7	Quality or Safety Concerns [13]	Product Recalls, Quality Failure, Product Defect, Slow/Uneconomical Product Development Cycle, Emerging Documentation Requirement
C8	Cybersecurity Threats [14]	Ransomware, DoS Attack, (Industrial) Espionage, Cyberattack, Data Loss, Data Breach

(continued)

Table 1. *(continued)*

No.	Causes with top cited paper	Keywords (synonyms, spellings and plural excluded)
C9	Globalization [15]	Increased Competition (e.g. Pricing Strategy, Margin Profits), Outsourcing, Offshoring, Counterfeiting, Piracy, Unauthorized Use of Intellectual Property, Interconnected, Unclear Supply Chain, International Sourcing
C10	Raw Material Shortages [16]	Natural Resource Constraints, Lack of Diversification, Environmental Concern, Poor Harvest
C11	Labor and Employees Issues [17]	Protest, Strike, Labor Regulation, (Minimum) Wage, Lack of Specialist, Shortage of Labor, Loss of Experienced Employees, Lack of Attractiveness for Employees, Dissatisfied and Unmotivated Employees, Individual Faults, Poor Management Decisions
C12	Technology and Infrastructure [18]	Malfunction, Power Outage, Failure of (Software) Systems, Lack of Realization of Emerging Technologies and Continuous Improvements (e.g. Artificial Intelligence, Digitalization, etc.), Maintenance and Downtimes of Machines/Systems
C13	Market and Demand Volatility [15]	(Market) Trend, Market Transform, Competitor, Seasonal Fluctuation, Lack of Reputation and Attractiveness, Increasing/Changing Customer Demands, Cultural Shifts
C14	Lack of Resilience and Redundancy [19]	Lack of Resilience, Lack of Redundancy, Lack of Diversification, No Alternative Suppliers, Single Sourcing, Inadequate Planning, Concentration of Critical Components or Processes in Specific Locations, Risk Management
C15	Increasing Sustainability Requirements [20]	Carbon Neutrality, Recycling, Reduce Waste and Pollution, Reduced Usage of Resources, Traceability, Transition to Renewable Energy, Emission Trading, Circular Economy

Note that causes can influence and condition each other. Therefore, the assignment of keywords to the main causes can vary depending on the individual situation or articles can be assigned to several causes. The results of the automated analysis out of 513 research articles (234 unmatched) and 476 news articles (280 unmatched) can be seen in Fig. 3.

It shows that the COVID-19 pandemic is the most relevant topic in research at present concerning disruptions in value chains. In news, political factors play an essential role, likely due to the war in Ukraine. The high number of unmatched articles is based on the short text excerpt with which the keywords are matched and shows the limit of the automated analysis.

The holistic, ChatGPT assisted cause identification forms the basis for the next sections. Levels of SDVN adaptivity are formulated offering explicit benefits in terms of resilience. Subsequently, requirements, current barriers, possible enablers, and research questions are derived in order to meet those benefits.

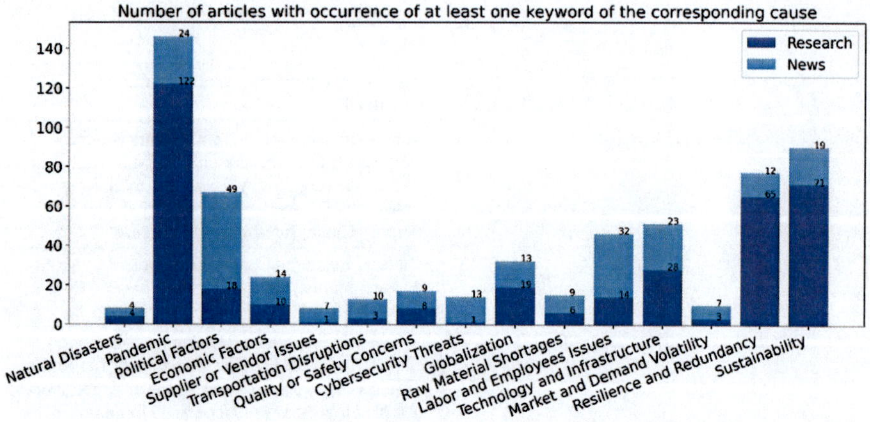

Fig. 3. Overview of the keyword matches to research papers and news articles.

3 Levels and Requirements of SDVN

The various causes that challenge the resilience of GVCs will be condensed to four global levels of SDVN adaptivity. These levels are formalized intervention strategies to influence an existing SDVN, enabling the network to adapt to the identified causes of disruption. By a software-defined, and thus to be achieved automatic action, benefits beyond the decision space of a single company are provided. As Fig. 4 shows, the four levels (Balancing, Adaption, Reconfiguration, Release) are arranged in a way to build on each other. While former levels act as enabler, latter levels may entail previous ones. The adaptivity strategy of SDVN, from disruption causes and their exemplary induced changes, to the adaptations on the four levels can be seen.

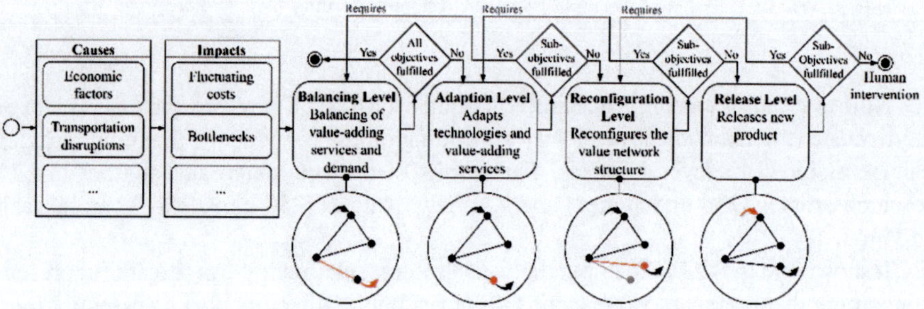

Fig. 4. Strategies in the case of changes regarding a software-defined value network.

Thus, e.g. a software-defined balancing of value-adding services and demand in a value network enables the going beyond adaptations of the production technology, the reconfiguration of the value-network structure or product releases. However, latter levels evoke changes in previous ones, such as the reconfiguration implies changes in balancing of the value network. The mentioned causes may affect multiple levels. Regarding the example of a substitution of a dropped-out supplier, primary its exclusion needs to be conducted. Consequently, however, it may be necessary to integrate a new supplier, change the technology and shift the needs for a new SDVN. If this flexibility does not provide enough success to resilience benefits, the superordinate level, in this case the release of the product, needs to be conducted. This represents a greater change, which is why their software-defined adaptation should be considered in a possible procedure in ascending order.

Besides the levels, respective notes, regarded causes, and requirements (R...) in their digitalization are stated in Table 2. Fundamental, application-oriented requirements in respective of each level are positioned. These are necessary properties for business processes in a value network to run in a software-defined manner and thus offer the aforementioned added value. They represent holistic principles for planning, realization, monitoring, and action, rather than a complete, detailed list. The levels and their requirements holistically cover the identified causes at the level of the value network.

4 From Software-Defined Manufacturing to Software-Defined Value Networks

Ongoing research and development in digitalization is already partially addressing, or transferable to the goal of achieving SDVN. Their contribution as enabler to raise software-defined manufacturing to SDVN in the preceding holistic context is introduced in the following. For this purpose, technological barriers (B...) to digitalization at the value creation level are identified and a brief overview of current enablers (E...) is given. In addition, remaining research questions (RQ...) with the requirements necessary for SDVN from the authors' point of view are discussed.

Data models serve as the blueprint for organizing and structuring data within value networks. They provide a standardized framework that allows for automation in the context of seamless integration, exchange, and analysis of data across different entities within the network. By defining the relationships, attributes, and rules governing the data, these models enable a common understanding and interpretation of information, ensuring consistency and accuracy throughout the network. One of the key advantages of data models in the digitalization of value networks is their ability to facilitate data-driven decision-making. Whilst companies develop product, process and resource (PPR) data models to uniformly describe these assets, **no uniform and standardized data models (B1)** exist to describe value networks, their participants with according skills and products with required capabilities. In this context, two enablers deserve special mention:

E1.1: Asset Administration Shells (AAS) are essential components in the realm of Industry 4.0 and for semantic modeling of lifecycle information of digital twins [21]. AAS provide a standardized framework for representing and managing physical and virtual assets in a digital ecosystem.

E1.2: OPC Unified Architecture (OPC UA) is a cross-platform, open-source, standard for data exchange, especially used on production and machine level. OPC UA provides the foundation for communication and, more importantly in this case, modeling through Companion Specifications.

Table 2. Levels and requirements of SDVN adaptivity.

Levels	According to e.g.	Causes	Requirements
Balancing: Balancing of value-adding services and demand by: **a)** Adaptation of production amount or **b)** Change of the cost structure	**1.** Production and transportation bottlenecks and capacities **2.** Fluctuating demand **3.** Resource availability **4.** Fluctuating costs	C1, C2, C4, C5, C6, C10, C11, C13, C14	**R1.1:** Monitoring of demand, resources and capable production equipment or transport as well as their costs **R1.2:** Identification of deficits regarding an optimal performance of the value network **R1.3:** Evaluation of suitable adaptation scenarios for a multi-criteria optimal selection **R1.4:** Realization of the recalculated scenario
Adaption: Adapts technologies and value-adding services for: **a)** Deployment of required technology or **b)** Integration of emerging technology	**1.** Quality or safety concerns **2.** Upcoming or expiring regulations **3.** Technology advantages	C7, C8, C10, C12, C15	**R2.1:** Overview of general, political and industrial, conditions such as policies and standardizations **R2.2:** Capability description, conformity and analysis of existing and emerging production technologies **R2.3:** Replanning of the manufacturing process **R2.4:** Verification of process capability
Reconfiguration: Reconfigures the value network structure: **a)** Exclusion of existing network participant or **b)** Integration of new network participant	**1.** Missing or reached fulfillment of requirements **2.** Drop out or join of supplier or transporter	C3, C5, C8, C9, C14	**R3.1:** Monitoring of performance and resilience of existing, used and unused, suppliers **R3.2:** Automatic identification of drop out **R3.3:** Knowledge about possible capable suppliers besides the current value network **R3.4:** (Un-)Authorizing data access to uniform models
Release: Adaptation of product in terms of: **a)** Requirements **b)** Design	**1.** Changing customer preferences **2.** Policies	C2, C3, C7, C13, C15	**R4.1:** Analysis of changing customer preferences and policies regarding product requirements **R4.2:** Documentation/data management for product adaptation **R4.3:** Design modification and evaluation

It is necessary that all participants agree on standardized data models. Many efforts are currently underway to define Companion Specifications (OPC UA) and Submodels (AAS).

RQ1.1: Appropriate standards for mapping holistic value networks are still lacking and need to be researched. Especially cross-company functions such as supply chain tracing and sustainability must be taken into account.

RQ1.2: Technological integration and the use of the above-mentioned tools are currently often challenging, especially for small and medium-sized enterprises. Therefore, these technologies as well as the architectures and interfaces necessary for their use must be developed and made accessible.

In order to share data and information among multiple stakeholders and to enable seamless collaboration, **Data Access and Exchange** mechanisms need to be established. Especially, elevating existing bilateral relationships into a network requires new methods for secure and controllable data exchange. Robust mechanisms for authentication and authorization must be implemented in the network's communication infrastructure and the underlying data models. Additionally, ensuring data validity and addressing potential concerns regarding secrecy and privacy, are crucial for enabling successful participation that nevertheless ensures the benefits of the levels.

Currently, there are **no ecosystems, frameworks, or tools for automated and secure data exchange (B2)** in the context of value networks, although several approaches are under development:

E2.1: Current initiatives, such as **Gaia-X, Catena-X, International Data Spaces (IDS) or Manufacturing-X** address cross-company, trust-based data infrastructure ecosystems. Their contribution is seen in the development of infrastructure, connectors, gateways, and federated services.

Data spaces play a pivotal role in the context of value networks, enabling seamless collaboration and the efficient exchange of information and insights among diverse stakeholders. Moreover, data spaces foster trust among participants by implementing robust data governance mechanisms. These mechanisms ensure data quality, enforce data usage policies, provide transparency around data access and usage rights, and enable a shift in the way organizations think about making their data available. By establishing trust, data spaces encourage greater participation and collaboration among stakeholders, unlocking the potential for innovative value creation.

RQ2.1: The application-oriented use of data spaces and the according tools, such as the eclipse data space connector, **is currently unclear** from the view of a single participant. Detailed application and usage scenarios have to be identified, demonstrated and the according architectures and implementations need to be provided.

RQ2.2: Access mechanisms guaranteeing e.g. the limited number of usages, time restrictions, or a secure computation regarding privacy of data [22] have to be included in order to ensure the correct use of the accessed data,

Establishing **independent orchestration mechanisms** becomes essential, as the reliance solely on original equipment manufacturers (OEMs) as pulling initiators of bilateral chains **limits the initiation and monitoring to the pull principle (B3).** Decentralized or as an orchestration unit, they build up on the required data models and data spaces, enabling (1) effective monitoring and (2) coordinating activities and thus the administration. For this purpose, no technological solutions are known by the authors. However, widely used mechanisms and architectures from software systems could be adopted, namely:

E3.1: Microservice architecture, which is an approach to building software systems that decompose applications into small, independent, and loosely coupled services. Each service is responsible for a specific business capability and can be developed, deployed, and scaled independently. Microservices enable flexibility, scalability, and resilience in large and complex systems. They promote agility and allow for the adoption of different technologies for each service based on its requirements.

E3.2: Observer design pattern (or publish-subscribe messaging pattern) establishes a one-to-many relationship between objects, where a subject (observable) notifies multiple observers when its state changes. The observers register with the subject and are automatically updated when changes occur. This pattern promotes loose coupling between objects, as the subject doesn't need to know the details of its observers. The observer pattern is commonly used in event-driven systems.

RQ3.1: An architecture and implementation for independent orchestration mechanisms needs to be provided to enable monitoring and activities of the network.

RQ3.2: Orchestration methods for equal opportunity of network partners and **trusted services** need to be developed to prevent abuse of the SDVN.

RQ3.3: The role of regulatory (e.g. government) **and standardization organizations** need to be developed, since the SDVN spans more participants than GVC,

Finally, the absence of effective **planning methods for value networks (B4)** presents a significant deficit that needs to be addressed. Without proper planning methods in the orchestration, value networks lack the necessary tools and strategies to self-organized optimize their operations and align their activities with the overall business objectives. Recent planning methods for production systems show suitable approaches:

E4.1: Skill-based manufacturing allows flexible matching of available resources and existing (product) requests described by capabilities. The possibility to compose and schedule workflows in complex, adaptive production systems could be raised to SDVN. Approaches of **iterative planning** to establish organizational and technological control loops for continuous improvement can also be transferred [23].

E4.2: Multicriteria optimization and evolutionary algorithms represent fundamental optimization methods to identify (stochastically) optimal solutions in complex systems. They could be used, especially in the balancing level, to determine optimal network production amounts.

RQ4.1: A uniform level of abstraction of the value chain is particularly important to consider, when researching the methods themselves. Regarding the whole capabilities of a participant, an aggregated version in terms of its interest needs to be considered in the planning methods.

RQ4.2: Efficient adaptivity mechanisms, regarding the variety and consequences of value chain disruptions, need to be developed to realize the strategies in Fig. 4.

The introduced barriers and open research questions highlight the challenges faced in achieving seamless integration and efficient operations in software-defined value networks. Looking again at Fig. 1, it is clear that the external world of the automation pyramid is virtually missing, and with it the interoperability of current GVCs, leading to the disruptions introduced. Each barrier highlights a specific area requiring attention, while the lack of effective planning methods underscores the importance of developing tools and strategies to optimize value network operations.

5 Conclusion and Outlook

This paper introduces and motivates SDVN, presents approaches and highlights ongoing research activities to enable SDVN. Therefore, a holistic view of global value chain disruptions has been set up with an automated literature analysis using an initial set of keywords from ChatGPT. Based on this, levels of SDVN adaptivity are developed, enabling flexibility benefits beyond the decision space of a single company. Global barriers to digitalization at the value-added level, the contribution of current enablers and open research questions are derived from the associated application-oriented requirements. Thus, a consistent and holistic approach for SDVN was established. SDVN leads to increased flexibility, value-added resilience through redundancy, scalability through standardized linkages, and data and time efficient exchange through direct linkages compared to GVC.

Enablers already under development and open research questions were already mentioned in the previous section. Addressing them will pave the way for SDVN and thus a successful and resilient production of the future in a VUCA world. Future work will develop a detailed concept of SDVN, showing the use of the mentioned enablers in a service-oriented software-defined architecture.

Acknowledgements. The authors would like to thank the Federal Ministry for Economic Affairs and Climate Action (BMWK) for funding the joint project: SDM4FZI as part of the "Future Investments in the Automotive Industry" funding program.

References

1. Hutter, C., Weber, E.: Russia-Ukraine War: Short-run Production and Labour Market Effects of the Energy Crisis. IAB-Discussion Paper (2022)
2. Schiffling, S., Valantasis Kanellos, N.: How the Ever Given exposed the fragility of global supply chains (2021)
3. Neubauer, M., Walker, M., Oechsle, S., et al.: Cloud-based evaluation platform for software-defined manufacturing: Cloud-basierte Evaluierungsplattform für Software-defined Manufacturing. at - Automatisierungstechnik **71**, 351–363 (2023)
4. OpenAI – ChatGPT. https://chat.openai.com/. Accessed 15 May 2023
5. SerpApi - Google Scholar API https://serpapi.com/google-scholar-api. Accessed 14 June 2023
6. NewsAPI https://newsapi.org. Accessed 14 June 2023
7. Carvalho, V.M., Nirei, M., Saito, Y.U., et al.: Supply chain disruptions: Evidence from the Great East Japan earthquake. Q. J. Econ. **136**, 1225–1321 (2021)
8. Zhang, D., Hu, M., Ji, Q.: Financial markets under the global pandemic of COVID-19. Financ. Res. Lett. **36**, 101528 (2020)

9. Lee, I., Shin, Y.J.: Fintech: Ecosystem, business models, investment decisions, and challenges. Bus. Horiz. **61**, 35–46 (2018)
10. Ozili, P.K., Arun, T.: Spillover of COVID-19. In: Wang, J., Akkucuk, U. (eds.) Managing Inflation and Supply Chain Disruptions in the Global Economy. IGI Global, pp. 41–61 (2022)
11. Garnett, P., Doherty, B., Heron, T.: Vulnerability of the United Kingdom's food supply chains exposed by COVID-19. Nat Food **1**, 315–318 (2020)
12. Scheibe, K.P., Blackhurst, J.: Supply chain disruption propagation: a systemic risk and normal accident theory perspective. Int. J. Prod. Res. **56**, 43–59 (2018)
13. Hastig, G.M., Sodhi, M.S.: Blockchain for supply chain traceability: business requirements and critical success factors. Prod. Oper. Manag. **29**, 935–954 (2020)
14. Masip-Bruin, X., Marín-Tordera, E., Ruiz, J., et al.: Cybersecurity in ICT supply chains: key challenges and a relevant architecture. Sensors (Basel) **21**, 6057 (2021)
15. Eggers, F.: Masters of disasters? Challenges and opportunities for SMEs in times of crisis. J. Bus. Res. **116**, 199–208 (2020)
16. Gereffi, G.: What does the COVID-19 pandemic teach us about global value chains? The case of medical supplies. J. Int. Bus. Policy **3**, 287–301 (2020)
17. Koberg, E., Longoni, A.: A systematic review of sustainable supply chain management in global supply chains. J. Clean. Prod. **207**, 1084–1098 (2019)
18. Ben-Daya, M., Hassini, E., Bahroun, Z.: Internet of things and supply chain management: a literature review. Int. J. Prod. Res. **57**, 4719–4742 (2019)
19. Ivanov, D., Dolgui, A.: A digital supply chain twin for managing the disruption risks and resilience in the era of Industry 4.0. Prod. Plan. Control **32**, 775–788 (2021)
20. Saberi, S., Kouhizadeh, M., Sarkis, J., et al.: Blockchain technology and its relationships to sustainable supply chain management. Int. J. Prod. Res. **57**, 2117–2135 (2019)
21. Bader, S., Barnstedt, E., Bedenbender, H., et al.: Details of the Asset Administration Shell - part 1: the exchange of information between partners in the value chain of Industrie 4.0 (Version 3.0RC02) Berlin (2022)
22. Yao, A.C.: Protocols for secure computations. In: 2013 IEEE 54th Annual Symposium on Foundations of Computer Science, pp 160–164. IEEE Computer Society, Los Alamitos, CA, USA
23. Dietrich, D., Neubauer, M., Lechler, A., et al.: Iterative Planning as a Holistic Framework for Production System-Wide Optimization Control Loops. In: Silva, F.J.G., Pereira, A.B., Campilho, R.D.S.G. (eds.) Flexible Automation and Intelligent Manufacturing: Establishing Bridges for More Sustainable Manufacturing Systems FAIM 2023. Lecture Notes in Mechanical Engineering, pp. 611–621 (2024)

Risk Management in Production Planning: A Systematic Literature Review

C. Kuhn[1](✉), P. Shah[1], J. Fulterer[1], S. Schmitz[1], and G. Schuh[1,2]

[1] Laboratory for Machine Tools and Production Engineering (WZL), Aachen, Germany
c.kuhn@wzl.rwth-aachen.de

[2] Fraunhofer Institute for Production Technology IPT, Aachen, Germany

Abstract. In recent years, unforeseen and rare events in the environment of manufacturing companies have been a dominating influence in production planning. Examples are dynamic product demand developments due to the COVID-19 pandemic and geopolitical conflicts or changes in production resource availabilities due to technological disruptions and shortages. In addressing these unforeseen events, large changes in the production systems can cause errors or disruptions raising the need for the inclusion of risk management in production planning. In order to insure applicability for unforeseen and rare events, risk management approaches in production planning require realistic and objective risk assessments. This paper conducts a literature review for risk management in production planning with the aim of assessing the state of the art in regards to realistic and objective risk assessment and concludes with the recommendation for further consideration of rare and unexpected events in future methodologies.

Keywords: Literature Review · Risk Management · Production Planning

1 Introduction

Production planning is increasingly affected by disruptions, such as environmental disasters or supply chain failures. The proportion of organisations which are reported to have experienced disruptions has reached its highest level in the last decade at ~88% in 2021 [1]. In the course of the COVID-19 pandemic, producing companies have experienced shortages in supply as well as unpredictable customer demand developments [2, 3]. The result of these events are disruptions in the planning of production systems. Examples are the short term adaption of the production plan to account for insufficient raw material supply and surging customer demand [2] or to account for labour-force shortages due to illness and lock-downs [4]. Given these developments, there is an increasing demand to anticipate and address risks in production planning [5], defined as the effects of uncertainties on the objectives in production planning [6].

With the aim of addressing risks, risk management comprises coordinated activities to direct and control an organisation [6]. Risk management balances different concerns such as profits, safety or reputation and evaluates suitable measures to address risks [7].

T. Bauernhansl et al. (Eds.): WGP 2023, LNPE, pp. 525–536, 2024.
https://doi.org/10.1007/978-3-031-47394-4_51

As such, the inclusion of risk management in production planning can assist in reducing the impact of risks on production systems.

This paper conducts a systematic literature review on methodologies for risk management in production planning. The paper is structured as follows. In Sect. 2, the background for approaches in risk management is discussed. In Sect. 3, the systematic literature review methodology is described, the results of which are presented in Sect. 4. In Sect. 5 the findings are discussed and recommendations for further research directions are given. Section 6 concludes this paper with a summary.

2 General Risk Management Process and Selection of Risk Treatment Measure Based on Knowledge Basis

The general risk management process in accord with ISO 31000 comprises the four steps *risk identification*, *risk analysis*, *risk evaluation* and *risk treatment* (see Fig. 1). *Risk identification* uses a range of techniques such as historical data or creative methods to find risks on the organisation. *Risk analysis* considers and quantifies, when possible, the likelihood of events and the magnitude of the consequences. *Risk evaluation* determines possible measures by comparing the results of the risk analysis with established criteria. *Risk treatment* is the process of selecting and implementing measures to address risks [6].

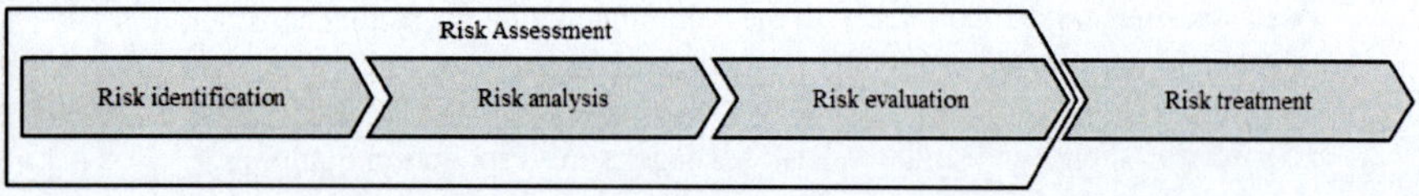

Fig. 1. General risk management process in accord with ISO 31000 [6]

An appropriate description of the risk in the risk assessment phase from Fig. 1 is a prerequisite for the objective evaluation of risk treatment measures [7]. Consequently, identifying the correct treatment measure is dependent on the knowledge basis available for the description of the assessed risk.

AVEN et al. describe decision guidelines for the choice of the risk treatment measure depending on the available knowledge basis of the risk [8]. Hereby, the available knowledge is distinguished between certain risk probabilities and uncertain risk probabilities. A possible source for uncertain risk probabilities is the rarity of the risk event such that established statistical methodologies are not suitable for the description of that event [9]. Rare risk events can bare a proportionally large impact on the overall organisation (e.g. natural disasters such as pandemics), such that the treatment of rare risk events is within the scope of overall risk management. The treatment of uncertainties, such as unknown risk probabilities, is regarded as a challenge for the field of risk management [7].

A risk knowledge basis with certain risk probabilities is suitable for the selections of risk-informed treatment measures. Risk-informed treatment measures have the aim of avoidance, reduction, transfer and retention of the effects from individual risks. Higher stakes and larger uncertainties lead to the application of precautionary measures. Precautionary measures aim at increasing the general robustness or resilience of a system, rather than addressing individual risks [8].

3 Methodology of the Systematic Literature Review

The systematic literature review procedure is based on BORREGO et al. [10]. The focus of the systematic literature review, in accordance with a framework provided by COOPER [11], is to present research outcomes and applications for risk management in production planning. As such, this paper aims to synthesise and integrate past research as well as identifying central issues for further research from a neutral perspective. A representative review scope is chosen with general scholars and practitioners as the audience.

The search for literature uses search strings created with combinations of synonyms as depicted in Fig. 2. The synonyms are organised within subject domains and methodological disciplines. A combination of the synonyms using "AND" and "OR" operators results in 108 search strings.

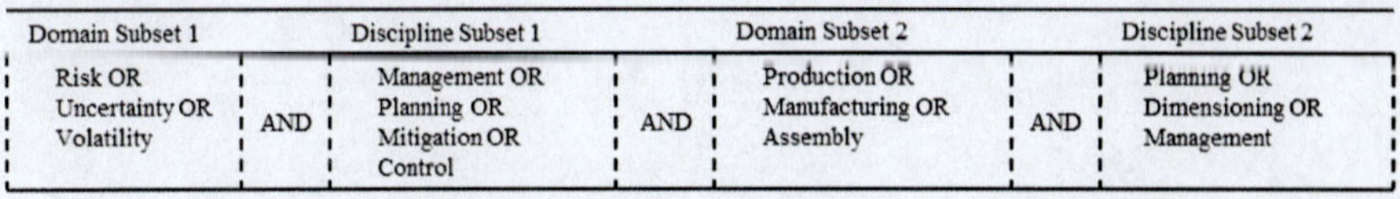

Fig. 2. Keyword combinations for the formulation of search strings.

In the review process, the search strings were applied to titles, keywords and abstracts in the databases ScienceDirect, Web of Science, Taylor and Francis, IEEE Xplore, Scopus and SpringerLink. Only results in the English language were considered. As the scope of this literature review is to identify approaches in production planning, the databases were further configured to only display results in the engineering field. Across all databases, a total of 21,423 publications were identified, which resulted in 17,433 publications after the removal of duplicates. A manual title screening further reduced the number of publications to 302. Figure 3 shows the subsequent review process. The review process methodology from BORREGO et al. [10] was complemented by a citation screening, such that only publications were included, which have been cited at least once since 2021 or were released after 2021. This additional step was included to ensure that the findings depict the current state of research. A filtering exclusively by publication date was not chosen, as older publications may still yield relevant findings. In the end, 47 publications remained for further analysis.

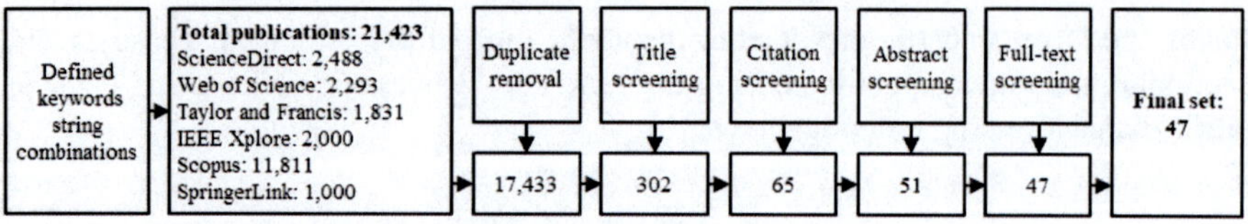

Fig. 3. Review process.

4 Results

The yearly distribution of the final set is depicted in Fig. 4. It is noticeable that the publications in the final set go back to 1984. On the one hand, this underlines the grounding for the field of research in established methodologies while on the other hand depicting the continued relevancy for further research.

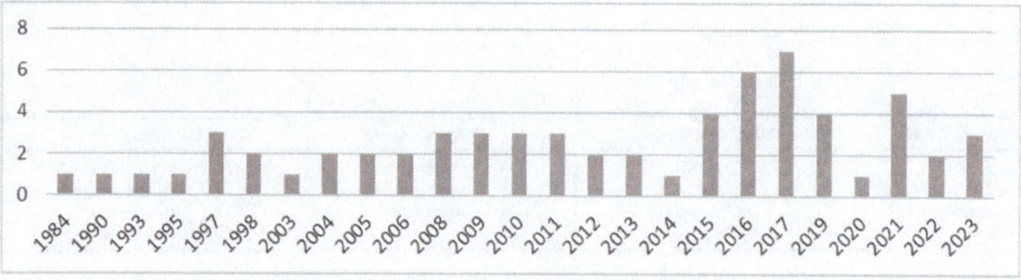

Fig. 4. Yearly distribution of the publications in the final set.

The final set of publications is presented in a framework, which orders the identified risk management methodologies for production planning. The categories of the framework are a direct derivation from the general risk management process (see Fig. 1). Most risk management methodologies identified in the literature review do not identify new risks but assume a type of risk to exist, which is then further analysed. As a result, the category *assessed risk* is derived from the risk management process step *risk identification.* As the risk management process steps *risk analysis, risk evaluation* and *risk treatment* are widely used in the literature, they are used as categories for the ordering of risk management methodologies.

For each of the categories, several elements are identified in the final set of publications. The publications are allocated to the elements, such that an overview of the identified risk management methodologies is given. Furthermore, the elements are grouped into thematically similar approaches and a heading is given to each group. Table 1 shows the results.

In total, 34 of 47 risk management methodologies for production planning include the entire general risk management process and contain one or more elements within each category. The remaining 13 risk management methodologies do not include the entire risk management process and are not allocated to an element in each category.

Table 1. Categorization and grouping of approaches for risk management.

Assessed risk	
General risks	
• Demand uncertainty [5, 12–34]	• Lead time uncertainty [5, 15, 20, 28, 33, 35–37]
• Supply uncertainty [5, 15, 20, 23, 27, 33, 38, 39]	
Specific risks	
• Machine breakdowns and plant failures [20, 31, 32, 36, 40–42]	• Customer due dates (rush jobs/ cancellations) [36, 38, 40]
• Machine yield [27]	• Resource suitability for production task [43]
• Production cost uncertainty [25]	• Risk FMEA [44–51]
Risk analysis	
Not suitable for analysis of rare risk events	
• Normal distribution [5, 12, 14, 16, 19, 20, 24, 27, 30, 36]	• Historical empirical data [13, 51]
• Fuzzy [15, 22, 23, 37–39, 43]	• Wear and tear modelling [41]
• Lognormal distribution [21]	• Risk FMEA [35, 44–50]
• Uniform distribution [23, 24, 37, 42]	
Compatible with analysis of rare risk events	
• Scenario formulation [36, 40]	• Compatible with any distribution [29, 31, 32]
Risk evaluation	
Qualitative approaches	
• Qualitative framework [17, 35, 52]	• If-then structures to codify expert knowledge [15, 20]
• Risk FMEA [44–50]	
Simulation approaches	
• Process simulation [5, 12, 14, 23–25, 36, 45, 53]	• Monte Carlo simulation [13, 16, 21, 53]
Algorithmic approaches	
• Deterministic algorithm [5, 19, 22, 29, 30, 34, 36, 38, 43, 54]	• Control engineering [28, 42]
• Linear programming [13, 16, 39]	
Artificial intelligence approaches	
• Genetic algorithm [24, 27, 31, 32, 37, 40]	• Neural network [25]

(*continued*)

Table 1. (*continued*)

Risk treatment	
Qualitative concepts	
• Postponement of decisions [17] • Effectuation [55]	• Risk FMEA [44–50]
Changes to the production system	
• Production schedule design/ adjustment [5, 13, 14, 16, 19, 22, 25, 27–29, 31, 32, 36–41, 56] • Safety stocks of resources [26, 30, 42] • Safety stocks of products [12, 14, 20, 23, 26, 42] • Safety lead times [20, 23, 33]	• Safety capacity [18, 33] • Production layout and process design [24, 34, 37] • Resource and production capacity allocation [21, 43, 54] • Maintenance planning [27, 31, 41, 46]
Changes to the production planning procedure	
• Increased frequency for re-planning [5] • Production schedule freeze/ time fencing [14, 33]	• Real options valuation of investment decisions [57]

5 Discussion

The discussion is structured such that first the individual categories in the framework of Table 1 are discussed, before a general discussion of the overall framework is held.

Before addressing the individual categories, it is noticeable that the element *Risk Failure Mode and Effect Analysis (FMEA)* occurs as an element in each category. This is due to the Risk FMEA being a risk management methodology, which covers the entire risk management process as depicted in Fig. 1. It refers to a methodology derived from the FMEA and applied to the subject of risk management.

Assessed Risks. The elements in the category *assessed risk* describe aspects in production planning, which undergo uncertain developments and are sources for risks as a result. The elements for the assessed risks can be split into two groups. The first group assesses only general risks in production planning, such as lead time uncertainty, without specifying the causes. The elements in the first group act as an aggregate measure of the effects of uncertainties from a variety of external and internal sources. The first group is better suited for the risk assessment of rare and unexpected events, as the exact causes of such events is often unknown in advance. The second group describes the effect of specific risks tied to specific causes, such as machine breakdowns. Both the first and the second group are well suited to assess specific risks with certain probabilities with the aim of deriving risk-informed measures.

Risk Analysis. Risk analysis elements can be categorised according to their suitability for analysis of rare risk events, as introduced in Sect. 2. This categorisation is displayed in the grouping of the elements. From the risk management approaches in the final set of publications, no approach specifically addresses the handling of uncertain information or the modelling of rare events for risk analysis.

The first group of elements includes risk analysis approaches, which are not suitable for the analysis of rare risk events. This is due to events in the tails of the probability distributions being explicitly not modelled, such as in the fuzzy or uniform distribution. Alternatively, such as with the normal distribution, events in the tails of the probability distributions are modelled, such that events sufficiently far away from the mean can be neglected in the long run through the sum of effects from repeated exposure [58]. The second group of elements includes risk analysis approaches, which do not include methods suitable for the analysis of rare risk events, but would be compatible with the inclusion of such an analysis. In case of *Scenario formulation*, extremal events can be defined in edge scenarios. Approaches of the element *Compatible with any distribution* are capable of introducing distributions suitable for the analysis of rare risk events.

Risk Evaluation. The risk evaluation methodologies display a range of existing approaches with varying mathematical and digital sophistication. Qualitative approaches include *Qualitative frameworks*, *Risk FMEA* or *If-then structures to codify expert knowledge*. Simulation approaches, including *Process simulation* and *Monte Carlo simulation* are the most common approaches in the final set of publications to evaluate the effects of risks and to identify correct risk treatments. Algorithmic approaches include *deterministic algorithms*, which are specifically created for the task of risk management in production planning. Although *Linear programming* and *Control engineering* can also be considered to be deterministic algorithms, they are listed as separate elements, as these algorithms have an established, separate background in literature. Lastly, approaches from the artificial intelligence background, namely *genetic algorithms* and *neural networks* have been applied to the problem of risk management in production planning.

Overall, some risk evaluation approaches can be executed in an analogue manner, especially qualitative methodologies. *Simulation* or *artificial intelligence* approaches are most suited for a digitised implementation of the risk management approach.

Risk Treatment. It is apparent, that a wide range of risk treatment measures is considered in the current state of research. As such, the elements in Table 1 are grouped into qualitative concepts, changes to the production system and changes to the production planning procedure.

Qualitative concepts include *the postponement of decisions* to meet decisions with more accurate information, as uncertainties are decreased with decreased forecasting horizons. This postponement leads to an increase in frequency of re-planning as decisions made over a shorter time horizon. A second example, *Effectuation*, is a qualitative concept, which aims to describe the behaviour of entrepreneurs as they operate in uncertain conditions. Rather than specific measures, effectuation gives guiding principles for decision making under uncertain conditions.

Changes to the production system are either executed by changes in the production schedule, including the planning of maintenance operations, or a long-term adjustment of the production process design, such as layout of safety stocks.

The third group of risk treatment approaches are changes to the production planning procedure. *Increased frequency of re-planning* addresses uncertainties of planning information by re-planning the production system as information changes or becomes more concrete. A downside of frequent re-planning is a frequent change of tasks and orders in the production system, called schedule nervousness. To address this, *production schedule freezing/time fencing* is employed to set the production schedule for a time frame or to reduce variations of a given production schedule [33]. *Real options valuation of investment decisions* is an investment valuation method, by which values for the return on investment are calculated for different future scenarios. The values are weighted by the estimated probabilities of the given scenarios and summed. This allows for the evaluation of investments for risk mitigation.

Overall Framework. As stated in Sect. 2, the appropriate application of risk mitigation measures depends on choosing the correct risk description. However, the risk analysis approaches for production planning identified in Table 1 do not address the challenge of uncertain probabilities for rare risk events. Instead, the current state of research primarily assumes certain risk probabilities and consequently only includes approaches appropriate for the application of risk-informed measures. The appropriate application of precautionary measures for the creation of a resilient or robust production system is conditional on the evaluation of risks with uncertain probability information, such as rare risk events.

A recommendation for further research would be the development of risk management approaches for production planning, which include risk analysis methodologies capable of statistical treatment for rare risk events. A possible approach to model rare events with high stakes, so called extremal events, can be found in the field of extreme value theory. Extreme value theory provides statistically optimal methods for the quantification of consequences of rare events in the tails of probability distributions [58]. Extreme value theory is employed in risk management to model the effects of catastrophic events such as flood levels [9] or earthquake insurance claims [59].

6 Summary and Conclusion

An increase in rare and unpredictable risk events in recent years has been a prevailing influence on production planning. To address the increasing need for the anticipation and addressing of risks in production planning, this paper conducts a systematic literature review for approaches for risk management in production planning. The databases ScienceDirect, Web of Science, Taylor and Francis, IEEE Xplore, Scopus and SpringerLink yielded a total of 21,423 results, which were narrowed down in a review process to a final set of 47 publications. The final set of publications is presented in a framework, which depicts the approaches for the individual steps in the general risk management process in accord with ISO 31000.

In comparison with the state of research for general risk management, the risk management in the field of production planning is well suited for the evaluation of individual risks with well-defined probabilities. However, approaches for risk management in production planning identified in this paper fail to consider risks with uncertain probabilities such as rare risk events. Consequently, the identified current approaches are not suitable for the design of precautionary measures with the aim of designing a more resilient and robust production system. A recommendation for further research is to consider uncertain information in risk probability evaluation as a result of rare risk events such as through the inclusion of extreme value theory.

Acknowledgement. The authors would like to thank the Federal Ministry for Economic Affairs and Climate Action for the support within the research project "TrendAuto2030plus".

The authors thank Jan Maetschke from the Laboratory for Machine Tools and Production Engineering (WZL) for fruitful discussions in the creation of formative ideas included in this paper.

References

1. Supply Chain Resilience Report 2021. Reading, UK (2021)
2. Paul, S.K., Chowdhury, P.: A production recovery plan in manufacturing supply chains for a high-demand item during COVID-19. Int. J. Phys. Distrib. Logist. Manage. **51**, 104–125 (2021)
3. Okorie, O., Subramoniam, R., Charnley, F., Patsavellas, J., Widdifield, D., Salonitis, K.: Manufacturing in the time of COVID-19: an assessment of barriers and enablers. IEEE Eng. Manage. Rev. **48**, 167–175 (2020)
4. Shen, W., Yang, C., Gao, L.: Address business crisis caused by COVID-19 with collaborative intelligent manufacturing technologies. IET Collab. Intell. Manuf. **2**, 96–99 (2020)
5. Ma, Q., Che, Y., Cheng, C., Wang, Z.: Characterizations and optimization for resilient manufacturing systems with considerations of process uncertainties. J. Comput. Inf. Sci. Eng. (2023)
6. International Standard Organisation: ISO. 31000: Risk management - Guidelines. ISO copyright office. Geneva (2018)
7. Aven, T.: Risk assessment and risk management: review of recent advances on their foundation. Eur. J. Oper. Res. **253**, 1–13 (2016)
8. Aven, T., et al.: Risk analysis: fundamental principles. Herndon, USA (2018).
9. Siffer, A., Fouque, P.-A., Termier, A., Largouet, C.: Anomaly detection in streams with extreme value theory. In: Proceedings of the 23rd ACM SIGKDD International Conference on Knowledge Discovery and Data Mining, pp. 1067–1075. ACM. New York (2017)
10. Borrego, M., Foster, M.J., Froyd, J.E.: Systematic literature reviews in engineering education and other developing interdisciplinary fields. J. Eng. Educ. **103**, 45–76 (2014)
11. Cooper, H.: Organizing knowledge syntheses: a taxonomy of literature reviews, pp. 104–126 (1988)
12. Schmitt, T.G.: Resolving uncertainty in manufacturing systems. J. Oper. Manage. **4**, 331–345 (1984)
13. Thompson, S.D., Davis, W.J.: An integrated approach for modeling uncertainty in aggregate production planning. IEEE Trans. Syst. Man Cybern. **20**, 1000–1012 (1990)
14. Kadipasaoglu, S.N., Sridharan, V.: Alternative approaches for reducing schedule instability in multistage manufacturing under demand uncertainty. J. Oper. Manage. **13**, 193–211 (1995)
15. Grabot, B., Geneste, L.: Management of imprecision and uncertainty for production activity control. J. Intell. Manuf. **9**, 431–446 (1998)
16. Uribe, A.M., Cochran, J.K., Shunk, D.L.: Two-stage simulation optimization for agile manufacturing capacity planning. Int. J. Prod. Res. **41**, 1181–1197 (2003)

17. Yang, B., Burns, N.D., Backhouse, C.J.: Management of uncertainty through postponement. Int. J. Prod. Res. **42**, 1049–1064 (2004)
18. Gottschalk, L., König, R.: Control of uncertainties in demand by flexible capacities, Bedarfsunsicherheiten durch flexible Kapazitäten beherrschen. Zeitschrift wirtschaftlichen Fabrikbetrieb 715–719 (2004)
19. Clark, Alistair R.: Rolling horizon heuristics for production planning and set-up scheduling with backlogs and error-prone demand forecasts. Prod. Plann. Control **16**, 81–97 (2005)
20. Koh, S.C.L., Gunasekaran, A.: A knowledge management approach for managing uncertainty in manufacturing. Industr. Manage. Data Syst. **106**, 439–459 (2006)
21. Huang, M.-G., Chang, P.-L., Chou, Y.-C.: Demand forecasting and smoothing capacity planning for products with high random demand volatility. Int. J. Prod. Res. **46**, 3223–3239 (2008)
22. Barba-Gutiérrez, Y., Adenso-Díaz, B.: Reverse MRP under uncertain and imprecise demand. Int. J. Adv. Manuf. Technol. **40**, 413–424 (2009)
23. van Kampen, T.J., van Donk, D.P., van der Zee, D.-J.: Safety stock or safety lead time: coping with unreliability in demand and supply. Int. J. Prod. Res. **48**, 7463–7481 (2010)
24. Jithavech, I., Krishnan, K.K.: A simulation-based approach for risk assessment of facility layout designs under stochastic product demands. Int. J. Adv. Manuf. Technol. **49**, 27–40 (2010)
25. Lan, Y., Zhao, R., Tang, W.: Minimum risk criterion for uncertain production planning problems. Comput. Industr. Eng. **61**, 591–599 (2011)
26. Radke, A.M., Tolio, T., Tseng, M.M., Urgo, M.: A risk management-based evaluation of inventory allocations for make-to-order production. CIRP Ann. **62**, 459–462 (2013)
27. Purohit, B.S., Lad, K.: Bhupesh: production and maintenance planning: an integrated approach under uncertainties. Int. J. Adv. Manuf. Technol. **86**, 3179–3191 (2016)
28. Abbou, R., Loiseau, J.J., Moussaoui, C.: Robust inventory control of production systems subject to uncertainties on demand and lead times. Int. J. Prod. Res. **55**, 2177–2196 (2017)
29. Kim, B.S., Chung, B.D.: Affinely adjustable robust model for multiperiod production planning under uncertainty. IEEE Trans. Eng. Manage. **64**, 505–514 (2017)
30. Korponai, J., Tóth, Á.B., Illés, B.: The effect of the safety stock on the occurrence probability of the stock shortage. Manage. Prod. Eng. Rev. **8**, 69–77 (2017)
31. Wang, Z., Pang, C.K., Ng, T.S.: Robust scheduling optimization for flexible manufacturing systems with replenishment under uncertain machine failure disruptions. Control Eng. Pract. **92**, 104094 (2019)
32. Wang, L., Lu, Z., Ren, Y.: Integrated production planning and condition-based maintenance considering uncertain demand and random failures. Proc. Inst. Mech. Eng. Part B: J. Eng. Manuf. **234**, 310–323 (2020)
33. Graves, S.C.: Uncertainty and production planning. In: Kempf, K., Keskinocak, P., Uzsoy, R. (eds.) Planning Production and Inventories in the Extended Enterprise. International Series in Operations Research & Management Science, vol. 151, pp. 83–101. Springer, New York (2011). https://doi.org/10.1007/978-1-4419-6485-4_5
34. Sakhaii, M., Tavakkoli-Moghaddam, R., Bagheri, M., Vatani, B.: A robust optimization approach for an integrated dynamic cellular manufacturing system and production planning with unreliable machines. Appl. Math. Modell. **40**, 169–191 (2016)
35. Wildemann, H.: Management of operational risks in production, Management leistungswirtschaftlicher Risiken in der Produktion. Zeitschrift wirtschaftlichen Fabrikbetrieb 187–191 (2005)
36. Lou, P., Liu, Q., Zhou, Z., Wang, H., Sun, S.X.: Multi-agent-based proactive–reactive scheduling for a job shop. Int. J. Adv. Manuf. Technol. **59**, 311–324 (2012)

37. Wang, W., Li, X., Gao, L., Wen, X., Wan, L.: Application of interval theory and genetic algorithm for uncertain integrated process planning and scheduling, pp. 2796–2801. IEEE (2013)
38. Wang, D., Fang, S.-C., Hodgson, T J.: A fuzzy due-date bargainer for the make-to-order manufacturing systems. IEEE Trans. Syst. Man Cybern. Part C (Appl. Rev.) **28**, 492–497 (1998)
39. Tan, R.R., et al.: A fuzzy linear programming enterprise input–output model for optimal crisis operations in industrial complexes. Int. J. Produ. Econ. **181**, 410–418 (2016)
40. Jain, A.K., Elmaraghy, H.A.: Production scheduling/rescheduling in flexible manufacturing. Int. J. Produ. Res. **35**, 281–309 (1997)
41. Sanmartí, E., Espuña, A., Puigjaner, L.: Batch production and preventive maintenance scheduling under equipment failure uncertainty. Comput. Chem. Eng. **21**, 1157–1168 (1997)
42. Azarskov, V.N., Zhiteckii, L.S., Solovchuk, K.Yu., Sushchenko, O.A., Lupoi, R.O.: Inventory control for a manufacturing system under uncertainty: adaptive approach. IFAC-PapersOnLine **50**, 10154–10159 (2017)
43. Coroiu, A.M.: Fuzzy methods in decision making process - a particular approach in manufacturing systems. In: IOP Conference Series: Materials Science and Engineering, p. 12154 (2015)
44. Islam, M.A., Tedford, J.D., Haemmerle, E.: Managing operational risks in small- and medium-sized enterprises (SMEs) engaged in manufacturing – an integrated approach. Int. J. Technol. Policy Manage. **8**, 420 (2008)
45. Naghab, A.P., Siadat, A., Tavakkoli-Moghaddam, R., Jolai, F.: An integrated approach for risk-assessment analysis in a manufacturing process using FMEA and DES, pp. 366–370. IEEE (2011)
46. Kiran, S., Prajeeth Kumar, K.P., Sreejith, B., Muralidharan, M.: Reliability evaluation and risk based maintenance in a process plant. Procedia Technol. **24**, 576–583 (2016)
47. Shah, L.A., Etienne, A., Siadat, A., Vernadat, F.: Process-oriented risk assessment methodology for manufacturing process evaluation. Int. J. Prod. Res. **55**, 4516–4529 (2017)
48. Klöber-Koch, J., Braunreuther, S., Reinhart, G.: Predictive production planning considering the operative risk in a manufacturing system. Procedia CIRP **63**, 360–365 (2017)
49. Djapan, M., Macuzic, I., Tadic, D., Baldissone, G.: An innovative prognostic risk assessment tool for manufacturing sector based on the management of the human, organizational and technical/technological factors. Saf. Sci. **119**, 280–291 (2019)
50. Burduk, A.: Risk assessment of a production system with series-parallel reliability structure. In: Batako, A., Burduk, A., Karyono, K., Chen, X., Wyczółkowski, R. (eds.) GCMM 2021. LNNS, vol. 335, pp. 656–666. Springer, Cham (2022). https://doi.org/10.1007/978-3-030-90532-3_49
51. Sousa, S., Nunes, E., Lopes, I.: Measuring and managing operational risk in industrial processes. FME Trans. **43**, 295–302 (2015)
52. Vargas, A., et al.: Decision-making system and operational risk framework for hierarchical production planning, pp. 72–81 (2016)
53. de Landtsheer, R., et al.: A discrete event simulation approach for quantifying risks in manufacturing processes. In: Proceedings of 5th the International Conference on Operations Research and Enterprise Systems, pp. 313–322. SCITEPRESS - Science and Technology Publications (2016)
54. Denkena, B., Dittrich, M.-A., Münch, G.V.: Ontology-based production planning under the consideration of system robustness. Procedia CIRP **104**, 547–552 (2021)
55. Brettel, M., Bendig, D., Keller, M., Friederichsen, N., Rosenberg, M.: Effectuation in manufacturing: how entrepreneurial decision-making techniques can be used to deal with uncertainty in manufacturing. Procedia CIRP **17**, 611–616 (2014)

56. Tometzki, T., Engell, S.: Risk management in production planning under uncertainty by multi-objective hybrid evolutionary algorithms. In: 20th European Symposium on Computer Aided Process Engineering, pp. 151–156. Elsevier (2010)
57. Milberg, J., Möller, N.: Valuation of changeable production systems. Prod. Eng. **2**, 417–424 (2008)
58. Embrechts, P.: Klüppelberg, Claudia, Mikosch, Thomas: Modelling Extremal Events. Springer, Heidelberg (1997). https://doi.org/10.1007/978-3-642-33483-2
59. Embrechts, P., Resnick, S.I., Samorodnitsky, G.: Extreme value theory as a risk management tool. North Am. Actuarial J. **3**, 30–41 (1999)

Environmentally Neutral Production

Data Mining-Enabled Temperature Control for Sustainable Production in Rubber Extrusion Lines: An Artificial Neural Network-Based Approach

M. Lukas[1(✉)], S. Leineweber[1], B. Reitz[1], L. Overmeyer[1], A. Aschemann[2], B. Klie[2], and U. Giese[2]

[1] Institute of Transport and Automation Technology, Leibniz University Hannover, 30823 Garbsen, Germany
Marco.Lukas@ita.uni-hannover.de

[2] Deutsches Institut für Kautschuktechnologie e. V., 30519 Hannover, Germany

Abstract. Rubber extrusion processes involve a high number of control parameters and batch-dependent material fluctuations, leading to lengthy and error-prone testing cycles for new products. To address this challenge and enable a more sustainable production, an artificial neural network-based data mining algorithm is developed that identifies correlations between process parameters and product characteristics. This approach allows for consideration of a larger quantity of non-linear correlations that improve with each new application, enabling more precise temperature control of rubber production. Real-world manufacturing data are used to validate the model, which incorporate batch- and recipe-dependent material variations. The results provide new insights into the complex relationships between process parameters and product characteristics, highlighting the potential of data mining algorithms to drive sustainable production in the rubber extrusion industry. This paper presents the iterative development of the data mining algorithm for rubber extrusion lines and demonstrates its practical application in the industry.

Keywords: Rubber Extrusion · Sustainable Production · Artificial Neural Network

1 Introduction

Rubber is an essential material in many industrial applications due to its wide range of properties, including high elasticity, durability, viscoelastic damping and impact resistance, making it suitable for fields ranging from aerospace, automotive to medical industries [1, 8, 13]. However, the properties of rubber can be affected by various factors, including the material basis, composites, manufacturing process parameters and environmental conditions. Temperature is one of the major process parameters that significantly influences quality of rubber extruded products [14]. Therefore, controlling

T. Bauernhansl et al. (Eds.): WGP 2023, LNPE, pp. 539–549, 2024.
https://doi.org/10.1007/978-3-031-47394-4_52

the temperature during the extrusion process is critical to ensuring consistent quality and reliable properties of the final product [2]. However, the high number of control parameters and batch-dependent material fluctuations involved in the temperature control process can lead to lengthy and error-prone testing cycles for new products [6]. These challenges can result in significant waste, high costs and have a notable impact on the energy consumption [14]. As a result, there is an increasing need for more sustainable rubber production methods [3, 20].

To address these challenges, statistical methods and artificial intelligence-based algorithms have been proposed as effective approaches for conducting a detailed analysis of production processes [15–17]. Prior studies have demonstrated the successful application of machine learning algorithms, such as artificial neural networks, for revealing correlations between different aspects of rubber production. For instance, Ghaffarian et al. [4] proposed a method for optimizing the process of designing rubber compounds using artificial neural networks and genetic algorithms. The approach reduces designing time by modeling the behavior of the rubber compound and using quick search technique to determine the values of raw materials. Another study by Kopal et al. [1] presents a generalized regression neural network for predicting curing characteristics of rubber blends with variation of carbon black filler at different temperatures. Park et al. [5] developed a machine learning-based method for simulating recipes for synthetic rubber production during the kneading process. The presented method predicts temperature changes and the synthetic rubber quality during the kneading operation using a prediction walk model and cross-updating prediction. Another study by Ahangar-Asr et al. [7] applied an evolutionary polynomial regression to develop models for predicting the mechanical properties of rubber concrete. The data mining model provides a representation of the materials behavior in the form of a polynomial function. Kojima et al. [18] combine computer simulation and machine learning to analyze the material properties of filled rubber. The approach uses a surrogate convolutional neural network model trained on simulation results to achieve more accurate predictions, identifying the dominant fillers influencing the mechanical properties.

Despite the wide usage of rubber in various industrial applications and studies in different areas, there is a lack of research in implementing machine learning-based data mining techniques to identify and analyze correlations between process parameters and product characteristics in the rubber extrusion process. This represents a gap in the literature, given the potential benefits of using such an approach to improve process efficiency and reduce scrap. In this paper, we address this gap by developing an artificial neural network-based data mining algorithm for rubber extrusion lines as a basis for an automatic temperature control and evaluate its performance in real-world manufacturing settings.

2 Experimental

This chapter focuses on the experimental setup and data analysis used to develop the proposed artificial neural network-based data mining algorithm. The following section covers the extruder specifications, data analysis process, data preprocessing, model selection, hyperparameter optimization, training and validation process (see Fig. 1).

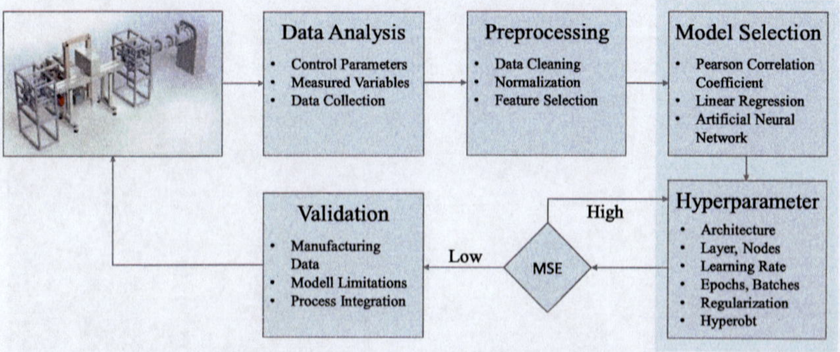

Fig. 1. Procedure Flow Chart: Data Mining development

2.1 Experimental Setup

Extruder Specifications

Figure 2 shows the extruder and the measurement technology used. To monitor the temperature profile along the screw channel, three specific locations are measured: the initial area (T_{S1}), the middle section (T_{S2}), and the final area (T_{S3}). In addition, a measuring blade of FOS Messtechnik GmbH in front of the screw tip records a radial temperature profile within the extruded compound. Whereby seven measured values (T_1 to T_7) are recorded in the rubber compound itself and one sensor measures the die temperature (T_8). In addition, the die pressure (p_S) is determined. To control the geometric stability of the extruded profile and to detect surface defects an *iProfilControl Dual Vision (iPCDV) 90* inspection unit of Pixargus GmbH was installed at the end of the extrusion unit. Furthermore, the figure shows the entry of the feeding strip and the conveyor belt for the removal of the extrudate. The extruder has an adjustable screw speed and five temperature-controlled zones as control values, which are described together with the measured values in Sect. 2.2.

To store all control and measured values of the extruder, an industrial PC with integrated time series database was incorporated into the process chain. The control and measured values can be exported so that the data mining algorithm can be used decentral to identify cause-effect relationships. The rubber used in this study was an EPDM (ethylene propylene diene monomer) based compound to produce windshield gaskets.

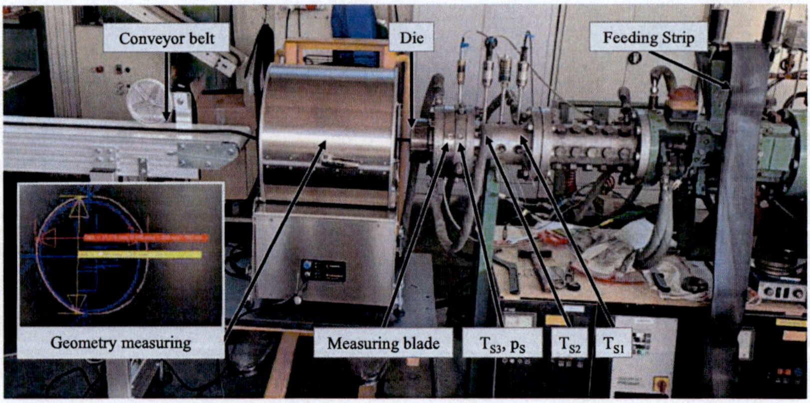

Fig. 2. Extruder and the measurement technology

2.2 Artificial Neural Network-Based Data Mining

Data Collection and Preprocessing

In order to provide a comprehensive understanding of the collected data, an overview of the control and measured parameters is presented in Table 1. The data collection process involved the measurement of various process parameters, such as screw speed, temperature, pressure and geometry measurements to gather information on process and product characteristics. The analysis incorporates a dataset consisting of 14,923 measurement sets that exclusively capture the extrusion process. All measurement sets were captured at one-second intervals.

Table 1. Overview of Control and Measured Parameters in the Data Collection Process

Control Parameters (Input)	Unit	Measured Parameters (Output)	Unit
Screw speed n	rpm	Radial temperature profile in the screw channel T_1-T_8	°C
Tempering screw T_S	°C		
Tempering feed roll T_F	°C	Near-surface mass temperature T_{S1}, T_{S2}, T_{S3}	°C
Tempering intake T_I	°C		
Tempering cylinder T_C	°C	Die pressure p_S	bar
Tempering die T_T	°C	Geometry measurements X, Y	mm

Control Parameters (Input): During the test sequence, the screw speed n was varied between 10 and 30 rpm, while the tempering screw T_S, tempering feed roll T_F and tempering intake T_I were adjusted simultaneously to the same value. As such, only tempering screw T_S was included in the modeling to avoid overfitting. Similarly, the tempering die T_T and the tempering cylinder T_C were held constant throughout the test and thus were not included in the modeling to prevent unnecessary complexity. The control parameter progression after changeover was not measured during the experiments due to a lack of appropriate measuring equipment. As a result, the model assumes that each temperature change occurs instantaneously, which does not reflect the physical reality of the systems high inertia. Based on this, the largest deviation from the actual process behavior is present immediately after a temperature change and subsequently decreases over time.

Measured Parameters (Output): The temperature distribution across the measuring sword is symmetrical, with the lowest temperatures observed at the edges T_1, T_7 and at T_8. In contrast, the center temperature T_4 exhibits an average temperature of 88.87 °C, which is comparatively lower than the average temperatures observed at the remaining measurement points T_2, T_3, T_5, and T_6. This temperature difference can be attributed to the heat conduction process occurring within the rubber extrusion system. The remaining temperature points display higher values of up to 106.56 °C. While temperature shows a lower mean standard deviation of 6.04, the die pressure p_S exhibits a much higher one of 34.62, with a mean value of 94.62 bar.

Preprocessing: Before training the model, the collected data was preprocessed to ensure that it is in a suitable format. First, any missing values and outliers were removed. To analyze the feature correlations and simplify the selection of the machine learning model architecture, statistical methods such as the Pearson correlation analysis were applied [21]. Additionally, the data was scaled to a range between 0 and 1 using a normalization function to improve the convergence rate of the models during training [19]. Finally, the data was split into training and testing sets, with a ratio of 80:20, respectively. The training set was used to fit the models, while the testing set was used to evaluate their performance.

Artificial Neural Network Model

Model Selection: In order to select a model type for the regression problem, the Pearson correlation coefficient was calculated to analyze the correlations between control and measured parameters. The analysis revealed a high linear correlation of the measured variables due to their symmetrical arrangement. However, there was a low linear correlation between the measured parameters and control parameters. Given the low linear correlation, it is likely that a nonlinear correlation exists between them. In addition, it was examined that nonlinear models outperform linear models when predicting temperature using time series characteristics [9]. Artificial neural networks have been extensively studied and applied for the prediction of nonlinear time series [10–12]. Due to the low linear correlation between the measured parameters and control parameters, we selected Artificial neural networks as basis for the modeling of the regression problem.

Artificial Neural Network Model Architecture and Training: To develop the artificial neural network model, an iterative development approach was used to optimize the hyperparameters. The performance of the model prediction and the hyperparameter optimizing process were evaluated using mean squared error (MSE) and convergence rate as metrics. In addition to manual hyperparameter tuning, we employed an optimization technique called hyperopt to further enhance the model performance. This sequential model-based optimization method, also known as Bayesian optimization, offers efficient algorithms and parallelization infrastructure for conducting model selection in Python [22]. By utilizing hyperopt, we were able to systematically search and identify various combination of hyperparameters, allowing us to improve the overall accuracy of the neural network model. Figure 3 presents the iterative development process of the model for the example of the measured temperature profile variable T_3, which leverages 80% of the actual measured process data (depicted in black) as training data (represented by orange predictions) to predict the remaining 20% of the measured process data (represented by blue predictions) used for validation. This approach enables model evaluation, ensuring reliable predictions for the unobserved portion of the data. Through the combination of the iterative approach and hyperopt, a significant improvement regarding the prediction accuracy was achieved (represented by prediction delta in red), reducing the mean squared error (MSE) of the unoptimized model (left graph) from 24.43 to 0.49 (right graph). Section 3 provides a comprehensive discussion of all obtained prediction results for every measured parameter.

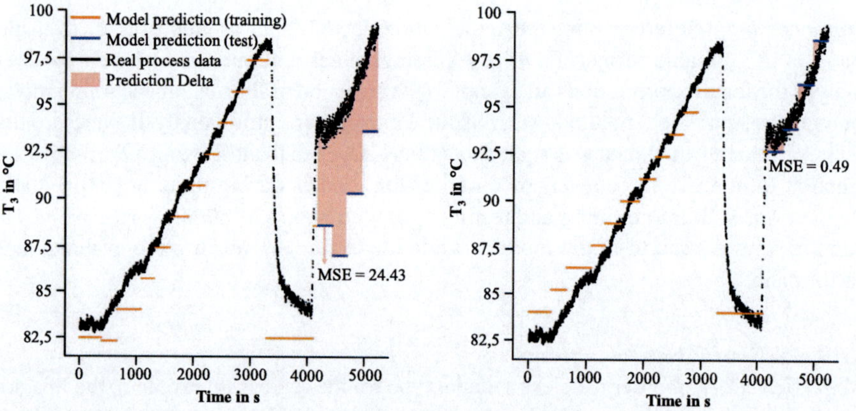

Fig. 3. Application of the ANN to predict T_3 – (left) raw model, (right) optimized model

During the iterative development of our model, various architectures were explored. It was found that a feedforward neural network with an increased number of multiple hidden layers was particularly effective. As a result, we implemented a feedforward neural network architecture in Python using TensorFlow 2.0, utilizing the tf.keras.layers.Dense class to define fully connected layers. The network architecture comprises four layers, including three hidden layers with 128 neurons each and one output layer with 12 neurons. The selection of the number of layers and neurons was based on a combination of trial and error as well as the utilization of hyperopt during the iterative development process. The hidden layers used rectified linear unit (ReLU) activation functions, with various types being applied, while the output layer employs a linear activation function. The model is trained using the mean squared error (MSE) loss function and was optimized with the Adam optimizer. The training process involved 200 epochs with a batch size of 64. We also achieved further improvements compared to the unoptimized model by increasing the learning rate to 0.003691 during training. Overall, the hyperparameter choices were made based on a combination of prior knowledge and empirical experimentation with the utilization of hyperopt to optimize the model performance on the specific dataset.

3 Results and Discussion

In this section, the experimental results for the prediction of rubber extrusion temperatures and pressures are analyzed and compared.

3.1 Experimental Results

Model Accuracy: The developed model demonstrated a high level of accuracy in predicting the measured temperature, with a mean squared error (MSE) of less than 0.93 for all measured variables. With this accuracy, the model achieves a maximum mean average deviation (MAE) of less than 0.71 for all temperature variables. This indicates a high level of precision for rubber extrusion applications, with a minimal temperature deviation of less than 1 °C for the prediction of real system behavior across most operating temperatures. The slightly higher deviation for T_8, as indicated by a MAE of 0.7, can be attributed to its position within the metal matrix, which is located furthest away from the rubber. Consequently, it experiences different temperature dynamics compared to the other measuring points in the measuring sword. In comparison to the temperatures, the prediction accuracy of the pressure p_S is due to its high variance lower, achieving a MSE of 37.40 and a MAE of 5.89. These results highlight the model ability to effectively capture the underlying patterns in the data and accurately predict the measured parameters of the extruded rubber material. The results in Table 2 show the accuracy of the model for each prediction variable. The predicted values in Table 2 represent the last 20% of the measurement series in which the temperature increment, representing the rate of adaptation achieved through manual adjustment of input parameters, was reduced to 2 °C (see Sect. 3.1 Limitations). When predicting the entire series of 2 °C temperature measurements, considering the remaining test series as training data (approximately 65% of the entire data), the average MSE for the temperatures T_1 to T_8 rises to 22.03, accompanied by an average MAE of 3.21. This underscores the impact of the reduced temperature increment on the predictive accuracy of the model and emphasizes the need to include further test series for the training of the model.

In addition to the reduced MSE, optimizing the hyperparameters also resulted in a significant improvement in the convergence rate of the model. The model demonstrated a 70% faster convergence rate compared to the unoptimized version, indicating that the hyperparameter adjustment led to a more efficient training process. To further evaluated the accuracy of the model, the coefficient of determination (R^2) was calculated for each prediction variable. The calculated R^2 values, which measure the proportion of variance explained by the model [23, 24], demonstrate a considerable degree of variability during both the training and prediction phases. This variability, characterized by horizontal prediction patterns within each increment, can be primarily attributed to the unmeasured control variables change trajectory for the respective increment, as illustrated in Fig. 3 and

Table 2. Overview of the MSE and MAE of the measured variable prediction

Measuring sword	MSE	MAE	Measured Parameters	MSE	MAE
Radial temperature profile T_1	0.50	0.59	Near-surface mass temperature T_{S1}	0.77	0.69
Radial temperature profile T_2	0.24	0.38	Near-surface mass temperature T_{S2}	0.93	0.71
Radial temperature profile T_3	0.49	0.56	Near-surface mass temperature T_{S3}	0.49	0.58
Radial temperature profile T_4	0.84	0.65	Die pressure p_S	37.40	5.89
Radial temperature profile T_5	0.82	0.66			
Radial temperature profile T_6	0.24	0.38			
Radial temperature profile T_7	0.59	0.65			
Radial temperature profile T_8	0.69	0.70			

Fig. 4 (represented by the orange and blue prediction patterns). Therefore, considering the specific characteristics of our model, the R2 values are not utilized for evaluating the model performance at this development stage.

Limitations: One major limitation of the model respectively input data is the lack of direct measurement of the control variables changeover, which results in the assumption that input variables change instantaneously. However, the real system behavior exhibits inertia and a temperature change can take in dependence of the parameter difference several minutes to occur. This discrepancy between the model and the real system behavior led to inaccuracies in the predictions and highlights the need for further research on incorporating control variable measurements or alternative simulation methods into the model. To address the effect of the control variable not being measured, we made adjustments to the variable changeover during the data collection process by reducing temperature increments from 10 °C to 2 °C. This change resulted in a higher model prediction accuracy and further reduction of MSE. In addition, we explored the possibility of using only steady-state data for training the model. For this we assumed that steady state is achieved in the last 20% of the respective temperature increment. Figure 4 presents a comparison of both approaches for the example of T_{S1}. The right graph represents the approach that utilizes only steady-state training data, while the left graph represents the approach where 80% of the available data (as described in Sect. 2.2) is used for training. For this specific example, using the approach that solely relies on steady-state training data leads to a slightly decrease in prediction accuracy, resulting in an increase of the MSE from 0.77 to 1.49.

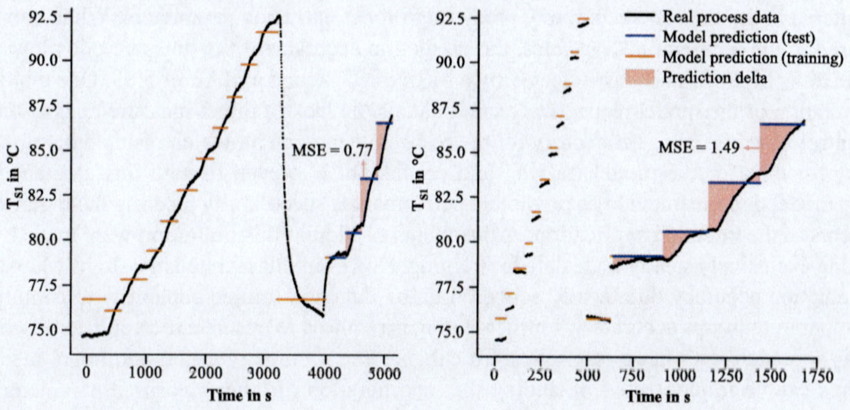

Fig. 4. Comparison of training approaches: using 80% of the whole dataset (left graph) and using exclusively steady-states as training data (right graph) to predict T_{S1}

The results indicate that using only steady-state data in training of the model resulted in higher MSE values for most variables compared to using 80% of the whole dataset. This outcome can be attributed to the reduced amount of the available training data (80% less) and the limited capability of the steady-state assumption to capture the true system behavior regarding temperature jumps.

4 Conclusion

This study presents the development of an artificial neural network-based data mining algorithm for rubber extrusion lines. The developed algorithm effectively identifies correlations between process parameters and product characteristics, resulting in accurate temperature predictions by capturing non-linear correlations.

The selection of the artificial neural network model was based on the low linear correlation between the measured parameters and control parameters, as indicated by the calculation of the Pearson correlation coefficient. Artificial neural networks have been extensively studied for the prediction of non-linear time series, making them suitable for modeling the regression problem in this study.

During the iterative development of the model, various architectures were explored and it was found that a feedforward neural network with multiple hidden layers was particularly effective. The model architecture consisted of three hidden layers with 128 neurons each and one output layer with 12 neurons. Hyperparameter optimization techniques, including manual iterative tuning and the use of Bayesian optimization through the hyperopt library, were employed to enhance the model performance.

The experimental results showed that the developed model achieved a high level of accuracy in predicting the measured temperature variables, with a MSE of less than 0.93 and a MAE of less than 0.71. This model accuracy results in a minimal temperature deviation of less than 1 °C for predicting the real system behavior across a wide range of operating temperatures, highlighting the model effectiveness in capturing underlying patterns in the data and accurately predicting rubber extrusion parameters. When compared to the temperature variables, the prediction accuracy of the pressure p_S is lower due to its high variance, as reflected by a MSE of 37.40 and a MAE of 5.89. One major limitation of the model respectively input data is the lack of direct measurement of the control variable change trajectory for the respective increment, which resulted in minor inaccuracies for the model training and prediction. However, despite this challenge, the model demonstrated high predictive performance, successfully meeting the requirements of the intended application. Adjustments to address this limitation were made by using exclusively steady-state data for training. However, this resulted in a slightly lower prediction accuracy due to the reduced training data and limited capability to capture temperature jumps accurately. Further testing is required to evaluate the applicability of this approach with more test series. To enhance the accuracy further, additional measures can be implemented, including the incorporation of other measured parameters such as geometric data of the extrudate and the utilization of simulation techniques for the control variable change trajectory.

Overall, the results of this study highlight the potential of artificial neural network-based data mining algorithms in accurately predicting rubber extrusion parameters, enabling a basis for an automated process control. By providing insights into the intricate relationships between process parameters and product characteristics, the developed algorithm can be used to reduce testing cycles of new rubber extruded products. Future research should focus on the incorporation of additional test series to further enhance the model prediction accuracy and address the current limitations by integrating control variable measurements or alternative simulation methods. The results of this study provide a foundation for developing a process control system that aligns with specific tolerance limits. By implementing a control system that builds upon the insights gained from this study, significant enhancements in operational efficiency and sustainability of rubber extrusion lines can be achieved.

Acknowledgments. The authors gratefully acknowledge the financial support of the German Federal Ministry for Education and Research (BMBF) within the project Digit Rubber (13XP5126A).

References

1. Kopal, I., et al.: A generalized regression neural network model for predicting the curing characteristics of carbon black-filled rubber blends. Polymers **14**(4), 653 (2022)
2. Du, Z., et al.: Effects of mixing temperature on the extrusion rheological behaviors of rubber-based compounds. RSC Adv. **11**(56), 35703–35710 (2021)
3. Warren-Thomas, E., Dolman, P.M., Edwards, D.P.: Increasing demand for natural rubber necessitates a robust sustainability initiative to mitigate impacts on tropical biodiversity. Conserv. Lett. **8**(4), 230–241 (2015)
4. Ghaffarian, N., Hamedi, M.: Optimization of rubber compound design process using artificial neural network and genetic algorithm. Int. J. Eng. **33**, 2319–2326 (2020)
5. Park, K., Park, H., Bae, H.: A data-driven recipe simulation for synthetic rubber production. IEEE Access **10**, 129408–129418 (2022)

6. Lukas, M., Leineweber, S.: Künstliche intelligenz und data-mining in der kautschukextrusion. In: phi – Produktionstechnik Hannover informiert, Newsletter Nr. 39 (2023)
7. Ahangar-Asr, A., Faramarzi, A., Javadi, A.A., Giustolisi, O.: Modelling mechanical behaviour of rubber concrete using evolutionary polynomial regression. Eng. Comput. **28**(4), 492–507 (2011)
8. Shit, S.C., Shah, P.: A review on silicone rubber. Natl. Acad. Sci. Lett. **36**(4), 355–365 (2013)
9. Mateo, F., Carrasco, J.J., Sellami, A., Millán-Giraldo, M., Domínguez, M., Soria-Olivas, E.: Machine learning methods to forecast temperature in buildings. Expert Syst. Appl. **40**(4), 1061–1068 (2013)
10. Egrioglu, E., Yolcu, U., Aladag, C.H., Bas, E.: Recurrent multiplicative neuron model artificial neural network for non-linear time series forecasting. Neural. Process. Lett. **41**(2), 249–258 (2015)
11. Khandelwal, I., Adhikari, R., Verma, G.: Time series forecasting using hybrid ARIMA and ANN models based on DWT decomposition. Procedia Comput. Sci. **48**, 173–179 (2015)
12. Deb, C., Zhang, F., Yang, J., Lee, S., Wei, S.: A review on time series forecastingtechniques for building energy consumption. Renew. Sustain. Energy Rev. **74**, 902–924 (2017)
13. Alvino, A., Borsini, S.: Testing of rubber nanocomposites for aerospace, automotive and oil and gas applications. In: High-Performance Elastomeric Materials Reinforced by Nano-Carbons, pp. 177–191. Elsevier (2020)
14. Silva, A., Silva, F.J., Campilho, R.D., Neves, P.M.: A new approach to temperature control in the extrusion process of composite tire products. J. Manuf. Process. **65**, 80–96 (2021)
15. Weichert, D., Link, P., Stoll, A., Rüping, S., Ihlenfeldt, S., Wrobel, S.: A review of machine learning for the optimization of production processes. Int. J. Adv. Manuf. Technol. **104**(5–8), 1889–1902 (2019)
16. Thiede, S., Turetskyy, A., Loellhoeffel, T., Kwade, A., Kara, S., Herrmann, C.: Machine learning approach for systematic analysis of energy efficiency potentials in manufacturing processes: a case of battery production. CIRP Ann. **69**(1), 21–24 (2020)
17. Moldovan, D., Cioara, T., Anghel, I., Salomie, I.: Machine learning for sensor-based manufacturing processes. In: 13th IEEE International Conference on Intelligent Computer Communication and Processing (ICCP), pp. 147–154 (2017)
18. Kojima, T., Washio, T., Hara, S., Koishi, M.: Synthesis of computer simulation and machine learning for achieving the best material properties of filled rubber. Sci. Rep. **10**(1), 18127 (2020)
19. Jo, J.-M.: Effectiveness of normalization pre-processing of big data to the machine learning performance. J. Korea Inst. Electron. Commun. Sci. **14**, 547–552 (2019)
20. Otten, F., Hein, J., Bondy, H., Faust, H.: Deconstructing sustainable rubber production: contesting narratives in rural Sumatra. J. Land Use Sci. **15**(2–3), 306–326 (2020)
21. Li, G., Zhang, A., Zhang, Q., Wu, D., Zhan, C.: Pearson correlation coefficient-based performance enhancement of broad learning system for stock price prediction. IEEE Trans. Circuits Syst. II **69**(5), 2413–2417 (2022)
22. Bergstra, J., Komer, B., Eliasmith, C., Yamins, D., Cox, D.D.: Hyperopt: a Python library for model selection and hyperparameter optimization. Comput. Sci. Disc. **8**(1), 14008 (2015)
23. Di Bucchianico, A.: Coefficient of determination (R^2). In: Ruggeri, F., Kenett, R. (eds.): Encyclopedia of Statistics in Quality and Reliability. Wiley, Chichester (2008)
24. Renaud, O., Victoria-Feser, M.-P.: A robust coefficient of determination for regression. J. Stat. Plann. Inference **140**(7), 1852–1862 (2010)

Material Flow of an Agile Battery Cell Production System Based on Diffusion-Tight Transport Boxes and Driverless Transport Systems

S. Henschel(✉), F. Kößler, and J. Fleischer

wbk –Institute of Production Science at Karlsruhe Institute of Technology (KIT), Kaiserstraße 12, 76131 Karlsruhe, Germany

sebastian.henschel@kit.edu

Abstract. The growing demand for lithium-ion battery cells, driven by increasing electrification in daily life, is currently met by inflexible mass production plants using rigidly linked manufacturing lines. These are set up in large dry rooms to protect moisture-sensitive battery materials. Since up to 30% of the total energy demand is caused by the operation of these dry rooms, the agile and customized battery cell production is based on compact local dry rooms, also called microenvironments. This publication presents the development and validation of a diffusion-tight transport box that enables the transportation of moisture-sensitive intermediate products between these microenvironments. The flexible and configurable material flow can incorporate alternative process routes. Furthermore, a modular expansion of the production system is enabled. By combining reusable diffusion-tight, automatically handled transport boxes and driverless transport systems, this approach enables an agile material flow in battery cell production.

Keywords: Agile Production System · Battery Cell Production · Material Flow · Intralogistics · Microenvironment · Dry Room · Transport Box

1 Motivation

The increasing electrification of everyday life is largely based on lithium-ion battery cells, which were originally developed for consumer electronics and are now also being used as energy storage devices in power tools and especially in electric vehicles. Compared to other battery cells available on the market, lithium-ion battery cells are characterized by their high volumetric and gravimetric energy density. Since the market launch in 1991 by Sony, both the product and the production have been continuously developed and optimized [1]. With regard to the cell chemistry used, a trend towards more moisture-sensitive materials that allow increases in energy density and use less critical raw materials such as cobalt has become apparent in recent years, particularly as a result of the increasing applications within electromobility. To protect these materials, the moisture-critical production process steps take place in a fully air-conditioned

T. Bauernhansl et al. (Eds.): WGP 2023, LNPE, pp. 550–558, 2024.
https://doi.org/10.1007/978-3-031-47394-4_53

dry-room atmosphere at a temperature of approx. 20 °C. The very low moisture content of this atmosphere is usually specified as the dew point temperature and represents the temperature at which the air humidity would condense. In the field of battery cell production, dew point temperatures of −20° to −50 °C are used. This corresponds to a relative humidity of well below 1% and an absolute moisture concentration of less than 100 parts per million by volume [2]. Since drying the air is very energy-intensive, a trend towards small local dry rooms, so-called microenvironments can be observed, particularly in research. These microenvironments reduce the volume of air which has to be dried and thus help reduce the economic impact of dry rooms. In order to flexibly interconnect these microenvironments, a material flow based on diffusion-tight transport boxes is required, which is explained in the course of this work.

2 State of the Art

2.1 Battery Cell Production

Battery cell production is divided into electrode production, cell assembly and cell finishing. In electrode production, the solvent-mixed battery slurry consisting of active material, conductive carbon black and binder is coated onto metal substrate foils, dried, compacted to a target porosity and post-dried. In cell assembly, the electrodes are separated and stacked or wound into battery cells and then partially packaged. In cell finishing, the battery cells are filled with electrolyte in a vacuum and fully sealed, then charged and discharged (formed) the first few times outside a dry room and subjected to end-of-line testing [3]. Until the battery cell is fully sealed, production must take place in dry rooms. This is due to the irreversible loss of cell capacity which occurs when especially the cathode active material comes into contact with ambient air. The severity of the degradation and therefore the care that needs to be taken when choosing the correct dry room atmosphere varies from the cell chemistry which is being used. While lithium-iron-phosphate (LFP) does not react as fast, especially nickel-manganese-cobalt (NMC) chemistries are very sensitive. This is especially true for active materials with a high cobalt content, such as NMC 622 or NMC 811 (stoichiometric ratio of 8:1:1) [4].

On an industrial scale, battery cells for electric cars in particular are produced in so-called gigafactories, which are characterized by their annual production output in the GWh range. Overall, due to the increased demand for lithium-ion battery cells, production capacities are being expanded worldwide and new factories are being planned. The market is currently dominated by large corporations that want to expand their market share and focus on the cost-effective mass production of common cell types. Individual factories are therefore often designed for the production of one cell type and optimized in terms of unit costs. Flexibility with regard to the producibility of further product variants plays a subordinate role in gigafactories due to the high scrap rates, currently faced during production start-up. At the plant level, this often means rigidly linked, highly productive special machines adapted to a product, which are set up in large dry rooms. Battery cells for niche applications, often require customized cells to maximize the benefit of the end user, for example in terms of achieved energy density within a given special restriction. These cells are not currently covered by the product portfolio of the large cell manufacturers. They are produced in small production plants, which usually have a labor intensive and thus expensive manual laboratory cell production for very small quantities. Usually, several production lines for different process steps are operated together at one dew point in one dry room [5, 6].

2.2 Dry Rooms

The operation of dry rooms is very energy-intensive and can account for up to 50% of the total energy demand of cell production [7]. During the operation, the dehumidification of the air generally takes place in two steps. First, the air is dehumidified by means of a condensation dryer, in which the air is cooled, so that it can bind less moisture. Part of the entrained humidity condenses and is removed. In order to remove as much moisture as possible, this requires process cooling ideally in the single-digit °C temperature range (inlet and outlet temperature). Afterwards, the pre-dehumidified air is dehumidified to the desired dew point by means of an absorption dryer. In this process, the moist air flows onto a drying wheel coated with a drying agent such as silica gel. This drying agent absorbs the moisture so that the air is dried. To prevent saturation of the drying wheel, it must be regenerated with warm air after moisture absorption. This regeneration air extracts the moisture from the desiccant and is normally drawn in from the environment, heated by means of heat input and finally discharged into the environment. During operation of the dry room, the air must be circulated and dehumidified constantly so that there is no local accumulation of moisture in the dry room and moisture inputs from employees or leaks are compensated. Due to the high energy demand for the drying operations, much of the air is circulated for cost reasons, as it is more energy efficient to dehumidify pre-dried air. Due to unavoidable leaks in a dry room and the oxygen supply necessary for employees, a continuous influx of freshly dehumidified air is nevertheless required to maintain the required air exchange rate [8]. A reduction of the drying room volume offers great potential in terms of energy savings, which is why the concept of agile battery cell production based on local and compact drying rooms, so-called microenvironments, is presented below.

2.3 Agile Manufacturing Concepts

In contrast to the plant concept of a gigafactory, which is based on rigidly linked special machines, an agile battery cell production is currently being set up at the wbk – Institute of Production Science. It is based upon a modular and freely scalable structure [9, 10]. The aim is to be able to produce pouch cells according to customer requirements quickly and highly automated. Flexibility can be achieved in terms of the used materials, the number of manufactured units and the geometric cell format. Thus, the current gap between manual laboratory production and mass production in gigafactories on the production side can be closed. The core element of the agile battery cell production are uniform robotic cells, designed as microenvironments whose dew points can be individually adjusted by decentralized air-conditioning technology placed on top of these microenvironment. Within the environments, flexible production modules with standardized interfaces are placed for carrying out the individual process steps, such as calandering for example [11]. The physical system at the wbk consists of four individual microenvironments which incorporate the production steps from the electrode coating to the cell assembly, as can be seen in Fig. 1. Scaling effects are mapped in a digital twin of the production system. A key feature of this agile production system is the high degree of automation, so that employees only need to enter the microenvironments for maintenance purposes. This enables the reproducible cell production and protects employees from working under dry room conditions, which are not without risk due to possible drying out of the mucous membranes. A key component and fundamental requirement for the agile battery cell production is the concept for an automated flow of material within and between the microenvironments.

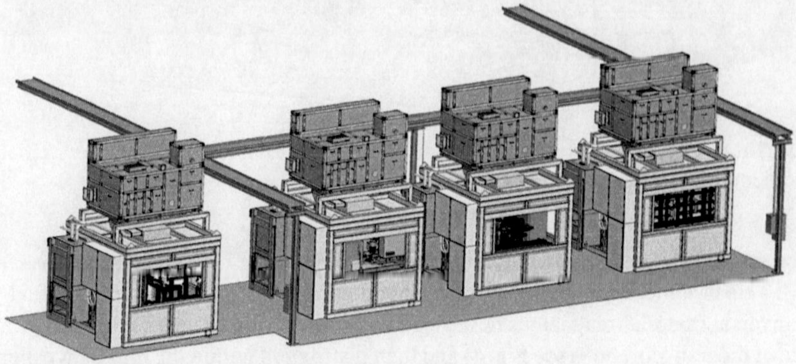

Fig. 1. The four microenvironments being set up at the wbk – Institute of Production Science [own illustration]

3 Material Flow in the Agile Battery Cell Production

Within the microenvironments, the material flow is automated by a centrally positioned 6-axis KUKA KR 22 R1610-2 Robot, as seen in Fig. 2. By equipping the robot with a quick-change coupling system SWK-011 from Schunk, the robot can automatically execute multiple tasks. These include the opening and closing of the transport boxes, the movement of uniform material magazines within each microenvironment, as well as material handling operations within different production processes, such as the calandering for example.

Each microenvironment is supplied with input materials and intermediate products via diffusion-tight transport boxes. A Herbie XL driverless transport system from Carrybots is used to transport them between the microenvironments. For material transfer, the driverless transport system drives from the outside to the material lock of the microenvironment and transfers the transport box to the material lock opened from the outside. For this purpose, driven roller conveyors are mounted on the automated guided vehicle and in the airlock. After the transfer of the transport box, the airlock is closed and an air exchange with dry air takes place. The transport box is then transported inside the microenvironment. It is positioned in a reproducible manner, so that the transport box

Fig. 2. The calandering microenvironment with the central handling robot, and the material lock on the side [own illustration]

can be opened by the central material handling robot in the next step, as can be seen in Fig. 3. This is achieved by usage of the quick couplings described earlier. As a next step, the universal material magazines developed within the project can be retrieved from the inside of the transport box (see Fig. 4) and then distributed within the microenvironment. Once all necessary production processes are finished, the material is repacked into the transport box, the lid is positioned and closed and the material is discharged through the material lock onto the driverless transport system again.

Fig. 3. The central material handling robot opening the transport box by utilizing the quick exchange coupling [image of the author]

Fig. 4. Retrieval of a electrode magazine from within the transport box [image of the author]

The transport box is a proprietary design, since no diffusion-tight reusable transport boxes for an automated logistics process have been available on the market to date. Current industrial battery cell production takes place in large interconnected dry rooms, so the only need for external material transport arises in case of dry room shutdowns, or when shipping moisture sensitive products to an external plant for further processing. Since dry rooms are normally operated around the clock, such cases tend to be an exception and therefor allow for manual packaging operations. For this purpose, the material, such as electrode coils, are packed in aluminum composite hoods, rinsed with dry air or inert gas, provided with desiccant bags if necessary, and then sealed with heat-sealing tongs. To protect against mechanical damage, the aluminum composite hood can be transported and stored in a cardboard box or crate. The difficulty to automate said process can be primarily found in the bendable composite film. Another disadvantage of this procedure is the limited number of times, the aluminum composite hood can be reused. This is because the sealing seam must be cut off, in order to open the hood at its destination.

For these reasons, aluminum composite hoods in combination with a transport box are not suitable for highly automated agile battery cell production, where moisture-sensitive intermediate products are transported on flexible routes under normal atmospheric conditions between the microenvironments. The transport box developed at wbk - Institute of Production Science is characterized by its airtightness and the automatable opening mechanism. It is constructed of bend aluminum sheet metal with fully welded seams. The lid is equipped with a Schunk quick coupling SWA-011 with integrated pneumatic passages. These can be used to activate four single acting cylinders within the box, used to compress a 3mm EPDM seal in between the lid and the box itself. By means of iteration, several techniques for applying the compressive force have been tested. The first generation of the box (see Figs. 3 and 4) allowed for a lot of play between the lid and the box itself. It was even possible to adjust the placement of the lid by hand, since it was primarily held in place by its own force of weight. The next generation of the transport box will feature four wedge type closures within the corners of the box. They will interlock with their counterpart placed within the box, as can be seen in Fig. 5 and thus allow for an optimal sealing effect.

Fig. 5. Explosive view of the finalized diffusion tight transport box with the wedge type closures and their counterparts in the corner of the box [own illustration]

Air tightness tests were carried out to validate the functionality of the first generation of the box. For this purpose, the opened transport box was stored in a dry room at a dew point temperature of −65 °C. A data logger (Rotronic BL-1D) was positioned in the transport box to record humidity, air pressure and temperature. After 2.5 h, the transport box along with the data logger inside was closed. Another data logger recording humidity and temperature (Rotronic HL-1D) was placed on top of the box. The closed transport box including the data logger on the outside was moved from the dry room to a normal atmospheric environment. After 10 days, the transport box was opened and the data from the two data loggers was compared. This time period was chosen, to allow for a sufficient intermediate storage capability within the transport box in case of shifting production volumes. A secondary test was carried out with an available industrial reference box. In contrast to the proprietary transport box, it is composed of triple folded and not fully welded seams. However, it does create a tight seal in between the lid and the box by utilizing manually operated compression buckles.

The data logger outside the box jumped to a relative humidity of 30% as soon as it left the dry room. The data logger within the transport box required three days to arrive at this level of humidity. The reference box, while being subjected to a longer exposure to the dry room atmosphere took less than 24 h to reach a moisture level comparable to standard environments. It was therefore shown, that the current sealing concept can protect the moisture-sensitive materials from moisture for a short period of time and in a better fashion than store bought alternatives. However, an improvement of the sealing concept is recommended, especially with regards to the storage of the

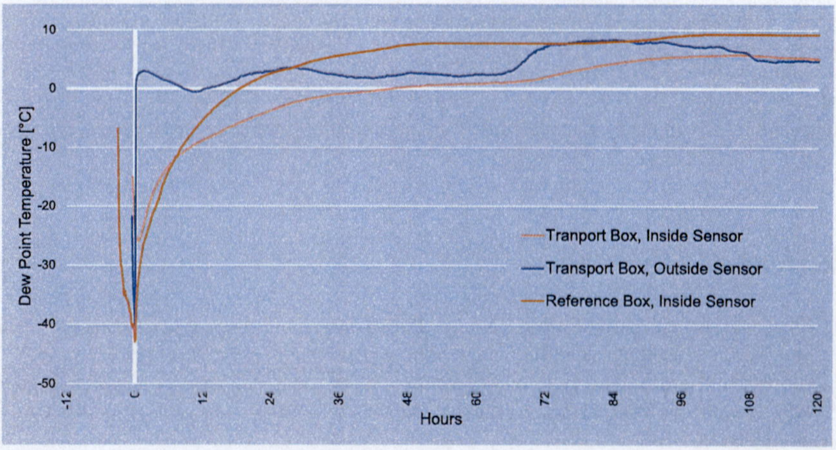

Fig. 6. Dew point temperature measurements over time for the transport box and a reference box

electrodes. Furthermore, calculations of the production relevant dew point show that the minimum dew point temperature inside the box was −26 °C, as can be seen in Fig. 6. The dew point temperature of the dry room was −65 °C. It is therefore advised, to ventilate the transport box by means of a fan as to minimize the buildup of moisture pockets and their entrapment within the closed transport box, when conducting further validation tests. In addition, the relatively inexpensive data loggers should be calibrated with a dew point mirror, which has a negligible measured value drift due to its design.

As mentioned earlier, the second generation of the transport box has been optimized by implementing wedge type closures within the four corners of the lid. This is a promising improvement, since in addition to a counterpart mounted within the transport box itself, the point of attack for the sealing force is now redirected to the corners for maximum contact area. Tightness tests to validate this concept are still pending.

4 Conclusion and Outlook

To enable agile battery cell production based on microenvironments, a logistics system for transporting intermediate products within and in between different manufacturing setups has to be developed. This paper shows an approach based upon reusable, diffusion tight transport boxes, which can be introduced into the microenvironments automatically. They are capable of being opened by a centralized material handling robot and protect intermediary products from external moisture as needed. The first generation of a self-developed transport box still shows potential for optimization with regards to tightness during a validation test. For further optimization of this transport box, it is imperative to involve colleagues from the field of cell chemistry in order to set clear requirements for the tightness of these boxes. The essential question in this context will be how long and with how much moisture the moisture-sensitive materials may come into contact in order to avoid any loss of quality in the finished battery cell. With regards to the design of the transport box, the current inlay concept may be expanded in the future in order to also enable the transportation of entire electrode coils instead of cut sheets.

Acknowledgements. The presented research is funded by the *German Federal Ministry of Education and Research* and the *Ministry of Science, Research and the Arts Baden-Württemberg* under the funding code 03XP0369A (Projects: *AgiloBat1* and *AgiloBat2*), the *Center for Electrochemical Energy Storage Ulm | Karlsruhe (CELEST)* and the *Battery Technology Center of KIT*. The authors would like to thank *Maurice Müller* and *Runqui Yao* for their contributions to the subject matter.

References

1. Korthauer, R. (ed.): Lithium-Ion Batteries Basics and Applications. Springer eBook Collection. Springer, Heidelberg (2018). https://doi.org/10.1007/978-3-662-53071-9
2. Ahmed, S., Nelson, P.A., Dees, D.W.: Study of a dry room in a battery manufacturing plant using a process model. J. Power Sources **326**, 490–497 (2016). https://doi.org/10.1016/j.jpowsour.2016.06.107
3. Bockholt, H., Indrikova, M., Netz, A., Golks, F., Kwade, A.: The interaction of consecutive process steps in the manufacturing of lithium-ion battery electrodes with regard to structural and electrochemical properties. J. Power Sources **325**, 140–151 (2016). https://doi.org/10.1016/j.jpowsour.2016.05.127
4. Armand, M., et al.: Lithium-ion batteries – Current state of the art and anticipated developments. J. Power Sources **479**, 228708 (2020). https://doi.org/10.1016/j.jpowsour.2020.228708
5. Orangi, S., Strømman, A.: A techno-economic model for benchmarking the production cost of lithium-ion battery cells. Batteries **8**(8), 83 (2022). https://doi.org/10.3390/batteries8080083
6. Ciez, R.E., Whitacre, J.F.: Comparison between cylindrical and prismatic lithium-ion cell costs using a process based cost model. J. Power Sources **340**, 273–281 (2017). https://doi.org/10.1016/j.jpowsour.2016.11.054
7. Erakca, M., et al.: Energy flow analysis of laboratory scale lithium-ion battery cell production. iScience **24**(5), 102437 (2021). https://doi.org/10.1016/j.isci.2021.102437
8. Vogt, M., Koch, K., Turetskyy, A., Cerdas, F., Thiede, S., Herrmann, C.: Model-based energy analysis of a dry room HVAC system in battery cell production. Procedia CIRP **98**, 157–162 (2021). https://doi.org/10.1016/j.procir.2021.01.023
9. Fleischer, J., Fraider, F., Kößler, F., Mayer, D., Wirth, F.: Agile production systems for electric mobility. Procedia CIRP **107**, 1251–1256 (2022). https://doi.org/10.1016/j.procir.2022.05.140
10. Ruhland, J., et al.: Development of a parallel product-production co-design for an agile battery cell production system. In: Anderson, A.L., et al. (eds.) CARV MCPC 2021. Lecture Notes in Mechanical Engineering, pp. 96–104. Springer, Cham (2022). https://doi.org/10.1007/978-3-030-90700-6_10
11. Kößler, F., Mayer, D., Fleischer, J.: Plant concept for highly automated and agile calendering of battery electrodes; [agile produktionssysteme in der batteriezellfertigung]. WT Werkstattstech. **112**(7–8), 492–495 (2022)

Power Forecast of Photovoltaic Systems-An Approach for Improving Energy Management of DC-Supplied Production Plants

B. Gutwald[1](✉), M. Barth[1], O. Mönius[1], B. Zeilmann[2], and J. Franke[1]

[1] Institute for Factory Automation and Production Systems (FAPS), Friedrich-Alexander-Universität Erlangen-Nürnberg, Egerlandstr. 7-9, 91058 Erlangen, Germany
Benjamin.Gutwald@faps.fau.de

[2] Richter R&W Steuerungstechnik GmbH, Körzendorf 52, 95491 Ahorntal, Germany

Abstract. Regeneratively fed Direct Current (DC) grids play a key role in the energy transition. The promising supply architecture for industrial plants is not only materially efficient in its topology, but it also distributes energy without large conversion losses, especially for electric drives. While the architecture already offers a particularly efficient integration for renewable energies, the natural fluctuation must be mastered in energy management. Depending on the orientation, dimensioning and geographical conditions, this paper describes an approach and implementation of power forecasting for photovoltaic (PV) systems based on numerical weather predictions. Furthermore, the associated energy management scenarios for storage and loads in DC-supplied production plants are presented. The advantages of energy forecasting lie in the optimal design of the DC-Grid components, the improvement of peak shaving and the early energy optimisation of production control through connection to a Manufacturing Execution System (MES).

Keywords: photovoltaic forecast for grid serving operation · energy management · energy optimised manufacturing execution system

1 Introduction

1.1 Motivation of DC-Grids

The rivalry between Thomas Alva Edison and George Westinghouse in the electricity industry sparked a competition between DC and alternating current (AC) for transmitting energy. The invention of the transformer in 1881 allowed for more efficient energy transportation over long distances, leading to the widespread adoption of AC networks worldwide. As a result, electrical devices have been designed to operate on the AC voltages used in each country.

However, decentralised power grids for industrial applications are now incorporating DC-based technologies, such as renewable energy sources and storage systems. Many internal components of devices already use DC power. Converting and feeding energy

T. Bauernhansl et al. (Eds.): WGP 2023, LNPE, pp. 559–568, 2024.
https://doi.org/10.1007/978-3-031-47394-4_54

from renewable sources and batteries into the AC grid is inefficient and leads to losses [1]. The need for conversion and associated costs can be eliminated by utilising DC networks. In the context of frequency conversion, the existing rectification and inversion methods employed by high-capacity frequency converters can be streamlined to a singular inverter operating within a DC network. This optimisation leads to a reduction in both energy losses and material expenses. Additionally, DC networks with storage and consumers allow for the utilisation of electrical braking energy that is typically wasted as heat through braking resistors [2–4]. Implementing DC grids in industries can save around 10 to 12% of total electrical energy consumption annually [5].

Apart from energy savings, the adoption of DC grids also reduces the requirement for electrically conductive materials. The topology of DC grids, with fewer conductors compared to three-phase AC grids, can lead to savings of up to 40% in conductive substances [6]. Moreover, the elimination of multiple rectifiers and converters in equipment results in further material savings. This results in the generation of components characterised by diminished production expenses and a concomitant decrease in embodied energy [7, 8]. Furthermore, controlling DC grids is less complex compared to AC grids since reactive power provision is no longer necessary [9, 10]. With the increasing use of renewable energy sources and material reduction, the CO_2 footprint is significantly reduced [3, 11, 12]. A simplified grid structure is shown in Fig. 1.

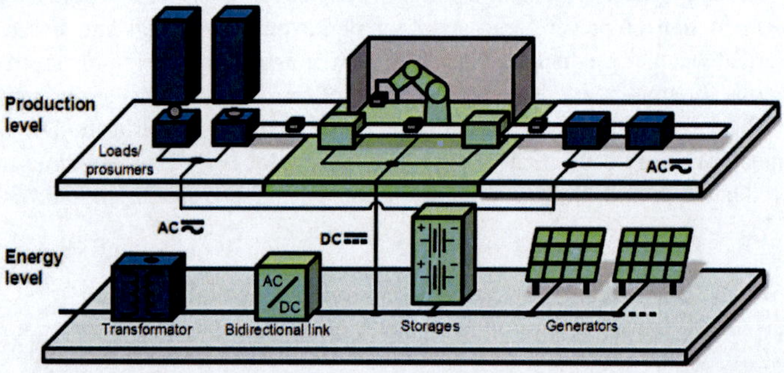

Fig. 1. Simplified grid structure for the presented grid architecture

1.2 State of Research: Energy Management in DC-Grids

Although the existing architecture already provides a notably efficient integration for renewable energies, the inherent variability in energy generation must be effectively controlled and managed in energy systems. The most popular European projects DC-Industry 1 and DC-Industry 2 as well as the resulting Open Direct Current Alliance (ODCA) provide for the use of different droop control mechanisms for the grids power exchange [13, 14]. In the DC|hyPASim project, the simulation of energy flow scenarios is predicated, among other things, on characteristic-based control [15]. In this control strategies, the DC grid voltage encodes the information for the operating mode of the Active Infeed Converter (AIC) and the storage units in specified voltage bands. The converters act as voltage controlled current sources [16]. When the grid voltage rises, the storage units are inactive until the AIC also feeds back into the AC grid when either the generation power is no longer buffered because the storage unit is fully charged or the charging power is at its limit. A decreasing grid voltage causes storage to discharge and, if necessary, also the AIC to enter the so-called source mode. The characteristic curves (droop curves) stored in the converters determine the corresponding power level of the energy flows between the storage unit and the grid [17]. The control concept sensibly provides for maximum power point (MPP) operation of the photovoltaic (PV) system [18]. Overall, the aim of the concepts is to optimise the self-consumption of the energy generated by solar power in the DC branch of the grid [19]. Since commercially available products usually do not offer droop characteristic parameterisation, programmable logic controllers (PLCs) generally are used for event-based power control of the first DC-supplied systems.

Compared to conventional self-consumption optimisation of AC-grid-connected PV systems with storage for households and commercial buildings, the control objective of self-consumption optimisation is quite similar. The notable difference lies in how surplus energy is utilized: it is initially consumed on the DC side, then on the AC side, and only afterward is it fed into the utility grid [10].

1.3 Need for Action and Methodology of Research Work

The energy management strategies presented here work automatically and regulate the optimal use of storage only on the basis of momentary power flow data. From the perspective of fluctuating electricity prices, energy-diverse production processes and with a view to the volatility of renewable energies, energy forecasting brings significant added value to energy management. Section 2 provides an overview of the current research progress in predicting PV power, along with a practical demonstration achieved using a software framework that involves simulation and validation on a specific solar power plant. Furthermore, a solution is proposed to deploy the power forecast data into an industrial energy management control system of a DC grid. Finally, profitable application scenarios of the forecast in the promising supply structure for production plants are presented.

2 PV Power Forecasts

2.1 Previous Work

Previous research has described implementations based on numerical weather prognosis (NWP) models of the Deutscher Wetterdienst (DWD) to achieve targeted forecasts for PV systems located in Germany. These are limited to the application of outdated and now decommissioned NWP systems of the DWD [20, 21]. In general, a distinction is made between offline and online forecasting methods. Offline forecasts use statistical models based on historical data from PV systems and the environment, without the need for real-time data. These forecasts achieve acceptable accuracy over a yearly period but cannot provide precise hourly predictions [22]. Online forecasts rely on NWPs and satellite-based radiation predictions, requiring continuous data availability. A recent study by the

Zentrum für Energietechnik (ZET) at the University of Bayreuth developed an online PV power forecast using the COSMO-D2 model of DWD providing intraday and day-ahead predictions up to +45 h. A significant concurrence between the prediction and the measurement data from five PV systems was observed in the long-term average. However, a substantial degree of variability is evident in the data for individual days, implying a wide scattering. [23] The COSMO-D2 ensemble system was deactivated and replaced by the ICON-D2 NWP in early 2021. An implementation based on the operational successor NWP is an evident research approach as a continuation of previous research work. Furthermore, a supplementary aspect to be considered is the incorporation of DWD's ICON-EU model, encompassing a forecast horizon of up to five days ahead. Due to the lower resolution in time and space compared to ICON-D2, greater deviation can be expected but with the added advantage of a larger combined prediction period [24].

2.2 Procedure of PV Power Forecast

The physical parameters of the modules and inverters form the basis for modeling a PV system. The required parameters follow either from the inverter and module databases of the California Energy Commission (CEC) or by user-defined acquisition and transfer of the parameters saved in a JavaScript Object Notation (JSON) file, which can be obtained from technical data sheets of the devices [25]. Together with the information on the tilt and azimuth angle of the solar cells, the simulation model is fully defined. The simulation is done by the open-source software pvlib python developed by the Sandia National Laboratories (SNL), a research facility of the U.S. Department of Energy [26]. The input parameters are the PV system object, irradiation and temperature data provided by the NWPs. The DC output power of the solar modules is calculated by the CEC Single Diode Equivalent Circuit Model [27]. The values serve as input variables to the CEC PV Watts inverter model [28]. The result of the simulation model chain, shown in Fig. 2, is the prediction of the time-variable power output of the converter.

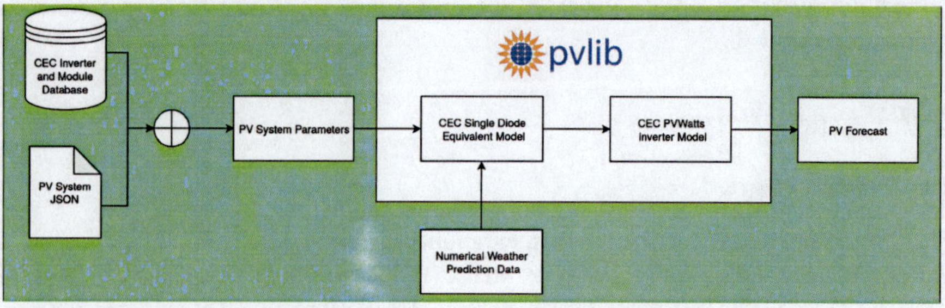

Fig. 2. Procedure of PV power forecast

The entire PV forecast framework is written in the Python programming language and is designed to be executed on a UNIX system. The Raspberry Pi 4 Model B is selected as the target device for the decentralised deployment of the software framework.

2.3 Concept Coupling PLC/REST-API/energy Management

The provision of the PV power forecast data is realised with a REST-API, which offers a common implementation of a software architecture for the stateless exchange of data between a client and server. The client can generally be of any type, provided it supports and knows the specification of the API. The advantages of a REST-API include the inherent flexibility, robustness, and simplicity it offers for exchanging data over the Internet or within an Ethernet-based local network [29]. The supply of data via the interface is conceivable for a wide range of applications. In the case of this publication the REST-API forms the direct communication to a PLC (Siemens S7-1500 series) that takes over the energy management of a real DC-powered production with a similar supply architecture to Fig. 1. The PLC controls the loads in the form of production process control. Also, it forms the setpoint specification for the charging currents of the storage in combination with the battery management system as well as regulating the feeding back into the AC grid. Furthermore, it is possible to influence droop control-based energy managements. The developed REST API is designed not only for integration with a PLC but also with other clients, such as simulation or direct end users. Specifically, the communication interface allows the forecast to be incorporated into a planning tool used for DC grids and their corresponding energy management [30].

2.4 Validation

The validation of the forecast results was performed on a 19.825 kW_P pitched roof system located in the north of Bavaria during the winter period from February 1, 2023, to February 25, 2023. The PV plant is affected by shading losses caused by the roof structure, trees, and a chimney. The short-term forecast, spanning up to 48 h, was performed utilising the higher time and space resolution ICON-D2 model. This forecast was subsequently merged with data obtained from the ICON-EU model, which covers a longer forecast horizon of up to 120 h. It should be noted that the average yield of a PV system in February at a German site is less than 5% of the total annual production. In addition, the relative share of the much more complex to predict diffuse radiation is higher than in the summer months [31]. The forecasts were calculated daily at 8:45 UTC+1 in the investigated period. Thereby the parameter data of the 6:00 UTC+0 run of ICON-D2 and the 0:00 UTC+0 run of ICON-EU were used.

Especially shading-related losses and the soiling of the solar cells are not considered due to the complexity of the calculation. Figure 3 illustrates the forecast of February 19th compared to the measurements of the converter of the validation plant. The forecast provided reliable data despite complex cloud situations in the high-resolution time domain up to 48 h of ICON-D2. A greater deviation can be observed on the third day after the beginning of the forecast in the time range of ICON-EU. A decline in the expected power generation was accurately predicted for the 5th day since the start of the forecast.

In order to evaluate the reliability and functionality of the PV forecast, the forecast data (F) and measured data (M) as well as the absolute error (E) are statistically processed in Table 1. The data are normalised to the peak power of the PV system and only consider the times between 7 a.m. and 5 p.m. in order to eliminate falsification by zero data points. The absolute errors of the forecast, which result from the absolute difference between forecast and measured data, are shown as histograms in Fig. 4.

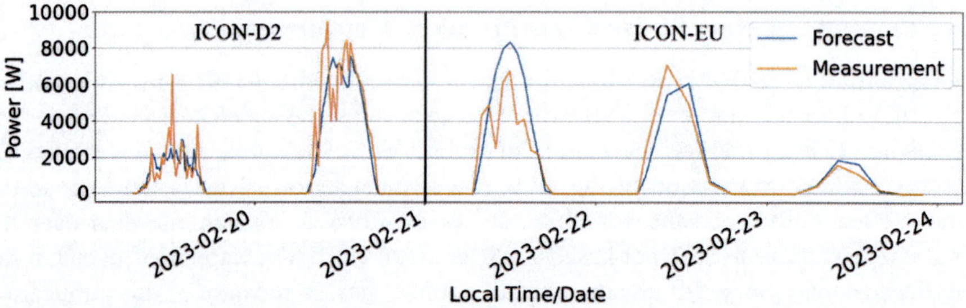

Fig. 3. Exemplary 5 days PV power forecast from 19th of February compared to the real measurement

Table 1. Combined statistical key figures of the forecast data (F), measured data (M) and the absolute error (E)

	Mean [kW/kW$_p$]			Standard deviation [kW/kW$_p$]			Max [kW/kW$_p$]		
	F	M	E	F	M	E	F	M	E
Combined	0.13077	0.11780	0.05559	0.12297	0.12543	0.06122	0.43527	0.47398	0.31587
ICON-D2	0.13216	0.12133	0.05492	0.12418	0.12823	0.06073	0.43527	0.47398	0.31394
ICON-EU	0.12430	0.10139	0.05869	0.11707	0.11016	0.06343	0.43423	0.43917	0.31587

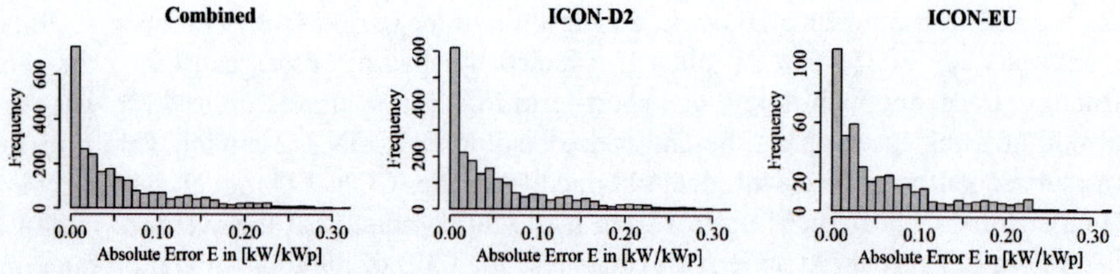

Fig. 4. Histograms of the absolute errors of the NWPs

As the statistics and the example forecast show, the forecast already provides a good prediction of energy production under the given conditions. However, it does not consider systematic errors such as shading by the environment or the effects of parameter deviations, which leads to a relatively high scatter. Further work will aim to reduce this error by extending the framework using self-learning black-box models to calculate the systematic error using additional time-based arguments.

3 Application Scenarios of PV Power Forecasts in DC Grids

There are various application scenarios that can be considered from both an electrical engineering perspective, which includes component sizing and operation, and a process perspective, which involves considering economic and environmental factors.

3.1 Application Scenarios for Economic Benefits

An important advantage lies in the implementation of **peak shaving** strategies. Through the acquisition of load profiles, which can be obtained by monitoring the energy consumption of production processes, efficient management of energy storage systems can be achieved by employing energy forecasting techniques.

An additional benefit is the incorporation of **MES to facilitate load shifting and CO_2 optimisation.** When the production planner is provided with a power forecast that indicates the anticipated output of PV systems, energy-intensive production processes can prioritise the utilisation of solar energy based on their specific energy needs. This strategy reduces dependency on expensive grid electricity and effectively diminishes the carbon footprint associated with certain products.

The knowledge of **variable electricity prices** allows for **economic optimisation**, enabling informed decision-making regarding the timing of storage facility charging or discharging and production sequence adjustments. In situations of high electricity prices, particularly when accompanied by a significant expected PV system output, it is advisable to fully charge the storage facility early on. Conversely, during periods of low solar power generation, it is also feasible to utilise electricity from the energy supplier to replenish the storage when electricity prices rise. In scenarios with flexible electricity prices, PV systems can even export electricity to the grid despite depleted storage tanks, due to the prevailing high electricity prices at that time.

3.2 PV Power Forecast to Optimise the Operation of Power Supply

A PV power forecast can offer substantial support in optimising the operation of grid components in various ways. One such application is the **preparation for emergency power operation**. In circumstances involving scheduled maintenance or expected substantial fluctuations in the general supply grid, the application of PV power forecasting aids in evaluating the feasibility of transitioning a particular grid branch to emergency power or island grid operation using only storage systems and PV power as support.

Considering sustainability and the lifespan of components, storage systems such as batteries can be operated in a charging mode optimised for maximising their overall lifetime. Higher charging current results in a reduced service life. Typically, limited information is accessible regarding the duration of a power surplus in the system. In cases where there is an ample surplus, charging is performed using the maximum charging current to store as much energy as possible. By combining the generation forecast with the known load curve, it becomes feasible to estimate the surplus generation and determine the optimal charging power for the battery, ensuring its longevity.

Another promising use case is **the enhancement of system efficiency by shifting the operation points of grid components that incur losses.** Research conducted in [32] demonstrates that the efficiency of components in the DC grid is heavily influenced by the load as shown in Fig. 5. Typically, energy-converting components exhibit the highest efficiency near their nominal operating points. By utilising PV forecasting, it becomes possible to efficiently control the power flows in the grid, ensuring that components like the central AIC are operated at optimal operating points, thereby maximising their efficiency.

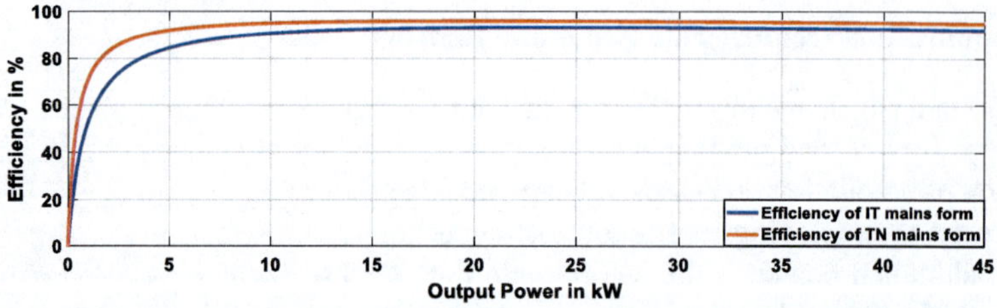

Fig. 5. Example for the grid efficiency of two design variants of DC grids according to [32]

3.3 Management of (Bidirectional) Vehicle Charging Infrastructure

Advancements in battery technology within modern electric vehicles not only result in significantly extended ranges and improved performance but also pave the way for their potential use as energy storage units in private households and businesses. In industrial fleets, electric vehicles have the potential to contribute greatly to grid stability by temporarily storing renewable energy and compensating for peak loads generated by production processes. DC grids represent the most energy-flexible, efficient, and environmentally friendly charging architecture. In the research project DC|hyPASim [15], researchers are developing bidirectional charging stations for industrial DC grids. By incorporating a PV power forecast, the fleet charging architecture can be seamlessly integrated with industrial processes, allowing vehicles to become part of energy management as flexible grid storage. The PV forecast plays a crucial role in ensuring the consistent availability of vehicle ranges.

4 Conclusion

This publication introduces a software framework based on models to predict the performance of PV systems. The framework utilises weather prognosis data provided by the DWD and generates a power forecast for a period of 5 days by employing module and converter models along with their alignments. The accuracy of the forecast was validated using over 5000 forecast data points collected from a real plant located in northern Bavaria during a winter month. The results demonstrate a useful accuracy and resolution for the proposed applications. For data transmission a flexible communication interface in the form of a REST-API is provided for energy management systems to facilitate automated usage. The study also presents various value-added application scenarios for the forecast within a DC-supplied supply architecture for production plants, focusing on a prime example of PLC-based energy management. These scenarios highlight the potential for economically optimised, environmentally friendly, and CO_2-neutral production. While the application scenarios primarily revolve around DC grids, they can also be adapted to conventional AC-supplied production facilities in many cases.

Acknowledgement. This publication was funded by the Federal Ministry for Economic Affairs and Climate Action through the project "DC|hyPASim" (Project no.: 50 LN/51 LN).

References

1. Dragicevic, T., Lu, X., Vasquez, J.C.: DC Microgrids—part II: a review of power architectures, applications, and standardization issues. IEEE Trans. Power Electron 3528–3549. https://doi.org/10.1109/TPEL.2015.2464277 (2016)
2. ZVEI e. V. Energieeffizienz mit elektrischen Antrieben. https://www.zvei.org/fileadmin/user_upload/Presse_und_Medien/Publikationen/2017/September/Fachbereich_Elektrische_Antriebe/Fachbereich_Elektrische_Antriebe_2017.pdf. Accessed 01 Aug 2023
3. Gritzner, D., Knochelmann, E., Greenyer, J.: Specifying and synthesizing energy-efficient production system controllers that exploit braking energy recuperation, pp. 408–413 (2018). https://doi.org/10.1109/COASE.2018.8560544
4. VDE DKE Gleichstrom im Niederspannungsbereich: Deutsche Normungs-Roadmap Version 2. https://www.dke.de/de/arbeitsfelder/energy/normungsroadmap-gleichstrom-im-niederspannungsbereich. Accessed 01 Aug 2023
5. Sauer, A.: Die Gleichstromfabrik: Energieeffizient. Robust. Zukunftsweisend (2020). ISBN 9783446465817
6. Blank, F., Körner, W., Puls, S. Die energieverteilung und antriebstechnik im gleichstromnetz. https://www.computer-automation.de/feldebene/antriebe/artikel/164614/4/. Accessed 01 Aug 2023
7. AlLee, G., Tschudi, W.: Edison Redux: 380 Vdc brings reliability and efficiency to sustainable data centers. IEEE Power Energy Mag. 50–59 (2012)
8. Maerz, M., Wunder, B., Ott, L.: LVDC-netze – herausforderungen und perspektiven: 7. In: ETG Conference (2017)
9. Sanjeev, P., Padhy, N.P., Agarwal, P.: Peak energy management using renewable integrated DC microgrid. IEEE Trans. Smart Grid 4906–4917 (2018)
10. Olivares, D.E., Mehrizi-Sani, A., Etemadi, A.H.: Trends in microgrid control. IEEE Trans. Smart Grid 1905–1919 (2014)
11. Weckmann, S., Sauer, A.: DC micro grid for energy efficient and flexible production. Procedia Manuf. 655–664 (2019)
12. Schaab, D.A., Weckmann, S., Kuhlmann, T.: Simulative analysis of a flexible, robust and sustainable energy supply through industrial smart-DC-Grid with distributed grid management. Procedia CIRP 366–370 (2018)
13. Stern, S. M. ODCA Summary. https://odca.zvei.org/fileadmin/odca/Dateien/2023-04-14_DC-ZVEI-Seiter_Gleichstrom-energieeffiziente-Fabrik.pdf. Accessed 15 Jun 2023
14. ZVEI DC-INDUSTRIE2 Vorstellung. https://dc-industrie.zvei.org/publikationen/dc-industrie2-projektvorstellung. Accessed 11 May 2022
15. WGP Gleichstromnetze sparen Energie. https://wgp.de/de/gleichstromnetze-bergen-hohes-einsparpotenzial/. Accessed 02 Jun 2023
16. Ott, L.: Modellbasierte Stabilitätsuntersuchungen in Niederspannungs-DC-Verteilnetzen (2019). ISBN 9783843940825
17. Mhankale, S.E., Thorat, A.R.: Droop control strategies of DC microgrid: a review, pp. 372–376 (2018). https://doi.org/10.1109/ICCTCT.2018.8550854
18. Gagic, M., Huang, K., Qin, Z.: Modular multilevel photovoltaic interfaced converter with low voltage energy integration for DC systems, pp. 1–6 (2019)
19. Hartwig Stammberger, Dr. Johann Austermann, Prof. Dr.-Ing. Holger Borcherding Systemkonzept DC-INDUSTRIE2. https://dc-industrie.zvei.org/publikationen/systemkonzept-fuer-dc-industrie2. Accessed 01 Aug 2023
20. Lorenz, E., Kühnert, J., Heinemann, D.: Short term forecasting of solar irradiance by combining satellite data and numerical weather predictions. In: 27th European Photovoltaic Solar Energy Conference and Exhibition, pp. 4401–4405 (2012). https://doi.org/10.4229/27thEUPVSEC2012-6DO.12.1

21. Meilinger, S., Herman-Czezuch, A., Kimiaie, N.: Development of innovative satellite-based methods for improved PV yield prediction on different time scales for distribution grid level applications (MetPVNet) (2021)
22. Starosta, A., Jhaveri, P., Munzke, N.: Wirtschaftlicher einfluss von PV-prognose-ungenauigkeiten auf eine ladeinfrastruktur für unternehmensparkplätze. In: 27. Symposium Photovoltaische Solarenergie, Bad Staffelstein (2022)
23. Kull, T.: Fischerauer gerhard simulation in den umwelt- und geowissenschaften: workshop 20203 (2020). ISBN 978-3-8440-7579-
24. Reinert, D., Frank, H., Prill, H.: DWD database reference for the global and regional ICON and ICON-EPS forecasting system: R&D at DWD (2022)
25. California Energy Commission CEC Modules Database. https://www.energy.ca.gov/sites/default/files/2021-10/PV_Module_List_Full_Data_ADA.xlsx. Accessed 01 Aug 2023
26. Holmgren, W.F., Hansen, C.W., Mikofski, M.A.: pvlib python: a python package for modeling solar energy systems. JOSS 884 (2018). https://doi.org/10.21105/joss.00884
27. Dobos, A.P.: An improved coefficient calculator for the California energy commission 6 parameter PV module model. J. Solar Energy (2012)
28. Dobos, A.P.: PVWatts version 5 manual (2014)
29. Doglio, F.: REST API development with node.js: manage and understand the full capabilities of REST development (2018). ISBN 978-1-4842-3714-4
30. Barth, M., Gutwald, B., Russwurm, E.: Simulation-based planning and design of hybrid AC/DC energy grids for production systems: a holistic approach. In: ASIM Fachtagung Simulation in Produktion und Logistik (2023)
31. Mertens, C.: Photovoltaik Lehrbuch zu Grundlagen, Technologie und Praxis: 5. aktualisierte Auflage (2020). ISBN 978-3-446-44232-0
32. Gutwald, B., Ndjiemeni, F.N., Barth, M.: Simulation-based efficiency comparison of different mains configurations for DC grid branches for supplying production plants based on a rule-compliant design, pp. 440–448 (2023). https://doi.org/10.1007/978-3-031-28839-5_49

T-Profile-Rolling of Stringers with Tailored Thicknesses – Challenges and Approaches

F. Aign(✉) and P. Groche

Institute for Production Engineering and Forming Machines, Technical University of Darmstadt, Otto-Berndt-Straße 2, 64287 Darmstadt, Germany
franziska.aign@ptu.tu-darmstadt.de

Abstract. Stringers in aircrafts have a longitudinally variable, load-adapted material thickness, which has been realized to date by chemical milling. A new flexible rolling process for T-profiles is being developed to enable more environmentally friendly and resource-efficient production of stringers in the future. The tailored distribution of wall thicknesses in the T-profiles in the longitudinal direction requires geometries of rolls that enable uniform rolling of the web and flange. This can be achieved by rolls with varying radii. However, due to the varying roll radii, varying longitudinal strains are introduced across the profile cross-section, which in turn lead to unwanted profile bowing.

This paper presents the basic idea of a new flexible rolling process for T-profiles - Flexible T-Profile-Rolling. Numerical investigations with different parameters of the roll circumferential speed and the roll gap are performed. The results show that accurate and straight T-profiles with longitudinally variable thicknesses can be produced by the proposed process.

Keywords: flexible rolling · T-profiles · straightness deviation

1 Introduction

10% of global CO2 emissions are caused by the production of steel and aluminum, and this share is expected to double by 2050 [1]. An important contribution to reducing emissions is first and foremost the saving of material in the production of new goods. Carruth et al. show that lightweight construction can reduce material consumption by 25–30% [1]. Particularly in the transport sector, but also in machine tools and architecture, lightweight construction has top priority in order to meet the demands of climate protection and increased performance [2]. One successful lightweight design principle is the use of load-adapted components, such as tailor-made blanks (TMB) [2–4]. These include individual gradients of material thickness, which lead to further weight and resource reductions by reducing the number of components and joints [3]. Thickness-variable TMB can be produced by flexible rolling, adhesive bonding, welding, milling or chemical milling [4].

Flexible load-adapted profile components are also used in aircraft design [3]. Stringers as stiffening profiles in aircraft fuselages have a variable material thickness

T. Bauernhansl et al. (Eds.): WGP 2023, LNPE, pp. 569–578, 2024.
https://doi.org/10.1007/978-3-031-47394-4_55

in the longitudinal direction to reduce weight. Air traffic is expected to grow by 4.3% per year until 2038, which, including the replacement of old aircrafts, corresponds to a demand for 39,210 new aircrafts [5]. The new process could therefore make an important contribution to resource efficiency in stringer production. At the same time, flexible T-profile-rolling is in line with the trend towards individual component geometries on a single line without tool changes as part of Industry 4.0. Flexible forming processes enable a rapid response to changing customer requirements [6].

A typical form of stringers is the T-section [7]. According to the current state of the art, thickness variations are introduced into the extruded aluminum profiles by chemical milling [8]. Chemical milling is used because it is so far the only process for producing TMB that allows the production of continuous thickness transitions for complex geometries such as T-profiles. This avoids stress peaks due to sudden thickness transitions and allows the profiles to be well formed [3, 4]. In chemical milling, thickness reduction is achieved by immersing the extrusion profiles in a chemical etchant. By masking or varying the dwell time in the etchant, the material thickness can be reduced in specific areas [8]. Etching produces both chemical and material waste, which is why it is classified as an environmentally critical process.

In order to meet the requirements of resource efficiency and sustainability, a new process is therefore needed to produce the T-stringers. A suitable alternative manufacturing process for TMB capable of producing continuous thickness transitions is flexible rolling.

In flexible rolling according to Kopp et al. [9], the roll gap is adjusted inline during cold strip production in order to introduce the variable thickness in the rolling direction. Number, position and shape of the thickness transitions are arbitrary. Since flexible rolling is a continuous forming process, high material utilization and a high production speed are attainable. However, flexible rolling is limited to flat sheets and cannot be used to roll branched profiles such as T-sections. Therefore, it is necessary to develop a new flexible rolling process for T-sections.

This paper thus presents a new process for the production of T-sections with variable thickness in the longitudinal direction by flexible T-profile-rolling. Challenges in the design of processes and equipment arise in particular due to the limited installation space and the unwanted profile bowing. Possibilities for overcoming these challenges are presented. The new process will be a cold rolling process, as it offers better surface finishes and thickness tolerances compared to hot rolling [10]. As the aluminum alloy Al2024 is of great importance for the aerospace industry [11], it is used for the process simulation.

2 Conceptual Design

2.1 Roll Arrangement

In the new T-profile-rolling process, an extruded T-profile is rolled into a thinner profile with variable sheet thickness in the longitudinal direction. The entire profile cross-section, i.e. web, flange and corner areas, is rolled simultaneously with the same roll gap height in order to avoid bowing due to different longitudinal strains in web and flange. This results in a new roll geometry with two skew rolls arranged at an angle of 45° and an additional cylindrical roll, see Fig. 1. The skew rolls are tapered and have a varying radius, that touches one side of the T-profile. The corner of the skew roll and that of the T-profile are rounded with the same radius of curvature.

The thickness change is introduced by the translational radial infeed of the skew rolls to minimize the number of drives required. To achieve the same infeed depth in the web and flange, the vertical infeed in the flange must be twice as large as the horizontal infeed in the web. All rolls are driven in rotation to achieve uniform sheet feed. The radii of the cylindrical and the skew roll in the centre of the flange half are the same. Similar to flexible rolling, the pitch at the transition between two sheet thicknesses is assumed to be 1:100 (1 mm thickness change per 100 mm length) [7].

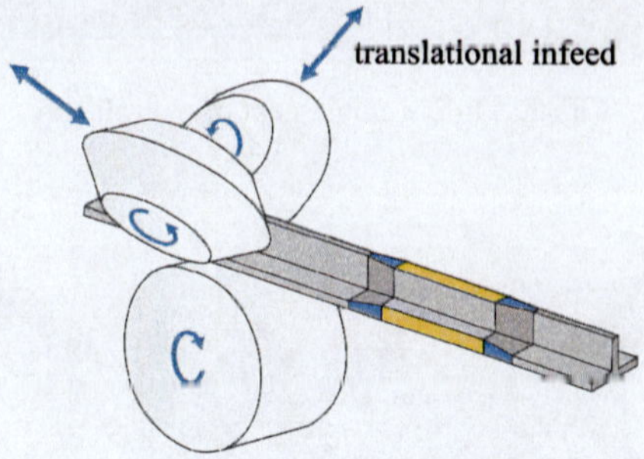

Fig. 1. Roll arrangement in flexible T-profile-rolling [12]

2.2 Implementation of the Rolling Stand

The adjustment of the skew rolls during the rolling process largely determines the geometry of the roll stand. It is performed by driven screw jacks. Since the cylinder roll is not fed inline and the same total infeed depth Δh must be achieved in the web and flange, the individual skew rolls are fed twice as deep (Δh) in the flange as in the web ($\Delta h/2$). The web is rolled by both skew rolls and thus achieves the same infeed depth as in the flange. To achieve this feed ratio the screw jacks, i.e. the axes of the adjustment movements with the infeed distance s, are inclined by 63.43° with respect to the horizontal, see Fig. 2. The movement is guided by linear roller guides. The cylindrical roll is not adjusted online and can therefore only be adjusted manually for assembly and straightening by means of a wedge gear. This makes it possible to roll T-sections with different initial web and flange thicknesses. In the future, however, different flexible infeed depths in the web and flange can only be achieved by driving the wedge gear. The skew rolls are internally supported by tapered roller bearings in an O-arrangement and driven by a spur gear stage. This avoids space problems caused by overlapping cardan shafts or bearing blocks. Shafts and axles are kept as short as possible for improved rigidity.

The rotary drive of the rolls is provided by cardan shafts in a Z-arrangement, which can compensate for the adjustment movement. An O-frame was chosen to increase the rigidity of the rolling stand. It consists of a plate-welded construction of two continuous frames connected by plates with guiding functions of the components. This can be seen in Fig. 3.

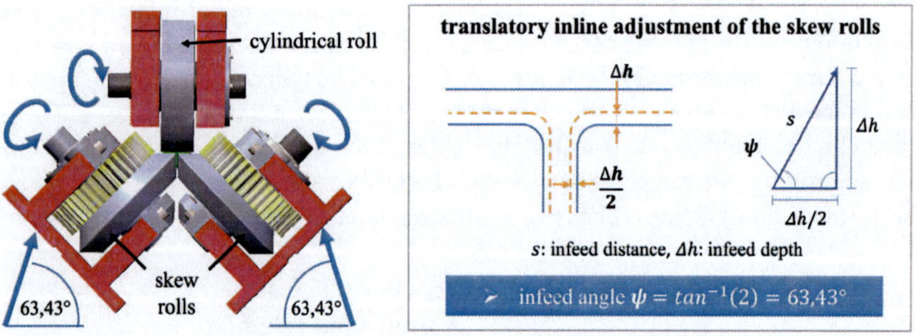

Fig. 2. Online adjustment of the roll gap

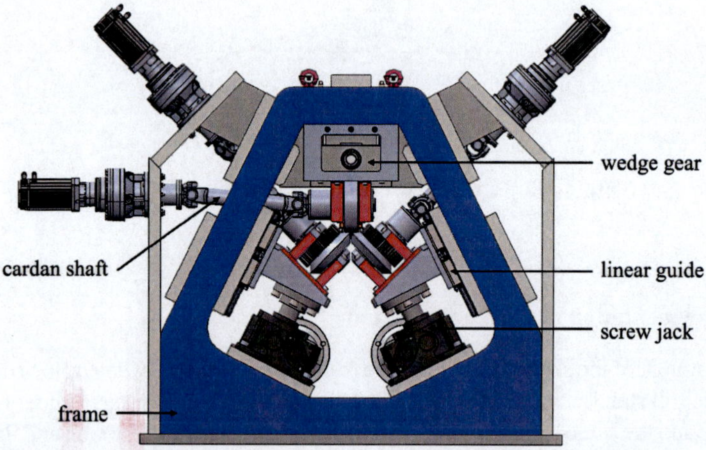

Fig. 3. Design concept roll stand

2.3 Control and Sensor System

A profile guide is placed in front of the roll stand. To control the adjustment movement of the skew rolls, the time of thickness change of the profile entering the roll gap must be known. For this purpose, the profile speed in the profile guide is measured by a friction wheel at a known distance from the roll stand. In addition, a laser distance sensor in the profile guide measures the profile thickness and thus the time of the profile thickness change. This allows the roll control system to calculate the position of the thickness transition in the roll gap and, at the same time, the start time of the translatory infeed movement. This makes it possible to roll multiple passes in sequence and achieve greater pass depths for the application. The infeed speed is a function of the sheet feed rate, the angle change and the roll length of the infeed ramp. To ensure the desired thickness distribution along the profile, the roll gap itself is monitored by laser sensors. In addition, linear potentiometers are mounted on the linear guide of the lifting device.

3 Numerical Model

The numerical model was built with the software package Marc Mentat. The rolls are simulated as rigid bodies and the sheet metal as a deformable part.

The radius of the cylindrical roll is 136 mm and the max. Radius of the skew roll in the corner (without rounding) is 145 mm. The initial angular velocity of both rolls is 0.25 rad/s. The initial geometry of the T-profile with a length of 350 mm is shown in Fig. 4. To save computation time, a symmetry plane is used and only half of the T-profile is simulated. The thickness reduction is 0.5 mm over a length of 50 mm, which corresponds to an infeed of the skew roll of 0.5 mm in y-direction and 0.25 mm in x-direction for the half profile. The sheet is modelled as a solid using element type 7 as "8-node hexagonal full integration". The size of the elements was determined using convergence analysis. The elements are 1.125 mm long, 0.875 mm wide and 4 elements are arranged in thickness direction.

The material used is Al2024 O and the flow curve from flat crush tests according to Becker [13] is implemented. Elastic-plastic isotropic material behaviour is assumed with a Young's modulus of 73 GPa, a density of 2.78 g/cm^3 and a Poisson's ratio of 0.3. The constant dilatation option is used to prevent the locking effect and the assumed strain formulation enables an improved modelling of the shear behaviour. Von Mises' yield criterion is applied.

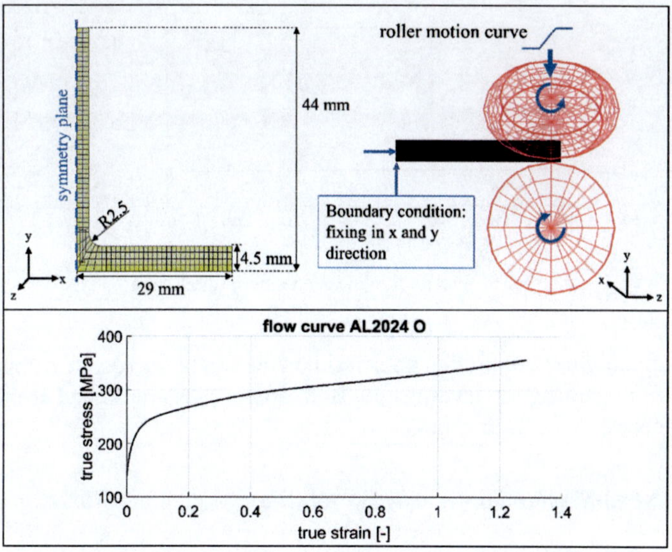

Fig. 4. Numerical model of flexible T-profile-rolling

At the start of the simulation, the profile is already in the roll gap, as shown in Fig. 4. The rolls start to rotate and due to the implemented Coulomb's friction coefficient of 0.1 the sheet is advanced. After 20 mm of advance, the skew rolls begin to infeed at a pitch of 1:100 until the respective infeed depth is reached and then continuously roll out the reached infeed depth. This is given by a stored linear motion curve. The rest of the profile is rolled out at a constant thickness. At the left end of the profile, the profile is fixed to the nodes in the centre of the sheet by a boundary condition in x- and y-direction, which is released as soon as the end of the profile enters the roll gap. Thus, the profile exits the roll gap without boundary conditions and can therefore deform freely.

4 Results and Discussion

4.1 Profile Dimensional Accuracy

For the thinner, rolled section of the profile in the stationary area, see Fig. 6, the average profile thickness in the flange $h_{flange,avg}$ is 3.97 mm, which is a deviation of 0.75% from the nominal profile thickness. The total length of the rolled profile is 378.29 mm, see Fig. 5.

The average width deviation of the profile flange in this area in the x-direction is 0.23%, since the material flow in the width direction is hindered by the friction between the rolled material and the rolls. For the rolling stock width b and the rolling stock height h, in this case the profile thickness h_{flange} or h_{web}, the width deviation can therefore be neglected if $b/h \geq 20$, analogous to flat rolling [14]. Web and flange must be considered separately.

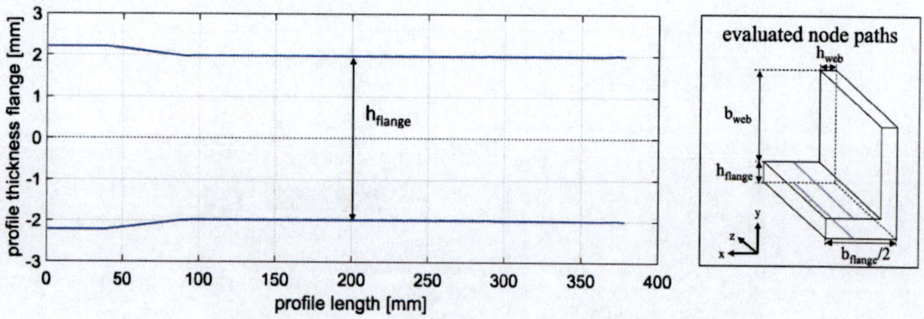

Fig. 5. Profile thickness change in the flange

Based on the simulated profile, a vertical bowing can be seen, so two approaches to compensate this bowing are investigated. Both approaches are limited at this point to the area of constant profile thickness.

4.2 Straightening Methods for Vertical Bowing

According to Groche et al. [15], an inhomogeneous longitudinal strain distribution leads to profile bowing and partial rolling can influence the bowing. It can be observed that the unstraightened bowing is negative. This suggests that the longitudinal strains in the corner must be larger than at the top of the web. This longitudinal strain difference could be attributed to the different radii of the skew rolls. With increasing radius, the pressed area during rolling increases and the deformation zone becomes larger. From this consideration, the first method to straighten the profile is to use a different roll adjustment in the web and flange. The longitudinal strain difference between the bottom of the flange and the top of the web can be adjusted by increasing the thickness reduction in the web. Takashima and Hiruta [16] show, that it is possible to straighten hot rolled T-profiles with different web and flange thickness reductions.

Figure 6 depicts the longitudinal strain of the profile in the centre of gravity G over the profile length. In the range from 40 mm to 90 mm, the increase in longitudinal strain as a function of the roll gap infeed can be observed. The range from 110 mm to 240 mm is selected as the steady state range for evaluation of bowing. Due to the constant conditions during rolling, the effects of the straightening methods can be better evaluated.

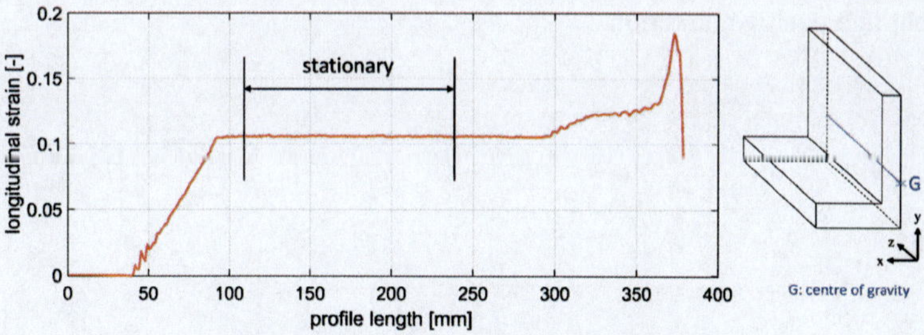

Fig. 6. Longitudinal strain in the flange

To test the first straightening method, the total infeed depth in the web was increased iteratively from 0.5 mm to 0.525 mm and 0.55 mm, while the infeed depth in the flange remained the same. It can be seen in Fig. 7 that the bowing initially decreases as the infeed depth in the web increases. At 0.525 mm, an almost straight profile is obtained with a maximum bow of 0.0115 mm, which corresponds to 2.94% of the unstraightened bow. With increasing infeed depth, the longitudinal strain level in the flange is exceeded and the profile bows in a positive direction. The evaluation of bowing is performed in the centre of gravity G.

According to Yekta et al. and Liang et al. [17, 18], bowing during rolling is caused by shear stresses above the plate thickness. Asymmetric rolling conditions of the skew and cylindrical rolls, such as different roll radii, different roll circumferential speeds or different friction conditions, cause a shear zone in the roll gap. The rolled material always curves in the direction of the roll with the lower peripheral velocity [17, 19]. This leads to the second method for straightening the profile. By reducing the peripheral velocity of the cylindrical roll ω_{ZW}, a straight profile can be produced by counterbending.

Figure 8 shows the vertical bow of the profile at the center of gravity when varying the peripheral velocity of the cylindrical roll ω_{ZW} in relation to the peripheral velocity of the skew roll ω_{SW}. It can be seen that as the ratio of the peripheral velocities decreases, the bowing also decreases initially until an almost straight profile is achieved at a ratio

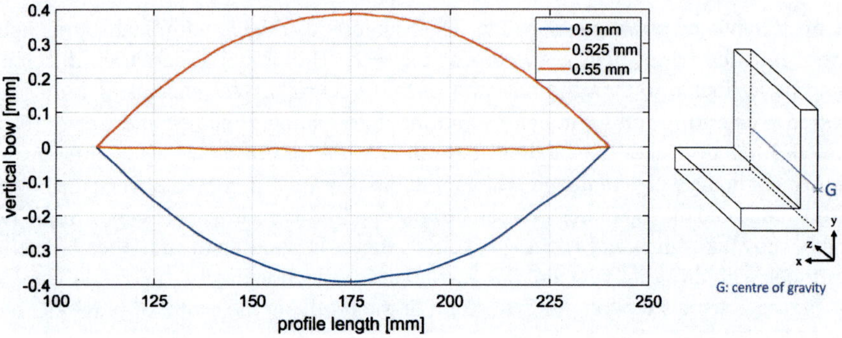

Fig. 7. Straightening by different roll adjustment in web and flange

of 0.9. For the velocity ratio of 0.9, the maximum bow is 0.0281 mm, which is 7.18% of the bow for a ratio of 1. As the speed ratio decreases further, the bowing increases again, but in a positive direction.

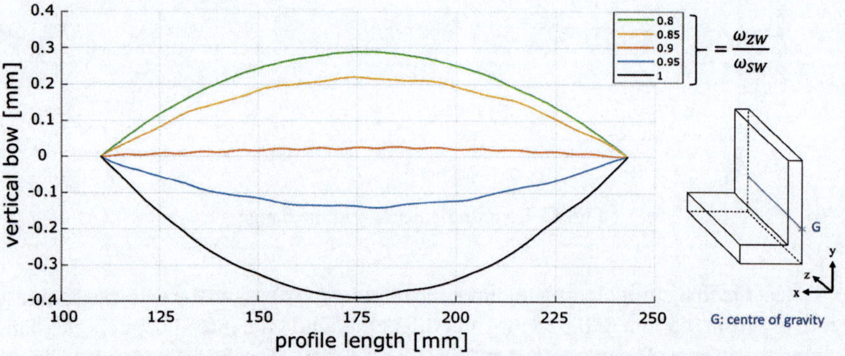

Fig. 8. Straightening by varying the peripheral velocity of the cylindrical roll

As expected, the vertical bowing can be influenced by varying the roll peripheral velocity, thus straightening the profile. This can also be seen in the longitudinal strain in the web, see Fig. 9. For the unstraightened profile rolled at a speed ratio of 1, the longitudinal strain decreases with increasing web height due to the decreasing radius of the skew roll. For both straightened profiles, the level of longitudinal strain remains almost constant. This confirms the assumption that there is a longitudinal strain difference across the web in the unstraightened profile, leading to vertical bowing of the profile.

In conclusion, both methods can be used to straighten the profiles. It remains to be seen which method will prove successful in practice. Varying the infeed ratio of web and flange will produce different profile thicknesses in the web and flange. In addition, the path differences to be set for roll gap adjustment are very small and difficult to achieve from a control standpoint. On the other hand, the speed variation method can cause slippage and surface defects.

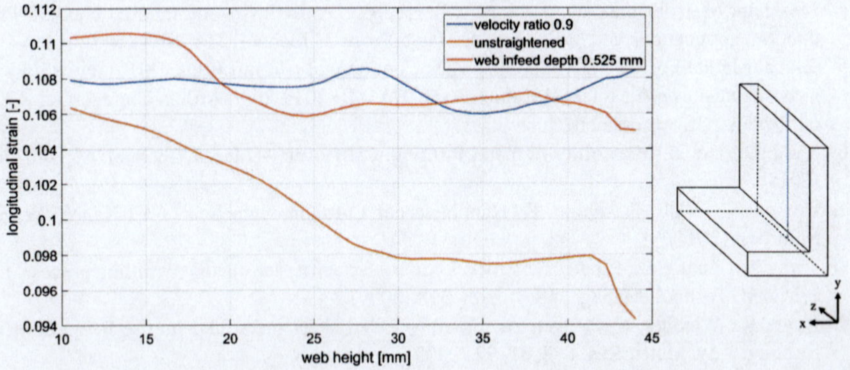

Fig. 9. Longitudinal strain in the web – straightened vs. unstraightened profile

5 Conclusion

In the future, flexible T-profile-rolling could be used to produce dimensionally accurate T-profiles from Al2024 O with variable section thicknesses over the length. A plant concept has been presented in this work. The challenge in implementing the process is the vertical bowing of the profile caused by the varying radius of the skew roll and the asymmetry of the profile shape. Simulations have shown that the vertical bowing can be straightened by reducing the peripheral velocity of the cylindrical roll or by varying the infeed depth in the web and flange. Implementation of the roll stand and experimental verification of the new technology, including multiple rolling passes, are planned. It is expected that the bowing will change as a function of profile size and infeed depth. This relationship will also need to be investigated in the future in order to produce dimensionally stable, flexible T-sections in different sizes. Furthermore, it is important to investigate whether the relationship between infeed depth and vertical bowing can also be applied to the behaviour in the thickness transition area.

Acknowledgements. The research presented here is part of the project "MeGro - Flexibler Walzprozess zum Einbringen variabler Wanddickenverläufe in Stringer". It is part of the Lower Saxony aviation research program and is funded by the Investitions- und Förderbank Niedersachsen (NBank). Furthermore, we would like to thank all industry partners, particularly AIRBUS Aerostructures GmbH, for their support in the research project.

References

1. Carruth, M., Allwood, J., Moynihan, M.: The technical potential for reducing metal requirements through lightweight product design. Resour. Conserv. Recycl. **57**, 48–60 (2011)
2. Kleiner, M., Geiger, M., Klaus, A.: Manufacturing of lightweight components by metal forming. CIRP Ann. **52**(2), 521–542 (2003)
3. Sinke, J., Zadpoor, A.A., Benedictus, R.: Tailor made blanks for the aerospace industry. Tailor Welded Blanks for Advanced Manufactoring, 1st edn. Woodhead Publishing (2011)

4. Merklein, M., Johannes, M., Lechner, M., Kuppert, A.: A review on tailored blanks—Production, applications and evaluation. J. Mater. Process. Technol. **214**, 151–164 (2014)
5. Global Market Forecast 2019–2038 Airbus Commercial Aircraft Book (2019). https://www.airbus.com/sites/g/files/jlcbta136/files/2021-07/GMF-2019-2038-Airbus-Commercial-Aircraft-book.pdf. Accessed 02 June 2023
6. Yang, D.Y., et al.: Flexibility in metal forming. CIRP Ann. – Manuf. Technol. **67**, 743–765 (2018)
7. Rossow, C., Wolf, K., Horst, P.: Handbuch der Luftfahrzeugtechnik. Carl Hanser Verlag, München (2014)
8. Torng, C., Huang, C., Chang, H.: Process control for aerospace chemical milling process. Int. J. Manuf. Technol. Manage. **18**(3), 308–318 (2009)
9. Kopp, R., Wiedner, C., Meyer, A.: Flexibly rolled sheet metal and its use in sheet metal forming. Adv. Mater. Res. **6–8**, 81–92 (2005)
10. Black, J.T., Kohser, J.: Degarmo's Materials and Processes in Manufacturing, 10th edn. John Wiley & Sons, Hoboken (2007)
11. Lenczowski, B.: New innovative lightweight materials and processes for aerospace application. 31. Aachener Stahlkolloquium Umformtechnik, Aachen (2016)
12. Aign, F., Groche, P., Wang, T.: Flexibles Auswalzen von T-Profilen zur Anwendung im Flugzeugbau. Blechnet **4**, 54–55 (2021)
13. Becker, N.: Weiterentwicklung von Verfahren zur Aufnahme von Fließkurven im Bereich hoher Umformgrade. In: IFU - Berichte aus dem Institut für Umformtechnik der Universität Stuttgart, vol. 122. Springer-Verlag, Berlin (1994)
14. Kugler, H.: Umformtechnik – Umformen metallischer Konstruktionswerkstoffe. Fachbuchverlag Leipzig im Carl Hanser Verlag (2009)
15. Groche, P., Güngör, B., Kilz, J.: Straight roll formed profiles through partial rolling. CIRP Ann.- Manuf. Technol. **00**, 1–4 (2023)
16. Takashima, Y., Hiruta, T.: T-bar universal rolling and its deformation properties. ISIJ Int. **52**(7), 1328–1334 (2012)
17. Yekta, R., Zohoor, H., Motekallem, A.: Analysis of asymmetric cold rolling using finite element method to create flat sheet. Sci. Iranica B **25**(6), 3133–3140 (2018)
18. Liang, H., Di, H., Gong, D.: Analysis of sheet curvature in asymmetrical cold rolling. J. Iron. Steel Res. Int. **20**(5), 34–37 (2013)
19. Johnson, W., Needham, G.: An experimental study of asymmetrical rolling. In: Applied Mechanics Convention, Institute of Mechanical Engineers, vol. 180, no. 9, Cambridge, UK (1966)

Studies on Manufacturing Polymer-Metal Hybrid Components Through Integrated Bulk Forming

Diego R. Alba[1](✉), Philip Reitinger[2], Calvin Ebert[2], Christian Bonten[2], and Mathias Liewald[1]

[1] Institut für Umformtechnik, University of Stuttgart, Holzgartenstraße 17, 70174 Stuttgart, Germany
diego.alba@ifu.uni-stuttgart.de

[2] Institut für Kunststofftechnik, University of Stuttgart, Pfaffenwaldring 32, 70569 Stuttgart, Germany

Abstract. The development of components with tailored properties is one of the ways to increase efficiency in areas such as the automotive sector. Hybrid components, in particular plastic-metal composites, meet this requirement by combining different material properties. In this context, this paper presents preliminary investigations regarding the use of fibre-reinforced plastics as an active medium for forming metallic materials. Several challenges arise for such a processing method, such as, the compressive strength of the plastics depending on the fibre fraction, -length and -orientation. Furthermore, the dissimilar elastic behavior of the two materials has a decisive influence on the composite- and bond strength. Therefore, in a first step, a material selection was performed, followed by material characterization. In addition, this paper presents a preliminary numerical analysis showing the feasibility of manufacturing a hybrid component using as example a spur gear with a polymer core made of glass fibre reinforced polypropylene and an outer geometry made of aluminum alloy AA6082.

Keywords: Hybrid bulk forming · polymer-metal composites · numerical simulation

1 Introduction

Sustainability is an issue of concern across all industries. In order to succeed, companies need to address new challenges and opportunities, and concurrently, while accelerating the pace of hybrid products development and release. In this context, there are numerous studies addressing the topic of forming and joining by plastic deformation techniques and hybrid manufacturing [1–5]. The use of mechanical joining techniques is a promising method to fabricate hybrid components with tailored properties made of polymers and metals [4].

T. Bauernhansl et al. (Eds.): WGP 2023, LNPE, pp. 579–588, 2024.
https://doi.org/10.1007/978-3-031-47394-4_56

Regarding polymer-metal joining, there are various processing techniques available [6–12]. However, the production of such hybrid components causes several challenges. For example, dissimilar chemical and physical properties, e.g. Young's Modulus and thermal expansion coefficients, could interfere with springback effect during and after forming, which could potentially compromise the joint strength. Moreover, the necessary forming pressures to form the metallic material could be very high if the forming of bulk materials is considered. As a result, these constraints directly impact the geometric intricacy of the final product.

On the scope of hybrid component manufacturing, this paper presents a preliminary analysis of using solid-phase polymers as an active medium for metal forming. The study conducted material selection and characterization, followed by a numerical analysis of the forming process of a hybrid component. The chosen component geometry is a spur gear with a polymeric core made of glass fibre reinforced polypropylene and an outer geometry made of aluminum alloy AA6082. The process principle is presented in Fig. 1.

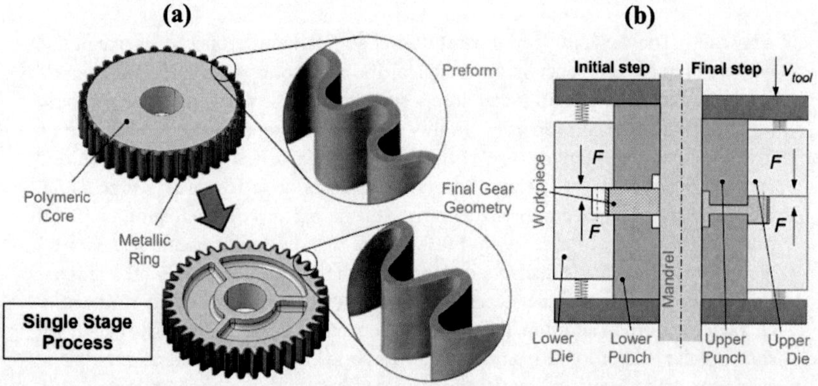

Fig. 1. Process principle for hybrid manufacturing of spur gear. (a) Targeted component; (b) Process kinematics.

2 Methodology

2.1 Polymer Mechanical Characterization

In order to evaluate the feasibility of using solid phase polymers as an active medium for metal forming, a material selection should first be made. To achieve this, a glass fibre reinforced polypropylene (PP) was chosen, as the glass fibers have a positive effect on the mechanical properties such as stiffness or strength and potentially reduce the springback effect because of less elastic deformation. As polymer matrix, PP is a thermoplastic polymer that has gained popularity across industries due to its properties and versatility. Known for its high strength-to-weight ratio, chemical resistance, and excellent processability, polypropylene has become a material of choice in various sectors. As reinforcement material, glass fibers have been identified as having a significant positive impact on the mechanical properties. This includes enhancing tensile strength, impact resistance, flexural strength and stiffness, and improving dimensional stability. As metallic material, the widely used aluminum alloy AA6082 was chosen due to its higher formability and lower required stresses for forming, compared to steel. The use of aluminum alloys in gear manufacturing has gained considerable attention in recent years due to their distinct properties and advantages. Aluminum alloys provide several benefits, such as lightweight construction, high strength-to-weight ratio, and excellent corrosion resistance, making them an attractive choice for gear applications.

Since the evaluation of polymer behavior during elastic-plastic deformation is the main focus of this processing routine, compression tests were carried out on the polymeric material. In accordance with the DIN EN ISO 604:2003 standard, the testing was performed at room temperature using a Zwick Roell 1445 Universal Testing Machine (Ulm, Germany), equipped with a load cell capable of measuring forces up to 20 kN. To perform the compression tests, a long fibre reinforced polypropylene (PP-LGF) from TechnoCompound GmbH, Bad Sobernheim, Germany, was chosen with fibre mass fractions of 20 M.-%, 30 M.-% 40 M.-%, and 60 M.-% of the product type TechnoFiber. The test specimens are injection-molded plates with a height of 4 mm, from which samples of size 10 mm × 10 mm were milled. The traverse speed of the machine is set at 1.0 mm/min, ensuring a quasi-static controlled and consistent application of load. The test results allow for assessing the performance and suitability of the PP composite with varying glass fibre content under compressive loads.

2.2 Hybrid Forming FEM Analysis

In order to investigate the feasibility of the process, in this study, a finite element analysis (FEA) simulation using DEFORM 3D software was conducted to analyze the forming process of a spur gear. The simulated gear with a module of 3, consisting of 28 teeth and a height of 15 mm, was formed using a polymer inner billet and a metallic outer ring. To minimize computational effort, symmetry planes were employed and the simulated section focused on half of a tooth, which represents an angle of 6.4287°. Tetrahedral elements were used for the simulation, with a total of 72,000 elements for the polymeric billet and 84,000 elements for the metallic ring. Figure 2a presents the FEM model with the above assumptions and simplifications. To investigate the effects of different metallic outer ring geometries, three geometric variants were considered (Fig. 2b).

The friction between the bodies was modeled using the Tresca model with a friction coefficient of $m = 0.12$. The tool velocity during the forming process was set to 20 mm/s, while the ambient temperature was maintained at 20 °C. Heat conduction was accounted by a thermal conductivity value of 11000 $W/m^2 \cdot K$, and heat convection was included with a heat transfer coefficient of 11 $W/m^2 \cdot K$.

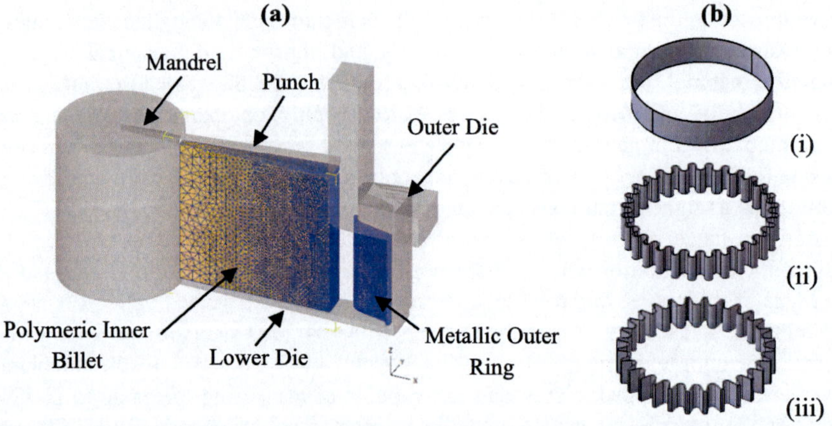

Fig. 2. (a) FEM model for hybrid forming process for a spur gear; (b) Metallic Outer ring geometries for preliminary feasibility analysis of the process.

3 Results and Discussions

3.1 Polymer Mechanical Characterization

As described in Sect. 2.1, compression tests were performed with glass fibre reinforced PP with varying fibre mass fractions. Figure 3 presents the tested samples (10 mm × 10 mm × 4 mm) with their corresponding identification of the fibre content. Based on Fig. 3, it can be verified that the adding of glass fibre reinforcement unsurprisingly resulted in a more brittle fracture behavior in the samples, compared to the pure PP with a more ductile behavior. This behavior becomes more pronounced with increasing fibre content. For example, the unreinforced samples did not exhibit this characteristic and show a smooth surface without broken layers.

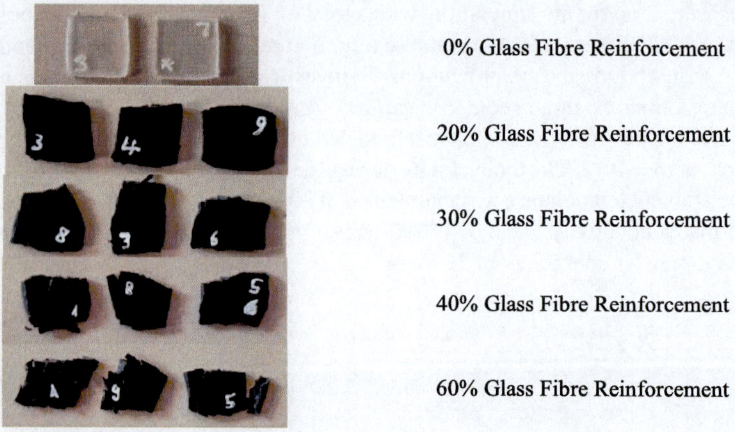

Fig. 3. Compression test samples with different glass fibre contents after experiments.

Additionally, it is possible to plot the engineering stress/strain curve of the performed tests. Figure 4 presents such curves for the tested samples. Nevertheless, based on these plots it is possible to define important compressive mechanical properties of the materials combinations, such as Young's modulus, fracture stress, which can be observed as the first maximum in the graph, and compression at break.

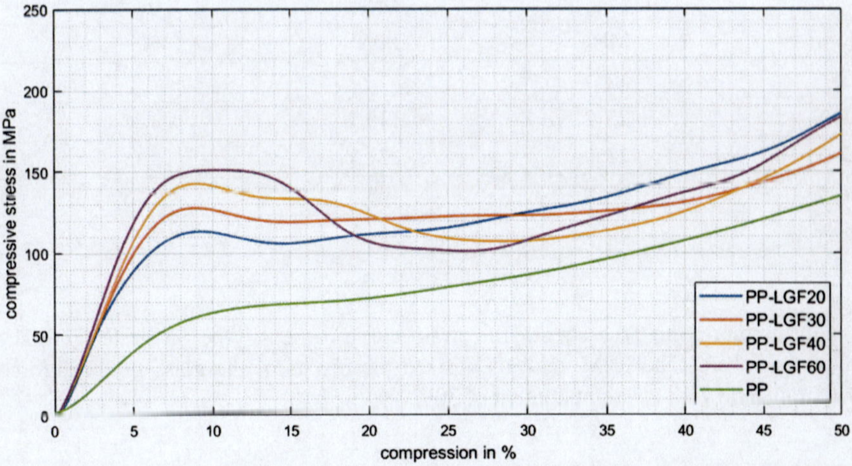

Fig. 4. True Stress/True compression curve for PP with several contents of glass fibre reinforcement.

In the graph, the major difference between reinforced and unreinforced PP is already evident in the elastic range. Thus, all reinforced PP shows an explicit fracture point at which maximum compressive stress is measured before it undergoes plastic deformation, whereas the unreinforced PP undergoes purely ductile deformation.

The determination of Young's Modulus and fracture stress also shows differences in the various tests. For Young's modulus, which is defined as the gradient in the linear elastic region, the differences are relatively small for the reinforced variants, but generally significantly higher than for unreinforced PP. Fracture stress, on the other hand, also increases significantly with increasing fibre content. Furthermore, for high fibre contents, a drop in stress in the plastic range can be observed. Afterwards, the compressive stress increases again quickly and reaches similar values for high compression as for low fibre contents.

Table 1 summarizes the results regarding the main mechanical properties of the tested samples. Trends are observable based on the data. The same behavior as seen in Fig. 4 can be observed considering the values of mechanical properties. Another point of consideration is that just an 20% addition of glass fibre increases in 75% the stress values at 8% compression, which is the area of fracture of the reinforced materials, compared to the not reinforced material. A further increase of glass fibre content has a less significant impact. On the other hand, no trend is visible considering the Young's modulus as well as the values of compression at fracture. Due to the ductile material behavior of the unreinforced PP, no values were determined for the fracture stress of this material.

Table 1. Mechanical properties of PP for different glass fibre reinforcements content.

Glass Fibre Content	Compressive Young Modulus	Compressive fracture stress	Compression at fracture stress
in wt	MPa	MPa	%
0%	1267	–	–
20%	2572	115,7	8,35
30%	2750	131,3	8,13
40%	2659	146,3	8,15
60%	3070	154,2	8,27

Furthermore, the True stress/True strain curve can easily be integrated into simulation software. The FEM analysis utilized the results of the glass fibre-reinforced polypropylene with a weight content of 30%.

3.2 Hybrid Forming FEM Analysis

Numerical simulations were performed in order to preliminarily analyze the hybrid forming feasibility of a spur gear. To gain insights regarding the process, a comparison was made between the different outer ring geometries. The optimization of the outer ring is necessary to achieve full filling of the die cavities and, consequently, components within the geometrical specifications. Figure 5 presents the results of the cylindrical outer ring geometry (Fig. 2b (i)) after 76% of the intended process duration. As indicated by the high values of the dimensionless DEFORM 3D damage indicator, it is clear that the outer ring has fractured. This default indicator of the software shows potentially fracture-prone areas on the component. The fracture starts at 64% of the simulation when the tensile stresses reach their higher values. The strains required to form the cylindrical outer ring in order to completely fill the die cavities are very high, which the aluminum

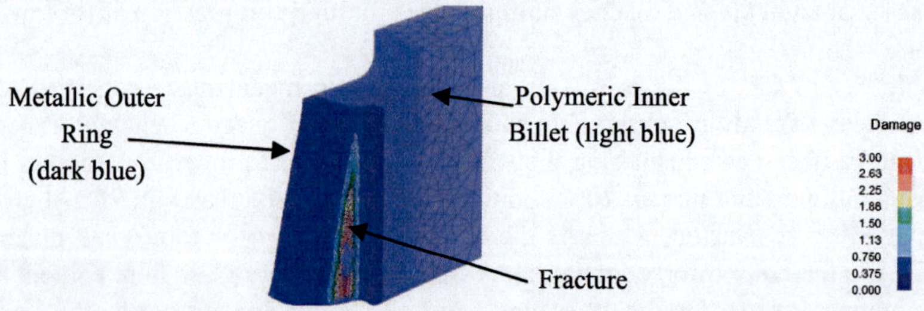

Fig. 5. Fracture of cylindrical outer ring geometry at 76% of process duration.

alloy cannot withstand. Hence, fracture starts and propagates perpendicularly along all the outer ring geometry. These outcomes demonstrate the need for alternative outer ring geometries.

Based on the results of the cylindrical outer ring geometries, two different geometries were simulated. The final simulation-step of the outer ring geometry similar to a spline (Fig. 2b (ii)) is shown in Fig. 6. The outer ring geometry presented a satisfactory behavior without considerable thinning or fracture. However, this geometry does not completely fill the cavities, as shown in detail in Fig. 6. Such an underfilled region can have a detrimental effect on the performance of a gear in future applications.

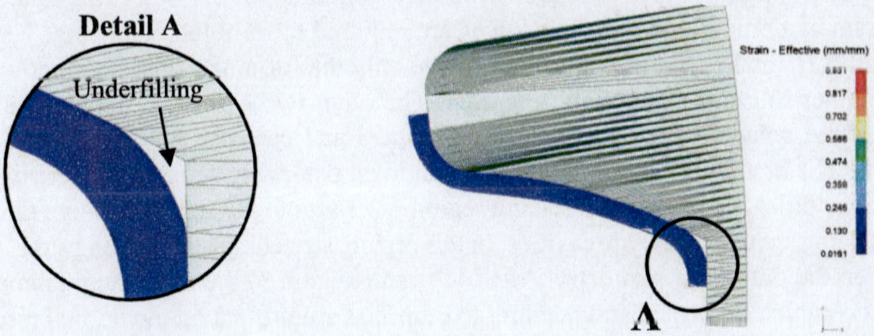

Fig. 6. FEM model for the spline outer ring geometry at the end of the process; Detail A presents the underfilled region.

Hence, a new geometry was proposed for the outer ring with the aim to address this underfilled region. This geometry is similar to a gear tooth, but with a shape easy to produce (Fig. 2b (iii)). The results of the final step of the simulated geometry are showed in Fig. 7. Although a small amount of underfilling occurs in a distinct area, the filling conditions could already be improved. This outer ring geometry is suitable for further experimental tests to manufacture a hybrid gear with a polymer core and an outer ring of aluminum. Based on these results, it is possible to conclude that the optimization of the outer ring geometry is a crucial aspect of the hybrid forming of geometrical complex

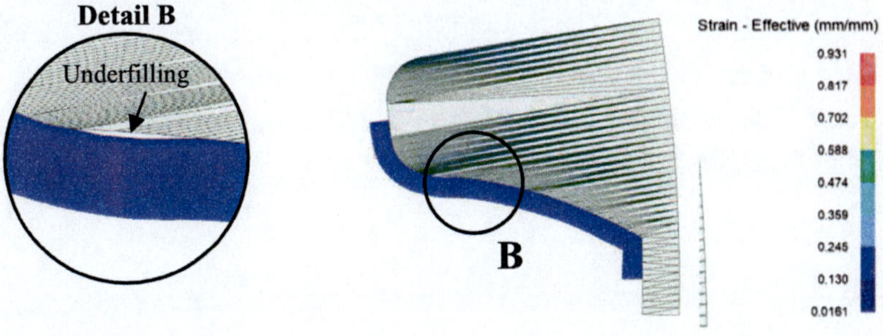

Fig. 7. FEM model for the optimized outer ring geometry at the end of the process; Detail B presents the underfilled region.

parts. Further studies should be carried out in order to optimize the geometry for this application more notably.

Lastly, another important aspect to consider in the hybrid formation process is the required force. Figure 8 displays the *Force vs. Punch Stroke* diagram of the three different simulated cases. The cylindrical outer ring has a different profile, compared to the other two geometries, because fracture occurs during the process. The *Force vs. Stroke* behavior can be classified into three distinct regions. The first region – from 0 mm to 2.3 mm – represents only the polymeric billet forming. The second region – from 2.3 mm to 4.6 mm – represents the region where the outer ring is being formed until fracture. It is notable that the force is higher in comparison to the other two cases. This could be explained by higher levels of strain, which are dominant on the outer ring and due to the strain hardening effect, greater forces are required to form the outer ring. The final region – 4.6 mm to 6 mm – represents, again, only the forming of the polymeric billet, as the outer ring has completely fractured. The other two simulated geometries show comparable behavior during the forming process and can also be divided into three regions. The first region is similar to the cylindrical outer ring geometry where only the polymeric billet is formed. The second region – 2.3 mm to 4.7 mm – is where the outer ring is being shaped on the die cavities. In this region, stretching and bending stress states occur on the outer ring geometry. After 4.7 mm (Region 3) the outer ring geometry is almost completely formed and compressive stresses are present on the normal direction of the outer ring in order to ultimately form the outer ring on the cavities. Lastly, since the process is considered as a precision-forming process, i.e. without flashing, a sudden increase in forces indicates that the die cavities are adequately filled and there is hardly any space for the materials to flow.

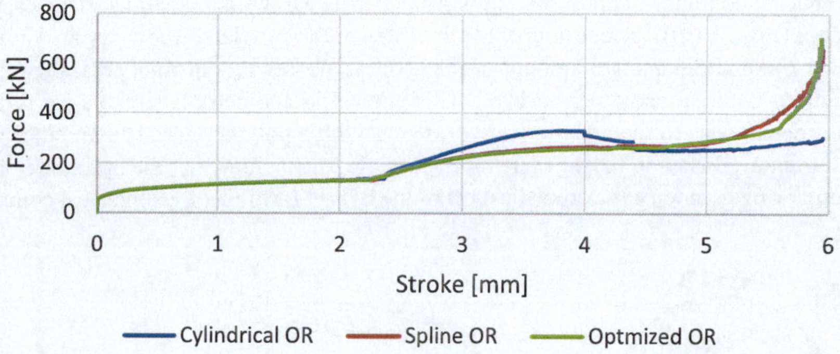

Fig. 8. Force vs. Stroke curve for the three different outer ring simulated geometries for hybrid forming process of a spur gear.

4 Conclusions and Outlook

A Preliminary feasibility analysis was performed to evaluate the possibility of producing hybrid polymer-metal components using forming technologies. In a first step, materials selection was carried out, followed by polymer characterization and FEM analysis. The following conclusions can be addressed from this study:

- Glass fibre reinforced polypropylene is a suitable polymeric material for such a manufacturing process. In terms of metallic materials, initial testing with high-strength aluminum alloy AA6082 was executed.
- The addition of glass fibre has a significant impact on the mechanical properties of the material. With higher glass fibre fractions, higher values of Young's Modulus and yield stress were present. The yielding point of the samples does not exhibit a clear trend.
- At room temperature, the glass fibre reinforced PP presents a low strain-to-yield value. Therefore, additional research is necessary to evaluate higher processing temperatures.
- To gain a better understanding of the processes involved in forming the material ring, compression tests of the polymer materials should be carried out under isochoric conditions. In these tests the influence of the fibre content on the bulk modulus can be investigated more in detail and will help to improve the accuracy of the material model for the numerical simulations.
- The Outer Ring geometry plays a decisive role in determining the success of the hybrid forming process as seen in the example of the shown spur gear.
- The Outer Ring Geometry was optimized through FEM simulations, resulting in satisfactory filling of die cavities. Additional optimization efforts could be pursued to eliminate underfilled areas in the final product.
- Additional research is needed to examine the joint pressure and joint strength of the polymer, in order to achieve a reliable connection.

References

1. Groche, P., et al.: Joining by forming – a review on joint mechanisms, applications and future trends. J. Mater. Process. Technol. **214**, 1972–1994 (2014)
2. Salamati, M., et al.: Processing and tooling considerations in joining by forming technologies; part A – mechanical joining. Int. J. Adv. Manuf. Technol. **101**, 261–315 (2019)
3. Meschut, G., et al.: Review on mechanical joining by plastic deformation. J. Adv. Join. Process **5**(100113), 1–27 (2022)
4. Weber, F., et al.: Joining by forming – a selective review. J. Adv. Join. Process **3**(1000054), 1–18 (2021)
5. Mori, K., et al.: Joining by plastic deformation. CIRP Ann. – Manuf. Technol. **62**, 673–694 (2013)
6. Sankaranarayanan, R., et al.: Review of research on friction riveting of polymer/metal light weight multi-material structures. Trans. Indian Inst. Met. **74**(11), 2541–2553 (2021)
7. Galinska, A., Galinski, C.: Mechanical joining of fibre reinforced polymer composites to metals—a review. Part II: riveting, clinching, non-adhesive form-locked joints, pin and loop joining. Polymers **12**, 1681 (2020)

8. Schmachtenberg, E., Schuck, M., Kühnert, I.: Urformen und Montieren in einem Prozess. Kunststoffe **4**, 24–31 (2007)
9. Müller, T.: Methodik zur Entwicklung von Hybridstrukturen auf Basis faserverstärkter Thermoplaste. Erlangen, Universität, Lehrstuhl für Kunststofftechnik, Diss. (2011)
10. Rauscher, B.: Formschlüssig verbundene Metall-Kunststoff-Hybridbauteile durch Integration von Blechumformung und Spritzgießen. Dissertation, TU Dortmund University (2011)
11. Hussain, M.M.; Rauscher, B.; Tekkaya, A.E.: Wirkmedienbasierte Herstellung hybrider Metall-Kunststoff-Verbundbauteile mit Kunststoffschmelzen als Druckmedium. Mat.wiss. u. Werkstofftech. **9**, 627–632 (2008)
12. Michaeli, W., Mäsing, R.: Metalldekorierte Bauteile in einem Schritt – Hinterspritzen und integriertes Tiefziehen. DIF-Tagung "Kunststoff-Metall-Verbundtechnik", 9./10. März 2010, Bad Herrenalb, S., pp. 1–16 (2010)

Influence of Process Parameters on the Product Quality in Active Media-Based Forming of Paper

C. Brunk(✉) and P. Groche

Insitute for Production Engineering and Forming Machines, Technical University of Darmstadt, Otto-Berndt-Str. 2, 64287 Darmstadt, Germany
cedric.brunk@ptu.tu-darmstadt.de

Abstract. Reducing plastic waste is an important challenge in the consumer goods industry´s transformation towards greater sustainability. Natural fibre-based materials such as paper can already partially substitute plastic-based packaging. Shells with wrinkle-free sealing areas, as needed for gastight packaging, can be achieved by active media-based forming. Nevertheless, the specific material properties of paper, such as limited plastic elongation or anisotropy, pose challenges regarding the products quality. The evaluation of the geometric characteristics of formed paper products and how they can be adjusted by appropriate selection of process parameters has been mostly empirical.

This paper presents an approach for the quantitative evaluation of the product quality in the active media-based forming of paper. The methodology is applied to different materials, material conditions and process parameters to evaluate the factors influencing the products geometry. A systematic evaluation of the experiments reveals that using material-adapted process control enables a significant increase in geometric flexibility.

Keywords: active media-based forming · forming quality · packaging · paper forming · wrinkle-free · process window

1 Introduction

Single-use plastic products account for a significant proportion of the plastic entering the environment each year and are a major ecological problem [1]. Without legally binding action, the plastic consumption in the 20 leading industrialised and emerging economies (G20) is expected to double by mid-century [1]. Legislators have recognised this problem and, through the European Union Directive on "Reducing the Impact of Certain Plastic Products on the Environment", have banned the use of certain single-use plastic products in Germany from July 2021 [2]. As a result, the demand for alternative materials and manufacturing processes to produce disposable tableware, to-go cups or fast food packaging sustainably and economically is growing at the same time. In Germany alone, around 60 kt of plastic packaging is expected to be replaced by fibre-based products between 2020 and 2025 [3].

In addition to the increasing demand, which requires high quantities and production rates, the high quality requirements for a successful product substitution must be

T. Bauernhansl et al. (Eds.): WGP 2023, LNPE, pp. 589–598, 2024.
https://doi.org/10.1007/978-3-031-47394-4_57

maintained. The requirements for complex product geometries, which are common in plastic packaging, should also be met when substituting with sustainable materials. In this context, the forming process of natural fibre-based products such as paper offers the possibility to meet these challenges.

Paper-based packaging can be produced by a number of processes, including fibre moulding and forming of paperboard. Fibre moulding offers a high degree of geometric freedom, but its main disadvantages are surface quality, complex and therefore expensive tools and high energy consumption. Folding and wrapping are variants of two-dimensional forming processes, where the paperboard is formed along a straight line. While satisfactory surface quality and production time can be achieved, the geometrical freedom is highly restricted, presenting a challenge for the marketing and recognition of product packaging. Reasonable geometric freedom and economical production can be achieved by using three-dimensional forming techniques such as deep drawing [4]. Due to the resulting tangential compressive stresses and the limited yield of paper, excessive wrinkling is a characteristic visual appearance of products produced by deep drawing with rigid tools. Another process variant is active media-based forming, which allows forming with a high degree of geometric freedom. With well-chosen process parameters and low drawing depth, a wrinkle-free flange can be achieved, as required for a gas-tight seal. Active media-based forming of paper has been studied by different research groups [5, 6]. In this process, a paper blank is formed into a die using an active medium.

Different types of paperboard have significantly different forming properties, which are further influenced by forming parameters. Geometric shape reproduction is a key factor in assessing quality. Until now, there has been no uniform systematic description of the geometric shape of formed parts and no investigations of the influencing process parameters have been performed in the active media-based forming of paper. The introduction of a methodology for the quantitative evaluation of forming quality and the systematic investigation of different materials, material conditions and process parameters will provide a valuable guidance to maximise the forming quality achieved, paving the way for paper-based packaging.

2 Approach

2.1 Paper as a Material in Forming Technology

Paper is a material made from natural fibres. On an industrial scale, paper is produced by drying a water-fibre mixture on a moving web. Due to the movement of the web, the fibres are mainly oriented in the direction of the movement, also known as the machine direction (MD). This directional fibre orientation, together with a non-uniform contraction limit during the drying process, results in different material properties in the machine and the transverse direction (cross direction, CD) [7]. This anisotropic material behaviour is evident in the stress-strain diagram, as shown in Fig. 1. In MD, paper has a higher tensile strength and lower strain, while in CD, paper is more ductile. The fibres can be either gained from plants or from recycled fibre materials, resulting in virgin-based paper or recycled paper, respectively.

The inter-fibre connection of dried paper is achieved by hydrogen bonds. The addition of water leads to an increase in extensibility and a decrease in tensile strength, as hydrogen bonds are dissolved by water molecules. As a hygroscopic material, the behaviour of paper can therefore be influenced by specific moistening, as shown in Fig. 1. The moisture content of paper w_{H_2O} is defined as follows [8]:

$$w_{H_2O} = \frac{m_{water}}{m_{total}} = \frac{m_{total} - m_{dry}}{m_{total}}, \tag{1}$$

where m_{total} is the total mass of the humid paper which can be split into m_{water}, the mass of water in the paper, and m_{dry}, the mass of the paper at 0% humidity. Most papers have their maximum elongation at a moisture content of about 15%, which is desirable for satisfactory forming properties [9].

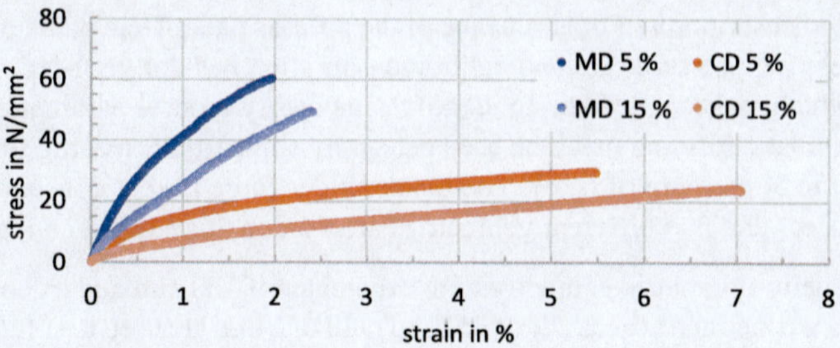

Fig. 1. Stress-strain diagram for a virgin-fibre based paper with 190 g/m^2 in machine and cross direction at different humidities

Paper also exhibits a time-dependent behaviour with respect to load. This viscoelasticity influences the forming process and has been further described by Haslach [10].

Within this study three different types of paperboards are used. A virgin fibre board with a three-layer fibre structure is used at two different grammages of 190 g/m^2 and 350 g/m^2. In addition, a recycled paperboard, with a two-layer fibre construction and a grammage of 200 g/m^2 is used.

2.2 Active Media-Based Forming of Paper

Active media-based forming is a variant of the deep drawing of paper in which an active medium is used to form a paper blank into a die. The process is shown in Fig. 2 (left). A blank holder force can be used to prevent the formation of wrinkles caused by tangential compressive stresses. Compressed air can be utilised as a forming medium if it is separated from the blank by an airtight membrane, as the paper itself is too porous to generate sufficient forming pressure.

The usage of active media-based forming of paper was first described by Huttel et al. and Östlund et al. [5, 6, 11], both giving first insights into the influence of process parameters on the forming result. In the investigation of Östlund et al., the blank holder force was kept constant over all the materials tested and the forming pressure was only considered at three different pressure levels, so that a deeper understanding of these two process parameters is not possible. By using four blank holder force levels, an influence of the blank holder force and applied pressure becomes apparent in the study of Groche, Huttel et al. [11]. The blank holder force must be chosen correctly to avoid both wrinkles and cracks.

The use of a higher moisture content was shown to be beneficial for an increased formability and was further investigated by Linvill et al. [12]. At higher moisture contents, the increased ductility leads to a better form filling within the materials used. With the proposed FE method of a moisture- and temperature-dependent constitutive model, the influence of die temperature and friction between die and paper was shown to be of high relevance. By assuming a coefficient of friction lower by a factor of four, more than twice the volume was achieved in the simulation. In addition, the extent of drying effects after the completion of the forming process was analysed and shown to have a significant effect on the springback.

A detailed description of the influence of the process parameters blank holder force, forming pressure, pressure rate and calibration duration has not yet been presented in literature, which makes it difficult to identify satisfactory process parameters.

The pneumatic forming machine used can apply a linearly increasing forming pressure up to a total pressure of 6 bar. The blank holder force is kept constant during the forming process and two different geometries are used in this study (see Fig. 2 right):

1. A rotationally symmetric geometry with a diameter of 100 mm and a depth of 8 mm. The sides are inclined at an angle of 45° and radii of 2 mm are used to soften the edges. Assuming that there is no draw-in during the forming process, a minimum strain to rupture of the material of 5.75% would be required to achieve complete filling of the cross section in the axis of symmetry. This can be estimated by calculating the area increase required. This estimate can be considered conservative, as draw-in can be expected.
2. A rectangular geometry 100 mm long, 80 mm wide and 10 mm deep. The corners are rounded with radii of 20 and 30 mm. The angle between the wall and bottom is set to 60°. Again, the edges are softened with 2 mm radii. Assuming no draw in, a minimum extensibility of 9.39% and 11.61% would be required along and transverse the long side of the rectangle, respectively.

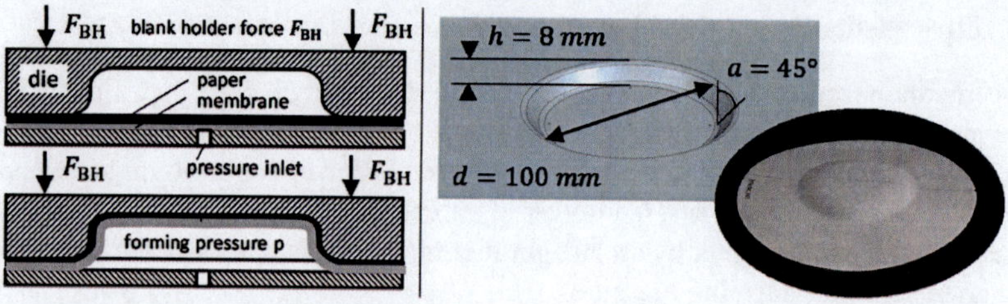

Fig. 2. Principle of active media-based forming (left, adopted from [13]); used conical forming geometry and picture of formed paperboard (right)

2.3 Determination of Geometric Forming Quality

The quality of formed paper parts can be described in terms of geometric aspects that reflect the formation achieved after removal and springback. Other possible quality measures could be the stability or air tightness after sealing. In this research, only the geometric forming quality is used and further optimized. The evaluation of the geometric forming quality was mainly done by visual comparison of sections of formed and scanned samples so far.

A quantitative evaluation of the geometric forming quality was briefly described by Östlund et al. who estimated the enclosed volume of the formed parts to quantify the forming quality of the isotropic paper sheets used after forming [12]. The volume was estimated by the integral of a 180° rotation of a 2 D profile scan of the rotationally symmetric form.

As part of this evaluation, the formed parts are scanned with an industrial scanning system, resulting in a point cloud that can be further analysed. As the paperboards used are industrial grade and therefore anisotropic, sections in the axis of symmetry are taken in MD and CD to analyse the effects on the directional dependent forming quality. The enclosed area of the sections is calculated and divided by the enclosed area of the die to give a quantitative measure of the form filling in both directions. For high form fillings, the straight (bottom and wall) and curved (bottom radius) components of the profile can be distinguished from each other by further processing of the first and second order gradients.

3 Results

3.1 Influence of Material and Humidity

The geometry of the form used and a section in CD of two exemplary paperboards for the conical geometry are shown in Fig. 3 (left). As can be seen from the section, the formed parts do not completely fill the die. This can be explained by the limited extensibility of the paper. Furthermore, the strain during forming can be divided into elastic and plastic strain. As the geometry is scanned after removal, only the plastic strain remains, so the area filling cannot be expected to be 100%. The virgin fibre paperboard at 15% moisture achieves a higher filling compared to the recycle fibre paperboard at 5% moisture, which can be explained by the higher elongation at break and reduced yield stress, resulting in a higher elongation at the same forming pressure.

An individual setting of forming pressure and blank holder force was determined for each material and humidity combination. The increased humidity of 15% leads to an increase in extensibility and a decrease in yield stress, which in turn leads to a better form filling for all the papers used. The area filling (orange area in Fig. 3 (left)) for all paperboards at both humidity levels is shown in Fig. 3 (right). The scans of the paperboards with high moisture content were taken after 24 h of storage at room conditions to ensure comparability. Individual samples measured twice (immediately after forming and after 24 h storage) showed that the drying process reduced the area filling by about 6%. This is in accordance with findings from Linvill et al. [12].

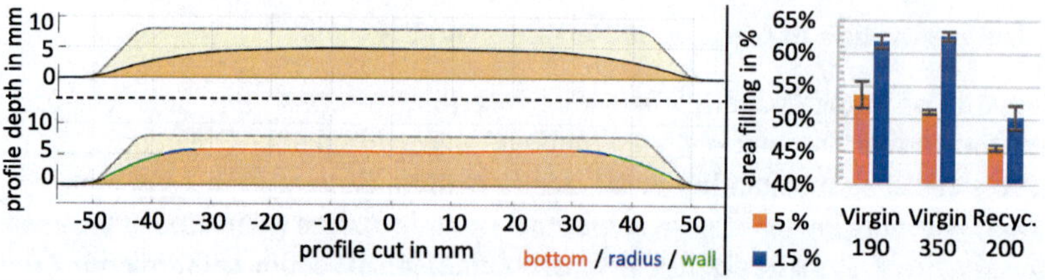

Fig. 3. Section in CD for recycled paperboard at 5% (left upper) and virgin fibre paperboard at a moisture level of 15% for the conical geometry with distinguished geometric features (left lower); calculated area filling for all paperboards and humidities (right); n = 5

The recycled fibre material shows the lowest area filling. During the recycling process, the fibres get shortened, leading to a reduced elongation at break and tensile strength. For both virgin fibre-based papers, the area filling is at a higher level. The increase in area filling due to the increased moisture is visible for all the paperboards, as expected from uniaxial testing. For the virgin fibre paper at the grammage of 350 g/m^2, the increase is around 11.5%, making it the highest area filling for all three materials.

The process window in active media-based forming is influenced by the applied pressure and the blank holder force. If the blank holder force is sufficiently high, no draw-in is possible. At this point, die filling is only possible by straining of the material. When the extensibility of the paper is reached with increasing pressure, an almost horizontal crack boundary is visible. Due to the low extensibility of the paper, only limited drawing ratios are possible. If the blank holder force is reduced, the paper can be drawn in. If set too low, the compressive stresses in the tangential direction will cause wrinkles to form. As no wrinkles are intended, this will result in a wrinkle boundary. If neither wrinkling nor cracks are visible, the forming is considered as successful. An example process window is shown in Fig. 4.

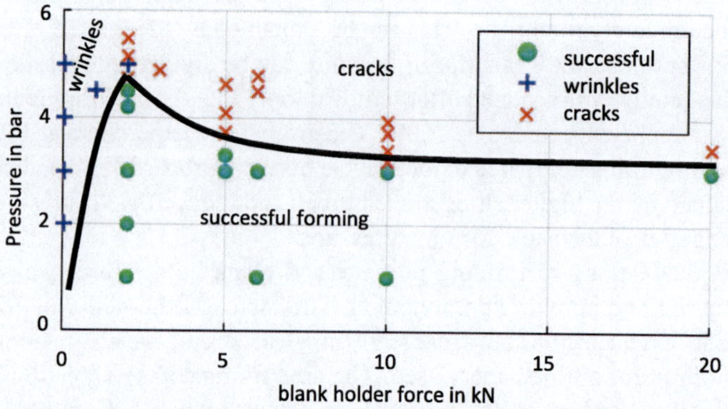

Fig. 4. Process window for virgin-fibre material with 190 g/m^2 at a humidity of 5% and conical forming geometry

The pressure, at which cracks occur is influenced by uncertainties, leading to somewhat blurred boundaries of the process window. This fact can be explained by inhomogeneities within the structure of the natural fibre material paper.

Further investigations on influencing factors are carried out on the 350 g/m^2 virgin-fibre paper at 15% humidity. This level of humidity allows for the best form filling. A comparison of the influence of the blank holder force and the forming pressure on the calculated area filling in MD shows that the applied pressure is significant, see Fig. 5. With the aim of maximising the total area filling, the forming pressure should be maximised until cracks or wrinkles occur, see Fig. 4. The influence of the blank holder force is of minor importance. A higher extension of the fibre network and therefore a higher proportion of plastic strain could explain the slight increase in area filling.

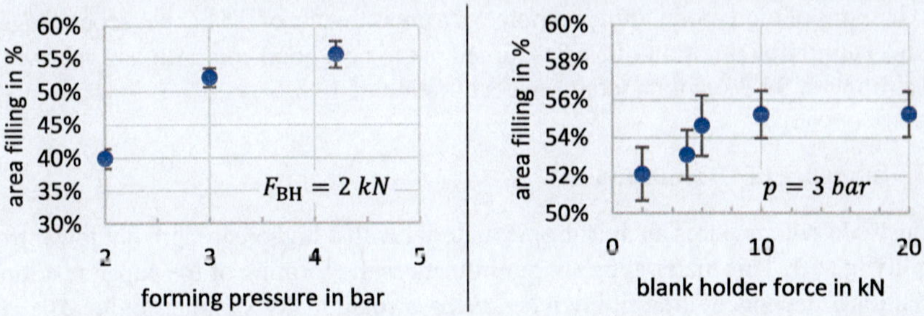

Fig. 5. Influence of forming pressure (left) at constant blank holder force and blank holder force (right) at constant forming pressure on area filling in MD for virgin-fibre material with 190 g/m^2 at a humidity of 5% and conical forming geometry; n = 5

3.2 Influence of Forming Speeds and Calibration Duration

In addition to the maximum value of the linearly increasing applied pressure, the pressure rate is another parameter of interest, as it determines the process time and thus the productivity in an industrial application. The pressure rate was set to increase at a rate of 0.05, 0.1 and 0.5 bar per second up to 4 bar, giving a total forming time of 80, 40 and 8 s, respectively (see Fig. 6).

Another variation of the pressure appliance is to add a constant pressure after linearly reaching the maximum value. A subsequent calibration duration of 0, 1, 5 and 15 s at a pressure of 4 bar was investigated for a pressure rate of 0.1 bar per second, resulting in a total forming time of 40, 41, 45 and 55 s, respectively.

An increase in area filling can be achieved at a reduced pressure rate. This is in accordance with the rate-dependent material properties of paper, as described by Haslach [10]. At the same time, the increase in pressure rate affects the scatter, resulting in a higher standard deviation.

With increasing duration of the calibration phase, an increase in area filling can be observed until saturation becomes visible after approx. 5 s. The continuation of the maximum pressure allows the viscoelastic material to be strained further, with the disadvantage of increasing the total forming time.

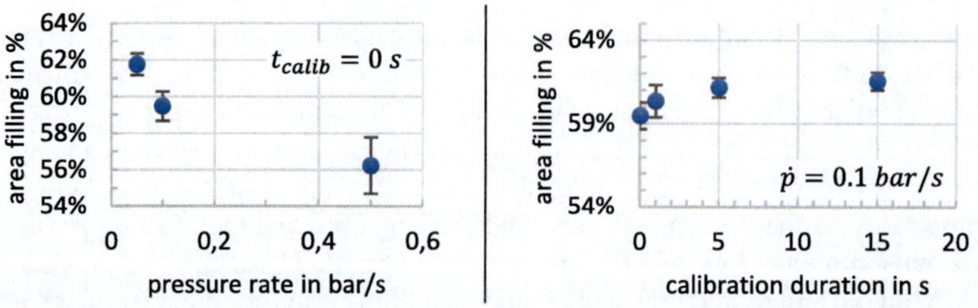

Fig. 6. Area filling in MD for different pressure rates with no calibration duration (left) and at different calibration durations with a pressure rate of 0.1 bar per seconds (right). All data for virgin-fibre paper at 350 g/m^2, 15% humidity and conical forming geometry; n = 5

In terms of high productivity, a choice of a pressure rate of 0.5 bar per second together with a calibration duration of 5 s has turned out to be a good compromise, achieving an area filling of 59.2% with a total process duration of 13 s.

3.3 Influence of Orientation

Industrial grade paper is an anisotropic material with a higher strength but lower extensibility in MD. This material behaviour influences the forming of the paper, resulting in directional dependent area filling when using a rotationally symmetric die. The effect on the profile sections in MD and CD using the conical geometry is shown in Fig. 7 for the virgin fibre paperboard with a grammage of 350 g/m^2 at 15% humidity.

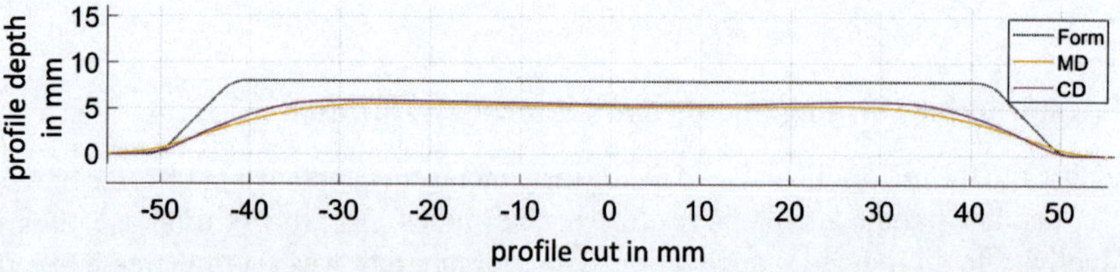

Fig. 7. Sections in MD and CD for virgin fibre paper at 15% moisture for the conical geometry

The area filling in MD is 62.0%, while it reaches 66.4% in CD. The difference in area filling is mainly influenced by the bottom height and the extension of the bottom radius.

While the effect of anisotropy is most evident in rotationally symmetric geometries, the material orientation can be used as a process parameter when non-rotationally symmetric geometries are used. For this purpose, a rectangular die with one longer and one shorter side is considered. The resulting sections in both directions are shown in Fig. 8.

Aligning MD parallel to the shorter side of the forming geometry and having CD parallel to the longer side results in maximum area filling along the longer side, but a minimum area filling along the shorter side. This minimum is due to the fact that the shorter side would require a higher extensibility to achieve the same forming depth over the shorter side length. Therefore, orienting the paper in MD with lower possible strain significantly reduces the area filling achieved.

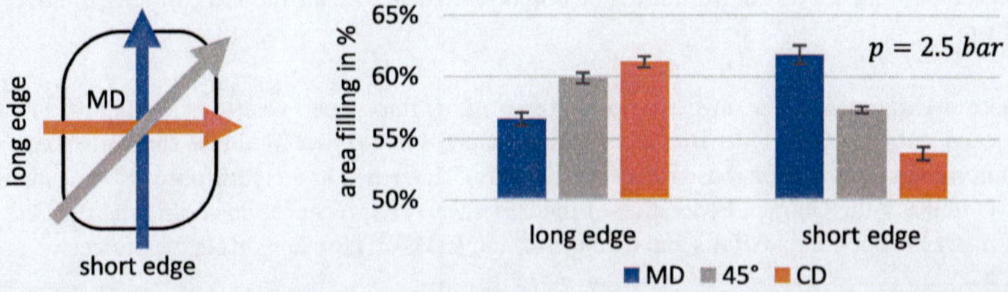

Fig. 8. Rectangular forming geometry and material orientation (left); area filling in MD and CD for different material orientations for virgin fibre paper at a moisture level of 15%; n = 5

By orienting MD along the longer side, a more even area filling in both directions can be achieved. The higher extensibility in CD compensates for the higher extensibility required along the shorter side. When oriented at an angle of 45°, the area filling of the longer and shorter sides lies between the results for the above-mentioned orientations. In order to achieve the most homogeneous area filling, an orientation between MD and 45° would be required for this particular geometry.

It should be noted that the orientation chosen influences the maximum forming pressure achievable, which in turn influences the form filling. Whereas the CD and 45° orientation could only withstand a forming pressure of 2.5 bar, the MD orientation could withstand 3.5 bar. Using the higher forming pressure in MD, an area filling in MD of about 63% could be achieved and approximately 68% in CD. With the aim of maximising the total volume of the formed product, as required for the packaging of goods, it is preferable to orient MD along the longer edge.

4 Conclusion

It has been shown that the formability of paperboard in active media-based forming is influenced by a number of parameters. The choice of the paper type and an increased moisture content of 15% have the greatest impact when a product with maximum area filling is desired. The influence of the blank holder force and forming pressure is described by the presentation of the process window. The limits of the process window are indicated by the appearance of wrinkles if the blank holder force is too low and cracks if the forming pressure is too high. By selecting the correct blank holder force and forming pressure, it is possible to maximise the area filling. However, due to elastic elongation, 100% area filling cannot be achieved. The latter process variable was found to be of greater importance for the area filling than the blank holder force and was therefore further modified. The pressure rate of the linearly increasing forming pressure and a subsequent calibration by constant pressure influence the area filling. Overall, longer process times increase the area filling, so a trade-off between high productivity and area filling must be made. Using a pressure rate of 0.5 bar per second with a subsequent calibration time of 5 s resulted in a reasonable compromise.

The anisotropic material behaviour of industrially produced paperboard leads to a directional dependent form filling when using rotationally symmetric geometries. When using forming geometries with different required strains in two directions, the correct choice of material orientation must be considered with the aim of maximising the overall filling.

Acknowledgement. The investigations presented in this paper were carried out within the research project IGF-21513 BG "Entwicklung einer universellen Methode zur Vorhersage der Umformbarkeit von papierbasierten Materialien im Tiefziehen und Hydroforming". The authors wish to thank the German Federation of Industrial Research Associations (AiF) and the Federal Ministry for Economic Affairs and Climate Action (BMWK) for supporting the project.

References

1. Economist Impact and the Nippon Foundation. Plastics Management Index: Evaluating Effective Management and Sustainable Use of Plastics (2021)
2. Europäische Union. Richtlinie (EU) 2019/ des Europäischen Parlaments und des Rates vom 5. Juni 2019 über die Verringerung der Auswirkungen bestimmter Kunststoffprodukte auf die Umwelt: 2019/904 (2019)
3. Gesellschaft für Verpackungsmarktforschung. Substitution von Kunststoffverpackungen durch papierbasierte Verbunde: Endbericht. Mainz (2021)
4. Hauptmann, M., Kaulfürst, S., Majschak, J.-P.: Advances on geometrical limits in the deep drawing process of paperboard. BioResources **11**(4), 10042–10056 (2016)
5. Östlund, M., Borodulina, S., Östlund, S.: Influence of paperboard structure and processing conditions on forming of complex paperboard structures. Packag. Technol. Sci. **24**(6), 331–341 (2011)
6. Huttel, D., Post, P.-P., Schabel, S., Groche, P.: The stress strain behaviour of paperboard in tensile and bulge tests. In: Proceedings of the 10th International Conference on Technology of Plasticity, pp. 811–816 (2011)
7. Blechschmidt, J., Naujock, H,-J.: Taschenbuch der Papiertechnik. Hanser, München (2021)
8. Deutsches Institut für Normung e. V. Papier und Pappe – Bestimmung des Feuchtegehaltes eines Lieferpostens – Wärmeschrankverfahren, 2018; (287: 2017)
9. Jessen, N., Mushövel, J.: Papier als nachhaltige Alternative in der Verpackungsindustrie. VDI Technik und Mensch (4), 13–14 (2020)
10. Haslach, J.H.W.: The moisture and rate-dependent mechanical properties of paper: a review. Mech. Time-Depend. Mater. **4**(3), 169–210 (2000)
11. Groche, P., Huttel, D., Post, P.-P., et al.: Experimental and numerical investigation of the hydroforming behavior of paperboard. Prod. Eng. Res. Dev. **6**(3), 229–236 (2012)
12. Linvill, E., Östlund, S.: Parametric study of hydroforming of paper materials using the explicit finite element method with a moisture-dependent and temperature-dependent constitutive model. Packag. Technol. Sci. **29**(3), 145–160 (2016)
13. Stein, P.: Individualisierte Formgebung papierbasierter Halbzeuge [Dissertation]. Darmstadt: Technische Universität Darmstadt (2019)

Effects of Modifying a Hot Stamping Process of 22MnB5 Thick Sheets in Terms of Fatigue Strength

T. Fünfkirchler[1(✉)], M. Jensen[2], S. Hübner[1], R. Masendorf[2], A. Esderts[2], and B. -A. Behrens[1]

[1] Institute of Forming Technology and Machines, Leibniz University Hannover, An der Universität 2, 30823 Garbsen, Germany
fuenfkirchler@ifum.uni-hannover.de

[2] Institute for Plant Engineering and Fatigue Analysis, TU Clausthal, Leibnizstraße 32, 38678 Clausthal-Zellerfeld, Germany

Abstract. Hot-stamped ultra-high-strength steels, as established in automotive body-in-white, have already been considered for add-on parts of commercial vehicle chassis frames such as rear and front underride guards. Replacing thinner hot-formed steels for the fine-grained structural steels with plate thicknesses of up to 9 mm usually used in this area offers enormous potential for weight-savings. In order to fully exploit these advantages, the substitution must be extended to other components, such as load-bearing structures. For this purpose, the material has to meet high fatigue strength requirements. Therefore, a modified hot forming process is presented which allows the adjustment of the advantageous fatigue properties of the manganese-boron steel 22MnB5. In this paper, different heat treatments are investigated to produce favorable reference manufacturing parameters in terms of fatigue strength. The material samples produced by different heat treatments are tested for their fatigue strength. In addition, the influence of sandblasting on the fatigue strength is investigated.

Keywords: Hot stamping · 22MnB5 · Fatigue strength

1 Introduction

Today, hot stamping is an established process for the industrial production of ultra-high-strength, near-net-shape components. With the aid of suitable process routes, the forming and heat treatment of a sheet can be combined in one process step [1]. The advantages of this process are high forming accuracy and the absence of springback effects. Furthermore, this forming process can be used to produce ultrahigh-strength parts with tensile strengths of up to 1500 MPa, which are used in particular for crash-relevant applications in the body structure. At the same time, however, the elongation at break of the material decreases from over 20% in the as-delivered condition to about 6% in the as-hardened condition [2].

T. Bauernhansl et al. (Eds.): WGP 2023, LNPE, pp. 599–608, 2024.
https://doi.org/10.1007/978-3-031-47394-4_58

Heavy-duty vehicle (HDV) chassis frames are another potential area of application for hot-stamped components. In this specific application, it is necessary to use materials and structures that can withstand several times the forces and payloads of passenger cars. In addition, the structures must achieve a significantly longer service life, measured in terms of mileage. The hot stamping material 22MnB5 has the potential to deliver these properties [3, 4]. Compared with normal passenger car structures, however, higher sheet thicknesses must be used here to meet the requirements. The fatigue behavior of the material is also important for HDV [5]. In recent years, several studies on the fatigue strength of 22MnB5 have been carried out in this respect. The majority of the investigations consider the influence of defects introduced by punching and trimming of blanks [6, 7] and Al-Si coating [8]. A general result is that 22MnB5 is very sensitive to these defects [9]. As mentioned in Lara et al. [7], this sensitivity of 22MnB5 is significantly higher than for cold-formable steels. In addition, the influence of the microstructure on the fatigue strength was investigated [10, 11]. Picas et al. found that the sensitivity to pre-existing defects decreases with softer microstructure, as does the fatigue strength [10]. For hot forming of thick plates, studies about an additional tempering step with a positive effect on the fatigue strength have been carried out by Sunderkötter et al. [12]. The special case of HDV structures offers further research potential due to the higher sheet thicknesses used, which has not yet been sufficiently investigated. In particular, graded material properties, i.e. properties that are distributed differently over the cross-section, could be generated here by means of an adapted hot stamping process route. In the optimum case, these have a positive effect on the fatigue strength of the material and thus improve the service life of the components produced in this way. The objective of this study is to adapt the process route, including heat treatment and post-treatment, in such a way that properties with a positive influence on the fatigue strength are achieved during the process.

2 Experimental Procedure for Hot Stamping Tests

2.1 Test Material

In this study the heat treatable steel 22MnB5 in sheet thickness s_0 of 6.0 mm was used. The material was provided without coating. The chemical compositions of the material have been analyzed by spark spectroscopy and the results are presented in Table 1. For the tests, sheets measuring 295 × 155 mm were cut out of a 22MnB5 sheet using a waterjet cutting machine. These were then subjected to heat treatment and hardened in a plate tool.

Table 1. Chemical composition (mass contents in %) of investigated 22MnB5 sheet

Material	C	Si	Mn	B	Al	Ti	Cr	Mo
22MnB5 ($s_0 = 6.0$ mm)	0.22	0.21	1.16	0.004	0.036	0.04	0.34	0.071

2.2 Experimental Setup for Hot Stamping

The process route for hot stamping of the components investigated in this work is shown schematically in Fig. 1. The conventional process route, as usually used in industry, is highlighted in grey. Here, the precut sheets are first heated in a furnace to the desired austenitizing temperature. The sheet is then transferred to a plate tool in a forming press, where the actual hot stamping process takes place. After cooling, the component is removed. In this work, the process route was adapted respectively extended at various points in order to improve the properties of the manufactured component in terms of fatigue strength. These adaptions are highlighted in green in Fig. 1. For example, resistance heating was investigated as an alternative heating method and a post-treatment in form of sandblasting was validated. Furthermore, different cooling routes were investigated as part of the experimental matrix.

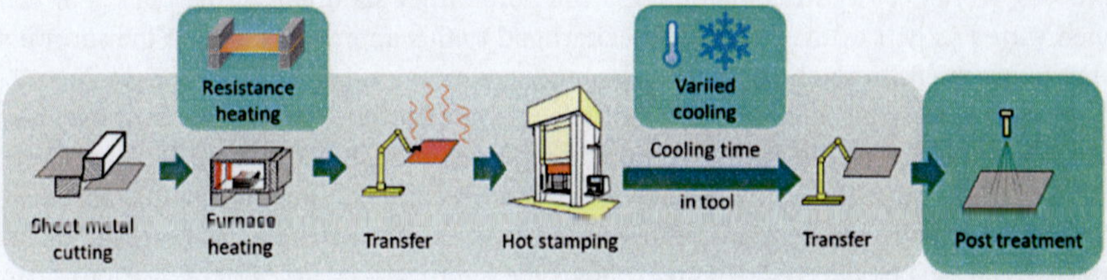

Fig. 1. Schematic process route for hot stamping (grey: conventional route, green: adapted process route)

For the first series of tests, the sheets were heated in the two ways described, and then compared with each other in terms of the properties obtained. The target temperature in both cases was 950 °C, which corresponds to the austenitizing temperature of the material required for hot stamping. Figure 2 shows the two heating systems from the experimental field at IFUM (chamber furnace on the left and resistance heating system in the center). Both heating systems allow reproducible heating of the sheets by means of intelligent temperature control. After heating, the sheets were then transferred into the plate tool of the forming press for hot stamping at IFUM (Fig. 2, right). In order to successfully hot-stamp sheet metal components, a martensitic microstructure must be established. This is done by rapid cooling above the critical cooling rate of 27 K/s. In order to generate these cooling rates, a special plate tool equipped with cooling channels below the surface was installed. The tool thus enables rapid cooling in combination with a high surface pressure, which is also required for successful hot stamping. A surface pressure of 30 MPa was used for all the tests described here. This value was determined in preliminary tests and was found to be the most favorable in terms of the resulting material hardness and strength. This value was determined in preliminary tests and was found to be the most favorable in terms of the resulting material hardness and strength. The cooling time of the resistance-heated specimens was varied as part of the test matrix. The precise press control ensures reproducible tests under variation of the process parameters. Due to the precise press control, reproducible tests under variation of the process parameters can be ensured.

Fig. 2. Chamber furnace (left), Resistance heating system (center) and plate tool for hot stamping process (right) at IFUM

To investigate the impact of the various cooling times, a sheath thermocouple was put into the sheet for the resistance-heated specimens to measure the temperature of the sheet during the testing. This allows the temperature to be recorded over the entire process, from heating to hot stamping. The time of hot stamping in the plate tool was then varied as part of the tests. The background to this approach was that the short hot stamping time followed by cooling in air produces a component with graded microstructure properties, which exhibits different strength and hardness properties over the cross section of the sheet. Ideally, these graded properties should lead to a positive influence on the fatigue strength. It is expected that a hard core in combination with a soft edge layer will provide the best properties in terms of fatigue strength, as cracking at the surface can thus be delayed. Figure 3 shows the temperature curves of the various tests. All tests shown here were performed with samples heated by resistance heating. The figure shows the different stages of the process. After a heating period of 12 s to reach the target austinitization temperature of 950 °C, this is held for a further 15 s to ensure a homogeneous temperature distribution in the sheet. This is followed by the transfer to the press, during which the temperature decreases minimally. The next step is the hot stamping process, in which the die moves together for a defined time in order to apply pressure with simultaneous cooling. After this time, the die opens again and the sheet is further cooling down in the air. The reference is a component that was hot stamped for 30 s and thus formed a fully martensitic microstructure. As a variation, the hot stamping time was reduced to 5 s respectively 3 s. The temperature curves show that the samples produced in this way only cool down to approx. 350 °C and 400 °C during the hot stamping process, while the temperature of the reference sample drops to below 200 °C. In this way, an attempt was made to avoid complete martensite formation and to produce a heterogeneous microstructure. Ideally, this should contain more martensite in the edge region than in the core of the sheet. The components produced in this way were then examined for their hardness and their fatigue strength. The results of these tests are presented in Sects. 3 and 4.

In order to analyze the influence of post-treatment on the fatigue behavior, selected specimens were subjected to post-treatment in the form of sandblasting after hot stamping. The blasting material used was high-grade corundum with a grain size of F30 (500–710 μm). Sandblasting was carried out manually on a suction blasting cabin at the IFUM.

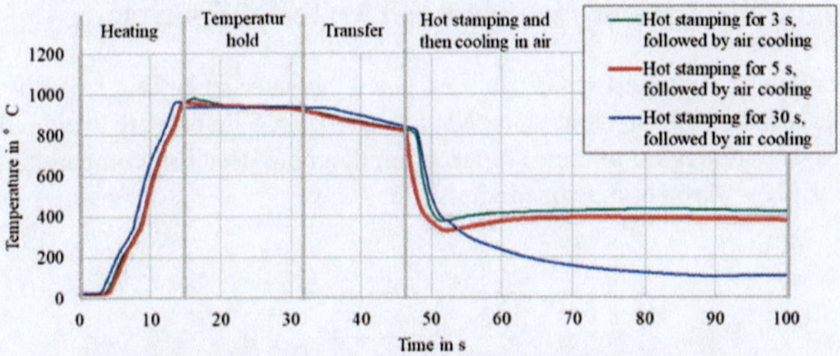

Fig. 3. Different cooling routes of resistance heated specimen

3 Hardness Properties of Hot Stamped Material Specimens with Different Cooling Routes

In order to check whether graded properties can be set by modifying the cooling route, hardness measurements were first carried out on selected specimens across the sheet cross-section. For this purpose, the 6 mm thick reference component was first examined and then compared with a component of the same thickness, in which the cooling in the press was interrupted after 5 s and the remaining cooling followed in air. Figure 4 shows a comparison of the hardness profiles of the two components.

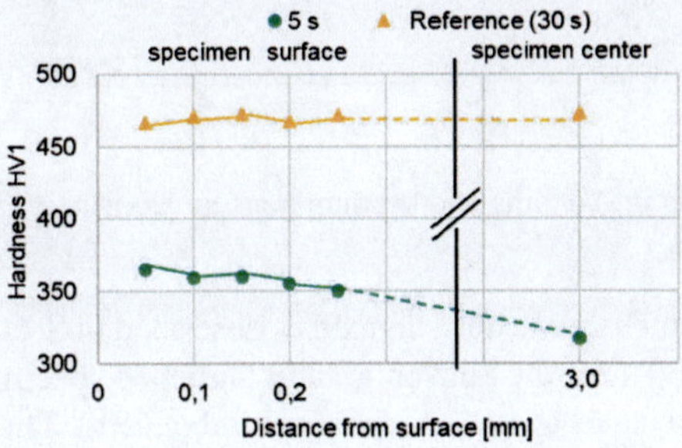

Fig. 4. Hardness characteristics over the cross-section with different cooling routes

While the hardness of the reference component shows an almost constant hardness profile at approx. 470 HV over the sheet cross-section, the hardness of the component whose cooling in the press was interrupted after 5 s changes over the sheet cross-section. It can be seen that the hardness, which is about 360 HV at the edge, decreases significantly toward the center of the specimen. Here it reaches about 310 HV and is thus about 15% lower than at the edge. This shows that graded properties, which change over the cross-section, can basically be generated with the modified cooling route. However, it should also be mentioned that the overall hardness of the part produced in this way is also lower than that of the reference part.

4 Fatigue Strength of Hot Stamped Material Specimens

For the fatigue strength test, specimens are taken by water jet cutting from the conventionally or resistance-heated hot-stamped plates, see Fig. 5. In order to avoid influencing the cut edges with regard to their fatigue strength and to establish comparability of the specimens, they were not machined afterwards.

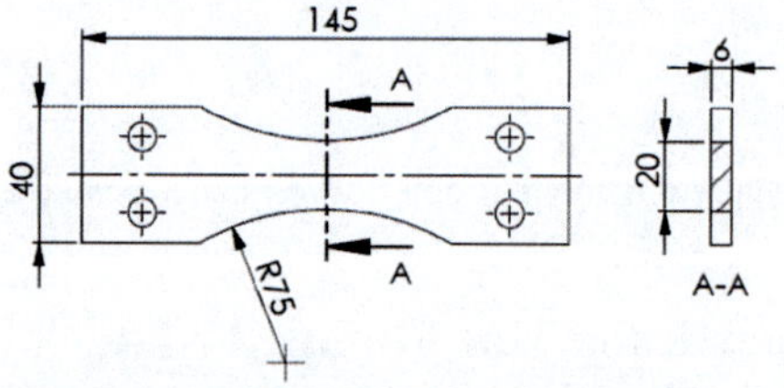

Fig. 5. Specimen shape for the fatigue tests

For the fatigue tests, a 4-point bending device is used, which generates a constant bending moment over the test cross section (Fig. 6).

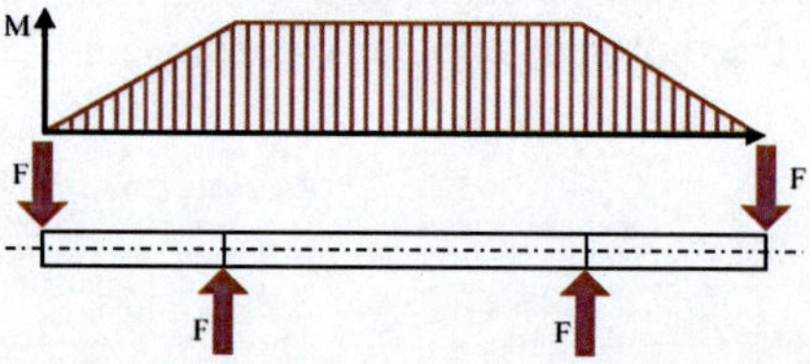

Fig. 6. 4-Point-bending and bending moment along specimen length

The material specimens were hot-stamped as described in Sect. 2.2 before being cut. The heating time of 330 s for the conventionally hardened specimens was significantly longer than for the resistance-heated specimens (total of 27 s). The applied surface pressure of 30 MPa was the same in both cases. The manufacturing parameters investigated are shown in Table 2. In the fatigue strength test, the bending stress is applied to the material specimen with a stress-ratio of $R = -1$. The tests are carried out in the high-cycle-fatigue range with the pearl string method according to DIN 50100:2016 with the crack as damage criterion.

4.1 Comparison Between the Heating Methods

The investigated material specimens V01 and V02 are completely through-hardened material specimens. The resulting S-N curves from the 4-point bending show a steep drop in strength, as is typical for components with cracks, Fig. 7 (left). The resistance-heated material specimens show higher fatigue strength than those produced in the chamber furnace. In both heat treatment processes, numerous incipient cracks coalesce

Table 2. Investigated manufacturing parameters for the hot stamping of 22MnB5

Variant	Heating	Temperature	Heating time	Cooling route	Post-treatment
V01	Chamber furnace	950 °C	330 s	conventional	–
V02	Resistance heating	950 °C	27 s	conventional	–
V03	Resistance heating	950 °C	27 s	conventional	Corundum F30
V04	Resistance heating	950 °C	27 s	Termination after 5 s	–
V05	Resistance heating	950 °C	27 s	Termination after 3 s	–

due to the oscillating stress, Fig. 7. (right). The number of cracks appears to be lower in resistance-heated sheets. This may be due to the lower scale formation during the heating process.

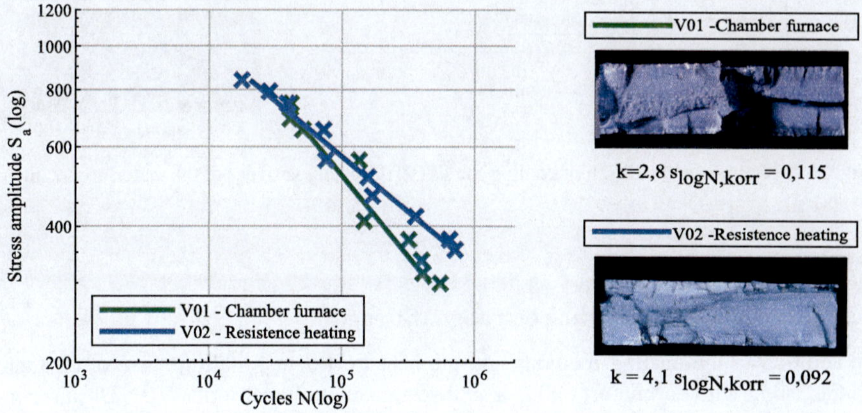

Fig. 7. S-N curves of the different heating methods (left) fracture surfaces of the material specimens (right)

4.2 Comparison of the Cooling Routes

Different cooling routes were also investigated. The quenching process was briefly interrupted in test series V04 and V05, see Table 2. The intention behind this was to create a non-full-martensitic microstructure by interrupting the hot stamping process, which gives the plate graded properties across the cross-section. These can influence the fatigue strength. Test series V02 was quenched completely without interruption. The resulting

S-N curves are shown in Fig. 8 (left). All the S-N curves shown have a steep slope k, which indicates a crack propagation S-N curve. Both S-N curves with an interruption of the cooling process show a lower fatigue strength. Looking at the fracture surfaces, Fig. 8 (right), it is also evident here that small surface cracks coalesce. However, the proportion of coalescing cracks is lowest in the fully hardened material specimen. It can be seen that the brief interruption of the cooling process leads to a lower fatigue strength and higher crack formation on the surface.

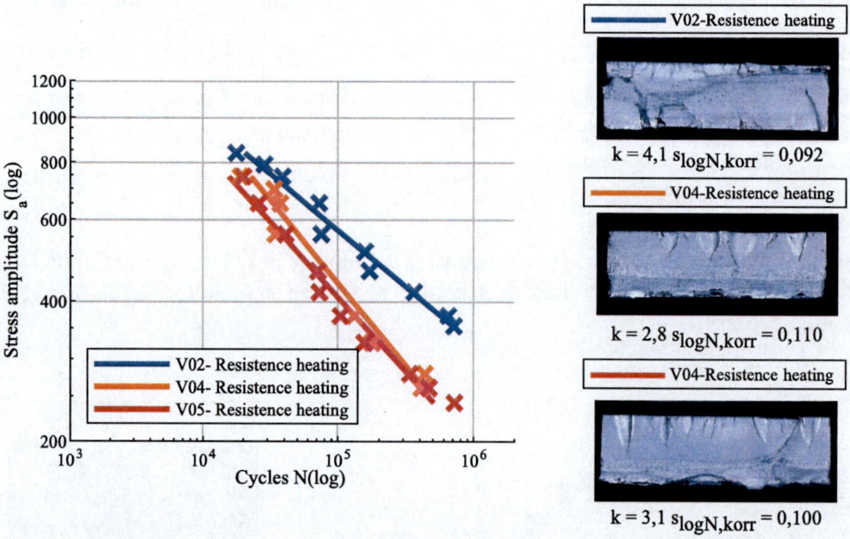

Fig. 8. S-N curves of the different cooling routes (left) fracture surfaces of the material specimens (right)

4.3 Influence of Sandblasting on Fatigue Strength

In addition to the heating methods and the heat treatments, the influence of cleaning sandblasting with corundum F30 is also investigated with test series V03. During conventional heating in a chamber furnace and during resistance heating of hot-stamped material samples, a scale layer forms on the surface. If this is removed by sandblasting, it has an influence on the fatigue strength, see Fig. 9 (left). It can be seen that after blasting with Corundum F30, there is an increased fatigue strength and a flatter S-N curve slope k, which indicates that fewer incipient cracks are effective in the material sample before testing. Looking at the fracture surfaces it can be seen that significantly fewer cracks coalesce at the surface of the material specimen in cleaning-blasted material specimens than is the case in the non-blasted specimen, see Fig. 9 (right). This can probably be attributed to the development of residual stresses on the surface as a result of post-treatment, which in this case have a positive effect on the fatigue strength. Further investigation of crack formation and crack propagation is planned in AIF project 22081N.

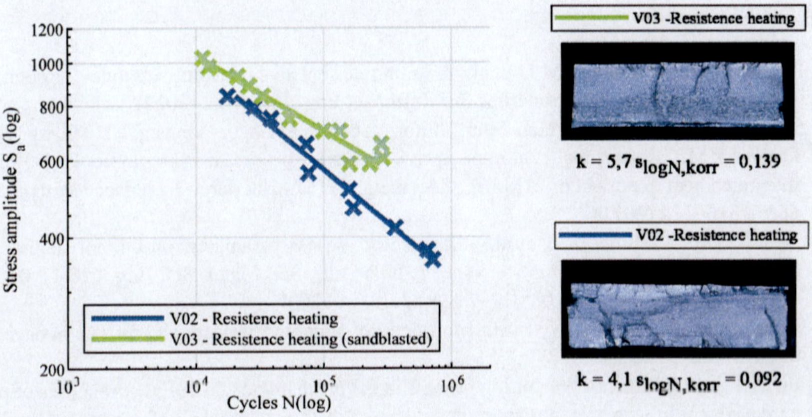

Fig. 9. S-N curves of the blasted and non-blasted material specimens (left), fracture surfaces of the material specimens (right).

5 Summary and Outlook

In the study presented here, sheet metal components made of the material 22MnB5 with a sheet thickness of 6.0 mm were hot-stamped in various ways and partially posttreated. The aim was to investigate the influence of the variation of various parameters in the process chain with regard to the fatigue strength of the components. The study showed that the heating method has a significant influence on the fatigue strength. Conductively heated material samples show a higher fatigue strength at faster process times than those produced in the chamber furnace. Furthermore, in both cases the post-treatment of the hot-stamped components by cleaning sandblasting has a significantly positive effect on the fatigue strength. More in-depth investigations are planned for the future in order to further classify the influence. In this work, it was further demonstrated that by changing the cooling route through a short cooling phase in the tool, followed by cooling in air, graded properties with respect to component hardness across the sheet cross-section can be generated. Overall, however, these graded properties have so far still led to a reduction in fatigue strength. In this regard, further tests with an intermediate cooling step are planned in order to obtain graded qualities that are suitable for fatigue strength. Furthermore, microstructural analyses are planned in order to determine the composition of the microstructure via the sheet thickness. With regard to cracking, a softer surface in combination with a hard core seems to be a desirable combination.

Acknowledgements. The research project "Formhärteverfahren zur Beeinflussung der Schwingfestigkeit von Blechen dickerer Wandstärke" from the Research Association for Steel Application (Forschungsvereinigung Stahlanwendung e. V., FOSTA) was supported under grant number 22081N by the Federal Ministry of Economic Affairs and Climate Action (BMWK) through the German Federation of Industrial Research Associations (AiF) as part of the program for promoting industrial cooperative research (IGF) on the basis of a decision by the German Bundestag.

References

1. Bardelcik, A., Worswick, M.J., et al.: A strain rate sensitive constitutive model for quenched boron steel with tailored Properties. Int. J. Impact Eng. **50**, 49–62 (2012)
2. Doege, E., Behrens B.-A.: Handbuch Umformtechnik. Springer- Vieweg, Heidelberg (2018)
3. Diekamp, M., Hübner, et al.: Optimised press-hardening process using spray cooling - Process integrated heat treatment of 22MnB5 sheet metal. HTM - Haerterei-Technische Mitteilungen **66**(6), 316–322 (2011)
4. Behrens, B.-A., Hübner, S., et al.: Influence of process parameters on the hot stamping of carbon-martensitic chromium steel sheets. IOP Conf. Ser. Mater. Sci. Eng. **418**(1), 012007 (2018). https://doi.org/10.1088/1757-899X/418/1/012007
5. Behrens, B.-A., Esderts, A., et al.: Modified hot stamping process and fatigue behavior of thick 22MnB5 sheets (2022)
6. Picas, I., et al.: Effect of the cutting process in the mechanical and fatigue properties of press hardened 22MnB5 steel. In: Proceedings of the 3rd International Conference on Hot Sheet Metal Forming of High Performance Steel, pp. 85–92 (2011)
7. Lara, A., Picas, I., Casellas, D.: Effect of the cutting process on the fatigue performance of press hardened and high strength dual phase steels. J. Mater. Process. Technol. **213**, 1908–1919 (2013)
8. Albracht, L., Hübner, S., Holländer, U., Behrens, B.-A.: Coating materials under oxygen-free silane atmosphere for hot stamping. In: Behrens, B.-A., Brosius, A., Drossel, W.-G., Hintze, W., Ihlenfeldt, S., Nyhuis, P. (eds.) WGP 2021. LNPE, pp. 3–10. Springer, Cham (2021). https://doi.org/10.1007/978-3-030-78424-9_1
9. Cangwang, Y., et al.: Effect of hierarchical martensitic microstructure on fatigue crack growth behavior of ultra-high strength hot stamping steel. Mater. Character. **174**, 111041 (2021)
10. Picas, I., et al.: Fatigue performance of press hardened steels with tailored material properties: a compromise between strength and damage tolerant microstructures. In: 4th International Conference on Hot Sheet Metal Forming of High Performance Steel, pp. 251–258 (2013)
11. Parareda, S., et al.: Fatigue resistance of press hardened 22MnB5 steels. Int. J. Fatigue **130**, 105262 (2020)
12. Sunderkötter, C., et al.: Press hardening of truck chassis components. In: 5th International Conference on Steel in Cars and Trucks (2017)

Numerical Modelling of Transformation-Induced Plasticity in Hot Forging Simulations

N. Mohnfeld(✉), J. Uhe, H. Wester, C. Kock, and B.-A. Behrens

Institute of Forming Technology and Machines, Leibniz University Hannover, 30823 Garbsen, Germany
mohnfeld@ifum.uni-hannover.de

Abstract. Hot forming processes involve several interactive material phenomena. This includes transformation-induced plasticity (TRIP), which significantly affects the resulting deformations and residual stresses of the workpieces. The magnitude and orientation of TRIP strains are dependent on stress states that occur during cooling, making it essential to consider TRIP effects in process design. Material parameters required for modelling this were previously recorded and validated through experiments on AISI 52100. In this study, a material model based on this data is used to investigate a forged component by predicting the distortion occurring over the course of die forging, deflashing, and cooling. By comparing simulations with and without TRIP effects, the study demonstrates the potential for resource-efficient design of forging process routes, minimising scrap due to distortion and maximising material savings, up to the limit of a distortion-free component.

Keywords: Hot forging · FE-simulation · phase transformation · transformation-induced plasticity · AISI 52100

1 Introduction

Forging processes are highly advantageous for the production of high-loaded components with complex geometry [1]. By using elevated forming temperatures of about 1,250 °C, the yield stress decreases, facilitating improved material flow and reduced forming forces [2], thereby enabling the production of complex parts [3]. Additionally, forging results in a refined microstructure [4], enhancing the strength and dynamic load capacity of the final product [5]. During hot forging various material phenomena arise which significantly influence the material behaviour during and after forming as well as the performance of the final component [6]. The occurring effects are significantly influenced by the locally present plastic deformations as well as the locally varying cooling rates within the material. Apart from inhomogeneous elastic and plastic deformations, rapid cooling from 1,250 °C can induce crucial thermal shrinkage strains. Furthermore, lattice structure changes resulting from microstructure transformations lead

T. Bauernhansl et al. (Eds.): WGP 2023, LNPE, pp. 609–618, 2024.
https://doi.org/10.1007/978-3-031-47394-4_59

to transformation-related deformations. As shown in Fig. 1 the occurring effects are driven by thermal, mechanical and metallurgical changes within the material. However, the specific effects interact with each other and can lead to varying local process conditions and transient stress states resulting in intricate distortion of forged and quenched components.

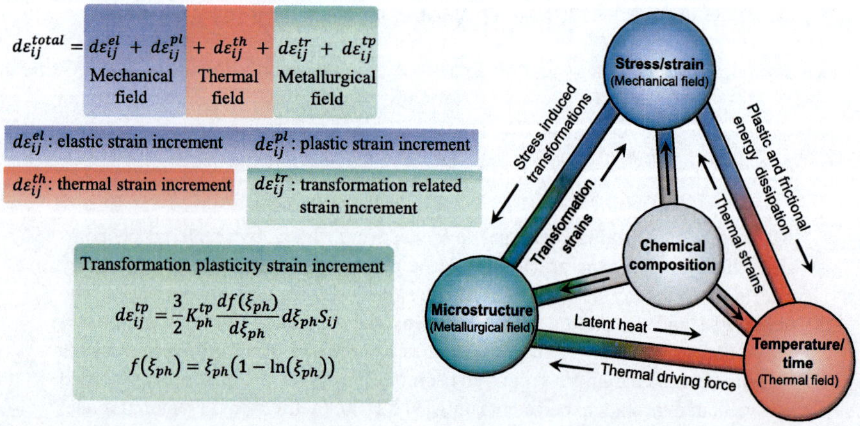

Fig. 1. Interrelated thermo-mechanical-metallurgical material phenomena in the thermomechanical processing of steel and basic principles of the additive strain reduction method as well as the equation for the transformation plasticity strain increment according to [7].

Consequently, predicting final distortions in hot-forged steel components presents a challenge, often leading to increased scrap production. In this regard, process simulation utilising the finite element (FE) method provides a high potential for efficient process design, thereby saving time and costly trial-and-error iterations [8]. However, for a realistic prediction of resulting distortions the occurring and interacting thermo-mechanical-metallurgical material phenomena need to be taken into account during numerical simulation [9]. In this context, the additive strain decomposition method, which is based on separately consideration of the five elastic, plastic, thermal, transformation related, and transformation plastic strains was applied by Behrens et al. [10]. This methodology allows for the individual strain increments to be determined, taking into account the local thermal, mechanical and metallurgical conditions present. Subsequently, the strain increments are summed up to the total strain increment as depicted in Fig. 1.

In this context, transformation-related strains represent volumetric strains which occur in polymorphic materials during phase transformations, such as the conversion of austenite into microstructural constituents like ferrite, pearlite, bainite, or martensite. Anisotropic transformation stresses arise when mechanical stresses coincide with phase transformations [11]. This phenomenon is described by transformation plasticity strains. Material parameters for calculating transformation-induced plasticity (TRIP) are not general available yet [12]. Previous works have addressed the consideration of the TRIP effect. However, they typically involve a mixing effect without phase-specific consideration [13]. In the work of Behrens et al. the diffusion-less transformation to martensite was investigated for AISI 4137 in a forging process with subsequent cooling in water. A good agreement between the simulated microstructure as well as the geometry in comparison to the experimentally produced component could be achieved. Distortions could not be detected in the simulation or in the real process, since the flash had a distortion-stabilising effect. A comparison of the residual stresses showed that the calculated values were always above the measured values. The reason given by the authors is that the flow properties and the modeling of the TRIP effect were not phase-specific [13]. Consequently, in numerical designs, phase-specific TRIP coefficients must be determined. This was accomplished in a previous study, which involved determining the transformation behaviour for AISI 52100 and subsequently recording the K_{ph}^{tp} factors as a function of phases, loading direction, and temperature. These factors were then validated through comparison with experimental findings [10]. Within the work the calculated TRIP coefficients were validated by comparison with experimental test data. From the state of the art and precious research work of the autors, it can be deduced that phase-specific material modelling with regard to the flow curves, transformation diagrams and the TRIP effect is necessary in order to numerically predict the distortions and residual stresses of forged components throughout the entire process chain. In particular, deflashing must be taken into account, as the flash has a distortion-stabilising effect. Therefore, phase-specific modelling is to be used to check whether the TRIP effect numerically leads to distortions in the component after deflashing and slow cooling. The slow cooling serves to validate the diffusion-controlled phase transformation, e.g. from austenite to bainite.

Within the presented study a phase-specific material model, which takes the thermo-mechanical and metallurgical effects into account, was developed in order to demonstrate the influence of the TRIP effect on the resulting distortion in a hot forging process chain. For this purpose, phase-specific flow curves for AISI 52100 were determined by experimental tests and implemented into the FE software Simufact Forming. To determine the influence of the TRIP effect, a process chain was numerically depicted. The considered process chain is illustrated in Fig. 2 and includes heating, forming, deflashing and cooling. In addition, the transfer times are also taken into account in order to ensure a consistent thermo-mechanical representation of a real process chain. Finally, the part, which is susceptible to distortion, is cooled to room temperature. The considered process chain was simulated with and without taking into account the TRIP effect, to determine its influence on the prediction of the resulting distortion.

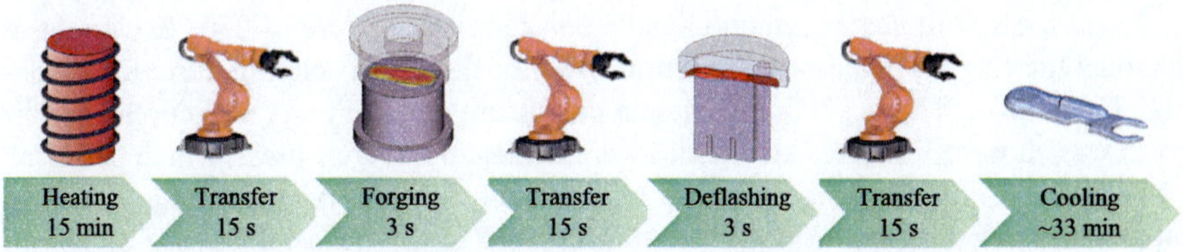

Fig. 2. Schematic representation of the numerically investigated process chain.

2 Material and Methods

The determination of phase-specific flow curves requires a high number of experimental tests. Therefore, phase-specific flow curves were first calculated with the software J-MatPro for the material AISI 52100. Subsequently, flow curves for the austenitic and pearlitic phases were determined by means of experimental tests. These were used to validate the calculated flow curves. For the experimental determination of the phase-specific flow curves, notched round tensile specimens with a diameter of 8 mm were used and tested on the Gleeble 3800-GTC forming simulator. In order to be able to record phase-specific flow curves, the transformation behaviour has to be characterised. For this purpose, the time-temperature-transformation (TTT) diagram of AISI 52100 was used, which was previously determined [10]. To analyse the flow behaviour of the pure austenitic phase the specimens were austenitised at 1,000 °C and then rapidly cooled with 30 K/s to the test temperature and immediately tested before a partial transformation of the microstructure occurred. The pearlitic specimens were heat treated according to the TTT diagram and subsequently heated to the test temperature. Figure 3 shows the TTT diagram with the time-temperature profiles for the respective phase-specific tensile tests. Each test temperature was repeated three times at a strain rate of $\dot{\varphi} = 1s^{-1}$. Slower strain rates lead to partial phase transformations occurring, especially in the case of the supercooled austenite, and thus no pure austenite flow curves can be recorded. Higher test temperatures for austenite, such as those encountered in forging, were not investigated, as these are purely austenitic and are sufficiently present for AISI 52100. The test temperatures for pearlites were chosen to determine pure pearlitic flow curves, since a further increase in temperature would lead to partial austenitisation. Flow curves for fully martensitic specimens could not be recorded. When heated in the oven and quenched in water, the specimens are very brittle, so that they crack spontaneously at the start of the test. A partial martensite formation in the tensile specimen center with the smallest cross-section, causes the specimen to crack at the clamping on the screw thread of the specimen. The experimentally recorded flow stresses were finally compared with the phase-specific flow stresses calculated by J-MatPro.

The numerical process chain was modelled within the FE software Simufact Forming 16.0 SP1. A semi-finished product with a diameter of 20 mm and a length of 125 mm was discretised with hexahedral elements with an element edge length of 0.5 mm and calculated as a full 3D model, as there are no symmetries. After the heating simulation to a temperature of 1,200 °C, the transfer step from the induction unit to the press is represented via a cooling simulation. This results in an inhomogeneous temperature distribution that better corresponds to the real process conditions.

Rigid tools with heat conduction and a constant temperature of 250 °C were used to model the forging process. The friction was modelled by a combined friction model from Coulumb and Tresca. For the friction coefficient $\mu = 0.3$ and the friction factor m = 0.5 was chosen. The press kinematics were defined by a screw press, which performs a stroke of 18 mm resulting in a flash thickness of 3.5 mm. After forging, a cooling simulation is again used to model the transfer step to the deflashing process. Within this process the flash is removed by a shear cutting simulation. First a cutting path of 3.0 mm is calculated and then the residual flash is separated from the workpiece using a boolean operation. The cutting path is used to calculate the stresses and the degrees of

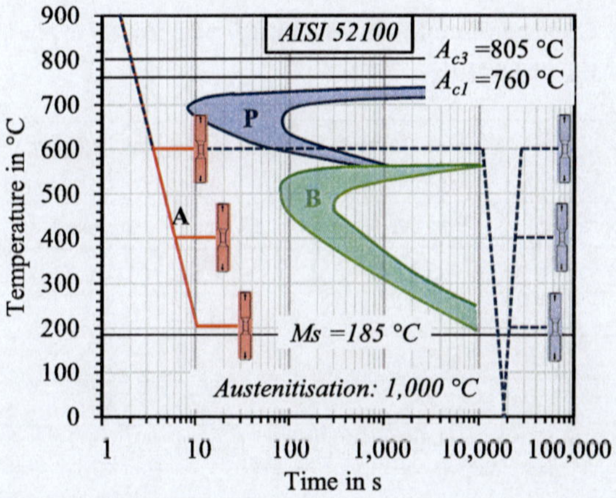

	Pearlite	Austenite
	200 °C	200 °C
1 s^{-1}	400 °C	400 °C
	600 °C	600 °C

Fig. 3. TTT-diagram with the time-temperature-route of the tensile specimens for testing the phase-specific flow curves as well as the test parameters.

deformation at the cutting edges of the workpiece, which have a significant influence on the final distortion. With the boolean subtraction, a significant reduction of the calculation time is achieved compared to conventional mesh separation algorithms such as element deletion or twinning. The final step is a transfer step from the deflashing die to the final cooling of the workpiece in air. Workpiece cooling in still air is done so that the diffusion-controlled transformations can be validated, as diffusion-less transformation has already been experimentally and numerically investigated in [13]. The required heat transfer coefficients were taken from the work presented in [14].

To consider the metallurgical field within the FE simulation, the transformation related strain increment $d\varepsilon_{ij}^{tr}$ and the transformation plasticity strain increment $d\varepsilon_{ij}^{tp}$ were calculated according to [8] via a subroutine. The considered equation for calculation of the transformation plasticity strain increment $d\varepsilon_{ij}^{tp}$ is shown in Fig. 1. For the function of transformation evolution $f(\xi_{ph})$, the approach according to Leblond et al. was used [15]. The phase-specifics parameter K_{ph}^{tp} was recorded load- and temperature-dependent as shown in [10] for AISI 52100 and validated based of the occurring lengthwise and the crosswise strain. Within the subroutine the current value of K_{ph}^{tp} is calculated under consideration of current microstructure, temperature and stress state. Table 1 lists the determined values for K_{ph}^{tp} from the experimental investigations from [10]. In the subroutine, the K_{ph}^{tp} value was linearly interpolated between the investigated temperatures. Above and below the temperature, the K_{ph}^{tp} value is constant for the respective phase and load type. In addition to the flow curves, other material parameters were also calculated by JMatPro as a function of phase and temperature, such as thermal conductivity, thermal expansion and Young's modulus. In order to take the phase transformation into account in the FE simulation, the TTT diagram determined in [10] was integrated into the material model.

Table 1. Summary of the coefficients of the transformation plasticity K_{ph}^{tp} with respect to phase type, temperature and stress type for AISI 52100 [10].

		*Transformation plasticity coefficient $K_{ph}{}^{tp}$ [10^{-5} *MPa^{-1}]*	
Phase type	Cooling rate/ Temperature	Positive stresses	Negative stresses
Martensite	10 K/s	4.32	6.96
Bainite	400 °C	3.04	3.76
	500 °C	4.56	3.60
Pearlite	600 °C	12.46	7.70
	700 °C	13.44	13.60

3 Results and Discussion

The comparison of the phase-specific flow curves for AISI 52100 from the experimental tensile tests with the calculated flow curves from JMatPro is shown in Fig. 4. For the pearlitic flow curves in Fig. 4(a) a good agreement can be observed. The experimental determined flow stress is slightly above the calculated at all temperatures investigated. However, the calculated flow curves represent the same characteristics, a work hardening at 200 °C and 400 °C as well as a constant course at 600 °C. The flow curves for austenite are also in good agreement with the calculated flow curves, Fig. 4 (b). Compared to the pearlitic phase, the flow stress at 200 °C and 400 °C is higher. For 600 °C the behaviour changes and the flow stress of the austenitic phase is much lower than for the pearlitic phase. The course of the calculated flow curve for austenite at 400 °C shows a beginning softening starting at a degree of deformation of 0.4. This effect could not be represented experimentally due to the early necking of the tensile specimen. In summary, it can be stated that the experimental and calculated flow curves show a sufficiently accurate agreement for pearlite and austenite. Due to the good agreement, the calculated flow curves from JMatPro are used. Furthermore, flow curves are calculated for the other phases such as martensite and bainite as well as for an extended temperature range from 25 °C to 1,200 °C and strain rates between 0.1 s^{-1} and 1,000 s^{-1}, which in part can only be determined experimentally with very high effort.

The thermo-mechanical-metallurgical coupled material model was used to simulate the process chain shown in Fig. 2 with and without taking into account the phase-specific TRIP effect. In Fig. 5 a top view of the deflashed workpiece at the end of the final cooling in air is shown. The shape deviation is determined between the geometry at the beginning of the cooling simulation, shown in transparent, and superimposed with the geometry at the end of the cooling. The colour gradient shows the difference between these two geometries.

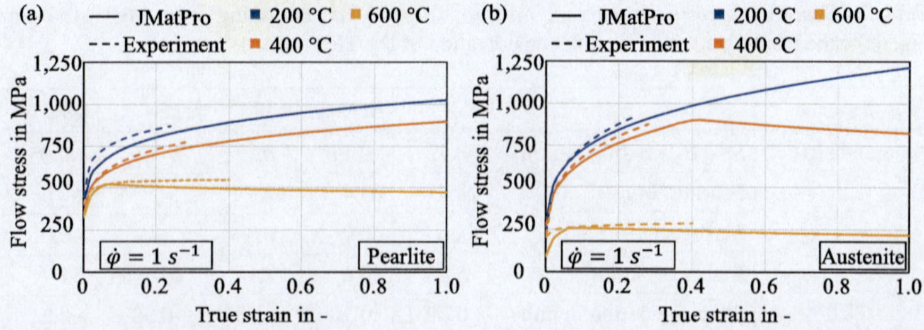

Fig. 4. Comparison of the experimentally recorded phase-specific and calculated flow curves for AISI 52100 (a) pearlite and (b) austenite.

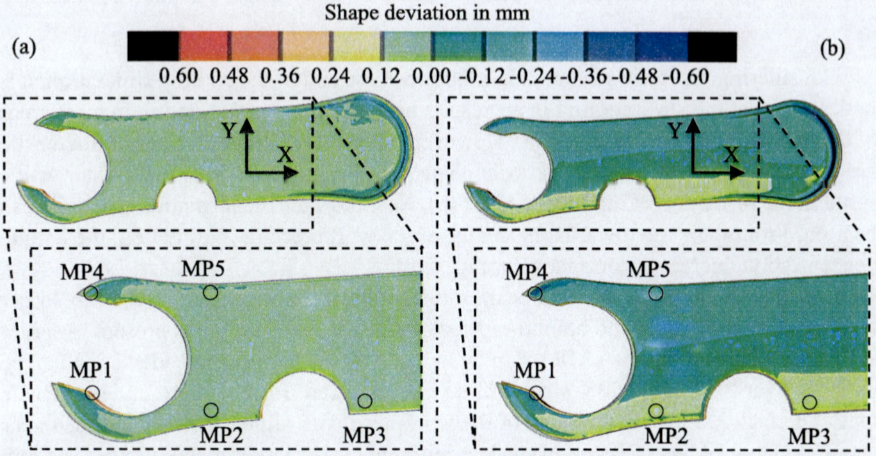

Fig. 5. Comparison of the shape deviation between start geometry and end geometry of the cooling simulation (a) without TRIP effect and (b) with TRIP effect.

In Fig. 5 (a) the shape deviation for the process chain without consideration of the TRIP effect and in Fig. 5 (b) the shape deviation taking the TRIP effect into account are presented. In order to analyse the influence of the TRIP effect in more detail, five measuring points (MP) were defined, which are compared in Table 2. The three measuring points (MP1-MP3) at the lower edge of the workpiece show 0.08 mm increased average shape deviation when the TRIP effect is taken into account. On the opposite side MP4 and MP5 show an average difference of 0.27 mm. The fact that the difference is smaller on the lower side could be due to the low thickness of the component in this area and thermal shrinkage dominates the evolution of distortion. Considering the phase fractions at the respective measuring points, it is noticeable that especially at the measuring points MP 4 a larger martensite fraction has developed due to the consideration of the TRIP effect. This also explains the larger shape deviation there.

Table 2. Shape deviation and microstructure proportions of the measuring points MP1-MP5 from Fig. 5 for the results without and with consideration of the TRIP effect.

		MP1	MP2	MP3	MP4	MP5
Without TRIP	Shape deviation in mm	0.72	0.11	0.15	−0.09	0.05
	Pearlite %-vol	23.4	63.4	92.9	29.2	68.6
	Martensite %-vol	62.6	0.0	0.0	57.2	0.0
	Bainite %-vol	0.0	36.6	7.1	0.0	31.4
With TRIP	Shape deviation in mm	0.80	0.19	0.23	−0.38	−0.21
	Pearlite %-vol	22.3	62.1	92.8	26.5	72.1
	Martensite %-vol	63.5	0.0	0.0	60.0	0.0
	Bainite %-vol	0.0	37.9	7.2	0.0	27.9

Considering the microstructure distribution at the end of the cooling simulation, it is noticeable that the slow cooling in air results in a pearlitic microstructure in most areas of the final component, see Fig. 6 (a). However, within claw larger portions of martensite are visible. Due to the low material volume in this area, higher cooling rates occur which result in the formation of martensite Fig. 6(b). Near the claw the formation of bainite can be seen. From the claw towards the centre of the workpiece in X-direction, the bainite concentration decreases continuously, Fig. 6 (c).

Considering the K_{ph}^{tp} factors of pearlite in Table 1, it is noticeable that they are larger than those of martensite and bainite especially when tensile loading is present. A larger K_{ph}^{tp} value leads to a larger TRIP strain, which is why the largest shape deviation can be explained for MP5 and MP5, since primary pearlites are formed there under a tensile load. The fact that in the right area of the workpiece with almost 100% pearlite, a very small influence of the TRIP strain can be seen is due to the larger material cross-section, which has a stabilising effect against distortion.

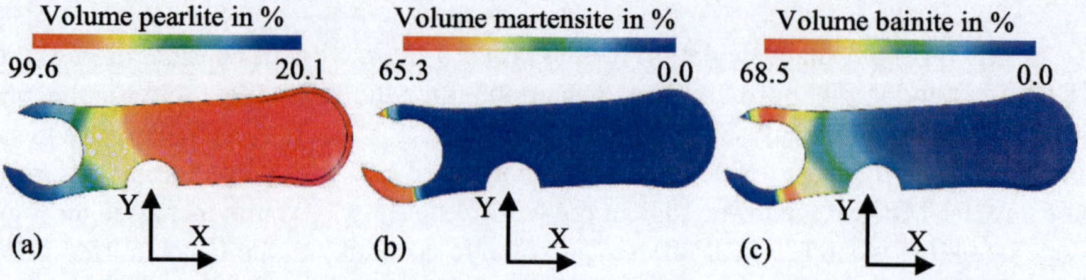

Fig. 6. Numerical calculated microstructure distribution after cooling with consideration of the TRIP effect (a) pearlite, (b) martensite and (c) bainite.

4 Conclusion and Outlook

Within the scope of this study, the influence of the consideration of the TRIP effect on the resulting distortion in the FE simulation of forged components was investigated. For the numerical simulation, phase-specific flow curves for the material AISI 52100 were calculated using JMatPro. The calculated flow curves were validated by experimental flow curves for pure austenitic and pearlitic microstructure at different temperature. TRIP effect was taken into account using temperature, microstructure and load dependent K_{ph}^{tp} coefficients. The shape deviation induced by the TRIP effect varies between 0.08 and 0.27 mm depending on the phase transformation and the type of loading. Shape deviation seems to be within a reasonable range for slow cooling respectively diffusion-controlled phase transformation. However, faster cooling can be expected to have a greater influence on distortion, as assumed in [13]. There is a significant effect, even if it is a slow cooling in air. The slow cooling leads to a predominantly pearlitic structure where the TRIP effect is greater, especially under tensile load in areas with thin material cross-sections. At higher cooling rates, e.g. in water or oil, greater shape deviations are to be expected, which are not predominantly due to the TRIP effect, but rather to the second strain increment in the metallurgical field, the transformation related strain increment. Transformation related strain increment describes the strains that occur due to the change in volume of the lattice structure of the individual phases. There, a martensitic transformation shows the greatest increase [10]. The results show, that the consideration of TRIP effect has a significant effect on predicted distortions and should therefore be taken into account for during numerical process design. A more realistic prediction of distortions of the final component allows for efficient process optimisation with regard to material usage and reduction of necessary post production steps.

In future work, the numerically designed process chain will be technologically implemented and the distortion-prone component presented here will be manufactured. Experimental data will be used to validated the developed thermo-mechanical-metallurgical coupled material model. In addition, optimisations are carried out to reduce the amount of material used while at the same time maintaining the dimensional accuracy. This can be done, for example, through targeted inhomogeneous cooling, which is first designed numerically. Furthermore, the influence of the back stress effect on the resulting distortions is to be investigated. For this purpose, the developed subroutine can be extended according to [16].

Acknowledgements. This study was funded by the German Research Foundation (DFG, Deutsche Forschungsgemeinschaft) - 212963651 (BE 1691/142–2). The authors gratefully acknowledge the German Research Foundation for their financial support of this project.

References

1. Semiatin, S.L.: Introduction to Forming and Forging Processes. In: Metals Handbook, vol. 15, Forming and Forging, Ohio (1988)
2. Magri, M.L., Diniz, A.E., Button, S.T.: Influence of surface topography on the wear of hot forging dies. Int. J. Adv. Manuf. Technol. **65**, 459–471 (2013)

3. AL-Kareem Ahmed, I.A., Mohammed, A.I., Allow, M.A.: Improvement of forging die life by failure mechanism analysis. J. Mech. Behav. Mater. **30**(1), 309–317 (2022)
4. Huang, W., Zhong, H., Lei, L., Fang, G.: Microstructure and mechanical properties of multi-pass forged and annealed 42CrMo steel. Mater. Sci. Eng., A **831**, 142191 (2022)
5. Jahazi, M., Eghbali, B.: The influence of hot forging conditions on the microstructure and mechanical properties of two microalloyed steels. J. Mater. Process. Technol. **113**(1–3), 594–598 (2001)
6. Verlinden, B., Driver, J., Samajdar, I., Doherty, R.D.: Thermo-Mechanical Processing of Metallic Materials. Elsevier, Amsterdam (2007)
7. Behrens, B.-A., Bouguecha, A., Bonk, C., Chugreev, A.: Numerical and experimental investigations of the anisotropic transformation strains during martensitic transformation in a low alloy Cr-Mo steel 42CrMo4. Procedia Eng. **207**, 1815–1820 (2017)
8. Simsir, C.: 3D finite element simulation of steel quenching in order to determine the microstructure and residual stresses. Ph. D. thesis, Middle East Technical University (2008)
9. Behrens, B.-A., et al.: Experimental and numerical investigations of the development of residual stresses in thermo-mechanically processed Cr-Alloyed Steel 1.3505. Metals **9**, 480 (2019)
10. Behrens, B.A., Brunotte, K., Wester, H., Kock, C.: Methodology to investigate the transformation plasticity for numerical modelling of hot forging processes. Key Eng. Mater. **926**, 547–558 (2022)
11. Videau, J.C., Cailletaud, G., Pineau, A.: Experimental study of the transformation-induced plasticity in a Cr-Ni-Mo-Al-Ti steel. Le Journal de Physique IV **6**(C1), C1-465 (1996)
12. Wolff, M., Böhm, M., Dalgic, M., Löwisch, G., Lysenko, N., Rath, J.: Parameter identification for a TRIP model with backstress. Comput. Mater. Sci. **37**(1–2), 37–41 (2006)
13. Behrens, B. A., Chugreeva, A., Chugreev, A.: FE-simulation of hot forging with an integrated heat treatment with the objective of residual stress prediction. In: AIP Conference Proceedings, AIP Publishing LLC, vol. 1960, no. 1, p. 040003 (2018)
14. Behrens, B., Brunotte, K., Wester, H., Kock, C.: Targeted adjustment of residual stresses in hot-formed components by means of process design based on finite element simulation. Arch. Appl. Mech. **91**, 3579–3602 (2021)
15. Leblond, J.B., Devaux, J., Devaux, J.C.: Mathematical modelling of transformation plasticity in steels I: case of ideal-plastic phases. Int. J. Plast **5**(6), 551–572 (1989)
16. Wolff, M., Böhm, M., Dalgic, M., Löwisch, G., Rath, J.: Validation of a TP model with backstress for the pearlitic transformation of the steel 100Cr6 under step-wise loads. Comput. Mater. Sci. **39**(1), 49–54 (2007)

Production of Fine-Grained Ti-6Al-4V ELI for Medical Implants Using Equal-Channel-Angular-Swaging

S. O. Kraus(✉), L. Kluy, and P. Groche

Institute for Production Engineering and Forming Machines, TU Darmstadt, Darmstadt, Germany

stefan.kraus@ptu.tu-darmstadt.de

Abstract. Multi-resistant bacteria pose a global threat, as infections can lead to life threatening conditions. In orthopaedics, implant surfaces are critical for infection prevention. Implants with a fine-grained surface are a promising way to achieve antibacterial properties by inhibiting biofilm formation.

In this paper, the continuous manufacturing process for fine-grained materials called Equal-Channel-Angular-Swaging (ECAS) is applied to the most commonly used medical titanium alloy Ti-6Al-4V ELI. First, the homogeneity of the shear strain (which is crucial for grain refinement) is investigated in a numerical simulation under different process parameters (e.g. feedrate, counter pressure, number of passes). In the ECAS process, the material fails due to shear bands. Using the process parameters found in the simulation, damage free forming is possible. The material properties are discussed using microstructural analyses, hardness measurements and tensile tests.

Keywords: medical implants · fine-grained titanium · SPD processes

1 Introduction

Medical implants, such as dental or hip implants, contribute significantly to our quality of life. However, the ageing population is demanding that implants last longer to avoid age-related replacements and additional surgeries. In addition, biological incompatibilities and bacterial infections can lead to complications that can result in bone resorption, implant loosening, loss of the entire implant, or even life-threatening conditions [1, 2]. In particular, antibiotic-resistant bacteria cause approximately 670,000 invasive infections and approximately 33,000 deaths in Europe each year as a direct result [3]. The use of ultrafine-grained titanium as an advanced implant material demonstrates excellent biological and mechanical properties to address these challenges. The fine-grained structure stimulates interactions between the implant and the surrounding bone tissue, which in turn activates bone healing and prevents implant loosening [4]. Stress shielding and the associated regression of bone tissue associated with the use of coarse-grained titanium implants with a high Young's modulus can also be avoided by using ultrafine-grained titanium with a bone-matched Young's modulus [12]. However, the production of ultrafine-grained titanium on an industrial scale remains a challenge [5].

T. Bauernhansl et al. (Eds.): WGP 2023, LNPE, pp. 619–627, 2024.
https://doi.org/10.1007/978-3-031-47394-4_60

One way to achieve the desired fine grain size below 1 μm is to refine the microstructure by severe plastic deformation (SPD) processes [4]. A special feature of the SPD process is that the shape of the workpiece remains unchanged, allowing the process to be repeated several times and very large strains to be achieved incrementally. Of all SPD processes, Equal Channel Angular Pressing (ECAP) is the best researched process for machining bar-shaped workpieces. Some studies show that the ECAP process can be used to produce ultrafine-grained (UFG) titanium (Ti) alloys. These studies also show that the antibacterial properties, bone healing and corrosion resistance are improved [5]. Although significant progress has been made in recent years in introducing a continuous process, the industrial production of UFG materials is still hampered by high process forces and friction, which, together with discontinuous process control, result in low productivity [5].

The Equal-Channel-Angular-Swaging (ECAS) process was developed to address these issues [6]. The forming mechanism of the ECAS process is similar to that of the ECAP process, but the tool is divided into two oscillating halves to enable lower friction and axial forces than ECAP. Previous studies have shown that ECAS enables the continuous production of UFG materials by grain refinement from conventional copper, steel, FeCo alloys and CP titanium [6, 7].

In the class of titanium alloys, Ti-6Al-4V (Ti64) and its medical representative Ti-6Al-4V ELI (Ti64 ELI) are considered to be the most widely used titanium materials, accounting for more than 50% of the world's titanium production. Ti-6Al-4V is used in high-stress aerospace applications and medical implants due to its excellent strength properties. The alloy belongs to the $(\alpha + \beta)$ titanium alloys and consists of phase fractions with different crystal structures, α-phase in hexagonal close-packed (HCP) structure and β-phase in body-centred cubic (BCC) structure [8–11]. However, the poor ductility of titanium makes damage-free forming difficult, as discussed in our previous studies [12, 13].

In this work, the ECAS process is adapted for the production of ultrafine-grained medical Ti-6Al-4V ELI. It is analyzed which process conditions lead to the desired strain state required for grain refinement. The resulting microstructure and mechanical properties are investigated and the current process limits are shown.

2 Approach with Equal-Channel-Angular-Swaging

2.1 Experimental Setup

Figure 1 shows the experimental set-up for ECAS forming. In the initial state, Ti-6Al-4V ELI is present as a bar with a coarse-grained structure. This is fed to the ECAS tools (f), which oscillate (s). During the forming process, the bar is formed by the tool jaws and thereby subjected to a counter pressure (p) against the feed direction. Due to the channel geometry (angle Φ), high shear strains are introduced into the material, increasing the dislocation density in the grains and enabling grain refinement. The end product is an ECAS-formed titanium bar with increased dislocation density at an unchanged diameter, allowing multiple passes. The ECAS process can be carried out on a variety of machines such as high speed stamping presses or rotary swaging machines.

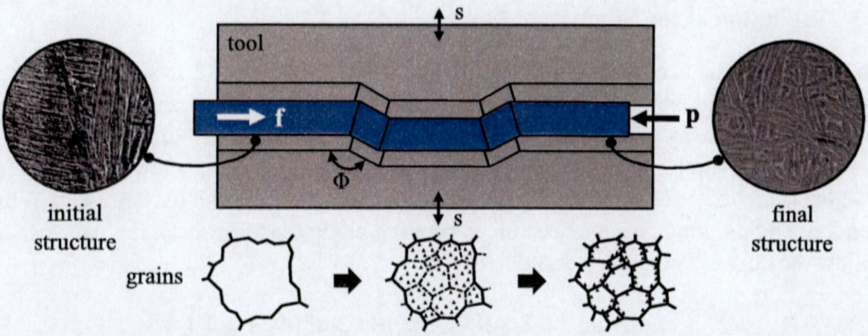

Fig. 1. Equal-Channel-Angular-Swaging (ECAS) process and grain refinement.

In this study, the ECAS process is carried out on a rotary swaging machine at an elevated process temperature. Preliminary studies with titanium have shown that forming at room temperature leads to material failure, in particular shear bands, while temperatures above 300 °C favor dynamic grain growth [14]. For this purpose, the tool jaws are heated to 150 °C using heating cartridges and, in addition, the Ti64 ELI bar is heated to 175 °C by inductive heating immediately prior to forming. The process parameters used are a feed rate of 400 mm/min, a swaging frequency of 30 Hz, a tool stroke of 0.7 mm and a maximum radial force of 1000 kN. The channel angle Φ of the tools is 150°. Hardness measurements are carried out using a hardness testing machine of the type Struers DuraScan 20 for HV1. Tensile tests according to DIN 50125 are performed on a Zwick Roell 100 combined tensile and compression testing machine. Three tensile tests are carried out for each material condition.

2.2 Finite Element Simulation

In order to investigate parameter combinations for the production of ultrafine grained Ti64 ELI using ECAS, the process is modelled using the finite element simulation software Simufact Forming 2021 with an implicit solver. This is done as a 3D simulation in order to observe the homogeneity of the strain distribution and the associated microstructural refinement in the material cross section. The dimensions of the ECAS tools and the process parameters correspond to the experimental setup. The tools are modelled as rigid bodies. Heat transfer between the workpiece and the tools is not simulated.

The isotropic elastic-plastic material behavior is represented by flow curves determined by El-Magd [15] for temperatures from 20–800 °C, divided into nine steps, and strain rates from 0.001–10000 1/s, divided into eight steps. The length of the simulated bar is 150 mm with a diameter of 8 mm. The coefficient of friction employed in the model is 0.12, determined by a sliding compression test. The counter pressure is varied in steps of 0 MPa, 5 MPa and 10 MPa. To reproduce continuous process conditions, the process is simulated over a feed path of 100 mm.

2.3 Validation of the Simulation Model

To investigate the accuracy of the simulation, it is validated for 2 Pass ECAS without back pressure. Equation 1 can be used to calculate the degree of deformation ε from the number of shear planes (n = 4) and the angles 2Φ (inner bending angle) and 2φ (outer bending angle) at each shear plane of the deformed bar, as shown in Fig. 2 [6]. Due to elastic material behaviour and different stress states without back pressure (inner bending angle compressive stress, outer bending angle tensile stress), the bar does not follow the channel geometry exactly.

$$\varepsilon = \sum\nolimits_{k=1}^{n} \left[\frac{2\cot(\Phi_n + \varphi_n) + \varphi_n csc(\Phi_n + \varphi_n)}{\sqrt{3}} \right] \qquad (1)$$

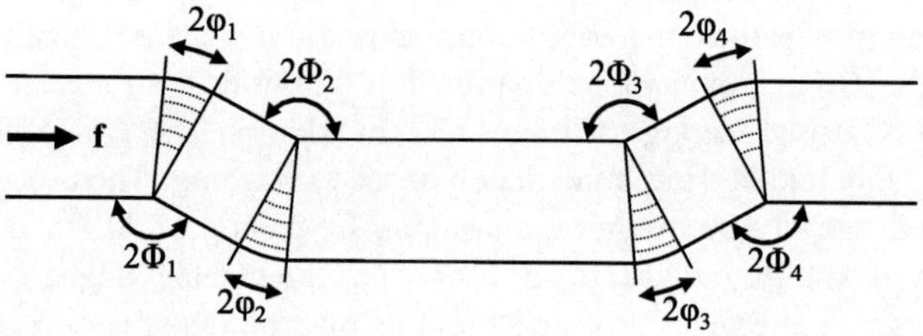

n	$2\Phi_1$	$2\varphi_1$	$2\Phi_2$	$2\varphi_2$	$2\Phi_3$	$2\varphi_3$	$2\Phi_4$	$2\varphi_4$
4	149°	12°	148°	8°	147,5°	16°	146,5°	10°

Fig. 2. Geometry of the ECAS processed bar for analytical validation

This approach leads to ε = 1,08 for the real bar geometry, which corresponds to a deviation of 9% from the simulated degree of deformation (ε = 1,18). In addition, the surface geometry of the simulated and the processed bar is compared by a 3D scan using GOM ATOS 5. The maximum deviation of the surface geometries is 5% of the bar diameter.

Both the analytical and geometric validation show that the simulation and the real experiment correlate with satisfactory accuracy.

3 Results and Discussion

The results of the simulation and experiments are discussed with respect to the production of a homogeneous nanostructure.

3.1 Simulation of Process Parameters

In the simulation, the counter pressure is varied and the homogeneity of the shear strain distribution in the cross section is evaluated, which is essential for the formation of the ultrafine-grained microstructure.

Counter Pressure and Number of Passes

The effect of counter pressure on the shear strain distribution in the bar cross section after 1 Pass and 2 Pass ECAS forming is shown in Fig. 3. The top row of the figure shows that the shear strain distribution after 1 Pass forming with and without counter pressure is inhomogeneous due to insufficient tool filling, in the form of the occurring rounding instead of a sharp bend at the shear planes and superimposed bending stress. Improved homogeneity of the shear strain distribution is achieved after 2 Pass ECAS forming. The advantage of 2 Pass forming (Fig. 1 and Fig. 2) is that the bar is sheared first in one and then in the opposite direction, which improves the homogeneity and straightness of the bar after the process. A counter-pressure of 10 MPa results in significantly improved tool filling, so that a symmetrical shear strain distribution can be observed with almost no positive strains (which in turn can favour the formation of shear bands). However, this simulation shows buckling of the bar in front of the ECAS tools, which would result in a damaged bar in the real process.

Based on the simulation results, the 2 Pass ECAS forming process with a counter pressure of 5 MPa is selected for the following investigations.

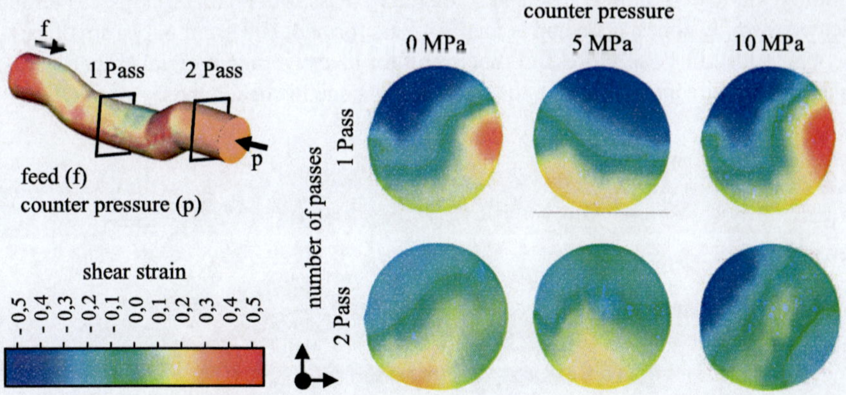

Fig. 3. Contour plots of the shear strain distribution in the cross-section as a function of the counter pressure and the number of passes

3.2 Experimental Investigations

Based on the experimental investigations, process limits with regard to formability and mechanical properties (e.g. hardness distribution with regard to microstructural homogeneity) are discussed.

Material Conditions of Ti-6Al-4V ELI

At the beginning of the experimental investigations, the material properties of Ti64 ELI are of crucial importance. Ti64 ELI is an (α + β) titanium alloy whose microstructure is stabilised by the alloying elements aluminium (Al) and vanadium (V) and whose excellent strength values can be well adjusted by heat treatment processes [8]. In this study, two material states of Ti64 ELI are distinguished. In the first state, prior to ECAS forming, Ti64 ELI is present in a lamellar microstructure. In order to adjust this initial lamellar microstructure, the as-received material with a globular microstructure is first subjected to a heat treatment (980 °C / 15 min / AC). The second condition considered is called ECAS Ti64 ELI, where the energy stored in the material by forming and the hardness are gradually increased by several ECAS passes.

Process Limits

The process limits are shown in Fig. 4. The 2 Pass ECAS forming of the material with a counter pressure of 5 MPa results in a slight thickening of the bar diameter ahead of the forming zones (A). Contrary to the expectations of the simulation, damage free 2 Pass forming (B) is only possible in reality without counter pressure. On the one hand, the counter pressure should lead to an improved forming capability due to a hydrostatic compressive stress state. On the other hand, process control is complicated by the two-sided clamping of the brittle material. Therefore, it is assumed that the damage is caused by the process control. Analogous to the simulation, insufficient tool filling can be seen without counter pressure. In (C) 4 Pass formed material is visible. 4 Pass ECAS forming is based on the same tool geometry as 2 Pass forming. However, immediately after 2 Pass forming, the feed motion is reversed so that the 2 Pass formed material passes through the two forming zones again and is finally 4 Pass formed. However, only a small part of the 4 Pass forming was successful (but useful for investigating material properties such as microstructure and hardness), the material fails due to shear bands.

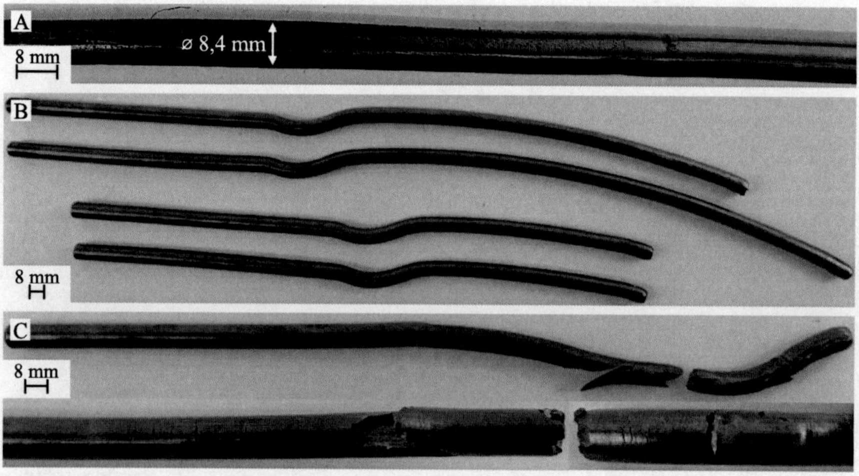

Fig. 4. Process limits of Ti64 ELI in the ECAS process. (A) diameter enlargement by counter pressure (2 Pass, 5 MPa counter pressure), (B) successful forming (2 Pass, 0 MPa counter pressure), (C) failure due to shear bands (4 Pass, 0 MPa counter pressure)

Homogeneity of the Microstructure

Once the process limits are known, the homogeneity of the microstructure is examined by means of hardness distribution in the cross section. For this purpose, the hardness is determined at nearly 70 points and visualised in contour plots as shown in Fig. 5. The initial lamellar structure shows an almost homogeneous hardness distribution of 330 HV1 across the cross section. After 2 Pass ECAS, an inhomogeneous hardness distribution can be seen. After 4 Pass ECAS the homogeneity of the hardness distribution improves.

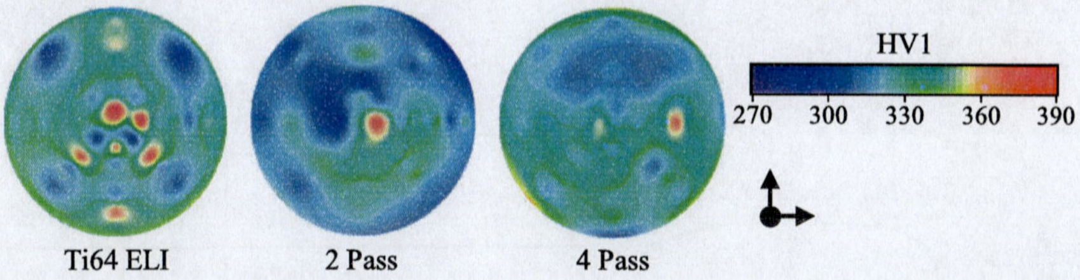

Fig. 5. Contour plots of the hardness distribution in the cross-section as a function of the material condition and the number of passes.

The reduced hardness after 2 Pass indicates that the hardening of the initial heat treatment is reduced by the forming. Further ECAS forming up to 4 Pass increases the hardness values due to the increasing dislocation density and associated grain refinement.

Mechanical Properties and Microstructure

The initial lamellar state of Ti64 ELI (A) with the associated mechanical properties, high strength and low elongation at break, is shown in Fig. 6 as a microscope image and in diagram and table form. This condition is not suitable for use as a medical implant material. The ASTM F136–13 standard requires a tensile strength of > 825 MPa and an elongation at break of > 8%, Young's modulus requirements aren't given in the standard, but to prevent stress shielding a bone-matched Young's modulus of 70 - 80 GPa (*) is desired [13, 16]. After 2 Pass ECAS forming (B) of lamellar Ti64 ELI, an increased strength and a significantly reduced Young's modulus at a lower elongation at break are observed. The corresponding microstructure image shows a microstructure that has been very strongly deformed. This microstructure has stored a very large amount of forming energy and is ideally suited for subsequent heat treatment to achieve the desired microstructure and material properties. In our previous study [13] it was shown for a different titanium alloy that the required ASTM specifications as well as a nano-scale grain size could be achieved by a subsequent heat treatment.

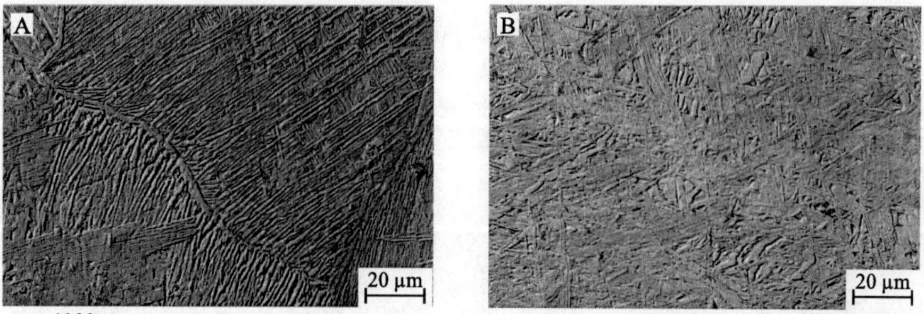

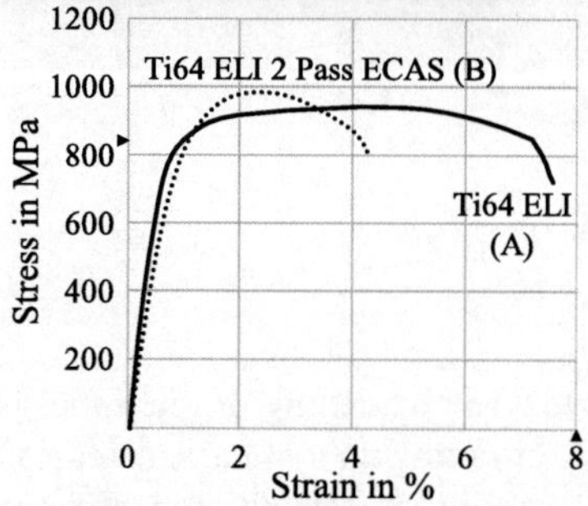

	E in GPa	*UTS* in MPa	*A* in %
goal acc. ASTM	*70-80* (*)	> 825	> 8
Ti64 ELI (A)	118	940	7
Ti64 ELI 2 Pass ECAS (B)	103	990	4

Fig. 6. Microstructure and mechanical properties of Ti-6Al-4V ELI: initial lamellar structure before ECAS (previously adjusted by heat treatment) (A), 2 Pass ECAS (B) and desired properties (▶). Three tensile tests according to DIN 50125 are carried out for each material condition.

4 Summary and Outlook

This study investigates the production of ultra-fine grained Ti64 ELI by Equal-Channel-Angular-Swaging (ECAS). In order to achieve damage-free forming of a homogeneous and fine grained microstructure, the influence of multiple passes and back pressure is investigated. Damage-free forming without shear bands is possible when (1) a lamellar microstructure is used in the initial state, (2) no counter pressure is applied, (3) 2 Pass ECAS is performed and (4) an elevated process temperature is set.

Future work will investigate subsequent heat treatments to achieve improved mechanical properties for use as an implant material.

Acknowledgements. The authors would like to thank the German Federal Ministry of Education and Research (BMBF) for funding the IdentiTI project, the German Federal Ministry for Economic Affairs and Climate Action (BMWK) for funding the DIAMOND project, Carsten Siemers, Lina Klinge, Elias Merz and Benjamin Grote, the Joachim Herz Foundation Hamburg, ADVANTIQX Dr. Johannes Scherer, Schweizer Feinwerktechnik GmbH and GFM GmbH.

References

1. Schwarz, F., et al.: Peri-implantitis. J. Clin. Periodontol. **45**, 246–266 (2018)
2. Murray, C.J.L., et al.: Global burden of bacterial antimicrobial resistance in 2019: a systematic analysis. The Lancet **399**, 629–655 (2022)

3. European Centre for Disease Prevention and Control & World Health Organization: Antimicrobial resistance surveillance in Europe 2022–2020 data. WHO Regional Office for Europe, Copenhagen (2022)
4. Elias, C.N., et al.: Ultrafine grained titanium for biomedical applications: an overview of performance. J. Markt. Res. **4**, 340–350 (2013)
5. Stráský, J.: Advanced Titanium Alloys for Medical Applications. Habilitation Thesis, Charles University, Prague (2020)
6. Görtan, M.O.: Severe plastic deformation of metallic materials by equal channel angular swaging: Theory, experiment and numerical simulation. Dissertation, Technische Universität Darmstadt, Darmstadt (2014)
7. Kluy, L., Chi, F., Groche, P.: Equal-Channel-Angular-swaging for the production of medical implants made of fine-grained titanium. In: WGP Production at the Leading Edge of Technology (2021)
8. Navarro, M., et al.: Biomaterials in orthopaedics. J. R. Soc. Interface **2008**, 1137–1158 (2008)
9. Peters, M., Clemens, H.: Titan, Titanlegierungen und Titanaluminide – Basis für innovative Anwendungen. BHM Berg-Hüttenmännische Monatshefte **155**, 402–408 (2010)
10. Agarwal, K.M., et al.: Simulated analysis of Ti-6Al-4V processed through equal channel angular pressing for biomedical applications. Mater. Sci. Energy Technol. 4, 290–295 (2021)
11. Oshida, Y.: Bioscience and Bioengineering of Titanium Materials, 2nd edn. Elsevier, Amsterdam (2013)
12. Lim, S.-S., et al.: Flow stress Ti-6Al-4V during hot deformation: decision tree modeling. Metals **10**, 739–749 (2020)
13. Klinge, L., et al.: Nanostructured Ti-13Nb-13Zr for dental implant applications produced by severe plastic deformation. J. Mater. Res. **37**, 2581–2588 (2022)
14. Klinge, L., et al.: Nanocrystalline Ti-13Nb-13Zr for dental implant applications. In: Proceedings of the 60th Conference of Metallurgists (2021)
15. El-Magd, E., et al.: Experimentelle und numerische Untersuchung zum thermomechanischen Stoffverhalten. In: Tönshoff, H.K., Hollmann, C. (eds.) Hochgeschwindigkeitsspanen metallischer Werkstoffe, pp. 183–206. Wiley-VCH Verlag GmbH & Co. KGaA, Weinheim (2005)
16. ASTM International: Designation, F136-13 (2021): Standard Specification for Wrought Titanium-6Aluminium-4Vanadium ELI (Extra Low Interstitial) Alloy for Surgical Implant Applications (UNS R56401). ASTM International, West Conshohocken, PA, USA (2021)

Reduction of Thermally Induced Wear on a Forging Tool by Heatpipes

R. Laeger(✉), J. Peddinghaus, D. Rosenbusch, and B.-A. Behrens

Institute of Forming Technology and Machines, Leibniz University Hannover, An der Universität 2, 30823 Garbsen, Germany
laeger@ifum.uni-hannover.de

Abstract. During forging, tool wear occurs as a result of thermomechanical stress. In addition, the deterioration of material behavior due to an overheated surface zone leads to reduced tool service life resulting in higher production costs. Heatpipes can be used to dissipate heat from the loaded tool surface area and thus optimize the material behavior of the tool. The performance of this method is strongly influenced by the heat conductivity between tool and heatpipe. To evaluate this novel method, the influence of the force-fit connection between heatpipe and tool on the thermal load is investigated during a forging process. In addition, the influence of the connection surface finish has been investigated by varying the roughness of the contact surface. Reference tools without heatpipes and with loose connection of the heatpipes were used for comparison. All tools were tested with 1,000 strokes in a fully automated forging process. The die temperatures were recorded to evaluate the resulting wear behavior of the tools. Based on the tests, reduced wear was observed using the heatpipes applied.

Keywords: wear reduction · heatpipes · forging tools

1 Introduction

During hot forming, mechanical and thermal stresses lead to tool geometry-influencing wear phenomena. The extent depends on the intensity and orientation of the stress. Different approaches are used to prevent different wear phenomena. Mechanical load usually leads to abrasion, which is counteracted by the high hardness of the die surfaces, e.g. by nitriding [1]. Adhesion, which is favored by high temperatures, can be prevented using boundary layer treatments that cause a low surface energy [2]. Nevertheless, higher temperatures also reduce the surface hardness of a forging tool. As a result of the cyclic mechanical stress and the lowered hardness, plastic deformation of the forming tools occurs, which usually leads to tool failure. By increasing the near-surface hardness and applying suitable cooling strategies, tool failure can be prevented [3].

In die forging, the lubricant is usually applied by spray application directly after the part has been ejected. A disadvantage of this application is an increased cooling rate, which leads to large temperature gradients near the tool surface. These in turn lead to internal stress differences resulting in a crack network on the surface [4]. Further, the use

T. Bauernhansl et al. (Eds.): WGP 2023, LNPE, pp. 628–637, 2024.
https://doi.org/10.1007/978-3-031-47394-4_61

of cooling lubrication requires the consideration of many influences like the direction of application or the appropriate application system [5].

A promising approach to avoiding the wear phenomena caused by spray lubrication and to ensuring the required cooling is the integration of heatpipes. Compared to conventional spray cooling, heatpipes offer the advantage of continuous heat dissipation, as this prevents the high cooling rates and consequently reduces wear due to thermal shock. In preliminary investigations, a clear influence of the type of connection between the die and the heatpipe has already been demonstrated in model tests [6]. A force-fit and form-fit connection proved to be the most effective strategy for heat conduction. In this study, the application behavior will be investigated and the best possible connection between heatpipe and forging die will be determined.

2 Experimental Setup

In the following subsections, the function of heatpipes, the design of the tools and the parameters of the forging process are explained. Furthermore, the methods of data acquisition are presented.

2.1 Basic Characteristics of Heatpipes

Heatpipes are closed tubes usually made of copper or stainless steel. In electrical engineering, the heatpipe is an established means of heat dissipation due to its small installation size and high effectiveness [7]. Its material and the filling medium vary depending on the application and are chosen according to the application temperature.

For the present investigations, copper heatpipes filled with water were used. They offer the advantage of high thermal conductivity compared to steel and are suitable for application temperatures up to 250 °C. This corresponds to the typical basic tool temperatures of forging tools [8, 9]. The schematic principle of heatpipes is shown in Fig. 1. On the right side, heat energy is supplied, the medium evaporates and the steam is transported to the left. There it condenses due to the release of heat energy and flows back to the right side thus causing a circulation inside the heatpipe. The negative pressure inside is used to shift the triple point and thus enable a more effective heat transportation.

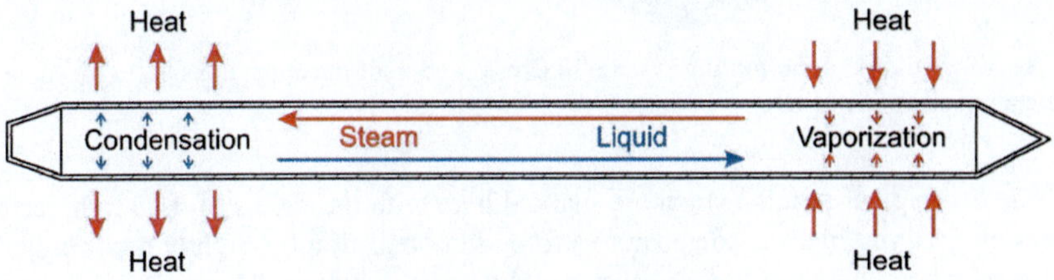

Fig. 1. Principle of a Heatpipe

The internal structure of heatpipes also has an influence on their application behavior. Groove, mesh and sinter structures are conventionally used. However, groove heatpipes lose their effectiveness with even small changes in shape and are therefore not used in this project. When the heatpipe diameter is increased, the effectiveness of mesh structures increases only slightly. Consequently, only sintered structures were used for this research. [3].

2.2 Integration in the Forging Die

The type of connection between the heatpipe and the forming tool is decisive for the effectiveness of heat transfer. In particular, the real contact area, mainly influenced by roughness, influences the heat transfer through the cavities and contact surfaces [10]. The connection by means of a material bond has already been tested in preliminary tests [6]. However, it was not possible to produce a sufficient connection through soldering since the high temperatures required would damage the heatpipe.

For the variants investigated here, force-fit connections were used to generate an increased real contact area. The clamping pressure used causes minor deformation of the heatpipe but is sufficiently small not to restrict its effectiveness.

In the tests, a reference without heatpipes, a variant with plugged heatpipes and two force-fit tooling systems were compared. The force-fit tools are divided into different surfaces roughnesses of Rz = 3 µm and Rz = 12 µm in the contact surface. The surface with different roughnesses is located at the connection area between heatpipes and tools (heatpipe channels, Fig. 2). For each variant, two tools were manufactured from hot-work tool steel 1.2343 and tempered to a hardness of 48 HRC.

Figure 2 shows the geometry of the upper die with mounted heatpipes and fins (a) and the investigated upper dies with heatpipe channels in cross-section (b). The division into two parts was not necessary for the plugged tool and the reference without heatpipe. Figure 2, b) also shows the areas A, B and the groove, which are examined in terms of wear.

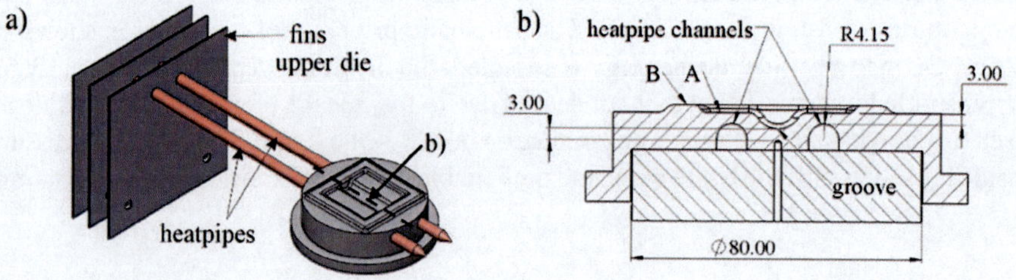

Fig. 2. a) 3D figure of the cooling system, b) Cross-section of the upper die system for force-fit connection, all values in mm

The heatpipe channels, which are marked here with the radius of 4.15 mm, have a height of 7.15 mm in order to generate a force-fit connection by slightly deforming the heatpipes when clamping the tool system ($Ø_{heatpipes}$ = 8 mm). The tool was fixed via the outermost shoulder by using a clamping ring. The plugged version features holes with diameters of 8.1 mm. The small distance of 3 mm to the engraving surface allows the heat energy to be dissipated close to the surface. This prevents the heat energy from being stored in the die.

2.3 Forging Process

The forming tests were performed in a fully automated forging cell with a cycle time of 10 s. For each variant, one die was tested with 1,000 cycles. An eccentric press of the type SP30d by Eumuco was used. Industrial robots are used to feed and remove the billets in order to guarantee a reproducible process sequence.

The billets were heated to a billet temperature of 1250 °C by means of an induction pusher furnace. The semi-finished material used was 42CrMo4 (1.7225) with a length of 40 mm and a diameter of 30 mm.

After removal, a short blast of air was blown onto the tool surface to remove scale. This was parameterized as low as possible to prevent cooling.

A hydraulic cylinder with an application head was moved to the press chamber after scale removal to apply the dry lubricant boron nitride by means of electrostatic charging. The use of a dry lubricant prevents abrupt cooling of the die surface and allows functional separation of cooling and lubrication, as well as an evaluation of the cooling performance of the heatpipe concept.

The upper tools were preheated to a base temperature of 180 °C by heating cartridges in order to set a tool temperature almost corresponding to the typical base temperature for forging dies. It is lower compared to the base temperature of 250 °C of [8] and [9] to prevent an overheating after the first strokes. After a holding time of 20 min, an equilibrium was established between the heat energy dissipated by the heatpipe systems and the heat energy supplied by the die heating system. Shortly before the tests started, the heating was switched off so that no further heat energy was supplied.

2.4 Measurement

During the forming process, temperatures were measured 3 mm below the tool surface. For this purpose, type K thermocouples were inserted in the center of the tool (drill hole at the bottom, shown in Fig. 2). To improve the heat transfer to the thermocouples, thermal paste was used.

All tools were measured before and after the tests using a Keyence VR-3200 profilometer. With the help of the geometry comparison before and after forging, the geometrical deviation due to wear was determined. This illustrates the build-up and removal as a result of tool use.

To identify larger cracks, adhesion and deformation of the tools, macroscopic photographs were taken. Focus stacking made it possible to obtain an overall image in depth of field by varying the focal length ranges.

The identification of microscopic cracks, microstructural states and hardness was carried out with the aid of metallographic examinations of cross-sections of the transition areas from B to A and from A to the base (see Fig. 2).

3 Results and Discussion

3.1 Temperature Curves

Figure 3 shows the temperature curves during forging. A section of 2000 – 6000 s of the process duration was selected, since this range is particularly suitable for presenting the results due to its continuity. Interruptions in the test series led to temperature decreases, as for example in the case of variant Rz12 in Fig. 3 at about 3000 s.

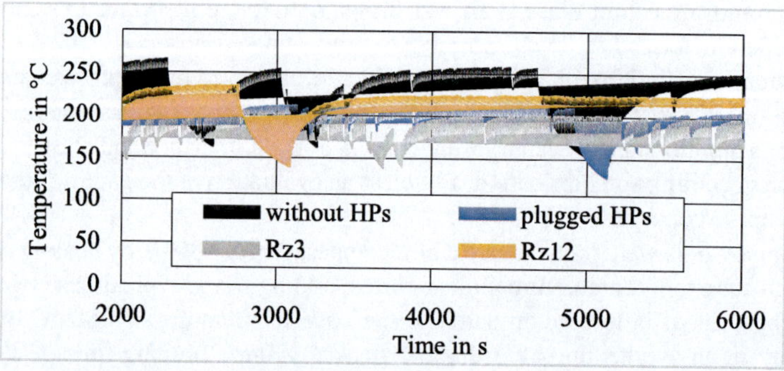

Fig. 3. Part of the temperature curves of the process with 1000 strokes

After a certain time however, a plateau was always reached, with a period of small amplitudes around one temperature. This approximately stationary state differs in its level depending on the fitting variant. Without heatpipes, a high plateau of approx. 250 °C is reached, resulting from the lack of targeted heat dissipation through appropriate cooling strategies. In contrast, the variant with plugged heatpipes has a low plateau at approx. 200 °C, since the heatpipes expanded as a result of the heat and effective heat dissipation took place. The force-fitted variants Rz3 and Rz12 were also able to generate low plateaus, which did not reach the initial level again despite interruptions. The variant Rz12 shows a constant but higher plateau compared to Rz3.

A strong influence of the force-fit-joined tools could clearly be demonstrated based on the temperature curves. It can be stated that a connection by means of a force-fit results in an improved and more constant temperature control. The connection of the supposedly plugged connected variant also proved to be force-fitted, as it expanded as a result of the heat. Since the bore of the plugged heatpipe has a similar surface finish to Rz3, the temperature profile was similar. The roughness of Rz12 resulted in a smaller real contact area and thus in lower heat transfer.

3.2 Height Difference

Temperature measurements are a suitable measure for evaluating supplied and dissipated thermal energy. However, the local surface temperature during forging is different from the base die temperature due to component contact times and tribological effects. These occur cyclically and can increase significantly for a short time as a result of the forging strokes. A difference to the measurement 3 mm below the surface cannot be avoided. For this reason, the wear on forming tools is investigated more in detail and examined for correlations with the measured temperatures. For this, Fig. 4 shows sections of the height differences between 0 and 1,000 cycles of the different variants.

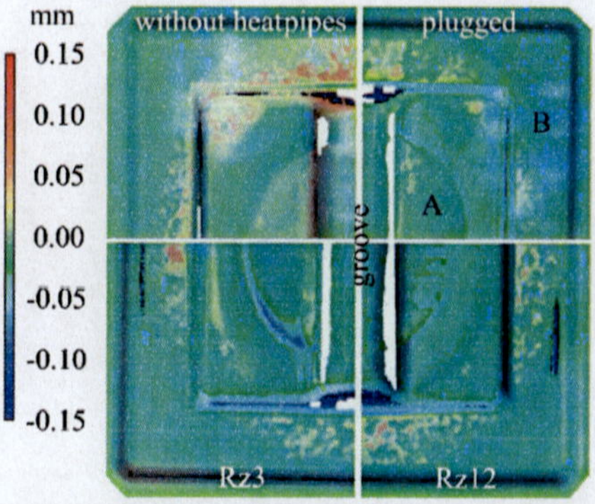

Fig. 4. Height differences of relevant areas

Compared to the variant without heatpipes and Rz12, the areas B of the plugged and Rz3 variants show lower build-up. This material build-up could be due to the adhesion of scale, billet material or lubricant.

The height step from B to A shows significant wear in both the plugged variant and in Rz12. In the variant without heatpipes, this only occurs locally (not shown here). Both the adhesion of the variant without heatpipes and the wear at Rz12 indicate increased temperatures, as already shown in Fig. 3. Higher temperatures lead to softening of the material and thus favor wear, which can be seen here in the form of abrasion. Adhesion is also favored by higher temperatures and consequently increased surface energies [2, 11].

The area A differs for each variant. In the variant without heatpipes and the plugged variant, a material build-up can be seen in the outer corners of A in each variant. The structure is inhomogeneous, similar to B, and thus indicates adhesion. The proximity of the heatpipes to the surface leads to direct heat dissipation on A of Rz3, minimizing temperature and thus adhesion. In the case of the plugged variant and the variant without heatpipes, clear build-ups can also be seen in the contact zones with the heated billet (circular markings centered on A) and on the curve of the groove to A.

All variants except Rz3 show a negative geometry change on A from the edge of the billet contact zone to B (even without heatpipes, but not visible in Fig. 4).

Rz3, on the other hand, has a deep indentation, especially on the left side of A. The improved cooling at Rz3 probably leads to a superficial cooling of the billets before forging, rendering them harder than in the warm state, consequently favoring abrasive wear. The dissipated heat from the billets may additionally have favored this phenomenon by softening the tool surface.

3.3 Photographic Documentation

Figure 5 shows an example of a section of the tool without heatpipes. Each tool has grooves at the transition from A to B, but of different intensity. This abrasive wear is typical for areas of flash formation as a result of the relative movement and high contact normal stress between tool and semi-finished product.

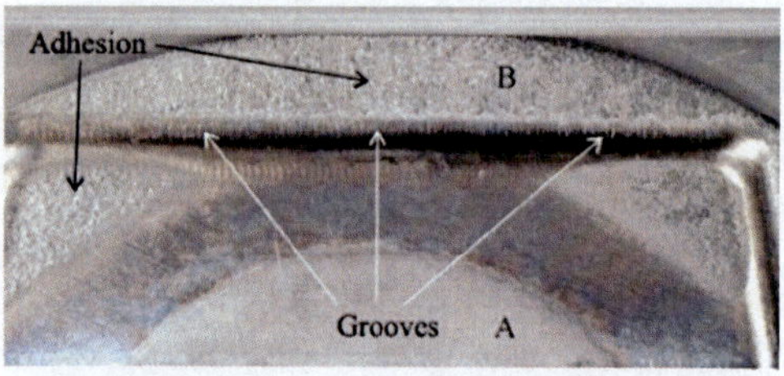

Fig. 5. Cutout of the variant without heatpipes after 1,000 strokes

Furthermore, material buildup with an undirected structure can clearly be seen on B and in the corners of A. This wear phenomenon can also be seen with different intensity on all tools, but is most visible on the variant without heatpipes. Since adhesion is favored by temperature, these results were to be expected. The results correlate with the temperature curves of Fig. 3 and the analyses of Fig. 4.

As already discussed regarding the height differences, Rz3 has a clear groove at the edge of the billet contact zone. This can clearly be seen in Fig. 6. In addition, abrasion marks can be seen in the groove, which due to the direction (outwards) confirm the assumption that the billets cool down too much on contact with the tool and the increased hardness favors abrasive wear. Furthermore, adhesion can be seen, which is nearly invisible on the pictures of the height differences (Fig. 4) due to the low height.

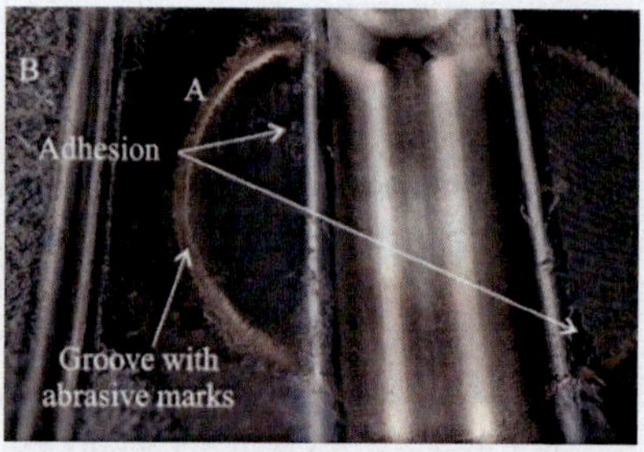

Fig. 6. Wear at Rz3 after 1,000 strokes

It can be summarized that the previously named wear phenomena could be validated on the basis of the photos. Higher temperatures resulted in more adhesion. Further, because of the low temperature and the higher hardness, abrasive wear occurred at the dies which were cooled by heatpipes. In addition, some signs of wear were so small in volume that they only became visible through the photographs.

3.4 Metallographic Examination

To visualize the microstructural changes and the microscopic wear, the transition from B to A and from A to the groove was prepared metallographically for each tool and a light-microscope image was taken. Cross-sections of the transition from B to A for each tool are shown in Fig. 7.

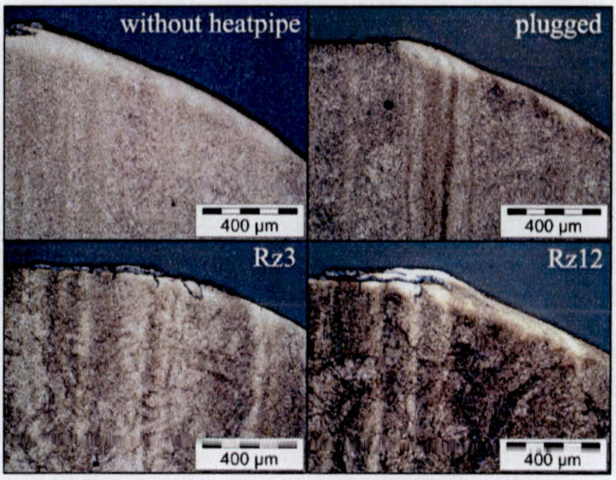

Fig. 7. Cross-section of the transition from B to A

Increased surface temperatures, which then experience strong cooling, can locally lead to the formation of white layers (martensitic phase). Bright areas (possibly white layers) are visible on all specimens to a minor extent. An influence due to the differently inserted heatpipes is not discernible. In each case, the lighter phase can be seen particularly in the upper area of the rounding.

The variant without heatpipes, the plugged variant and the variant Rz12 also show strong deformation in the area depicted (Fig. 7). In addition, these variants exhibit plastic deformation of the tool material via an adhesive layer at the upper area of the curve. This phenomenon is most noticeably developed in Rz12. Rz3 has the lowest average temperatures (see Fig. 3) and consequently the microstructure softens less and deforms very little. A correlation of the tool temperatures and the deformation is clearly evident, since the heatpipes lead to lower temperatures and consequently to less thermally induced damage to the tools.

All specimens also have small cracks at various points of their surfaces, but this is only clearly visible here in Rz3 at the upper specimen edge. These small cracks are typical wear phenomena as a result of alternating thermomechanical stress. With a larger temperature gradient in the outer layer, as is the case with Rz3, the tool is subjected to increased stress due to thermal shock, resulting in an increased tendency to crack.

Furthermore, the cross-sections of the height difference from B into the groove show that there is a thin adhesive layer in each variant. In this case, a direct correlation to test parameters is not discernible. However, since the longest billet contact time is here, adhesion could form due to the high surface energy as a result of the temperature.

The hardness measurements of the height difference from B to A showed low hardness of approx. 300 HV 0.1 near the surface compared to the basic hardness of approx. 500 HV 1 set by tempering. This is a typical phenomenon in forging dies as a result of softening by increased temperatures. It favors plastic deformation, as also demonstrated in Fig. 4 and Fig. 7.

Also noticeable were locally higher near- surface hardnesses (lighter areas) on Rz12 of >500 HV 0.1 down to a depth of 2.1 mm. Due to this, further hardness tests were performed in the white areas on Rz12, which showed significantly higher hardness (>600 HV 0.1). This higher hardness indicates a martensitic microstructure, which also exhibits the typical light coloration as seen on the variant without heatpipes in Fig. 7.

The light coloration in Fig. 7 and the hardness measurement correlate with the temperature curves of Fig. 3 and the optical analyses of Sect. 3.2 and 3.3. Rz3 and the variant with plugged heatpipes have less built-up material due to lower temperatures.. Depending on the application, the deterioration of adhesive wear can lead to tool failure. Furthermore, the applied material has a higher hardness and thus suggests to be martensitic.

4 Conclusion

It can be stated that an increased heat dissipation due to force-fitted heatpipes could be determined. A low, constant tool base temperature, less adhesion and lower plastic deformation show the potential of a pressed heatpipe with high real contact area (Rz3 and unexpected before the plugged variant).

From the temperature profiles in Fig. 3, it can be deduced that a force-fit connection results in constant and lower temperature plateaus compared to a reference without heatpipes. The increased surface quality (lower roughness) also leads to improved heat transfer due to the larger real contact area between heatpipe and die.

For the design of forming tools, the temperature field must be known or estimated so that suitable distances of the heatpipes to the surface can be selected. Too high temperature gradients from surface to base material showed a slightly increased cracking tendency due to thermal shock in the Rz3 variant (Fig. 7).

In order to prevent abrasive wear, e.g. the groove in Fig. 6, the tool should undergo a near-surface strength increase by surface modification, e.g. by nitriding or coating, in addition to temperature control by heatpipes.

Acknowledgements. The results presented were developed while working on the project "Targeted die temperature control by integrating heatpipes in hot forging tools", project number 189451423. The authors thank the German Research Foundation DFG (Deutsche Forschungsgemeinschaft) for the financial support.

References

1. Gronostajski, Z., Kaszuba, M., Widomski, P., Smolik, J., Ziemba, J., Hawryluk, M.: Analysis of wear mechanisms of hot forging tools protected with hybrid layers performed by nitriding and PVD coatings deposition. In: Wear, pp. 269–280 (2019)

2. Groche, P., Nitzsche, G.: Temperatureinfluss auf den Adhäsionsverschleiß beim Umformen von Aluminiumblechen. Materialwiss. Werkstofftech. **35**(7), 461–466 (2004). https://doi.org/10.1002/mawe.200400774
3. Puppa, J., Behrens, B.-A.: Optimization of cooling and lubrication for nitrided and ceramic-coated hot forging dies. Appl. Mech. Mater. **794**, 97–104 (2015)
4. Schneidt, A., Mahnken, R.: Experimentelle Untersuchungen zum Thermoschock beim Hybridumformprozess. Proc. Appl. Math. Mech. **7**(1), 4030013–4030014 (2007). https://doi.org/10.1002/pamm.200700074
5. Hawryluk, M., Ziemba, J., Dworzak, Ł., Kaczyński, P., Kasprzak, M.: Wear analysis of forging tools used in the hot forging processes using 3D reverse scanning techniques and cooling-lubricating system. Int. J. Adv. Manuf. Technol. **97**(5–8), 2009–2018 (2018). https://doi.org/10.1007/s00170-018-2066-y
6. Behrens, B.-A., Brunotte, K., Peddinghaus, J., Laeger, R.: Influence of the connection between forming die and heatpipe on the heat transfer. In: METAL 2022 Conference Proceedings. METAL 2022, 18.05.2022 - 19.05.2022: TANGER Ltd (METAL Conference Proceedings), pp. 227–232 (2022)
7. Chen, X., Ye, H., Fan, X., Ren, T., Zhang, G.: A review of small heat pipes for electronics. Appl. Therm. Eng. **96**, 1–17 (2016). https://doi.org/10.1016/j.applthermaleng.2015.11.048
8. Hawryluk, M., Jakubik, J.: Analysis of forging defects for selected industrial die forging processes. Eng. Fail. Anal. **59**, 396–409 (2016). https://doi.org/10.1016/j.engfailanal.2015.11.008
9. Gronostajski, Z., Kaszuba, M., Polak, S., Zwierzchowski, M., Niechajowicz, A., Hawryluk, M.: The failure mechanisms of hot forging dies. Mater. Sci. Eng. A **657**, 147–160 (2016). https://doi.org/10.1016/j.msea.2016.01.030
10. Burghold, E.M.: Experimental determination of transient heat transfer between solid bodies during dynamic loading. Dissertation. RWTH Aachen University (2016)
11. Woska, R., Barbehön, J.: Metallische Adhäsion unter trockener Reibung. Materialwiss. Werkstofftech. **13**(10), 348–355 (1982). https://doi.org/10.1002/mawe.19820131005

Characterisation and Qualification of Hot Working Tool Steels and Inconel for the Production of a Hybrid Forging Die

J. Siring(✉), M. Till, H. Wester, J. Uhe, and K. Brunotte

Institute of Forming Technology and Machines (LUH), An der Universität 2, 30823 Garbsen, Germany
siring@ifum.uni-hannover.de

Abstract. In forging, premature die failure caused by high cyclic thermo-mechanical loads leads to higher production costs. The die life can be increased by using high strength materials like nickel-based alloys (Inconel). However, the application of Inconel as die material is challenging due to its high costs and poor machinability compared to conventional tool steels. Therefore, an approach to produce hybrid dies consisting of hot working tool steels and Inconel according to the Tailored Forming concept is presented. The dies are manufactured by forging pre-joined semi-finished workpieces. To resolve challenges like different flow behaviours, first a material characterisation of two hot working tool steels (1.2343/1.2367) and two Inconel alloys (625/718) was carried out using uniaxial compression tests. Based on the determined mechanical properties suitable process windows were investigated and FE-simulations for the hot forming of hybrid Inconel and tool steel semi-finished workpieces were carried out to investigate the resulting material distribution.

Keywords: Material characterisation · Tailored Forming · Hybrid forging dies

1 Introduction

Dies in hot forging processes are exposed to high thermo-cyclic loads, which lead to wear and failure [1]. The replacement of a failed die and the additional production downtime result in high costs [2]. These costs can be reduced by increasing the service life of the dies. On the one hand, this can be achieved by adapting the geometry [3] or on the other hand by using wear-resistant surface adaptations [4] or materials such as Inconel [5].

The nickel-based alloy Inconel offers in contrast to conventional tool steels high strength properties even at elevated temperature. However, it is more expensive and difficult to manufacture [6]. Therefore, new approaches for the application of Inconel as die material are necessary. In order to minimise the amount of material used and the amount of machining required, only near-surface layers of Inconel should be used and thus a near-net-shape forming is desirable [7]. The combination of different materials to reduce wear were investigated by Landgrebe et al. [8]. Here, a nickel-based alloy was applied

T. Bauernhansl et al. (Eds.): WGP 2023, LNPE, pp. 638–647, 2024.
https://doi.org/10.1007/978-3-031-47394-4_62

locally in order to increase thermal and corrosion resistance. To combine the different materials a selective laser melting (SLM) process was used. However, appropriate die geometries can also be produced by hot forging. In the study of Behrens et al. [9], the production of forging dies by forming was investigated. For the production, an adapted process route for the tool steel was developed and tested. Overall, a higher hardness in the surface area of the die was achieved compared to dies produced conventionally by machining. For the production of hybrid bulk metal components, the Tailored Forming concept is a possible solution, in which pre-joined hybrid semi-finished products are formed at elevated temperatures. The use of Tailored Forming was already successful demonstrated for the production of high loadable hybrid parts like shafts [10] or bearing bushings [11] consisting of different materials such as steel and aluminium. The basic idea of Tailored Forming is the use of geometrically simple pre-joined semi-finished products which are subsequently formed. By means of the subsequent forming process, a tailored material distribution and thermo-mechanical modification of the joining zone can be achieved. For example, using the occurring thermo-mechanical loads, the coarse and brittle microstructure of deposition welded layers can be transformed into a fine grained and ductile microstructure. A detailed description of the Tailored Forming process chain can be found in the work of Behrens and Uhe [10].

In this context the Tailored Forming concept is a promising approach for the production of near-net-shaped hybrid dies consisting of a conventional tool steel as base material and an Inconel surface layer. In the Tailored Forming process chain considered in this work, a general hot working tool steel and an Inconel alloy are first joined by friction welding and then the dies are forged to produce a near-net-shape geometry.

However, due to the different materials and resulting material properties, the design of such a combined forging process is challenging and FE simulation has great potential for an efficient process design. [2, 5]. Therefore, this work focuses on the determination of the temperature and strain-rate dependent flow behaviour of the considered materials Inconel 625/718 and tool steel AISI H11/H10 (1.2343/1.2367). The general flow behaviour of these materials was analysed with regard to a combined forming. Furthermore, the data is used to parametrise material models for numerical simulations of the Tailored-Forming process. The FE-model was initially validated by comparison with experimental results of an abstracted hot forging process. Finally, the influence of modified process conditions like forging temperature, material combination and Inconel layer thickness were investigated numerical.

2 Determination of Material Parameters

To analyse the temperature and strain rate dependent material behaviour of Inconel 625, 718, tool steel 1.2376 and 1.2343, uniaxial cylinder compression tests were conducted on the Gleeble 3800-GTC forming simulator. The tests were carried out at 900 °C, 1,050 °C and 1,200 °C using cylindrical specimens with a length of 9 mm and a diameter of 6 mm. This temperature range was chosen in order to keep the flow stress of the Inconel alloys as low as possible and meet the temperature range of conventional hot forging process. The strain rate was varied between 1 1/s, 10 1/s and 50 1/s. Each test was repeated three times. Based on the uniaxial cylinder compression tests, the flow curves were calculated and used to determine the material specific coefficients ($n_{1/2}$ = static strain hardening coefficient; $m_{1/2}$ = kinematic strain hardening coefficient; $C_{1/2}$ = proportionality factor; $l_{1/2}$ = scaling factor) of the analytic GMT flow curve model (Table 1). The equation for the GMT flow curve model is given in Eq. (1).

$$k_f = C_1 \cdot e^{C_2 \cdot T} \cdot \varphi^{n_1 \cdot T + n_2} \cdot e^{\frac{l_1 \cdot T + l_2}{\varphi}} \cdot \dot{\varphi}^{m_1 \cdot T + m_2} \quad (1)$$

With the help of the GMT flow curve model, it is possible to consistently describe the flow stresses k_f as a function of strain φ, strain rate $\dot{\varphi}$ and temperature T in the numerical simulation.

Table 1. The individual parameters for the analytic GMT model of the different materials.

Material	C_1	C_2	n_1	n_2
1.2343	3,177.96	−2.82e-3	2.33e-4	−0.26173
1.2367	6,210.03	−3.50e-3	−6.40e-5	−1.25e-3
625	77,557.78	−5.41e-3	−5.90e-4	0.436919
718	101,383.30	−5.49e-3	5.77e-4	−0.64934
	l_1	l_2	m_1	m_2
1.2343	2.74e-5	−4.411e-2	1.15e-4	−0.02132
1.2367	−1.60e-5	7.04e-4	2.38e-4	−0.13157
625	3.82e-5	−0.06329	4.72e-4	−0.36352
718	1.04e-4	−0.11935	9.09e-4	−0.83406

In Fig. 1 the calculated flow curves using the parameterised GMT model for the Inconel alloys 718/625 (a, c) and the tool steels 1.2343/1.2367 (b, d) for the three temperatures (900 °C, 1,000 °C, 1,100 °C) (a, b) and strain rates of 1 1/s and 10 1/s (c, d) are shown. A comparison of the considered tool steels shows small differences regarding their flow behaviour (Fig. 1b, d). In contrast to the tool steels, the Inconel alloys show a stronger dependence on temperature and strain rate considering the absolute values (Fig. 1a, b). Both Inconel alloys show a strong reduction of flow stress with increasing temperature. However, in percentage terms, tool steels and Inconel alloys show a 50–60% increase in flow stress with a decrease in temperature (1,100–900 °C). Both Inconel alloys show a strong reduction of flow stress with increasing temperature. At the highest investigated temperature of 1,100 °C, the closest convergence of flow stresses between the tool steels and Inconel alloys can be observe. Comparison of different strain rates shows for both Inconel alloys a decrease of flow stress with increasing strain rate (Fig. 1c). The absolute values of flow stress differ more from each other for the two Inconel alloys than for the tool steels.

In order to enable a homogeneous forming of the two materials and reduce the occurrence of critical stresses in the joining zone the flow stresses of the used materials should be equalised [10]. The comparison of the flow properties at 900 °C, 1,000 °C and 1,100 °C shows that inhomogeneous heating, with an Inconel layer that is as hot as possible and a cold tool steel part, should be aimed to equalise the flow stresses. Such inhomogeneous heating has already been implemented in the Tailored Forming process chain by inductive heating of pre-joined steel aluminium semi-finished products [11]. However, due to the small layer dimensions of the Inconel targeted within this project, inhomogeneous heating is challenging. Therefore, in this study a homogeneous heating was carried out as a first approach to proof a simple implementation concept for further experimental tests.

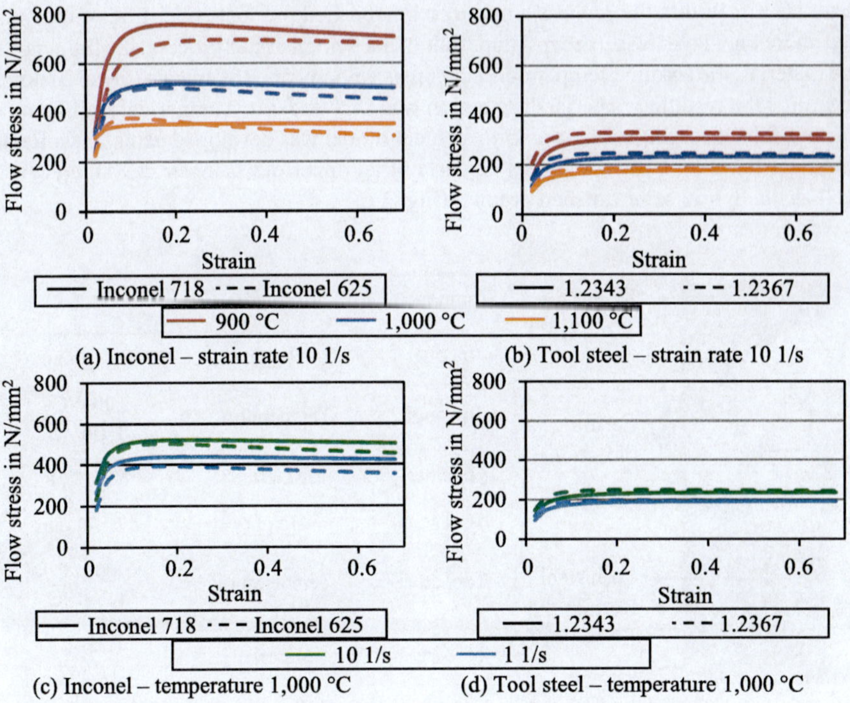

Fig. 1. Calculated flow curves using GMT of the two Inconel alloys (a) and tool steels (b) at different temperatures with strain rates of 10 1/s; Flow curves with different strain rates at 1,000 °C for Inconel (c) and tool steels (d).

Flow curves indicate that high forming temperatures should be applied to the Inconel as the yield stress difference is reduced with increasing temperature. However, performed upset tests for the Inconel alloys showed increased brittleness for temperatures of 1,200 °C, shown by cracking of the specimens at the edge. Therefore, a forming temperature of 1,200 °C is not considered further in the following. In addition to the brittleness with increasing temperature, the resulting hardness of the Inconel should also be taken into account. An increase of the forming temperature reduces the resulting hardness of the Inconel [5, 12]. Therefore, a balance must be made between high temperatures to reduce flow stress and low temperatures to avoid hardness loss. Accordingly, temperatures of 1,000 °C and 1,100 °C are preferred in the numerical investigations.

3 Numerical Simulation Model

For a first investigation the forming of hybrid semi-finished workpieces in an abstracted hot forging process was considered. The demonstrator process is based on an existing tool system, which was already successfully used for the hot forging of dies from mono-material [9]. Within the process a hybrid semi-finished product consisting of tool steel and an Inconel layer is formed to a cup. Using this demonstrator process, the influence of the material, the forming temperature, the press velocity and the Inconel layer thickness regarding the resulting material distribution was analysed. For a numerical simulation of this hybrid hot forging process a finite element model was developed using the software Simufact Forming v16. The model consists of an upper die, a lower die, an ejector as well as the hybrid semi-finished product (Fig. 2 (a)).

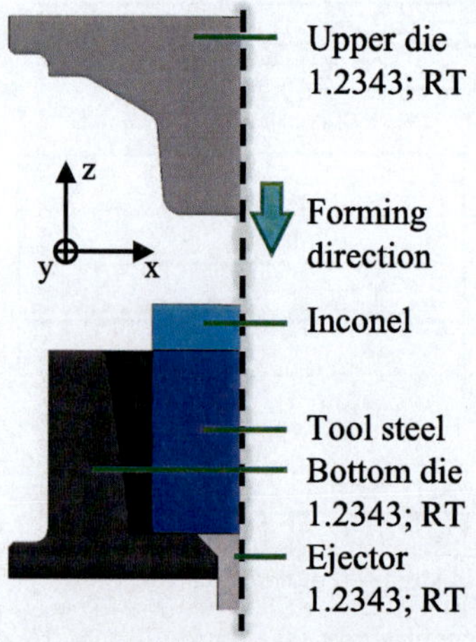

(a) Simulation model

Component	Parameter	Value
Inconel	Height	3 / 10 mm
Inconel	Temperature	1,000 °C / 1,100 °C
Inconel	Material	625 / 718
Tool steel	Height	47 / 40 mm
Tool steel	Temperature	1,000 °C / 1,100 °C
Tool steel	Material	1.2343 / 1.2367
Press	Forming speed	5 / 15 mm/s

(b) Varied parameters

Fig. 2. a) Simulation model with designations for the different components; b) Listing of the varied parameters within the simulation (RT = room temperature).

The total height of the semi-finished product was 50 mm, resulting in a height of the steel part of 40 mm when using an Inconel layer of 10 mm or a height of the steel part of 47 mm when using an Inconel layer of 3 mm. The diameter of the semi-finished product was 40 mm. Within the study, initial temperatures of 1,000 °C and 1,100 °C for the semi-finished product were considered. Initially, homogeneous heating of the tool steel and the Inconel layer was assumed. Furthermore, the influence of the press velocity was investigated, taking into account a punch speed of 15 mm/s and 5 mm/s.

Analogous to the experimental tests, a hydraulic press was assumed. The stroke was adjusted to achieve a final component thickness of 17 mm in the middle of the formed product. An overview over considered process parameters (full factorial variation) is given in Fig. 2 (b). The friction between the semi-finished hybrid product and the tools were modelled by means of combined friction model with a friction coefficient μ of 0.15 and a friction factor m of 0.3 [3]. The dies were assumed to be rigid with heat conduction and an initial temperature of 20 °C were defined. Since pre-joined hybrid semi-finished products are to be used in accordance with the Tailored Forming concept, the contact between tool steel and Inconel is modelled as tied contact. In this case, no relative movements or separation between the bodies are possible [8]. If the distortion of the Inconel or tool steel was too strong, the programme remeshed the parts. In a first step, the comparison between simulation and experimental tests was used to verify the determined material models. In a second step, the boundary conditions such as starting temperature or press speed were varied to analyse the effect on Inconel layer distribution and investigate possible process windows for further experimental trials.

4 Numerical Simulation Results

4.1 Comparison of Simulation and Experiment

To validate the material data and the numerical model, the demonstrator process was carried out experimentally. First, the material combination tool steel 1.2343 and Inconel 625 was used. The layer thickness of the Inconel was 3 mm. Heating was carried out homogeneously to 1,100 °C and the subsequent forming was carried out at a press speed of 15 mm/s. The tests were carried out on the hydraulic press.

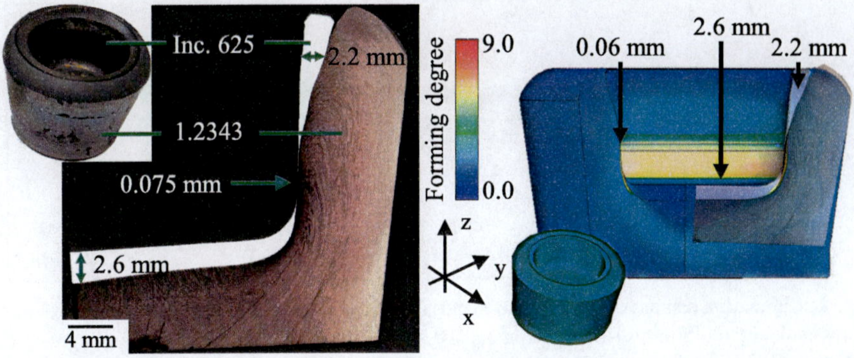

Fig. 3. Result overview and microsection (tool steel was cut down) of the experimental abstracted hot forging test (1,000 °C, 15 mm/s) of 1.2343 (47 mm height) with Inconel 625 (3 mm height, Inc. 625) and comparison between simulation and experimental result.

Experimental results indicate that forming of pre-joined semi-finished products consisting of tool steel and an Inconel alloy as layer is possible. The hybrid semi-finished product was successfully formed and no separation of the materials or tearing off of the

Inconel layer can be seen in the micrograph (Fig. 3). However, a significantly inhomogeneous layer distribution can be observed. Nevertheless, no cracking of the layer was identified. In the bottom area of the cup, the layer thickness is 2.6 mm. In the side area, however, the Inconel is considerably thinner. The layer thickness is reduced to as little as 0.075 mm. In the upper area of the cup, the layer thickness increases again.

For validation, the numerically calculated material distribution was compared with the experimental results in Fig. 3. Overall, a good agreement was achieved. Only in the side area a small deviation of 0.015 mm can be seen. Small differences between simulation and experiment may have been caused by non-centred insertion of the semi-finished products into the press during the experimental tests.

4.2 Results of Numerical Sensitivity Study

The simulations were carried out for all material combinations of Inconel and tool steel. First, the results considering a starting temperature of 1,000 °C and 1,100 °C with a press velocity of 15 mm/s are discussed (Fig. 4).

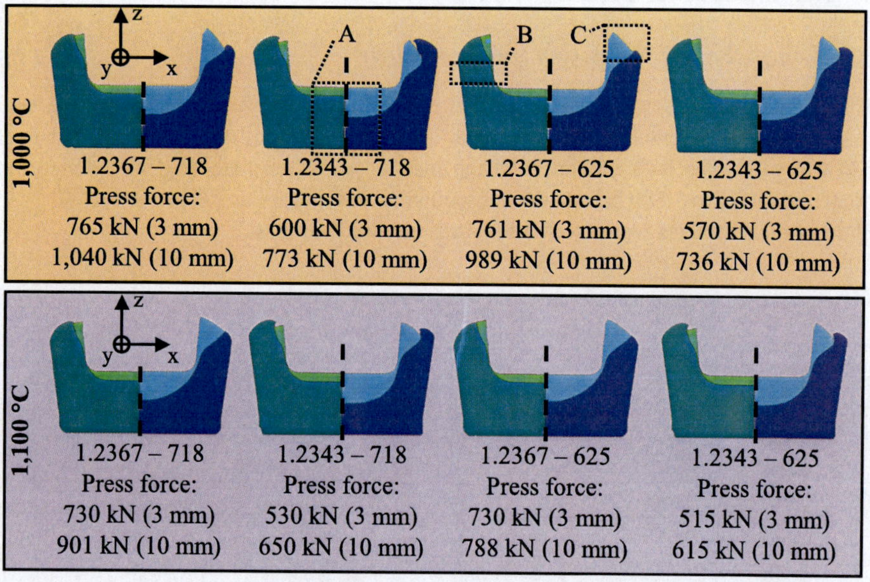

Fig. 4. Simulation results for the different material combinations at two starting temperatures and a press velocity of 15 mm/s (initial Inconel = 10 mm in blue; initial Inconel = 3 mm in green; A = bottom area; B = side area; C = upper part).

The chosen material combination had a minor influence on the component geometry with regard to the formation of the minimum layer thickness or the distribution of the two materials in the formed component (Fig. 4). In the case of the material combination 1.2367 with Inconel 625 for example, the upper part (C) of both materials ends evenly. In comparison, the material combination of 1.2343 and Inconel 718 results in a higher difference between the height of tool steel and Inconel (C). This observation could be made at both investigated temperatures and can be explained with the help of the flow curves as well as the selected geometry (Fig. 5). Due to the arrangement of the materials and the choice of process, the material initially flows radially before the material flow occurs against the direction of the press. As a result of the lower flow stress of the tool steel, it is always first plasticised and radially displaced. With a thicker Inconel layer, the cross-section for the tool steel becomes smaller in order to flow against the press direction after the radial form filling (Fig. 5(a)). As a result, the Inconel layer is formed into a convex area (A), which increases the press force see Fig. 4. Whereas with a thin Inconel layer the press force is lower, because the cross-section to the upward movement of the tool steel is larger (Fig. 5(b)). This allows the Inconel layer to be formed with a constant thickness in area (A). Regardless of the thickness, a relatively strong thinning of the Inconel layer for both thicknesses can be observed in the side area (B).

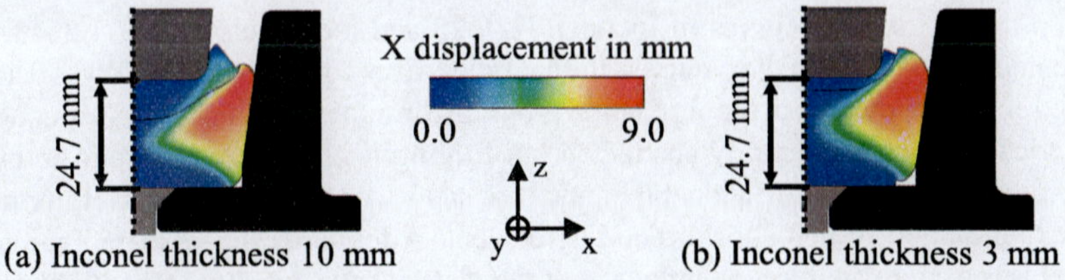

Fig. 5. The X displacement of 1.2367 with Inconel 718 at a press speed of 5 mm/s (a) Inconel 3 mm; b) Inconel 10 mm).

The variation of the Inconel alloy and the temperature shows an influence regarding the required press force. This becomes particularly clear when comparing 1.2367 with Inconel 718 (1,100 °C: 901 kN) and 1.2367 with Inconel 625 (1,100 °C: 788 kN), each with a 10 mm Inconel thickness. Furthermore, a difference in the tool steel used is also evident, as for example in the comparison with Inconel 718 with a height of 10 mm (1.2367: 1,040 kN; 1.2343: 773 kN). The higher material influence due to the change in tool steel can be explained by the larger volume ratio compared to Inconel. The highest influence on the press force can be seen for a variation of the Inconel layer thickness.

Subsequently the press speed was reduced to 5 mm/s for starting temperatures of 1,000 °C and 1,100 °C (Fig. 6). The results show the same tendencies as for the press speed of 15 mm/s. Again, the use of the Inconel alloy 625, the tool steel 1.2343 as well as a reduction of the thickness layer or the increasing of the starting temperature result in lower press forces. When comparing the simulation results regarding the two press speeds, an influence on the required press force can be seen (Fig. 4, Fig. 6). Although higher strain rates initially lead to higher flow stresses, the press force is reduced due to lower heat loss during the process. The lowest press force could be determined for the material combination 1.2343 and Inconel 625 (3 mm thickness) at an initial temperature of 1,100 °C.

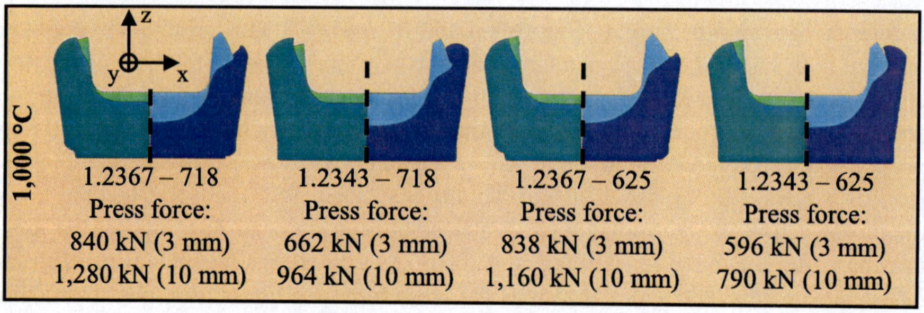

Fig. 6. Simulation results for the different material combinations at two starting temperatures and a press velocity of 5 mm/s (initial Inconel = 10 mm in blue; initial Inconel = 3 mm in green).

5 Discussion and Outlook

Within this work, flow curves of Inconel 718/625 and tool steels 1.2367/1.2343 were presented. Based on these flow curves, the analytical flow curve model GMT was parameterised and implemented in the simulation program Simufact Forming v16. Subsequently, numerical and experimentally abstracted hot forging tests were carried out on hybrid semi-finished products of Inconel 625 and tool steel 1.2343. Comparison of numerical and experimental results show a good agreement. Afterwards the validated numerical model was used to analyse the influence of the material combination, layer thickness as well as process conditions such as starting temperature and press speed. The comparison had shown that the selected Inconel layer thickness has an influence on both the resulting material distribution in the side area and especially in the bottom area and the required press force. In contrast, the press speed had only a minimal influence on the final geometry. The final geometry in the upper outlet area of the press was mainly affected by the choice of material and the initial Inconel thickness. For example, for an initial thickness of 10 mm the Inconel has flowed beyond the tool steel. In comparison, same hight in the upper area (C) was achieved when Inconel was used at a lower thickness. This becomes particularly clear when comparing the identical material combinations. An influence of the temperature on this effect cannot be determined.

Results show, that due to significantly differing flow properties of considered tool steels and Inconel alloys an adapted process design is necessary for combined hot forging. Furthermore, the influence of temperature on the formability of Inconel needs to be taken into account. As high temperatures result in an embrittlement the temperature should be as low as possible. On the other hand, lower temperature increases the flow stress differences between tool steel and Inconel and results in higher press forces.

Overall, an influence on the required press force can be seen for all considered parameters. Since the Inconel layer is to be used as a tool protection layer, the focus is on the material distribution and layer thickness. Both the experimental investigations and the further numerical sensitivity studies showed a clearly inhomogeneous distribution of the layer thickness. In the side area of the cup, for example, there is a strong thinning of the material. A similar percentage maximum thinning can also be achieved with a significant increase in the thickness of the initial layer (98% at 3 mm; 98.4% at 10 mm). It should be emphasised that despite thinning, no cracking of the Inconel layer could be observed. The strong thinning is due to the tool geometry used. With regard to a later application for the production of hybrid tools, the material flow of both materials has to be investigated and controlled by a suitable process control and tool design. In general, a higher starting temperature (approx. 1,100 °C) and a lower Inconel layer thickness should be selected.

In future work, a hot forging process for a more complex mandrel die geometry will be numerically designed. Subsequently, hybrid mandrel dies will be produced and tested in experimental hot forging tests to analyse the potential regarding an increase of die service life.

Acknowledgement. The results presented were obtained in the research project "High-temperature-resistant tool edge layer by applying the Tailored Forming technology to hot forging tools" (T03) funded by the Deutsche Forschungsgemeinschaft (DFG, German Research Foundation) – Collaborative Research Centre 1153 "Process chain to produce hybrid high performance components by Tailored Forming" - 252662854. The authors would like to thank the German Research Foundation for the financial support.

References

1. Vollrath, K.: Simulation in the forging industry, forging industry information service – Special Edition, Hagen (2013)
2. Bouguecha, A., Behrens, B.-A., Bonk, C., Rosenbusch, D., Kazhai, M.: Numerical die life estimation of a crack susceptible industrial hot forging process. In: AIP Conference Proceedings, vol. 1896, p. 190012 (2017)
3. Behrens, B.-A., Rosenbusch, D., Wester, H., Siring, J.: Numerical investigations on stresses and temperature development of tool dies during hot forging. Key Eng. Mater. **926**, 559–568 (2022)
4. Klocke, F., et al.: Wear protection of deep drawing tools by systematic optimization of highly stressed surfaces. Adv. Mater. Res. **907**, 439–453 (2014)
5. Renhof, L.: Mikrostruktur und mechanische Eigenschaften der Nickellegierung IN 718 (2007)
6. Davis, J.R.: ASM Speciality Handbook Heat-Resistant Materials. ASM International, Novelty (1997)
7. Klocke, F.: Zerspanung mit geometrisch bestimmter Schneide. 9. Auflage. Springer, Berlin (2018)
8. Landgrebe, D., Krüger, L., Schubert, N., Jentsch, E., Lehnert, T.: Resource-efficient development of thermally highly resistant engine components of hybrid metal composites – experiments and numerical analysis. Procedia Eng. **207**, 884–889 (2017)
9. Behrens, B.-A., Brunotte, K., Petersen, T., Relge, R., Till, M.: Transfer of ausforming parameters to scaled forging tools. In: METAL 2021 Conference Proceedings (2021)
10. Behrens, B.-A., Uhe, J.: Introduction to tailored forming. Prod. Eng. Res. Dev. **14**(2), 133–136 (2021)
11. Kruse, J., et al.: Numerical simulation and experimental validation of the cladding material distribution of hybrid semi-finished products produced by deposition welding and cross-wedge rolling. Metals **10**, 1336 (2020)
12. Lalam, S.V., Reddy, G.M., Mohandas, G., Kamaraj, M., Murty, B.S.: Continuous drive friction welding of Inconel 718 and EN24 dissimilar metal combination. Mater. Sci. Technol. **25**(7), 851–861 (2009)

Numerical Analysis of Tools for Hot Forming of Sheet Metal with Consideration of Mechanical and Thermal Process Loads for Structural Optimisation

S. Vakulenko[1(✉)], V. Kräusel[1], M. Dix[1], and R. Neugebauer[2]

[1] Fraunhofer Institute for Machine Tools and Forming Technology, Reichenhainer Straße 88, 09126 Chemnitz, Germany
Serhii.Vakulenko@iwu.fraunhofer.de
[2] Headquarters of the Fraunhofer-Gesellschaft, Hansastraße 27C, 80686 Munich, Germany

Abstract. An accurate understanding of the loads on the active parts of highly loaded forming tools is crucial for the robust design and subsequent optimisation of the tools. This publication presents a numerical simulation-based methodology for reliably predicting the load and structural behaviour of hot forming tools, addressing the challenge of process-related mechanical-thermal loads.

In this approach, the first step is to determine the mechanical and thermal loads acting on the active tool parts during the forming process by means of process simulations. Subsequently, the effects of these mechanical and thermal process loads on the deformations and stress distributions in the forming tool are investigated by means of a coupled thermomechanical simulation. The simulation results are validated using real tests. The developed methodology based on the numerical approach shows good agreement with the results of the real experiments and can be used for topology optimisation and design of the cooling system.

Keywords: Finite Element Analysis · Hot Sheet Metal Forming Tool · Load Distribution

1 Introduction

Hot sheet metal forming (SMF) is a process used in industry to produce complex high-strength parts made of boron-manganese steel. By combining of forming and heat treatment (quenching) in one process step, the formability and mechanical properties of these parts can be improved [1]. Accurate prediction of the complex mechanical and thermal stresses is essential for reliable tool design. However, the design of hot forming tools, such as dies and punches, is currently based on empirical knowledge [2] and standards [3], resulting in excessive tool weight and potential design limitations [4]. Advanced numerical analysis techniques are required to predict the mechanical and thermal stresses in hot forming processes, leading to optimised tool designs with improved performance and reduced weight.

T. Bauernhansl et al. (Eds.): WGP 2023, LNPE, pp. 648–659, 2024.
https://doi.org/10.1007/978-3-031-47394-4_63

Finite element (FE) analysis is commonly used to predict tool deformation and stress distribution in SMF processes [5] and for optimising cooling channels in hot forming tools [6]. Previous research has used FE methods to investigate the prediction and influence of contact forces on tool behaviour in SMF.

For example, the authors in [7] introduced a numerical method for determining the contact pressure during the stamping process, enabling the characterisation of time-dependent contact conditions between the tool and the workpiece surfaces. Sah and Gao [8] used a Bayesian framework to combine the results of FE analysis with experimental measurements from sensors. Their objective was to estimate the pressure distribution embedded in the die, similar to the approach outlined in [9].

Numerical methods are commonly used to predict contact forces in SMF tool optimisation via topology optimisation (TO). Much of the previous research [10] aimed to integrate structural analysis and forming simulation in a single FE model, but this was computationally intensive, modelling is complex and limited for large industrial tool designs [11]. This study proposes a separate approach with independent forming simulation and transfer of contact force results for structural analysis of the tool component.

Some studies have explored a similar approach. For example, in [12] LS-DYNA® software was used to simulate SMF and optimise the die structure by obtaining the loads during the production process. The active parts of the tool were approximated as rigid and the contact forces were projected onto the 3D die geometry for TO using the Optistruct software. Similar methods have recently been proposed for the structural analysis and weight reduction of SMF tools for the production of automotive panels [13] and advanced high-strength steel parts for automobiles [14].

Temperature stresses have been used in previous studies to facilitate the design and optimisation of cooling systems in hot SMF dies. For example, in [15] the hot SMF was simulated to validate the effectiveness of conformal die cooling channels and in [16], die channels produced using additive manufacturing techniques were analysed to assess their efficiency.

However, the content of the previously mentioned work focused on cold SMF, considering the behaviour of a die under mechanical loads, or the efficiency of the design of a new cooling system for a hot SMF die. In this paper, a numerical simulation-based method is proposed for the analysis of highly stressed forming tools used in hot SMF processes. The proposed approach aims to accurately predict the mechanical and thermal loads on the active parts of the tool and to study their impact on the structural behaviour of the tool.

2 Methodology

Figure 1 illustrates the proposed methodology for the numerical analysis of tools used in hot SMF applications, considering both mechanical and thermal process loads.

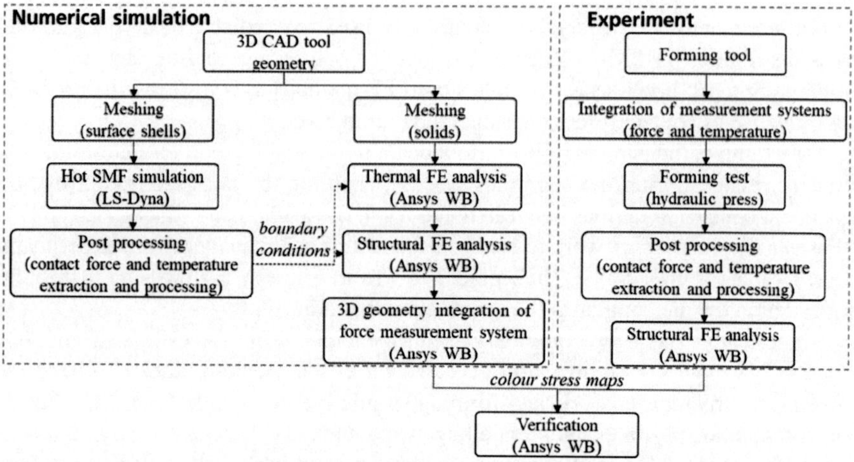

Fig. 1. Proposed methodology for the FE analysis.

To demonstrate the proposed methodology, a press hardening tool for manufacturing the base of a B-pillar of a passenger car was chosen as the object of study. The tool consists of three active components: the die, the punch, and the blank holder. Although this paper focuses on the die, it is important to note that the methodology can be applied to evaluate other active components of the tool.

The methodology is based on FE methods using LS-DYNA® to simulate the press hardening process. The active tool parts were represented in the simulation as 2D rigid surfaces without deformation. The resulting simulation provided contact forces and temperatures at each node of the thin shell elements during SMF, which were then used as boundary conditions for the thermomechanical analysis in ANSYS Workbench (WB).

To validate the numerical model, real tests were carried out to record the forces and temperatures during the press hardening process. The simulated results were compared with the experimental data using colour resultant stress maps projected onto the force measurement system within ANSYS WB. This visual representation allowed a direct comparison and assessment of the agreement between the simulated and experimental results.

3 Numerical Simulation

3.1 Modelling of the Forming Tool

To validate the proposed methodology, a forming tool was used to manufacture a crash-relevant press-hardened structural component, specifically a B-pillar base of an automobile. The active parts of the hot forming tool (Fig. 2) responsible for forming the B-pillar base are made of nodular cast iron (EN-GJS-700–2). In particular, the punch and die have cooling channels with different configurations.

To perform the numerical simulation of press hardening using 2D shells, the first step in constructing the simulation model was to generate them based on the 3D computer-aided design (CAD) geometry of the active tool components. This process aimed to

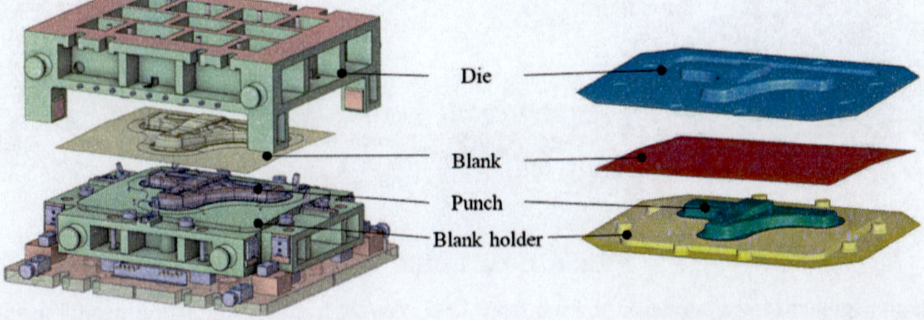

Fig. 2. B-pillar base tool (left); surface shells of active tool parts for SMF simulation (right).

accurately replicate the surface geometry of the tool, as shown in Fig. 2. The FE model used in this study consisted of 332,500 elements and 331,000 nodes, with the largest element size of 10 mm located mainly behind the sheet deformation zone and the smallest element size set at 0.5 mm for the small radii of the active tool parts.

3.2 FE Simulation of the Hot Sheet Metal Forming Process

The SMF process was simulated using LS-DYNA® LS-PrePost v4.7 software. To enable a realistic prediction of the material behaviour during forming, the blank was modelled in the simulation as an elastic-plastic material using the material card *MAT_UHS_STEEL [17]. The parameters used in this material card are based on experimental data for the boron-alloyed quenched and tempered steel 22MnB5 of 1.5 mm thickness. They include not only the chamfer transformations in the material but also the temperature-dependent flow curves.

In the simulation setup, the punch was constrained to allow movement and rotation in all directions based on the global coordinate system. The die and blank holder were allowed to move freely in the punching direction. The simulated hot SMF was divided into five distinct and sequential operations that define the kinematics of the calculated simulation and accurately capture the different operations (OP) of the press hardening process: OP 10 – heating of the parts up to a temperature of 950 °C; OP 20 – loading the parts into the die (by gravity); OP 30 – closing the tool; OP 40 – forming; OP 50 – quenching.

During the last two operations, the die movement was controlled by applying a force of 1,000 kN, corresponding to the actual tests discussed in Chapter 4. The cooling process was excluded during the OP 50 for simplification purposes. The highest tool contact loads (forces and temperatures) were recorded during the last two operations and these data were exported from LS-DYNA® and converted to CSV (comma-separated values) format for seamless integration into the subsequent thermomechanical analysis. The distribution of forces at each location is shown visually in Fig. 3, providing a clear representation of the resulting behaviour.

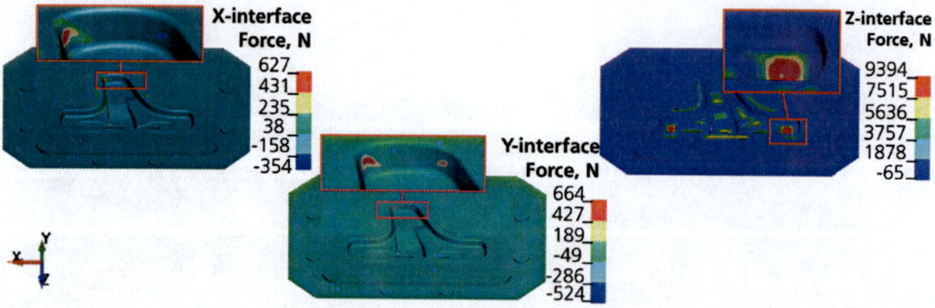

Fig. 3. Contact forces acting on the die from LS-DYNA® forming simulation results in three directions.

3.3 Thermomechanical Analysis

The next step was to carefully prepare the 3D CAD tool geometry required for reliable FE analysis. To evaluate the thermal loading, a thermal analysis is first necessary to provide information on the temperature distribution in the mechanical assembly. This methodology used a steady-state thermal analysis approach that required the inclusion of thermal conductivity as a primary material input parameter. Table 1 below shows the material parameters used in the thermomechanical simulation.

Table 1. Properties of the die material [18].

Material name	Young modulus (GPa)	Poisson's ratio	Yield strength (MPa)	Ultimate strength (MPa)	Thermal conductivity W/(m·K)	Specific heat J/(kg·K)
EN-GJS-700-2	176	0.275	440	700	31.1	515

In this paper, a tetrahedral mesh was used due to the complex geometry of the tool. The mesh generated was of high quality, as confirmed by an internal mesh quality check, with a minimum element quality of 0.21 and a minimum Jacobian ratio (corner nodes) of 0.5. The mesh had a minimum element size of 1 mm and a maximum of 10 mm. The total number of elements was 2,091,820, with 1,403,879 nodes. The subsequent step involves importing the node data obtained from the three X, Y, and Z coordinates, containing the interface node force and node temperature data, into the deformable solid elements created explicitly for linear thermal-structural analysis. The data, formatted as CSV files, are directly imported and used as a boundary condition within the thermo-mechanical analysis. Nodal temperatures obtained from the SMF simulation using quadratic shell mesh elements were effectively projected onto a tetrahedral solid element type. Despite the different mesh types, the mesh density at locations with radii was carefully controlled to achieve mesh identity between the two models. This meticulous approach ensured a seamless mapping of the data onto the surface of the tool, taking into account its precise position along the coordinate axis.

Once the thermal analysis is complete, the structural analysis integrates the material data, geometry and/or model level branches from the thermal and structural models. This integration eliminates the need for mesh reconstruction and other user intervention in the thermal analysis, resulting in potential time savings and reduced errors [19]. For the mechanical analysis, the boundary conditions are derived from the thermal analysis results and the nodal forces from the forming simulation are applied to the die surface. Constraints were applied in all directions on the opposite side of the die to prevent any movement, representing the actual contact with the top of the press.

The evaluation of the effect of the forces acting on the die was based on the comparative stresses using the Mises criterion and displacement. Figure 4 shows the stress distribution in the die under both thermal and mechanical loads during OP 50.

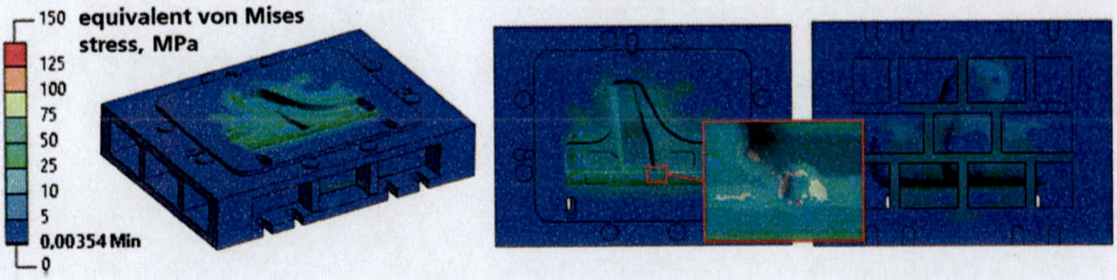

Fig. 4. Equivalent von Mises stress values isometric (left), top (centre) and (right) back view of the die during OP 50.

The results show that the stresses are mainly concentrated under the formed sheet with limited propagation along the tool. Stress levels remain within the elastic limit of the material and range from (50 – 100) MPa with intermittent peaks up to 150 MPa. The simulation results also showed maximum stresses of up to 150 MPa associated with singularities.

By including both thermal and mechanical influences in the FE analysis, a comprehensive assessment of their effects on the tool structure has been achieved. This provides a more accurate understanding of the tool's response to loading conditions during hot SMF.

4 Experiments

To verify the accuracy of the FE simulation results, experiments were carried out using a hydraulic double-column press with a nominal press force of 250 t. A force measurement system, eight type K thermocouples (four in the die and four in the punch) and a string potentiometer sensor to monitor the time-motion curve were used to collect data during the forming process. The purpose of placing thermocouples at selected locations on the die (Fig. 5) was to gain a better understanding and control of the temperature profile during the hot SMF process at various points of contact between the die and the sheet (edges, radii, deepest point) and to evaluate the efficiency of the cooling system. The force measurement system consisted of 14 measurement modules with four measurement points embedded between two plates at pre-determined positions. To ensure optimum performance, it was determined through experimentation that each module worked best with a maximum load of approximately 100 kN, giving a total of 1,400 kN. The tests were carried out using a press force of 1,000 kN. The measurement system was positioned between the ram and the die as shown in Fig. 5.

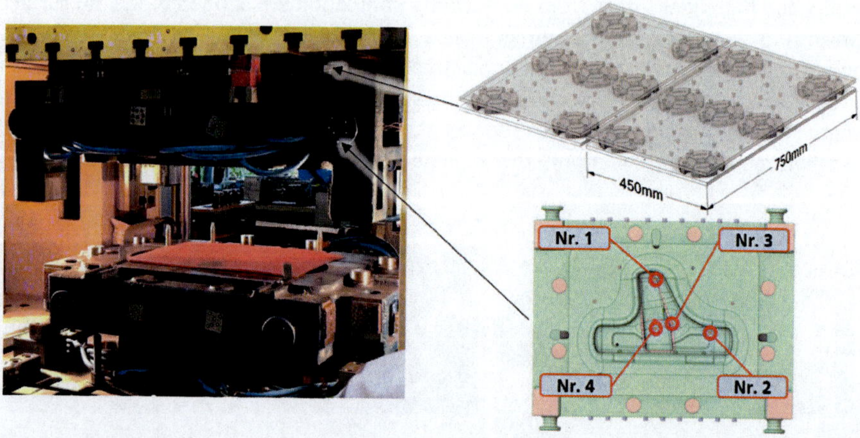

Fig. 5. Experimental set-up: B-pillar base tool installed in press (left) with force measurement system (top right) and thermocouples in the die (bottom right).

The data obtained during the forming process is presented as force and temperature values plotted against time at each measurement point. For additional validation, this data is superimposed on the time-displacement curve obtained from the string potentiometer sensor. This allows the identification of specific times or positions within the press hardening process for later verification of the simulation model.

5 Results and Discussion

Prior to the comparative evaluation of the accuracy of load prediction using the developed methodology, the temperature prediction at the die and its transfer from the hot SMF simulation was thoroughly evaluated. This involved comparing the predicted temperatures with thermocouple measurements in the die. Steady-state thermal analysis was used to determine the maximum temperature corresponding to the highest thermal stresses in the die. Figure 6 shows the maximum temperature recorded at the thermocouple locations corresponding to the end of quenching, which is of considerable interest for thermal stress analysis.

The results of the thermal analysis showed good agreement with the actual data obtained from the experiment. The largest deviation of 6 K was observed at thermocouple No. 2. This deviation may be due to a slight displacement of the formed blank, as thermocouple No. 2 is located under the edge of the blank. The results indicate a reliable temperature transfer from hot SMF simulation using the proposed method and can be further used to predict thermal stresses in combined thermomechanical simulation.

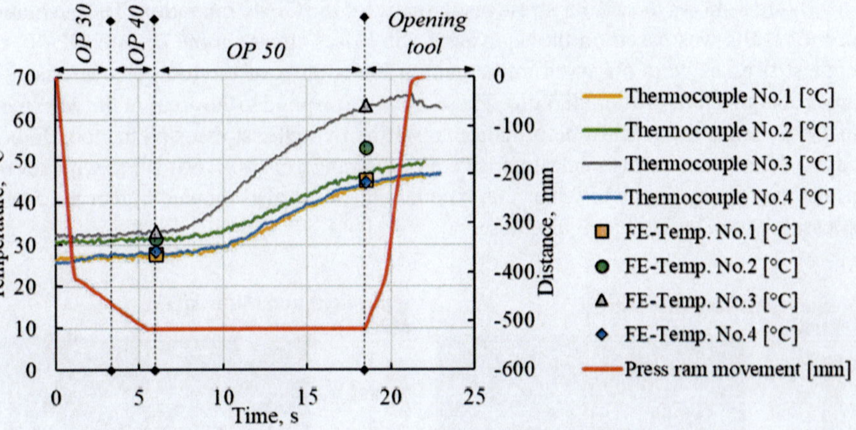

Fig. 6. Comparison of experimental data from thermocouples in the die with simulation data.

To compare the results of the tool load mapping in the numerical analysis with the real experiment, the collected data were projected onto a 3D model of the force measurement system. The force measurement system was integrated into the thermomechanical analysis of the numerical model. In addition, a separate numerical model was created for the experimental data, which included the measurement modules mounted on the base plate of the force measurement system. The experimental data collected was applied to the surfaces of the measurement modules in contact with the top plate where forces were experienced. The base plate of the force measurement system was fixed in all coordinates. The stress distribution on the base plate of the measurement system is visualised by colour stress maps. Figure 7 illustrates the implementation of this approach.

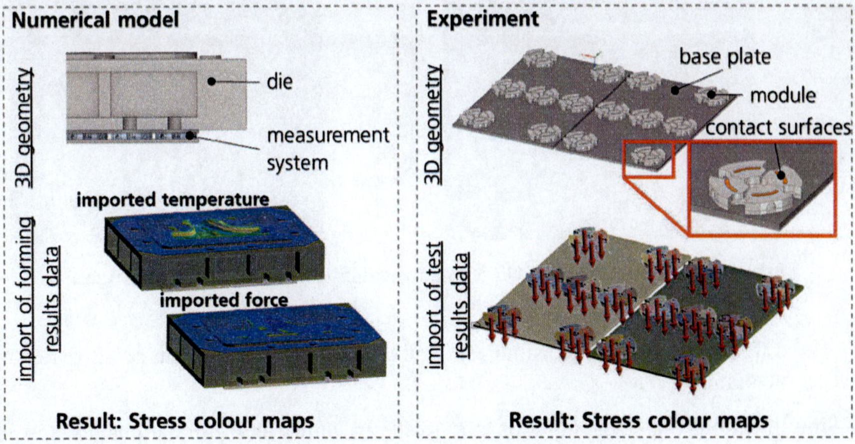

Fig. 7. Integration of a force measurement system to record the stresses resulting from press hardening.

Figure 8 shows the colour stress result maps for the OP 40 operation. The evaluation of the results was based on the equivalent von Mises stress. In the case of OP 40, the highest stresses were observed in the central lower parts of the tool, characterised by areas of high curvature (small radii). These areas correspond to the areas of the workpiece that experience the greatest deformation, resulting in higher stresses on the tool. In both cases, the stresses in the central part were in the range of (40 – 60) MPa, with several peaks varying by about 15 MPa. The simulation results also showed higher maximum stresses associated with singularities.

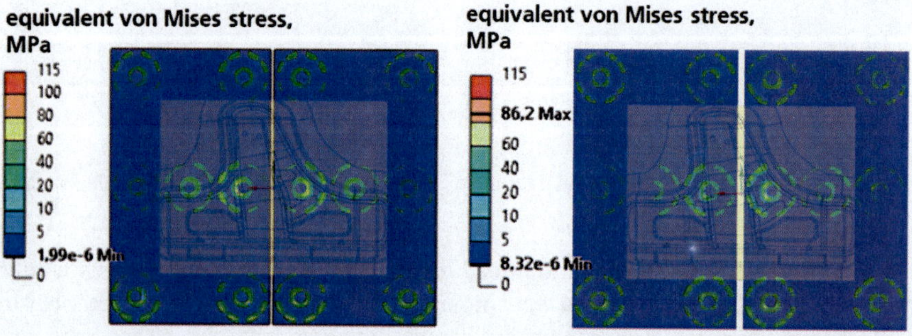

Fig. 8. Colour stress maps of OP 40: numerical model (left); experiment (right).

Similar observations can be made for OP 50, where the stress concentration in the numerical model is more pronounced in the center and at the bottom of the force measurement system and less pronounced at the edges (Fig. 9).

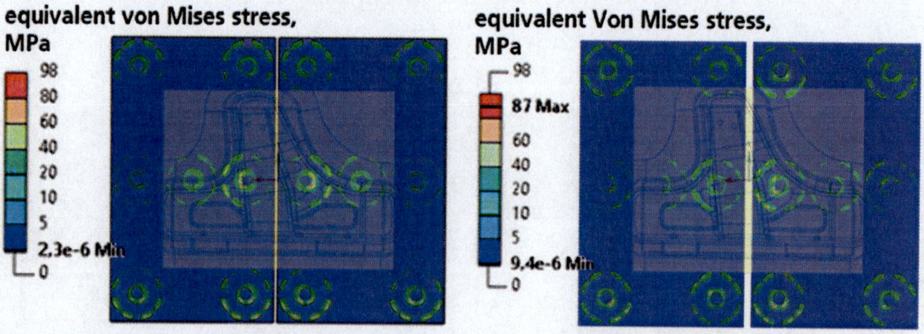

Fig. 9. Colour stress maps of OP 50: numerical model (left); experiment (right).

The differences between the simulated and experimental results can be attributed to several factors:

- Simplifications and assumptions are made in numerical simulations to ensure computational feasibility, but these can lead to deviations from real observations.
- The positioning of the force measurement system in relation to the die can significantly affect the recorded force values during the forming process.

- Misalignment of the die can have a significant effect on both the stress distribution in the die and the recorded forces on the measurement system.
- The recorded force results can be significantly affected by even a slight tilt of the press ram.

However, it should be emphasised that the numerical model shows good agreement with the experimental data. This is evidenced by the consistent trend of the resulting stress distribution on the base plate of the measurement system and the small deviation between the simulated and experimental data.

6 Conclusions

This paper presents a method, based on numerical simulations, for accurately predicting the thermal and mechanical loads as well as the structural behaviour of the highly loaded hot forming tool. The numerical model developed in this study includes thermo-mechanical loads, which distinguishes it from previous research. The effectiveness of the model has been validated by comparison with experimental data.

- The method of registering contact forces and temperatures at the surface nodes of the active tool parts, considering them as rigid shells, allows data to be extracted from the forming simulation. These data can then be used as boundary conditions for the thermomechanical analysis of the tool structure. This approach offers several advantages, including good agreement with real practice, reduced model complexity, simplified adjustment of contact conditions and time savings.
- The developed methodology can be applied to the topology optimisation and cooling system design of the hot forming tool. Furthermore, this approach can be extended to predict thermal stress distribution and assess structural strength in other industrial scenarios such as plastic moulding, aluminium casting and forging.
- Several potential fields for further improvement of the approach have been identified but were not explored in this study. These areas include conducting mesh studies to assess the effect of element size, number and shape on model accuracy; refining the model by eliminating irrelevant contact force and temperature data to increase its simplicity; investigating the effect of maximum press loads to identify critical regions and validate numerical accuracy; and integrating a press model into the numerical analysis to account for all factors influencing stress distribution in the die. These avenues offer promising opportunities for future research and development in this area.

References

1. Hoffmann, H., Neugebauer, R., Spur, G. (eds.): Handbuch Umformen, 2nd edn. Edition Handbuch der Fertigungstechnik / hrsg. von Günter Spur, vol. 2. Hanser, München (2012)
2. Saal, P., Meier, L.: Reduction of residual stresses in forming tools – stress reduction in topology optimized cast part by well chosen geometric parameters, München (2010)

3. Verein Deutscher Ingeniere e.V.: Gusswanddicken in Stanzerei-Großwerkzeugen – Konstruktionsempfehlungen für Werkzeuge mit mehr als 1000 mm Seitenlänge. Beuth Verlag GmbH, Berlin 25.120.10(3120:2015–06). https://www.beuth.de/de/technische-regel/vdi-3120/233326623 (2015)
4. Kruschke, G.: Autobauer Audi entwickelt neue Werkzeuggeneration. Produktion Online, 12 March 2016. https://www.produktion.de/wirtschaft/audi-entwickelt-neue-werkzeuggeneration-311.html?page=2. Accessed 12 Jun 2023
5. Wang, H., Xie, H., Liu, Q., Shen, Y., Wang, P., Zhao, L.: Structural topology optimization of a stamping die made from high-strength steel sheet metal based on load mapping. Struct. Multidisc. Optim. (2018). https://doi.org/10.1007/s00158-018-1899-1
6. Muvunzi, R.: Application of additive manufacturing for improved thermal management of hot sheet metal forming tools. Dissertation, Stellenbosch University (2020)
7. Pereira, M.P., Yan, W., Rolfe, B.F.: Sliding distance, contact pressure and wear in sheet metal stamping. Wear (2010). https://doi.org/10.1016/j.wear.2010.01.020
8. Sah, S., Gao, R.X.: An experimental study of contact pressure distribution in panel stamping operations. Int. J. Adv. Manuf. Technol. (2011). https://doi.org/10.1007/s00170-010-3050-3
9. Sah, S., Mahayotsanun, N., Peshkin, M., Cao, J., Gao, R.X.: Pressure and draw-in maps for stamping process monitoring. J. Manuf. Sci. Eng. **138**, 091005 (2016). https://doi.org/10.1115/1.4033039
10. Xie, Y., et al.: Topology optimization of blank holders in nonisothermal stamping of magnesium alloys based on discrete loads. Int. J. Adv. Manuf. Technol. **106**, 671–681 (2020). https://doi.org/10.1007/s00170-019-04467-7
11. Pilthammar, J., Sigvant, M., Kao-Walter, S.: Introduction of elastic die deformations in sheet metal forming simulations. Int. J. Solids Struct. (2018). https://doi.org/10.1016/j.ijsolstr.2017.05.009
12. Nilsson, A., Birath, F.: Topology optimization of a stamping die. In: AIP Conference Proceedings. Materials Processing and Design; Modeling, Simulation and Applications; NUMIFORM '07; Proceedings of the 9th International Conference on Numerical Methods in Industrial Forming Processes, Porto (Portugal), 17–21 June 2007, pp. 449–454. AIP (2007). https://doi.org/10.1063/1.2740852
13. Su, T., He, T., Yang, R., Li, M.: Topology optimization and lightweight design of stamping dies for forming automobile panels, vol. 121 (2022)
14. Xu, T., Wu, H., Xue, F., Guo, J., Ran, J., Gong, F.: Structural design of stamping die of advanced high-strength steel part for automobile based on topology optimization with variable density method, vol. **121** (2022)
15. Muvunzi, R., Hagedorn-Hansen, D., Matope, S., Madyibi, X., Swart, C.B., Nagel, M.: Industry case study: process chain for manufacturing of a large hybrid hot stamping tool with conformal cooling channels. Int. J. Adv. Manuf. Technol. **110**, 1723–1730 (2020). https://doi.org/10.1007/s00170-020-05992-6
16. Cortina, M., Arrizubieta, J., Calleja, A., Ukar, E., Alberdi, A.: Case study to illustrate the potential of conformal cooling channels for hot stamping dies manufactured using hybrid process of laser metal deposition (LMD) and milling. Metals **8**, 102 (2018). https://doi.org/10.3390/met8020102
17. LS-DYNA®: Keyword user's manual. Volume II Material Models.: Livermore Software Technology Corporation (LSTC), California (2014)

18. Giesserei Hegi AG: Technisches Merkblatt Gusseisen mit Kugelgraphit nach DIN EN 1563. https://www.meuselwitz-guss.de/fileadmin/daten/Dateien/pdf/werkstoffe/Werkstoffkenndaten_Kugelgraphit_getrennt_gegossen.pdf
19. Rosendo, F., Guillermo, L., Paul, P.L., Herbert, Y.: A one way coupled thermo-mechanical model to determine residual stresses and deformations in butt welding of two ASTM A36 steel plates. In: VII International Conference on Computational Methods for Coupled Problems in Science and Engineering (2017)

Investigation of the Process Limits for Hybrid Parts Made of Fiber-Reinforced Thermoplastics and Metallic Sheets Produced with a Two-Stage Isothermal Forming and Consolidating Tool

J. Wehmeyer[1(✉)], T. Fünfkirchler[1], S. Hübner[1], C. Schumann[2], A. Raatz[2], and B.-A. Behrens[1]

[1] Institute of Forming Technology and Machines (IFUM), Leibniz University Hannover, Garbsen, Germany
wehmeyer@ifum.uni-hannover.de
[2] Institute of Assembly Technology (Match), Leibniz University Hannover, Garbsen, Germany

Abstract. Multi-material structures are promising in the automotive industry for achieving lightweight design body construction and functionalization due to their favorable mechanical properties and low structural weight. These structures typically combine metal and plastic materials to create final components with enhanced properties compared to mono-material structures.

The focus of this paper lays on the creation of a manufacturing cell that combines joint forming and heat-assisted press joining techniques to produce components made from steels and continuous fiber-reinforced thermo-plastics in the form of unidirectional carbon-fiber tapes. To improve production efficiency and reduce costs, a manufacturing cell was created and tested, which utilized two robots for automated handling and an isothermal, two-stage forming tool concept to shorten cycle times. The resulting composite components were tested for mechanical performance. Compared to pure steel components, all composite components exhibited a higher specific load capacity. Cycle times in well under 60 s were achieved. The new manufacturing cell led to a significant reduction in process time compared to variothermal tool concepts.

Keywords: Advanced materials · multi-material concepts/solutions · lightweight structures

1 Introduction

Given the growing significance of resource efficiency, driven by both environmental concerns and competitive factors, the effective utilization of resources is increasingly crucial. Numerous strategies can be employed to enhance resource efficiency, such as more efficient product design, reducing the use of raw materials in manufacturing, and facilitating material separation and recycling.

T. Bauernhansl et al. (Eds.): WGP 2023, LNPE, pp. 660–669, 2024.
https://doi.org/10.1007/978-3-031-47394-4_64

One approach to leverage the benefits of both metal and fiber-reinforced plastics and maximize the potential of lightweight construction is through the combination of these materials into hybrid components. This allows for synergistic effects and optimal utilization of lightweight construction capabilities. However, the main obstacle to implementing such material concepts often relies in the lack of cost-effectiveness associated with the manufacturing processes involved. The production of fiber-reinforced plastic-metal hybrid components necessitates tailored handling and forming processes that accommodate the distinct material behaviors, as well as a means of effectively connecting the two components on a large scale.

There are two primary approaches for achieving such connections. Firstly, the two joining partners can be bonded together using a suitable adhesive. The application of such bonding agents enhances the bond strength, but it also introduces significant cost implications and manufacturing challenges. The use of adhesives requires additional process steps, increased safety measures for workers, and prolonged process times due to the curing of the adhesive.

Press joining with variothermic tools is a common method for creating large-area joints between metals and fiber-reinforced plastics. This process involves heating the materials within the tool, melting the thermoplastic, and pressing it at once with the metallic component. Subsequently, the tool is cooled, allowing the thermoplastic to solidify. However, the industrial application of these heated-cooled tools in series production poses challenges in terms of economic feasibility, technical viability, and energy efficiency. The long process times and high thermal alternating loads make it impractical for efficient and sustainable production [1].

Most of the process variants developed thus far have primarily focused on basic feasibility or meeting specific requirements, as confirmed by initial results. The next phase aims to prioritize cost reduction in terms of process complexity and time. The challenge relies in developing processes and process sequences that ensure complete functionality while also providing economic incentives for mass production.

To address this objective, a process chain based on a two-stage, isothermal tool concept with automated handling technology was developed and implemented in this research for producing load-adapted components. Initially, a sample component based on a structural automotive part was defined. The component's load-bearing capacity is achieved through reinforced areas, accomplished by applying thin stiffeners made of fiber-reinforced plastic tapes.

Following commissioning, the optimal process window for fiber impregnation and component consolidation was determined by varying process parameters such as forming force, holding time, and tool temperature, while analyzing their effects on component stability and quality. A comparison between a pure steel component and load-compliant hybrid components will highlight the lightweight construction potential of the final products.

2 State of the Art

2.1 Fiber-Reinforced Thermoplastics

Fiber-reinforced thermoplastics (FRP) offer a promising solution to address the challenges mentioned in [1]. The utilization of FRP has been steadily increasing in recent years and has become a crucial element in achieving future lightweight construction goals. Semi-finished FRP products encompass various types, including long fiber reinforced thermoplastics (LFT), glass mat reinforced thermoplastics (GMT), fiber reinforced unidirectional tapes (UD tapes), and organic sheets [2].

Stamp forming processes can be employed to shape thermoplastics when exposed to high temperatures, enabling their use in large-scale production, depending on the matrix material [3, 4]. Additionally, other joining techniques such as clinching can be utilized at elevated temperatures to consolidate the matrix materials [5].

While FRPs are considered a distinct class of materials, they essentially consist of a composite structure comprising multiple components. Within this composite structure, the individual components have distinct roles: the high-strength fibers bear the mechanical compressive forces and loads, while the matrix material in which the fibers are embedded secures them in place, provides support, and transmits mechanical stresses to the fibers [6].

2.2 Heat-Assisted Press Joining of Hybrid FRP Steel Components

Heat-assisted press joining is a technique that enables the bonding of thermoplastic FRPs and metals, offering the potential for reduced process times and streamlined process chains. In this process, the thermoplastic matrix material of the fiber composite is melted, allowing for a fusion bonding effect with the metallic joining partner [7]. Heat-assisted press joining eliminates the need for separate bonding steps, and with appropriate process control and surface pretreatment, it can achieve structural strengths [8]. The thermoplastic is melted through heat conduction resulting from the contact between the metallic workpiece and the tool in the joining zone [9]. The temperature at which the metal surface is wetted by the molten plastic is crucial for bond formation. For instance, an optimal joining temperature range of 260 °C to 270 °C was determined for a combination of polyamide 66 and steel [10], while a promising bond between polyamide 6 and steel was achieved at a joining temperature of 250 °C [11]. Other subprojects explored the use of inductive heating of the tool [12] and conductive heating of the metallic joining component to reduce process time [13]. The heat-assisted press joining process involves several substeps. Firstly, both materials must be heated to the required joining temperature, which is monitored using thermocouples. The heating occurs superficially in the joining zone area, eliminating the need for complete consolidation of the organic sheet. Once both materials have reached the desired joining temperatures, the metal sheet is pressed onto the organic sheet with a defined force. During this process, the molten thermoplastic fills the undercut geometries of the metal surface, creating a form-fit joint. Additionally, material bonds are formed between the thermoplastic and the metal surface [15]. Subsequently, the cooling process begins. Once the thermoplastic of the organic sheet has solidified, the test specimen can be removed.

Key process parameters for heat-assisted press joining include the joining temperatures of both partners and the joining force, which have been extensively studied [14].

Mechanical pretreatment methods such as sandblasting [9, 16], imprinting of structures [7, 17] and different chemical pretreatments [18] have been considered in further works. These pretreatments generate an activated and partially altered surface structure. Extensive research has also been conducted in the field of optimized surface structuring with undercuts. Figure 1 provides examples of approaches such as CMT pin welding [12], HyTensile [18], laser structuring [17], and electron beam structuring [12].

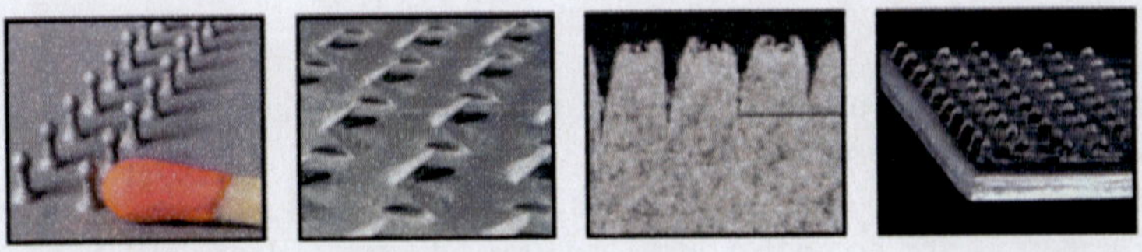

Fig. 1. Surface structuring processes by CMT pin welding [12], HyTensile [18], laser structuring [7] and electron beam structuring [12]

In [19] a variothermal forming tool was developed for producing hat sections comprising carbon fiber reinforced thermoplastic tapes and metal sheets with varying frame angles. Once the tool was operational, hybrid hat profiles were manufactured, and the properties of the components were evaluated. The forming tool has a maximum heating power of 12.4 kW and is heated to a temperature of 250 °C. It can be rapidly cooled down to a removal temperature of 150 °C within 45 s using cooling channels that are filled with continuously circulating water and are positioned near the contour of the tool.

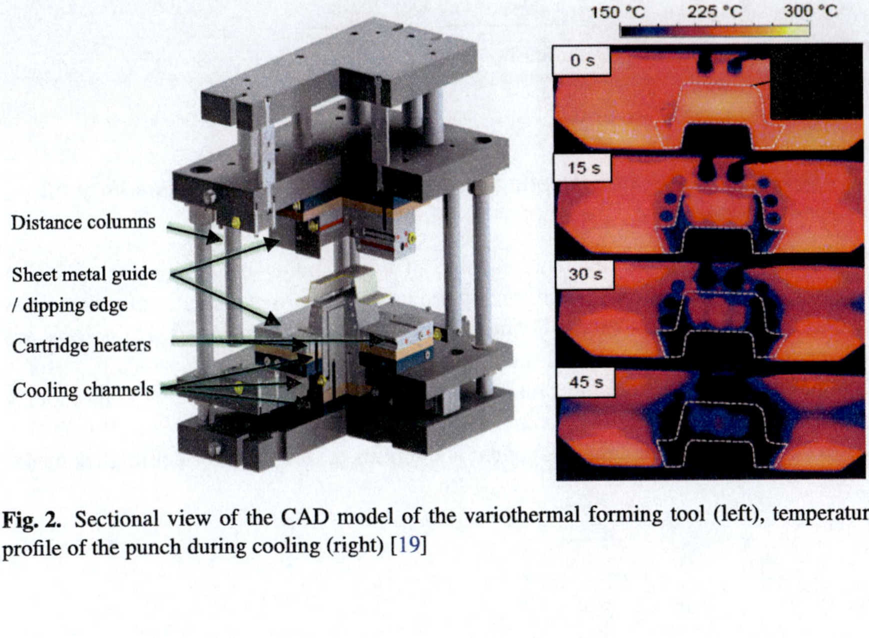

Fig. 2. Sectional view of the CAD model of the variothermal forming tool (left), temperature profile of the punch during cooling (right) [19]

Figure 2 depicts a sectional view of the CAD model representing the forming tool. It should be mentioned that the tool requires several minutes for reheating. This aspect served as the primary motivation for the research discussed in this paper, aiming to establish a process chain that enables economically viable production times for FRP-metal hybrids. An additional objective is to conserve energy by eliminating the necessity of reheating the mold after each part has been produced.

3 Experimental Setup

3.1 Work Piece

The developed component is a hybrid sandwich component featuring specifically reinforced regions. It comprises metallic cover sheets with PA6 (Polyamide 6) films placed as adhesion promoters. No additional adhesives are required for achieving a permanent bond. The core of the component consists of a full-surface layer of UD-CFRP tape (unidirectional carbon fiber reinforced plastic tape). Narrow strips of UD-CFRP tape are further applied to reinforce the specific areas of the reinforcing structures. Figure 3 illustrates the schematic layer structure of the semi-finished product as well as an example of the manufactured component.

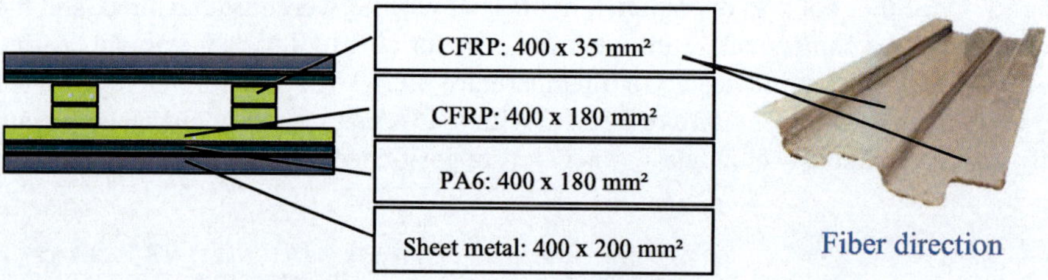

Fig. 3. Layer structure of the semi-finished products (left), manufactured component (right) [20]

3.2 Two-Stage Isothermal Heating/Impregnating and Consolidation/Forming Tool

To produce sandwich components capable of withstanding bending loads, reducing energy costs, and significantly minimizing production time, a two-stage isothermal heating/impregnation tool and consolidation/forming tool were developed. These tools are designed to facilitate automated handling technology. Unlike the variothermal tools mentioned in existing literature, this concept involves both stages being continuously heated at different temperatures using heating coils. This eliminates the energy-intensive heating process for each new component and eliminates the need for cooling channels.

In the first stage of the tooling process, the stacked semi-finished products are heated, and the loosely arranged carbon fibers are impregnated into the PA6 plastic through a force of 325 kN, resulting in thermal joining. The temperature of this tooling stage is set at 270 °C, which is 10 °C below the decomposition temperature of PA6. This temperature is selected to ensure the melting of PA6 and the creation of an optimal heat cushion for transferring to the second tooling stage, while preventing premature solidification of the plastic before the actual forming process. The contours within the mold's die ensure that the additional layers of CFRP-UD tape for the reinforcement contour remain securely in place and do not shift.

In the second stage of the tooling process, the actual forming of the component occurs at a temperature of 120 °C, taking into account the deeper contours of the die. The temperature is chosen to prevent premature solidification of the plastic during the forming process. Subsequently, cooling occurs during an average holding time of 25 s within the mold. Once this time has elapsed, the consolidation of the composite is completed, and the component, which is still warm but fully solidified, can be safely removed.

3.3 Manufacturing Cell with Automated Handling Technology

In order to further reduce production time and enhance the consistency of the tests, an automated system was developed for the stacking of semi-finished products and their transfer into the tools. Two robots were used to carry out these tasks effectively. For the stacking process of the semi-finished products, a Kuka KR6 robot was utilized. One of the key challenges was to design a gripper that could reliably handle and place various components such as rigid metal sheets, CFRP UD tapes of different sizes, and flexible PA6 films in their unprocessed state. To address this challenge, a vacuum suction cup configuration consisting of two arrays, each equipped with four suction cups, was chosen. Additionally, the gripper was equipped with two small heated punches that could be utilized to create thermal bonds between specific points of the semi-finished product layers. This bonding process ensured that the layers remained securely in place during the transfer to the mold. Once the stacking process was completed, the assembled stacks of semi-finished products were deposited at a predefined location for further handling.

For the transfer of the stacked semi-finished products to different stages of the production process, including the heating stage, forming stage, and deposit location, a Kuka KR60 robot was employed. The KR60 robot utilized a magnetic gripper to securely hold the stacks of semi-finished products. The stacking process and the transfer process were carefully synchronized to optimize production time. With a heating or forming time of 20 s (plus 12 s for opening and closing) and a transfer time of 38 s, the total production time per component amounted to 70 s. Since both stages of the tool were loaded in each stroke, this value could be divided by 2, resulting in a theoretical cycle time of 35 s. Figure 4 provides an overview of the complete manufacturing cell.

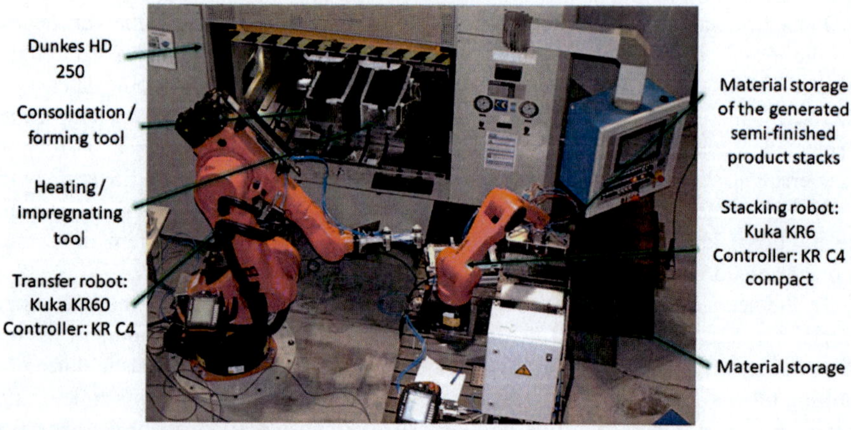

Fig. 4. Production cell consisting of Kuka KR60, Kuka KR6, two-stage isothermal tool and Dunkes 250 forming press [20]

4 Conducting Experiments

The research conducted in this study involved various production parameters, including forming force, holding time in the tool, and temperature in the forming stage. For the metallic cover sheets DC03 and TH340 were used. The average surface roughness (Ra), measured with a confocal LED microscope (Table 1), for DC03 was 0.1401 µm, while for ZM70, it was 0.5711 µm. Additionally, monolithic reference components made from 1 mm thick DC03 were manufactured to serve as references for assessing the impact of the reinforcement structures designed to withstand the load. A special adhesion promoter film was also tested. To demonstrate the potential of the hybrid components in lightweight construction compared to the monolithic components, destructive component tests in the form of three-point bending tests were conducted.

Table 1. Test matrix for the production of the components

Material		Metallography images	
DC03	TH340	DC03	TH340
Tensile strength [MPa]		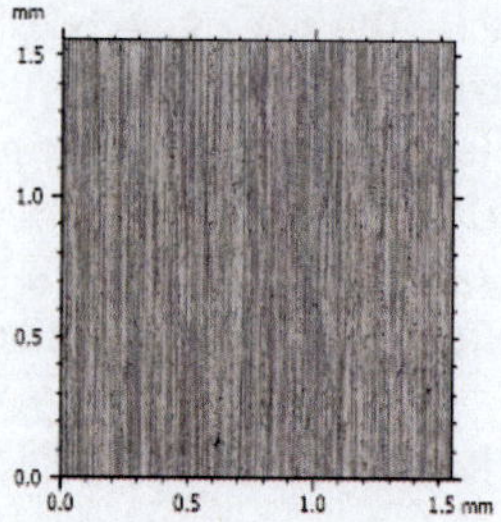	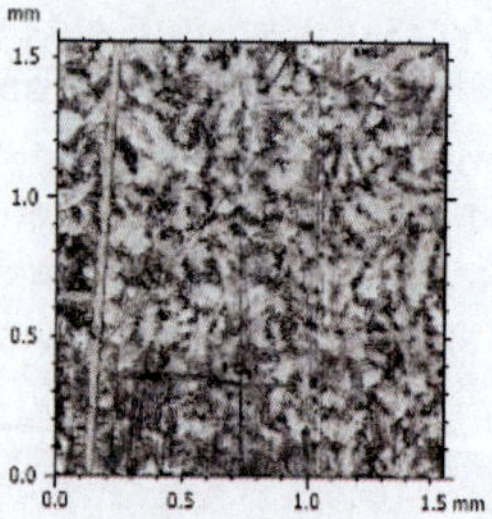
270-370	340-395		
Arithmetic average deviation Ra [µm]			
0,1401	0,5711		
Maximum height of the roughness profile [µm]			
1,534	4,567		

5 Results

The double-hat profiles, produced using the test setup described in chapters 3 and 4, underwent testing through three-point bending tests according to DIN 14130. The specific load capacity of the composites was determined by calculating the ratio between the weight of the composites and the maximum average forces obtained from the three-point bending tests conducted in each test series. Different weight values were obtained depending on the combination. The specific load value can be calculated using the following formula:

$$Max.spec.load\left[\frac{N}{g}\right] = \frac{Force}{Weight\ of\ the\ unit}$$

The test results of the three-point bending tests are shown in Table 2.

Table 2. Parameter combination for generating the maximum specific load capacity per test series.

Test series	Steel	Forming force [kN]	Holding time [s]	Temp. Forming stage [°C]	Max. Spec. Load [N/g]
A	DC03	400	20	100	3.66
B	ZM70	325	20	100	4.72
C*	ZM70	250	20	100	4.08
D*	DC03	250	20	100	4.23

*: Additional layer of adhesion promoter film

All test series exhibit a higher specific load capacity in comparison to the monolithic component. In test series A, the impact of holding time in the mold was investigated by varying it from 20 to 35 s.

It was determined that the influence of holding time was negligible since PA6 completely melted within 20 s. Furthermore, a lower temperature of 100 °C in the forming stage proved to be beneficial across all test series. This can be attributed to the fact that the actual forming process is completed in approximately 2 s, allowing for better solidification of the material composite thereafter.

The highest quality bond was achieved with ZM70 using a forming force of 325 kN, as the rough surface of the steel provides better conditions for thermal press joining compared to the smoother DC03. In this case, the specific load capacity was 56.8% higher than that of the 1 mm thick DC03 sheet. In test series C* and D*, an additional adhesion promoter film was utilized. While it had a positive impact on the DC03 components, it had a negative effect on the ZM components. This is attributed to the fact that the tool was not designed to accommodate the additional layers on the top and bottom sheet metal. The increased volume of material caused the PA6 to be squeezed out, resulting in delamination of the composite.

6 Summary and Outlook

This paper presents a novel two-stage isothermal tool concept with automated handling technology, utilizing two robots. This tooling system enables the production of stacks of semi-finished products composed of steel, PA6 film, and unidirectionally reinforced carbon fiber tapes. In this process, both stages of the tool are continuously occupied, and new stacks of semi-finished products are stacked by the first robot. The production time per component is 70 s, considering a heating or forming time of 20 s (+12 s opening and closing) and a transfer time of 38 s. Since both stages of the tool are always loaded in each stroke, this value can be divided by 2, resulting in a theoretical cycle time of 35 s. Compared to a variothermic tool, the two-stage isothermal tool concept reduces the process time by a factor of 8.5 for this particular component, demonstrating its significant time-saving potential.

Following the manufacturing process, a thorough examination of all produced components was conducted using 3-point bending tests. Four test series were analyzed, involving different metals and an additional bonding agent. All test series demonstrated higher specific load capacities, with improvements of up to 56.8% when compared to a single 1 mm thick DC03 sheet.

Within the scope of this project, a process was developed that enables the production of components from loosely stacked semi-finished products consisting of sheet metal and UD tapes, utilizing a two-part isothermal heating and impregnation stage, as well as a consolidation and forming stage. Due to the individual assembly of the component from individual layers and the comparatively short cycle time, this production technology offers the potential for economically manufacturing components capable of withstanding loads in series production. Future investigations will focus on further validation tests with lower temperatures in the forming stage, as well as additional component testing to address the alpha-delta problem and fatigue strength.

Acknowledgements. The authors thank the Industrial Collaborative Research (IGF) and the European Research Association for Sheet Metal Working (EFB) for the financial support of the research project "Economical production of load-compliant FRP/metal composites" (IGF-Code: 19603N).

References

1. Behrens, B.-A., Neumann, A.: Ermittlung der Prozessgrenzen bei der Umformung von Musterbauteilen aus Faser-Kunststoff/MetallVerbunden. In: Schlussbericht zu dem IGF-Vorhaben DFG-AiF-Cluster Großserientaugliche Prozessketten für hochintegrierte Bauteile aus hybriden FaserKunststoff/Metall-Verbunden Teilprojekt 4 (2016)
2. Friedrich, K., Almajid, A.A.: Manufacturing of advanced polymer composites for automotive applications. Appl. Compos. Mater. **20**, 107–128 (2013)
3. Behrens, B.-A., Hübner, S., Bonk, C., Bohne, F., Micke-Camuz, M.: Development of a combined process of organic sheet forming and GMT compression molding. In: Procedia Engineering 207, Cambridge, pp. 101–106 (2017)
4. Behrens, B.-A., Hübner, S., Neumann, A.: Forming sheets of metal and fibre-reinforced plastics to hybrid parts in one deep drawing process. Procedia Eng. **81**, 1608–1613 (2014)

5. Behrens, B.-A., Hübner, S., et al.: Forming and joining of carbon-fiber-reinforced thermoplastics and sheet metal in one step. Procedia Eng. **183**, 227–232 (2017)
6. Schürmann, H.: Konstruieren mit Faser-Kunststoff-Verbunden. Springer-Verlag, Berlin, ISBN 9783540721895 (2007)
7. Ageorges, C., Ye, L., Hou, M.: Advances in fusion bonding techniques for joining thermoplastic matrix composites: a review. Compos.: Part A Bd **32**, 839–857 (2001)
8. Heckert, A., Zaeh, M.F.: Induction heated joining of aluminum and carbon fiber reinforced nylon 66. J. Laser Appl. Bd. **27**, Nr. S2 (2015)
9. Brecher, C.: Integrative Produktionstechnik für Hochlohnländer. Springer-Verlag, Berlin Heidelberg (2011)
10. Flock, D.: Wärmeleitungsfügen hybrider Kunststoff-Metall-Verbindungen, Norderstedt (2011)
11. Reig, F., Steinhilper, R.: Handbuch Konstruktion, 2nd edn. Hanserverlag, München (2018)
12. Scheik, S., Schleser, M., Reisgen, U.: Thermisches Direktfügen von Metall und Kunststoff – Eine Alternative zur Klebtechnik?, Adhäsion (2012)
13. Schoene, J., Marx, B., Schönberger, J.: Fügen von Metall-Kunststoff-Verbunden. In: ISF Direkt 49 (2014)
14. Lippky, K., Hartwig, S., Blass, D., Dilger, K.: Bonding performance after aging of fusion bonded hybrid joints. Int. J. Adhes. Adhes. **93**, 102331 (2019)
15. Üzüm, M.: Metal polymer hybrids – Multiscale adhesion behavior and polymer dynamics. Technische Universität Berlin, Dissertation (2015)
16. Bergmann, J.P., Stambke, M.: Potential of laser-manufactured polymer-metal hybrid joints. Phys. Procedia **39**, 84–91 (2012)
17. Matsuzaki, R., Tsukamoto, N., Taniguchi, J.: Mechanical interlocking by imprinting of undercut micropatterns for improving adhesive strength of polypropylene. Int. J. Adhes. Adhes. **68**, 124–132 (2016)
18. Bräunling, S., Staiger, E.: Textil-Blech-Verbund Hybride auf Basis von Kohlenstofffasern und Thermoplast durch umformende Verbundherstellung. In: EFB Projekt, 3. Zwischenbericht AK-Sitzung “Hybride Strukturen” (2015)
19. Behrens, B.-A., Neumann, A.: Großserientaugliche Prozessketten für hochintegrierte Bauteile aus hybriden Faser-Kunststoff/Metall-Verbunden – Teilprojekt 4: Ermittlung der Prozessgrenzen bei der Umformung von Musterbauteilen aus Faser-Kunststoff/Metall-Verbunden. In: AiF-Abschlussbericht 17689BR, Hannover (2016)
20. Behrens, B.-A., Raatz, A., Wehmeyer, J., Bohne, F., Lorenz, R., Schumann, C.: Wirtschaftliche Fertigung belastungsgerechter FVK/Metall-Verbunde. In: IGF-Vorhaben 19603 N, Schlussbericht (2021)

Investigation of the Influence of Slightly Increased Process Temperatures on the Extrusion of Functional Aluminum Components in Sheet-Bulk Metal Forming

T. Wild(✉) and M. Merklein

Institute of Manufacturing Technology, Friedrich-Alexander University Erlangen-Nuremberg, Egerlandstraße 13, 91058 Erlangen, Germany
thomas.wild@fau.de

Abstract. By using high-strength aluminum, weight savings can be achieved. However, limited formability of these alloys at room temperature is a challenging aspect regarding the manufacturing of components with filigree functional elements within sheet-bulk metal forming. So as to derive detailed process knowledge and to improve the achievable process results, forward extrusion of geared components made of EN AW 6082 sheets at room temperature is investigated within this work. Furthermore, robust large-scale industrial cold forming is a demanding task, as tool temperatures increase over time due to forming and frictional heat, leading to varying material and process properties. In order to determine measures for the compensation of these fluctuations, additional experiments are carried out with slightly increased tool temperatures. By analyzing workpiece and process related parameters for both sets of experiments, potential approaches to control the outcome of forming operations under variable conditions are discussed.

Keywords: Sheet-Bulk Metal Forming · Extrusion · Aluminum

1 Introduction

Ecological and economic challenges define the requirements for modern manufacturing of components in the industrial environment [1]. In various domains, this leads to the fact, that high volume manufacturing processes become more complex and more demanding in terms of a robust implementation [2]. On the one hand, it is increasingly important to produce near net-shape and functionally integrated components, which then leads to difficulties in forming these workpieces [3]. On the other hand, this challenging setting is further intensified by the fact, that high-strength lightweight materials like precipitation hardenable aluminum are increasingly used [4]. These are characterized by low formability at room temperature. However, cold forming of these alloys is beneficial as process chains can be shortened, as preliminary and subsequent energy-intensive heat treatment can be avoided. Furthermore, good mechanical properties of the components can be achieved due to strain hardening in cold forming [4]. Another main aspect of

T. Bauernhansl et al. (Eds.): WGP 2023, LNPE, pp. 670–679, 2024.
https://doi.org/10.1007/978-3-031-47394-4_65

modern manufacturing is the development of a robust production chains with the lowest possible scrap rates across different production batches [2]. However, transient process conditions can occur during the run-in phase in cold forming. As a consequence of forming and frictional heat, the temperature of the used tool set increases [5]. During the manufacturing of the first 100 components tool temperature rises [6] and may reach values up to 80 °C [7] and thus altering tribological conditions in this timeline [5]. In the course of this study, it is the objective to investigate the effects of these fluctuations on an extrusion process of geared components within sheet-bulk metal forming (SBMF). Characteristic for this process class is the application of bulk forming operations on sheet metal in order to manufacture filigree functional elements. This results in complex process conditions, leading to restrictions with regard to the geometric accuracy of components and the overall achievable die filling of functional elements [8]. In previous studies, the local adaption of the tribological system presented an effective approach to control the material flow, therefore improving the process results [9]. For this reason, the influence of fluctuating mechanical and tribological conditions as a consequence of tool heating in high volume manufacturing and possible compensation measures for this behavior will be investigated in the following.

2 Experimental Setup and Methodology

In order to analyze the impact of tool heating as a consequence of frictional and forming heat in high volume manufacturing on an extrusion process within sheet-bulk metal forming, both the implication on the mechanical forming and on the tribological conditions are investigated within this study. For this purpose, experiments are conducted at room temperature (RT), 50 °C and 80 °C, which represents the temperature range cold forging tools can reach [7].

2.1 Materials

As workpiece material for the investigation the precipitation hardenable aluminum alloy EN AW 6082 in the artificially aged T6 condition and an initial sheet thickness of $t_0 = 2$ mm is chosen. Due to its lightweight potential, which is a consequence of its high mechanical properties, application areas for this alloy are for example forgings and structural components in the automotive sector [10]. For the mechanical characterization of the material layer compression tests in accordance to DIN 50106 [11] were conducted on a thermomechanical simulator Gleeble 3500 by Dynamic Systems Inc. With a constant upsetting speed of 5 mm/min. This machine enables the controlled heating of the specimen during the experiment through resistance heating. At room temperature (RT) the material is characterized by an initial yield stress of $k_{f0,RT} = 252.4$ MPa. When increasing the temperature, the initial yield stress is reduced to values of $k_{f0,50\,°C} = 240.5$ MPa at 50 °C and $k_{f0,80\,°C} = 233.4$ MPa at 80 °C, which correlates to a total reduction of 7.5%.

For the realization of the forming experiments, the wax-based and water-soluble Lubricant Beruforge 152DL (BF 152DL) is used. The application of the lubricant is possible without the necessity of a conversion layer on the workpiece [12], thus providing shorter process chains and ecological benefits. Furthermore, BF152DL was characterized as suitable for the demanding process requirements of SBMF as it provides advantageous tribological conditions with low friction factors during forming [13]. All used active tool components are made of the powder metallurgical cold working steel K890 microclean by voestalpine Böhler, which is hardened to 63 + 2 HRC. The resulting surfaces of the active tool set is obtained by hard-machining operations.

2.2 Forming Experiments

Tribological Charactization. For the tribological characterization ring compression tests (RCT) according to Male and Cockroft [14] are carried out at the chosen temperatures in order to inversely derive the corresponding friction factors m. The inner diameter of the ring-shaped specimen is evaluated after upsetting, since the predominant friction conditions have a significant influence on this characteristic value. The adaption of this test procedure to the process conditions of SBMF was conducted by Vierzigmann [13]. For the characterization of sheet metal, the specimen geometry is machined by wire electrical discharge machining to an inner diameter of 9 mm and an outer diameter of 15 mm. The upsetting of the specimens to 50% of the initial sheet thickness is performed on a hydraulic press Lasco TSP100SO with a forming speed of 5 mm/s using cylindrical and parallel anvils, which are also equipped with ceramic insulated heating bands. Subsequently, the resulting inner diameters are measured on a coordinate measuring machine. For the experiments, the lubricant amount is set to 10 g/mm2. The numerical modeling of the RCT is carried out in the simulation software Simufact. forming 2021.1, in order to inversely determine the resulting friction factors. For this purpose, calibration curves in dependence of the inner specimen diameter are interpolated with a rational model in Matlab for each forming temperature.

Experimental Analysis of the SBMF Process. The experimental analysis of an SBMF forward extrusion process is carried out on a tool setup, which can be seen in Fig. 1. This setup is used in a triple acting hydraulic Press Lasco TZP400/3 at a forming speed of 5 mm/s. The active tool set consists of the forming punch and die, which can be heated up by ceramic insulated heating bands, and a counterholder. To control the forming stroke of the process, mechanical stops are integrated in this tool setup, which can be adjusted with steel foils in a minimum of 0.05 mm increments. In order to achieve a uniform temperature distribution in the components for the forming experiments at elevated temperatures, the experiments are started minimum 50 s after the workpiece have been placed into the tool system.

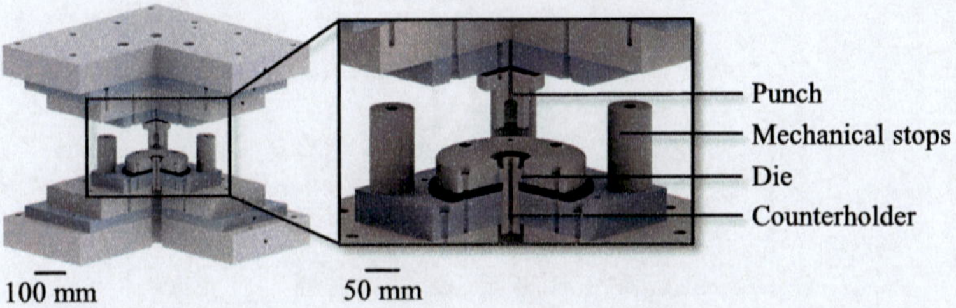

Fig. 1. Tool setup for the forward extrusion of geared functional elements

The investigation geometry for the forming experiments within this study is a planar and geared workpiece (see Fig. 2) with 40 abstracted locking teeth on its outer diameter. These functional elements are formed on ring-shaped blanks with an inner diameter of 25 mm and an outer diameter of 50 mm. The performed forming stroke is set to a reference value of s = 0.6 mm, which in theory results in a complete die filling of the functional elements. This is controlled by the measurement of the residual material thickness in the displacement zone around the functional elements with a micrometer.

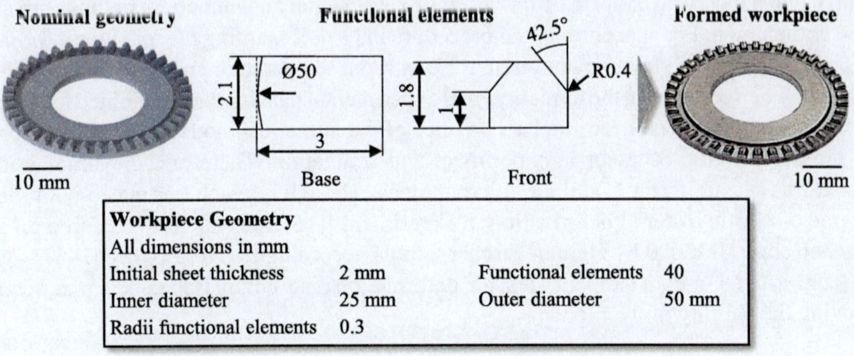

Fig. 2. Workpiece geometry of the investigated functional component

In Fig. 3a schematic depiction of the forward extrusion process is given. At the beginning, the blank is placed on the hydraulically actuated counterholder, which applies a constant pressure of $\sigma_{CH} = 120$ MPa in the workpiece center, preventing a crowning of the blank during forming. The punch movement then displaces the counterholder until the forming die is reached and the functional elements are formed onto the workpiece. Due to the locally varying forming loads, which are characteristic for SBMF, unintended material flow towards the workpiece center occurs, resulting in incomplete die filling of the functional elements [9]. In order to analyze this behavior for the conducted experiments, the complete geometry of the formed parts is optically measured with the system ATOS by Carl Zeiss GOM Metrology GmbH. Subsequently, the volume of only the functional elements is derived in the OpenSource software CloudCompare by boolean operations.

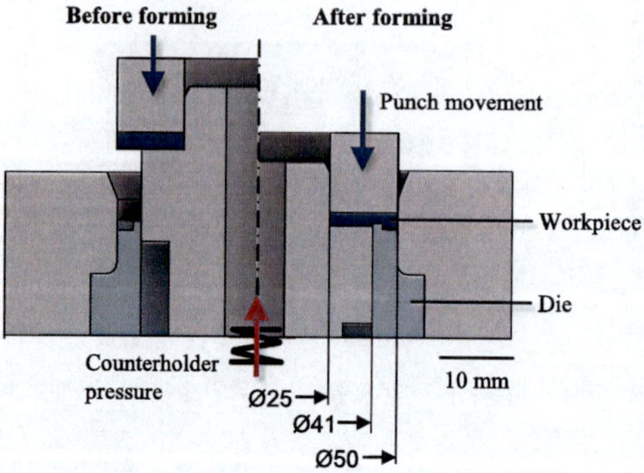

Fig. 3. Forming procedure for the forward extrusion of geared components

3 Results and Discussion

A significant influencing factor for the forming results in SBMF is the resulting tribological condition during forming [13]. For this reason, temperature dependent friction factors are inversely determined in RCTs. In order to verify the validity of the simulation model and the derived calibration curves, the experimental and numerical findings are at first compared in Fig. 4 according the procedure in [15]. Regarding the maximum forming force F_{max} (Fig. 4a) a deviation of 1.2% at room temperature and 2.0% at 80 °C is evident, when comparing the numerical and experimental data for the determined friction factor. Another indicator for good accordance of the numerical model is the correlation of the deformation behavior in experiment and simulation. Therefore, the simulation based true plastic strain φ and the microhardness HV 0.05, which correlates with the degree of plastic material deformation in experiments [16] and which is measured on a Fischerscope HM2000 by Helmut Fischer GmbH according to DIN EN ISO 14577–1, are contrasted. For both temperatures, the deformation distribution is in good agreement, proving the validity of the model.

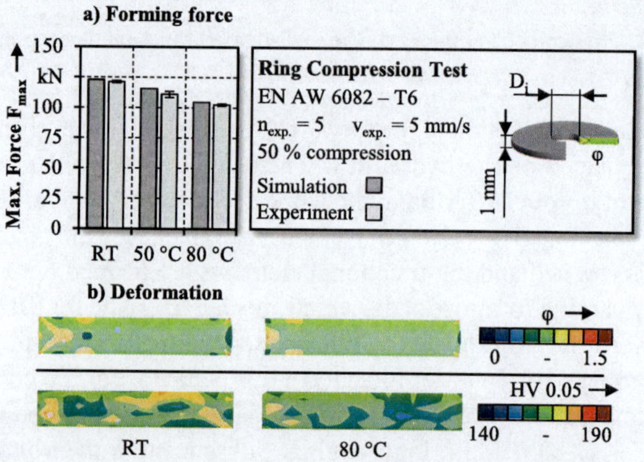

Fig. 4. Comparison of the resulting a) forming forces and b) deformation distribution in the numerical analysis and experimental data of ring compression test

The influence of the forming temperature on the resulting friction factors of EN AW 6082 when using the lubricant BF152DL in RCTs is shown in Fig. 5. At RT, the friction factor of m = 0.08 ± 0.02 is determined as a reference. As a consequence of this relatively low value, it can be concluded that the lubricant provides a sufficient tribological systems for the requirements of SBMF, as stated in former studies [13]. When the forming temperatures are increased, a reduction of the friction factors up to m = 0.06 ± 0.01 at 80 °C is measured. This is can be explained by two overlapping influencing parameters, since increasing the forming temperature changes the mechanical flow behavior of the material on the one hand and the lubricant properties on the other. Similar results were also reported by Bay [6] in the bulk forming of steels, where the coefficients of friction decreases up to minimum at a forming temperature of 130 °C.

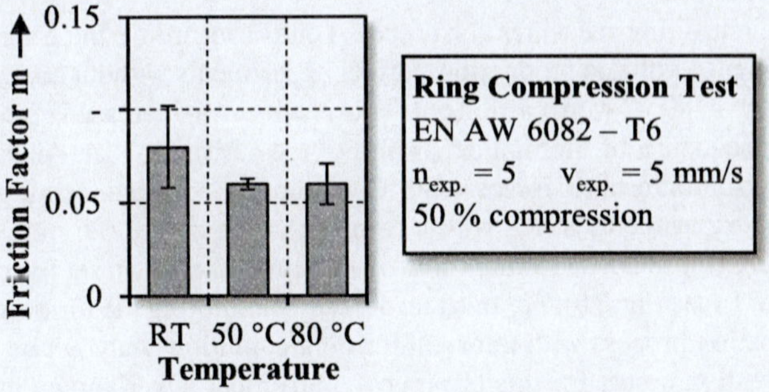

Fig. 5. Resulting friction factors at RT, 50 °C and 80 °C from RCT

In the next step, it will be investigated how these tribological conditions at slightly increased forming temperatures affect the extrusion of functional thin-walled workpieces. The results for the forming experiments with a constant forming stroke of s = 0.6 mm are given in Fig. 6. In order to compensate thermal expansion of the active tool set and differing forming conditions, adjustments to the mechanical stops are necessary to maintain the desired resulting forming stroke, which is the case for the experiments at 80 °C. Additional 0.05 mm have to be added to the mechanical stops at this forming temperature. Due to the reduced mechanical properties and frictional forces during forming, a linear decrease of the maximum forming force from F_{max} = 634.1 ± 8.0 kN at RT to F_{max} = 591.4 ± 2.7 kN at 80 °C is measured in the experiments (see Fig. 6a). Furthermore, the resulting die filling of the functional elements increases for the experiments at 50 °C and 80 °C, also as consequence of the reduced mechanical properties of the material (see Fig. 6b).. This influence apparently exceeds the effect of friction, since a reduced friction factor is accompanied by lower die filling in SBMF [9].

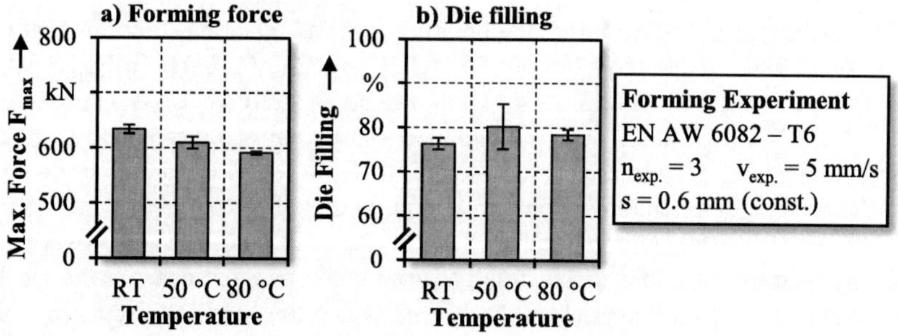

Fig. 6. Influence of the forming temperature on the a) maximum forming forces F_{max} and the b) resulting die filling at a constant forming stroke of $s = 0.6$ mm

To derive controlling measures for process control regarding the transient forming temperatures in high volume production, further experiments without adjustments to the mechanical stops at 80 °C were carried out. This results in an increased effective forming stroke as a consequence of thermal expansion of the active tool set. As can be seen in Fig. 7a, the maximum forming forces at 80 °C without the adjusted forming stroke reach the level of the experiments at RT with a forming stroke of $s = 0.6$ mm while the die filling approximately increases to the value of experiments with an set forming stroke of $s = 0.7$ mm at RT (see Fig. 7b). For this reason, only monitoring the forming forces of an ongoing production process with transient forming conditions may not be sufficient, to identify intervention points for in-situ process adjustment. Overlapping effects, in this case the increasing die filling, may mask the changes in the resulting forming force, so that the geometric result can deviate unnoticed over time.

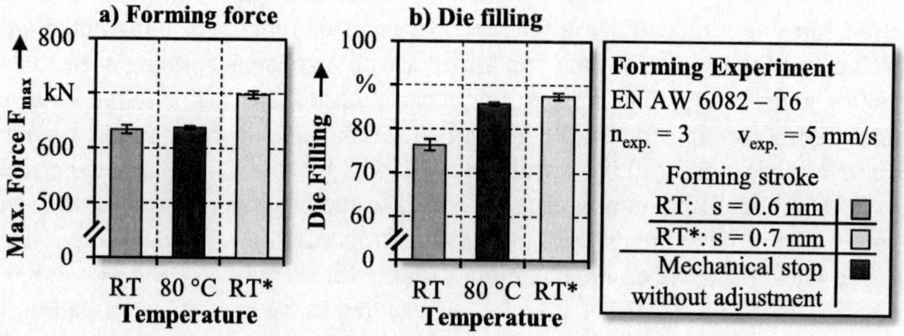

Fig. 7. Resulting a) maximum forming forces F_{max} and b) die filling without the adjustment of the mechanical stop due to thermal effects

Furthermore, the workpiece properties have to be analyzed, in order to evaluate the impact of differing forming temperatures on the manufactured components properties. For this reason, the microhardness distribution of the formed components is depicted in Fig. 8. In experiments with a constant forming stroke of $s = 0.6$ mm no significant differences are visible regarding the different forming temperatures RT, 50 °C and 80 °C. In general, an increase of the microhardness values can be seen within the displacement zone and the gearing elements. Maximum values occur in the radii of this area, as the main deflection of the material is located there. In the workpiece center, no significant material deformation takes place and therefore the measured microhardnesses remains at the level of the initial hardness of the base material. As discussed earlier, without the adjustment of the mechanical stops at the forming temperature of 80 °C, die filling of the functional elements is increased and therefore comparable to the results at a higher forming stroke s = 0.7 mm at RT. For this reason, the microhardness plots of these two variants are discussed in the following. To facilitate the increased die filling, a higher degree of material deformations is necessary, which then leads to overall higher microhardness values. Furthermore, more material is also displaced into the workpiece center unintentionally, which explains the higher microhardness values in these areas compared to the experiment with a forming stroke of s = 0.6 m. However, when comparing the last two variants, lower microhardness values are measured for the workpiece formed at 80 °C (without adjusted mechanical stops), indicating the influence of the raised forming temperature on reduced strain hardening of the material. This shows the challenging aspect of an in-situ adjustment of process parameters in high volume manufacturing, since interactions between a geometry-based process control and the resulting component properties are in part contrary.

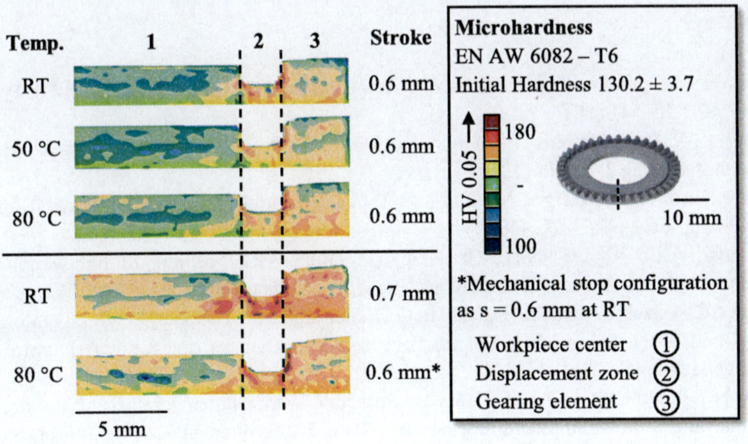

Fig. 8. Influence of the process temperature on the microhardness distributions of the formed parts

4 Summary and Outlook

Within this study, the influence of slightly increased forming temperatures as a consequence of frictional and forming heat in high volume manufacturing in a SBMF extrusion process was analyzed. These transient forming conditions were replicated by conducting experiments with a heated tool system. The results of these investigations show an increase in the die filling of functional elements, when the forming temperature is increased. If this requires an adjustment of the process parameters to guarantee a robust forming outcome, resulting forming forces are of limited suitability to define intervention points for the process control, as this parameter can partially mask changes in the resulting die filling. Further investigations should focus on the influence of transient forming conditions on the operational behavior of workpieces in order to derive holistic measures for robust in-situ process control incorporating the workpiece geometry and properties. Thereby, waste reduction in the run-in phase of high volume production can be achieved. Furthermore, interdependencies of transient forming conditions with local tool-sided surface modifications should be investigated, as these present a valid measure for improving the forming results in the field of SBMF. Additionally, transmission of these findings to coil-based forming operations is necessary in order to generate a fundamental and transferable understanding of the impact of transient forming conditions on the high volume forming of filigree functional components.

Acknowledgements. This work was supported by the German Research Foundation (DFG) within the scope of the Project "Improvement of the geometrical accuracy of parts by material flow-optimized coil layouts when extruding functional components" under the grant number 492930781.

References

1. Stock, T., Seliger, G.: Opportunities of sustainable manufacturing in industry 4.0. Procedia CIRP **40**, 536–541 (2016)
2. Wang, K.-S.: Towards zero-defect manufacturing (ZDM) - a data mining approach. Advances in Manufacturing **1**, 62–74 (2013)
3. Lange, K., Cser, L., Geiger, M., Kals, J.: Tool life and tool quality in bulk metal forming. CIRP Ann. **41**(2), 667–675 (1992)
4. Tekkaya, A.E., Khalifa, N.B., Grzancic, G., Hölker, R.: Forming of lightweight metal components: need for new technologies. Procedia Engineering **81**, 28–37 (2014)
5. Müller, C., Filzek, J., Groche, P., Oehler, O., Scherzinger, P., Twickler, M.: Temperaturentstehung und die tribologischen Folgen bei Produktionsbeginn der Kaltmassivumformung. Schmiede Journal (2014)
6. Bay, N., Wibom, O., Nielsen, J.A.: Einfluss der Werkzeugtemperatur auf Reibung und Schmierung beim Kaltmassivumformen von Stahl. Umformtechnik **29**(2), 106–112 (1995)
7. Lange, K.: Umformtechnik Handbuch für Industrie und Wissenschaft - Band 4: Sonderverfahren, Prozeßsimulation, Werkzeugtechnik, Produktion. 2nd edn. Springer, Berlin (1993)
8. Merklein, M., et al.: Bulk forming of sheet metal. CIRP Ann. **61**(2), 725–745 (2012)
9. Merklein, M., Löffler, M., Schneider, T.: Plastic flow and its control in sheet–bulk metal forming of thin-walled functional components. CIRP Ann. **64**(1), 245–248 (2015)
10. Ostermann, F.: Anwendungstechnologie Aluminium, 3rd edn. Springer, Berlin (2014)
11. DIN 50106: Prüfung metallischer Werkstoffe - Druckversuch bei Raumtemperatur (2016)
12. Carl Bechem GmbH: Technische Produktinformation Beruforge 152 DL (2019)
13. Vierzigmann, H.U.: Beitrag zur Untersuchung der tribologischen Bedingungen in der Blechmassivumformung - Bereitstellung von tribologischen Modellversuchen und Realisierung von Tailored Surfaces. Dissertation, Meisenbach (2016)

14. Male, A.T., Cockroft, M.G.: A Method for the determination of the coefficient of friction of metals under conditions of bulk plastic deformation. J. Inst. Met. **93**, 38–46 (1964)
15. Tekkaya, A.E.: A guide for validation of FE-simulations in bulk metal forming. Arab. J. Sci. Eng. **30**(1), 113–116 (2005)
16. Tekkaya, A.E.: Improved relationship between Vickers hardness and yield stress for cold formed materials. Steel Research **72**(8), 304–310 (2001)

Macro-structured Deep Drawing of Steel Alloys - Influenceability on Thinning and Drawing Forces

A. Wolf(✉), M. Tulke, and A. Brosius

TU Dresden University of Technology, 01069 Dresden, Germany
alexander.wolf3@tu-dresden.de

Abstract. Deep-drawn components are found across all industries. With respect to the challenges of manufacturing towards a more climate-friendly and resource-efficient industry, various approaches such as lubricant-free forming, are being investigated. Deep drawing with macro-structured tools is a key approach in this regard. In addition to the expansion of the process window, other benefits can be achieved, such as the improvement in resource efficiency resulting from the reduction of induced residual stresses and the saving of lubricant.

The investigation of the sheet thinning and drawing force in connection with the macro-structure with the intention to elaborate the sensitivity of the parameters is the subject of this work. Within a systematic investigation, the influence of a changing wavelength λ along with the associated immersion depth δ of the macro-structure is examined. Concluding, a contribution to the design methodology of the macro-structured tools based on the usable process window is presented.

Keywords: Deep Drawing · Macro-structured tools · Thinning

1 Introduction

Deep drawing allows complex designs and seamless structures to be created with high dimensional accuracy in a single step. However, like other manufacturing processes, deep drawing also has its process limitations. The main challenge is to achieve the desired shape without causing defects such as wrinkling, tearing or excessive thinning of the material.

In recent years, there has been an increasing importance on sustainable manufacturing practices and resource efficiency. The consumption of energy, the use of sometimes polluting lubricant, as well as the generation of non-useable parts through tryout and badly controlled processes, are some of the concerns associated with the traditional process. To address these worries, researchers and industry professionals have been exploring innovative approaches to enhance the sustainability and resource efficiency of deep drawing process [1]. One major approach is the usage of lightweight materials which reduces the overall weight of the components for the use cases. Another way is the development of eco-friendly lubricants or the complete elimination of them [2]. Furthermore, advancements in the usage of the Finite Element Method (FEM) allows

T. Bauernhansl et al. (Eds.): WGP 2023, LNPE, pp. 680–687, 2024.
https://doi.org/10.1007/978-3-031-47394-4_66

the virtual testing and optimization of the process prior to the tool production. This option enables manufacturers to minimize trial-and-error iterations and optimize their tool designs and blank shapes to reduce waste and make a contribution to sustainability. The usage of macro-structured tool offers a contribution to these major challenges by reducing or even eliminating the use of lubricants and increasing the energy efficiency.

2 Macro-structured Deep Drawing

This innovative approach, first presented by Mousavi and Brosius [3] aims to enhance the formability of sheet metal by introducing macro-scale structures on the surface of the tools (see Fig. 1). These structures can influence the deformation behavior and allow an improved stretchability, reduced wrinkling and enhanced material flow during the deep drawing process. Macro-structured deep drawing has shown promising results in terms of overcoming the traditional limitations of deep drawing, such as material thinning and cracks and opens up a larger process window. Brosius and Tulke used the reduction of the contact area and reduced heat transfer due to the macro-structure to enable cryogenic deep drawing of common aluminum alloys without additional tool cooling [4]. The process can achieve higher deformation capabilities and a higher process stability with less induced residual stresses [5]. One of the key advantages of macro-structured deep drawing is its potential for material and energy savings. The enhanced formability enables the usage of thinner sheets, which reduces material consumption while maintaining the desired structural integrity.

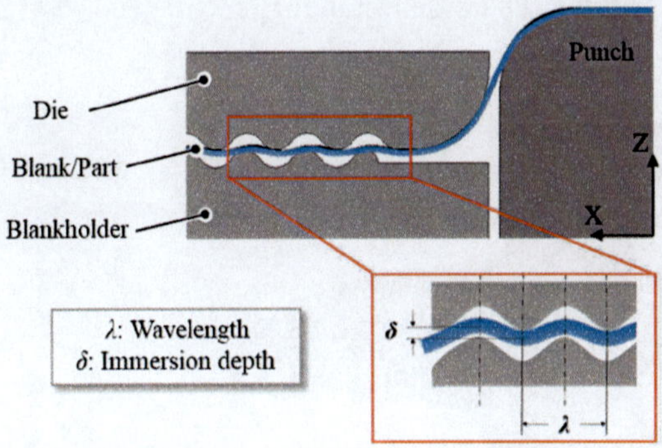

Fig. 1. Tool with macro-structured die and blankholder, the schematic sketch of the parts and parameters for the deep drawing process [5]

There were various other approaches, which aim for a larger process window and a higher drawing ratio. Some researchers aim for an improvement of the blankholder to control the draw in of the blank by varying the force during the process [6] or design a segmented blankholder [7] while others go for the die and optimize the surface through coatings [8] or adding ultrasonic vibration during the process [9].

Due to the geometry and the resulting deformations of the deep drawn part various stress states occur during the process. Mousavi et al. have shown by tools with a macro-structured surface, springback in deep drawn components can be reduced due to the induced restraining forces [10]. Next to less thinning, lower drawing forces are also a main advantage in the usage of macro-structured tool, aiming for less energy consumption in the process. This will have a major impact on the sustainability and efficiency of future deep drawing application. Therefore, a comparison of the drawing forces will be included in the discussion.

3 Approach and Methodology

3.1 Objective of This Paper

The aim of this work is to examine the influence of different wavelengths, immersion depths and structure radii on the resulting components. In particular, the thinning and the resulting drawing forces in the component are investigated and evaluated. This will help to develop a design methodology for the versatile usage of macro-structures for the deep drawing by improving the understanding of the process and the mechanisms during the process. In contrast to Tulke et al. [11] the paper focus on the thinning of the drawn part and addresses DC06 as steel instead of aluminum alloys.

3.2 Setup for the Evaluation Thinning and Resulting Strain

The experiments were performed mainly virtual with the usage of the FEM. The setup and the dimensions for the cup are shown in Fig. 2.

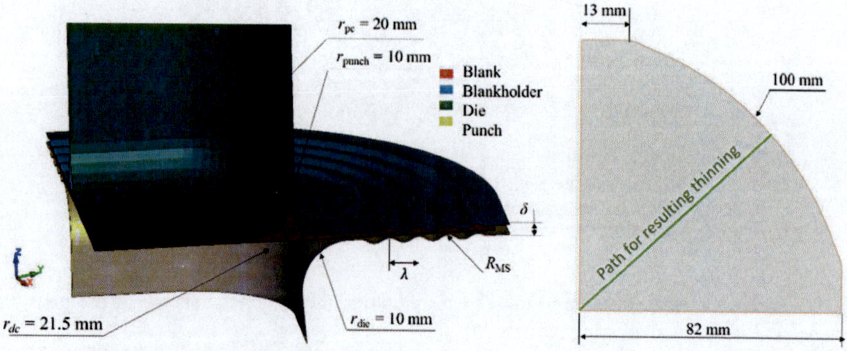

Fig. 2. Virtual setup with the main dimensions for the tools and the blank (right)

The chosen rectangular cup was reduced to a quarter to reduce computing time and power. The calculations were performed explicit in LS-DYNA R12.0.0. The preparations and the postprocessing was done in LS-PrePost V4.9.8. The blank was meshed with an element length of 1 mm in Hypermesh and transferred into LS-PrePost. A fully integrated shell element was used (ELFORM16) with nine integration points through the shell thickness. The initial thickness of the blank was $s_0 = 1$ mm. The material model for describing the forming behavior during the process was MAT133, which based on the Yield2000-2D by Barlat et al. [12].

Figure 2 shows the reduced setup with the parts included. The tools were calculated as rigid. The process is parted, first the blankholder sets the immersion depth, afterwards the deep drawing process takes place. The main parameters remain constant during the tests. The punch is 82 mm in total which reduces the side length in this model by a half. The punch radius r_{punch} with 10 mm and the corner radius r_{pc} with 20 mm defining the punch. The die as counterpart got a die corner radius r_{dc} of 21.5 mm and a drawing radius r_{die} of 10 mm. The friction coefficient $\mu = 0.15$ between all surfaces. The drawing depth was set to 45 mm. For a comparison of the calculated values to the conventional deep drawing, the usage of a conventional drawing ring-blankholder combination is also calculated for two blankholder forces.

3.3 Tryout Parameter and Postprocessing

For the design of the experiments (DoE) on the influence on resulting thickness and the intrinsic strains, the immersion depth, the wavelength and the wave radius were altered in a specific range. Therefore, findings from the circular and a T-shaped cup were used for initial suggestion for the parameter range [2]. The parameters are shown in Table 1..

Table 1. Parameter set for the evaluation

Parameters	Level	1	2	3
Wavelength in mm	λ	6	8	10
Immersion depth in mm	δ	-0.1	-0.2	-0.3
Radius macro-structure in mm	R_{MS}	1	2.5	4

Due to the possible 27 variations which are possible, a fractional-factorial design based on Taguchi [14] was chosen for balanced comparison of the process parameters to reduce the energy consumption for full-factorial calculations. The reduced orthogonal array is shown in Table 2.

For the postprocessing on the results for the development of the thickness and the drawing force was extracted along the sketched path from the center to the flange area through the corner of the rectangular cup (see Fig. 2). The results for the minimal blank thickness and the force were used for an evaluation to outline the influence of the different parameters. The distance from the center is divided into the different sections of the part wall.

Table 2. Array for the reduced parameter set

Level			
Exp.	λ	δ	R_{MS}
1	1	1	1
2	1	2	2
3	1	3	3
4	2	1	2
5	2	2	3
6	2	3	1
7	3	1	3
8	3	2	1
9	3	3	2

4 Results and Discussion

4.1 Resulting Sheet Thickness Along the Defined Path

Table 3. shows the calculated resulting thickness and the drawing force for the performed virtual experiments. Whereas wrinkling is clearly visible, without a FLC the crack is only an assumption based on a major reduction of the thickness.

Table 3. Resulting minimum sheet thickness and maximum drawing force for the virtual experiments

Run	Thickness t in mm	Thickness reduction in %	Max drawing force F_{punch} in kN	Comment
1	0.853	15%	90.12	good
2	0	100%	110.64	crack
3	0	100%	103.29	crack
4	0.94	6%	69.03	wrinkling
5	0.877	12%	88.85	wrinkling
6	0.824	18%	100.39	good
7	0.945	6%	63.69	good
8	0.935	6%	69.15	wrinkling
9	0.915	9%	78.83	wrinkling

The data was extracted and plotted in the following charts (see Figs. 3 and 4). Figure 3 shows the progress of the sheet thickness in the rectangular cup, beginning from the center. The different areas of the cup were marked at the bottom of the chart. The initial sheet thickness of $t_0 = 1$ mm is shown by the dashed line. Figure 4 displays the drawing forces F_{punch} for the different runs.

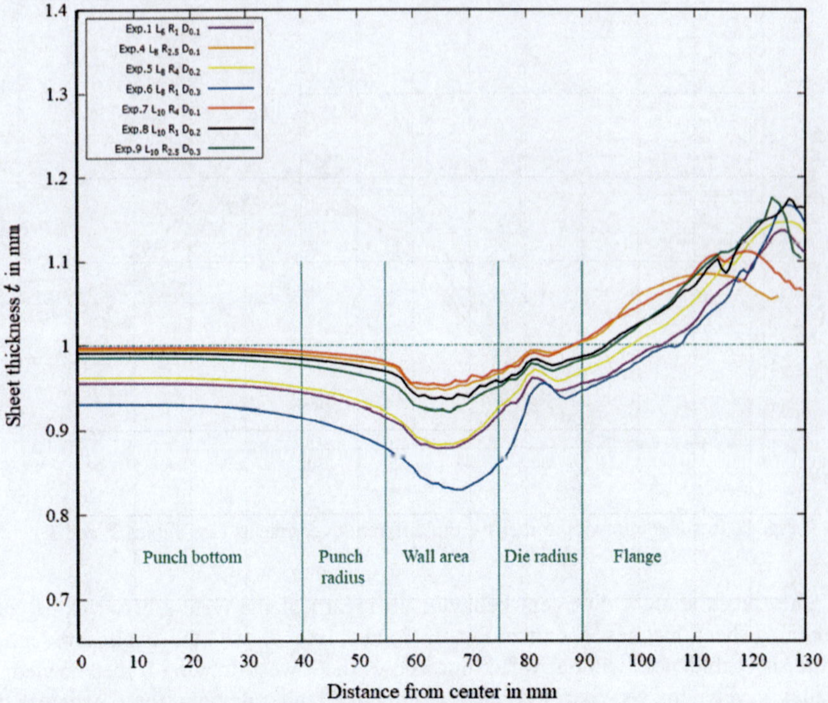

Fig. 3. Resultant thickness in different areas of the cup for the virtual experiments (see Table 2.)

It's obvious, that there is minimal and constant change of the thickness in the bottom area due to the biaxial stress state. Beginning with the punch radius the reduction starts whereas the minimum is, as expected, similar in all cups right above the punch radius and the corner of the cup in the side wall. Afterwards the thickness increases and exceeds the initial thickness in the flange area due to the tangential stresses during the drawing resulting in a thickening and sometimes wrinkling. All shown cups will result in drawable parts with minimal risk of bottom cracks, whereas wrinkling could be observed in the FEM.

The strong interaction between the parameters is visible in the experimental runs with the highest reduction in the thickness. The blue line (Exp. 6) represents the wavelength of $\lambda = 8$ mm. Due to the immersion depth δ of 0.3 mm combined with a small wavelength, the restraining forces were high. This results in an increased thinning but also a reduced wrinkling in the flange, even though the minimum contact area is reduced to a minimum due to the smallest wave radius R_{ms}. Choosing a reduced immersion depth of $\delta = 0.1$ mm results in the smallest reduction of the sheet thickness (see Exp. 4 and 7). Whereas both were calculated with the same immersion depth, the wavelength and the wave radius vary. The differences occur in the flange area, where Exp. 7 shows a higher thickening than Exp. 4., but no visible wrinkling in the calculation. Therefore, it can be assumed, that a higher contact area suppresses wrinkling better than a smaller wavelength.

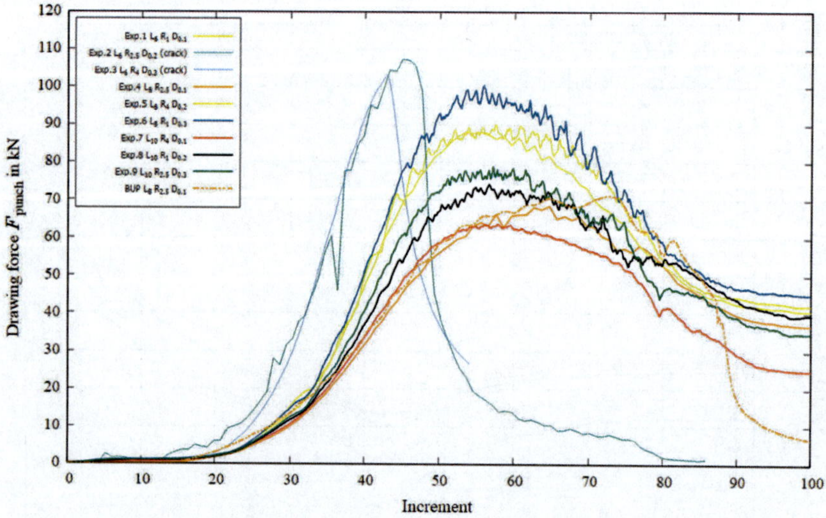

Fig. 4. Resultant drawing forces for the different experiments (see Tables 2. and 3.)

The forces show a divergent behavior, in regard of the results of wrinkling and remaining sheet thickness shown in Fig. 3. Hereby, Exp. 7., with the smallest reduction of the sheet thickness without wrinkling, shows the lowest drawing forces needed. A smaller wavelength, see Exp. 1, 5 and 6 results in a higher drawing force, whereas the wave radius plays a minor role. Exp. 1 and 6 have the smallest wave radius, but Exp. 5 with R_{ms} of 4 mm shows the highest drawing force in this comparison. Knowing that only Exp. 1, 6 and 7 performed with no wrinkling, the interaction between the drawing depth and the wavelength for a robust process can be shown. A higher immersion depth leads to a suppression of wrinkling and the need of higher needed drawing forces.

The forces of the experiments on the existing tool set on the BUP600 were included in the chart (see Fig. 4). For $\lambda = 8$ mm and R_{ms} of 2.5 mm an immersion depth δ of 0.2 mm was set with the use of distancing rings on the outside of the tool. The drawing depth was set with a potentiometer to reach 45 mm drawing depth as used in the FEA. A good agreement between simulation and experiment is visible. It can be recognized that with the correctly selected parameter set, a lower energy consumption with simultaneous better distribution of the sheet thickness in the component is possible.

5 Conclusion and Outlook

The paper investigates the effects of different parameter sets on the thinning and drawing force of a rectangular cup, using an altering wavelength, immersion depth and wave radii. The results in the thinning of the sheet show a strong interaction between the altered parameters, whereas the immersion depth can reduce the wrinkling by superimposing the tangential stress in the flange area but also leads to more thinning in the sheet. As a better option with less thinning, a well-chosen design of wavelength and wave radius results in good parts with a minimum in thinning and a reduction in the needed drawing force. As the useful option in this comparison, Exp. 7 with a wavelength $\lambda = 10$ mm and R_{ms} of 4 mm showed the best outcome for the specific cup with a drawing depth of 45 mm. Therefore, it can be concluded, that for this wavelength and immersion depth, a minimum thinning with a reduced drawing force is reached for DC06 and the presented rectangular cup geometry. It was shown, that the process is suitable to reach a reduction in the drawing force and therefore the energy consumption, which is a major contribution to the sustainability and resource efficiency.

References

1. Ingarao, G., Di Lorenzo, R., Micari, F.: Sustainability issues in sheet metal forming processes: an overview. Journal of Cleaner Production **19**(4), 337–347 (2011). https://doi.org/10.1016/j.jclepro.2010.10.005
2. Mousavi, A.: A novel approach towards a lubricant-free deep drawing process via macro-structured tools. In: TU Dresden (2019)
3. Brosius, A., Mousavi, A.: Lubricant free deep drawing process by macro structured tools. CIRP Annals **65**, 253–256 (2016). https://doi.org/10.1016/j.cirp.2016.04.060
4. Brosius, A., Tulke, M.: Forming of aluminium alloys with macro-structured tools at cryogenic temperature. CIRP Annals (2023). https://doi.org/10.1016/j.cirp.2023.03.018
5. Wolf, A., Tulke, M., Brosius, A.: Deep Drawing with Macro-structured Tools and Their Influence on Residual Stresses, pp. 94–102 (2022). https://doi.org/10.1007/978-3-030-78424-9_11
6. Palmieri, M.E., Tricarico, L.: Control of draw-in in the deep-drawing process by regulating the force on the blank holder. In: IOP Conf. Ser.: Mater. Sci. Eng. **1238**(1) (2022). https://doi.org/10.1088/1757-899X/1238/1/012071
7. Zhang, H., Qin, S., Meng, L.: Research on deep drawing with multi-ring blank holder technique. (in En;en). Int J Adv Manuf Technol **122**(9–10), 4115–4126 (2022). https://doi.org/10.1007/s00170-022-10142-1
8. Taube, K.: Carbon-based coatings for dry sheet-metal working. Surface and Coatings Technology **1–3**, 976–984 (1998). https://doi.org/10.1016/S0257-8972(97)00178-3
9. Malekipour, E., Sharifi, E., Shahbazi Majd, N., Heidary, H.: Observation on the behavior of ultrasonic micro-hammer and its effects on the deep drawing process: Numerical simulation and experimental study. Ultrasonics **119**, 106566 (2022). https://doi.org/10.1016/j.ultras.2021.106566
10. Mousavi, A., Brosius, A.: Improving the springback behavior of deep drawn parts by macro-structured tools. In: IOP Conference Series: Materials Science and Engineering, vol. 418, pp. 12–18 (2018). https://doi.org/10.1088/1757-899x/418/1/012105
11. Tulke, M., Lafarge, R., Wolf, A., Brosius, A.: Contact conditions and temperature distribution during cryogenic deep drawing with macro-structured tools: [accepted manuscript]. In: ICTP 2023. Mandelieu-La Napoule (2023)
12. Barlat, F., et al.: Plane stress yield function for aluminum alloy sheets—part 1: theory. International Journal of Plasticity, pp. 1297–1319 (2003). https://doi.org/10.1016/s0749-6419(02)00019-0

Investigation of the High Dynamic Wood Processing Regarding Tool Imbalance and Surface-Quality

Jonas Duntschew(✉), Thomas Stehle, and Hans-Christian Möhring

Institute for Machine Tools (IfW), University of Stuttgart, Holzgartenstr. 17, 70174 Stuttgart, Germany
jonas.duntschew@ifw.uni-stuttgart.de

Abstract. High dynamic wood processing is manufacturing of sustainable resources at its limits. Facing challenges to determine the influence of minor tool and clamping imbalance regarding the surface quality of wood machining and structural behavior of the machine tool, this article is presenting an approach to evaluate and classify the impact of tool imbalance at high speed rotation.

Therefore, a variety of process scenarios and tool configuration are investigated. To quantify the surface-quality several state of the art measurement systems and evaluation methods are used.

Finally, it is possible to connect the structural surface information with the balance factor and implement routines to measure natural massive wood surfaces.

Keywords: Woodworking · Surface Analysis · Tool Imbalance

1 Introduction

The balance factor of rotational tools describes the resulting centrifugal forces due to the mass and revolutions per minute [1]. The effect of a high balance factor (hereafter described as imbalance) results in vibrations and displacements of the tool center point, that ultimately effects the surface quality [2–4]. Approaches to determine measured wood surface quality values mostly focused on surface roughness [5]. Current in-process investigations focused on algorithm to support the manufacturing of wood and wood based material [6].

High dynamic wood processing in particular leads to a significant influence of imbalanced tools. The mathematical correlations and equations to define the limits for the permitted imbalance are described in DIN 69888, ISO 15642, EN 847 and ISO 1940–1 [7–10].

Previous investigations at the Institute for Machine Tools (IfW) regarding the influence of imbalance focused on the induced displacements of the tool centre point [2] and described, inter alia, an approach to measure the dynamic tool displacements. Based on that IGF project a follow up project funded by the German Federal Ministry for Economic Affairs and Climate Action (BMWK) investigates the limits of current imbalance

T. Bauernhansl et al. (Eds.): WGP 2023, LNPE, pp. 688–696, 2024.
https://doi.org/10.1007/978-3-031-47394-4_67

recommendations and the resulting effects on surface quality and structural machine integrity to derive a balance prediction model [11].

Proceeding the research project several investigations towards the influence of balancing effects were performed. This paper presents parts of the results identified to be the most noticeable for the audience in our ongoing journey. It intends to sensitize researcher in the area of woodworking regarding the influence of imbalance and to present our unconventional fast approach to measure the surface quality of large probes.

2 Experimental Set-Up

In the following chapter the details of the performing process und the used equipment is presented. The investigation focused on two measuring locations to detect the excitation of the tool forced by the imbalance:

- Directly at the moving tool (WEINIG Powermat 500)
- The workpiece surface after process

The chapter is referring to the requirements of the two measuring locations and will maintain that logic throughout the result section to keep a clear understanding between action and reaction.

2.1 Planing and Profiling Machine – WEINIG Powermat 500 and the Applied In-Line Sensors

The plan-milling experiments were performed on the WEINIG Powermat 500 (see Fig. 1: A-D). The overall tool configurations, process parameters and workpiece (see Fig. 1: E) information are stated in Table 1.

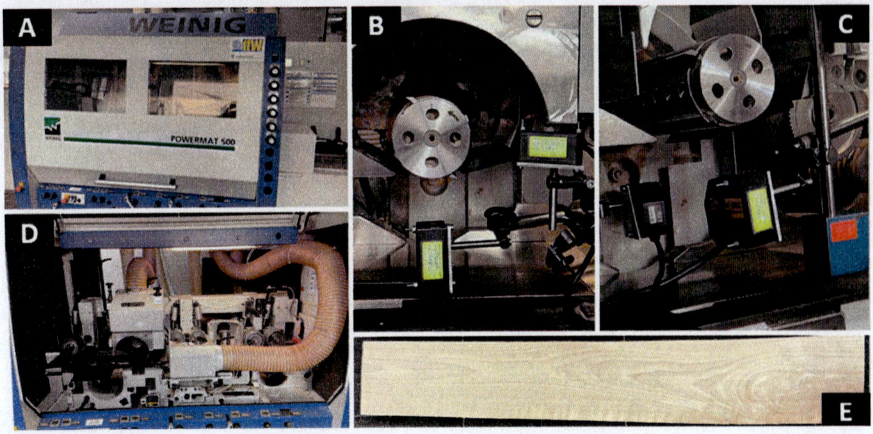

Fig. 1. WEINIG Powermat 500 – cutting test performing through feed planning machine (A, D), laser triangulation sensors to measure the tool-face-movements (B), laser triangulation sensors to evaluate the overall movement alongside the cutting edge (C) at the main spindle and the workpiece (E)

To investigate the highest dynamic impact, the processing limits of the machine has been chosen (rpm: 12.000 1/min and feedspeed: 30 m/min). Two KEYENCE laser triangulation sensors (Type LK-H052) measured the dynamic deviations of the planning tool, using two different strategies (see Fig. 2) at a frequency of 392 kHz. For the First strategy the sensors are positioned at the tool face (shown in Fig. 1: B and Fig. 2: left).

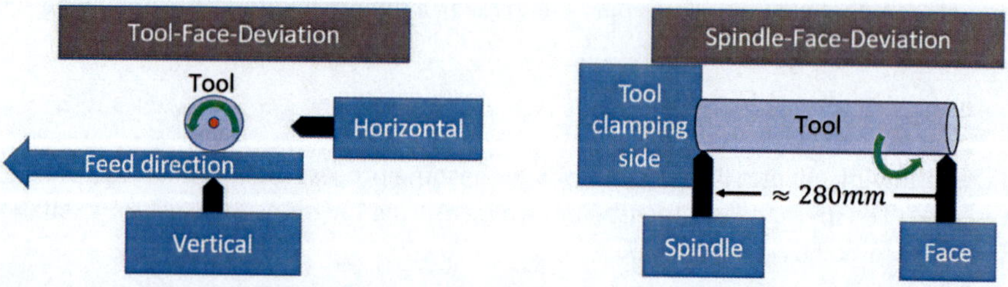

Fig. 2. Measuring strategies of the laser triangulation sensor heads regarding the tool and spindle regarding the two approaches

Second strategy measured the excitation between the spindle and the tool-face alongside the cutting blades (see Fig. 1: C and Fig. 2: right).

Table 1. Tool, workpiece and process Parameters

Tool		Workpiece		Process Parameters	
Length	240 mm	Length	1400 mm	Rpm	12000 min−1
Diameter	90 mm	Width	220 mm	Feed rate	30 m/min
Weight	12.35 kg	Type	beech	Depth of cut	1.5 mm
Tooth-No	2, 4			Cutting speed	3396 m/min (up-milling)
Cutting angle	20° deg				
Variable balancing class	G1, G6.8, G16, G40				

To constantly apply the variable balancing class presented in Table 1 the tools are prepared to be mounted with defined masses. Figure 3 shows the two-blade version of our testing tool with the marked bolting spot to apply the imbalance mass (Fig. 3: middle).

Fig. 3. Details of the cutter head tool

The experiments were performed with the two blade cutter head at the maximum rotational and feed speed the machine can operate, resulting in the stated cutting speed of 3396 m/min specified in Table 1. In addition, the 4-blade tool version has been applied to investigate the effect of higher dynamics taking increasing tooth passing frequency into account. To simplify the reading experience, the presented results are performed with the two blade tool while not specifically stated otherwise.

2.2 Surface Quality Evaluation Techniques

The measurement and quantification of the workpiece surface structure were performed by Bruker Alicona Compact Cobot (Fig. 4: A) and Wenzel LH87 (Fig. 4: B) systems. The selection is based on the objective to find the most suitable concept to analyze the effect of tool imbalance reflected upon the workpiece.

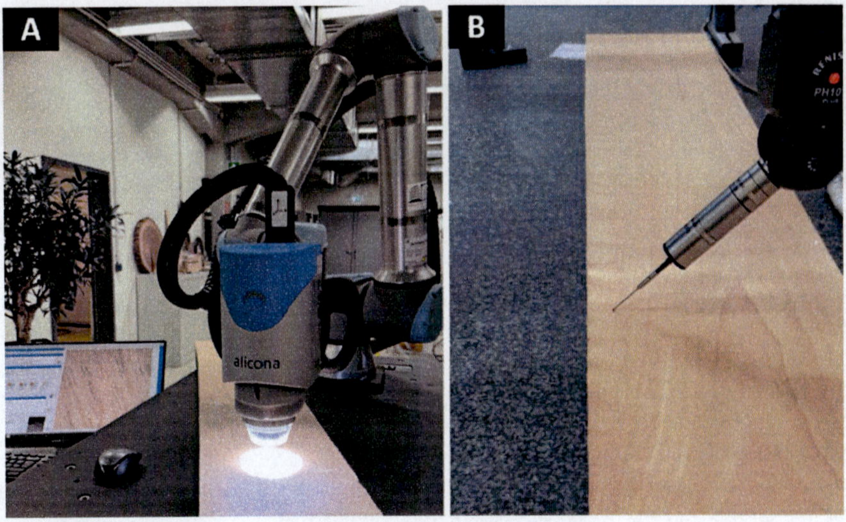

Fig. 4. Bruker Alicona Compact Cobot (A) and Wenzel LH87 (B) performing the surface measurement operations

As listed in Table 2 and discernable in Fig. 4 the main difference is the employed scanning technique and realized measuring area. Both suited the demand to process the large workpiece. Not disputable is the overall resolution of the optical Alicona system.

Table 2. Overview - Surface measurement equipment: Used settings/approaches

Bruker Alicona Compact Cobot		Wenzel LH87	
Scanning technology	Confocal focus (optical / 10x lens), Area scanning	Scanning technology	Coordinate measuring machines (tactile), Line scanning
Measured area	40 mm x 2 mm	Measured area	160 mm x 210 mm
Measuring points	80 MP	Measuring points	100,000 P
Accuracy	< 1 µm	Accuracy	< 1 µm
Approach	Alicona Software: Gaussian planar filter, surface roughness	Approach	Using the probe head diameter (2 mm) as passive filtering, MATLAB

Theoretically it can match the area of the Wenzel system. Ahead of the results it has to be noted, that concomitant examinations lead to the conclusion that increasing Alicona measuring areas result in no further informational improvement regarding this investigation. The amount of data will reach a point the analysis software is unable to process.

3 Results

Based on the sectioning during the introduction of the investigated measuring scenarios the results will be presented regarding the tool balance class and its physical effects. Therefore, the results are split into the dynamic tool displacement and workpiece surface to evaluate the major effects.

3.1 Dynamic Tool Displacement

As previous investigation indicated the imbalance directly effects the tool deviation while idle running. Characteristic effects occur which have machine dynamic effects. Increasing centrifugal forces cause the spindle bearings or other effective surface pairings to move relative to each other. Spindle axes can be forced to resonate at specific frequency. The overall machine structure can be forced to excitation due to resonating responses to the processing parameters.

To avoid processing in resonate frequencies a modal analysis has been performed. It approved the stated process parameters to be not critical.

The introduced displacement measuring of the tool face deviation showed a significant increase of deviation after the balance class switch from G16 to G40 (see Fig. 5). The vertical deviation increases from 0.04 mm to 0.28 mm.

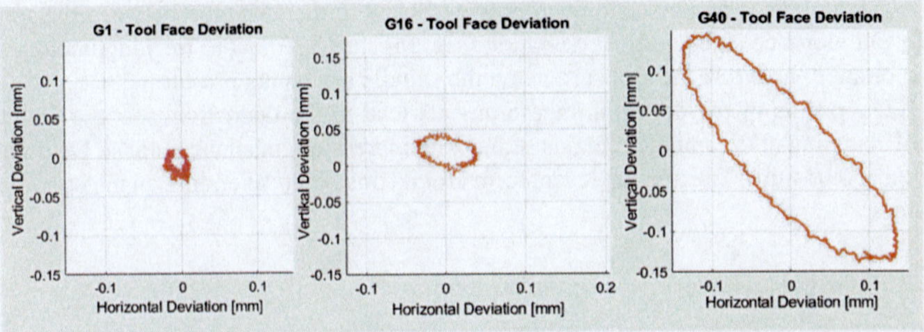

Fig. 5. Effect of increasing imbalance to the tool face deviation

In comparison the deviation between G1, G6.8 and G16 was insignificant. Based on this knowledge ongoing investigations either chose to compere G1 to G40 or G16 to G40 balance classes.

The same results where implied by triaxial acceleration sensor covering the cutting process to document the overall vibrations of the machine structure.

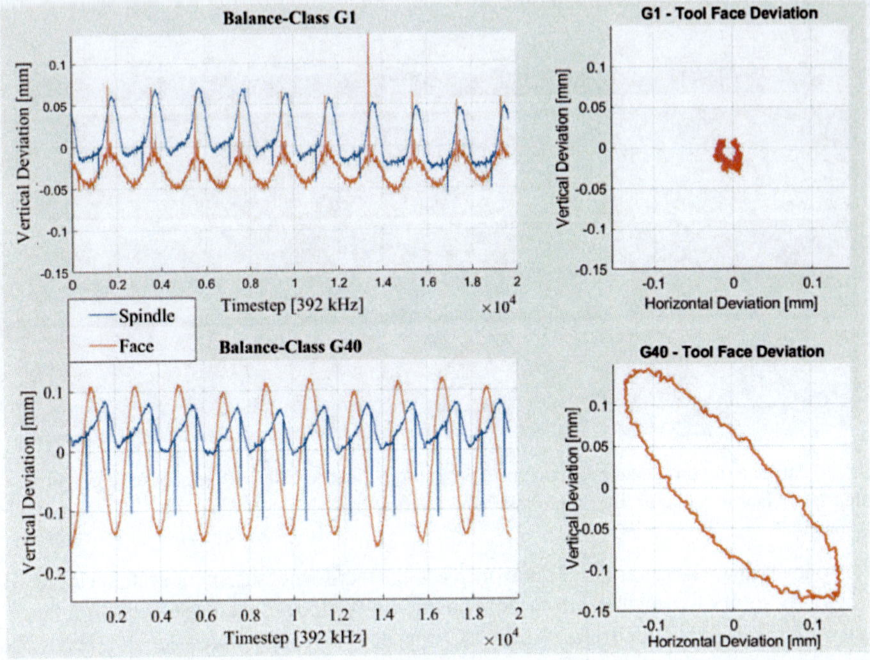

Fig. 6. Movements of the cutter head –Alongside the cutting edges (Spindle-Face-Deviation, left) and face-sided (Tool-Face-Deviation, right)

The tool face movements thereafter were linked to the deviation between the spindle and tool face sided vertical deviation (see. Figure 6). To avoid misjudgments, it is important to note that the tool surface on the spindle side is not pre-turned.

The results of the deviation measurements lead to the deduction, that the area of tool clamping at the spindle remains stable with increasing imbalance (const. Deviation delta of 0.08 mm). The moving levers form around this point depending on the operating forces.

3.2 Workpiece Surface

To evaluate the effects of changing imbalance regarding the workpiece quality we started measurements with the Alicona Cobot system. To investigate the highest possible dynamics these tests were performed using the 4-bladed tool. To interpret the high resolution 3D surface, the software integrated surface texture analysis was used to apply integrated Gaussian planar filter methods (see Fig. 7).

The natural inhomogeneity of solid timber can lead to unpredictable microscopic effects. Therefore, it is difficult to classify the whole process by observing a relatively small area. The processed results show that a specifically tooth engagement becomes present at G40. The imbalance forces the tool to a dominant single cut finish (16 visible cuts resulting in a 2.5 mm tooth engagement distance). Usually it is a mix due to minimal differences of the radial tooth height.

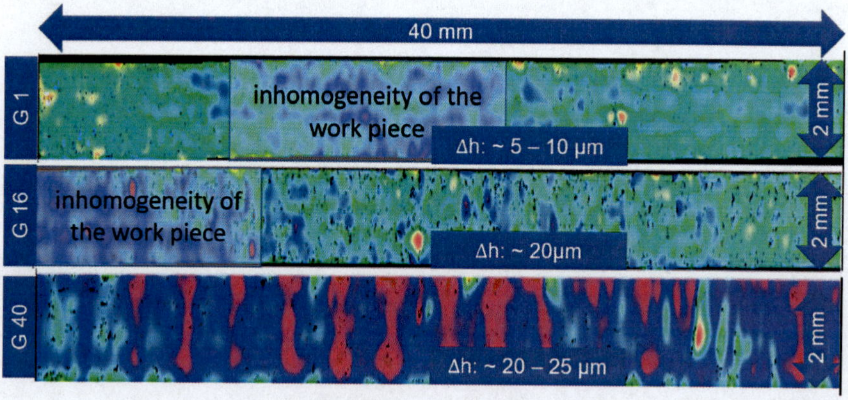

Fig. 7. Surface texture analysis performed by the Alicona Compact Cobot using software integrated Gaussian planar filter methods (4-blade-tool)

To investigate larger areas of the workpiece the tactile measurement system (Wenzel) provided a solution to evaluate the global positioning of the processed surface orientation und roughness. Since the Bruker Alicona Compact Cobot is not able to detect the global plane orientation the Wenzel System was engaged to discover the indicated vertical deviation alongside the tool axis measured in Fig. 6.

It was possible to detect the effect of the measured tool deviation. It results in an increasing inclined plane (see Fig. 8) in vertical probe direction (vertical surface deviation).

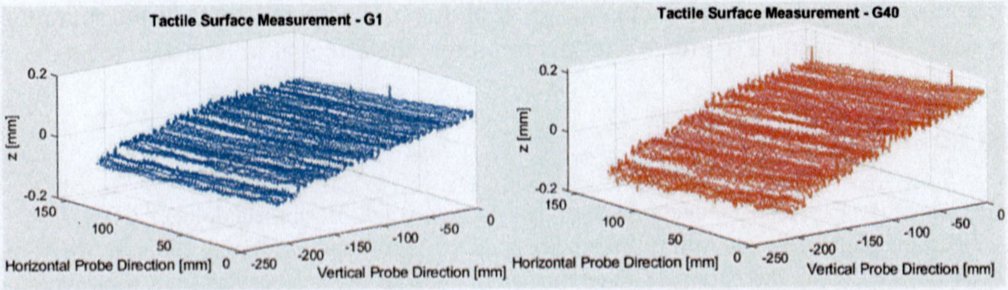

Fig. 8. Tactile Measurements: Large surface structure results of the beech surface regarding the balance factor

Extracting the values, the G1 tool configuration lead to a vertical surface deviation of 0.17 mm in comparison to 0.29 mm with the G40 tool configuration alongside the measured vertical workpiece area of 210 mm. This confirms the idle measured maximum tool face deviations are present during the cutting process.

It has to be stated, that the vertical and horizontal number of data points of the Wenzel machine is limited. The accuracy of the positional values itself are state of the art and exceed the needs of the investigation.

One significant advantage of the tactile approach is the independence regarding changing surface light interactions and other optical interferences. In addition, the probe head diameter can be used as a physical passive filter to adapt naturally caused local workpiece inhomogeneity. In comparison and regarding the two individual X-Y-regions the Alicona System can process performing with 0.15 $\frac{mm^2}{s}$ and the Wenzel System reached 32 $\frac{mm^2}{s}$.

4 Conclusion and Outlook

Several investigations to evaluate and classify the impact of the tool balance class at high speed rotation of a massive slicer were conducted. Linking the tool imbalance, tool deviation and surface structure it is to conclude, that the quality of wood surfaces and tool excitations begin to increase significant after exceeding the G16 balance classification. Besides the changes of the surface quality, concomitant examinations detected significant structural machine excitation while processing with the G40 tool configuration and even more critical while testing higher imbalance configuration. A recommendation to not exceed the G16 balance class can be given to ensure an optimal process window.

Regarding the underlying optical and tactile wood workpiece surface investigations the microscopic and macroscopic effects of increasing tool imbalance are presented. The technical advantages of each system helped to create different approaches to validate the surface quality of wooden surfaces. Results indicate that increasing imbalance forces the tool to perform a dominant single cut finish and results in oscillations alongside the rotation axis leading to an overall vertical surface deviation.

Further investigations will determine the effects using wood milling tools at a CNC-machining center with diameter smaller 20 mm. A configuration facing challenges to balance the tool during production and a significant influence of the tool clamping, that relativize the single balancing class of the tool due to the overall mass of the rotational system. All performed to create a modal to predict the effect of imbalanced tools.

References

1. Schneider, H.: Auswuchttechnik, Springer-Vieweg, vol. 8. Berlin, Heidelberg (2013)
2. Kimmelmann, M.: Measuring imbalance-induced vibrations in rotating tools. In: MATEC Web Conf. - 8th International Conference on Manufacturing Science and Education (Volume 121), pp. 9–110 (2017)
3. Schulz, H., Wuerz, T.: CIRP Annals-Manuf. Techn. **47**(1), 321–324 (1998)
4. Heisel, U., Krondorfer, H.: Proceed.of the 13th Internat, pp. 253–264. Wood Machining Seminar, Vancouver (1997)
5. Costes, J.: Towards high cutting speed in wood milling. Annals of Forest Science, Springer Verlag/EDP Sciences **59**(8), 857–865 (2002)
6. Eschelbacher, S., Duntschew, J.: Recognition of wood and wood-based materials during machining using acoustic emission. In: Wulfsberg, J.P., Hintze, W., Behrens, BA. (eds.) Production at the leading edge of technology. Springer Vieweg, Berlin, Heidelberg (2019)
7. DIN 69888: Auswuchtanforderungen an rotierende Werkzeugsysteme. Beuth-Verlag, Berlin (2008)
8. ISO 15641: Milling cutters for high speed machining - Safety requirements. Beuth-Verlag, Berlin (2001)
9. DIN EN 847-1: Tools for woodworking - Safety requirements - Part 1: Milling tools, circular saw blades. Beuth-Verlag, Berlin (2013)
10. ISO 1940-1: Mechanical vibration - Balance quality requirements for rotors in a constant (rigid) state - Part 1: Specification and verification of balance tolerances. Beuth-Verlag, Berlin (2004)
11. Institute for Machine Tools - Research projects of Wood and Composites Machining: Determination of balancing strategies and technically reasonable balance qualities with regard to the machine dynamics for tools in wood machining, LINK: https://www.ifw.uni-stuttgart.de/en/research/wood-and-composites-machining/, Stuttgart (10.05.2023)

Automated Determination of Suitable Finishing Operation Types for Milling of Surfaces

M. Erler(✉) and A. Brosius

Chair of Forming and Machining Processes, Technichal University of Dresden, 01062 Dresden, Germany
martin.erler@tu-dresden.de

Abstract. Milling is one of the most productive manufacturing processes for creating high quality surfaces. This process is divided into different types of milling operations, which differ in terms of cutting geometry and process kinematics. For NC-programming, the planner must take all of these into account in order to select the most appropriate type of operation. Most of the restrictions are linked to the topology of the surface and the shape of the overall component. Choosing the right type of finishing is crucial for automated planning of milling operations. This paper presents an approach for this. By considering the local geometry and global accessibility, it determines the technically possible types of finishing operations for any point on a surface. This creates the informational basis for the automated determination of processing types for entire surfaces and components. Conceivable applications are automated process planning (CAPP) or automated NC-programming.

Keywords: Cutter selection · CAPP · Milling

1 Introduction

Finishing is the last operation in the milling process when machining a component. It is used to achieve the required surface quality and form and position tolerances. During finishing, the surface of the workpiece is machined with the end face of the tool, the lateral surface or both. The relative movement of the tool to the workpiece in conjunction with the tool used results in the most important process variants for finishing: Face milling, end milling, peripheral milling and free-form milling. End milling and circumferential milling differ in terms of the position of the tool relative to the workpiece. In contrast to end milling, the face side of peripheral milling is usually not in contact with the workpiece. The cutting portion of the end face of the end mill can be made usable for machining different surfaces by geometric variation. Primarily, three types of end mills are distinguished (Fig. 1). Depending on the required surface properties and machining times, one process type is usually clearly superior.

However, the selection of a tool is subject to further restrictions. The most important are the global accessibility and the surface geometry. While for flat surfaces only the accessibility has a limiting effect, for curved surfaces usually several restrictions are

T. Bauernhansl et al. (Eds.): WGP 2023, LNPE, pp. 697–706, 2024.
https://doi.org/10.1007/978-3-031-47394-4_68

given. In addition, the restrictions are not constant, but can change depending on the position on the surface. In particular, the global accessibility depends on the workpiece geometry, which changes during machining.

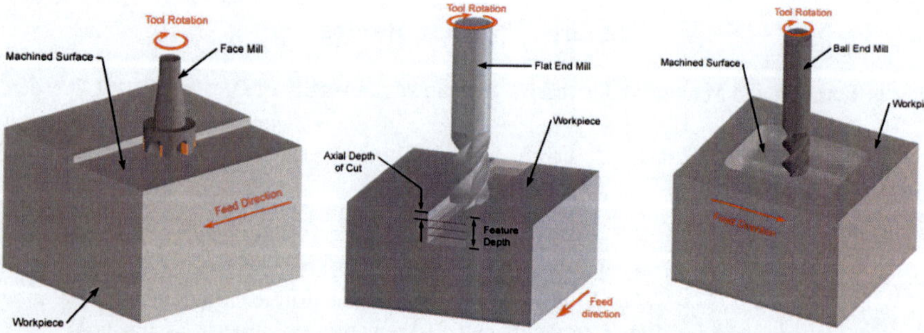

Fig. 1. Operation types: face milling (l.), end milling (m.), peripherical milling (r.) [1]

For an automatic selection of the finishing tool, the required surface properties, the surface geometry and the local and global accessibility must therefore be taken into account.

2 Related Work

Since the process type is very closely coupled to the tool, no approach dealing with the selection of the process type can be found until about 2010. Rather, technologies suitable for this purpose are subsumed under the approaches to automated tool selection or path planning. Both research areas are completely limited to 2.5D cases until the early 1990s [2]. In 1996, Lee et al. present a first approach for the sizing of freeform surfaces [3]. However, the focus remains on prismatic machining situations with some exceptions [4–6]. A differentiation of the application between finishing and roughing is also often missing before 2000 [7, 8]. With the advancement of NC technology and CAM systems, research is increasingly intensifying and differentiating. As of today, 4 areas can be identified around the automated tool and process type selection for milling:

1. Tool type selection
2. Optimal tool usage planning
3. Optimal tool diameter determination during roughing
4. Accessibility analysis during roughing and finishing
5. Selection of valid tool (geometries) when finishing free-form surfaces

2.1 Tool Type Selection

For a long time, the tool type was firmly linked to the feature to be machined. Until about 2000, "cutter selection" meant the determination of the correct cutter diameter. [9]. Thus the selection of the tool type already took place implicitly by the feature classification

[10]. With the advent of expert systems and the closely related notion of features, rule-based systems gained in importance [11]. Even today, the tool type selection is still coupled to the feature definition [12] and typically takes place through firmly defined rules [13]. However, there are also approaches to consider the more advanced parameters - e.g. by means of fuzzy [14] or Artificial Neural Networks [15]. However, this has limited success because the processing areas are already defined in this approach and the solution spaces are thus considerably restricted.

2.2 Optimal Tool Usage Planning

For roughing, the larger the tool diameter, the larger the metal removal rate. However, this only applies if the entire tool is engaged. Tool change also takes time. The division of a feature into machining areas thus represents an optimization problem. While explicit solutions can be found for geometrically trivial cases [6], the general case is np hard [16]. Approaches to solve the problem are primarily in the field of heuristics [8]. These initially used contour descriptions [9, 10] or 2-dimensional surface models [8]. However, methods for 3-dimensional surfaces soon followed [17, 18]. Approaches also exist for deployment planning across multiple components [19]. Almost all approaches are based on geometric decomposition of the given feature (Fig. 2). Mainly offset-based [20] and area model based approaches [21] are used. The problem is closely related to optimal tool path planning. The use of meta-heuristics has also been observed recently [16].

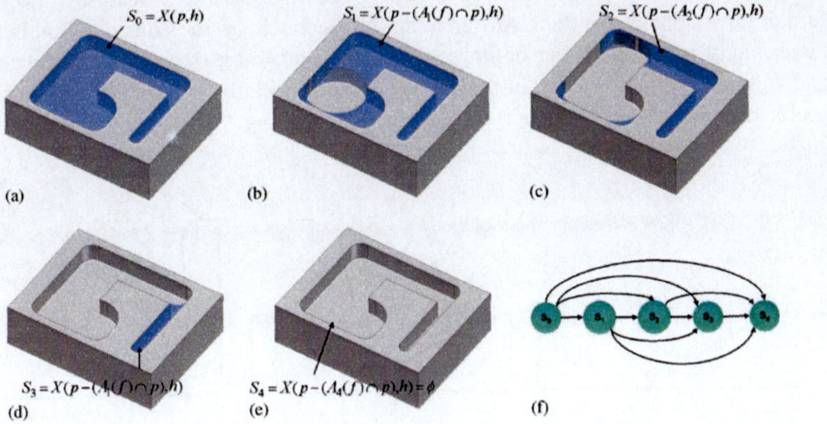

Fig. 2. Segmentation of a pocket and graph representation of operation sequence [22]

2.3 Accessibility Analysis

The accessibility to a feature or machining location basically limits the tool selection. A distinction is made between local and global accessibility. This problem has been researched since the early 1990s [23]. The still current approach for this is to use a Gauss map to project the accessibility onto a sphere (Fig. 3) [24].

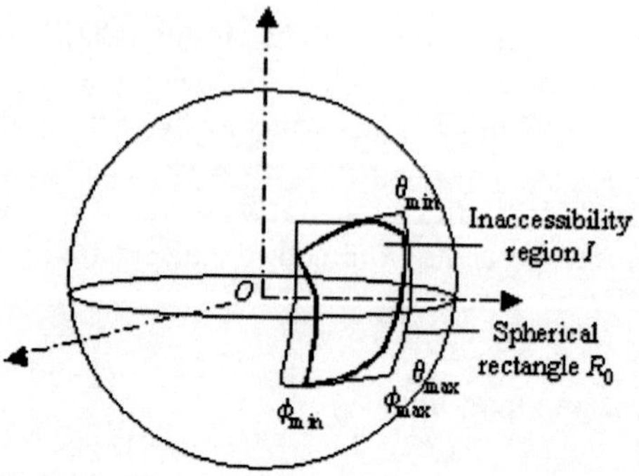

Fig. 3. Representation of inaccessible region and enclosing spherical rectangle [25]

There are different calculation methods for this [25]. This approach is applied either to entire features or only areas of them [26]. A rapid approach to global visibility using area segments instead of explicit formulas is presented by Suthunyatanakit et al. [27].

2.4 Selection of Valid Tool (Geometries)

Basically, tool types are only suitable for certain types of machining. Initially, this aspect was largely ignored and limited to 2.5D machining with end mills. The diameter was decisive here, since accessibility is always given with prismatic features [28]. With further developments in the CAM area and NC technology, the consideration became increasingly mixed with the optimization of path planning (see 2.3). The focus was and still is on roughing [29]. In the 3-D case, but especially in 5-axis machining, new problems occured (Fig. 4) which are relevant for finishing.

Fig. 4. Three types of gouging in 5-axis machining. (a) Local gouging. (b) Rear gouging. (c) Global gouging (collision) [30]

Thus, the focus shifted to finishing with ball end mills, where there are numerous methods. These are generally based on geometrical/mathematical [31], rule-based [32] and/or/oder ANN-based [33] approaches. Surface models are almost always used exclusively.

3 Description of Method

In order to avoid the limitations described above, all aspects that are important for the process selection must be taken into account. In principle, this can be done by a cutting test of the tool geometry at all machining points. However, this approach is not suitable for generating a solution, since the number of possible tools, machining locations and orientations is almost infinite.

The presented method (Fig. 5) calculates properties for discrete surface segments from which the suitability for certain process types can be concluded. For tools suitable for this purpose, their accessibility to the surface segment is checked. Invalid tools and poses are dismissed. For the remaining combinations of tool and pose, the largest possible diameter is determined using our approach [26]. Finally, any required process-specific properties of the surface (e.g. freedom from runout for face milling) are calculated. From the information now available for a surface segment, a rule-based list of possible process types is generated.

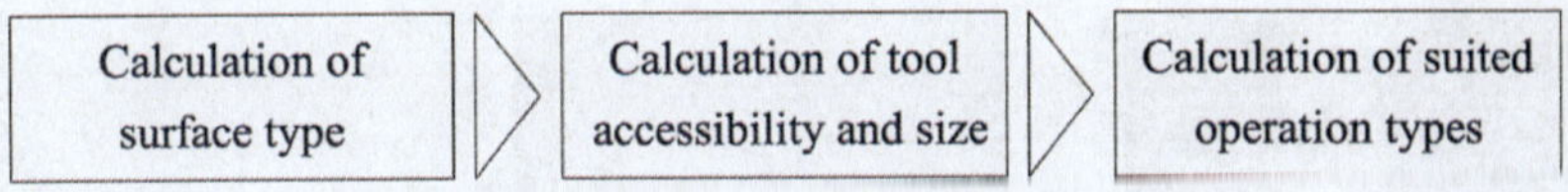

Fig. 5. Structure of proposed method

Calculation of Surface Type

The basis is formed by the area segments, which are determined by means of classical segmentation methods. Each area segment is then evaluated using the scheme shown in Fig. 6 and assigned to a type. The tests for planarity, symmetry and cylindricity are performed using the angles between the patches of a surface segment and geometrical tests if the surface fits into the desired shape.

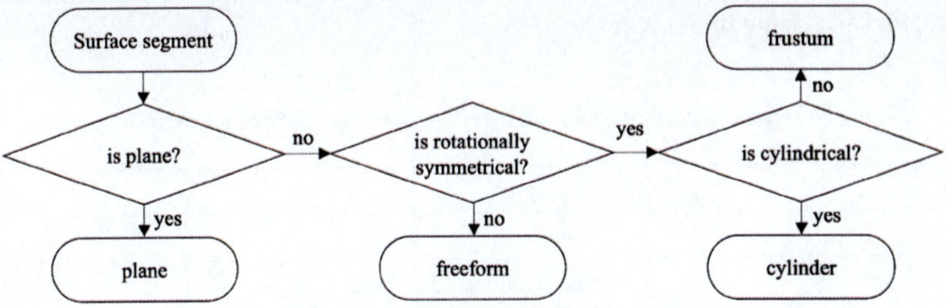

Fig. 6. Algorithm to determine surface type

Calculation of Tool Accessibility and Size

For each surface, based on its type, the possible tools are determined (Table 1). At selected

Table 1. Surface types and assigned tool types

	plane	cylinder	frustum	freeform
End-Mill	■	■	■	
Bullnose-Mill	■	■	■	■
Ballnose-Mill	■	■	■	■
Face-Mill	■			

points of the surface segment, the accessibility is then calculated for the suitable tool types. Discretization of the tested directions makes them comparable.

The largest possible opening angle of the visibility cone can also be calculated from the accessible directions (Fig. 7).

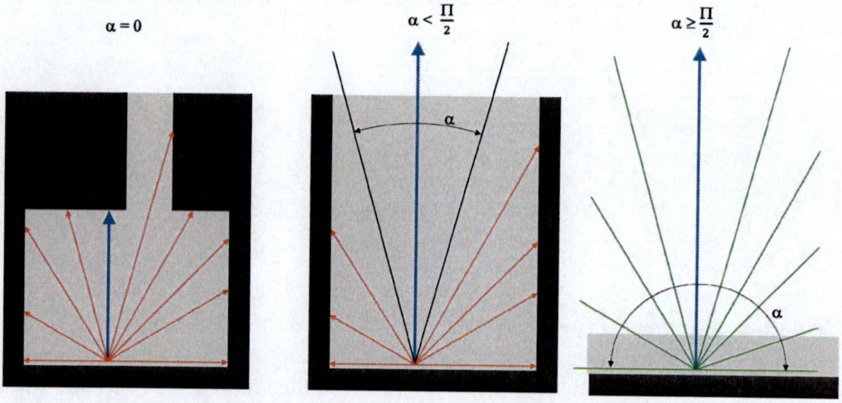

Fig. 7. Calculation of biggest opening angle for visibility cone

The tool type is determined by the biggest opening angle of the accessibility cone as shown in Table 2. Finally, for each direction, the largest possible tool diameter is calculated according to our approach [26].

Table 2. Determination of tool type based on opening angle of accessibility cone

α	Tool type
≤ 0	None
$< \frac{\Pi}{2}$	End, Bullnose, Ballnose
$\geq \frac{\Pi}{2}$	End, Bullnose, Ballnose, Face

Calculation of Suited Operation Types

Tool types are differently suitable for different types of machining. For example, face milling cutters are not at all suitable for circumferential milling and ball nose cutters are only suitable for face milling with very great restrictions. These manufacturing-technological conditions are fundamentally dependent on many other factors, but for the time being they will be used generally as shown in Table 3 for the sake of simplicity.

Table 3. Evaluation scheme for assigning operation types to tool types

	face milling	end milling	peripheral milling	freeform milling
End-Mill	0,9	1,0	1,0	0,1
Bullnose-Mill	0,5	1,0	0,9	0,8
Ballnose-Mill	0,1	1,0	0,8	1,0
Face-Mill	1,0	0,0	0,0	0,0

A weight ω of 0 means that machining with this tool is not possible - a 1 means that this is the best possible tool for this machining type. The values estimated by the author provided good results, but they require further evaluation.

The accessibility and diameter information can be used to determine the possible machining types for each surface segment and thus also for each surface. A machining type for a direction is considered valid if all surface segments in this direction can be machined with at least one valid tool.

By using a fitting function (1), a score *s* can be assigned finally to the different tools (*tool*) and directions (*dir*). The used fitting function also considers the surface area of each surface segment *k* and the tool diameter *d* to the power of a factor f_{mrr} as simple representation for the material removal rate achievable with that tool. Feasible values for f_{mrr} were found between 1,5 and 1,8 for diameters between 1 and 20 mm.

$$s_{tool,dir} = \sum\nolimits_{k=0k}^{n} A_k d_k^{f_{mrr}} \quad (1)$$

In this way, the many possibilities can be weighted and evaluated.

4 Results and Discussion

The presented approach provides valid and evaluated machining options for finishing. By using discrete surface segments, the granularity can be adapted to the needs (resolution vs computation time). A level-of-detail approach would be very interesting here. The tool types considered correspond to the most common types for finishing. The method reliably detects valid and good machining types depending on tool type and accessibility (Fig. 8). It can be seen that the method is able to distinguish between plane faces, which can be machined using a face mill (blue) and those which not due to bordering faces, that prevent face milling tool usage for the whole face (e.g. flat grey face in right figure with red drilling holes surrounded by green freeform face). A disadvantage of the approach is the discretization, which leads to inaccuracies and information loss. This can be reduced, but not eliminated.

However, the information obtained is sufficiently detailed to provide, for example, the information needed for path planning (Fig. 9).

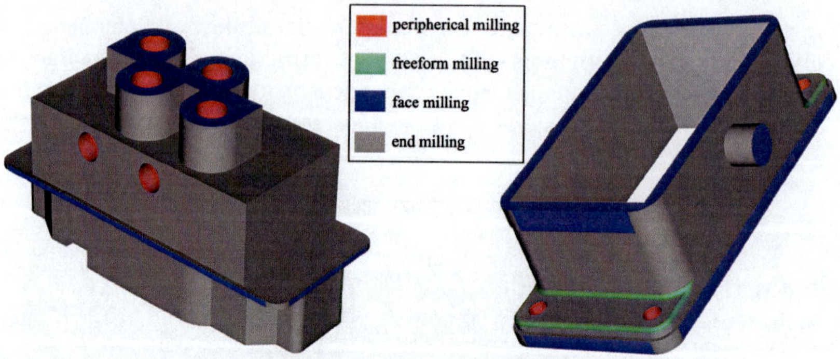

Fig. 8. Determined operation types

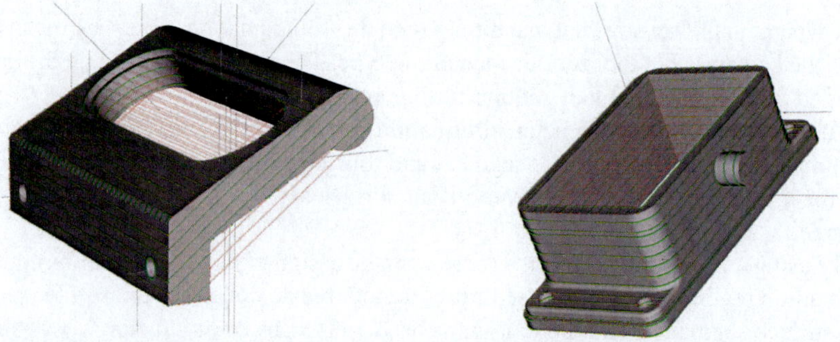

Fig. 9. Automatically generated tool paths as example application of proposed method

Other possible applications within the scope of CAPP include automated tool selection, dynamic operation sequence planning, generation of planning data and finally automated NC programming.

References

1. Machinery, H.: Technology Milling. https://www.hoston.cc/news-events/technologymilling.html, zugegriffen 18 August 2023
2. Bala, M., Chang, T.-C.: Automatic cutter selection and optimal cutter path generation for prismatic parts. Int. J. Prod. Res. **29**(11), 2163–2176 (1991). https://doi.org/10.1080/00207549108948076
3. Lee, Y.-S., Chang, T.-C.: Automatic cutter selection for 5-axis sculptured surface machining. Int. J. Prod. Res. **34**(4), 977–998 (1996). https://doi.org/10.1080/00207549608904946. Apr.
4. Elber, G., Fish, R.: 5-axis freeform surface milling using piecewise ruled surface approximation. J. Manuf. Sci. Eng. **119**(3), 383 (1997). https://doi.org/10.1115/1.2831117
5. Glaeser, G., Wallner, J., Pottmann, H.: Collision-free 3-axis milling and selection of cutting tools. CAD Comput. Aided Des. **31**(3), 225–232 (1999). https://doi.org/10.1016/S0010-4485(99)00018-4

6. Chen, Y.-H., Lee, Y.-S., Fang, S.-C.: Optimal cutter selection and machining plane determination for process planning and NC machining of complex surfaces. J. Manuf. Syst. **17**(5), 371–388 (1998). https://doi.org/10.1016/S0278-6125(98)80004-6
7. Jha, N.K.: A set-theoretic automatic selection of cutting tools in manufacturing system. J. Mater. Process. Technol. **65**(1–3), 99–106 (1997). https://doi.org/10.1016/0924-0136(95)02248-1
8. Lee, Y.S., Daftari, D.M.: Feature-composition approach to planning and machining of generic virtual pockets. Comput. Ind. **31**(2), 99–128 (1996). https://doi.org/10.1016/0166-3615(96)00027-9
9. Kyoung, Y.M., Cho, K.K., Jun, C.S.: Optimal tool selection for pocket machining in process planning. Comput. Ind. Eng. **33**(3–4), 505–508 (Dez. 1997). https://doi.org/10.1016/S0360-8352(97)00179-4
10. Lee, Y.S., Chang, T.C., Choi, B.K.: Cut distribution and cutter selection for sculptured surface cavity machining. Int. J. Prod. Res. **30**(6), 1447–1470 (1992). https://doi.org/10.1080/00207549208942966
11. Carpenter, I.D., Maropoulos, P.G.: Flexible tool selection decision support system for milling operations. J. Mater. Process. Technol. **107**(1–3), 143–152 (2000). https://doi.org/10.1016/S0924-0136(00)00707-X
12. Zhou, M., Zheng, G., Chen, Z.C.: An automated CNC programming approach to machining pocket with complex islands and boundaries by using multiple cutters in hybrid tool path patterns. Int. J. Adv. Manuf. Technol. **83**(1–4), 407–420 (2016). https://doi.org/10.1007/s00170 015 7506 3
13. Ji, W., Wang, L., Haghighi, A., Givehchi, M., Liu, X.: An enriched machining feature based approach to cutting tool selection. Int. J. Comput. Integr. Manuf. **31**(1), 1 (2018). https://doi.org/10.1080/0951192X.2017.1356472
14. Phung, X.L., Truong, H.S., Bui, N.T.: Expert system based on integrated fuzzy AHP for automatic cutting tool selection. Appl. Sci. Switz. **9**(20) (2019). https://doi.org/10.3390/app9204308
15. Zhou, G., Yang, X., Zhang, C., Li, Z., Xiao, Z.: Deep learning enabled cutting tool selection for special-shaped machining features of complex products. Adv. Eng. Softw. **133**, 1–11 (Juli 2019). https://doi.org/10.1016/j.advengsoft.2019.04.007
16. Langula, S., Erler, M., Brosius, A.: An efficient method for automated machining sequence planning using an approximation algorithm. In: Liewald, M., Verl, A., Bauernhansl, T., Möhring, H.-C. (eds.) Production at the Leading Edge of Technology, in Lecture Notes in Production Engineering, pp. 727–736. Springer International Publishing, Cham (2023). https://doi.org/10.1007/978-3-031-18318-8_72
17. Lee, Y.S., Chang, T.C.: CASCAM-An automated system for sculptured surface cavity machining. Comput. Ind. **16**(4), 321–342 (1991). https://doi.org/10.1016/0166-3615(91)90073-I
18. Hao, M., Mizugaki, Y., Sakamoto, M.: Optimal tool selection based on genetic algorithm in a geometric cutting simulation. J. Jpn. Soc. Precis. Eng. **60**(9), 1237–1241 (1994). https://doi.org/10.2493/jjspe.60.1237
19. Yao, Z., Gupta, S.K., Nau, D.S.: Algorithms for selecting cutters in multi-part milling problems. CAD Comput. Aided Des. **35**(9), 825–839 (2003). https://doi.org/10.1016/S0010-4485(02)00110-0
20. D'Souza, R.M.: Tool sequence selection for 2.5D pockets with uneven stock. J. Comput. Inf. Sci. Eng. **6**(1), 33–39 (2006). https://doi.org/10.1115/1.2161228
21. D'Souza, R.M., Sequin, C., Wright, P.K.: Automated tool sequence selection for 3-axis machining of free-form pockets. CAD Comput. Aided Des. **36**(7), 595–605 (2004). https://doi.org/10.1016/S0010-4485(03)00137-4

22. D'Souza, R.M.: On setup level tool sequence selection for 2.5-D pocket machining. Robot. Comput.-Integr. Manuf. **22**(3), 256–266 (2006). https://doi.org/10.1016/j.rcim.2005.06.001
23. Woo, T.C.: Visibility maps and spherical algorithms. Comput.-Aided Des. **26**(1), 6–16 (1994). https://doi.org/10.1016/0010-4485(94)90003-5
24. Elber, G., Cohen, E.: Arbitrarily Precise Computation of Gauss Maps and Visibility Sets for Freeform Surfaces, 271–280 (1995)
25. Dhaliwal, S., Gupta, S.K., Huang, J., Priyadarshi, A.: Algorithms for computing global accessibility cones. J. Comput. Inf. Sci. Eng. **3**(3), 200 (2003). https://doi.org/10.1115/1.1606475
26. Erler, M., Koch, A., Brosius, A.: Determination of largest possible cutter diameter of end mills for arbitrarily shaped 3-axis milling features. In: Liewald, M., Verl, A., Bauernhansl, T., Möhring, H.-C. (eds.) Production at the Leading Edge of Technology, in Lecture Notes in Production Engineering, pp. 228–237. Springer International Publishing, Cham (2023). https://doi.org/10.1007/978-3-031-18318-8_24
27. Suthunyatanakit, K., Bohez, E.L.J., Annanon, K.: A new global accessibility algorithm for a polyhedral model with convex polygonal facets. CAD Comput. Aided Des. **41**(12), 1020–1033 (2009). https://doi.org/10.1016/j.cad.2009.08.002
28. Veeramani, D., Gau, Y.-S.: Selection of an optimal set of cutting-tool sizes for 2D pocket machining. Comput.-Aided Des. **29**(12), 869–877 (Dez. 1997). https://doi.org/10.1016/S0010-4485(97)00042-0
29. Balasubramaniam, M.: Tool selection and path planning for 3-axis rough machining. Nr. **1997**, 1–77 (1999)
30. Lasemi, A., Xue, D., Gu, P.: Recent development in CNC machining of freeform surfaces: A state-of-the-art review. Comput.-Aided Des. **42**(7), 641–654 (Juli 2010). https://doi.org/10.1016/j.cad.2010.04.002
31. Meng, F.J., Chen, Z.T., Xu, R.F., Li, X.: Optimal barrel cutter selection for the CNC machining of blisk. CAD Comput. Aided Des. **53**, 36–45 (2014). https://doi.org/10.1016/j.cad.2014.03.006
32. Lin, T., Lee, J.W., Bohez, E.L.J.: A new accurate curvature matching and optimal tool based five-axis machining algorithm. J. Mech. Sci. Technol. **23**(10), 2624–2634 (2009). https://doi.org/10.1007/s12206-009-0724-6
33. Rojek, I.: Technological process planning by the use of neural networks. Artif. Intell. Eng. Des. Anal. Manuf. **31**(1), 1–15 (2017). https://doi.org/10.1017/S0890060416000147. Feb.

Advanced Process Analysis for High Performance Tapping Technology in Aluminum Cast Alloy AlSi10Mg

N. Felinks(✉), E. Krumme, and D. Biermann

Institute of Machining Technology (ISF), TU Dortmund University, Baroper Str. 303, 44227 Dortmund, Germany

nils.felinks@tu-dortmund.de

Abstract. In context of industrial mass production, innovative machining technologies offer a potential for significant gain of efficiency and sustainability. Whereas in conventional tapping processes the tool moves in and out in sync with the thread pitch, the kinematics of the novel helical thread forming technology allows to reduce the process time by half. This paper is about an extended process analysis of its specific process characteristics in contrast to conventional tapping. By high-speed-microscopy and advanced analysis of mechanical tool loads, it is possible to evaluate specific differences in the forming process of the thread profile. In context of sustainable production, aluminum cast alloy AlSi10Mg – which represents a recyclable material for near-net-shape components in mobility sector – is regarded in this study. In addition, the resulting quality of the produced M6 threads is evaluated in context of decreased production time and aspects of sustainability.

Keywords: Helical Thread Forming · High Performance Tapping · Aluminum Cast Alloy

1 Introduction

In this study the aluminum cast alloy AlSi10Mg is regarded in context of high performance tapping. This material is typically used for casting parts in automotive industry. Innovative manufacturing technologies for AlSi components are required to face high energy and labor cost in industrial production of Europe. Especially, after electrical energy costs have reached a peak in 2022, a sustainable and productive value chain is important to face challenges in industrial mass production [1]. The production of aluminum raw material is linked to high amounts of energy required. In contrast novel approaches of recycling routes make aluminum alloys potentially more sustainable in the future to provide for example 'Green Alumina' [2, 3].

Due to its high strength and corrosion resistance this material is suitable for crash-relevant components as well as for motor housings [4]. For example automotive components are designed weight-optimized and with complex structures, which increase

T. Bauernhansl et al. (Eds.): WGP 2023, LNPE, pp. 707–716, 2024.
https://doi.org/10.1007/978-3-031-47394-4_69

usability and specific weight. Additionally the mechanical properties can be approved by adapted heat treatment [5], which makes Al alloys worth applying in mobility sector.

In context of manufacturing, aluminum cast alloys are regarded as well machinable. High tool life and feed speed lead to productive machining processes in general [6]. Regarding high performance manufacturing technology, tapping methods are in focus because of their high percentage of non-productive process time. There are different approaches to increase the productivity of tapping in Al alloys, for example by increasing rotation speed, tool technology or lubrication [7, 8]. Furthermore the innovative technology of helical thread forming was developed, which minimizes the tool path for manufacturing of internal threads significantly [9]. In comparison to conventional tapping methods for AlSi alloys, the thread profile is interrupted by a helical groove. Additionally significantly higher tool loads - depending on the core hole diameter - occur in helical thread forming [10, 11].

This study deals with an enhanced analysis of the specific aspects of the forming process in comparison to conventional thread forming. In detail it contains an enhanced analysis of the forming forces and process by novel methods.

2 Material and Methods

The aluminum alloy AlSi10Mg is regarded in this study as an industrially common used material for die casting components. By a T6-heat treatment a hardness of 110 HV10 and elongation at break of 5% is given to the samples. Especially in context of thread forming a elongation of 5…8% is essential for sufficient formability [12]. Figure 1 shows the microstructure as well as an excerpt of mechanical properties.

Material: AlSi10Mg (3.2381; EN AC-43000)

Microstructure:

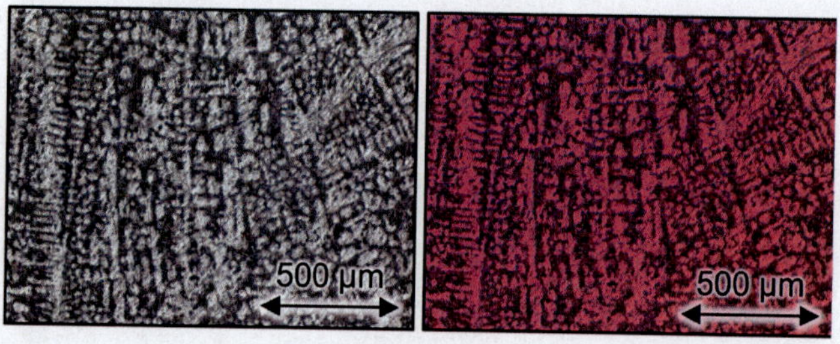

Heat treatment: T6 **Yield stress:** $R_{p0,2}$ = 280 MPa

Hardness: 110 ±3,5 HV10 **Elongation:** A = 5 %

%	Si	Fe	Cu	Mn	Mg	Zn	Ti	Ni	Pb	Sn	rest	Al
min.	9.0			0.2	0.2							
max.	11.0	0.55	0.05	0.45	0.45	0.1	0.15	0.05	0.05	0.05	0.15	bal.

Fig. 1. Microstructure, mechanical properties and chemical composition of regarded material AlSi10Mg

The performed tests on the two regarded thread forming technologies required a through-hole, which was chamfered afterwards. A standard 4-component dynamometer type *Kistler 9272* in combination with a special measuring setup, which contains a *Kistler 9067B* force ring, were applied to record the mechanical tool loads. The setup was designed to infer the forces working on the forming tooth within the so called 'double-ring-test' (DRT). The high-speed camera system type *Keyence VHX9000* was used to analyze the thread formation in a specially primed sample. Afterwards the thread's accuracy was checked by a thread gage. In order to evaluate the macroscopic shape a light microscope (*Keyence VHX5000*) was utilized to take pictures of the halved samples. In post-processing, the data of DRT were prepared to total force values. The complete structure of methodology is presented in Fig. 2.

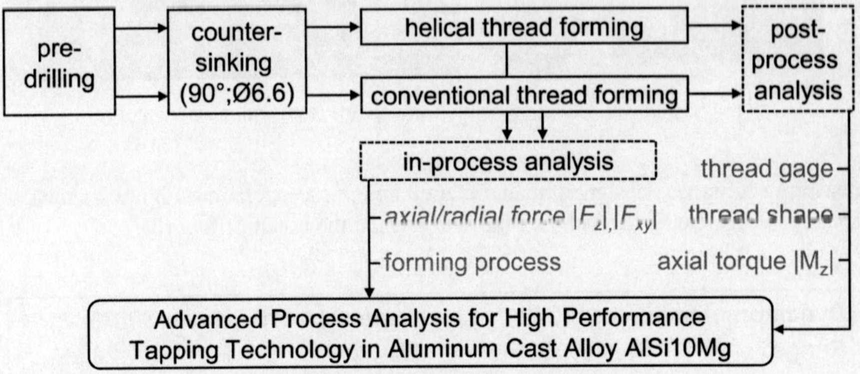

Fig. 2. Methods and structure of process chain

The experimental tests this study based on were performed on a high performance machining center type *GROB BZ600* in the lab of *Institute of Machining Technology* (ISF) at *TU Dortmund University*. This machine tool is equipped with a special NC-cycle and strengthened spindle bearings, which are adapted on kinematics and working forces of helical thread forming. The spindle clamped an inflexible HSK-A80 tool holder for helical thread forming. It carries the specific TiN-coated tapping tool for M6 threads, which has two helical rows of forming teeth. When initially the tool punches into the core hole, a helical groove is plowed, which enables the tool to retract in high speed. After that, the thread profile is basically formed by a synchronized tool movement of a half turn and extraction of a half thread pitch ($\Delta c = 180°$; $\Delta z = 0.5$ mm).

The M6 forming tap is clamped in a compensating collet holder and forms the thread conventionally by screwing in and out of the bore hole. This tool path is programmed by standard *Siemens 840D* 'Cycle 84'. For increased lubrication, the tool (type *InnoForm 1-GAL/E-SN-IKZ-PM-TICN* from *EMUGE)* carries oil grooves and internal coolant supply. Especially designed for AlSi cast alloys, the cold forming tap features lead tapper form E and a TiCN coating, as Fig. 3-c shows. The machine tool is equipped with cooling emulsion by *Quaker Houghton* type *Hocut4646* with a concentration of 9%.

As mentioned in the chapter methodology above, this study aims to evaluate the forming progress of thread forming in detail. For this purpose, the 'double-ring-test'

Machine tool:	GROB BZ600	**Drilling tool:**	twist-drill ø5.55.; TiAlN
Tool holder:	HSK-A80	**Tapping tools:**	
Spindle speed:	$n_{max} \leq 10000$ rpm		a) GAL-SN-E-IKZ M6; TiAlN
Lubricant:	Emulsion 9 %		b) PunchTap© M6; TiN

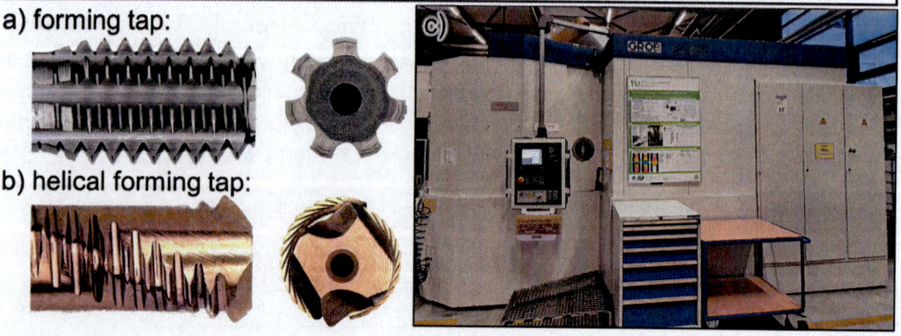

Fig. 3. Technical data of regarded forming taps (a-b) and machine tool (c)

measuring setup was designed, manufactured and put to operation. Figure 4 shows the designed CAD-model (a) and the final setup inside the machine tool (b).

Dynamometer:	Kistler 9272	**Force sensor:**	Kistler 9067B
Range:	F_z = 0…20 kN	**Range:**	$F_{x/y}$ = 0…20 kN
Sensitivity:	F_z = -3,554 pC/N	**Sensitivity:**	$F_{x/y}$ = -8,13 pC/N
Frequency:	f = 50 kHz	**Frequency:**	f = 50 kHz

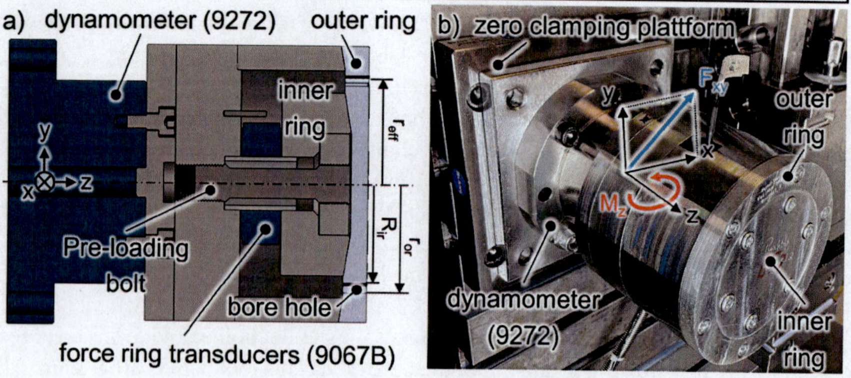

Fig. 4. Experimental setup for analysis of the forming forces

In detail, it carries an outer ring (-or), which is made of regarded material AlSi10Mg and fixed to the base. The inner ring (-ir) is mounted onto a 50 kN pre-loaded *Kistler 9067B* force ring, which measures forces triaxially. By tapping in the slot between inner and outer ring (width 0.3 ± 0.05 mm), the radial forces during thread forming are measured by the inner ring. The programmed radius is calculated by:

$$r_{eff} = \frac{(r_{or} + R_{ir})}{2} \tag{1}$$

In post-processing of the forces measured as described, the raw data were extracted and y-offset was compensated for preparing the calculation of axial torque. From the plane force component canals $F_x(t)$ and $F_y(t)$ the total values were calculated and added as vector sums to $|F_{xy}|(t)$. After that the calculated torque $|M_z|(t)$ was accounted by multiply $|F_{xy}|(t)$ with effective tool radius $r_{tool,eff}$:

$$r_{tool,eff} = \frac{(d_{core} - d_{tip})}{2} \tag{2}$$

$$M_z(t) = r_{tool,eff} \cdot F_{xy}(t) \tag{3}$$

Finally the radial forces $|F_{xy}|(t)$ and calculated torque $|M_z|(t)$ were plotted in charts. In order to evaluate the calculations, reference tests were performed in solid samples with coaxial attached dynamometer type *Kistler 9272*. The post-processing method used for these data is comparable studies previously published [10]. FE simulations of the helical thread forming process will be validated using the data obtained in this study.

In fact the thread profile is formed by a 180° tool rotation in helical thread forming; in contrast to conventional tapping, where the thread tips are formed successively in about three rotations. The setup in Fig. 5 was designed to make the forming process visible. By the aim of focusing a longitudinal section of the bore wall, one quarter of the cross section was removed. To balance between tool support and visibility of forming process, the center of the pilot hole was relocated by $\Delta x = \Delta y = 0.2$ mm from xyz-plane-section.

Camera:	*Keyence VHX9000*	**Magnification:**	50x
Lens:	*Keyence VH-Z50L*	**Frame rate:**	f = 12000 fps

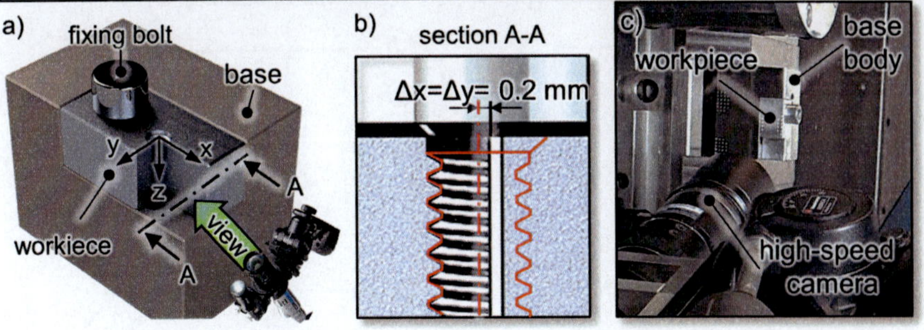

Fig. 5. Experimental setup for visualization of the forming process

3 Results and Discussion

In the following, the advanced process analysis for high performance tapping technologies is presented. Specific aspects of thread forming process are regarded by the example of manufacturing tests on M6 threads ($t_{M6} = 2 \times D$) in AlSi10Mg material.

3.1 Mechanical Tool Loads

In reference, a maximum amount $M_z = 5$ Nm was measured in retract end extension phase for conventional thread forming, as Fig. 6 shows. Both peaks rise and fall to the start respectively end tapping process. The total value of torque measured in DRT is 40% lower, but the qualitative progression of the $|M_z|(t)$ follows the reference. There is no axial torque measured in reference tests when the tool movement has stopped at the bore's exit at $t_{M6} = 10$ mm. In contrast, deformed burr material generates a force fit between the inner and outer ring in the slot. After the tool starts to unscrew, the radial force goes to zero.

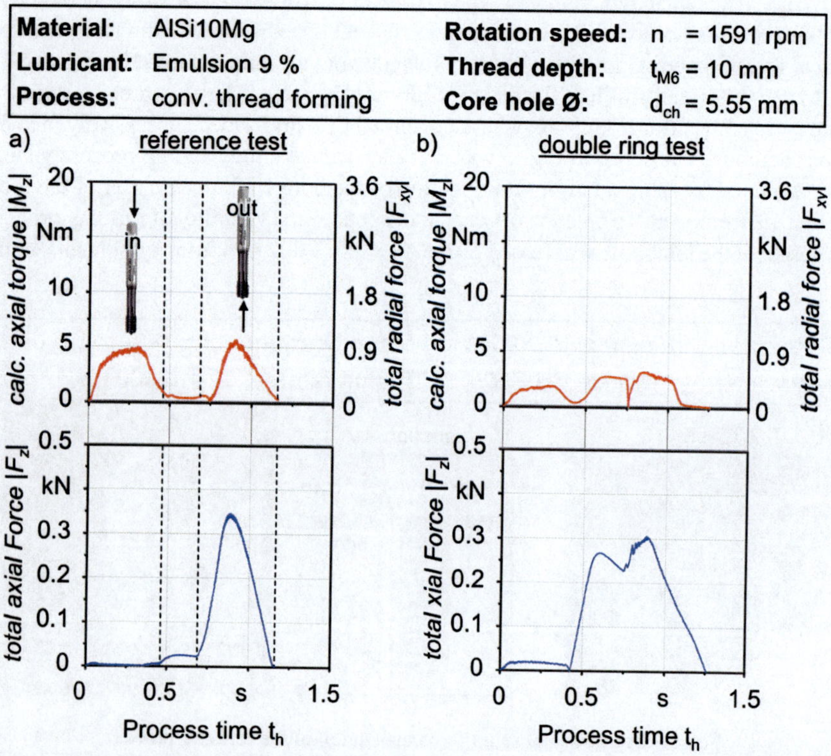

Fig. 6. Absolute values of axial torque $|M_Z|(t)$, axial force $|F_Z|(t)$ and radial force $|F_{xy}|(t)$ in reference process (a) and double ring test (b) in conventional thread forming

This observation is to be reasoned with traction between inner and outer ring by radial deformed material. In this phase the axial force increases simultaneously, so it can said that the inner ring braces the outer ring.

In Fig. 7-a) both total values of axial force and torque in reference test follow the results of former studies [11]: increased axial forces occur in punching in the core hole (t = 0…0.05 s), helical deburring (t = 0.2…0.25 s) and finally retract phase (t = 0.4…0.45 s). In contrast the axial torque $|M_z|(t)$ increase in both forming stages (t = 0.1…0.2 s, t = 0.3…0.4 s) as a result of the formation of the thread shape. In contrast, the total values in DRT (see Fig. 7-b) are 33% higher. This can be explained by the additionally deformed material blocked in the gap. By this phenomena the axial force decreases in discontinuous motion of helical thread forming. In consequence of the interrupted profile, burr material sheers of more regularly than in reference process.

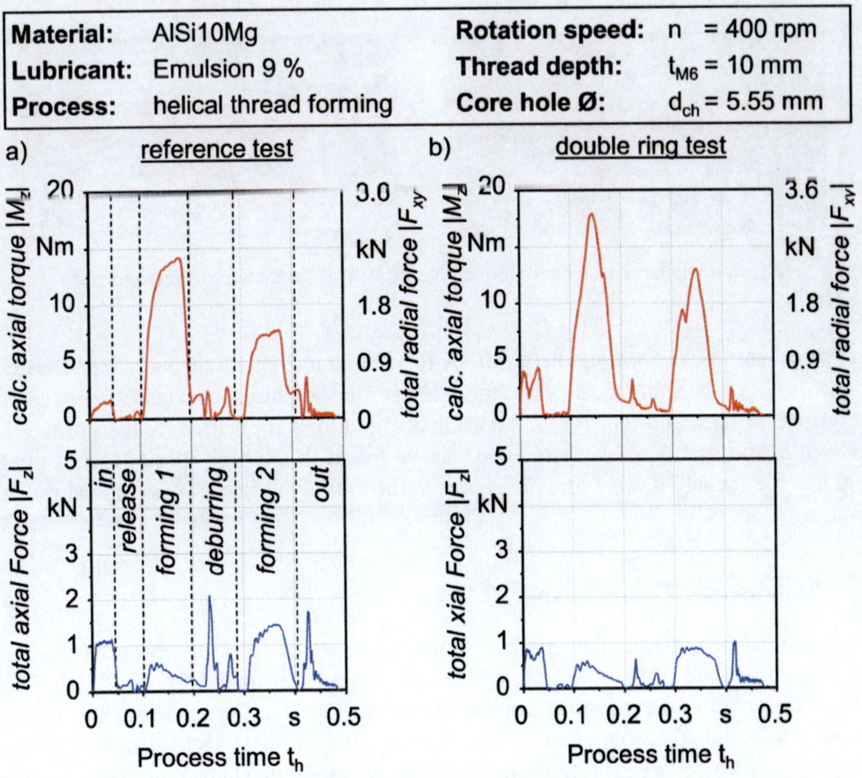

Fig. 7. Absolute values of axial torque $|M_Z|$ (t), axial force $|F_Z|(t)$ and radial force $|F_{xy}|(t)$ in reference process (a) and double ring test (b) of helical thread forming

After comparing the calculated forces and reference tests, it can be said that the qualitative progression can be measured by DRT. Concerning the total values, there are deviations depending on the regarded forming stage.

3.2 Thread Forming Process

After the mechanical tool loads were analyzed by a novel method, the forming process was visualized by the experimental setup explained in Fig. 5. The freezed images of most relevant phases for conventional (a) and helical thread forming (b) are shown in Fig. 8. In (a) the thread tips are on the one hand formed discontinuously by the polygon profile of the cold forming tap. On the other hand, the tool's conical tip lead to increasing deformation of the bore wall.

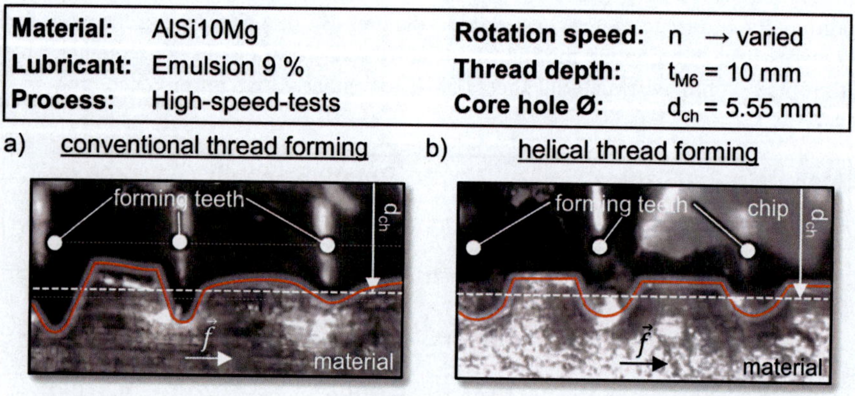

Fig. 8. Comparison of conventional (a) and helical (b) thread forming process

In helical thread forming the profile is formed almost simultaneously. The freezed frame in Fig. 8-b) show the similar stages of profile formation. This leads to increased deformation rates and tool loads. Additionally, a chip is formed while the grooves is plowed in the first process phase. This can be found in the high-speed-video as well and is to be named as one more difference to the entirely chipless conventional thread forming process.

3.3 Macroscopic Thread Shape

Furthermore the thread's shape produced differs in macroscopic view in some aspects. As known in state of the research, the thread tips split in thread forming [12]. This can be observed in the halved reference samples in Fig. 9-a. Consequently this thread shape results in higher tolerances, which confirm DIN 13–50 [DIN13–50]. The profile of helically formed threads differs significantly because of two helical grooves. Allover flank surfaces and profile are comparable to conventional thread forming like formally known [10].

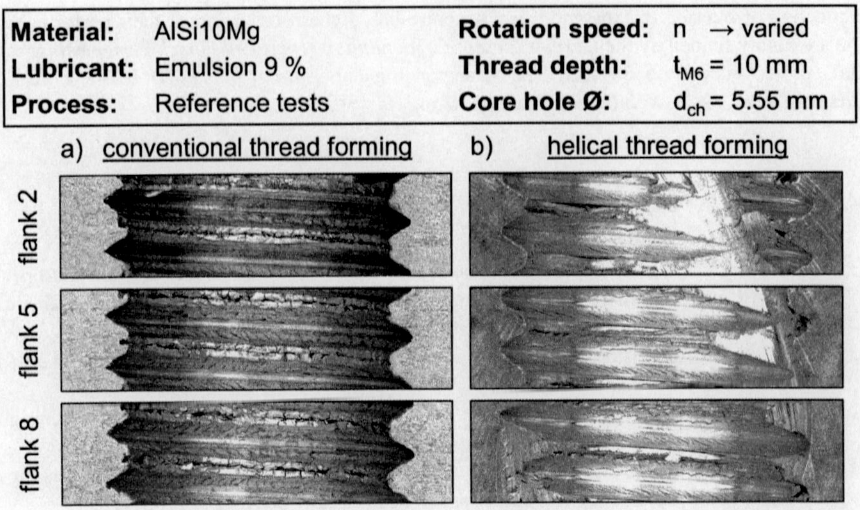

Fig. 9. Resulted thread quality of conventional (a) and helical (b) thread forming

4 Conclusion and Outlook

Finally this study deals with extended methods for detailed analysis of tapping processes. Two specific experimental setups were designed in order to evaluate radial forces and visualize the forming process by itself. In general the designed double-ring-test is a valid method to analyze radial and axial forces. These data are the data basis for the FE simulation of helical thread forming and are used in the following research.

In determining increased forces in axial direction and high level torque, the total values differ from reference tests. The gap between the inner and outer ring decrease by 33% because the lower the amount of material to be deformed. Values of axial force in deburring process stage differ by 167%. Due to the interrupted profile, less burr are formed which are to be sheered of.

The high-speed analysis of thread formation is valid to visualize the different forming methods of regarded tapping technologies. As the kinematics of helical thread forming specifies, the deformation rate is much higher as in conventional thread forming. In this case profile is formed in three rotations in $\Delta t = 0.1$ s. Moreover in helical thread forming the productive time is mostly similar. The process by itself mainly differs in the intensity of deformation method of thread profile.

Further research could focus on modelling the quantification of measured force. In general, this study proves that the setup is valid for enhanced analysis of the mechanical tool loads. In order to get precise measuring data, an algorithm for calculation of total values due to reference tests is to be developed. Furthermore enveloped methodology could be used for simulation input data as well as studies on influence of lubrication effect in tapping technology.

Acknowledgements. The investigations on high performance tapping presented are part of the research project 'Basic technological investigation of a new high-performance process for the production of internal threads and microstructure-based characterization of their performance' and are kindly funded by the *German Research Foundation (Deutsche Forschungsgemeinschaft, DFG,* project number 455887485). The corresponding author also thanks *EMUGE-Werk Richard Glimpel GmbH & Co. KG* for technical consulting.

References

1. Ember: Average monthly electricity wholesale prices in selected countries in the European Union (EU) from January 2020 to March 2023 (2023). https://www.statista.com/statistics/1267500/eu-monthly-wholesale-electricity-price-country/
2. OECD: Measuring distortions in international markets: the aluminium value chain (2019). https://doi.org/10.1787/c82911ab-en
3. Scarsella, A., Gasafi, E.: Green Alumina: A Technological Roadmap. In: Eskin, D. (ed.) Light Metals 2022, pp. 3–10. Springer International Publishing, Cham (2022)
4. DIN Deutsches Institut für Normung e. V.: Aluminium and aluminium alloys – Castings: Chemical composition and mechanical properties 77.150.1 0(DIN EN 1706) (2021)
5. Ružbarský, J.: Al-Si Alloys Casts by Die Casting: A Case Study, 1st edn. SpringerBriefs in Applied Sciences and Technology. Springer, Cham
6. Köhler, J.: Machinability of Aluminum and Magnesium Alloys. In: Laperrière, L., Reinhart, G. (eds.) CIRP Encyclopedia of Production Engineering, pp. 770–776. Springer, Berlin Heidelberg, Berlin, Heidelberg (2014)
7. Oliveira, J., Ribeiro Filho, S., Brandão, L.: Investigation of the influence of coating and the tapered entry in the internal forming tapping process. The Int. J. Adv. Manuf. Technol. **101**, 1051–1063 (2019). https://doi.org/10.1007/s00170-018-3011-9
8. Bhowmick, S., Lukitsch, M.J., Alpas, A.T.: Tapping of Al–Si alloys with diamond-like carbon coated tools and minimum quantity lubrication. J. Mater. Process. Technol. **210**, 2142–2153 (2010). https://doi.org/10.1016/j.jmatprotec.2010.07.032
9. Kopton, P., Hechtle, D., Biermann, D., et al. (eds.): Innovative Gewindefertigung im Automobilbau: Ein Beitrag zur energieeffizienten Produktion. Spanende Fertigung, vol. 6. Vulkan-Verlag, Essen (2012)
10. Felinks, N., Krumme, E., Beer, C., et al.: Comparison of modern tapping technologies for lightweight applications. Procedia CIRP **118**, 420–424 (2023). https://doi.org/10.1016/j.procir.2023.06.072
11. Felinks, N., Krumme, E., Beer, C., et al.: Influence of core hole diameter on thread quality for a high-performance tapping technology in alsi10mnmg cast alloys. Procedia CIRP **115**, 42–47 (2022). https://doi.org/10.1016/j.procir.2022.10.047
12. VDI-Richtlinie Machining of internal threads (VDI 3334)

Influence of Minimum Quantity Lubrication on Tool Temperature and Wear in Wood Machining

André Jaquemod(✉), Kamil Güzel, and Hans-Christian Möhring

Institute for Machine Tools (IfW), University of Stuttgart, Holzgartenstr. 17, 70174 Stuttgart, Germany

andre.jaquemod@ifw.uni-stuttgart.de

Abstract. The machining of wood and wood-based materials is becoming increasingly interesting in view of the progressive application of this class of materials in various industrial sectors, such as furniture, doors, windows, interior fittings and sustainable construction in the building sector. In order to promote the establishment and further development of wood construction in particular, it is necessary to research and develop approaches to solutions in the field of machinery, processes and tools in order to optimize manufacturing processes. In the industrial environment, wood and wood-based materials are usually processed dry. Unfavorable process parameters often lead to thermal problems, which can have a negative impact on the machining qualities. In contrast to dry cutting, there is hardly any scientific knowledge available for wood machining using minimum quantity lubrication (MQL). In this paper, the results on the influence of the use of MQL when sawing solid wood are discussed. Within the framework of experimental investigations, various process parameters were varied and the effects on tool temperature and tool wear were analyzed.

Keywords: Wood Machining · Sawing · MQL

1 Introduction

The current climate goals as well as new legislative drafts on sustainability and resource conservation demonstrate the increasing importance of these issues for politics and society. On the path to a climate-neutral future, the use of wood is particularly essential. Wood, as a naturally renewable construction and raw material, is an alternative to finite fossil resources based on petroleum [1]. In addition, wood has other properties that make it an attractive building material. During its growth, wood absorbs the greenhouse gas CO_2, which remains bound in the building material [2]. Only during thermal utilization or biological decomposition, it will be released again [3]. Furthermore, during its lifetime, a tree converts CO_2 into oxygen, which is crucial for other living organisms [4]. These aspects have led to wood now being perceived as a good alternative. Already, one in five houses in Germany are built using wood construction methods [5].

T. Bauernhansl et al. (Eds.): WGP 2023, LNPE, pp. 717–726, 2024.
https://doi.org/10.1007/978-3-031-47394-4_70

Compared to other materials such as steel, the machining of wood requires less energy [6]. However, serial production must be further optimized in order to increase the economic efficiency. Currently, traditional house construction is still more cost-effective than a comparable timber construction [7]. Consequently, there are increasing initiatives to develop the potential for productivity enhancement in the manufacturing of solid wood elements [8].

In wood machining, dry cutting is typically used due to the hygroscopic material properties. A new approach for this is the use of minimum quantity lubrication (MQL) to achieve a new tribological state without significant interaction between the lubricant and the wood. The influence of MQL on metal and fibre-reinforced polymer composite machining has already been analyzed in several publications [9–11]. For example, investigations on the impact of MQL on wear have been conducted in this field. [12] and [13] have shown that both wear and tool temperature decrease with the use of MQL. Thus, precise and economical use of the cooling lubricant quantity enables resource-saving and at the same time production-increasing manufacturing. For wood machining, there are only previous studies related to the content of this publication [14], which demonstrate the influence of MQL on the material surface. The use of MQL reduced the area covered with burn marks. For further analysis, this publication presents a study on the influence of temperature in conjunction with the forces and tool wear during sawing of solid wood with circular saw blades.

2 Experimental Approach

The cutting experiments were performed based on the investigations in [14] on a 5-axis machining centre (MAKA PE 170). Circular saw blades manufactured by Leuco AG with a diameter of 350 mm, a cutting width of 3.5 mm, and 24 teeth were used. The entire experimental setup for force measurement is illustrated in Fig. 1 for visualization purposes.

Fig. 1. Experimental setup for force measurement

Two identical saw blades were tested, and each blade was assigned for a specific machining method (dry/MQL). The individual saw teeth were measured using an Alicona InfiniteFocus G5 (Bruker Alicona). The cutting edge radius and cutting edge offset were determined for feed distances up to 350 m. In addition to the cutting edge measurements, force measurements were also performed for these feed distances. Rectangular material samples with the dimensions of [100 × 100 × 200] mm made of spruce were mounted on a force measurement platform (Kistler Type 9272). Grooves measuring 100 mm in length and 80 mm in depth were cut longitudinally along the fibre of the samples. During the process cutting forces were measured on workpiece side with a sampling rate of 1000 Hz. The data were recorded, stored and analyzed using a LabView program.

In order to implement a MQL in the cutting process, compressed air is used to supply the lubricant to the cutting edge. For this purpose, the HTR spray head (by HPM Technologies) is used, which can be installed in the machine using a magnetic arm. The orientation of the spray head is set at an angle of 60° between the nozzle and the circular saw blade and frontal to the circular saw blade. This ensures sufficient lubricant supply and good wetting of both the minor cutting edges. In particular, friction on the minor cutting edges is intended to be reduced. By placing the nozzle close to the cutting edge entry, it is intended to ensure the lubricant remains at the cutting edge until the cutting process. The Survos Standard (by HPM Technologies) is used as the lubricant, which exhibits evaporation of up to 100%.

For investigation of the influence of wear, a constant speed of 5000 min^{-1} and a feed rate of 10 m/min were maintained for all experiments. All experiments were performed in cutting along the fibre and at a cutting depth of 80 mm.

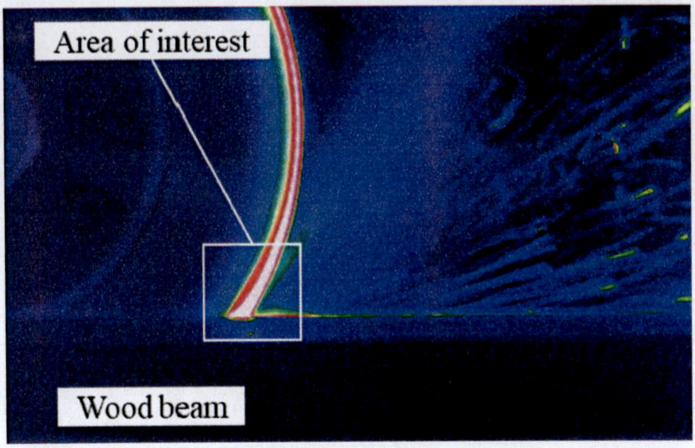

Fig. 2. Thermal image while sawing

In addition, the influence of wear on forces and temperature of the saw tooth were investigated. For this purpose, after reaching a feed distance of 24 m, the temperature was recorded using a thermal imaging camera during a 70 cm long bar cutting process. The VarioCAM hr head camera (by InfraTec) records 200 images at a frequency of 60 Hz. To avoid thermal fluctuations, especially at the beginning of the cutting process, only the temperatures of the middle section of the bar (30–50 cm) are considered for the analysis. The focus of the camera is directed towards the exiting cutting edge. Figure 2 shows a picture of the thermal imaging camera recordings. To perform repeatable measurements, the measurement is triggered using an inductive sensor.

3 Results

In the following, cutting force, normal cutting force and passive force are analysed in relation to wear. Figure 3 shows the forces during dry cutting along the feed distance. Additionally, a root regression analysis was conducted and visualised as well.

Particularly in the case of the normal cutting force, the characteristic trend of wear can be observed both in the regression curve and in the data points. Initially, the normal cutting force increases rapidly from 8.6 N to 20.27 N at 100 m of feed distance and then continues to rise to 22.86 N. A similar increase in forces can be observed for the cutting force (blue). However, the values exhibit significantly greater variation, leading to outliers. For example, the cutting force decreases to 19.29 N at a feed distance of 300 m, after previously being around 25 N.

There is no increase along the feed distance for the passive forces, indicating wear has no influence in this regard. It is noticeable the passive forces exhibit a wide range from 2.72 N to 10.81 N.

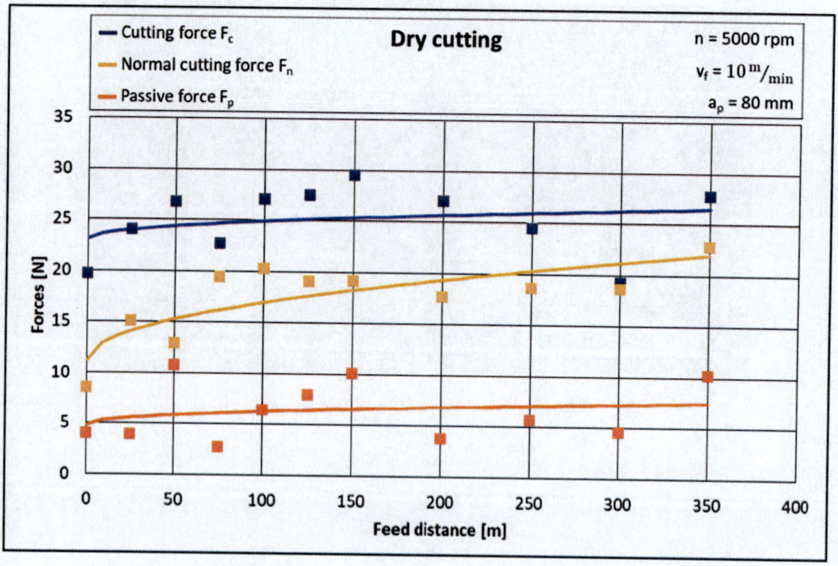

Fig. 3. Forces during dry cutting process

During the analysis of forces during MQL cutting (Fig. 4), both the cutting force and the normal cutting force show an increase of approximately 10 N along the feed distance. The trend of the regression curves also resembles the expected pattern. At the beginning, there is a rapid increase in the cutting force from 20.71 N to 25.17 N over a feed distance of 100 m. Over the next 200 m, the cutting force increases to 28.75 N. A similar trend can be observed for the normal cutting force.

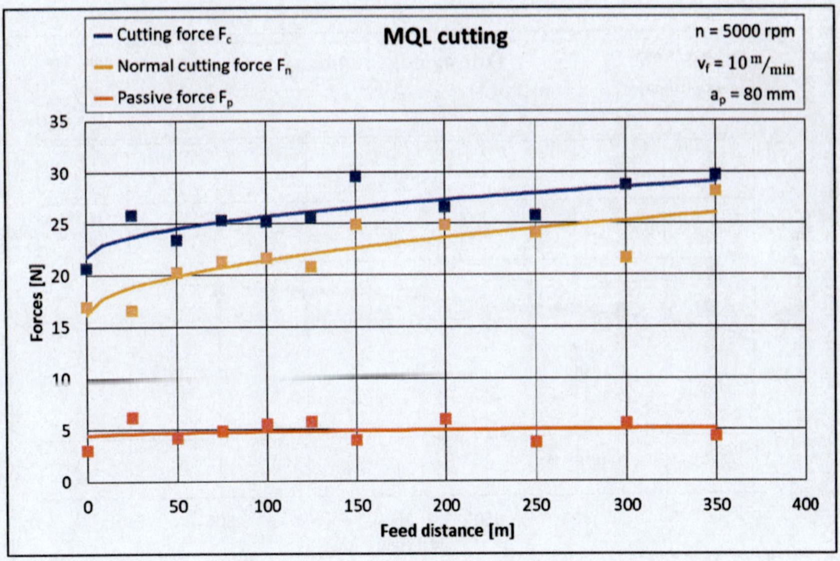

Fig. 4. Forces during MQL cutting process

The curves of the normal cutting force for both machining methods show an initial increase followed by a flattening. The cutting forces exhibit a similar trend, although the regression curve for dry cutting is flatter. When considering the individual force measurements, those from the lubricated machining align more closely with the regression curve. A significant difference is particularly evident in the passive force. In the lubricated machining, the values range from 3.01 N to 5.99 N, while in the dry cutting, the passive force ranges from 2.72 N to 10.81 N.

The smaller range of passive forces in magnitude can be attributed to reduced friction. Specifically, reducing friction at the minor cutting edge could lead to a decrease in lateral blade deflection. The wide variation in values can also be attributed to the heterogeneity of the wood. Knots, irregular grain patterns, and internal stresses in the wood have a significant impact on cutting forces. Therefore, no significant difference can be observed in this comparison.

Figure 5 displays the edge radius r_β of the two circular saw blades. The "dry blade" has an average radius of 9.93 μm, while the "MQL blade" already has a radius of 11.12 μm. After a feed distance of 25 m, the values increase to 15.58 μm for MQL and 13.45 μm for dry cutting. It is notable the values converge despite the lower initial rounding of the dry blade. Beyond a feed distance of 250 m, the cutting edge radius is larger for dry cutting compared to MQL. The trend line indicates wear is more pronounced in dry cutting. This can be attributed to higher friction and potential chemical influences.

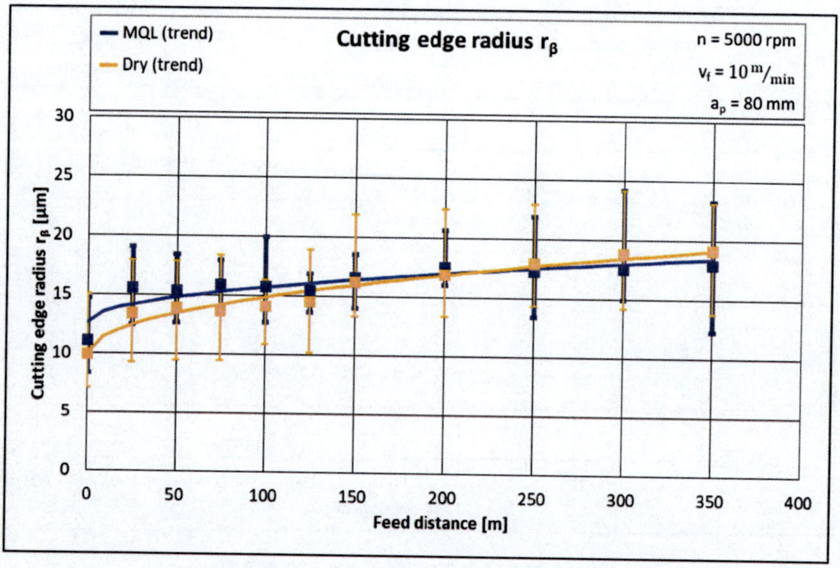

Fig. 5. Cutting edge radius during dry and MQL cutting

In Fig. 6, the results for the profile flattening are presented. The profile flattening for the dry blade is lower in the initial state at 10.98 μm compared to 12.62 μm for the "MQL blade". However, over the observed feed distance, it increases more significantly for dry cutting compared to the use of MQL. The use of MQL suggests a reduction in abrasive effects on the cutting edge. The special influences of wood (knots, irregular grain patterns, inclusions) must also be considered, as they can affect the wear behavior. Nevertheless, the basic trend of cutting forces can be correlated with the wear condition. As wear progresses and the edges round off, the fibers are no longer cleanly severed, leading to an increase in forces. According to this assumption, the forces in MQL cutting should decrease with increasing feed distance. However, it has been shown that this is not clearly the case. It should be noted only the working range in terms of wear level was considered here. For a solid statement and to determine a definitive trend, the saw blades would need to be tested up to the end of their feed distance. Tools in woodworking can reach several kilometers of cutting distance, which could not be realized within the scope of these investigations.

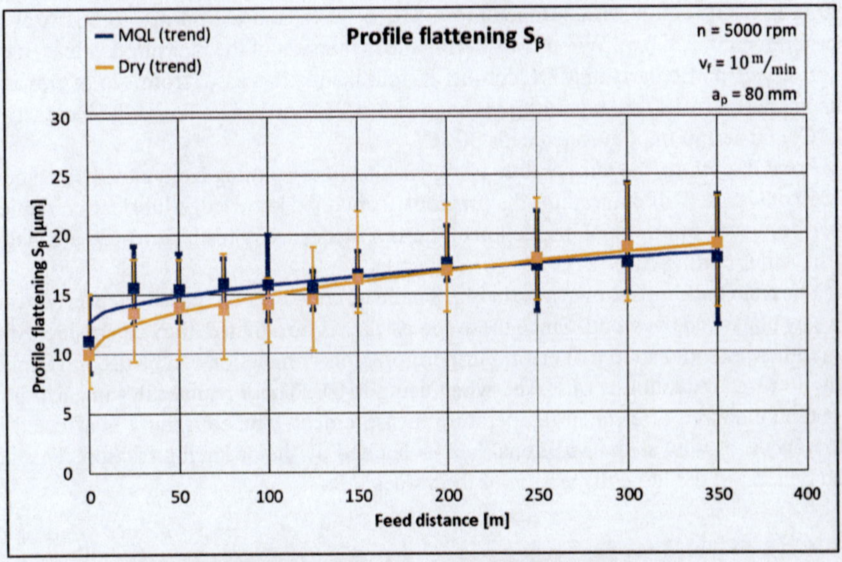

Fig. 6. Profile flattening during dry and MQL cutting

Figure 7 shows the average temperatures across the feed distance for both machining variants. Up to a feed distance of 100 m, the temperatures remain relatively constant at around 40 °C. However, a gradual temperature increase with occasional strong fluctuations can be observed towards the end of the analyzed feed distance. For example, at

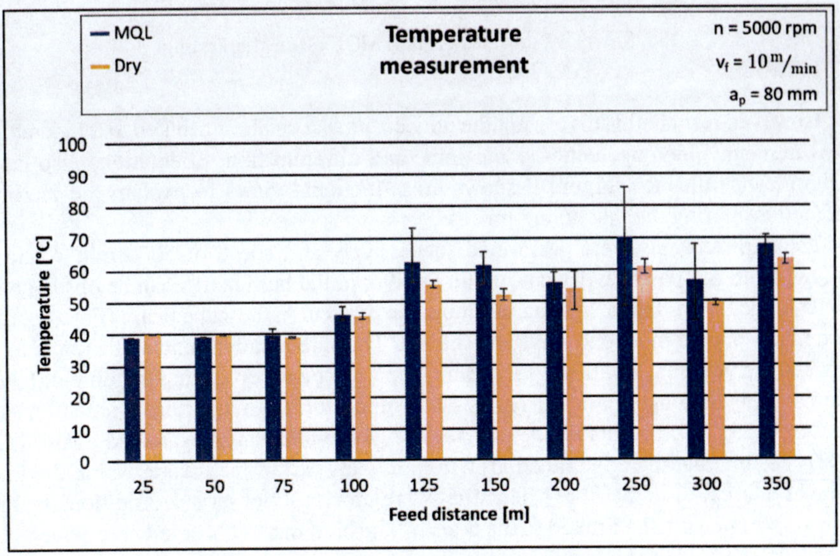

Fig. 7. Course of temperature during dry and MQL cutting

250 m in MQL cutting, the temperature rises to 70 °C and then drops to 56 °C at the measurement for 300 m. When considering the dispersion of the measured values, it can be observed, particularly in MQL cutting, it significantly increases from 125 m onwards. The values up to 100 m stay within a range of 3 °C. From 125 m, the range increases to 12 °C, and at 250 m, it even exceeds 30 °C.

In MQL cutting, the temperature is higher than in dry cutting from a feed distance of 125 m onwards. This contradicts the previously obtained knowledge lubrication reduces wear. However, upon visual inspection of the saw blade, heavy resin build-ups are visible in the MQL cutting case.

The resin build-ups shown here in Fig. 8 indicate contact or increased friction between the saw blade and the wood. Since these forces cannot be isolated, they can be identified as another potential source of error. Furthermore, this connection can be used to explain why no force reduction is observed when using MQL. These resin build-ups also pose a potential source of error in temperature measurement. The emissivity is affected by adherences, as well as the additional heat generated by the occurring friction. This can also be seen in the large dispersion of the values.

Fig. 8. a) Dry saw blade and b) MQL blade after testing

However, resin build-ups are not the only errors that can be attributed to temperature measurement. Inhomogeneities, inclusions, and different fibre orientations also have a significant influence. Figure 9 shows cross-sectional views to explain the thermal interactions during the machining process.

The depicted cuts were performed successively at a speed of 5000 min^{-1} and a cutting depth of 40 mm in dry cutting. In Fig. 9b, initial burn marks can be observed at the left edge of the image. When examining the next cut in the same beam (Fig. 9c), the burn marks appear more extensive and intense. They are mainly found at the transitions between the wood layers. In the second image, it can be observed that the fibre or layer orientation is no longer parallel. The cause of this anomaly in the fibre orientation can be seen in Fig. 9d). In this image, the surface of the next cut is examined again. It is clearly visible that there is branch growth in the left part of the image, where the first signs of anomalies in the fibre orientation were previously observed. Additionally, the burn marks appear to be limited to the branch. The burn marks observed here serve only as a visualization of the occurring temperatures.

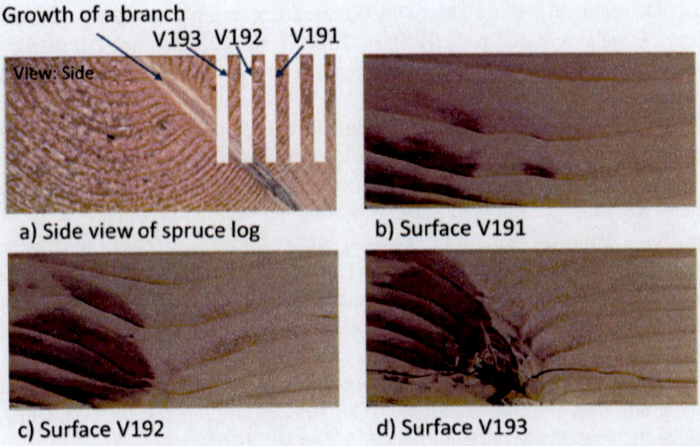

Fig. 9. Influence of anomalies in wood fibre on consecutive surfaces

4 Summary

The investigations have shown that the use of MQL leads to a reduction in wear. This can be observed through the profile flattening and the cutting edge radius. In conjunction with wear and tool bluntness, an increase in forces along the cutting path and a temperature rise are visible in the results. However, no improvement through the lubricant is evident in these two variables. This can be partly attributed to the properties of the wood. In particular, the influence of anomalies in the fibre orientation could be demonstrated.

Furthermore, when using MQL there is an interaction between lubricant, saw blade and wood, which leads to resin build-up on the tool body. This probably offsets the positive effects of the lubricant on forces and temperature development. Nevertheless, the results on reduced wear in particular show potential for the use of MQL in wood machining which needs to be further explored. A new series of experiments should confirm the results obtained here. In future studies, care should be taken to reduce the influence of resin build-ups by cleaning the saw blade and enable a direct comparison. Furthermore, the influence of the spray positioning on the results should be investigated.

References

1. Westkämper, E., Warnecke, H.-J.: Einführung in die Fertigungstechnik, 5. Teubner Verlag, Aufl (2002)
2. Walz, A., Taverna, R., Stöckli, V.: Effektiver Klimaschutz durch den Wald - Holz nutzen ist wirksamer als Vorräte anhäufen. Wald und Holz **4**(10), 37–40
3. Bundesministerium für Umwelt: Naturschutz, nukleare Sicherheit und Verbraucherschutz: Heizen mit Holz (2022). https://www.bmuv.de/heizen-mit-holz/umwelt/klimaauswirkungen-von-heizen-mit-holz. Last Access: 11 June 2023
4. Technische Universität München- Holzforschung München: Bauen mit Holz = aktiver Klimaschutz. TU München (2020)

5. Holzbau Deutschland - Bund Deutscher Zimmermeister im Zentralverband des deutschen Gewerbes: Lagebericht Zimmerer/Holzbau 2023 Stand 2023. Internet: https://www.holzbau-deutschland.de/fileadmin/user_upload/eingebundene_Downloads/Lagebericht_2023_mit_Statistiken.pdf. Last Access: 09 June 2023
6. Gottlöber, C.: Zerspanung von Holz und Holzwerkstoffen: Grundlagen–Systematik–Modellierung–Prozessgestaltung. Carl Hanser Verlag GmbH Co KG (2014)
7. Gieseke, W., Klameth, N., Herrmann, J.: Nachhaltiges Bauen mit Holz im Geschosswohnungsbau. Nachhaltiger Konsum Springer Gabler, pp. 537–554 (2021)
8. ZukunftBau Forschungsförderung: Kostengünstige, hochdämmende, schichtenreduzierte, sortenreine, klebstofffreie, digital gefertigte Holzmassivbauweise: Herstellungs- und Fügungsoptimierung für mehrgeschossige Außenwandkonstruktionen aus geschlitzten Kantvollhölzern. https://www.zukunftbau.de/projekte/forschungsfoerderung/1008187-2054. Last Access: 15 February 2023
9. Nagaraj, A., Uysal, A., Jawahir, I.S.: An Investigation of process performance when drilling carbon fiber reinforced polymer (CFRP) composite under dry, cryogenic and MQL environments. Procedia Manufacturing **43**, 551–558 (2020)
10. Iskandar, Y., Damir, A., Attia, M.H., Hendrick, P.: On the effect of MQL parameters on machining quality of CFRP. In: ICCM International Conferences on Composite Materials, Vol. 2013, pp. 3281–3290 (2013)
11. Ekinovic, S., Prcanovic, H., Begovic, E.: Investigation of influence of MQL machining parameters on cutting forces during MQL turning of carbon steel St52-3. Procedia engineering **132**, 608–614 (2015)
12. Dhar, N.R., Kamruzzaman, M., Ahmed, M.: Effect of minimum quatinty lubrication (MQL) on tool wear and surface roughness in turning AISI-4340 steel. Elsevier J. Mat. Proc. Technol. **172**, 299–304 (2006)
13. Salur, E., Kuntoglu, M., Aslan, A., Pimenov, D.Y.: The Effects of MQL and Dry Enviroments on Tool Wear, Cutting Temperature, and Power Consumption during End Milling of AISI 1040 Steel. MDPI Metals **11**(11) (2021)
14. Jaquemod, A., Güzel, K., Möhring, H.-C.: Investigation of the Effect of Minimum Quantity Lubrication on the Machining of Wood. In: Production at the Leading Edge of Technology, pp. 238–245. Springer International Publishing, Cham (2023). (Lecture Notes in Production Engineering)

Effect of Wetting Behavior of Shot Peened Surfaces on the Tribological System During Cup Backward Extrusion of Aluminum Alloy EN AW 6082

Niklas Schwark[1], Peter Herrmann[1(✉)], Martina Müller[1], Tim Herrig[1], and Thomas Bergs[1,2]

[1] Laboratory for Machine Tools and Production Engineering (WZL), RWTH Aachen University, Campus Boulevard 30, 52074 Aachen, Germany
p.herrmann@wzl-tf.rwth-aachen.de

[2] Fraunhofer-Institute for Production Technology (IPT), Steinbachstraße 17, 52074 Aachen, Germany

Abstract. In the context of resource efficient production, forming technologies such as cup backward extrusion become increasingly more important. The substitution of conventional materials by materials with similar mechanical properties and high lightweight potential, for instance the aluminum alloy EN AW 6082, leads to new challenges in the automotive industry. In order to avoid high economic damage caused by the adhesion tendency of aluminum alloys, which subsequently leads to increased tool wear, the tribological system must be optimized.

Through drop shape analysis, the influence of the wetting behavior of the lubricants GARDOMER L6261 and GARDOLUBE L301/1 on workpieces, which were shot peened using different materials, on the punch force during cup backward extrusion was investigated. In addition, visioplasticity tests were employed in order to understand the change in flow behavior in different tribological systems. It was shown that both suitable lubrication as well as workpiece structuring positively influence the tribological system.

Keywords: cup backward extrusion · visioplasticity · tribology

1 Introduction

The aim to produce energy efficient vehicles leads to an increased popularity of lightweight components in the automotive industry [1]. Due to a low density alongside high strength and corrosion resistance, aluminum alloys are highly suitable for lightweight applications. Aluminum alloys such as EN AW 6082 can be processed by cold impact extrusion processes [2]. Impact extrusion processes offer a high surface quality, shape accuracy, strength and material utilization [3], which is particularly advantageous due to the high material price of aluminum alloys [4]. A frequently used process variant of cold impact extrusion processes is the cup backward extrusion process.

T. Bauernhansl et al. (Eds.): WGP 2023, LNPE, pp. 727–736, 2024.
https://doi.org/10.1007/978-3-031-47394-4_71

However, the main disadvantage of cold forming processes are the high punch forces F_{max} that are required, which result in high local stresses and thus typically cause high tool wear [5]. During the cold impact extrusion of aluminum, tool wear is further amplified through the high adhesion tendency of aluminum alloys [6].

Tool wear can however be reduced by an adapted tribological system [7]. The tribological system is largely influenced by the friction conditions between workpiece and tool surface. On the one hand, a reduction in friction can be accomplished by using a suitable lubrication system. On the other hand, the introduction of targeted structures to the tool or workpiece surface can also contribute to a reduction in friction, e.g. through the formation of lubricant pockets [8]. During cup backward extrusion, the lubricant film thins out considerably due to the process-induced high surface enlargement. Lubricant pockets help to prevent the lubrication film from tearing off by increasing the lubricant storage capacity [9].

For a targeted adaptation of the friction conditions during the cup backward extrusion of aluminum alloys, the influence of different parameters on the tribological system has to be considered. In the fields of sheet metal forming and bulk forming of steel, previous studies have already shown that the use of lubricant pockets led to a decrease in friction between tool and workpiece surface [10]. Additionally, different surface structures led to different friction conditions [11].

However, the influence of different peening structures and lubricants on the tribological system as well as the change in material flow during cup backward extrusion of EN AW 6082 have not yet been systematically analyzed.

Thus, the aim of this study was to determine the influence of different surface structures and lubricants on the required punch forces F_{max} and material flow behavior. Lower punch forces F_{max} during the extrusion process result in a lower load for the forming tools and are thus an indicator for a possible wear reduction.

During this study, two typical lubricants for the cold forming of aluminum and three different, shot-peened workpiece surfaces were investigated. In order to gain a better understanding of the relations between the use of different lubricants in combination with different workpiece structures and the tribological system, the wetting behavior of the tested lubricants on the structured surfaces was examined using a drop shape analysis. Afterwards, cup backward extrusion tests were performed. During the experiments, all combinations between the lubricants and surface structures used during this study were investigated. To evaluate those tests, the maximum punch force F_{max} was measured during each experiment. Finally, through visioplasticity tests, areas with the highest true strain were identified and compared between different surface structures.

2 Materials and Methods

The material used during this study was the hardenable wrought alloy EN AW 6082 (3.2315), which is a frequently used aluminum alloy in the automotive and construction sector due to its excellent formability and machinability [12]. In preparation to the experiments, the specimens were soft annealed prior to the tests to ensure optimal forming properties. Subsequently, the specimens were structured by shot peening with different peening materials. As peening materials, corundum and glass (P1), steel (P2) and glass (P3) grains were used. The surface of the shot peened samples were then examined using a VK X 100 Laser Scanning Microscope from Keyence according to DIN EN ISO 25178–2 [13]. The specimens as well as their corresponding surface parameters are listed in Fig. 1.

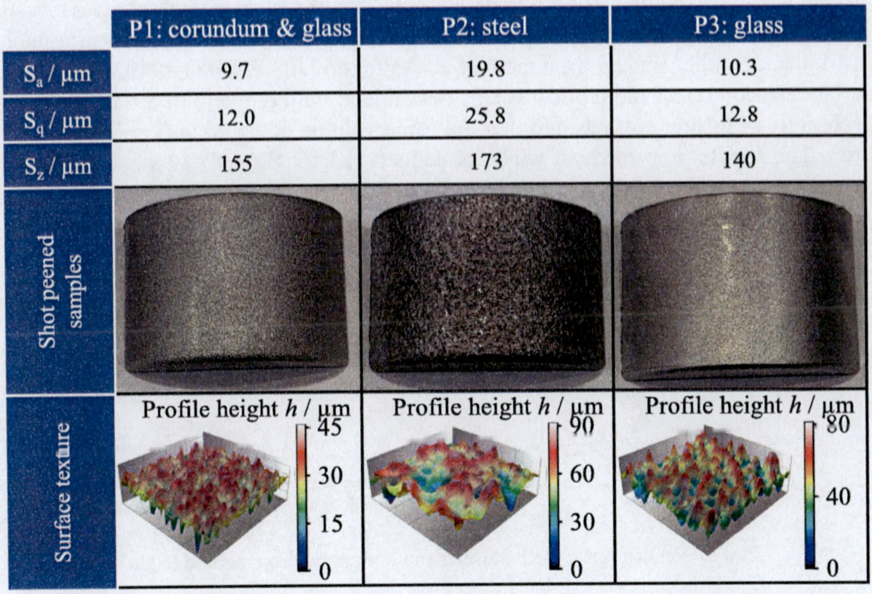

Fig. 1. Shot-peened surfaces and roughness values according to DIN EN ISO 25178–2

Different lubricants were tested during this study. For the lubricant application, a lubricant bath was heated to a temperature $T_1 = 45\ °C – 50\ °C$ and continuously mixed using a magnetic stirrer. The specimen remained in this lubricant bath twice for two minutes each. Between the baths, the specimen were dried in an environment with a temperature $T_d = 50\ °C$. The lubricants used during this study are listed in Table 1.

Table 1. Lubricants examined through drop shape analysis

Variant	Lubricant	Type of lubricant	Mixing ratio (lubricant/water)	Solids content in the mixture
L1	Gardolube L6301/1	Polymer dispersion	45/55	11.7%
L2	Gardomer L6261	Polymer dispersion	30/70	6.15%

During cup backward extrusion tests, all parameter combinations between the workpiece structures and the lubricants were considered. Each parameter combination was repeated three times to be able to identify outliers. During the impact extrusion process, punch force-time curves were recorded.

2.1 Drop Shape Analysis

To start with, the wettability of different lubricants on the structured surfaces was investigated. For this purpose, a drop shape analysis was performed on a DSA100 measurement instrument. By the pendant drop method according to DIN EN ISO 19403–3 [14], the surface tension σ_1 of the lubricants was determined with the help of the curvature of each drop. A dosing unit ensured that the same volume of liquid was present in each drop. The sessile drop method, which is defined in DIN EN ISO 19403–2 [15], then allowed the contact angle θ of the investigated lubricants on a reference surface of polydimethylsiloxane (PDMS) to be determined (Fig. 2).

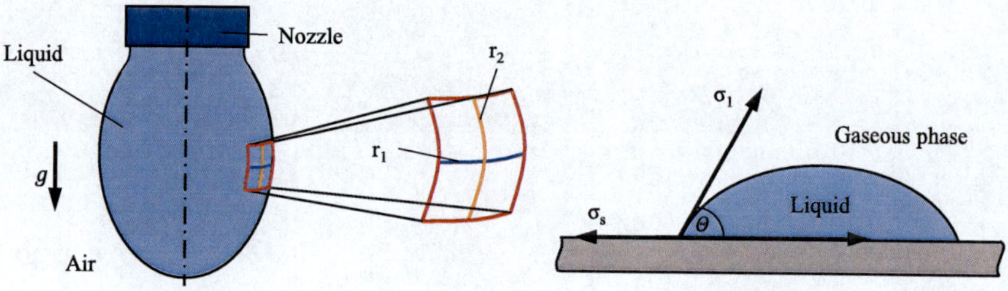

Fig. 2. Pendant drop method (left) and horizontal drop method (right)

The surface tension σ_1 was split into a dispersive and polar fraction. The same parameters were determined for the wetting behavior of the reference liquids ethylene glycol and distilled water on the structured surfaces P1-P3. The surface tension σ_s was then calculated from contact angles θ using the OWENS WENDT RABEL KAELBLE method and subsequently divided into a dispersive and polar fraction [16]. Afterwards, the time-dependent spreading pressure S_{sl} was calculated for the different workpiece and tool combinations in a time interval of $t = 240$ s. If the spreading pressure $S_{sl} >$ 0 mN/m, excellent wettability can be assumed. In a range of -10 mN/m $< S_{sl} <$ 0 mN/m, sufficient wetting is present.

Another possibility for checking the wetting behavior is the Wetting Envelope. For this purpose, a boundary line is determined with a reference liquid for a fixed surface, which results from the surface tension σ_1, and is plotted in a diagram for different wetting angles θ. The abscissa of the diagram represents the dispersive, and the ordinate the polar fraction of the surface tension. The dispersive and polar fractions of the surface tension σ_1 of the tested lubricants were then plotted in this diagram. Based on the position of these points in relation to the boundary lines in the Wetting Envelope, the wetting angle θ of each lubricant on a specific surface can be estimated.

2.2 Cup Backward Extrusion and Visioplasticity Tests

A SCHULER HPX 400 hydraulic press with a maximum punch force of $F_{max} = 4000$ kN was used to perform the experiments. The typical tool setup during cup backward extrusion is shown in Fig. 3 (a). A constant punch velocity $v = 15$ mm/s and a punch path of $s_p = 13$ mm/s were used during the tests. Before the forming process, the cylindrical specimen had an initial diameter $d_0 = 39.8$ mm and an initial height $h_0 = 25$ mm, as shown in Fig. 3 (b). After the cup backward extrusion, the workpiece had a wall thickness $s = 5.1$ mm, a bottom thickness $b = 12$ mm, an outer diameter $d_1 = 40$ mm and a cup height $h_1 = 40.66$ mm. The geometry of the workpiece after the impact extrusion process can be seen in Fig. 3 (c).

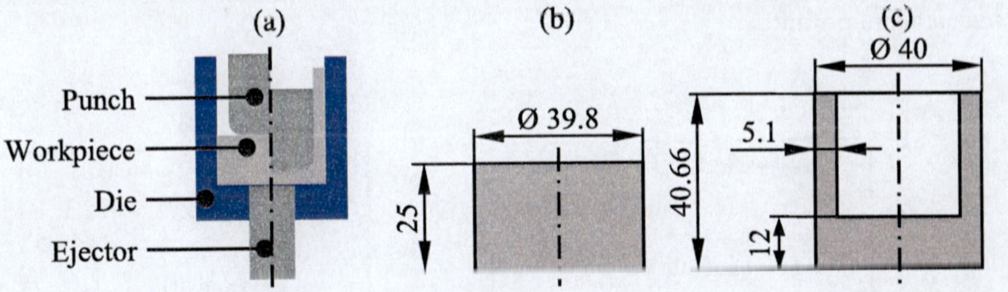

Fig. 3. Schematic cup backward extrusion (a) and geometry of the specimen before (a) and after (b) cup backward extrusion

Through visioplasticity tests, the material flow can be evaluated after the forming process [16]. To perform those tests, the cylindrical specimens were split in axial direction with a precise cut using wire cutting before the cup backward extrusion. A homogeneous dot pattern was then applied to the cut surface with a TRUMPH TRUMARK STATION 5000 LASER ENGRAVER. Each point had a diameter of $d_p = 1.5$ mm and a center point distance of $s_{pm} = 2.25$ mm. The software ARGUS by the company GOM was used to study the displacement and expansion of each point after the forming process (Fig. 4).

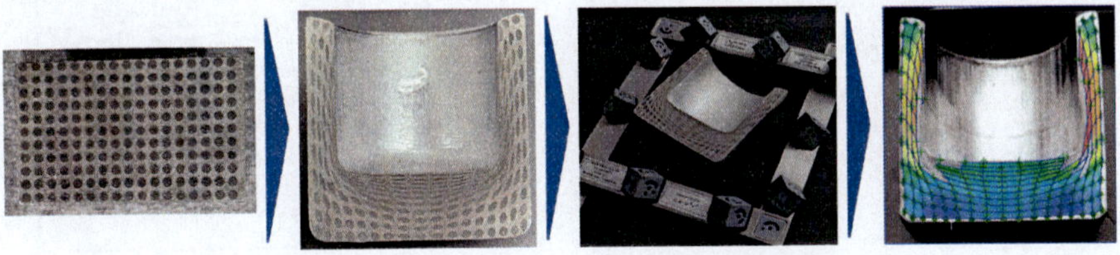

Fig. 4. Steps performed during visioplasticity tests

The object to be evaluated was placed between ten reference points. Through multiple images, the distorted ellipses were detected and connected with a mesh. The true strain was evaluated by comparing the deformation of the dot pattern in the experiments and the absolute displacement of each point.

3 Experimental Results

First, the wetting behavior of the lubricants on the tested surfaces was investigated. Afterwards, during cup backward extrusion tests, the influence of lubricant selection and workpiece structures on the tribological system, i.e. the friction behavior and consequently the required punch force and material flow behavior was determined.

3.1 Wetting Behavior on the Structured Surfaces

Figure 5 shows the time-dependent variation of the spreading pressure S_{sl} as well as the Wetting Envelope for the tested lubricants and surfaces. GARDOMER L6261 exhibited a lower spreading pressure S_{sl} than GARDOLUBE L6301/1 on all examined surfaces, independent of time.

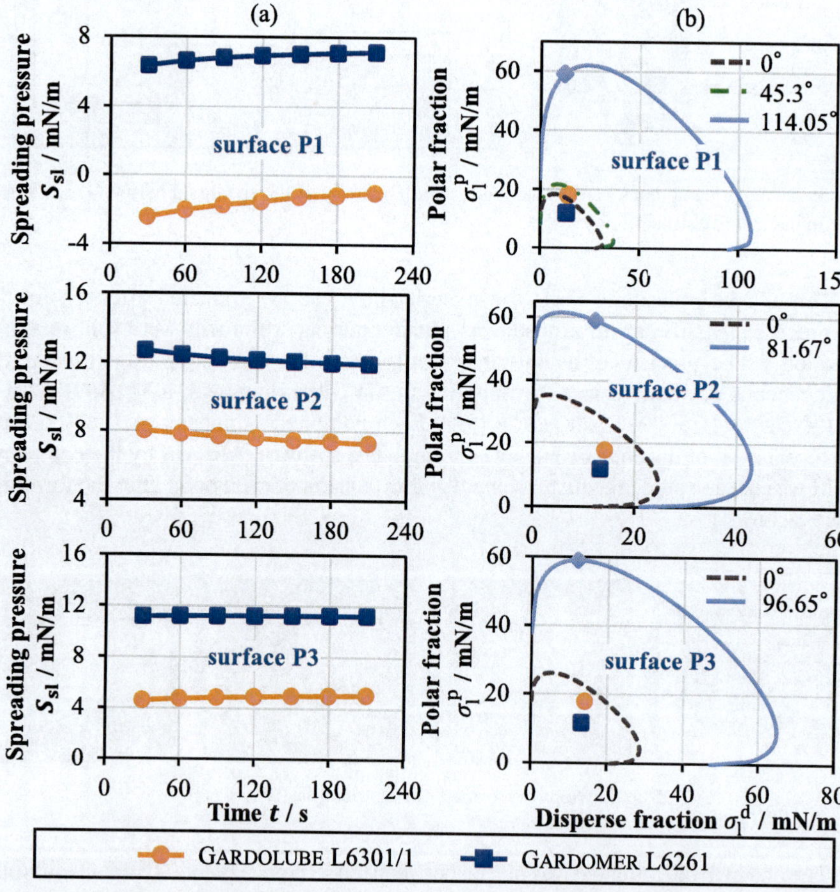

Fig. 5. Spreading pressure-time curve (a) and Wetting Envelopes (b) of the tested surfaces

Surface P2 showed the highest spreading pressure S_{sl}, which corresponds with the best wetting behavior, while the spreading pressure S_{sl} was lowest for surface P1. The combination of surface P1 with the lubricant GARDOLUBE L6261 even resulted in a negative spreading pressure S_{sl}, so that in this case the wetting behavior could only be described as sufficient. The investigation of the Wetting Envelope also showed that GARDOMER L6301/1 fell below the limit curve of the contact angle $\theta = 0°$, which did not occur using GARDOLUBE L6261. It is noticeable that the spreading pressure S_{sl} decreased over time only when using surface P1. Both considering the spreading pressure S_{sl} and the Wetting Envelope, GARDOMER L6301/1 showed improved wetting properties when compared to GARDOLUBE L6261 and was thus used for the following visioplasticity tests.

3.2 Punch Force During Cup Backward Extrusion Tests

The punch force-time curves of the different workpiece structures followed a very similar course. First, the punch force F_{St} increased until it reached a maximum force F_{max}, before it approached an approximately constant level due to the plastic flow of the material. Finally, the punch force F_{St} dropped with a high gradient before the initial value $F_{St} =$ 0 kN was reached again.

As shown in Fig. 6, by using the specimens that were shot peened with glass grains (P3), the maximum punch force F_{max} was reduced by about $\Delta F = 15$ kN compared to the surface that was shot peened with a combination of corundum and glass (P1). On the other hand, by using GARDOMER L6301/1 as a lubricant instead of GARDOLUBE L6261, a maximum punch force reduction of about $\Delta F = 25$–40 kN was reached. However, when GARDOMER L6301/1 was used, the difference between the required punch forces F_{max} for different workpiece structures was much lower and the effect of the different workpiece structures was inversed. To determine the data in Fig. 6, three tests of each setup were examined with regard to the punch force curve. The error bars show the lowest and highest maximum force determined.

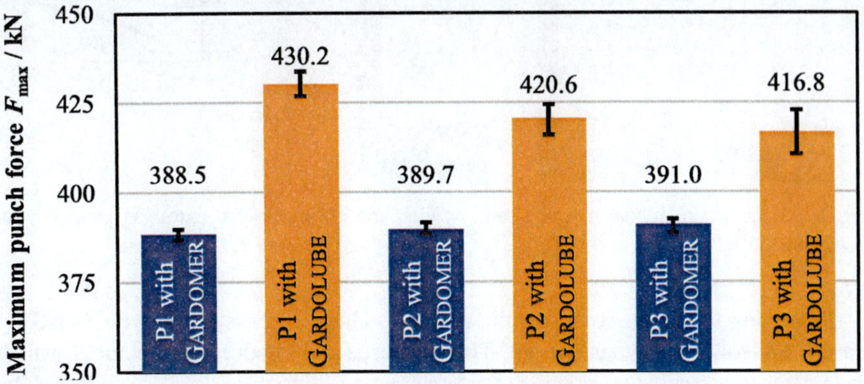

Fig. 6. Maximum punch force during cup backward extrusion using different surface textures and lubricants

The better wetting behavior of GARDOMER L6301/1 also manifested in lower maximum punch forces F_{max} during cup backward extrusion, which can be attributed to an improved material flow behavior. It was also noticeable that the workpiece surfaces which were textured with corundum and glass showed both the lowest spreading pressure S_{sl} and the lowest punch force F_{st} during the extrusion tests. Accordingly, regarding the surface structure, worse wettability of the lubricant on the respective workpiece surfaces resulted in a lower punch force F_{st} during the forming process.

3.3 Visioplasticity Tests

During the visioplasticity tests, a typical strain distribution pattern was characterized. The highest true strain values were reached on the inner wall of the extruded cup, while the lowest strain values were found in the upper and lower exterior edges of the cup. Figure 7 also shows that not all distortions were captured by the ARGUS software. This was particularly the case for the inner side of the cup. Zone 3 became largest using surface P3, and was least prevalent with surface P1.

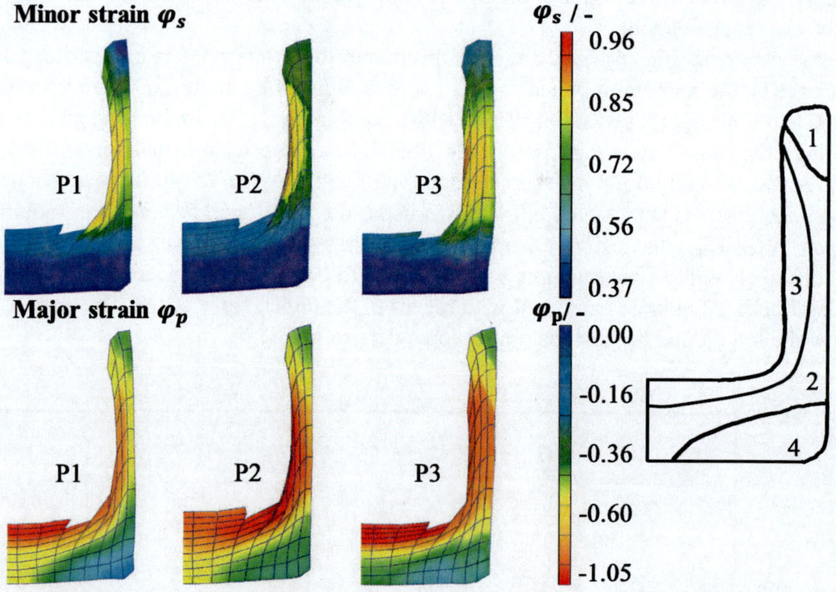

Fig. 7. Minor strain φ_s, and major strain φ_p after the cup backward extrusion process with GARDOMER L6261

Regarding the minor strain φ_s, the smallest values in zones 2 to 4 were found for surface P1, followed by surface P3. This was especially noticeable in zone 4, which underwent the lowest deformation. Surface P3 also offered the lowest values for the major strain φ_p in zones 1 to 3, while this effect was most visible in zone 1 (Table 2).

Table 2. Minor strain φ_s and major strain φ_p in different zones

Zone	Minor strain φ_s/%			Major strain φ_p/%		
	P1	P2	P3	P1	P2	P3
1	-0.36 to -0.47	-0.36 to -0.47	-0.36 to -0.47	0.37 to 0.44	0.67 to 0.72	0.56 to 0.72
2	-0.60 to -0.69	-0.69 to -0.80	-0.69 to -0.80	0.64 to 0.72	0.72 to 0.81	0.72 to 0.81
3	-0.69 to -0.80	-0.92 to -1.05	-0.80 to -0.92	0.79 to 0.85	0.81 to 0.88	0.81 to 0.88
4	-0.12 to -0.16	-0.36 to -0.47	-0.22 to -0.29	0.37 to 0.47	0.37 to 0.47	0.37 to 0.47

Surface P1 showed both the lowest true strain values φ as well as the lowest punch force F_{max}. However, surface P2 did not cause the highest required punch force F_{max}, although this structure led to both the highest major strain φ_p and minor strain φ_s values.

4 Summary and Outlook

In this study, a drop shape analysis was used to identify the wetting behavior of two lubricants that are typical for cup backward extrusion, GARDOMER L6261 and GARDOLUBE L301/1, on shot peened surfaces. It was shown that both lubricants enabled full wettability on surfaces that were shot peened with glass, steel, and a combination of corundum and glass grains. Of those two lubricants, GARDOMER L6261 offered better wetting behavior regarding the spreading pressure S_{sl} and the Wetting Envelope. With the aid of a cup backward extrusion test series using aluminum alloy EN AW 6082, it was shown that both the type of workpiece structuring as well as a suitable lubricant selection were able to reduce the required punch force F_{max} during the extrusion process. During this study, the use of different lubricants had a significantly larger effect than the use of different workpiece structures, and the effect of the different workpiece structures on the required punch force F_{max} was not uniform when using different lubricants.

Visioplasticity tests have shown that the resulting true strain was also influenced by the different workpiece structures. Better wetting behavior mostly decreased the required punch force F_{max} and the values for minor strain φ_s and major strain φ_p. When a higher maximum punch force F_{max} was required to be applied during the extrusion process, the resulting major strain φ_p, minor strain φ_s, and the areas experiencing high strain values also increased. The reduction in minor strain φ_s and major strain φ_p, that was enabled through the use of different workpiece structures and lubricants, is of particular interest, as lower strain values could counteract the adhesion mechanism that are prevalent in aluminum alloys. The highest true strain values φ occurred on the inner side of the cup, and thus highly affect the adhesion mechanism on the punch.

In future studies, the effects of different workpiece structures on the tribological system could be quantified with the aid of determining the coefficient of friction by tribometer tests [17]. This would enable accurate FE-simulations of the process, which could confirm the results that were found in the visioplasticity tests. Furthermore, through long-time tests, the actual effect of the different tribological systems on the tool wear could be investigated.

Acknowledgement. This project was funded by the Deutsche Forschungsgemeinschaft (DFG - project number 460819047 (TriboLub)).

References

1. Tisza, M., Czinege, I.: Comparative Study of the application of steels and aluminum in lightweight production of automotive parts. Int. J. Lightweight Mat. Manuf. **1**, 229–238 (2018)
2. Nadolski, D., et al.: Einfluss einer kaltmassivumformung und wärmebehandlung auf die Maß- und formänderungen. HTM J. Heat Treat. Mat. **69**, 295–305 (2016)
3. Dietrich, J.: Praxis der Umformtechnik. Umform- und Zerteilverfahren, Werkzeuge, Maschinen. 12th edn. Springer Vieweg, Wiesbaden (2018)
4. Kucharzewski, A.: Massivumformung – eine Prozesskette für den Leichtbau. Info-Reihe Massivumformung **2**, 34–40 (2010)
5. Lange, K., Kammerer, M., Pöhlandt, K., Schöck, J.: Fließpressen. Springer-Verlag, Berlin Heidelberg, Wirtschaftliche Fertigung metallischer Präzisionswerkstücke (2008)
6. Saha, P.: Thermodynamics and tribology in aluminum extrusion. Wear **218**, 179–190 (1998)
7. De Moraes Costa, A.L., Souza da Silva, U., Valberg, H.S.: On the friction conditions in FEM simulations of cold extrusions. In: 23rd International Conference on Material Forming, ESAFORM 2020, pp. 231–236. Cottbus (2020)
8. Hild, R.: Dry Full Forward Extrusion by Textured Workpieces and Self-Lubricating Tool Coatings, dissertation, RWTH Aachen University (2020)
9. Löffler, M., Engel, U., Willner, K., Beyer, F., Merklein, M.: Investigation of the tribological behaviour of microstructures for controlling the material flow in sheet-bulk metal forming. CIRP J. Manuf. Sci. Technol. **22**, 66–75 (2018)
10. Sobis, T., Engel, U., Geiger, M.: A theoretical study on wear simulation in metal forming processes. J. Mat. Proc. Technol. **34**, 233–240 (1992)
11. Costa, H.L., Hutchings, I.M.: Hydrodynamic lubrication of textured steel surfaces under reciprocating sliding conditions. Tribology International, pp. 1227–1238 (2007)
12. Leichtbau Aluminium Giesserei Hannover GmbH: Daten und Fakten für die Praxis. EN AW-6082. Die vielseitige Legierung, https://www.leichtmetall.eu/app/uploads/sites/2/Leichtmetall_Datenblatt_EN-AW-6082.pdf, last accessed 07 February 2023
13. DIN EN ISO 25178-2, Geometrische Produktspezifikation (GPS) - Oberflächenbeschaffenheit: Flächenhaft - Teil 2: Begriffe und Oberflächen-Kenngrößen. Deutsche Fassung (2012)
14. DIN EN ISO 19403-3, Beschichtungsstoffe - Benetzbarkeit - Teil 3: Bestimmung der Oberflächenspannung von Flüssigkeiten mit der Methode des hängenden Tropfens. Deutsche Fassung, Ausgabe Dezember (2011)
15. DIN EN ISO 19403-2, Beschichtungsstoffe - Benetzbarkeit - Teil 2: Bestimmung der freien Oberflächenenergie fester Oberflächen durch Messung des Kontaktwinkels. Deutsche Fassung (2011)
16. Chevalier, L., Dahan, N.: Metal-forming process analysis by the visioplasticity method: Application to stationary and non-stationary processes. Journal of Materials Processing technology **31** (1992)
17. Teller, M., et al.: Methodik zur Entwicklung von interaktionsminimierten Oberflächenschichten für das schmiermittelfreie Kaltfließpressen von Aluminium, Dry Metal Forming Open Access Journal, 34–41 (2015)

An Integrated Approach to Developing Modular Product Families in the Engineer-to-Order Sector

F. Girkes[1](✉), M. Günther[1], C. Rüger[1], J. P. Bergmann[1], and C. B. Töpfer-Kerst[2]

[1] Production Technology Group, Technical University of Illmenau, 98693 Ilmenau, Germany
florian.girkes@tu-ilmenau.de

[2] IWB Industrietechnik GmbH, Langenscheidtstraße 7, 99867 Gotha, Germany

Abstract. Nowadays, companies face challenges such as globalization, individualization and shorter product lifecycles, resulting in a constant stream of new product development processes (PDP). Modularized product families represent a powerful concept for reducing complexity and increasing resource efficiency in the PDP and beyond. Despite existing approaches and methods in the development of modular product families, there are deficits in the state of the art regarding their transfer and application to the engineer-to-order (ETO) sector, as well as for neutral indicator-based evaluation. Therefore, this paper derives a generic modularization procedure for the ETO sector and verifies it in an industrial use case. For this purpose, a heuristic swapping algorithm has been developed for grouping the components of a product family into clusters and enabling an objective mathematical evaluation. By integrating modular product structures into organizational processes, ETO manufacturers can strengthen their competitive position as well as increase their resource efficiency.

Keywords: modular product families · engineer-to-order · modularization

1 Introduction

Increasing globalisation is confronting industrial companies with completely new challenges. On the one hand, decreasing barriers to market entry are leading to increased competition, as companies are forced to reduce necessary production resources and shorten lead and delivery times. On the other hand, the desire for individualisation leads to a focus on smaller target groups with special requirements. This is also the case in mechanical and plant engineering. Here, products are predominantly offered as customised individual solutions [1]. Due to the high degree of individuality, the actual product development is typically only started after a customer order has been received. In the process, the design department primarily falls back on strategies and experiences that have already been worked out, but which are usually not processed in any adequate data collection. Under these conditions, product development is characterised by a high degree of complexity, which leads to uncertain resource requirements, lead times and delivery times. This complexity, as well as the product and process complexity, must

T. Bauernhansl et al. (Eds.): WGP 2023, LNPE, pp. 737–746, 2024.
https://doi.org/10.1007/978-3-031-47394-4_72

therefore be controlled and minimised [1]. To this end, the focus in recent decades has been set on the modularisation of product structures, whereby the existing product structure is broken down into its components and specifically combined into component groups, so-called modules [2]. These modules combine not only physical components and their relationships or interactions with each other, but also processes and technologies, so that once an order is received, the concept design turns from a pure design task into a selection task with appropriate adjustments. This reduces the iterations in the design. In general, this also creates the possibility of allocating more concrete resource, i.e. cost and capacity requirements, to modules [3]. Various methods have already been presented in research that can be used to generate a modular product structure [3, 4]. However, these modularisation methods focus on mass customisation, where products are often fully configurable and largely composed of predefined standard components. But this research does not sufficiently consider products that require additional order-specific development work and are not easily transferable to an environment whose main feature is customised product solutions [3]. Therefore, this paper presents a method for the modularisation of products for the mechanical and plant engineering industry.

2 State of the Art

Since the 1990s until today, the development of methodological approaches to modular product structuring has been a relevant research topic in the science of product development and design engineering. Since then, a large number of methods have been developed. In the context of this work, only those developed methods are considered whose application can be transferred to the products of a small and medium-sized enterprise (SME), since such enterprises are predominantly found in the ETO sector [4]. The ETO sector is characterised by highly individualised and customised products that are specifically designed and manufactured upon receipt of a customer order [5]. High complexity and uncertainty are typical. The reason for this is the long development and adaptation times that make up the majority of the cumulative lead time [6].

2.1 Methods for Developing Modular Product Families

A modularisation method specifically designed for the ETO sector must meet their development environment and requirements. The inclusion of customer requirements must therefore be a mandatory part of the method to be developed. It should also provide an easy-to-follow approach that can be applied with minimal training as well as an key figures- and rule-based evaluation, so that an objective basis for comparisons and decisions can be guaranteed, especially in the step of concept selection. In addition, above all, the highest possible degree of flexibility must be achieved. Although ETO companies inevitably lose some of their flexibility with modularisation, since their customers often introduce change requests during the development process, the method itself, but also the modularisation concept it creates, must leave room for adaptation. Furthermore, the method should not only consider technical-functional but also product-strategic module drivers in order to include the corporate strategy or corporate goals. For the development of modular product families, only those modularisation methods are generally suitable that take into account at least product-strategic, but ideally also technical-functional relationships, in order to allow a broad field of requirements to flow into the product structuring. Table 1 summarises the evaluation of the individual methods from the current state of the art with regard to these requirements. Although some of these methods already take the aforementioned requirements into account, none of the methods combines all the requirements.

Table 1. Evaluation of methods for the development of modular product families.

Evaluation criteria Methods for modular product structures	**Suitability development of modular product families**	**Transferability for modularisation in ETO sector**	**Consideration of product strategic relations**	**Consideration of technical-functional relations**	**Consideration of customer requirements**	**Comprehensibility & comprehensibility of the procedure**	**Key figures- and rule-based Evaluation**
Integration Analysis Methodology [7]	✕	(✔)	○	●	○	◑	○
Modular Function Deployment [8]	(✔)	✔	●	○	●	●	●
Merging DSM and MIM [9]	(✔)	✕	●	●	○	○	○
Modular Engineering [10]	(✔)	✕	●	●	●	○	◑
Modular Design Methodology [11]	✕	(✔)	○	●	●	●	○
Modularisation through component links [12]	(✔)	✕	●	●	○	○	○
Life phase modularisation [13]	✔	✕	●	●	○	◑	●

Applicable	Conditionally applicable	Not Applicable	fulfilled	Partially fulfilled	Not fulfilled
✔	(✔)	✕	●	◑	○

3 Methodology for the Modularisation of Product Families

The following chapter deals with the development of the methodology for the modularisation of product families in the ETO sector. The data for the use case considered here was delivered by a company based in Thuringia. In 2022, the company had over

60 employees and achieved a turnover of around 5.9 million euros. The company's core competences include the production of components for automation technology and customised probe machine construction. The application of the developed method explicitly refers to customised special machine construction, which in this case exclusively concerns different types of hopper conveyors. For the processing of the scenario, the corresponding customer order, the technical drawings, the bill of materials as well as the general cost drivers are available for each case. In addition, expert interviews have taken place with the company's designers. The methodology is illustrated in Fig. 1: In the ETO sector, individual customer wishes are translated into customised products [5]. Therefore, the starting point for this method should be the clarification of customer requirements. This is followed by the decomposition of the existing product structure, which should not be limited to the physical composition. Rather, it is about the reuse of knowledge, processes and calculations. The modular concept in ETO companies must therefore be broader. The technical-functional analysis was therefore complemented by product-strategic investigations and methodically supported with the help of DSM and MIM. To complete the preparation for the actual modularisation, the components are then analysed. This is followed by module identification. Subsequently, a module concept can be evaluated and selected. The last step is the revision of this concept and its transfer into a modular product structure.

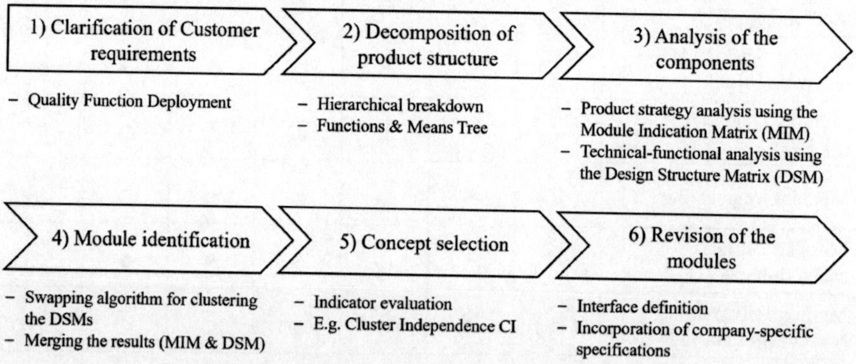

Fig. 1. Procedure of the modularisation methodology.

The further description of the methodology takes place with the creation of the DSM in point 3, as the preceding steps are already sufficiently described in the relevant literature. This publication also deliberately omits the MIM and point 6, as the individual implementation is highly company-specific.

3.1 Analysis of the Components

In this paper, the module drivers from Table 2 were used for evaluation. The evaluation of the affected relationships was carried out with the help of the evaluation scheme on a scale from -2 (harmful relationship), -1 (undesirable relationship), 0 (indifferent relationship), + 1 (desirable relationship) to + 2 (necessary relationship). In Table 2, this scheme is illustrated as an example for the component pair drive roller-motor and finally transferred to the Design Structure Matrix, see Fig. 2.

Table 2. Evaluation scheme for the Design Structure Matrix.

Driver Roll ↔ Motor					
Relation	The motor generates a torque which is transmitted to the drive roller by means of a shaft				
Evaluation	Physical proximity	Energy flow	Material flow	Process similarity	Recycling group
	+2	+2	0	0	0

		A	B	C	D	E
Driver Roll	A	0 0 0 0 0	0 0 0 0 1		2 2 0 0 0	2 2 0 0 0
Structures	B	0 0 0 0 1	0 0 0 0 0	2 0 2 0 0	2 0 -1 0 1	
Hopper	C		2 0 2 0 0	0 0 0 0 0	2 0 2 0 0	
Conveyor belt	D	2 2 0 0 0	2 0 -1 0 1	2 0 2 0 0	0 0 0 0 0	
Motor	E	2 2 0 0 0				0 0 0 0 0

Legend: Physical proximity R, Energy flow E, Material flow M, Process similarity P, Recycling group W

Physical proximity	R E	Energy flow
Material flow	M P	Process similarity
Recycling group	W	

Fig. 2. Section of the Design Structure Matrix for the hopper conveyor.

In order to prepare the assessment data for further processing, they are combined into a common value for each component combination. In this way, company-specific targets or strategic priorities can be better taken into account by weighting the types of interactions considered. This is done by forming the weighted mean $\bar{x}_W$ for each component combination acc. to Eq. 1:

$$\bar{x}_W = \frac{\sum_{i=1}^{n} w_i x_i}{\sum_{i=1}^{n} w_i} = \frac{w_1 x_1 + w_2 x_2 + \ldots + w_n x_n}{w_1 + w_2 + \ldots + w_n} \tag{1}$$

with: wi – Weighting, xi – Evaluation

Figure 3 shows an extract of the DSM for the weighted mean values of the hopper conveyor product family. The weighting was done by the project team as follows: Physical proximity 35%, energy flow 10%, material flow 20%, process similarity 25%, recycling group 10%.

		A	B	C	D	J
Driver Roll	A	0,0	0,1	0,0	0,9	0,9
Structures	B	0,1	0,0	1,2	0,6	0,0
Hopper	C	0,0	1,2	0,0	1,1	0,0
Conveyor belt	D	0,9	0,6	1,1	0,0	0,0
Motor	J	0,9	0,0	0,0	0,0	0,0

Fig. 3. Section of the DSM for the weighted mean values.

Since technical solutions are primarily dealt with in the special machine construction or ETO sector, only the weighted DSM is used for the module identification for the time being. The reason for this is the predominantly technical-functional analysis part compared to the MIM.

3.2 Swapping Algorithm for Identifying Module Candidates

The development of the swapping algorithm is based on the work of Pimmler and Eppinger [7]. The focus of the developed algorithm is therefore exclusively on the interactions of the components with each other. For the formation of the module candidates, no other variables have been used so far. The approach to clustering the matrix elements is based on the basic idea of collecting the positive elements of the matrix as close as possible to the main diagonal of the matrix in order to subsequently form blocks with strong internal relationships. The algorithm consists of two different functions: "Permute" (see Fig. 5) and "DSM Cluster" (see Fig. 4). "Permute" creates random permutations of the rows and columns of the matrix, calculates the sum of the elements in the first side diagonals of the permuted matrix and repeats this process for a given number of iterations. The "DSM Cluster" function forms clusters along the main diagonals of the permuted

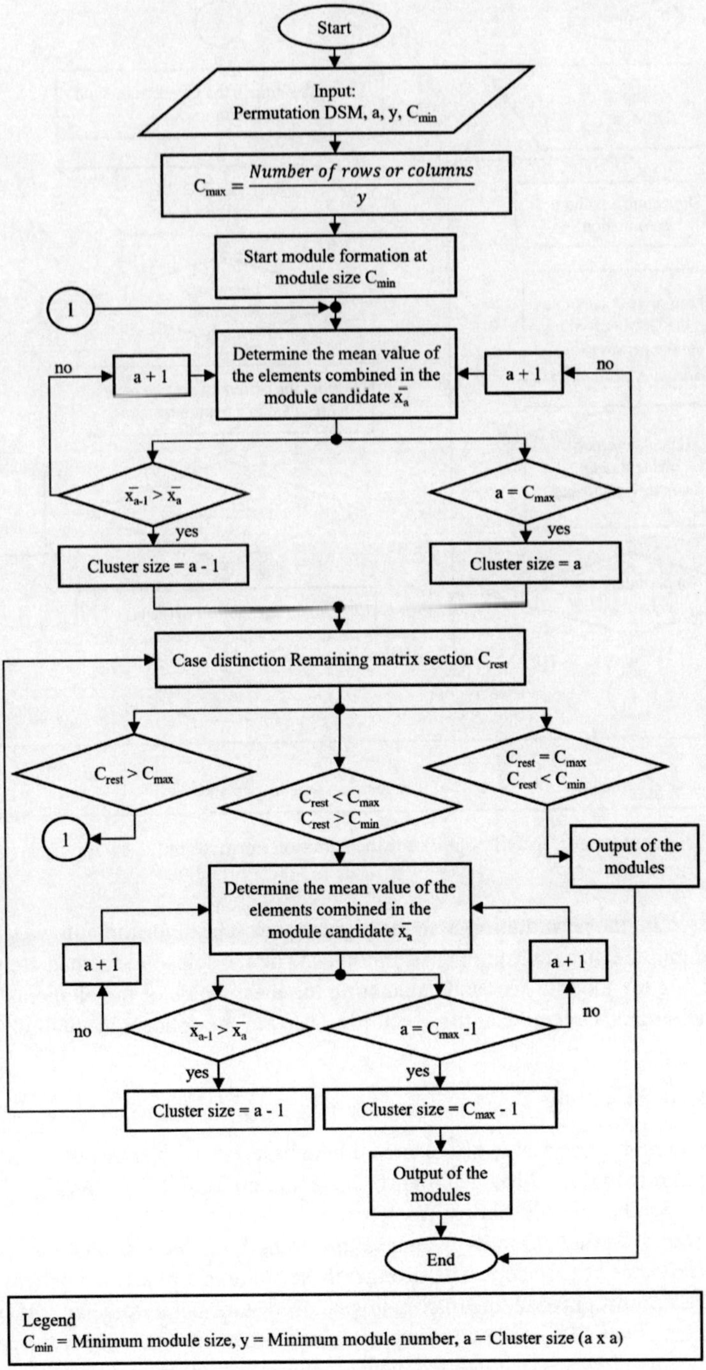

Fig. 4. Flow chart acc. to DIN 66001 for the function DSM-Cluster in the swapping algorithm.

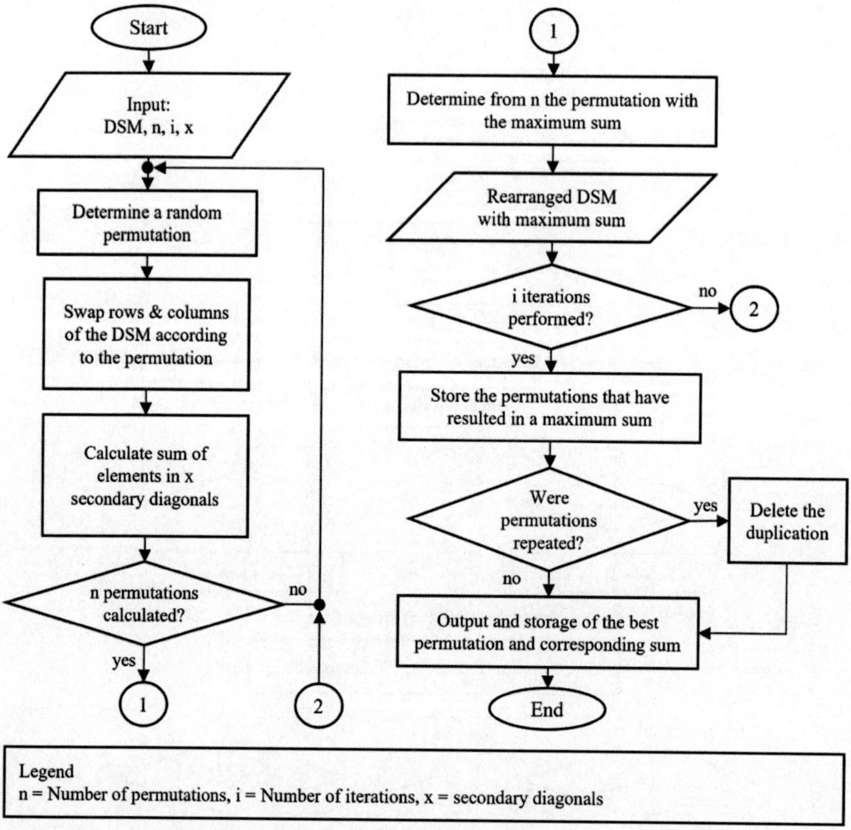

Fig. 5. Flow chart acc. to DIN 66001 for the function Permute in the swapping algorithm.

matrix based on the permutation with the highest sum. The algorithm allows setting the minimum module size and number of modules. The modules are formed step by step by expanding the module size and calculating the mean value of the elements within a block. The module concept can then be further used in the concept evaluation.

3.3 Concept Selection

The final module concept is to be selected on the basis of an objective evaluation using four key figures: Internal Module Quality Q_M, External Module Quality Q_{ex}, Interface Complexity S_k and the Reusability W.

The Internal Module Quality evaluates the strength of the relationships within the module candidates of a concept. The stronger these relationships are, the better the module is rated. For this purpose, the maximum achieved sums per module size $MaxSum_{axa}$ are determined over all considered concepts. In order to be able to classify the quality of the module, three intervals are defined in which the sum of all elements contained in the module Sum_M can be classified. For this purpose, the auxiliary variable x is calculated for all $x \in \mathbb{N}$ acc. to Eq. 2:

$$x = \frac{MaxSum_{axa}}{3} \tag{2}$$

The intervals then result acc. to formula 3 for a module size a > 1:

$$Sum_M \int \begin{cases} [0; x] & low\ (1\ Points). \\ [x; 2x] & moderate\ (3\ Points). \\ [2x; MaxSum_{axa}] & high\ (9\ Points). \end{cases} \tag{3}$$

On the basis of the evaluation of the individual modules in the concept, the resulting evaluation x_{Qi} assigned to the individual modules in the amount of 1 point (low quality), 3 points (moderate quality) or 9 points (high quality), depending on the quality level achieved. Thereupon the internal module quality is calculated acc. to Eq. 4:

$$Q_M = \frac{\sum_{i=1}^{n} x_{Qi}}{n} \tag{4}$$

x_{Qi} – Evaluation of the module candidates; n – Number of module candidates in the concept

The External Module Quality (Eq. 5) evaluates the quality of the individual modules in the concept based on their relationships to each other. The fewer relationships the module candidates have with each other, the more independent they are. To calculate the external module quality, the number of relationships between the modules x_{ex} is counted using the DSM and divided by the number of modules in the concept n. The lower the result, the higher the quality of the modules.

$$Q_{ex} = \frac{\sum x_{ex}}{n} \tag{5}$$

x_{ex} – Number of relationships between the modules, n – Number of modules in the concept

The Interface Complexity (Eq. 6) considers the number of different interfaces of the modules in the concept. Within the interface matrix, the number of different interfaces as well as all interfaces of the module candidates in the concept can be counted among each other.

$$S_k = \frac{x_{dS}}{s} \tag{6}$$

x_{dS} – Number of different interfaces, s – Number of all interfaces of the modules

Reusability acc. to Eq. 7 counts those module candidates in the concept that can be transferred to other product variants without further adaptation. The more reusable modules are formed, the better the evaluation for the respective concept.

$$W = Number\ of\ all\ transferable\ modules \tag{7}$$

The key figures are summarised in an overall rating by adding them up, taking into account their sign, and then normalising them. Thus, the module concept with the highest overall score can be selected.

4 Conclusion and Outlook

In this paper, a method for the modularisation of customer-specific products was developed by deriving tools from modularisation methods for series production and adapting them to the framework conditions of special machine construction. This supports the starting point of a resource-efficient product development and production. In accordance with the requirements of the ETO development environment, the method includes a possibility for recording and weighting customer requirements. To make the method as application-oriented as possible, tools were chosen that do not need extensive training. The module identification step is also supported by an algorithm to ensure an efficient and flawless solution process. In addition, a metrics-based evaluation procedure is provided to enable an objective selection of a suitable module concept. A critical aspect in connection with module identification is that a heuristic algorithm was developed that does not provide an optimal module solution. The extent to which the procedure allows adjustments by the customer even during the development phase could not be investigated within the scope of the work. Finally, the character of real modules must also be taken into account. Optimal modules do not exist in practice, but rather compromise solutions. One reason for this is the multitude of different and contradictory requirements.

References

1. Schönsleben, P.: Integrales Logistikmanagement, 8th edn. Springer, Berlin Heidelberg, Berlin, Heidelberg (2020)
2. Schuh, G., Riesener, M.: Produktkomplexität managen, 3rd edn. Carl Hanser Verlag GmbH & Co. KG, München (2017)
3. Pakkanen, J., Juuti, T., Lehtonen, T., Mämmelä, J.: Why to design modular products? Procedia CIRP **109**, 31–36 (2019)
4. Krause, D., Hartwich, T., Rennpferdt, C.: Produktentwicklung und Konstruktionstechnik, 19th edn. Springer, Berlin Heidelberg, Berlin, Heidelberg (2020)
5. Little, D., Rollins, R., Peck, M., Porter, J.: Integrated planning and scheduling in the engineer-to-order sector. Int. J. Comp. Integ. Manuf. **13**, 545–554 (2000)
6. Pandit, A., Zhu, Y.: An ontology-based approach to support decision-making for the design of ETO (Engineer-To-Order) products. Automation in Construction **16**, 759–770 (2007)
7. Pimmler, T., Eppinger, S.: Integration analysis of product decompositions. In: Alfred P. Sloan School of Management, Massachusetts Institute of Technology (1994)
8. Erixon, G.: Modular function deployment. A method for product modularisation. In: The Royal Institute of Technology, Stockholm (1998)
9. Lanner, M.: An Approach Towards Considering Technical and Economic Aspects in Product Architecture Design. 2nd WDK Workshop, Delft (1996)
10. Koppenhagen, F.: Systematische Ableitung modularer Produktarchitekturen. Komplexitätsreduzierung in der Konzeptphase. In: Hamburg-Harburg, Techn. Univ., Arbeitsbereich Produktionswirtschaft, Diss., Shaker, Aachen (2004)
11. Stone, R., Wood, K., Crawford, R.: A heuristic method for identifying modules for product architectures. Des. Stud. **21**, 5–31 (2000)
12. Koeppen, B., Kersten, W.: Systematic analyses of component couplings and their application for modularisation of complex products. In: IJMASSC, vol. 2, p. 303 (2008)
13. Blees, C.: A method for the development of modular product families. In: TUHH library, PKT (2011)

Potential of Scanning-Strategy Adaptations for Producing Homogenous Microlattices by PBF-LB

N. Ulff(✉), J. Schubert, and F. Zanger

Wbk Institute of Production Science, Karlsruhe Institute of Technology, Karlsruhe, Germany
nico.ulff@kit.edu

Abstract. For a sustainable production and increasing energy demands, physically smaller and more efficient heat exchangers are required. Hence, one challenge is to increase the surface area to volume ratio, therefore enabling the production of finer structures. Additive Manufacturing (AM) enables the efficient production of advanced heat exchangers. Especially Laser Powder Bed Fusion (PBF-LB/M) increases in industrial interest with continuously growing readiness level. The structure size in PBF-LB is limited by melt pool size and directional inhomogeneities, using equal strategies as for bulk solid components. To overcome this, the scalability potential and alteration of a single contour strategy was investigated on 316L to compensate geometric deviations during design phase. This also enables the possibility to explicitly adjust boundary parameters for compensating the directional overmelting of individual inclined struts. Thus, this approach showed a viably strategy to reduce geometric inaccuracies, enabling a predictable manufacturing of homo-genous microlattices.

Keywords: lattice structures · scan-strategy adaptations · homogenization · over-melting compensation

1 Introduction

To increase the efficiency and sustainability of thermally loaded processes, advanced heat exchangers are required. This becomes relevant especially in consideration of increasing global energy demands on power generation technologies requiring high thermal efficiency [1]. Therefore, the demand for heat exchangers is predicted to exceed a market size value from currently 18.19 billion USD to a revenue forecast in 2030 with 26.26 billion USD and a compound annual growth rate of 5.1% [2]. Heat exchangers enable the flow of thermal energy between two or more fluids at different temperatures, for an efficient heat transfer [3]. They are primarily used in power production, requiring optimized structures to increase efficiency. Furthermore, they are also relevant in chemical and food industries, electronics, manufacturing industries, refrigeration, air-conditioning and aeronautic applications [4–6].

Energy demands, decreasing space and increasing material requirements have led to the approach of manufacturing more efficient heat exchange devices for increasing

T. Bauernhansl et al. (Eds.): WGP 2023, LNPE, pp. 747–756, 2024.
https://doi.org/10.1007/978-3-031-47394-4_73

sustainability and reducing costs. The goal is to optimize the size regarding the required and given heat capacity and therefore reducing the physical size [7]. One approach is to investigate the possibilities of using the large design freedom of metal additive manufacturing and explicitly, as the most common, laser powder bed fusion (PBF-LB/M), for improving heat exchange characteristics [8]. This is achieved by improving the thermal performance in comparison to conventional manufacturing routes with novel designs including increased surface area to volume ratios. Therefore, the implementation of lattice structures is of major interest in advanced heat exchangers due to their high surface area to volume ratios, especially the use of triply periodic minimal surface lattices (TPMS) [9]. Lattice structures can ultimately be divided into strut- and skeletal- or sheet-based periodic types [10].

Van Bael et al. achieved an increase in effectiveness of conventional heat exchangers from 0.33 to 0.71 by using a novel TPMS-structured design and simultaneously mentioning the drawback of limited PBF-LB manufacturing capabilities for smaller structures (recommended structure thickness > 0.5 mm) [11]. Enabling the PBF-LB process for a reproducibly manufacturing of smaller lattice structures would lead to a better thermal performance of advanced heat exchangers to meet increasing energy demands.

PBF-LB is a direction-dependent process in which the process parameters significantly influence the final component, especially if the desired geometry is getting closer to the minimal melt pool or spot size of the laser beam [12]. Here, laser power P (in W), scanning speed v (in mm/s), layer thickness t (in μm) and hatch spacing h (in μm) can be summarized in a commonly used equation as volume energy density E_V (VED, in J/mm^3), affecting density and performance, given in (1) [13].

$$E_V = \frac{P}{v \times h \times t} \tag{1}$$

Furthermore, the melt pool size and shape in combination with the stair stepping effect in additive manufactured parts leads to deviations in the target geometry [12, 13]. For example, Hossain et al. showed an average strut diameter deviation of 11.9% for designed strut sizes of 250, 300 & 350 μm [14]. An effect that gets worse with an increasing inclination angle or smaller strut sizes, which can be explained by an increase in the amount of powder exposed by the laser instead of solid structures [15], illustrated in Fig. 1. The heat conduction through powder is much worse than compared to solid structures, leading to heat accumulation and therefore, to deeper melt pools and geometric inaccuracies [16, 17].

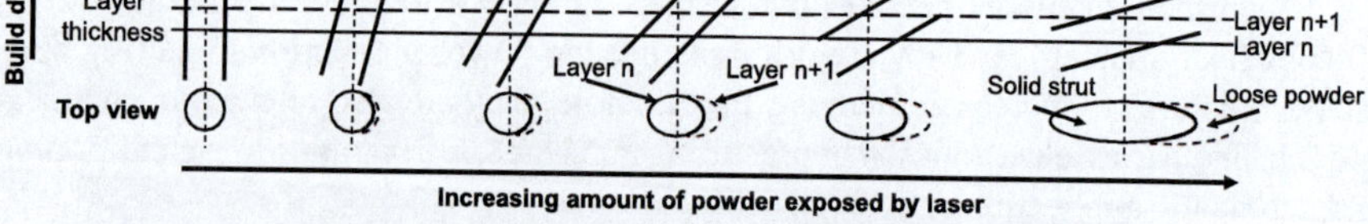

Fig. 1. Schematic increase of exposed powder for higher strut inclination angles

Since the volume energy density is not suitable as a scaling option for the strut geometry and the desired thickness, the resulting melt pool width and depth can be correlated with the absorbed energy density and the specific enthalpy ΔH, given in (2) [18, 19]. The specific enthalpy depends on absorptivity A, laser power P, scanning speed v, laser spot diameter Ø, as well as density ρ and thermal diffusivity D of the used powder.

$$\Delta H = \frac{P \times A}{\rho \times \sqrt{\pi \times D \times v \times \text{Ø}^3}} \tag{2}$$

Ghouse et al. used this approach for two different scanning strategies and were able to scale specific enthalpy with the resulting diameter. They also reported the issue of an increasing strut diameter for higher inclination angles for a set enthalpy [20]. A common issue, where adaptations have not been addressed sufficiently enough by literature.

Therefore, further possible strategy adaptations for improving the homogenous manufacturing of microlattices were investigated in this paper, including the aforementioned information.

On this basis, the possibility of scanning strategy adaptations and process parameter adjustments were considered. Regarding the general scanning strategy, three possible approaches can be adopted: The commonly used contour-hatch scanning strategy [21], the single contour scanning strategy [22] and the single-exposure strategies [23]. Somehow, contour hatch is not applicable for microlattices and the single-exposure strategy has limited possibilities for adaptation and is also higher in energy demand regarding the aimed cost reduction and sustainability. Therefore, the focus was set on single contour scanning. Here it is possible to either alter the diameter of the contour (single contour distance SC_d) or locally adapt the scanning parameters for down-skin surfaces, illustrated in Fig. 2, which should lead to a decrease in overmelting.

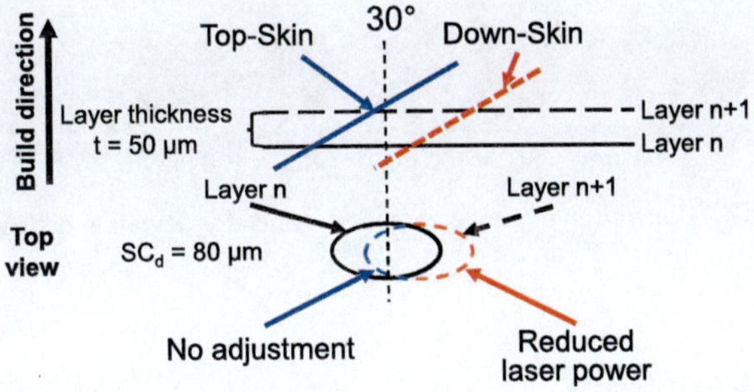

Fig. 2. Illustration of locally reduced laser power for down-skin strut regions

2 Materials and Methods

2.1 Materials and Test Specimen Manufacturing

For evaluating the effect of different SC_d values on the resulting strut diameter, including several inclination angles (90°, 75°, 60°, 45°, 30°, 15°, 0°), a test specimen was designed using Rhino 3D Version 7.0 [24] and exported as STL-file, shown in Fig. 3. The proposed

SC_d was decreased from 120 μm to 60 μm in 10 μm steps along the x-direction and was kept constant in y-direction for statistical validation. The scan-path of the laser were set equal to the edge of the geometry without off-set, illustrated in Fig. 4. The geometry was then placed on a 10 x 10 x 2 mm^3 baseplate, which was manu-factured with a standard parameter set (P = 235 W, v = 700 mm/s). All build jobs were prepared in materialise magics (version 23.01).

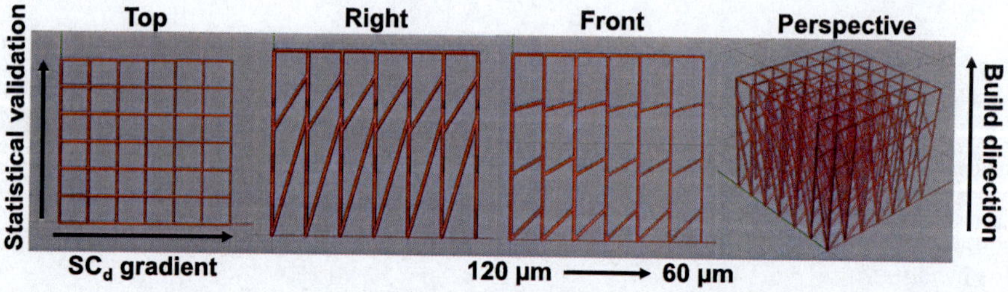

Fig. 3. 3D visualization of the designed test specimen in Rhino 3D

All test specimens were built on a SLM 280HL machine (SLM Solutions, Lübeck, Germany) with a fixed laser spot size of 80 μm in argon atmosphere using a pre-heated substrate plate at 100 °C and a 50 μm layer thickness of 316L stainless steel powder (20–63 μm). This material is of major interest for compact heat exchangers due to its good corrosion and pitting resistance, as well as being one of the most studied PBF-LB materials [25].

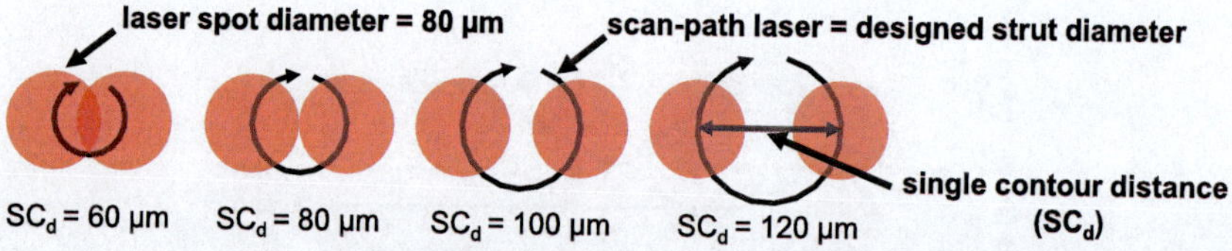

Fig. 4. Illustration of single contour distance for deviating distances

2.2 Design of Experiment

Since the aim of this work was to analyse the possibilities of compensating the described problem of the angle-dependent oversize, two approaches were pursued. On the one hand, the influence of the single contour distance was studied, which could enable the corresponding adjustments directly in the design phase with the help of a Rhino script. On the other hand, a local adjustment of the border parameters was studied, for locally reducing the melt pool depth. Based on literature information regarding the 316L VED processing window for higher densification (above 99% relative density), a wide range of parameter sets was chosen in the range of 50–150 J/mm^3 [26]. Since the laser power has the greatest influence on the melt pool geometry, it was varied between 50 W, 75 W,

100 W, 150 W, 200 W. The scanning speed was adjusted to maintain the process window as described above. An example is depicted in Fig. 5 showing exemplarily process values of interest for a laser power of 100 W.

The local adjustments were conducted for a laser power of 75 W and were adjusted to 50 W in the down-skin area. Furthermore, they were also evaluated for different SC_d values and inclination angles. Higher laser powers were not evaluated as initial results indicated an increase in porosity.

SC_D in µm	scanning speed in mm/s					laser power in W	layer thickness in µm
	200	300	400	500	600	100	50
60	166.67	111.11	83.33	66.67	55.56	-	-
70	142.86	95.24	71.43	57.14	47.62	-	-
80	125.00	83.33	62.50	50.00	41.67	SC_D = spot diameter	
90	111.11	74.07	55.56	44.44	37.04	-	-
100	100.00	66.67	50.00	40.00	33.33	-	-
110	90.91	60.61	45.45	36.36	30.30	-	-
120	83.33	55.56	41.67	33.33	27.78	-	-

Fig. 5. VED processing window for high densities ($P = 100$ W)

2.3 Test Specimen Preparation, CT Analysis and Process Mapping

After manufacturing the test specimens, they were cleaned of loose powder manually and separated from the substrate plate by a wire EDM machine Mitsubishi FA20S (Mitsubishi Electric Europe, Germany). Normally, full-density sample cubes are recommended to be scanned in angles between 10° and 45° for minimizing edge artefacts [27]. Since the struts are built on a 2 mm plate, this would distort the results of individual struts close to the platform. Therefore, a CT-holder was specially designed for this purpose and manufactured with a filament deposition printer, as depicted in Fig. 6.

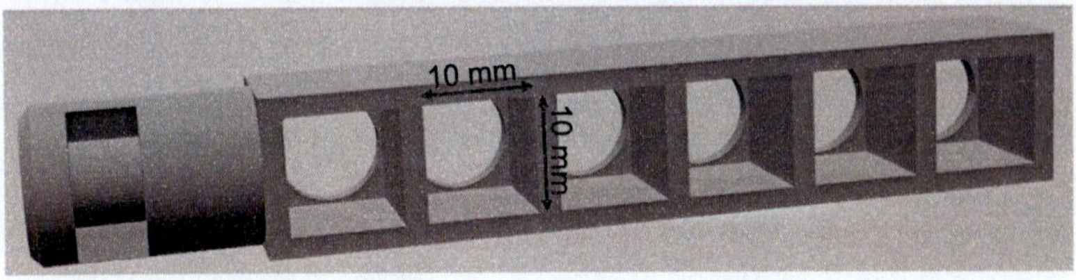

Fig. 6. Schematic illustration of the designed CT-holder for microlattices

The test specimens were scanned in a Metrotom 800 CT device by Zeiss IMT GmbH (Germany). They were scanned in a single 360° scan without stitching; the source voltage and current were set to 130 kV and 61 µA. A 0.25 mm Cu beam hardening filter was used to improve the image contrast (resulting voxelsize = 7.68 μm). The reconstruction was calculated on the CT device using its software Metrotom OS 3.2 by Zeiss IMT GmbH (Germany). VGStudioMax 3.2 by VGStudio (Germany) was used for analyzing the internal porosity (VGEasyPore algorithm) and resulting strut diameter (thickness measurement with sphere method). The results for the porosity measurement and resulting strut diameters were plotted in Matlab (version R2022a) in a range for several laser power and scan velocity values.

3 Results and Discussion

3.1 Single Contour Distance Approach

Before the influence of the variable contour distance is discussed, the process window was defined in which fully dense struts with a relative density > 99.8% were manufactured. Figure 7 depicts the results for $SC_d = 80$ μm, as the distance and inclination angle showed no significant influence. Although the process parameters were within the re-commended process window for full density, the SC_d strategy showed an increase in porosity towards laser powers > 75 W in regard to the tested scan velocity, limiting the region of interest. The measured sphericity value of the pores > 90% indicates mainly keyhole porosity, therefore a too high energy input. Similarly, higher laser powers limit the manufacturing of finer struts, because, as already mentioned, the laser power mainly influences the final geometry and the lower limit range is relevant for further adaptations.

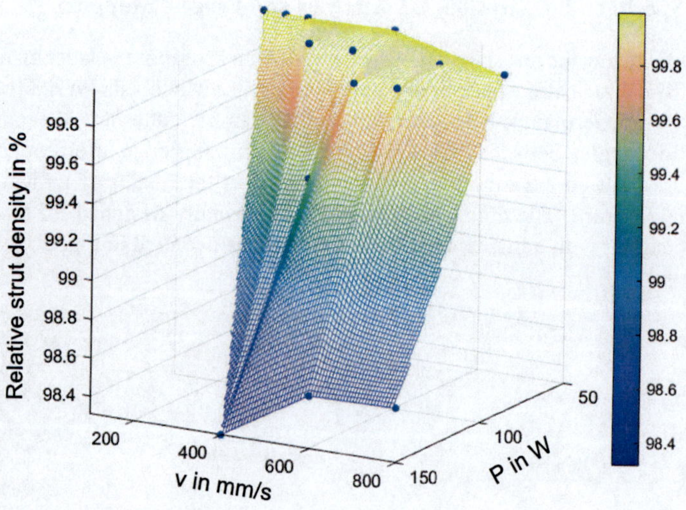

Fig. 7. Surface plot for relative strut density depending on laser power and scan velocity

Figure 8 shows the influence of laser power and inclination angle dependency on the resulting strut diameter for defined scan velocity and contour distance. With increasing inclination angle, the deviation from the 90° struts increases strongly for inclination angles < 60°, as the area of powder increases nonlinear. The diameter of 15° struts deviate from vertical struts in the range of 17.3 to 25.4%, demonstrating the importance of adaptations.

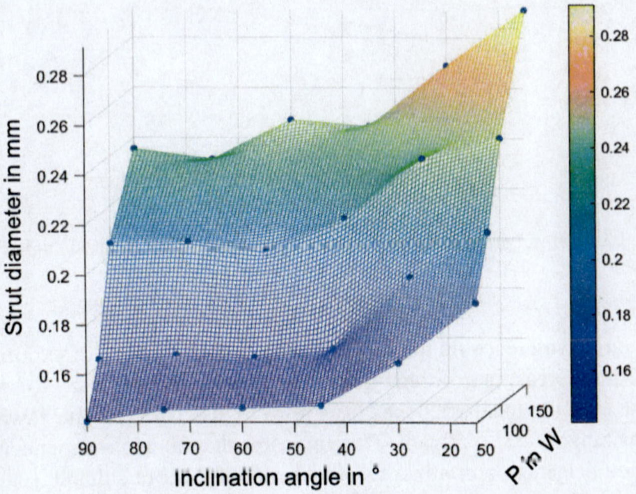

Fig. 8. Surface plot for relative strut diameter depending on laser power and inclination angle ($v = 400$ mm/s and $SC_d = 80$ μm)

One possibility for an adaptation is the variation of SC_d. In Fig. 9, the influence of different SC_d values on the resulting strut diameter including the standard deviation along the axis of the strut can be seen regarding the inclination angle. The grey dashed line includes the standard deviation for the average diameter value of six measured struts for each inclination angle. The angle-dependent oversizing is not affected, but e.g. for a strut diameter of 0.21 mm, an adjustment of the SC_d depending on the inclination angle could lead to a constant strut diameter for fixed process values. This is achievable by an adjustment in the design phase including a Rhino script, where the SC_d value would decrease with increasing overhang angle. However, as the VED is not suitable as a scaling option and general coherences were of interest initially, more suitable scaling options such as the specific enthalpy and a clear determination of the actually exposed powder area and correlating inclination angle will be used for homogenization laws in future investigations.

3.2 Approach of Down-Skin Parameter Adaptation

The evaluation of SC_d limited the region of interest for laser Powers < 75 W and also the down-skin laser power to 50 W. Figure 10 includes a comparison between no adaptations for 50 W and 75 W and also a combined approach for all inclination angles and the corresponding strut diameter. The indicated threshold of 60° inclination angle for oversizing was also visible for the adaptation, but with a major decrease of strut diameter and nearly fitted each course of line. For lower inclination angles, the melt pool

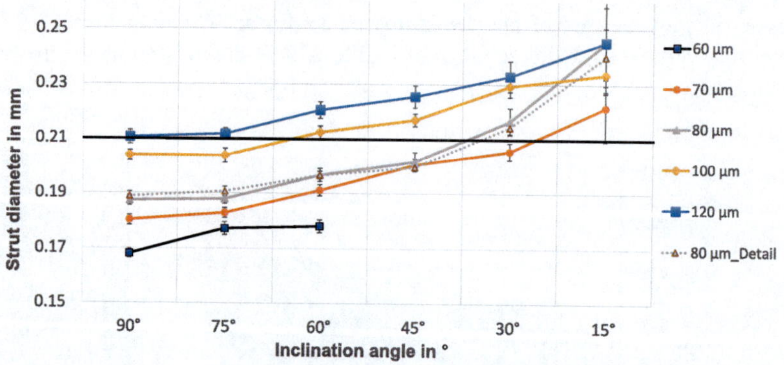

Fig. 9. Strut diameter in correlation to different SC_d values and inclination angle (P = 50 W and v = 100 mm/s)

width has a larger influence on the final strut geometry; the area of exposed powder is negligible. With increasing exposed powder, the melt pool depth starts to be the decisive factor and the effect of locally reduced laser power gets visible for the down-skin region with a downsizing by 41 to 60 μm. This indicates that an angle-dependent down-skin adaptation could lead to a continuous strut geometry. More suitable scaling options, such as the specific enthalpy or a correlation of the actually exposed powder area and the inclination angle, will be of major interest in further investigations.

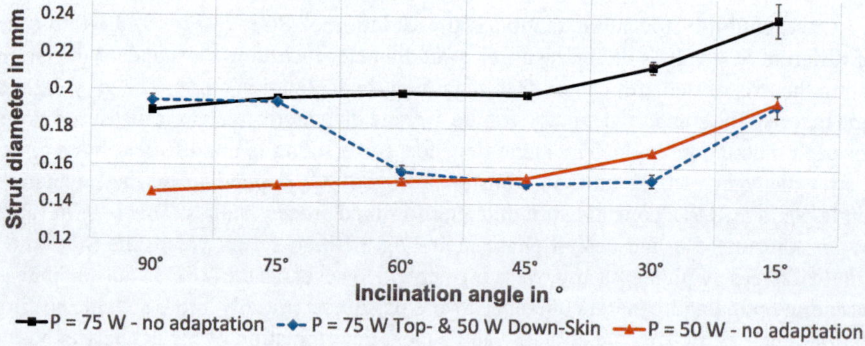

Fig. 10. Influence of locally adjusted laser power on resulting strut diameter (SC_d = 80 μm and v = 300 mm/s)

4 Conclusions

Two possible strategy-adaptations for microlattices regarding the angle-dependent oversizing have been evaluated, including the design of a test geometry reducing the time for manufacturing and required measurement methods. Both strategies, SC_d and down-skin adaptation, seem to be applicable, but include individual challenges that have to be addressed. Here, the implementation of scaling laws, e.g. the specific enthalpy and the specific amount of exposed powder, could lead to a solution.

References

1. Zohuri, B.: Compact heat exchangers. Springer (2017). ISBN 978-3-319-29834-4, https://doi.org/10.1007/978-3-319-29835-1
2. New York, March 09, 2023 (GLOBE NEWSWIRE) -- Reportlinker.com announces the release of the report "Heat Exchanger Market Size, Share & Trends Analysis Report By Product, By End-use, By Region And Segment Forecasts, 2023 - 2030" - https://www.reportlinker.com/p06277991/?utm_source=GNW
3. Shah, R.K., Sekulic, D.P.: Fundamentals of heat exchanger design. John Wiley & Sons (2003)
4. Dehkordi, K.S., Fazilati, M.A., Hajatzadeh, A.: Surface Scraped Heat Exchanger for cooling Newtonian fluids and enhancing its heat transfer characteristics, a review and a numerical approach. Appl. Therm. Eng. **87**, 56–65 (2015)
5. Kayaci, N., Demir, H., Kanbur, B.B., Atayilmaz, Ş.O., Agra, O., Acet, R.C., et al.: Experimental and numerical investigation of ground heat exchangers in the building foundation. Energy Convers Manage **188**, 162–76 (2019). LNCS Homepage, http://www.springer.com/lncs, last accessed 21 November 2016
6. Zheng, X., Wang, R.Z., Tu, Y.D.: Investigation on energy consumption of desiccant coated heat exchanger based heat pump: limitation of adsorption heat of desiccant. Energy Convers. Manage. **188**, 473–479 (2019)
7. Pordanjani, A.H., Aghakhani, S., Afrand, M., Mahmoudi, B., Mahian, O., Wongwises, S.: An updated review on application of nanofluids in heat exchangers for saving energy. Energy Conversion and Management **198**, 111886 (2019). ISSN 0196-8904, https://doi.org/10.1016/j.enconman.2019.111886
8. Thompson, S.M., Aspin, Z.S., Shamsaei, N., Elwany, A., Bian, L.: Additive manufacturing of heat exchangers: a case study on a multi-layered Ti–6Al–4V oscillating heat pipe. Addit. Manuf. **8**, 163–174 (2015). https://doi.org/10.1016/j.addma.2015.09.003
9. Kaur, I., Singh, P.: State-of-the-art in heat exchanger additive manufacturing. Int. J. Heat Mass Transf. **178**, 121600 (2021). https://doi.org/10.1016/j.ijheatmasstransfer.2021.121600
10. Al-Ketan, O., Rowshan, R., Al-Rub, R.K.A.: Topology-mechanical property relationship of 3D printed strut, skeletal and sheet based periodic metallic cellular materials. Addit. Manuf. **19**, 167–183 (2018)
11. Mahmoud, D., Tandel, S.R.S., Yakout, M., et al.: Enhancement of heat exchanger performance using additive manufacturing of gyroid lattice structures. Int. J. Adv. Manuf. Technol. **126**, 4021–4036 (2023). https://doi.org/10.1007/s00170-023-11362-9
12. Van Bael, S., Kerckhofs, G., Moesen, M., Pyka, G., Schrooten, J., Kruth, J.P.: Micro-CT-based improvement of geometrical and mechanical controllability of selective laser melted Ti6Al4V porous structures. Materials Science and Engineering: A **528**(24), 7423–7431 (2011). ISSN 0921–5093, https://doi.org/10.1016/j.msea.2011.06.045
13. Scime, L., Beuth, J.: Melt pool geometry and morphology variability for the Inconel 718 alloy in a laser powder bed fusion additive manufacturing process. Additive Manufacturing **29**, 100830 (2019). ISSN 2214-8604, https://doi.org/10.1016/j.addma.2019.100830
14. Hossain, U., Ghouse, S., Nai, K., Jeffers, J.R.T.: Mechanical and morphological properties of additively manufactured SS316L and Ti6Al4V micro-struts as a function of build angle. Additive Manufacturing **46**, 102050 (2021). ISSN 2214-8604, https://doi.org/10.1016/j.addma.2021.102050

15. Wang, D., Yang, Y., Yi, Z., et al.: Research on the fabricating quality optimization of the overhanging surface in SLM process. Int. J. Adv. Manuf. Technol. **65**, 1471–1484 (2013). https://doi.org/10.1007/s00170-012-4271-4
16. Khan, H.M., Dirikolu, M.H., Koç, E.: Parameters optimization for horizontally built circular profiles: Numerical and experimental investigation. Optik **174**, 521–529 (2018). ISSN 0030–4026, https://doi.org/10.1016/j.ijleo.2018.08.095
17. Hitzler, L., et al.: Direction and location dependency of selective laser melted AlSi10Mg specimens. J. Mat. Proc. Technol. **243**, 48–61 (2017). ISSN 0924-0136, https://doi.org/10.1016/j.jmatprotec.2016.11.029
18. Hann, D.B., Iammi, J., Folkes, J.: A simple methodology for predicting laser-weld properties from material and laser parameters. J. Phys. D Appl. Phys. **44**(44), 445401 (2011)
19. King, W.E., et al.: Observation of keyhole-mode laser melting in laser powder-bed fusion additive manufacturing. J. Mater. Process. Technol. **214**(12), 2915–2925 (2014)
20. Ghouse, S., Babu, S., Van Arkel, R.J., Nai, K., Hooper, P.A., Jonathan, R.T.J.: The influence of laser parameters and scanning strategies on the mechanical properties of a stochastic porous material. Materials & Design **131**, 498–508 (2017). ISSN 0264-1275, https://doi.org/10.1016/j.matdes.2017.06.041
21. Qiu, C., et al.: Influence of processing conditions on strut structure and compressive properties of cellular lattice structures fabricated by selective laser melting. Mater. Sci. Eng. A **628**, 188–197 (2015)
22. Cao, X., et al.: Optimisation of single contour strategy in selective laser melting of Ti-6Al-4V lattices. Rapid Prototyping Journal **28**(5), 907–915 (2022). https://doi.org/10.1108/RPJ-04-2021-0103
23. Tsopanos, S., et al.: The influence of processing parameters on the mechanical properties of selectively laser melted stainless steel microlattice structures. ASME. J. Manuf. Sci. Eng. **132**(4), 041011 (August 2010). (July 23, 2010), https://doi.org/10.1115/1.4001743
24. McNeel, R., et al.: Rhinoceros 3D, Version 7.0. Robert McNeel & Associates, Seattle, WA (2010)
25. da Silva, R.P.P., et al.: Thermal and hydrodynamic analysis of a compact heat exchanger produced by additive manufacturing. Applied Thermal Engineering **193**, 116973 (2021). ISSN 1359-4311, https://doi.org/10.1016/j.applthermaleng.2021.116973
26. Ahmed, N., Barsoum, I., Haidemenopoulos, G., Abu Al-Rub, R.K.: Process parameter selection and optimization of laser powder bed fusion for 316L stainless steel: A review. J. Manuf. Proc. **75**, 415–434 (2022). ISSN 1526-6125, https://doi.org/10.1016/j.jmapro.2021.12.064
27. du Plessis, A., et al.: Standard method for microCT-based additive manufacturing quality control 1: Porosity analysis. MethodsX **5**, 1102–1110 (2018). ISSN 2215-0161, https://doi.org/10.1016/j.mex.2018

Framework for Reinforcement Learning Production Control for One-Piece Flow Modular Concrete Structure Production with Incomplete Information

Susanne Lisa Piekarek(✉), Alex Maximilian Frey, Marvin Carl May, and Gisela Lanza

wbk – Institute of Production Science, Karlsruhe Institute of Technology (KIT), Kaiserstr. 12, 76131 Karlsruhe, Germany
susanne.piekarek@kit.edu

Abstract. Cement accounts for a large proportion of global CO_2 emissions, which can be reduced by prefabricating light weight modular concrete structures. For such modular construction methods with high accuracy requirements and rapid construction the separation of planning and production is futile due to occurring deviations in geometry, time, process and material. The processes must be connected in order to be able to compensate for deviations. For this purpose, a reinforcement learning agent can be used, which dynamically assigns modular concrete structures to a curing chamber and their processing time. In contrast to most approaches for RL in the literature, the considered problem is located at the interface between production control and process control. Thus, this paper presents a framework that outlines the use of RL in this problem domain by describing the simulation model, the actions of the RL agent as well as a possible state space and reward function.

Keywords: Reinforcement learning · production control · modular construction

1 Introduction

The construction industry is one of the largest sectors of the world‘s economy [1]. Cement manufacturing alone is responsible for 8 - 10% of the world’s CO_2 emissions [2]. As the world cement demand is increasing significantly, this leads to an increase in this industry’s absolute energy use and CO_2 emissions [3], it is crucial to save resources and reduce CO_2 emissions in the construction industry. Load-bearing structures from standardized concrete modules produced in flow production have great potential for resource conservation and efficient construction [4].

So far, structural design and production of modular concrete structures have been carried out sequentially and factually separately. The design defines the boundary conditions to be fulfilled by the production, such as tolerances, and thus the limits of acceptance or rejection. Deviations beyond the tolerance require reworking or excluding of a component. With modular construction methods combining many hundreds of modules,

T. Bauernhansl et al. (Eds.): WGP 2023, LNPE, pp. 757–764, 2024.
https://doi.org/10.1007/978-3-031-47394-4_74

precision requirements down to the sub-millimetre range, "just-in-time" production and rapid construction in which every single module is needed exactly on time and with a perfect fit, thinking separately in functional silos is hopeless - design, production and construction process must be developed into a unit with a "zero defect claim" (6 σ). This enables to intelligently and interactively compensate for the various deviations in geometry, time, process, material, etc. A component that is "too short" - previously still out of tolerance - can now be compensated for by one that is "too long". To achieve this goal, the production and thus the process must also be controlled accordingly.

Reinforcement Learning (RL) has recently shown the potential as a control system for complex production systems with limited domain knowledge [5]. Therefore, an RL agent is to be used in this problem case which is responsible for the production as well as the process control. Since this interface between production and process control has hardly been considered in the literature so far, this paper will develop a framework for the use of an RL agent in the problem under consideration.

This paper is structured as follows: Sect. 2 consists of a literature review which deals with the cases of application and the use of RL in the literature. In Sect. 3, the studied problem in this paper is described. The Framework for RL-based production and process control for one-piece flow modular concrete structure production with incomplete information is presented in Sect. 4. Finally, Sect. 5 provides a conclusion, limitations and ideas for future research.

2 Literature Review

Reinforcement Learning is one of the categories of machine learning, along with supervised and unsupervised learning [6]. Through interaction with the environment an RL agent learns which actions to choose to achieve the maximal reward [7]. In contrast to many approaches which only regard isolated subproblems, RL is therefore able to explicitly consider a problem as a whole [8]. Since supervision or a pre-defined set of data is not needed, it is especially considered in case of dynamic environments that require reactive decisions in real-time [9]. In the context of production systems, the literature provides plenty successful RL applications. Panzer and Bender (2022) present a recent literature review on the topic [9].

In the area of production control Kuhnle et al. (2021) for instance introduce and apply a methodology for the design of an adaptive RL-based production control system. The developed RL-agents exhibit effective adaptive control strategies and can be easily tailored to various objectives and application scenarios [5]. Waschneck et al. (2018) show that RL agents can optimize the production autonomously for different targets in an abstract semiconductor production facility use case [10]. Kuhnle et al. (2022) present an approach to increase the plausibility of RL-based control strategies. The combination of the advantages of high prediction accuracy and high explainability enables the design of an advanced RL-system including specific domain expertise [11].

In the area of process control Spielberg et al. (2017) for instance demonstrate that appropriately formulated reward hypothesis functions can be effectively employed for industrial process control [12]. Lillicrap et al. (2019) developed an RL-algorithm which consistently achieves robust solutions for over 20 simulated physics task finding competitive policies [13].

While RL has already been applied widely in production control as well as in process control of complex production systems, the interface between production and process control occurring in the problem at hand has hardly been considered in the literature so far.

3 Problem Description

The considered problem deals with the new modular production of concrete structures based on the flow production principle. In modular construction, all or individual modules of a structure are prefabricated and assembled according to the modular principle. Typical fields of application are skeleton construction, as is still common today, e.g. in hall construction. In contrast to monolithic construction, the industrial pre-production of the modules is intended to improve quality, reduce construction time and cut costs [14]. In the problem at hand, largely arbitrary, bar-shaped modules made of Ultra High Performance Concrete (UHPC) are considered in the sense of an optimised modular system.

A possible conceptual production system for the modules is depicted in Fig. 1. First the formwork is prepared in the process of shuttering. Reinforcement is used to increase the load bearing capacity. The concrete is poured into the formwork in the process of concreating. Subsequent heat treatment in a curing chamber is intended to reduce shrinkage deformations and influence the resulting dimensions of the modules while trying to enable a fast stripping of the formwork. After the stripping the module can be measured and transported to a construction site.

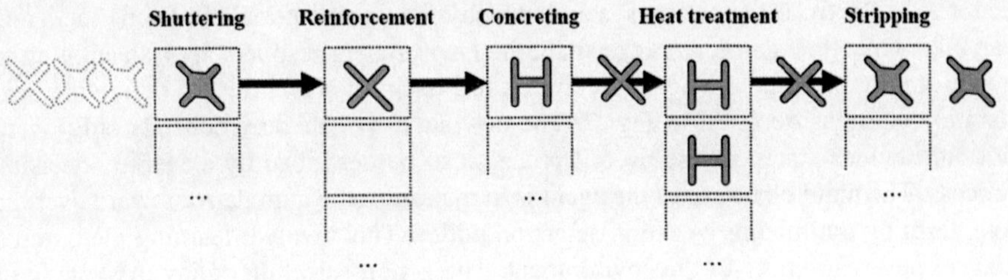

Fig. 1. Production system for concrete modules

The RL agent to be developed is to be used to intervene in the running process by removing modules from the heat treatment at certain points in time. In terms of process control, this influences the properties of the modules and in terms of production control, it influences the process time and the order of completion, which results in a problem that has not yet been considered in the literature.

In contrast to previous work, cf. [15], the heat treatment of concreted modules is started immediately without waiting for the measurement of previous lots, resulting in a lot-free flow production. However, this also means that the RL agent does not yet know the resulting dimensions of the modules already produced at this point, but their deviations are to be compensated for. This is why the heat treatment duration of a module should not be determined at the beginning of its heat treatment, but it should be decided dynamically on the basis of the system status, i.e. among other things on the basis of the curing chamber occupancy and the actual lengths of the modules already produced, whether a module is to be removed from the curing chamber. In this first approach no variants of modules are taken into account.

The described production system will be represented by a discrete-event process simulation model. The RL agent will be trained through the interaction with this environment.

4 Framework for RL-Based Production and Process Control

This section describes the framework for RL-based production and process control. Section 4.1 introduces the basic elements of RL-frameworks which are subsequently defined in the following sections: the state space in Sect. 4.2, the action space in Sect. 4.3 and the reward function in Sect. 4.4. Section 4.5 presents an overview of the entire developed framework.

4.1 Basic Elements of RL Frameworks

A RL framework consists of some basic elements which are explained in the following based on [5, 8]. The process starts with the agent observing the state of the environment, represented as $s_t \in S$, from a finite set of possible states at each time t. The state space S encompasses all possible observations the agent can make about the environment, providing information about the current situation to guide the agent in choosing actions that effectively address the control problem. On this basis the agent then selects an action $a_t \in A$ from the action space A which contains all possible actions the agent can take. Once the action a_t is executed, the environment responds by transitioning to a new state s_{t+1} while the agent receives a feedback, the reward r_{t+1}, before the next iteration starts as depicted in Fig. 2. The new state s_{t+1} thereby depends only on s_t, not on previous states, as the overall process can be described by a Markov decision process. The main objective of the agent is to maximize the cumulative reward over the long term by optimizing its action-selection policy. This involves learning the optimal control policy to apply to the environment. The action-selection policy or strategy of the agent is represented by a probability distribution for choosing a specific action in a given state.

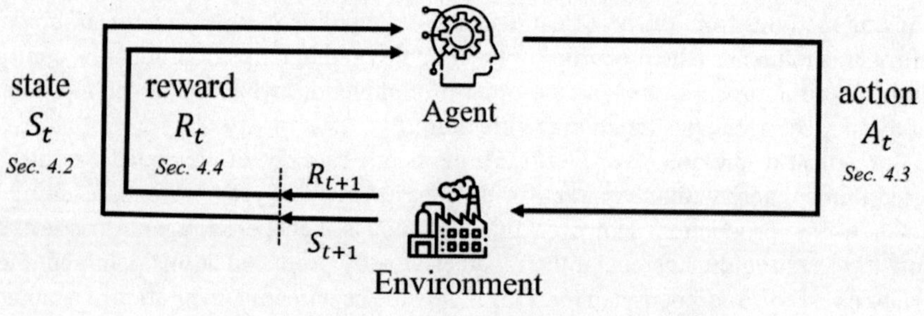

Fig. 2. Overview of the RL-framework based on [8]

4.2 State Space Modelling

The state space represents the relevant information provided to the agent as a foundation for making decisions regarding the selection of actions.

Since the RL agent has to decide on the processing time of all modules, one important information about the environment is the current duration of the heat treatment for all modules that are in the curing chamber at the considered moment. This results in a continuous state space which can be discretized. One reason for the discretization is that it might not be possible to release a module from the curing chamber at every time step and therefore it is more realistic to consider only certain time steps. Another reason is a better handling of the decision making since otherwise the agent is called at every time step of the simulation choosing to do nothing most of the time. However discretizing continuous states leads to several challenges. It is difficult to choose the right level of granularity since if it is too coarse it can lead to a poor policy representation and if the granularity is too fine it can result in a large state space. The biggest problem is that tabular methods do not generalize well across different states, even if two states are very close; the agent will need to learn their values independently. This leads to slow learning and poor generalization.

Another important state information that needs to be considered are the resulting dimensions of the already finished components in order to be able to compensate for deviations. However, it must be taken into account that these are not immediately available when a module is removed from the curing chamber, but have to be measured first.

4.3 Action Space Modelling

Whenever the agent is requested it has to decide whether one or more new modules are to be placed in the curing chamber and whether one or more modules currently in the curing chamber are to be removed. The curing chamber has only a limited number of slots which can either contain a module for heat treatment or are empty. Therefore, the agent has to decide for each slot if, in case the slot is empty, it wants to place a new module there, or, if the slot is already occupied, it wants to remove the placed module or to idle and do nothing. Only valid actions are executed, i.e. it is not possible to place a new module in an already occupied slot or to remove a module from an empty slot. Such invalid actions are not executed, penalized with a negative reward as feedback and the agent is requested to choose another action.

4.4 Reward Function Modelling

The reward function is the key element that directs the agent's behavior and determines the optimal policy. The main objective is to optimize the quality of the whole structural system which consists of the individual modules. At the same time the costs as well as the total lead time of the production should be minimized.

One approach to optimize the quality of the whole structural system is to minimize the standard deviation of the cumulative lengths of produced modules and make sure the modules are as close to the nominal value as possible. A possible reward could be $+1$ if the selected action results in an overall stronger approximation to the nominal value and - 1 if it leads to an overall stronger deviation from the nominal value. The selection of an invalid action could be penalized with a reward of – 2 while a reward of 0 is given for the idle action, i.e. doing nothing. However, since the resulting dimensions of the modules can only be measured during later processes the reward can only be calculated after some time has passed. Additionally, since heuristically the agent should do probably nothing most of the time, sparse rewards make the training process even slower.

4.5 Overview of the Entire Framework

The described basic elements of the framework for RL-based production and process control can be put together to the framework depicted in Fig. 3. By interacting with the environment, represented through a simulation model, the RL agent learns how to control the heat treatment duration to achieve the best reward.

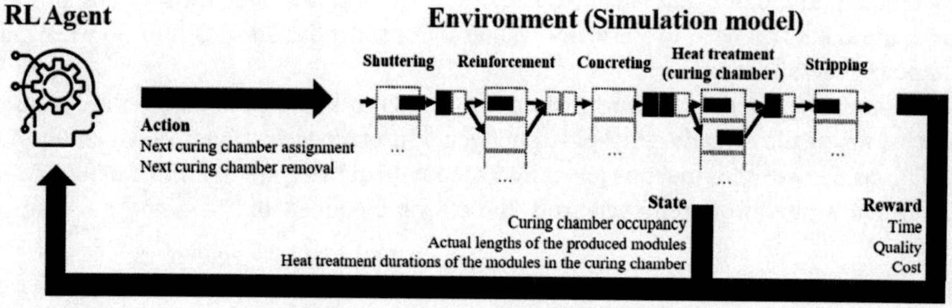

Fig. 3. Overview of the Framework for RL-based production and process control

5 Conclusion

In this paper a framework for the use of RL at the interface between production control and process control is presented. While RL has already been applied a lot in production control as well as in process control of complex production systems, the interface has hardly been considered in the literature so far. The problem under consideration regards a one-piece flow modular concrete structure production with incomplete information. With the aim of being able to compensate for occurring deviations, the RL agent should dynamically assign modular concrete structures to a curing chamber and determine their heat treatment duration. For the development of such an RL-based production and process control the simulation model of the production and the basic elements of RL-frameworks - the state space, the action space as well as a possible reward function - are described.

However, this conceptual framework will be implemented, tested and further adapted in future work, but yet has not been tested with field data. Moreover, one or more suitable optimization algorithms have to be tested and compared simulatively with regard to their results - i.e. the costs, the geometric quality of the supporting structure and the total throughput time. Furthermore, production control methods of industrial practice such as heuristics are to be modelled and compared with the results of the RL agent. This serves the purpose of identifying improvement potential and finding a solution superior to the state-of-the-art, in other words, to heuristics. In addition, an extension of the problem to consider variants and construction sites supplied in parallel shall also be considered in the future. Thereby the RL agent has to further decide on sequences and construction site assignments resulting in an even more complex problem.

References

1. Barbosa, F., et al.: Reinventing Construction: A Route To Higher Productivity. McKinsey & Company. https://www.mckinsey.com/de/~/media/McKinsey/Locations/Europe%20and%20Middle%20East/Deutschland/News/Presse/2017/2017-02-28/170228_mgi_construction.pdf (2017). Accessed 24 May 2023
2. Suhendro, B.: Toward green concrete for better sustainable environment. Procedia Engineering (2014). https://doi.org/10.1016/j.proeng.2014.12.190
3. Hasanbeigi, A., Price, L., Lin, E.: Emerging energy-efficiency and CO2 emission-reduction technologies for cement and concrete production: A technical review. Renew. Sustain. Energy Rev. (2012). https://doi.org/10.1016/j.rser.2012.07.019
4. Mark, P., et al.: Vom Handwerk zur individualisierten Serienfertigung. Bautechnik (2021). https://doi.org/10.1002/bate.202000110
5. Kuhnle, A., Kaiser, J.-P., Theiß, F., Stricker, N., Lanza, G.: Designing an adaptive production control system using reinforcement learning. J. Intell. Manuf. (2021). https://doi.org/10.1007/s10845-020-01612-y
6. van Otterlo, M., Wiering, M.: Reinforcement Learning and Markov Decision Processes. In: Wiering, M., van Otterlo, M. (eds.) Reinforcement learning. State-of-the-art, vol. 12. Adaptation, Learning, and Optimization, vol. 12, pp. 3–42. Springer, Berlin, New York (2012)
7. Russell, S., Norvig, P.: Artificial Intelligence. A modern approach, 3rd edn. Always learning. Pearson Education, Limited, Harlow, United Kingdom (2009)
8. Sutton, R.S., Barto, A.: Reinforcement learning. An introduction. Adaptive computation and machine learning. The MIT Press, Cambridge, Massachusetts, London, England (2018)
9. Panzer, M., Bender, B.: Deep reinforcement learning in production systems: a systematic literature review. Int. J. Prod. Res. (2022). https://doi.org/10.1080/00207543.2021.1973138
10. Waschneck, B., et al.: Deep reinforcement learning for semiconductor production scheduling. In: 2018 29th annual SEMI advanced semiconductor manufacturing conference (ASMC 2018). Saratoga Springs, New York, USA 30 April-3 May 2018. 2018 29th Annual SEMI Advanced Semiconductor Manufacturing Conference (ASMC), Saratoga Springs, NY, USA, 4/30/2018 - 5/3/2018, pp. 301–306. Institute of Electrical and Electronics Engineers, Piscataway, New Jersey (2018). https://doi.org/10.1109/ASMC.2018.8373191
11. Kuhnle, A., May, M.C., Schäfer, L., Lanza, G.: Explainable reinforcement learning in production control of job shop manufacturing system. Int. J. Prod. Res. (2022). https://doi.org/10.1080/00207543.2021.1972179
12. Spielberg, S.P.K., Gopaluni, R.B., Loewen, P.D.: Deep reinforcement learning approaches for process control. In: 2017 6th International Symposium on Advanced Control of Industrial Processes (AdCONIP). 2017 6th International Symposium on Advanced Control of Industrial Processes (AdCONIP), pp. 201–206 (2017). https://doi.org/10.1109/ADCONIP.2017.7983780
13. Lillicrap, T.P., et al.: Continuous control with deep reinforcement learning. https://arxiv.org/pdf/1509.02971 (2015)

14. Borchardt, A., Schwerm, D.: Modernes Bauen mit Betonfertigteilen. Stand und Entwicklungstendenzen im Hochbau. Beton- und Stahlbetonbau (2000). https://doi.org/10.1002/best.200001160
15. Frey, A.M., Lanza, G.: Adaptive Manufacturing Based on Active Sampling for Multicomponent Individual Assembly. In: Proceedings of the Changeable, Agile, Reconfigurable and Virtual Production Conference and the World Mass Customization & Personalization Conference, World Mass Customization & Personalization Conference, pp. 372–380. Springer, Cham (2022). https://doi.org/10.1007/978-3-030-90700-6_42

Author Index

T. Bauernhansl et al. (Eds.): WGP 2023, LNPE, pp. 765–768, 2024.
https://doi.org/10.1007/978-3-031-47394-4